Origin and Early Evolution of the Metazoa

TOPICS IN GEOBIOLOGY
Series Editors: F. G. Stehli, DOSECC, Inc., Gainesville, Florida
D. S. Jones, University of Florida, Gainesville, Florida

Volume 1 SKELETAL GROWTH OF AQUATIC ORGANISMS
Biological Records of Environmental Change
Edited by Donald C. Rhoads and Richard A. Lutz

Volume 2 ANIMAL–SEDIMENT RELATIONS
The Biogenic Alteration of Sediments
Edited by Peter L. McCall and Michael J. S. Tevesz

Volume 3 BIOTIC INTERACTIONS IN RECENT AND FOSSIL BENTHIC COMMUNITIES
Edited by Michael J. S. Tevesz and Peter L. McCall

Volume 4 THE GREAT AMERICAN BIOTIC INTERCHANGE
Edited by Francis G. Stehli and S. David Webb

Volume 5 MAGNETITE BIOMINERALIZATION AND MAGNETORECEPTION IN ORGANISMS
A New Biomagnetism
Edited by Joseph L. Kirschvink, Douglas S. Jones, and Bruce J. MacFadden

Volume 6 *NAUTILUS*
The Biology and Paleobiology of a Living Fossil
Edited by W. Bruce Saunders and Neil H. Landman

Volume 7 HETEROCHRONY IN EVOLUTION
A Multidisciplinary Approach
Edited by Michael L. McKinney

Volume 8 GALÁPAGOS MARINE INVERTEBRATES
Taxonomy, Biogeography, and Evolution in Darwin's Islands
Edited by Matthew J. James

Volume 9 TAPHONOMY
Releasing the Data Locked in the Fossil Record
Edited by Peter A. Allison and Derek E. G. Briggs

Volume 10 ORIGIN AND EARLY EVOLUTION OF THE METAZOA
Edited by Jere H. Lipps and Philip W. Signor

Origin and Early Evolution of the Metazoa

Edited by

Jere H. Lipps
University of California
Berkeley, California

and

Philip W. Signor
University of California
Davis, California

Plenum Press • New York and London

Library of Congress Cataloging-in-Publication Data

Origin and early evolution of the Metazoa / edited by Jere H. Lipps
and Philip W. Signor.
p. cm. -- (Topics in geobiology ; v. 10)
Includes bibliographical references and index.
ISBN 0-306-44067-9
1. Evolutionary paleobiology. 2. Metazoa. 3. Paleontology-
-Cambrian. .4. Paleontology--Precambrian. I. Lipps, Jere H., 1939-
. II. Signor, Philip W. III. Series.
QE721.2.E85075 1992
563--dc20 92-16461
CIP

ISBN 0-306-44067-9

A Division of Plenum Publishing Corporation
233 Spring Street, New York, N.Y. 10013

Printed in the United States of America

Contributors

Stefan Bengtson Institute of Palaeontology, S-752 22 Uppsala, Sweden

M. D. Brasier Department of Earth Sciences, Oxford University, Oxford OX1 3PR, England

Derek E. G. Briggs Department of Geology, University of Bristol, Bristol BS8 1RJ, England

Simon Conway Morris Department of Earth Sciences, Cambridge University, Cambridge CB2 3EQ, England

Thomas Peter Crimes Department of Earth Sciences, University of Liverpool, Liverpool L69 BBX, England

Françoise Debrenne Institute of Paleontology, 75005 Paris, France

M. A. Fedonkin Paleontological Institute of the Russian Academy of Sciences, Moscow, B-321, Russia

Richard A. Fortey Department of Palaeontology, Natural History Museum, London SW7 5BD, England

Richard J. F. Jenkins Department of Geology and Geophysics, University of Adelaide, Adelaide, South Australia 5000

Jiang Zhi-Wen Yunnan Institute of Geological Sciences, Kunming 650011, Yunnan, China

Andrew H. Knoll Botanical Museum, Harvard University, Cambridge, Massachusetts 02138

Ed Landing New York State Geological Survey, Albany, New York 11230

Jere H. Lipps Department of Integrative Biology and Museum of Paleontology, University of California, Berkeley, California 94720

Jeffrey F. Mount Department of Geology, University of California, Davis, California 95616; and University of California White Mountain Research Station, Bishop, California 93514

Leonid Ye. Popov Scientific Research Geological Institute, St. Petersburg, Russia

Alexei Yu. Rozanov Paleontological Institute of the Russian Academy of Sciences, Moscow, B-321, Russia

Philip W. Signor Department of Geology, University of California, Davis, California 95616

James Sprinkle Department of Geological Sciences, University of Texas, Austin, Texas 78713-7909

James W. Valentine Department of Integrative Biology and Museum of Paleontology, University of California, Berkeley, California 94720

Andrey Yu. Zhuravlev Paleontological Institute of the Russian Academy of Sciences, Moscow, B-321, Russia

Preface

Several years ago, we realized that the most prominent ideas that had been expressed about the origin and early evolution of the Metazoa seemed to have been developed chiefly by zoologists using evidence from modern species without reference to the fossil record. Paleontologists had, in fact, put forth their own ideas about the problem, but the zoological and the paleontological evidence were seldom considered together, especially by zoologists. We believed that the paleontological documentation of the first Metazoa was too scattered, too obscure to Western readers, and much of it too recent to have been readily available to our colleagues in zoology. Whether or not that was entirely true, we thought that a single volume reviewing the fossil record of the earliest Metazoa would be useful to many in both paleontology and zoology, especially since so much new information has been developed in the last few years. Some of this information has been summarized in general articles recently, but an overview of most of the field does not exist.

We therefore organized this book in five parts so that the evidence could be placed in perspective and summarized and inferences made from it. Part I introduces the previous hypotheses that have been proposed for the origin and early radiation of Metazoa. Part II consists of two summary chapters that set the sedimentological, geochemical, and biological background to the known radiations of Metazoa. Part III deals with the earliest known Metazoa, those of the Vendian Period, ranging in age from about 550 to 580 million years or so. These faunas are represented chiefly by soft-bodied or trace fossils. Part IV considers the early evolution of Metazoa from a geographical and systematic perspective. Three chapters summarize the occurrence of Early Cambrian fossils in three critical, well-documented areas of the world. Each of these areas is so important in a worldwide context that each was considered to be the area where the stratotype of the Precambrian–Cambrian boundary might be established. Of course, other areas exist where Precambrian to Early Cambrian fossils occur, and these three chapters make reference to the other critical regions as well. Four chapters are chiefly systematic, dealing with four of the major phyla that radiated in the Early Cambrian. Part V examines general hypotheses about the origin and radiation of Metazoa based on the fossil record. The origin of skeletons is a most critical issue in deciphering the early radiation of Metazoa because skeletons are more easily preserved in the geological record, they appear to have been acquired in many disparate groups at about the same time, and they may have conferred an evolutionary advantage that promoted the radiation itself.

Not all the chapters in this book present new information—in each case, they are summaries of a considerable body of detailed and dispersed knowledge. Some chapters include new information, and all provide new interpretations. So much information is available, in fact, that many other chapters could have been included in this book had space been available. Nevertheless, the authors have done a worthy job of presenting a general overview of the evidence. Moreover, they have provided references to further detailed documentation; readers will find the reference sections very helpful in this regard, particularly because many of these references are not widely known outside particular groups of specialists.

The fossil evidence for the origin and radiation of Metazoa is truly international, and the authors of these chapters work in all parts of the globe. Some represent views that are favored in particular regions and not in others. In part, we have selected contributors on that basis, and we have made no effort to constrain the views expressed, because there is plenty of room for a variety of viewpoints. Indeed, we believe it is essential that workers know the views of those in other parts of the world. Although some of our authors come from countries where English is not the mother tongue, they have written their chapters better than anyone else could. Their views and evidence remain unadulterated so that readers can make well-informed judgments of their own. We thank the authors for their excellent reviews.

Because the authors reside in many different countries, write in English with different degrees of surety, and have other duties, the chapters were submitted, edited, and revised at different times between mid-1987 and mid-1990. An opportunity to update manuscripts was available in November and December 1991.

Jere H. Lipps
Philip W. Signor

Contents

Part I • Introduction

Part II • The Late Proterozoic

Part III • Vendian Faunas of the World

Chapter 4 • Vendian Faunas and the Early Evolution of Metazoa

M. A. Fedonkin

Chapter 5 • Functional and Ecological Aspects of Ediacaran Assemblages

Richard J. F. Jenkins

Chapter 6 • The Record of Trace Fossils across the Proterozoic–Cambrian Boundary

Thomas Peter Crimes

Part IV • Radiations of the Cambrian Faunas

Chapter 7 • The Lower Cambrian Fossil Record of the Soviet Union

Alexei Yu. Rozanov and Andrey Yu. Zhuravlev

Chapter 8 • Lower Cambrian of Southeastern Newfoundland: Epeirogeny and Lazarus Faunas, Lithofacies–Biofacies Linkages, and the Myth of a Global Chronostratigraphy

Ed Landing

Chapter 9 • The Lower Cambrian Fossil Record of China

Jiang Zhi-Wen

Chapter 10 • The Early Cambrian Radiation of Arthropods

Derek E. G. Briggs and Richard A. Fortey

I

Introduction

Chapter 1

Origin and Early Radiation of the Metazoa

PHILIP W. SIGNOR and JERE H. LIPPS

1. Introduction

The complex of historical events encompassing the origin and early evolution of Metazoa is at once the salient feature and the most significant unresolved bio-historical phenomenon in the history of life. It has been the single most perplexing issue since paleontology emerged as a scientific discipline in the eighteenth and nineteenth centuries. Many of paleontology's heroic figures (W. Buckland, C. Lyell, C. R. Darwin, C. D. Walcott) and modern leaders (P. E. Cloud, B. Runnegar, S. M. Stanley, J. W. Valentine) have offered hypotheses or scenarios to explain or

PHILIP W. SIGNOR • Department of Geology, University of California, Davis, California 95616. JERE H. LIPPS • Department of Integrative Biology and Museum of Paleontology, University of California, Berkeley, California 94720.

Origin and Early Evolution of the Metazoa, edited by Jere H. Lipps and Philip W. Signor. Plenum Press, New York, 1992.

account for events of this critical juncture in the history of life. To date, none of these ideas are widely accepted.

Much of the difficulty stems from the enormous complexity of the phenomena involved. Within a geologically very brief period of time large numbers of Metazoa and, to a lesser extent, Parazoa and Protista appear in the fossil record. The early metazoans included representatives of most of the marine invertebrate phyla, classes, and orders recognized today (e.g., Valentine *et al.*, 1990; Valentine, this volume, Chapter 16; Bengtson and Conway Morris, this volume, Chapter 14). Skeletogenous protozoans also appeared for the first time and organic-walled phytoplankton diversified. The event was not *an* appearance or adaptive radiation; there were simultaneous events involving a large number of clades. Even if many clades had long, undetected Proterozoic histories of evolution, as some suggest, the simultaneous acquisition of skeletons among them must have been a complex event. Together with the evolutionary events of the time, whole new ecological associations formed and trophic relationships became more intricate. Perhaps no simple hypothesis can account for this complex sequence of events.

A second difficulty centers on the nature of the problem to be solved. The complexity of evolutionary events near the Precambrian–Cambrian boundary ensures that a variety of more or less discrete problems are available for potential solution. Many questions can be posed for each perceived problem. But the key question cannot be identified; where is our Gordian knot? There is no consensus on the nature of the fundamental questions. The hypotheses advanced to explain the metazoan radiation actually address several different questions, and the justifications supporting a particular hypothesis often focus more on the inferred importance of the particular question than on rigorous arguments supporting the hypothesis.

In this first chapter, we will briefly review the broad outlines of the metazoan radiation, discuss the outstanding questions regarding the origin of animals, and finally, review the more promising hypotheses in light of recent discoveries. This brief overview will serve as an introduction to the general problems of metazoan origins and diversification that are addressed in more detail, and from more specialized perspectives, in the subsequent chapters of this book.

2. Overview of the Radiation of Metazoa

2.1. Proterozoic and Early Cambrian Prokaryotes and Protists

A variety of prokaryotes and eukaryotic algae are known from the late Proterozoic (e.g., Knoll, this volume, Chapter 3). The first fossil evidence of prokaryotes is found in rocks dated at approximately 3.5 Ga, although this evidence has been questioned (Buick, 1991). The record of unicellular eukaryotes begins at about 1.4–1.7 Ga (Summons and Walter, 1990). Metaphytes are not common in the Proterozoic record, but generally accepted occurrences (e.g., *Grypania, Tawuia*) are known from the middle and late Proterozoic (Hofmann, 1987; Walter, 1987; Walter *et al.*, 1990; Knoll, this volume, Chapter 3). Together, these bacteria, algae, and other protists formed diverse communities in marine habitats.

Protistan phytoplankton are abundant and diverse throughout the later Proterozoic (Vidal and Knoll, 1983; Knoll, this volume, Chapter 3). They are preserved chiefly as cysts, the acritarchs, of unknown algal affinity. The floras are biogeographically differentiated, particularly between those onshore and offshore (Vidal and Knoll, 1983). From 1.4 Ga until near the end of the Proterozoic, the acritarchs increased in diversity. Their morphology also became increasingly complex in the later Proterozoic. Some, such as *Chuaria* and *Melanocyrillium* (Bloesser, 1985; Hofmann, 1987), are markedly different from other algae. At about 580 Ma, the same time as the appearance of the Vendian metazoans, phytoplankton declined in diversity and complexity, leaving only simple spherical forms. In the Early Cambrian, acritarchs radiated into many new types which were generally more highly ornamented and complex than the Proterozoic forms (Knoll, this volume, Chapter 3).

The first foraminifera are also found at the base of the Cambrian system (Glaessner, 1978; Lipps, in press). The foraminifera were simple agglutinated tubes, named *Platysolenites* [considered by Rozanov and Zhuravlev (this volume, Chapter 7) to be worm tubes]. *Platysolenites* is indistinguishable from Mesozoic to Recent *Bathysiphon* (Glaessner, 1978), including the type species of the genus (Gooday, 1988). Other forms are coiled tubes attibutable to *Ammodiscus* or *Spirosolenites* and hemispheres or spheres (Culver, 1991; Lipps, in press). Simple siliceous spheres, considered to be radiolaria by Nazarov (1975), also appear in the Early Cambrian. Rozanov and Zhuralev (this volume, Chapter 7) interpret these spheres as sponge spicules, although radiolaria are known from later in the Cambrian (Nazarov, 1975). Both foraminifera and radiolaria increased somewhat in diversity through the later Cambrian, becoming increasingly diverse in the Early Ordovician (Lipps, in press), a pattern resembling that of metazoans in general (Sepkoski, 1978).

The pattern of diversification of single-celled organisms thus shows a number of major changes at about the same time as the metazoans radiated in the late Proterozoic and early Phanerozoic (Knoll, this volume, Chapter 3). This indicates that the processes responsible for inducing the metazoan radiation had effects extending well beyond just the multicellular animals to probably the entire marine ecosystem.

2.2. Vendian Metazoa

The first fossil metazoans appear in the fossil record in marine sediments dated at approximately 580 Ma, although there is considerable variation in radiometric dates obtained for metazoan-bearing Vendian strata (Cowie and Harland, 1989). These rocks have been assigned to the Vendian, Sinian, Ediacaran, or Ediacarian periods by different workers, but each term appears to encompass rocks of the latest Proterozoic. Herein, we avoid the terminological controversy and employ the commonly accepted terms Vendian and Ediacaran as equivalents. This Vendian soft-bodied fauna is now known to have a global distribution (Cloud and Glaessner, 1982; Conway Morris, 1985b; Jenkins, this volume, Chapter 5). Vendian fossils have been reported from Australia (Sprigg, 1947; Glaessner, 1958), China (Wang *et al.*, 1980), the Soviet Union (Fedonkin, 1980, 1981, 1985, and this volume, Chapter 4), the Baltic Platform (Palij *et al.*, 1983), England (Ford, 1958, 1963,

1979, 1981), Africa (Gürich, 1933; Pflug, 1970a, b; Germs, 1973a, b), Newfoundland (Anderson and Misra, 1968; Misra, 1969, 1971; Anderson and Conway Morris, 1982), and eastern and western North America (Narbonne and Hofmann, 1987; Narbonne and Aitkin, 1990). Some are quite large (e.g., *Dickinsonia*), approaching 1 m in length (Runnegar, 1982c). The fauna is dominated by a variety of radially symmetric taxa, but a number of bilaterally symmetric forms and trace fossils are also known. These early, soft-bodied organisms display a number of different body plans and might include representatives of phyla extant today (Jenkins, 1985, and this volume, Chapter 5; Glaessner, 1984). Their architecture, however, suggests that these undisputable Metazoa could represent a suite of organisms unrelated to modern phyla, or perhaps a mix of the two (Seilacher, 1984, 1989; McMenamin, 1985b). Particularly troubling in this regard are morphological inconsistencies such as the failure of features interpreted as metameric segments to meet across the fossils' centerline on putative annelids or primitive arthropods (e.g., *Vendia*). Fedonkin (1986, and this volume, Chapter 4) interprets two of the better known genera, *Dickinsonia* and *Vendia*, as bilaterians unrelated to any subsequent forms. These inconsistencies led Seilacher (1984, 1989) to propose that the Vendian fauna, previously interpreted as evolutionary antecedents of our modern marine phyla, are a separate, extinct lineage (Vendozoa). McMenamin (1986b) expanded upon this theme, arguing that Vendian organisms likely possessed algal symbionts. Regardless of their affinities, the Vendian fauna quickly disappear from the fossil record. No convincing Ediacaran-aspect metazoan has been reported from Cambrian strata. [McMenamin (1986a) suggested that *Xenusion*, a fossil recovered from erratic boulders in Scandanavia, was a partially preserved soft-bodied metazoan fossil, but Dzik and Krumbiegel (1989), on the basis of new material, have reinterpreted *Xenusion* as an onycophoran (?) related to the Middle Cambrian genus *Aysheaia*.]

Two skeletal fossils, *Cloudina* and *Sinotubulites*, reportedly occur within strata containing a Vendian fauna (Germs, 1972; McMenamin, 1985). *Cloudina* occurs in late Proterozoic strata throughout the world and *Sinotubulites* occurs in the Proterozoic of China. Both are small (maximum diameter less than 1 cm), conical to tubular, irregular expanding forms. Their skeleton was apparently formed of calcium carbonate (Germs 1972; Glaessner, 1976; Grant, 1990). These two genera have also been reported from the Lower Cambrian (Yochelson and Herrera, 1974; McMenamin, 1985). The relationships of *Cloudina* and *Sinotubulites* are uncertain [e.g., Hahn and Pflug (1985); Grant (1990) regards *Sinotubulites* as a junior synonym of *Cloudina*). The significance of these two genera is that they are the first forms with mineralized tissues and they appear well before other skeletogenous taxa. Lined worm tubes, *Archaeichnium*, also appear first in the late Proterozoic of Namibia (Glaessner, 1984).

2.3. Trace Fossils

The first trace fossils appear in the fossil record at about the same level as the Vendian fauna. Older traces have been reported (e.g., Kauffman and Steidtmann, 1981), but these have not been generally accepted (Byers, 1982; Crimes, 1987, 1989, and this volume, Chapter 6; Bergstrom, 1990; Fedonkin and Runnegar,

1992). Initially, the earliest forms were simple surface traces, but burrows began to penetrate the substrate and became increasingly complex early in the Cambrian (Crimes, 1987, 1989, and this volume, Chapter 6; Fedonkin and Runnegar, 1992). Horizontal traces formed by burrowing organisms are present within Vendian sediments (e.g., *Gordia, Neonerities*) (Seilacher, 1985; Crimes, 1989). These traces imply the presence of coelomate-grade metazoans in the late Vendian, but no sooner (Seilacher, 1985; Bergstrom, 1990).

The diversification of traces may have been responsible, at least in part, for the disappearance of the Ediacaran fauna from the fossil record. Bioturbation could be responsible for disturbing sediments that entombed soft-bodied organisms, thus removing them from the fossil record (Stanley 1976a; Fedonkin, this volume, Chapter 4). The advent of bioturbation could also have had a major evolutionary impact on marine benthic communities, through the initiation of carbon recycling through sediment disturbance (Fischer, 1984). Bioturbation destroys sedimentary structures (while creating biogenic structures) and returns buried carbon to the carbon cycle.

Trace fossils are very important to our understanding of metazoan evolution. Organisms moving over or through sediment disturb the sediment, thereby leaving a record of their activity. Even organisms as small as foraminifera leave traces discernible to the unaided eye (e.g., Severin *et al.*, 1982). The apparent absence of such evidence in rocks older than the Vendian indicates that coelomate Metazoa have no extended period of hidden evolution, as suggested by some authors (e.g., Durham, 1978; Runnegar, 1982a, b), or that simple, small, flatwormlike animals might have roamed the surface sediments, interstitial sand, or firm substrates without leaving any late Proterozoic record.

Further evidence for the presence of Metazoa is provided by the decline of stromatolites in the late Proterozoic. The three-step decline of stromatolites at about 1.0 Ga, 0.7–0.8 Ga, and during the Cambrian has been attributed to the appearance and foraging of metazoans (Garrett, 1970; Awramik, 1971; Walter, 1987). The first decline in abundance and diversity may have resulted from the effects of small acoelomate metazoans, the second may be due to animal grazing, and the third due to the activities of large animals in the Cambrian (Walter, 1987). Although this episodic reduction in stromatolite diversity and corresponding set of hypotheses is reasonable, it is nevertheless still speculative and requires further study. Stromatolites continue to occur sporadically in the fossil record up to the Recent, but never in the abundance or diversity they attained in the Proterozoic.

2.4. Early Cambrian Metazoa

The sudden appearance of diverse metazoan skeletal fossils heralds the beginning of the Phanerozoic. The precise sequence of appearance of different clades varies from region to region [compare, for example, the chapters on the faunas of Siberia, Newfoundland, and Meishucun by Rozanov and Zhuravlev (Chapter 7), Landing (Chapter 8), and Jiang (Chapter 9) in this volume]. Nevertheless, there is little evidence that the capacity to form skeletons was acquired gradually or over a prolonged period (Stanley, 1976a; Brasier, 1979; Valentine *et al.*, 1990; but see Rozanov and Zhuravlev, this volume, Chapter 7). A wide variety of skeleton types and most of the major marine invertebrate clades appear suddenly in the fossil record (Stanley, 1976a; Sepkoski, 1978; Brasier, 1979, 1989; Glaessner, 1984; Jiang, this volume, Chapter 9; Bengtson and Conway Morris, this volume, Chapter 14). Of course, it is difficult to translate the stratigraphic record into a chronology. Gaps representing significant intervals of time are present in many (probably most) sections (e.g., Schindel, 1980; Sadler, 1981) and the evolution of skeletons might be less sudden than it appears. Such a gap, caused by a late Proterozoic regression, occurs below the first occurrence of skeletal fossils (Brasier, 1979, and this volume, Chapter 15). But radiometric dates from Proterozoic and Cambrian rocks (Cowie and Harland, 1989) indicate that the gap cannot be sufficiently long to alter the general pattern described above.

The ecological diversification of animals is equally dramatic. A wide variety of habitats were occupied by these biotas, from shallow to deep benthos and to the pelagic realm. Paleobiogeographic differences were also present among these biotas, with different continents each associated with a distinct fauna (Signor, 1991). Trophically, Vendian animals may have met their nutrient requirements passively, mostly as suspension feeders, a few as detrital feeders, and possibly some bearing endosymbionts, but few as active carnivores or herbivores, whereas the Early Cambrian biotas include a wide variety of active feeders. Photosynthesizing protists and algae were present in both biotas. Early Cambrian metazoan communities thus include most of the trophic roles found in modern marine communities. The diversification of skeletogenous organisms included the creation of a large number of new trophic roles.

The pattern of metazoan diversification itself has been the subject of considerable inquiry. Early investigations attempted to characterize the gradual or rapid nature of the radiation. More recently, the rapid but not instantaneous pattern of first occurrences in the fossil record (e.g., Stanley, 1976a; Durham, 1978) has been generally accepted, although disagreements about the actual timing of origins continue. Recent work has focused instead upon the precise pattern of diversification. Sepkoski (1978) suggested that the Proterozoic–Cambrian radiation was similar in form to an exponential diversification. Rates of first occurrence decline through the Cambrian and Ordovician, reaching a Paleozoic equilibrium in the Late Ordovician. The overall pattern is congruent with that predicted by a logistic model. More complex models have been employed to account for later evolutionary events and for the behavior of separate "faunas" in geological time (Sepkoski, 1979, 1984). Walker (1985) suggested a series of alternative quantitative models for diversification, but these remain to be compared, in detail, to Sepkoski's results.

Continuing studies of stratigraphic and evolutionary events of the late Proterozoic and Early Cambrian have tended to accentuate the apparent diversification. Recent radiometric dates from the Upper Proterozoic and Lower Cambrian indicate that the base of the Lower Cambrian could be as young as 530 Ma (e.g., Conway Morris, 1988), although Cowie and Harland (1989) prefer an age of 570 Ma, while the top of the Cambrian remains at 505–510 Ma. A young base of the Early Cambrian greatly shortens the duration of that interval and accelerates the rate of appearance of new taxa. Further work has also tended to produce older representatives of many major taxa and thus increase the apparent rate of evolution during this time (Valentine *et al.*, 1990).

3. Questions

Any attempt to define a set of specific questions about the Precambrian–Cambrian boundary will quickly yield far more (and difficult) questions than we have answers. Our goal here is not to be comprehensive, but to show that different authors have addressed quite different questions in their attempts to understand the origins of Metazoa. The individual authors have generally examined and emphasized the significance of their particular question, but it is not clear which of their questions might be the keystone of the problem (if, indeed, we dare hope for such good fortune). For example, Walcott (1910) accepted that animals must have evolved gradually over an extended period of time. Therefore, the problem was to explain the lack of any fossil record of developing or incipient Metazoa. This led to Walcott's identification of the supposed "Lipalian Interval", a time of worldwide regression when animals evolved. The Cambrian transgression restored the stratigraphic record and generated the well-known sudden first appearances of taxa in the fossil record (e.g., Brasier, 1979). Others (e.g., Brooks, 1894; Daly, 1907; Axelrod, 1958) have sought different means to obscure the early evolution of Metazoa. Stanley (1976a) and Sepkoski (1978), in contrast, accept the pattern as recorded in the fossil record and seek to identify processes that could generate the apparent pattern. The two interpretations differ wildly and lead to very different conclusions. Thus, the debate is poorly focused and awkward because the different research programs are based on different precepts and address very different questions.

A brief listing of some relevant questions, together with appropriate citations, is presented in Table I. Answers to some of these questions provide partial answers to others; our purpose is not to obscure or minimize that fact. But researchers have often presumed that an answer to one of these questions will resolve *all* those remaining. Afterall, the goal of most scientists is to find the simplest possible hypothesis to explain all available facts. Table I shows that the "problem" of the origin of Metazoa has been viewed in many different ways. Not surprisingly, no modern consensus emerges from this confusion: the same questions continue to arise. The large variety of hypotheses presented in the following section reflects in large measure this mixed perception of the questions surrounding the origin of animals.

Hypotheses invoked to explain the appearance of Metazoa generally belong to one of two classes. The theories either present a scenario for a presumed long, hidden history of animals followed by a sudden rapid appearance in the fossil record or they attempt to link the rapid evolution of animals to some key biological or physical event. The two classes of hypotheses presuppose very different chronologies and kinds of critical events in the early history of animal life. More fundamentally, this debate is a microcosm of the long-standing philosophical dichotomies in evolutionary paleontology.

Gould (1977) identified three dichotomies in paleobiological theory. He argued that these dichotomies remain unresolved and are essentially unsolvable; they are less questions of science than world views. Hence, Gould termed them "eternal metaphors." The first of these dichotomies is the contrast between a world at equilibrium and a world developing progressively through time. The second is a question of rate; do biological processes operate at steady rates or are there striking variations in rate through time? The final metaphor concerns the

TABLE I. Theoretical Questions Pertinent to the Emergence of Metazoa

1. "When" questions
   a. When did Metazoa first evolve? (Awramik, 1971; Durham, 1978; Runnegar, 1982a; Stanley, 1976b)
   b. How long was the delay between the appearance of monerans and the evolution of Metazoa?
   c. Why was the delay so extended? (Durham, 1978; Valentine *et al.*, 1990)
   d. Why then and not some other time? (Harland, 1964; Rudwick, 1964; Stanley, 1976a, b; Valentine *et al.*, 1990)
   e. When did the ability to form skeletons evolve? (Boaden, 1989; Brooks, 1894; Evans, 1912; Runnegar, 1982b; Signor, 1988, 1991)
   f. If the evolution of Metazoa and metazoan skeletons is not synchronous, why the delay? (Brooks, 1894; Cook and Shergold, 1984; Daly, 1907; Evans, 1912; Fox, 1975; Hutchinson, 1961; Kazmierczak *et al.*, 1985; Kempe and Degens, 1985; McMenamin, 1986a; Towe, 1970)

2. "Where" questions
   a. Where did Metazoa evolve? (Axelrod, 1958; Brooks, 1894; Goldring, 1965; Raymond, 1935)
   b. Could a long period of metazoan evolution be hidden in an unusual environment or obscured by nonpreservation? (Axelrod, 1958; Brooks, 1894; Clark, 1964, 1979; Goldring, 1965; Raymond, 1935; Walcott, 1910)

3. Process questions
   a. What happened to trigger the evolution of Metazoa? (Cloud, 1976; Cook and Shergold, 1984; Erwin and Valentine, 1984; Fischer, 1965; Nursall, 1959; Stanley, 1976b; Valentine and Campbell, 1975)
   b. Have patterns of sedimentation, preservation, or the paleoecology of early organisms combined to obscure a significant portion of Metazoan evolution? (Boaden, 1989; Nicol, 1966; Valentine, 1989; Walcott, 1910)
   c. Did physical (extrinsic) conditions on the early Earth prevent the evolution of Metazoa? (Berkner and Marshall, 1964a, b, 1965; Cloud, 1976; Fischer, 1965)
   d. Was the radiation of Metazoa triggered by some sort of adaptive (intrinsic) breakthrough? (Buss, 1987; Thomson, 1989; Valentine and Campbell, 1975; Valentine *et al.*, 1990)

4. Pattern questions
   a. What was the tempo of the radiation of Metazoa? (Darwin, 1859; Sepkoski, 1978; Stanley, 1976a)
   b. Was there a change in evolutionary tempo concurrent with the evolution of Metazoa? (Stanley, 1976b)
   c. Are Metazoa monophyletic?

driving forces behind change in the biosphere. Is the change driven externally by physical processes or is change controlled biologically, through processes originating within the biosphere?

The role of these metaphors in the debates surrounding the origins of Metazoa is striking. The differing views on the appearance of Metazoa, identified in the previous section, follow these metaphors closely. In particular, the dichotomies surrounding the questions of process rate and ultimate cause categorize the different hypotheses rather nicely. Some authors embrace extrinsic (physical) causation and constant, gradual rates of organic change, while others adopt intrinsic (biological) causation and strong variations in rate. All possible combinations of the foregoing pairs are found in the literature about the Proterozoic–Cambrian boundary. The better-known hypotheses for the origin of Metazoa are categorized in Table II according to this scheme. The third metaphor, steady state versus direc-

TABLE II. Hypotheses Proposed to Account for the Origin and Precambrian–Cambrian Radiation of Metazoans

Extrinsic factors/gradual rates of evolution

1. Absence of carbonate from Precambrian oceans (Daly, 1907)
2. Advent of sufficient oxygen to permit skeletonization (Towe, 1970)

Extrinsic factors/episodic rates of evolution

1. Accumulation of sufficient oxygen in the atmosphere and oceans allowed the development of larger organisms with complex musculature and metabolism (Nursall, 1959; Berkner and Marshall, 1964a, b, 1965, 1967; Fischer, 1965; Cloud, 1968, 1976; Runnegar, 1982c; Towe, 1970, 1981)
2. Emergence of continents would affect all groups through environmental changes caused by the new continental arrangements and elevations (Valentine and Moores, 1972; LaBarbara, 1978; Knoll, 1978)
3. Increased carbonate levels in the oceans permitted the development of skeletons, accounting for the second stage of the metazoan radiation (Daly, 1907; Kazmierczak *et al.*, 1985)
4. Changes in ocean chemistry other than carbonate, particularly phosphorous, account for the appearance of skeletons (Cook and Shergold, 1984)
5. Glaciation of the Earth changes paleoceanography and trophic resource supply so that all groups could radiate in concert (Harland, 1964; Rudwick, 1964)
6. Breakup of supercontinent (Valentine and Moores, 1972)
7. Variation in sea level (Brasier, 1979, 1982)

Intrinsic factors/gradual rates of evolution

1. Acquisition of sedentary habit with resulting competition (Brooks, 1894; Raymond, 1935)
2. Organisms invade habitats where preservation can occur (Brooks, 1894; Axelrod, 1958; Goldring, 1965)
3. Advent of predators, requiring skeletonization (Evans, 1912; Hutchinson, 1961)
4. Attainment of large size, permitting skeletonization (Nicol, 1966)

Intrinsic factors/episodic rates of evolution

1. Large size of metazoans permits skeletonization
2. Sedentary habit allows development of large organisms with complex morphology
3. Evolution of sex (Stanley, 1976b)
4. Advent of predators and/or croppers (Hutchinson, 1961; Stanley, 1973, 1976b; Edhorn, 1977)
5. Evolution of regulatory genes (Valentine and Campbell, 1975)
6. Evolution of metazoan development (Buss, 1987)
7. Evolution of genomic flexibility (Valentine and Erwin, 1987)

tional change through time, is not included for reasons of space and simplicity. But some, most notably Lyell (1830), have denied the existence of directional change through time. Lyell argued that animals had always been on earth but evidence for their presence was obscured by metamorphism, erosion, and poor preservation.

4. Hypotheses to Account for the Radiation of Metazoa

4.1. Hidden Evolution

The fossil record, as currently understood, indicates that hypotheses predicated upon a long, hidden history of Proterozoic evolution (hypotheses invoking both extrinsic and intrinsic gradual rates) are probably not supported by the available evidence. Three arguments have been advanced in support of these hypotheses. The first and second arguments invoke evidence from trace fossils and biochemistry. The last argument, that the existence of complex metazoan life in the Vendian and Cambrian requires prexisting, although somewhat less complex metazoans, is a theoretical argument lacking direct evidence.

The evidence from trace fossils is unequivocal: the diversification of trace fossils parallels but does not precede the diversification of shelly fossils (Byers, 1982; Crimes, 1987, 1989, and this volume, Chapter 6). As outlined in the preceding overview of the Precambrian–Cambrian fossil record, no trace fossils are known to support the contention that metazoans existed prior to the Vendian. Simple surface traces (*Cochlichnus, Planolites*) occur at approximately the same level as the Ediacaran fauna. Traces become more complex and attain a strong vertical component of development within the substrate at the base of the Cambrian (e.g., *Phycodes, Diplocraterion*). Arthropod-grade traces (e.g., *Cruziana, Monomorphichnus, Rusophycus*) appear above the Precambrian–Cambrian boundary but below the first trilobites. This pattern of sequential trace fossil appearances occurs in Upper Proterozoic and Lower Cambrian sections throughout the world (Brasier, 1979; Crimes, 1987, 1989, and this volume, Chapter 6; Fedonkin and Runnegar, 1992) and is particularly well displayed in Newfoundland (Narbonne *et al.*, 1987). The absence of older uncontrovertable traces in any ancient environments indicates an absence of coelomates and, possibly, other metazoan life.

Early metazoans might have been small, on the order of the size of meiofauna, and their small size might have allowed them to persist in the Proterozoic, literally without leaving a trace (Nicol, 1966; Runnegar, 1982b, c; Boaden, 1989). But even protozoa leave traces as they move through sediment, especially in silts and muds. Such traces, however, are small and indiscriminate and might be overlooked in older rocks. Small Metazoa inhabiting hard or coarse-grained substrates (e.g., stromatolites, sands) might not leave a fossil record at all. Perhaps Metazoa originated as meiofauna, a sort of Cope's Rule taken to the extreme (Nicol, 1966). But this hypothesis seems unlikely, given the larger size of the Metazoa that do appear first in the fossil record.

Runnegar (1982a, 1986) estimated magnitudes of molecular divergence (the molecular clock) for invertebrate hemoglobin. Based on comparisons of a- and b-hemoglobins of representative annelids, arthropods, phoronids, and vertebrates, he estimated that these groups last shared a common ancestor at approximately 0.9–1.0 Ga. If this analysis is correct, it would demonstrate an unknown, long Proterozoic history of Metazoa. However, Runnegar's conclusions have been challenged by Erwin (1989) on methodological grounds. At present, the molecular evidence for the age of the Metazoa is uncertain.

The final argument for older Proterozoic Metazoa is appealing in its simple logic. The first Metazoa that appear in the fossil record are complex and show considerable morphological divergence from their common ancestor, probably some type of flatworm. In turn, the ancestral metazoan must have evolved over some period of time. For example, fully developed trilobites appear throughout the world shortly after the beginning of the Cambrian. No primitive intermediate forms are known and no forms can be considered intermediate between primitive arthropods and their closest relatives. Similarly, Conway Morris (1985a) noted that the diverse arthropod fauna present in the Middle Cambrian Burgess Shale lacks any significant precursors in the fossil record [now with the exception of new soft-bodied Lower Cambrian assemblages from China (Zhang, 1985; Chen *et al.*, 1989) and Greenland (Conway Morris *et al.*, 1987; Conway Morris and Peel, 1990)]. A considerable history of arthropod evolution must have preceded that preserved in the fossil record. Similar arguments in support of the contention that other taxa have a long period of hidden evolution can be advanced.

A related argument centers upon the development of biogeographic provinces in the Early Cambrian (Signor, 1991). Provinces are biohistorical phenomena that develop through time as a consequence of dispersal and vicariance events. The presence of well-developed, distinct provinces at the beginning of the Cambrian strongly indicates a hidden prehistory of uncertain length (Signor, 1991). If plate tectonics is the primary control on the development of those provinces, the development of provinces could require an extended interval of geological time. In any event, Early Cambrian provinces remain intact throughout the remainder of the Cambrian.

The existence of some degree of hidden evolution is beyond dispute; even those who advocate a very rapid evolutionary origin for Metazoa accept this point. The dispute centers on the duration of time missing from the record.

4.2. Skeletons and Animals

The undisputed existence of soft-bodied animals prior to the appearance of skeletogenous animals is direct evidence that the evolution of metazoans was not linked to or predicated upon the evolution of skeletons. Nevertheless, persuasive arguments have been advanced in support of the contention that some kinds of animals must have possessed skeletons when they first evolved. In particular, the body plans of brachiopods and bivalve mollusks might not be viable without some type of skeletal support (Cloud, 1949; Stanley, 1976b; Valentine and Erwin, 1987). The brachiopod lophophore employed for feeding requires a mantle cavity (hence, a shell to create that cavity) to form and channel currents through the lophophore (Valentine and Erwin, 1987; Popov, this volume, Chapter 12). The evolution of the brachiopod required the evolution of the shell; the brachiopod did not precede the brachiopod shell (Cloud, 1949; Valentine and Erwin, 1987). Likewise, the presence of a shell is a prerequisite for the evolution of the molluscan classes (Stasek, 1972). For example, bivalve mollusks require a mineralized shell to burrow and to create a mantle cavity for the gills. A naked bivalve cannot function (Stasek, 1972; Stanley, 1976b).

Another argument suggesting that the Cambrian event was not simply the onset of biomineralization is found in the temporal distribution of agglutinated skeletons. Agglutinated skeletons require no biomineralization, and are often constructed by organisms lacking the capacity to form hard tissues. Yet the first agglutinated fossils occur at the base of the Cambrian, together with the advent of different modes of biomineralization (Brasier, 1979; Signor, 1988; Signor and McMenamin, 1988). At least three separate clades simultaneously acquire the capacity to construct agglutinated tests early in the Cambrian [*Platysolenites*,

Volborthella, and *Onuphionella* (Signor, 1988)]. If the organisms were present earlier but were unable to biomineralize, we should expect to find agglutinated skeletons preserved in the record. Thus, skeletonization was an integral component of the Proterozoic–Cambrian radiation that proceded concurrently with the evolution of major invertebrate clades. But the evolution of Metazoa preceded skeletonization by a geologically significant period of time.

4.3. Oxygen and Animals

Animals cannot persist for long without access to oxygen. In general, larger, complex, active, skeletogenous animals require oxygen in higher concentrations than do smaller, simpler, sedentary, and unskeletonized animals (e.g., Raff and Raff, 1970; Rhoads and Morse, 1971). This observation, combined with the discovery that the early Earth's atmosphere lacked free oxygen, led to the suggestion that the radiation of Metazoa followed the accumulation of sufficient free oxygen in the atmosphere to support animal life (Berkner and Marshall, 1964a, b, 1965, 1967; Fischer, 1965; Cloud, 1968, 1976). This scenario gained considerable support from interpretations of the Proterozoic sedimentary record. The presence of early to mid-Proterozoic detrital sulfides, uranites, and oxidized detrital iron (banded iron formations) was taken as evidence of an early atmosphere lacking in free oxygen. According to the scenario, molecular oxygen from photosynthetic activity accumulated in the atmosphere only after such chemical sinks were saturated with oxygen.

More recently, the interpretation of physical evidence for a reducing Proterozoic atmosphere has been questioned (e.g., Dimroth and Kimberley, 1976; Clemmey and Badham, 1982). Detrital sulfides are found in certain modern depositional environments (Clemmey and Badham, 1982). The genesis of banded iron formations is a contentious subject. Regardless, deposition of significant banded iron deposits ceased by about 1.7 Ga, well prior to the appearance of Metazoa (Clemmey and Badham, 1982). Furthermore, the presence of oxidized paleosols in lower Proterozoic sedimentary sequences and oxidation rinds on ancient pillow basalts (Dimroth and Lichtblau, 1978) suggests the presence of free molecular oxygen in the Earth's atmosphere well before the times predicted by Cloud (e.g., Cloud, 1976). Clemmey and Badham (1982) hypothesize an oxygenated atmosphere perhaps as early as 3.7 Ga. Others prefer younger dates (e.g., Schopf *et al.*, 1984). On the basis of geochemical analyses of Proterozoic paleosols from South Africa, Holland and Buekes (1990) estimate that atmospheric oxygen reached 15% of present atmospheric levels (PAL) by 2.2 to 1.9 Ga. Another analysis of a Proterozoic paleosol indicated that the concentration of atmospheric oxygen increased by about 1.8 Ga (Holland *et al.*, 1989). Regardless of the precise timing, oxygen apparently was present in the atmosphere well before the appearance of Metazoa. Therefore, it seems unlikely that oxygen for respiration was the proximal control on the evolution of animals, as observed in the fossil record.

Fossils have also been used to infer levels of atmospheric oxygen in the Proterozoic. The body plan of *Dickinsonia* was interpreted by Runnegar (1982c) as an evolutionary adaptation to low levels of oxygen. However, there are alternative interpretations for the thin, flattened morphology of that fossil (McMenamin, 1986a).

Towe (1970) proposed a mechanism to explain how insufficient free oxygen constrained the evolution of Metazoa (the collagen-priority hypothesis). Towe observed that collagen is a critical component in metazoan architecture, and that formation of collagen is an oxygen-intensive process. From this, he argued that the formation of collagen would have priority over the formation of any skeletal elements. Towe suggested that low levels of atmospheric oxygen might permit the evolution of soft-bodied animals, but not skeletonized forms. This hypothesis has the virtue of predicting (*post hoc*) the observed fossil sequence rather well, and it cannot be rejected with the data at hand. But there is no evidence that oxygen was limiting in the late Proterozoic (e.g., Schopf *et al.*, 1984).

4.4. Predators and Prey

A number of authors have proposed that the advent of metazoan predators triggered or accelerated the appearance of skeletons in the fossil record (Evans, 1912; Hutchinson, 1961; McMenamin, 1986a; Vermeij, 1989). In this scenario, the appearance of predators would place an enormous selective premium on the acquisition of skeletons. However, the hypothesis fails to address the actual origin of animals in any substantive manner, and it implies the existence of an older, unskeletonized, hidden fauna. As noted above, there is no evidence for such a presumption.

Stanley (1973) suggested that the advent of predation was the critical step in the evolution of animals, but had a very different impact than supposed by others. Stanley suggested that the late Proterozoic oceans' biota encompassed only bacteria and algae: in effect, a trophic structure of only one level. The appearance of heterotrophs (croppers) triggered the development of complex trophic relationships and, eventually, the evolution of animals. This hypothesis has considerable theoretical appeal, but it neglects the fact that the first organisms were almost certainly heterotrophs, and heterotrophs have surely persisted since. Perhaps these chemical heterotrophs were unable to consume cellular prey. In any event, Stanley's hypothesis only begs the question of the origin of croppers (in essence, the hypothesis substitutes "cropper" for "metazoans" in statements of the problem).

4.5. Evolution of Large Size

Some authors have accounted for the lack of early animals in the fossil record by suggesting that the oldest Metazoa were small and lacked skeletons (Nicol, 1966; Boaden, 1989). These animals only appeared in the fossil record when they reached large size and became skeletogenous. The processes that relate size and the capacity to form skeletons are unclear. Also, the lack of early bioturbation argues against this hypothesis, regardless of the size of the animals. In a sense, this hypothesis alters the nature of the question to be answered, but it does not explain the origin of animals.

4.6. Carbonate, Phosphate, and Ocean Chemistry

Clearly, the evolution of Metazoa and the evolution of skeletons are separate events, even through the evolution of skeletons is a prerequisite for the evolution of some clades (e.g., brachiopods). Several authors (e.g., Daly, 1907; Riding, 1982; Cook and Shergold, 1984; Kazmierczak *et al.*, 1985; Kempe and Degens, 1985) have argued that ocean chemistry first inhibited and then subsequently facilitated skeleton formation by Metazoa. Each author favors changes in various ions in the ocean (e.g., Ca^{2+}, P^{3-}) or atmosphere (e.g., CO_2) that would theoretically restrict or enhance biomineralization of particular compounds. For example, Riding (1982) suggests that changes in the partial pressure of CO_2 would alter the balance of Ca^{2+} and Mg^{+2} in the oceans, in turn affecting the ability of cyanophytes to calcify.

Certainly, oceanic chemistry has varied through the Phanerozoic (Wilkinson, 1980), and almost surely varied through the Proterozoic as well (Tucker, 1982; Brasier, 1986b). These changes have influenced patterns of accumulation of calcium carbonate, dolomites, phosphorites, and other sedimentary minerals. Likewise, patterns of biomineralization have probably been influenced as a result of atmospheric or oceanic chemistry. But it is difficult to attribute the onset of skeletonization to such physical processes. The simultaneous appearance in the fossil record of skeletons composed of calcium carbonate, calcium phosphate, biogenic silica, and other materials cannot be explicated by these hypotheses. Furthermore, the coeval appearance of agglutinated skeletons (constructed of detrital particles) in several unrelated taxa (e.g., *Platysolenites, Volborthella, Onuphionella*) suggests the Early Cambrian radiation of skeletogenous organisms was not controlled by simple changes in ambient conditions (but see Brasier, 1986b).

4.7. Glaciations, Sea Level, and Diversity

Rudwick (1964) proposed that the termination of the Varangian glaciations triggered the evolution of Metazoa (also see Harland, 1964; Hambry and Harland, 1985). The Varangian glaciation was probably the most severe in the history of the Earth, with continental glaciers reaching tropical latitudes (Harland, 1964, 1989; Embleton and Williams, 1986). Rudwick suggested that adverse climatological conditions and reduced areal extents of the continental shelves, caused by an increase in the volume of water sequestered in glaciers and a concurrent reduction in sea level, retarded the appearance of Metazoa. However, the mechanism producing such an effect is unknown. In fact, Runnegar (1982b) suggested an opposite effect for the glaciations; glaciations actually triggered the evolution of Metazoa.

4.8. Tectonics

Recent authors have suggested several ways that tectonic events might have influenced the evolution of animals. These include the evolution of continents, changes in the configuration of the continents, and changes in sea level. Hargraves (1976) adduced that continents did not form and become emergent until late in the Proterozoic. Chamberlain and Marland (1977) and LaBarbara (1978) linked the emergence of continents with the initial evolution of animals. However, Knoll (1978) summarized convincing evidence that continents must have appeared long before the evolution of animals.

Valentine and Moores (1972) related the appearance and radiation of animals to the breakup of a late Proterozoic supercontinent. The causative link between the rifting event and the evolution of animals is not explicit, and seems dubious. The configuration of the continents is widely acknowledged to be a major influence on species richness through time (Valentine and Moores, 1972; Signor, 1985, 1990), but that is an issue separate from the evolution of animals.

Brasier (1979, 1982, 1985, and this volume, Chapter 15) suggested that changes in sea level influenced the evolution of Metazoa, although he proposed no mechanism to account for such an effect. More likely, the patterns of sequential faunal appearances noted by Brasier resulted from the global, basal Cambrian transgression and are not the immediate result of evolutionary change (Mount and Signor, this volume, Chapter 2).

4.9. Genetic Mechanisms

Several authors have suggested that the key innovation in the evolution of Metazoa was the evolution of regulatory genes or the evolution of complex genetic mechanisms to regulate development (e.g., Valentine and Campbell, 1975; Arthur, 1982, 1988; Buss, 1987; Bonner, 1988). Certainly, metazoan development is a complex suite of adaptations. But these adaptations cannot be separated from the appearance of the Metazoa. To test for the existence of these genetic mechanisms is to test for the appearance of Metazoa.

4.10 Mechanical Efficiency

Cisne (1974) and Fox (1975) propose that the primary advantage of skeletons is mechanical efficiency. Skeletons permit organisms to support themselves or withstand currents without muscular exertion. In arthropods, skeletons promote efficient locomotion. Undoubtedly, skeletons provide a considerable energetic advantage for many organisms. But Metazoa almost certainly predate skeletons, so this hypothesis does not address the problem of the appearance of animals. Furthermore, the hypothesis does not provide insight into why skeletons appear relatively late in the history of life nor why they appear in protists.

5. Conclusions

The emergence of Metazoa remains the salient mystery in the history of life. The origin of skeletons and the origin of animals are separate phenomena, although the appearance of skeletons surely contributed to the subsequent, rapid diversification of skeletogenous animals in the Early Cambrian. The marine biota near the Proterozoic–Phanerozoic boundary shows rapid diversification in nearly all clades, and at different grades of organization, different ecologies, different skeletal types, and different trophic levels. This kind of simultaneous reorganization probably requires a process acting across ecosystems and influencing each biota. The fundamental question, yet to be identified, therefore must center on the physical or biological processes and phenomena of the time. Subsequent chapters in this book deal with aspects of this problem. They bring together much information on important taxa and broad geographic areas of occurrence, and summarize scenarios that could account for the origin and early radiation of Metazoa. But the solution to this mystery remains unknown.

ACKNOWLEDGMENTS. P. W. S. thanks S. M. Awramik, J. F. Mount, J. W. Valentine, and G. J. Vermeij for discussions of material presented here. Work by P. W. S. was supported by NSF EAR 88-04798 and work by J. H. L. was supported by NSF BSR 85-09301. This is Universtity of California Museum of Paleontology contribution 1554.

References

Anderson, M. M., and Conway Morris, S., 1982, A review, with descriptions of four unusual forms, of the soft-bodied fauna of the Conception and St. John's Groups (Late Precambrian), Avalon Peninsula, Newfoundland, in: *Proceedings. Third North American Paleontological Convention* (B. Mamet and M. J. Copeland, eds.), Vol. 1, pp. 1–8.

Anderson, M. M., and Misra, S. B., 1968, Fossils found in the Precambrian Conception Group in Southeastern Newfoundland, *Nature* **220:**680–681.

Arthur, W., 1982, A developmental approach to the problem of variation in evolutionary rates, *Biol. J. Linn. Soc.* **18:**243–261.

Arthur, W., 1988, *A Theory of the Evolution of Development,* Wiley, Chichester, England.

Awramik, S. M., 1971, Precambrian columnar stromatolite diversity: Reflection of metazoan appearance, *Science* **174:**825–827.

Axelrod, D. I., 1958, Early Cambrian marine fauna, *Science* **128:**7–9.

Bergstrom, J., 1990, Precambrian trace fossils and the rise of bilaterian animals, *Ichnos* **1:**3–13.

Berkner, L. V., and Marshall, L. C., 1964a, The history of oxygenic concentration in the earth's atmosphere, *Disc. Faraday Soc.* **37:**122–141.

Berkner, L. V., and Marshall, L. C., 1964b, The history of oxygenic concentration in the earth's atmosphere, in: *The Origin and Evolution of Atmospheres and Oceans* (C. J. Brancuzio and A. G. W. Cameron, eds.), Wiley, New York, pp. 102–126.

Berkner, L. V., and Marshall, L. C., 1965, History of major atmospheric components, *Proc. Natl. Acad. Sci. USA* **53:**1215–1225.

Berkner, L. V., and Marshall, L. C., 1967, The rise of oxygen in the earth's atmosphere with notes on the martian atmosphere, *Adv. Geophys.* **12:**309–331.

Bloesser, B., 1985, *Melanocryillium,* a new genus of structurally complex late Proterozoic microfossils from the Kwagunt Formation (Chuar Group), Grand Canyon, Arizona, *J. Paleontol.* **59:**741–765.

Boaden, P. J. S., 1989, Meiofauna and the origins of the Metazoa, *Zool. J. Linn. Soc.* **96:**217–227.

Bonner, J. T., 1988, *The Evolution of Complexity,* Princeton University Press, Princeton, New Jersey.

Brasier, M. D., 1979, The Cambrian radiation event, in: *The Origin of Major Invertebrate Groups* (M. R. House, ed.), Academic Press, New York, pp. 103–159.

Brasier, M. D., 1982, Sealevel changes, facies changes and the late Precambrian–Early Cambrian evolutionary explosion, *Precambrian Res.* **17:**105–123.

Brasier, M. D., 1986a, Precambrian–Cambrian boundary biotas and events, in: *Global Bio-Events* (O. Walliser, ed.), Springer-Verlag, Berlin, pp. 109–117.

Brasier, M. D., 1986b, Why do lower plants and animals biomineralize?, *Paleobiology* **12:**241–250.

Brasier, M. D., 1989, Toward a biostratigraphy of the earliest skeletal fossils, in: *The Precambrian–Cambrian Boundary* (J. W. Cowie and M. D. Brasier, eds.), Clarendon Press, Oxford, pp. 117–165.

Brooks, W. K., 1894, The origin of the oldest fossils and the discovery of the bottom of the ocean, *J. Geol.* **2:**455–479.

Buick, R., 1991, Microfossil recognition in Archean rocks: An appraisal of spheroids and filaments from a 3,500 m.y. chert-barite unit at North Pole, Australia, *Palaios* **5**:441–459.
Buss, L. W., 1987, *The Evolution of Individuality,* Princeton University Press, Princeton, New Jersey.
Byers, C. W., 1982, Geological significance of marine biogenic sedimentary structures, in: *Animal–Sediment Relationships* (P. L. McCall and M. J. S. Tevesz, eds.), Plenum Press, New York, pp. 221–256.
Chamberlain, W. M., and Marland, G., 1977, Precambrian evolution in a stratified global sea, *Nature* **265**:135–136.
Chen J.-Y., Hou W.-G., and Lu H.-Z., 1989, Early Cambrian netted scale-bearing worm-like sea animal, *Acta Palaeontol. Sinica* **28**:1–16 [in Chinese with English summary].
Cisne, J. L., 1974, Trilobites and the origin of arthropods, *Science***186**:13–18.
Clark, R. B., 1964, *Dynamics in Metazoan Evolution,* Clarendon Press, Oxford.
Clark, R. B., 1979, Radiation of the Metazoa, in: *The Origin of Major Invertebrate Groups* (M. R. House, ed.), Academic Press, New York, pp. 55–102.
Clemmey, H., and Badham, N., 1982, Oxygen in the Precambrian atmosphere: An evaluation of the geological evidence, *Geology* **10**:141–146.
Cloud, P. E., 1949, Some problems and patterns of evolution exemplified by fossil invertebrates, *Evolution* **2**:322–350.
Cloud, P. E., 1968, Pre-metazoan evolution and the origins of the Metazoa, in: *Evolution and Environment* (E. T. Drake, ed.), Yale University Press, New Haven, Connecticut, pp. 1–72.
Cloud, P. E., 1976, Beginnings of biospheric evolution and their biogeochemical consequences, *Paleobiology* **2**:351–387.
Cloud, P. E., and Glaessner, M. F., 1982, The Ediacarian Period and System: The Metazoa inherit the earth, *Science* **217**:783–792.
Conway Morris, S., 1985a, Non-skeletalized lower invertebrate fossils: A review, in: *The Origins and Relationships of Lower Invertebrates* (S. Conway Morris *et al.*, eds.), Clarendon Press, Oxford, pp. 343–359.
Conway Morris, S., 1985b, The Ediacaran biota and early metazoan evolution, *Geol. Mag.* **122**:77–81.
Conway Morris, S., 1988, Radiometric dating of the Precambrian–Cambrian boundary in the Avalon Zone, *N.Y. State Mus. Bull.* **463**:53–58.
Conway Morris, S., and Peel, J. S., 1990, Articulated halkieriids from the Lower Cambrian of North Greenland, *Nature* **345**:802–805.
Conway Morris, S., Peel, J. S., Higgins, A. K., Soper, N. J., and Davis, N. C., 1987, A Burgess Shale-like fauna from the Lower Cambrian of North Greenland, *Nature* **326**:181–183.
Cook, P. J., and Shergold, J. H., 1984, Phosphorus, phosphorites and skeletal evolution at the Precambrian–Cambrian boundary, *Nature* **308**:231–236.
Cowie, J. W., and Harland, W. B., 1989, Chronometry, in: *The Precambrian–Cambrian Boundary* (J. W. Cowie and M. D. Brasier, eds.), Clarendon Press, Oxford, pp. 186–198.
Crimes, T. P., 1974, Colonisation of the early ocean floor, *Nature* **248**:328–330.
Crimes, T. P., 1987, Trace fossils and the Precambrian–Cambrian boundary, *Geol. Mag.* **124**:97–119.
Crimes, T. P., 1989, Trace fossils, in: *The Precambrian–Cambrian Boundary* (J. W. Cowie and M. D. Brasier, eds.), Clarendon Press, Oxford, pp. 166–185.
Culver, S. J., 1991, Early Cambrian foraminifera from West Africa, *Science* **254**:689–691.
Daly, R. A., 1907, The limeless oceans of Pre-Cambrian time, *Am. J. Sci.* **23**:93–115.
Darwin, C., 1859, *On the Origin of Species,* John Murray, London.
Dimroth, E., and Kimberley, M. M., 1976, Precambrian atmospheric oxygen: Evidence in the sedimentary distributions of carbon, sulfur, uranium and iron, *Can. J. Earth Sci.* **13**:1161–1186.
Dimroth, E., and Lichtblau, A. P., 1978, Oxygen in the Archean ocean, comparison of ferric oxide crusts on Archean and Cenozoic pillow basalts, *Neues Jahrb. Mineral. Abh.* **133**:1–22.
Durham, J. W., 1978, The probable metazoan biota of the Precambrian as indicated by the subsequent record, *Annu. Rev. Earth Planet. Sci.* **6**:21–42.
Dzik, J., and Krumbiegel, G., 1989, The oldest "onychophoran" *Xenusion:* A link connecting phyla? *Lethaia* **22**:169–181.
Edhorn, A.-S., 1977, Early Cambrian algae croppers, *Can. J. Earth Sci.* **14**:1014–1020.
Embleton, B. J., and Williams, G. E., 1986, Low palaeolatitude of deposition for late Precambrian varvites in South Australia: Implications for palaeoclimatology, *Earth Planet. Sci. Lett.* **79**:419–430.
Erwin, D. H., 1989, Molecular clocks, molecular phylogenies and the origin of phyla, *Lethaia* **22**:251–257.

Erwin, D. H., and Valentine, J. W., 1984, "Hopeful monsters," transposons, and Metazoan radiation, *Proc. Natl. Acad. Sci. USA* **81**:5482–5483.

Evans, J. S., 1912, The sudden appearance of the Cambrian fauna, in: *International Geological Congress, 11th, Stockholm 1910. Compte Rendu*, Vol. 1, pp. 543–546.

Fedonkin, M. A., 1980, Early stages of evolution of Metazoa on the basis of paleontological data, *Zh. Obshch. Biol.* **41**:226–233 [in Russian].

Fedonkin, M. A., 1981, *White Sea Biota of the Vendian (Precambrian Non-Skeletal Fauna of the Russian Platform North)*, Nauka, Moscow [in Russian].

Fedonkin, M. A., 1985, Non-skeletal fauna of the Vendian: Promorphological analysis, in: *The Vendian System*, Vol. 1, *Paleontology*, (B. S. Sokolov and A. B. Ivanovskii, eds.), Nauka, Moscow, pp. 10–106, [in Russian].

Fedonkin, M. A., 1986, Precambrian problematic animals: Their body plan and phylogeny, in: *Problematic Fossil Taxa* (A. Hoffman and M. H. Nitecki, eds.), Oxford University Press, Oxford, pp. 59–67.

Fedonkin, M. A., and Runnegar, B. N., 1992, Proterozoic metazoan trace fossils, in *The Proterozoic Biosphere: A Multidisciplinary Study* (J. W. Schopf and C. Klein, eds.), Cambridge University Press, Cambridge, pp. 391–398.

Fischer, A. G., 1965, Fossils, early life and atmospheric history, *Proc. Natl. Acad. Sci. USA* **53**:1205–1213.

Fischer, A. G., 1984, Biological innovations and the sedimentary record, in: *Patterns of Change in Earth Evolution* (H. D. Holland and A. F. Trendall, eds.), Springer-Verlag, Berlin, pp. 145–157.

Ford, T. D., 1958, Precambrian fossils from Charnwood Forest, *Proc. Yorkshire Geol. Soc.* **3**:211–217.

Ford, T. D., 1963, The Pre-Cambrian fossils of Charnwood Forest, *Trans. Leicester Lit. Philos. Soc.* **57**:57–62.

Ford, T. D., 1979, Precambrian fossils and the origin of the Phanerozoic phyla, in: *The Origin of Major Invertebrate Groups* (M. R. House, ed.), Academic Press, New York, pp. 7–21.

Ford, T. D., 1981, The Edicaran fossils of Charnood Forest, Leicestershire, *Proc. Geol. Assoc.* **91**:81–83.

Fox, L. A. B., 1975, An adaptive model for the origin of invertebrate skeletons, M.S. thesis, University of California, Davis.

Garrett, P., 1970, Phanerozoic stromatolites: Non-competitive ecologic restriction by grazing and burrowing animals, *Science* **167**:171–173.

Germs, G. B. J., 1972, New shelly fossils from the Nama Group, South West Africa, *Am. J. Sci.* **272**:752–761.

Germs, G. B. J., 1973a, A reinterpretation of *Rangea schneiderhoehni* and the discovery of a related new fossil from the Nama Group, South West Africa, *Lethaia* **6**:1–10.

Germs, G. B. J., 1973b, Possible sprigginid worm and a new trace fossil from the Nama Group, South West Africa, *Geology* **1**:69–70.

Glaessner, M. F., 1958, New fossils from the base of the Cambrian in South Australia, *Trans. R. Soc. S. Australia* **81**:185–188.

Glaessner, M. F., 1966, Precambrian evolution, *Earth Sci. Rev.* **1**:29–50.

Glaessner, M. F., 1976, Early Phanerozoic annelid worms and the geological and biological significance, *J. Geol. Soc. Lond.* **132**:259–275.

Glaessner, M. F., 1978, The oldest foraminifera, *Aust. Bur. Miner. Resour. Geol. Geophys. Bull.* **192**:61–65.

Glaessner, M. F., 1984, *The Dawn of Animal Life*, Cambridge University Press, Cambridge.

Goldring, R., 1965, Sediments into rocks, *New Sci.* **26**:863–865.

Gooday, A. J., 1988, The genus *Bathysiphon* (Protista: Foraminiferida) in the north-east Atlantic: A neotype for *B. filiformis* G. O. & M. Sars, 1872, and the description of a new species, *J. Natl. Hist.* **22**:95–105.

Gould, S. J., 1977, Eternal metaphors of palaeontology, in: *Patterns of Evolution As Illustrated by the Fossil Record* (A. Hallam, ed.), Elsevier, Amsterdam, pp. 1–26.

Grant, S. W. F., 1990, Shell structure and distribution of *Cloudina*, a potential index fossil for the terminal Proterozoic, *Am. J. Sci.* **290A**:261–294.

Gürich, G., 1933, Die Kuibis-Fossilien der Nama-Formation von Sudwestafrika, *Palaeontol. Z.* **15**:137–154.

Hahn, G., and Pflug, H. D., 1985, Die Cloudinidae n. fam., Kalk-Rohren aus dem Vendium und Unter-Kambrium, *Senkenb. Lethaea* **65**:413–431.

Hambry, M. J., and Harland, W. B., 1985, The Late Proterozoic era, *Palaeogeogr. Palaeoclimatol. Palaeoecol.* **51:**255–272.
Hargraves, R. B., 1976, Precambrian geologic history, *Science* **193:**363–371.
Harland, W. B., 1964, Evidence of late Precambrian glaciation and its significance, in: *Problems in Palaeoclimatology* (A. E. M. Nairn, ed.), Interscience, New York, pp. 119–149.
Harland, W. B., 1989, Palaeoclimatology, in: *The Precambrian–Cambrian Boundary* (J. W. Cowie and M. D. Brasier, eds.), Clarendon Press, Oxford, pp. 199–204.
Hofmann, H. J., 1987, Precambrian biostratigraphy, *Geosci. Can.* **14:**135–154.
Holland, H. D., and Buekes, N. J., 1990, A paleoweathering profile from Griqualand West, South Africa: Evidence for a dramatic rise in atmosphere oxygen between 2.2 and 1.9 BYBP, *Am. J. Sci.* **290A:**1–34.
Holland, H. D., Feakes, C. R., and Zbinden, E. A., 1989, The Flin Flon paleosol and the composition of the atmosphere 1.8 bybp, *Am. J. Sci.* **289:**362–389.
Hutchinson, G. E., 1961, The biologist poses some problems, in: *Oceanography* (M. Sears, ed.), American Association for the Advancement of Science, Washington, D.C., pp. 85–94.
Jenkins, R. J. F., 1985, The enigmatic (late Precambrian) genus *Rangea* and related forms, *Paleobiology* **11:**336–355.
Kauffman, E. G., and Steidtmann, R., 1981, Are these the oldest metazoan trace fossils?, *J. Paleontol.* **55:**923–947.
Kazmierczak, J., Ittekkot, V., and Degens, E. T., 1985, Biocalcification through time: Environmental challenge and cellular response, *Palaeontol. Z.* **59:**15–33.
Kempe, S., and Degens, E. T., 1985, An early soda ocean?, *Chem. Geol.* **53:**95–108.
Knoll, A. H., 1978, Did emerging continents trigger metazoan evolution?, *Nature* **276:**701–703.
LaBarbara, M., 1978, Precambrian geological history and the origin of the Metazoa, *Nature* **273:**22–25.
Lipps, J. H., In press, Origin and early evolution of foraminifera, in: *Proceedings Benthos '90*, Tohoku University, Sendai, Japan.
Lyell, C., 1830, *Principles of Geology*, Vol. 1, John Murray, London [Facsimile ed., University of Chicago Press, Chicago (1990)].
McMenamin, M. A. S., 1985, Basal Cambrian small shell fossils from the La Cienega Formation, northwestern Sonora, Mexico, *J. Paleontol.* **59:**1414–1425.
McMenamin, M. A. S., 1986a, The garden of Ediacara, *Palaios* **1:**178–182.
McMenamin, M. A. S., 1986b, The emergence of animals, *Sci. Am.* **256:**94–102.
Misra, S. B., 1969, Precambrian (?) fossils from southeastern Newfoundland, *Geol. Soc. Am. Bull.* **80:**2133–2140.
Misra, S. B., 1971, Stratigraphy and depositional history of late Precambrian coelenterate-bearing rocks, southeastern Newfoundland, *Geol. Soc. Am. Bull.* **82:**979–987.
Narbonne, G. M., and Aitkin, J. D., 1990, Ediacaran fossils from the Sekwi Brook area, Mackenzie Mountains, Northwestern Canada, *Palaeontology* **33:**945–980.
Narbonne, G. M., and Hofmann, H. J., 1987, Ediacaran biota of the Werneke Mountains, Yukon, *Can. Palaeontol.* **30:**647–676.
Narbonne, G. M., Myrow, P. M., Landing, E., and Anderson, M. M., 1987, A candidate stratotype for the Precambrian–Cambrian boundary, Fortune Head, Burin Peninsula, southeastern Newfoundland, *Can. J. Earth Sci.* **24:**1277–1293.
Nazarov, B. B., 1975, Lower and Middle Paleozoic radiolarians of Kazakhstan, *Trudy Inst. Geol. Geofiz. Sibirsk. Otd. Akad. Nauk SSSR* **275:**1–203.
Nicol, D., 1966, Cope's rule and Precambrian and Cambrian invertebrates, *J. Paleontol.* **40:**1397–1399.
Nursall, J. R., 1959, Oxygen as a prerequisite to the origin of Metazoa, *Nature* **183:**1170–1172.
Nursall, J. R., 1962, On the origins of the major groups of animals, *Evolution* **16:**118–123.
Palij, V. M., Posti, E., and Fedonkin, M. A., 1983, Soft-bodied Metazoa in the Vendian and Lower Cambrian, in: *Upper Precambrian and Cambrian Paleontology of the East-European Platform* (A. Urbanek and A. Rozanov, eds.), Wydawnictwa Geologiczne, Warsaw, pp. 56–95.
Pflug, H. D., 1970a, Zur Fauna der Nama-Schichten in Sudwest-Afrika. I. Pteridinia. Bau und systematische Zugehorigkeit, *Palaeontogr. Abt. A* **134:**153–262.
Pflug, H. D., 1970b, Zur Fauna der Nama-Schichten in Sudwest-Africa. II. Rangeidae. Bau und systematische Zugehorigkeit, *Palaeontogr. Abt. A* **135:**198–231.
Raff, R. A., and Raff, E. C., 1970, Respiratory mechanisms and the metazoan fossil record, *Nature* **228:**1003–1005.

Raymond, P. E., 1935, Pre-Cambrian life, Bull. Geol. Soc. Am. **46:**375–392.

Rhoads, D. C., and Morse, J. W., 1971, Evolutionary and ecologic significance of oxygen-deficient marine basins, *Lethaia* **4:**413–428.

Riding, R., 1982, Cyanophyte calcification and changes in ocean chemistry, *Nature* **299:**814–815.

Rudwick, M. J. S., 1964, The infra-Cambrian glaciation and the origin of the Cambrian fauna, in: *Problems in Paleoclimatology* (A. E. M. Nairn, ed.), Interscience, New York, pp. 150–155.

Runnegar, B., 1982a, A molecular-clock date for the origin of the animal phyla, *Lethaia* **15:**199–205.

Runnegar, B., 1982b, The Cambrian explosion: Animals or fossils?, *J. Geol. Soc. Aust.* **29:**395–411.

Runnegar, B., 1982c, Oxygen requirements, biology and phylogenetic significance of the Late Precambrian worm *Dickinsonia*, and the evolution of the burrowing habit, *Alcheringa* **6:**223–239.

Runnegar, B., 1986, Molecular paleontology, *Palaeontology* **29:**1–24.

Sadler, P. M., 1981, Sediment accumulation rates and the completeness of stratigraphic sections, *J. Geol.* **89:**569–584.

Schopf, J. W., Hayes, J. M., and Walter, M. R., 1984, Evolution of Earth's earliest ecosystems: Recent progress and unsolved problems, in: *Earth's Earliest Biosphere* (J. W. Schopf, ed.), Princeton University Press, Princeton, New Jersey, pp. 361–384.

Seilacher, A., 1956, Der Beginn des kambriums als biologishe Wende, *Neues Jahrb. Geol. Palaeontol. Abh.* **103:**155–180.

Seilacher, A., 1984, Late Precambrian and Early Cambrian Metazoa: Preservational or real extinctions?, in: *Patterns of Change in Earth Evolution* (H. D. Holland and A. F. Trendall, eds.), Springer-Verlag, Berlin, pp. 159–168.

Seilacher, A., 1985, Discussion of Precambrian metazoans, *Philos. Trans. R. Soc. Lond. B* **311:**47–48.

Seilacher, A., 1989, Vendozoa: Organismic construction in the Proterozoic biosphere, *Lethaia* **22:**229–239.

Sepkoski, J. J., Jr., 1978, A kinetic model of Phanerozoic taxonomic diversity. I. Analysis of marine orders, *Paleobiology* **4:**223–251.

Sepkoski, J. J., Jr., 1979, A kinetic model of Phanerozoic taxonomic diversity. II. Early Phanerozoic families and multiple equilibria, *Paleobiology* **5:**222–251.

Sepkoski, J. J., Jr., 1984, A kinetic model of Phanerozoic taxonomic diversity. III. Post-Paleozoic families and mass extinctions, *Paleobiology* **10:**246–267.

Severin, K. P., Culver, S. J., and Blanpied, C., 1982, Burrows and trails produced by *Quinqueloculina impressa* Reuss, a benthic foraminifer, in fine-grained sediment, *Sedimentology* **29:**897–901.

Signor, P. W., 1985, Real and apparent trends in species richness through time, in: *Phanerozoic Diversity Patterns: Profiles in Macroevolution* (J. W. Valentine, ed.), Princeton University Press, Princeton, New Jersey.

Signor, P. W., 1988, The Precambrian–Cambrian metazoan radiation: Significance of earliest Cambrian agglutinated skeletons, *Geol. Soc. Am. Abstr. Prog.* **20:**A104.

Signor, P. W., 1990, The geologic history of diversity, *Annu. Rev. Ecol. Syst.* **21:**509–539.

Signor, P. W., 1991, Early Cambrian biogeography and the prehistory of early skeletogenous animals, in: *Paleozoic Paleogeography of the Western United States* (J. D. Cooper and C. A. Stevens, eds.), Pacific Section of the Society of Economic Paleontologists and Mineralogists, Los Angeles, California, pp. 801–810.

Signor, P. W., and McMenamin, M. A. S., 1988, The Early Cambrian worm tube *Onuphionella* from California and Nevada, *J. Paleontol.* **62:**233–240.

Sprigg, R. C., 1947, Early Cambrian (?) jellyfishes from the Flinders Ranges, South Australia, *Trans. R. Soc. S. Australia* **71:**212–224.

Stanley, S. M., 1973, An ecological theory for the sudden origin of multicellular life in the late Precambrian, *Proc. Natl. Acad. Sci. USA* **70:**1486–1489.

Stanley, S. M., 1976a, Fossil data and the Precambrian–Cambrian evolutionary transition, *Am. J. Sci.* **276:**56–76.

Stanley, S. M., 1976b, Ideas on the timing of metazoan diversification, *Paleobiology* **2:**209–219.

Stasek, C. R., 1972, The molluscan framework, *Chem. Zool.* **12:**1–44.

Summons, R. E., and Walter, M. R., 1990, Molecular fossils and microfossils of prokaryotes and protists from Proterozoic sediments, *Am. J. Sci.* **290A:**212–244.

Thomson, K. S., 1989, *Morphogenesis and Evolution*, Oxford University Press, Oxford.

Towe, K. M., 1970, Oxygen-collagen priority and the early metazoan fossil record, *Proc. Natl. Acad. Sci. USA* **65:**781–788.

Towe, K. M., 1981, Biochemical keys to the emergence of complex life, in: *Life in the Universe* (J. Billingham, ed.), Massachusetts Institute of Technology Press, Cambridge, Massachusetts, pp. 297–306.

Tucker, M., 1982, Precambrian dolomites: Petrographic and isotopic evidence that they differ from Phanerozoic dolomites, *Geology* **10:**7–12.

Valentine, J. W., 1989, Bilaterians of the Precambrian–Cambrian transition and the annelid–arthropod relationship, *Proc. Natl. Acad. Sci. USA* **86:**2272–2275.

Valentine, J. W., and Campbell, C. A., 1975, Genetic regulation and the fossil record, *Am. Sci.* **63:**673–680.

Valentine, J. W., and Erwin, D. H., 1987, Interpreting great developmental experiments: The fossil record, in: *Development as an Evolutionary Process,* Liss, New York, pp. 71–107.

Valentine, J. W., and Moores, E. M., 1972, Global tectonics and the fossil record, *J. Geol.* **80:**167–184.

Valentine, J. W., Awramik, S. M., Signor, P. W., and Sadler, P. M., 1990, The biological explosion at the Precambrian–Cambrian boundary, *Evol. Biol.* **25:**279–356.

Vermeij, G. J., 1989, The origin of skeletons, *Palaios* **4:**585–589.

Vidal, G., and Knoll, A. H., 1983, Proterozoic plankton, *Geol. Soc. Am. Mem.* **161:**265–277.

Walcott, C. D., 1910, Cambrian geology and paleontology II. No. 1. Abrupt appearance of the Cambrian fauna of the North American Continent, *Smithson. Inst. Misc. Collect.* **57:**1–16.

Walker. T. D., 1985, Diversification functions and the rate of taxonomic evolution, in: *Phanerozoic Diversity Patterns: Profiles in Macroevolution* (J. W. Valentine, ed.), Princeton University Press, Princeton, New Jersey, pp. 311–334.

Walter, M. R., 1987, The timing of major evolutionary innovations from the origins of life to the origins of Metaphyta and Metazoa: The geological evidence, in: *Rates of Evolution* (K. S. W. Campbell and M. F. Day, eds.), Allen and Unwin, London, pp. 15–38.

Walter, M. R., Rulin, D., and Horodyski, R. J., 1990, Coiled carbonaceous megafossils from the Middle Proterozoic of Jixian (Tianjin) and Montana, *Am. J. Sci.* **290A:**133–148.

Wang Y., Xing Y., Lin W., Zhang L., Lu Z., Gao Z., Ma G., and Lu S., 1980, Subdivision and correlation of the Upper Precambrian in China, in: *Research on Precambrian Geology, Sinian Suberathem in China,* Tianjin Science and Technology Press, Tianjin, China [in Chinese].

Wilkinson, B. H., 1980, Biomineralization, palaeoceanography, and evolution of calcareous marine organisms, *Geology* **8:**265–267.

Yochelson, E. L., and Herrera, H. E., 1974, Un fossil enigmatico del cambrico inferior de Argentina, *Ameghiniana* **11:**283–294.

Zhang W.-T., 1985, Preliminary notes on the occurrence of the unusual trilobite *Naraoia* in Asia, *Acta Palaeontol. Sinica* **24:**591–595.

II

The Late Proterozoic

Chapter 2

Faunas and Facies—Fact and Artifact

Paleoenvironmental Controls on the Distribution of Early Cambrian Faunas

JEFFREY F. MOUNT and PHILIP W. SIGNOR

1. Introduction

The fossil record of the Proterozoic–Phanerozoic transition appears to reflect the relatively sudden, global appearance and diversification of complex metazoan life. Despite assumptions and inferences of a long period of metazoan evolution prior to their preservation in the rock record, virtually all new fossil discoveries and stratigraphic sections only appear to reinforce the relative abruptness of this radiation event (Cloud, 1949; Valentine, 1973; Stanley, 1976; but see Durham, 1978; Runnegar, 1982a, b).

The burst of adaptive innovation that took place at the close of the Proterozoic

JEFFREY F. MOUNT • Department of Geology, University of California, Davis, California 95616; and University of California White Mountain Research Station, Bishop, California 93514. PHILIP W. SIGNOR • Department of Geology, University of California, Davis, California 95616.

Origin and Early Evolution of the Metazoa, edited by Jere H. Lipps and Philip W. Signor. Plenum Press, New York, 1992.

occurred during a period of fundamental change in global climate, continental configuration, ocean chemistry, and sea level (Brasier, this volume, Chapter 15; Knoll, this volume, Chapter 3). Given this association, it is tempting to conclude that extrinsic processes controlled and perhaps initiated the radiation of Metazoa (Signor and Lipps, this volume, Chapter 1). Despite ample and ingenious models that purport to explain this association, relatively few workers have attempted to evaluate the role that these extrinsic processes had on the fidelity of the record itself [notable exceptions include Brasier (1982) and Theokritoff (1979)]. Yet, in order to eventually unravel the causes of this event, it is imperative that artifact be separated from fact within its record.

In this chapter we evaluate the role that extrinsic processes, such as eustatic sea level changes, may have played in controlling the apparent record of the metazoan radiation event. Data for this analysis come primarily from the Lower Cambrian of the southwestern Great Basin, United States. The results of this work demonstrate that patterns of species first appearances are strongly influenced by the regional preservation of suitable habitats: an artifactual bias that exists in virtually all Lower Cambrian sections worldwide.

Although this chapter focuses largely on detecting artifact within the record, the data and analyses presented here may offer some insight into the paleoenvironmental setting of the Early Cambrian radiation event. The results of this study show that, as for most Paleozoic communities (Sepkoski and Miller, 1985) and clades (Bottjer and Jablonski, 1988), Early Cambrian evolutionary innovations appear to have taken place in shallow subtidal, inner shelf environments. However, we do not recognize any prediliction for onshore–offshore shifts in innovations or communities during the Early Cambrian (Jablonski *et al.*, 1983), nor are we able to detect any correlation between changes in habitat area and loci of origination.

2. Precambrian–Cambrian Transition in the Southwestern Great Basin

The focus of this study is the Lower Cambrian of the White-Inyo Region of eastern California and western Nevada (Figs. 1 and 2). Originally described by Walcott (1908, 1912), this section is one of the thickest, best-exposed, and most-studied of its kind in the world. The White-Inyo Region strata form part of a belt of uppermost Proterozoic and Cambrian units that crop out extensively throughout the Cordillera of western North America. These units, which form the base of the Cordilleran miogeocline, reflect the development of a rifted margin in the latest Proterozoic or earliest Cambrian (Stewart and Suczeck, 1977; Bond *et al.*, 1985, 1988). Limited paleomagnetic pole data for North America indicate that the western margin lay in an east–west orientation in a low-latitude tropical setting dominated by equatorial easterlies (Scotese *et al.*, 1979; Parrish *et al.*, 1986).

Like most Cambrian strata of the western Cordillera, the White-Inyo Region strata form large-scale alternations of siliciclastic-dominated and carbonate-dominated stratigraphic packages (Fig. 2b). These alternations, or “Grand Cycles”

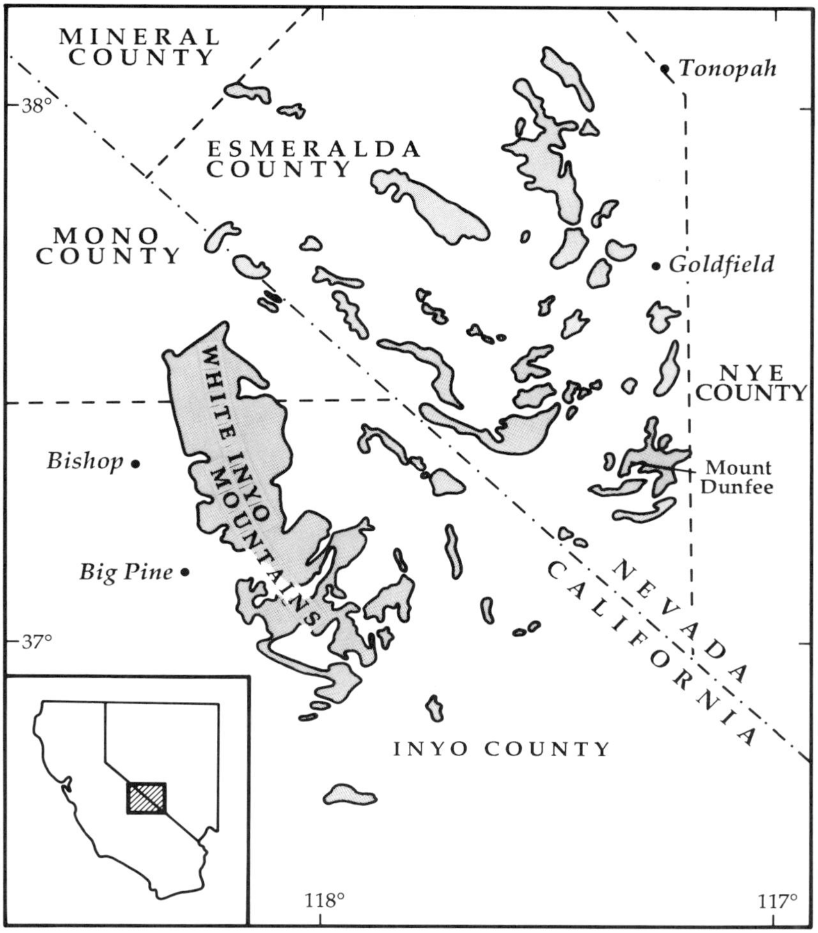

Figure 1. Outcrops of Lower Cambrian strata in the White-Inyo Region of eastern California and western Nevada. [Modified from Mount *et al.* (1983).]

(after Aitken, 1966, 1978), are unconformity-bounded, regionally-correlative depositional sequences that typically comprise one to two trilobite zones. Depending upon geographic position on the shelf, the cycles range from as little as 100 m to more than 1000 m in thickness. Each cycle consists of a lower siliciclastic-dominated or "shaley" half-cycle that is usually gradationally overlain by a carbonate-dominated half-cycle.

While variable in detail, the depositional history of each Grand Cycle appears to record three fundamental events: (1) initiation of the Grand Cycle through abrupt deepening and drowning of an emergent carbonate shelf coupled with the widespread deposition of siliciclastic material in dominantly subtidal settings; (2) shoaling of the shelf and development of a time-transgressive shift from siliciclastic- to carbonate-dominated deposition; and (3) landward expansion of

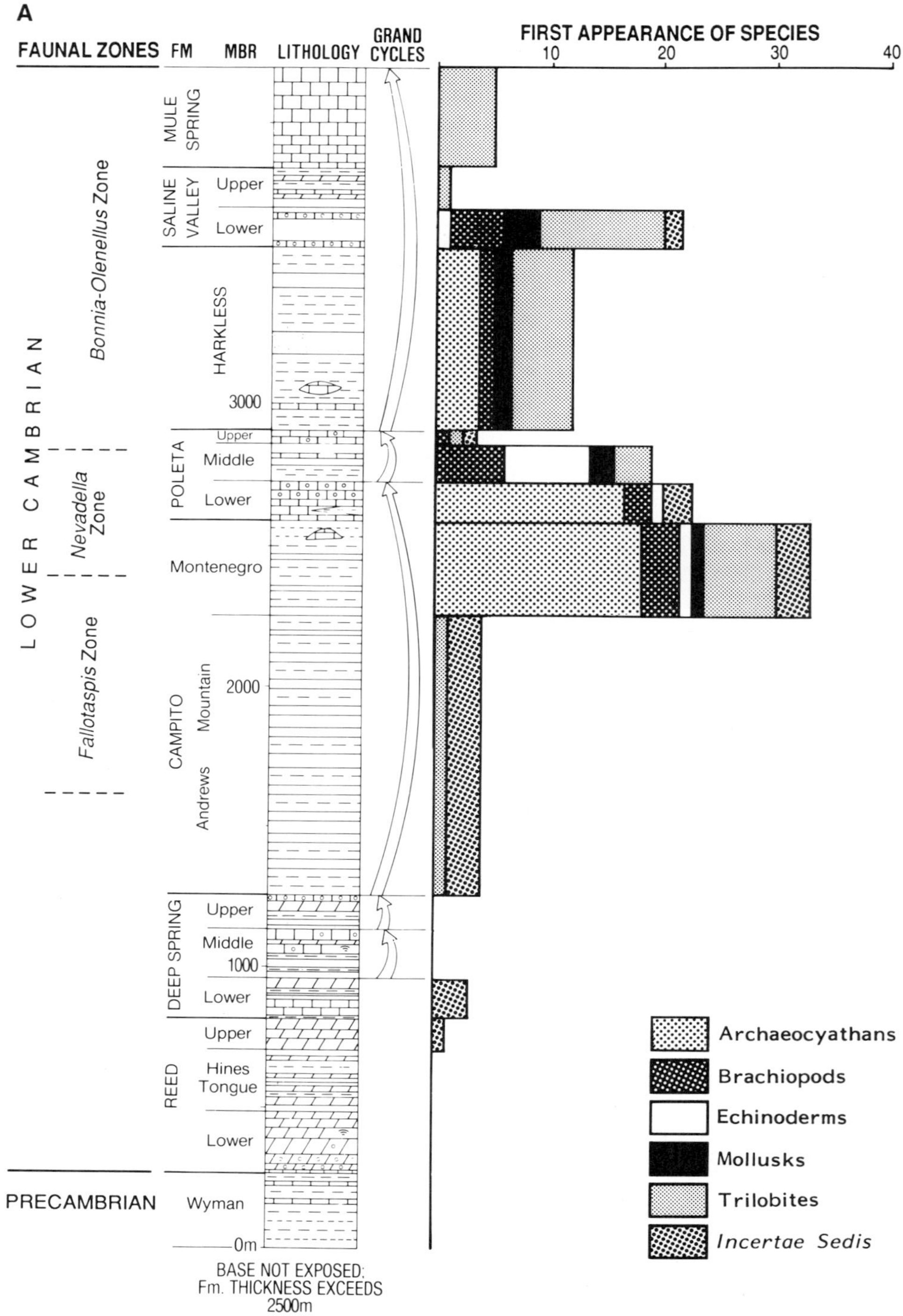

Figure 2. Lower Cambrian lithostratigraphy and first appearances of major taxa. (A) First occurrences of body fossil species. (B) First appearances of ichnospecies. Stratigraphy based on the work of Nelson (1962) with minor modifications by Stewart (1970). Trilobite zones for western North America were established by Fritz (1972). Biostratigraphy of the pretrilobite portion of the section and the placement of the Precambrian–Cambrian boundary are summarized in Signor and Mount (1986a, b). Faunal data

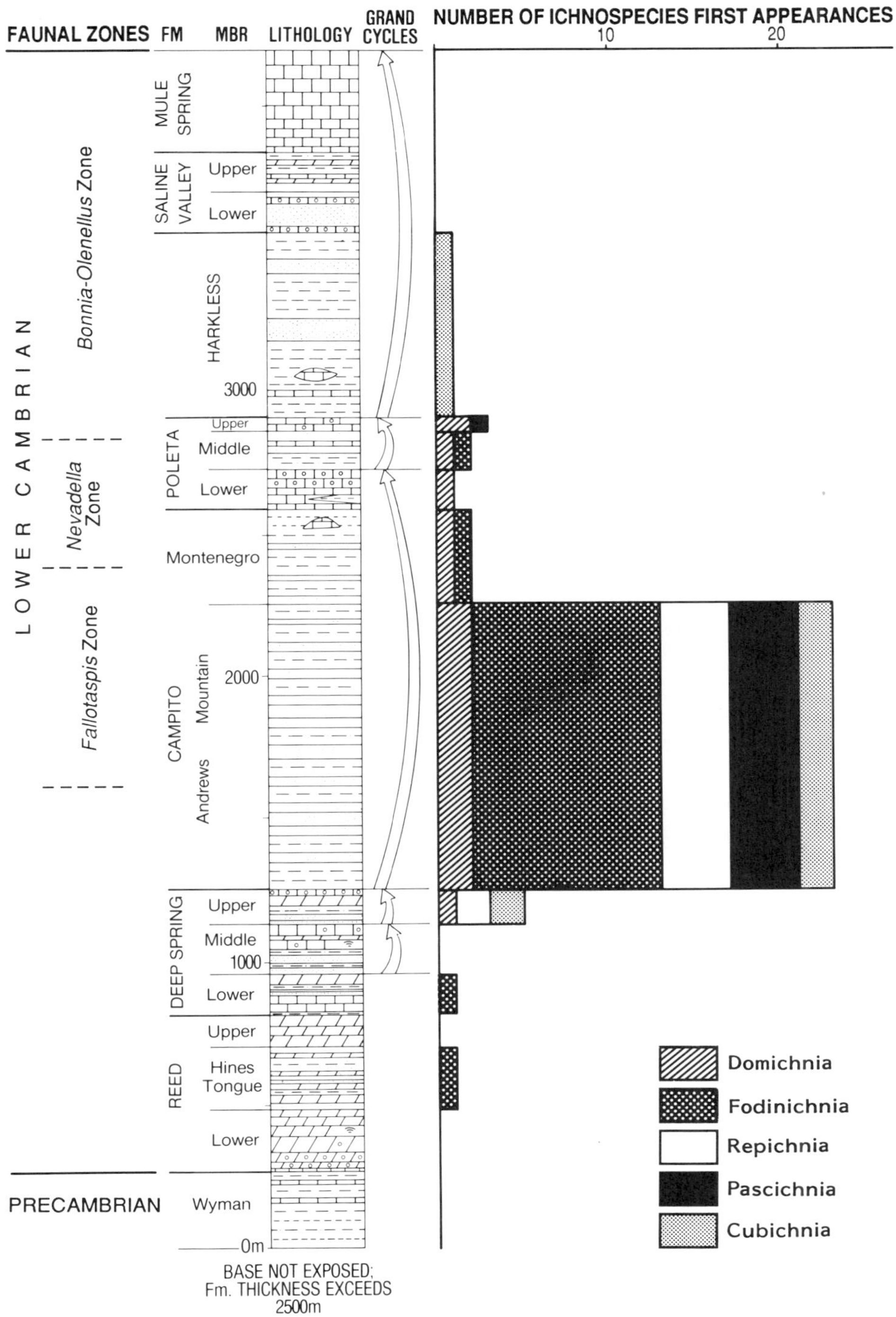

are summarized in Signor and Mount (1986a, b) and Onken and Signor (1988). Large arrows indicate shoaling-upward Grand Cycles. The three Grand Cycles that make up the Campito, Poleta, Harkless, Saline Valley, and Mule Spring Formations were originally described by Fritz (1975). The pretrilobite Grand Cycles have been described by Mount and Signor (1989a, b). See text for discussion.

peritidal carbonate deposition culminating with the development of regional emergence.

The mechanics of Grand Cycle deposition remain in dispute, although most workers accept that it reflects eustatic changes in sea level rather than cyclic tectonic or climatic changes (Aitken, 1978; Read *et al.*, 1986; Chow and James, 1987; Bond *et al.*, 1988). Their widespread preservation in North America appears to be associated with a relatively quiescent tectonic setting and the general persistence of eustatic sea level rise. Based on a study of Lower Cambrian Grand Cycles in the White-Inyo Region, Mount and Rowland (1981), Mount (1982a, b), and Mount, Hunt, Green, and Dienger (1991) have noted that the asymmetric nature of Grand Cycles reflects episodic changes in the rate of eustatic sea level rise during what Matthews and Cowie (1977) have termed the "Great Cambrian Transgression."

At the onset of Grand Cycle deposition, drowning of the emergent carbonate platform and widespread deposition of siliciclastic muds in subtidal environments occurred in response to an increase in the rate of sea level rise. Apparently, drowning of the platform and development of an open, storm-influenced shelf allowed reworking of siliciclastic sediments previously trapped in nearshore and marginal marine depocenters. Nucleation of a carbonate bank and formation of the gradational transition between the siliciclastic and carbonate half-cycle was associated with a sea level highstand and a decline in the rate of sea level rise. Landward expansion of the carbonate bank and shoaling of the shelf resulted in the widespread deposition of peritidal carbonates that form the bulk of the carbonate half-cycle. Termination of the carbonate half-cycle through emergence of the peritidal carbonate platform appears to have occurred during sea level stillstand or fall.

Although not unique to North America, Grand Cycles are not as well documented on other continents. Sections that contain Grand-Cycle-like successions are found in northern Iran (Brasier, 1989a), southern Kazakhstan (Missarzhevsky and Mambetov, 1981), and Mongolia (Voronin *et al.*, 1982), but at present "classic" Grand Cycle features have not been recorded in critical, often-cited localities such as the Siberian Platform, the North China and Yangtze Platforms, southeast Newfoundland, and the East European Platform. The lack of Grand Cycles is not surprising. Their formation appears to have been based on shifts in the balance between siliciclastic influx and carbonate production [see Mount (1984) for discussion] and on the presence of steady, regional thermotectonic subsidence of the North American passive margins. The eustatic sea level signature recorded in Grand Cycles is not likely to have been preserved in settings that were disrupted by tectonism, lay at relatively high-latitude settings, received no siliciclastic input, or had sedimentation rates that were too low or too high.

Regardless of the mechanism responsible for the development of Grand Cycles, they record episodic shifts in shelf depositional conditions in response to Cambrian sea level fluctuations. These shifts, which involved systematic changes in depth and sediment type, appear to have synchronously affected the entire Cordilleran margin and led to the accumulation of a predictable succession of lithofacies. The temporal and spatial distribution of these lithofacies may, in turn, have strongly influenced the apparent pattern of the metazoan radiation event.

2.1. Patterns of Diversification and Paleoenvironments

Lower Cambrian stratigraphic paleontology of the southwestern Great Basin has been extensively studied [see reviews in Signor and Mount (1986a, 1989) and Onken and Signor (1988)] (data sources for this study are listed in the Appendix). The record of the metazoan radiation event is grossly similar to records seen worldwide. Four sequential steps have been recognized: appearance of simple, horizontal traces in latest Proterozoic (Vendian) age strata (but see Lipps and Fedonkin 1988); appearance of pretrilobite, Tommotian-aspect small shelly fossils in units proximal to the Precambrian–Cambrian boundary (Signor *et al.*, 1987); appearance and diversification of complex traces; and appearance and diversification of trilobites and other complex shelly fossils.

As shown in Fig. 2, each recorded step of the metazoan radiation event appears to involve dramatic increases, followed by sharp declines, in number of first appearances. The fauna of each interval is as follows.

2.1.1. Pretrilobite Diversification

An impoverished shelly fauna occurs locally within the upper Reed Formation and the Lower Member of the Deep Spring Formation in both the White-Inyo Mountains and in Esmeralda County, Nevada (Mount *et al.*, 1983; Signor *et al.*, 1983, 1987). This fauna consists of calcitic and partially to wholly dolomitized tubes and cones. The first shelly fossil, *Wyattia* (Taylor, 1966), occurs in dolomites of the upper Reed Formation and in limestones of the lowermost Deep Spring Formation (Cloud and Nelson, 1966). The restricted geographic distribution and poor preservation of this taxon has minimized its biostratigraphic usefulness. The remainder of the fauna consists of *Nevadatubulus*, *Sinotubulites*, *Coleoloides*, the hyolith *Salanytheca*, and an undescribed small conical fossil (Signor *et al.*, 1987). Although the fauna has been collected throughout the local area its stratigraphic occurrence is limited to wackestones and packstones of the Lower Member Deep Spring Formation (Gevirtzman and Mount, 1986); no fossils were recovered elsewhere in the Deep Spring Formation despite extensive prospecting. With the exception of *Platysolenites* and *Onuphionella* (Signor and Mount, 1986a), there is a conspicuous absence of body fossils in the 1000-m-thick stratigraphic interval between the last appearance of pretrilobite fossils and the first appearance of the olenellid trilobite *Fallotaspis*.

2.1.2. *Nevadella* Zone Diversification

Following the first appearance of olenellid trilobites (*Fallotaspis*) in the Andrews Mountain Member of the Campito Formation, there is a dramatic increase in the number of first occurrences. This second local diversification in the uppermost *Fallotaspis* and *Nevadella* Zones accounts for the initial occurrence of over 60% of the total number of species reported from the Lower Cambrian of this region. When examined in detail, the diversification does not appear to be a synchronous event. Rather, it involves the stepwise and overlapping appearance of trilobites, followed by brachiopods and archaeocyathans, followed by echinoderms. A dramatic 90% decline in the number of first appearances occurs at the transition between the *Nevadella* Zone and the overlying *Bonnia–Olenellus* Zone. This decline appears to affect all groups, with the possible exception of brachiopods.

2.1.3. *Bonnia–Olenellus* Zone Diversification

In contrast to the first and second diversifications, the third diversification involves an increase in first appearances within established taxonomic groups. The radiation of olenellid trilobites in this interval accounts for the bulk of the diversification. Archaeocyathans show a consistent decline, while brachiopods reach the acme of their regional Lower Cambrian diversity. In the uppermost portions of the *Bonnia–Olenellus* Zone, but below the Lower–Middle Cambrian boundary, a depauperate fauna dominated by trilobites and mollusks occurs. By this time, the number of first appearances had declined by 95% from the peak of the *Bonnia–Olenellus* Zone diversification.

2.1.4. Trace Fossil Record

The appearance of diverse types of trace fossils appears to have occurred mostly during one brief episode (Fig. 2b). Over 60% of the total number of the first occurrences of Lower Cambrian ichnospecies known from the region are from the Andrews Mountain Member of the Campito Formation. Although this diversification is dominated by fodinichnid traces, it appears to have included a synchronous, but less dramatic diversification in all other ethological groups as well. Ichnospecies richness remains largely unchanged through most of the remainder of the Lower Cambrian in the Great Basin. In the middle to upper *Bonnia–Olenellus* Zone the number of reported ichnospecies decreases dramatically and reaches zero in the Mule Spring Formation.

The record of first appearances of trace fossils contrasts sharply with the record of body fossils. The first appearance of traces in the White-Inyo region occurs in the Proterozoic Wyman Formation (Alpert, 1977), well below the first appearance of shelly fossils. In addition, the peak in appearances of new traces is offset from the peaks in body fossil appearances. Finally, the pronounced decline in ichnospecies occurs primarily in the Lower Member of the Saline Valley Formation where there is a peak in the number of body fossil occurrences.

2.1.5. Bias within the Data Set

A variety of sampling biases might have influenced the patterns of diversification shown in Fig. 2. One concern is the possibility of unequal effort in collection and description of the various taxonomic groups. For example, Gangloff's (e.g., Gangloff, 1975, 1976; McKee and Gangloff, 1969) extensive taxonomic studies of archaeocyathans are the principal literature source of the dramatic rise in species richness and first appearances in the *Nevadella* Zone. In contrast, there are relatively few studies of the taxonomy of mollusks and echinoderms in the region. We are unable to determine with certainty if the low species richness of these groups reflects a depauperate fauna or a simple lack of attention. Our own field studies, while not conclusive, suggest that the general pattern of species richness illustrated here is not solely a product of monographic biases.

An additional bias may result from the way that the data are calculated. The number of first appearances for a stratigraphic interval might be affected by the thickness of the unit. The thicker the stratigraphic unit, the greater the chance of recovering species. The data presented in Fig. 2 are the numbers of first occurrences per stratigraphic unit (either a member or a formation). The unequal distribution of sampling intervals reflects the way that earlier workers defined the members and formations. In order to evaluate the effect of unequal sampling intervals, the species richness and number of first occurrences within a unit can be compared against its thickness. The resulting product-moment correlation ($r = 0.05$; $n = 15$) does not support a significant relationship between these variables. Unit thickness apparently does not exert any significant influence on the pattern of fossil occurrences.

In sum, although bias may result from inconsistent efforts at collection, taxonomic study, or unequal sampling units the general pattern of first and subsequent occurrences outlined above is a reasonable assessment of the Lower Cambrian fossil record in the southwestern Great Basin.

2.2. Fossil/Lithofacies Associations

As noted previously, the stratigraphic distribution of lithofacies may exert considerable influence on the character and completeness of the record of the metazoan radiation event. In order to evaluate the influence of facies controls on the Lower Cambrian fossil record, we have surveyed the range of Lower Cambrian lithofacies in rocks that crop out in the southwestern Great Basin. We recognize six principal depositional systems, which could be further subdivided into ten different facies (Table I). Recognition of these facies is based on standard techniques of lithofacies analysis, including lithofacies associations, bedding characteristics, and sedimentary structures [for examples, see Mount (1982a b), Mount and Signor (1985), and Gevirtzman and Mount (1986)]. In most cases we relied on published stratigraphic analyses that were verified and supplemented through our own unpublished field studies (Table II). We found considerable variation in the terminology and interpretations of the authors cited. Thus, the facies presented in Table I are standardized and may not match precisely with the published sources.

We have attempted to establish the lithofacies where each species of body and trace fossil make their first and subsequent appearances. We have used the following approach with varying degrees of success. Published descriptions (listed in the Appendix) were reviewed to determine, as precisely as possible, the locality, the stratigraphic horizon, and the lithofacies where each taxon was collected. Published reports with incomplete (or insufficient or scant) lithologic information were supplemented by our own stratigraphic studies. Some taxa that lacked sufficient lithologic data were not observed in our field studies. In those cases, we estimated their lithologic association by combining their locality data with published and unpublished stratigraphic descriptions.

From this information we developed a fossil/lithofacies assignment to the

TABLE I. Lower Cambrian Lithofacies of the Southwestern Great Basin[a]

	Siliciclastic-dominated and mixed siliciclastic–carbonate
1.1a.	Muddy tidal flat and sand-dominated tidal channel
1.1b.	Sand-dominated low intertidal, shoreface, foreshore
1.2a.	Sand-dominated, heterolithic, storm-influenced subtidal[b]
1.2b.	Mud-dominated, deep shelf[b]
1.3a.	Archaeocyathan/algal bioherms on muddy siliciclastic substrate
	Carbonate-dominated
2.1a.	Allochem-dominated peritidal shoal carbonates
2.1b.	Heterolithic intertidal and supratidal (evaporitic)
2.2a.	Sand-dominated, heterolithic, shallow subtidal (above storm wave base; includes normal marine, subtidal back-bank lagoon)
2.2b.	Mud-dominated deep subtidal (below storm wave base)
2.3a.	Archaeocythan/algal bioherms on lime mud substrates

[a]Paleoenvirommental determinations for lithofacies are based on standard techniques of lithofacies analysis, including lithofacies associations, bedding characteristics, and sedimentary structures. For examples see Mount (1982a, b), Mount and Signor (1985), and Gevirtzman and Mount (1986).
[b]Includes lithofacies indicative of siliciclastic-dominated deposition in subtidal, back-bank lagoon environments or intrashelf basin.

appropriate depositional system, and to a specific facies where possible, for the first appearance of each species. Our results, summarized in Fig. 3, indicate that the overwhelming majority of the Lower Cambrian taxa of the southwestern Great Basin are collected from a limited range of lithofacies. Over two-thirds of the first appearances in the White-Inyo region occur in environments below fairweather wave base, but usually above storm wave base. This includes bioherms established on siliclastic and carbonate mud-dominated substrates. The remaining one third of the first appearances are scattered between a range of environments from tidal flat to sandy subtidal depositional conditions. Only low intertidal siliciclastic-dominated environments lacked first appearances.

Although the concentration of first appearances in lithofacies representing shallow subtidal conditions is general, there is considerable variation between different taxonomic groups (Fig. 3).

The overwhelming majority of archaeocyathan first appearances are reported from bioherms that were established in muddy subtidal siliciclastic and carbonate environments. Despite the abundance of archaeocyathan/algal reefs within the Lower Cambrian of the White-Inyo Region (Rowland, 1984), surprisingly few archaeocyathan first occurrences are reported from reef or shoal environments. The erect, branching form *Vertocyathus* appears to be the only well-documented archaeocyathan that makes its initial appearance in reef facies.

The first occurrence of species of brachiopod, trilobite, and mollusk are concentrated in siliciclastic-dominated and mixed siliciclastic–carbonate, shallow subtidal, open shelf environments. No first occurrences are reported from bioherms and relatively few are reported from intertidal deposits. First occurrences of echinoderm species are reported only from muddy subtidal siliciclastic units. However, echinoderm plates have been reported from a wide range of carbonate and siliciclastic environments from as low as the base of the *Nevadella* Zone (Montenegro Member Campito Formation).

TABLE II. Estimated Volume of Lithofacies Preserved in Lower Cambrian Section of the Southwestern Great Basin[a]

Formation/ Member	Siliciclastics						Carbonates				Data sources[b]
	Tidal flat (1.1a)	Low intertidal (1.1b)	Sandy subtidal (1.2a)	Muddy subtidal (1.2b)	Bioherm (1.3a)	Shoal (2.1a)	Tidal flat (2.1b)	Sandy subtidal (2.2a)	Muddy subtidal (2.2b)	Bioherm (2.3a)	
Mule Spring	—	—	—	—	—	4.0	1.9	2.0	1.0	—	A,B,H,S,T
Saline Valley	1.1	0.8	1.3	1.1	—	1.1	1.9	0.5	0.8	—	A,B,H,S,T
Harkless	2.9	7.3	2.2	1.8	0.3	—	—	—	—	—	A,B,H,S,T
Upper Poleta	—	—	—	—	—	0.2	0.2	0.6	0.3	0.1	A,H,I,S,T
Middle Poleta	0.6	0.6	0.6	1.0	0.1	0.2	0.3	0.2	0.2	—	A,H,I,S,T
Lower Poleta	—	—	—	0.2	—	0.9	—	0.7	0.9	1.4	A,H,I,K,L,M,P,Q
Montengro Member	0.5	2.0	1.0	5.2	0.5	—	—	—	—	—	A,D,H,I,J,K,L,M
Andrews Mountain	—	0.8	10.8	12.7	—	—	—	—	—	—	A,H,L,M,N,O,S
Upper Deep Spring	—	—	1.2	0.4	—	0.9	—	0.6	—	—	A,G,H,L,M,N,P,S
Middle Deep Spring	0.6	0.4	0.6	0.6	0.1	1.0	0.5	0.5	0.1	—	A,C,G,H,L,M,S
Lower Deep Spring	0.3	0.2	0.7	0.6	—	0.4	0.5	0.9	0.3	—	A,E,F,H,L,M,S
Upper Reed	—	—	—	—	—	—	3.0	—	—	—	A,H,L,M,S,T
Hines Tongue	1.1	1.4	1.4	0.4	—	0.3	0.6	0.3	—	—	A,H,L,M,S,T
Lower Reed	—	—	—	—	—	4.0	1.3	—	—	0.1	A,H,L,M,S,T
Total %	7.1	13.5	19.8	24.1	1.0	13.0	10.0	6.3	3.6	1.6	

[a]Data expressed as an estimated percentage of the total measured stratigraphic section. See Table I for key to lithofacies (1.1a), etc.

[b]Citations in Appendix. (A) Albers and Stewart, 1972; (B) Crews, 1980; (C) Dienger, 1985; (D) Fuller, 1977; (E) Gevirtzman and Mount, 1986; (F) Gevirtzman, 1983; (G) Greene, 1986; (H) McKee and Moiola, 1962; (I) Moore, 1976; (J) Morgan, 1977; (K) Mount and Rowland, 1981; (L) Mount and Signor, 1985; (M) Mount and Signor, 1986; (N) Mount, 1980; (O) Mount, 1982a; (P) Mount, 1982b; (Q) Rowland, 1978; (R) Rowland, 1981; (S) Stewart, 1970; (T) unpublished field studies.

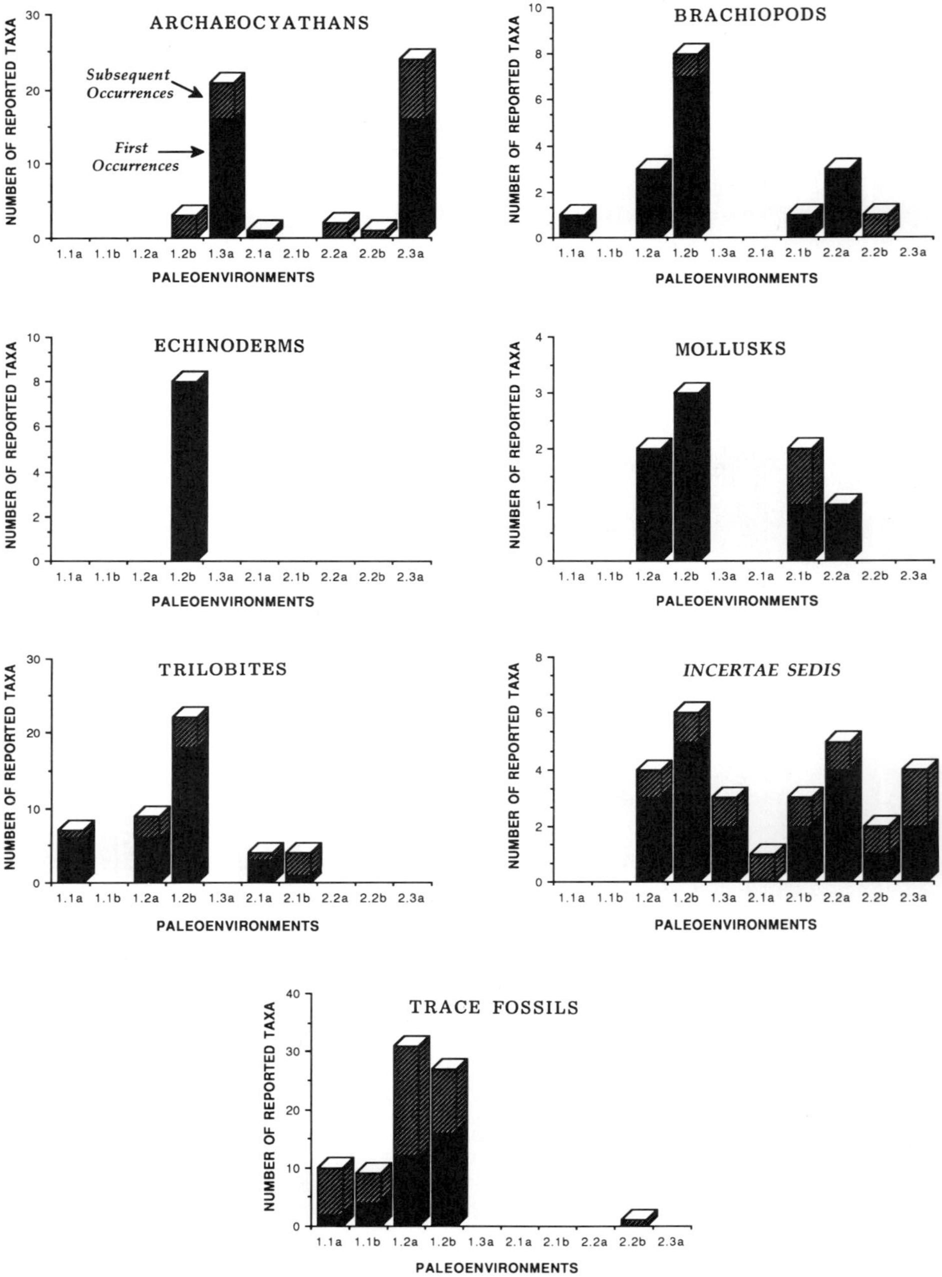

Figure 3. Lithofacies context of body and trace fossil first and subsequent appearances. See text for discussion and Table I for numerical designations of lithofacies. Data sources are summarized in Mount and Signor (1985, 1986, 1989a, b).

As might be expected for such a group of unrelated organisms, the *incertae sedis* make their initial and subsequent appearances in a broad range of carbonate and siliciclastic environments; only the intertidal siliciclastic units lack any reported *incertae sedis*. With the exception of the enigmatic fossil *Wyattia*, pre-trilobite or Tommotian-aspect small shelly fossils are restricted to shallow, subtidal, carbonate-dominated environments. In contrast, *incertae sedis* that appear to postdate the first occurrence of trilobites are much less facies-specific.

Virtually all trace fossil first occurrences are reported from siliciclastic-dominated depositional environments. Over 60% of the traces are from one unit, the Andrews Mountain Member of the Campito Formation (Fig. 2). The member is dominated by interstratified mudstones, siltstones, and fine sands deposited in storm-dominated open shelf environments: a lithologic setting ideal for the preservation of traces and the setting of the first occurrences of most Early Cambrian traces (Crimes, 1987).

2.2. Artifacts of the Record

Given the preference of trace and body fossil first appearances for shallow subtidal, subfairweather wavebase environments, it can be argued that the pattern of diversification shown in Fig. 2 is simply an artifact of Grand Cycle control of the stratigraphic distribution of lithofacies and only indirectly reflects global evolutionary processes. Alternatively, it could be argued that the record is reasonably accurate and that the fossil/lithofacies associations illustrated in Fig. 3 only record the relative proportion of different lithofacies in the Lower Cambrian section. For example, the high number of first occurrences in shallow subtidal siliciclastic units may simply be a product of the greater volume of that lithofacies (i.e., the larger the volume, the greater the opportunity for preserving first appearances within it).

One approach to test which relationship is most important is to compare lithofacies volumes to the numbers of first occurrences of body fossils (trace fossils are less suitable because over 60% make their debut in one single unit). Our null hypothesis was that first occurrences of taxa are randomly distributed and independent of lithofacies. One prediction from this null hypothesis is that the percentage of the total number of first appearances in each lithofacies will be approximately proportionate to the relative volume of that lithofacies. Using stratigraphic columns from sources listed in Table II, we estimated the relative volumes of lithofacies preserved in the entire Lower Cambrian section.

However, for the purposes of the test, we restricted the analysis to fossil occurrences and lithofacies in the *Nevadella* and *Bonnia–Olenellus* Zones. Empirical patterns of global fossil occurrences indicate that the diversity of earliest Cambrian faunas was low (e.g., Sepkoski, 1978, 1979; Brasier, 1979; Glaessner, 1984), and inclusion of these taxa and lithofacies might bias the result of our test. Therefore, we restricted our test to that portion of the section that should be most likely to show a correlation between first occurrences and lithofacies.

The lithofacies volumes are compared to the number of first occurrences in Fig. 4. From a chi-square contingency table analysis, we calculated that the deviation of the observed frequencies from the distribution expected from the null

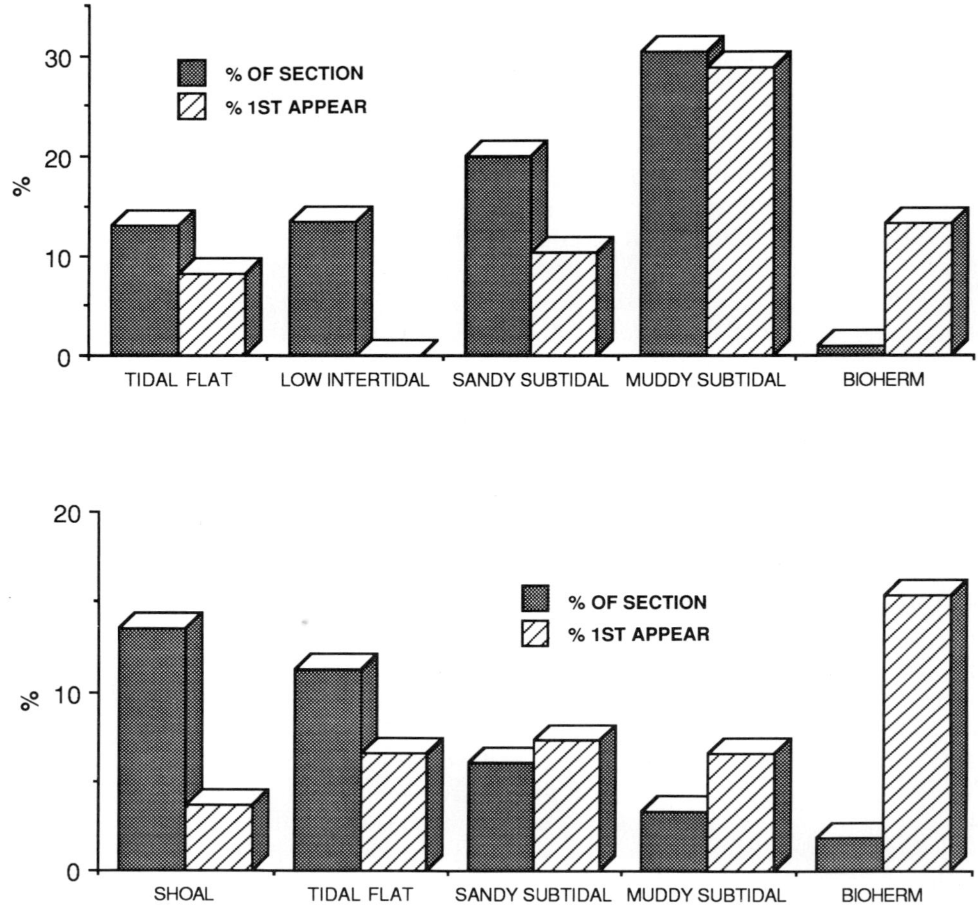

Figure 4. Comparison of total number of first appearances of body fossils with the volume of lithofacies preserved in the Lower Cambrian section. (A) Siliciclastic lithofacies, and (B) carbonate lithofacies. The volume of lithofacies preserved is an estimate based on a review of published and unpublished stratigraphic columns. Data sources are summarized in Mount and Signor (1985, 1986, 1989a, b).

hypothesis (a distribution random with respect to lithofacies) is vanishingly small ($p < 0.001$; Table III). Even if archaeocyathans, a clade restricted to and responsible for subtidal bioherms, are removed from the contingency table, the calculated result is still significantly different from random ($p < 0.01$).

The foregoing test supports the observation that first appearances within the Lower Cambrian of the southwestern Great Basin are tied to the widespread stratigraphic occurrence of lithofacies suitable for preservation and recovery of taxa. These lithofacies are predominantly carbonate or siliciclastic muds that accumulated in shallow subtidal environments below fairweather wave base, but above storm wave base. Since those lithofacies make up only about one-third of the entire Lower Cambrian section, it is unlikely that a reasonably complete record of the metazoan radiation is preserved. Rather, the eustatic changes that drove the

TABLE III. Calculated Chi-Square Statistic of Observed versus Expected Frequencies of First Appearances[a]

	Siliciclastics				Carbonates				
Taxon	1.1a	1.1b	1.2a	1.2b + 1.3a	2.1a	2.1b	2.2a	2.2b + 2.3a	χ^2
Archaeocyathans	0 (3.4)	0 (7.1)	0 (3.4)	16 (6.7)	1 (4.2)	0 (2.9)	0 (2.6)	16 (3.1)	87.32
Brachiopods + echinoderms	1 (2.4)	0 (4.9)	3 (0.2)	15 (22.7)	0 (2.9)	1 (2.0)	3 (1.8)	0 (2.2)	35.35
Trilobites	6 (3.5)	0 (7.3)	6 (3.5)	18 (6.9)	3 (4.4)	1 (2.9)	0 (2.7)	0 (3.2)	36.19
Incertae Sedis + mollusks	0 (2.7)	0 (5.6)	5 (2.7)	10 (5.3)	0 (3.3)	3 (2.2)	5 (2.1)	3 (2.4)	22.28
Total χ^2	1.97	24.82	0.40	52.79	7.93	2.49	0.18	6.02	96.60
Total χ^2 without archaeocyathans	0.26	17.76	3.61	40.14	5.47	0.64	0.28	2.95	71.11

[a]Chi-square matrix for fossil/lithofacies associations. See Table I for lithofacies key 1.1a, etc. Chi-square statistic calculated as $\chi^2 = \Sigma (f_o - f_e)^2/f_e$, where f_o is the number of observed first appearances in a given lithofacies and f_e is the (total number of observations of given taxonomic group) × (% of total Lower Cambrian section occupied by given lithofacies). The topmost of each pair of frequency values is f_o, the bottom is f_e. Chi-square values are not given for echinoderms and mollusks separately because more than one cell contained f_e values of <1.0. Data for bioherms in siliciclastic and carbonate depositional systems were combined with those for muddy subtidal siliciclastic and muddy subtidal carbonates, respectively, for analysis. The calculated chi-square statistic indicates that the probability that the fauna are randomly distributed with respect to lithofacies is <0.001. See text for discussion of results.

development of Grand Cycles have controlled much of the pattern and character of this record by dictating the types and amount of habitat preserved.

One casualty of the Grand Cycle-driven controls on fossil occurrences is likely to be the quality of biostratigraphic resolution of the Lower Cambrian. Since Early Cambrian radiometric dates are scarce and often contradictory and have large uncertainties, and because few sections yield useful magnetostratigraphic data, regional and interregional correlations are wholly dependent upon biostratigraphy. But, as demonstrated here, the occurrence of Lower Cambrian fossils is strongly controlled by the stratigraphic distribution of suitable lithofacies. Correlations based on first and last occurrences of such facies-restricted taxa will be strongly influenced by lithostratigraphy. The problem can be particularly acute when correlations depend upon only one taxonomic group, if that group has a tendency to occur within a narrow range of paleoenvironments.

Groups that evolve rapidly and are broadly distributed among paleoenvironments are going to be the most useful for Lower Cambrian biostratigraphic correlations. The data summarized in Fig. 3 indicate that there are considerable differences between various taxonomic groups. However, with the exception of the wide-ranging (and probably unrelated) *incertae sedis*, all of the Lower Cambrian taxa are highly facies-restricted. Archaeocyathans, which are used extensively by some Lower Cambrian stratigraphers, are largely restricted to subfairweather-wavebase, mud-dominated substrates. Almost all of the Early Cambrian traces appear to have originated in shallow subtidal siliciclastic environments, including traces that in younger strata are typically associated with deep-water deposits. Even trilobites, which show relatively high diversity and rapid turnover, are collected predominantly from subtidal siliciclastic units.

In sum, facies restrictions on virtually all Lower Cambrian taxa limit their biostratigraphic utility. Exclusive use of any one group for correlation (a common practice) will almost certainly lead to biases that are a product of the stratigraphic distribution of lithofacies.

3. Discussion

The fossil/lithofacies associations described herein illustrate the important role of environmental stratigraphy in biasing the record of the metazoan radiation. Although the concept of paleoecological controls on the distribution of fossils is certainly not new, remarkably little attention has been focused on paleoecological influences on this important interval of biological history. Our success at testing hypotheses about the nature and dynamics of the origin and early evolution of Metazoa depends directly upon our ability to analyze and interpret this record.

Recognition and understanding of fossil/lithofacies associations can contribute toward a variety of issues surrounding the Proterozoic/Cambrian transition and the radiation of Metazoa. It is beyond the scope of this chapter (and probably the reader's patience) to review all the ramifications of this analysis. However, the data presented here have a direct bearing on at least two interrelated issues: (1) the relationship between the pattern of the metazoan radiation event and the diversification of habitats during eustatic changes in sea level; and (2) the paleoenvironmental context of adaptive innovations during this radiation.

3.1. Habitat Diversity and the Metazoan Radiation

As noted in the introduction, the Proterozoic–Phanerozoic transition took place concurrently with fundamental changes in ocean chemistry, sea level, climate, and continental configuration. The tantalizing cooccurrence (albeit poorly constrained) of these changes and a revolution in the biosphere makes it tempting to seek a causal relationship. It is beyond the scope of this chapter to review all of the inferred models (Signor and Lipps, this volume, Chapter 1). However, the fossil/lithofacies associations described here do imply that some links may be more apparent than real.

Global rise in sea level has often been cited as a cause or trigger of the metazoan radiation event. The expansion and diversification of new habitats during the "Great Cambrian Transgression" has been touted by a number of workers as the primary control on the pattern of the event [see reviews in Brasier (1979, 1982, 1989b)]. In an analysis of taxonomic richness from six regions, Brasier (1982) found that Early Cambrian faunas tend to be unequally distributed among the lithofacies present in each region. He found few species in peritidal carbonate and coastal sand complexes and many in offshore limestones and shales. Brasier postulated that Sepkoski's (1978, 1979) asymptotic or sigmoidal diversity curve for Vendian (Ediacaran) through Middle Cambrian orders and families reflects an exponential growth process involving diversification within habitats that were expanding during the Cambrian transgression. Following the reasoning of Carr and Kitchell (1980) for Permian–Triassic extinctions, Brasier also suggested that declines in suitable habitat area caused by brief episodes of regression, during this extended transgression, led to catastrophic declines in "oversaturated" communities such as those dominated by the archaeocyathans. Finally, a prediction of the "transgression-diversity" hypothesis is that nearshore communities should develop and diversify before offshore and deeper-water communities. Brasier (1979) argued that because diversification of various Early Cambrian forms took place in a limited range of habitats, the common succession of *trace fossils → small shelly fossils → trilobites and complex shelly fossils* found in several regions reflects the sequential appearance and expansion of progressively offshore environments during the Cambrian transgression.

The "transgression-diversity" hypothesis offers an intriguing explanation of the origin of patterns of diversification during the initial radiation of metazoans. However, when evaluated in the light of discussions presented here, the hypothesis and its assumptions do not form a workable solution.

First, as reflected in North American Grand Cycles and stratigraphic successions worldwide, eustatic sea level rise during the Proterozoic–Cambrian transition was not steady and did not involve an orderly expansion of progressively offshore habitats. Intertidal, shallow subtidal, and deep subtidal habitats appear to have undergone considerable and unequal expansion and contraction numerous times prior to and during the radiation event. As the stratigraphic database of upper Proterozoic sections grows, it is increasingly clear that extensive shallow-to-deep marine environments persisted on most major continents well before the metazoan radiation event (e.g., Christie-Blick *et al.*, 1988). A Lipalian Interval when the continents were devoid of marine deposition is no longer viable. The association of metazoan fossils and transgressive deposits does not reflect a unique eustatic event during this interval of time. Rather, it appears to record the opportunity for preservation of fossil material during several sea level rises.

Second, an important assumption in the analysis that supports the "transgression-diversity" hypothesis is that the thickness of a given lithofacies in a stratigraphic column reflects the spatial distribution of the habitat. Because thickness at any locality is a function of subsidence, sedimentation rate, sea level change, and time, and not areal distribution, this logic can lead to incorrect results. The problem is exacerbated by use of a single stratigraphic column to represent a region. The predominance of a particular lithofacies in an interval of one column does not indicate that other lithofacies were not present on the shelf. Undoubtedly, shallower and deeper conditions existed elsewhere.

This has important implications for the relationship between habitat area and diversity. Without lithofacies maps of the various regions, no reliable estimate can be made for the association between habitat area and faunal diversity. We cannot determine if the rather small 16% increase in "unsuitable" lithofacies cited by Brasier as the cause of the global decline in archaeocyathans actually reflects a decrease in habitat area or is merely a product of a seaward shift in facies belts with no coincident decrease in actual habitat area (Jablonski, 1980). Moreover, it is not known whether the order of appearance of new forms is caused by the progressive formation and expansion of new habitats during transgression or if their ordering is only a consequence of the transgression-controlled stacking of lithofacies. In light of results presented here, a more parsimonious view would be that the relationship between increases and decreases in habitats and first appearances, diversifications, and declines of various Lower Cambrian taxonomic groups is more apparent than real.

3.2. Loci of Evolutionary Innovation

A number of studies of Phanerozoic benthic invertebrates have suggested that advanced or more modern communities appear first in nearshore environments and that archaic communities appear to inhabit more equable, offshore environments (Jablonski *et al.*, 1983; Jablonski and Bottjer, 1983; Sepkoski and Sheehan, 1983; Sepkoski and Miller, 1985; Bottjer and Jablonski, 1988). In a recent study of time–environment histories of crinoid, bivalve, and bryozoan clades, Bottjer and Jablonski (1988) have questioned the apparent onshore–offshore trends noted in Paleozoic communities. They argue that the trends are actually a product of separate, unrelated clade histories which, when viewed in aggregate on a coarse time scale, only appear to form a coherent pattern. At least two types of onshore–offshore migration are recognized within individual clades: (1) origination in the nearshore with subsequent expansion into the offshore, leaving behind representatives (termed expansion), or (2) movement of representatives into the offshore with loss of onshore representatives (termed retreat).

Although there has been considerable speculation about why certain environments appear to preferentially give rise to evolutionary innovations and what extrinsic or intrinsic processes may be responsible for onshore–offshore expansion or retreat, the causal links remain elusive [see review in Bottjer and Jablonski (1988)]. The uniqueness of the metazoan radiation event, including its polyphyletic expansion into largely empty ecospace, may provide a first-order test of at least some of the hypotheses put forward to explain this association (Sepkoski and Miller, 1985).

Although the data and analyses presented here have been used primarily to demonstrate artifactual bias within the record, they may also be used to evaluate the loci of evolutionary innovations and their assemblage into new communities during the Early Cambrian. While not accurately reflecting the timing of origination or appearance in a region, it is reasonable to expect that the paleoenvironmental context of a taxon's first appearance approximates the habitat of its origination. This is most likely to be true if the stratigraphic section represents a wide spectrum of lithofacies, as in the White-Inyo Region, thus lessening the prospect that first appearances might have taken place in paleoenvironments that were not recorded.

Based on the data provided here, it appears that the bulk of the metazoan radiation event took place in muddy siliciclastic and carbonate subtidal environments that lay below fairweather wave base but above storm wave base. This preference for first appearances in shallow subtidal habitats persists at all taxonomic levels (Mount and Signor, 1985) and confirms the widely recognized pattern of evolutionary innovations on inner parts of the shelf [see review in

Sepkoski and Miller (1985)]. However, the data do not support the onshore–offshore model of adaptive innovation suggested by Jablonski *et al.* (1983) and Bottjer and Jablonski (1988). With minor exceptions, the sites of evolutionary innovation and community development appear to have been largely static during the Early Cambrian and concentrated in inner-shelf, rather than nearshore settings.

These differences in pattern of origination and expansion between Early Cambrian and later Phanerozoic clades may offer some insight into processes responsible for the generation of adaptive innovations. The analyses of Bottjer and Jablonski (1988) argue against hypotheses that invoke extrinsic processes such as sea level changes or mass extinctions as controlling factors on origination and offshore expansion or retreat. They find that times of expansion and retreat of clades either do not coincide with or have longer durations than periods of mass extinctions or sea level change. For different reasons, the data from the southwestern Great Basin confirm this result. Mass extinction of offshore competitors or onshore members of individual clades is unlikely to have controlled the locus of radiation during the Early Cambrian. Little evidence exists for expansion or retreat of Early Cambrian communities or clades at the expense of preexisting members. Brasier (1979, 1982) and others have suggested that sea level-controlled expansion and contraction of habitats may have produced progressively offshore sites of evolutionary innovation. As noted above, on a global basis this may be more artifactual than real. Moreover there appears to be no systematic change in the loci of development of novelties.

Bottjer and Jablonski (1988; see references cited therein) point out that a broad range of hypotheses exist for onshore originations due to intrinsic processes. These include: (1) extinction bias, where onshore taxa may be more resistant to extinction than offshore taxa, with innovations appearing equally across a shelf, and nearshore taxa surviving to diversify; (2) probabilistic mechanisms, where originations in nearshore settings are either dependent upon standing diversity or rates of evolutionary turnover at lower taxonomic levels; and (3) environmental bias, where more innovations per speciation event are produced in nearshore environments than in offshore environments.

It seems unlikely that extinction bias controlled the loci of evolutionary innovations during the Early Cambrian. Based on the results of this study, it is clear that originations did not take place with equal frequency across the shelf. Preferential extinction of offshore taxa would not have enhanced the survival of novelties formed in inner-shelf or nearshore environments because there were few, if any innovations in offshore settings. Probabilistic mechanisms, such as diversity-dependent originations and turnover-dependent originations, are equally unlikely because of the near lack of offshore communities. Rather, it appears that the simplest, yet most difficult process to reconstruct, environmental bias in the origin of evolutionary novelties, may have been operating during the Early Cambrian radiation event.

4. Conclusions

The goal of this chapter has been to separate the artifactual from the factual aspects of the record of the Early Cambrian metazoan radiation. Based on the fossil/lithofacies data from the southwestern Great Basin, bursts and declines in first appearances of early metazoans may merely be a product of changes in the volume of suitable habitat preserved within any given stratigraphic interval. It is unlikely that the order of appearance of new forms was caused by the progressive formation and expansion of new habitats during sea level changes. Rather, much of their ordering is only a consequence of the transgression–regression-controlled stacking of lithofacies suitable for their preservation. In addition, no evidence exists on a global basis for a unique association between late Proterozoic/Early Cambrian sea level rise and the radiation of Metazoa. The factual aspect of the record is the consistent inner-shelf, shallow subtidal loci of evolutionary innovations with little expansion into offshore settings. Whether the difference between patterns of origination and expansion in Early Cambrian and later Phanerozoic clades reflects different processes responsible for evolutionary innovations remains to be seen.

Appendix: Data Sources

Albers, J. P., and Stewart, J. H., 1972, Geology and mineral deposits of Esmeralda County, Nevada, *Nevada Bur. Mines Geol. Bull.* **78:**1–80.

Alpert, S. P., 1973, *Bergaueria* Prantl (Cambrian and Ordovician), a probable actinian trace fossil, *J. Paleontol.* **47:**919–924.

Alpert, S. P., 1975, Planolites and Skolithos from the Precambrian–Lower Cambrian, White-Inyo Mountains, California, *J. Paleontol.* **50:**226–239.

Alpert, S. P., 1976a, Trilobite and star-like traces from the White-Inyo Mountains, California, *J. Paleontol.* **50:**226–239.

Alpert, S. P., 1976b, Trace fossils of the White-Inyo Mountains, in: *Depositional Environments of Lower Paleozoic Rocks in the White-Inyo Mountains, Inyo County, California* (J. N. Moore and A. E. Fritsche, eds.), Pacific Section, Society of Economic Paleontologists and Mineralogists Pacific Coast Paleogeography Field Guide 1, pp. 43–48.

Alpert, S. P., 1977, Trace fossils and the basal Cambrian boundary, in: Trace Fossils 2 (T. P. Crimes and J. C. Harper, eds.), *Spec. Issue* **9:**1–8.

Alpert, S. P., and Moore, J. N., 1975, Lower Cambrian fossil evidence for predation on trilobites, *Lethaia* **17:**307–329.

Crews, A. L., 1980, Sedimentology of a Lower Cambrian marine shelf sequence: Zabriske Quartzite, Saline Valley Formation and related strata, Ph.D. thesis, University of California, Los Angeles.

Dienger, J. L., 1985, Facies modelling of siliciclastic and carbonate sediments in the Lower Cambrian Middle Member Deep Spring Formation, White-Inyo Range, California, M. S. thesis, University of California, Davis.

Durham, J. W., 1967, Notes on the Helicoplacoidea and early echinoderms, *J. Paleontol.* **41:**97–102.

Durham, J. W., 1974, On the Precambrian–Early Cambrian trace fossil *Plagiogmus*, *Geol. Soc. Am. Abstr. Progr.* **6:**170.

Durham, J. W., 1978, The probable metazoan biota of the Precambrian as indicated by the subsequent record, *Ann. Rev. Earth Planet. Sci.* **6:**21–42.

Durham, J. W., and Caster, K. E., 1963, Helicoplacoidea: A new class of echinoderms, *Science* **140:**820–822.

Firby, J. B., and Durham, J. W., 1974, Molluscan radula from earliest Cambrian, *J. Paleontol.* **48:**1109–1119.

Fritz, W. H., 1975, Broad correlations of some Lower and Middle Cambrian strata in the North American Cordillera, *Geol. Surv. Can. Pap.* **75-1A:**145–159.

Fuller, R. D., 1977, Paleoenvironmental Analysis of an Early Cambrian Archaeocyathan Reef in the White-Inyo Mountains, CA, Idaho State University, unpublished manuscript.

Gangloff, R. A., 1976, Archaeocyatha of eastern California and western Nevada, in: *Depositional Environments of Lower Paleozoic Rocks in the White-Inyo Mountains, Inyo County, California* (J.

N. Moore and A. E. Fritsche, eds.), Pacific Section, Society of Economic Paleontologists and Mineralogists Pacific Coast Paleogeography Field Guide 1, pp. 19–30.

Gevirtzman, D. A., 1983, Paleoenvironments of an earliest Cambrian (Tommotian) shelly fauna in Esmeralda County, Nevada, M. S. thesis, University of California, Davis.

Gevirtzman, D. A., and Mount, J. F., 1986, Paleoenvironments of a Tommotian (Earliest Cambrian) Fauna in the Southwestern Great Basin, USA, *J. Sediment. Petrol.* **56:**412–421.

Greene, L. R., 1986, Cyclic sedimentation within the Upper Member of the Deep Spring Formation (Lower Cambrian), Eastern California and Western Nevada: The anatomy of a Grand Cycle, M. S. thesis, University of California, Davis.

McKee, E. H., and Gangloff, R. A., 1969, Stratigraphic distribution of archaeocyathids in the Silver Peak Range and the White and Inyo Mountains, western Nevada and eastern California, *J. Paleontol.* **43:**716–726.

McKee, E. H., and Moiola, R. J., 1962, Precambrian and Cambrian rocks of south-central Esmeralda County, Nevada, *Am. J. Sci.* **260:**530–538.

Moore, J. N., 1976, Depositional environments of the Lower Cambrian Poleta Formation and its stratigraphic equivalents, California and Nevada, *Brigham Young Univ. Geol. Stud.* **23:**23–38.

Morgan, N., 1977, The Montenegro bioherms: Their paleoecology, relation to other archaeocyathid bioherms and to Early Cambrian sedimentation in the White and Inyo Mountains, California, in: Depositional Environments of Lower Paleozoic Rocks in the White-Inyo Mountains, Inyo County, California, Pacific Coast Paleogeography Field Guide 1 (J. N. Moore and A. E. Fritsche, eds.), Pacific Section Society of Economic Paleontologists and Mineralogists, Los Angeles, pp. 13–18.

Mount, J. F., 1980, The environmental stratigraphy and depositional Systems of the Precambrian(?)–Cambrian Campito Formation, eastern California and western Nevada, Ph.D. thesis, University of California, Santa Cruz.

Mount, J. F., 1982a, Storm-surge-ebb origin of hummocky cross-stratified units of the Andrews Mountain Member, Campito Formation (Lower Cambrian), White-Inyo Mountains, Eastern California, *J. Sediment. Petrol.* **52:**941–958.

Mount, J. F., 1982b, Earliest Cambrian paleoenvironments of the southern Great Basin, in: *Geology of Selected Areas in the San Bernadino Mountains, Western Mojave Desert and Southern Great Basin, California* (J. D. Cooper, B. W. Troxel, and L. A. Wright, eds.), Geological Society of America Cordilleran Section Meeting Guidebook, pp. 187–193.

Mount, J. F., and Rowland, S. M., 1981, Grand Cycle A (Lower Cambrian) of the southern Great Basin: A product of differential rates of sea level rise, in: *Short Papers for the Second International Symposium on the Cambrian System* (M. E. Taylor, ed.), U. S. Geological Survey Open-File Report 81-743, pp. 143–146.

Mount, J. F., and Signor, P. W., 1985, Early Cambrian innovation in low-stress environments: Paleoenvironments of Early Cambrian shelly fossils, *Geology* **13:**730–733.

Mount, J. F., and Signor, P. W., 1986, The record of Cambrian Radiation Event in the White-Inyo region, eastern California and western Nevada, in: *Nello Pace Symposium Volume, White Mountain Research Station, August, 1985*, University of California Press, Berkeley, California.

Mount, J. F., and Signor, P. W., 1989a, Paleoenvironmental context of the metazoan radiation event and its impact on the placement of the Precambrian–Cambrian boundary: Examples from the southwestern Great Basin, U.S.A., in: *Late Proterozoic and Cambrian Tectonics, Sedimentation, and Record of Metazoan Radiation in the Western United States* (N. Christie-Blick and M. Levy, eds.), 28th International Geological Congress Field Trip Guidebook T331, American Geophysical Union, pp. 39–46.

Mount, J. F., and Signor, P. W., 1989b, Description of Field Trip Stops, in: *Late Proterozoic and Cambrian Tectonics, Sedimentation, and Record of Metazoan Radiation in the Western United States* (N. Christie-Blick and M. Levy, eds.), 28th International Geological Congress Field Trip Guidebook T331, American Geophysical Union, pp. 88–99.

Mount, J. F., Gevirtzman, D. A., and Signor III, P. W., 1983, Precambrian–Cambrian transition problem in western North America: Part I. Tommotian fauna in the southwestern Great Basin and its implications for the base of the Cambrian System, *Geology* **11:**224–226.

Mount, J. F., Dienger, J. L., and Greene, L. R., 1986, Large-scale cyclic sedimentation in shallow marine sediments and the production of mixed siliciclastic and carbonate sediments: Examples from the Lower Cambrian of the southern Great Basin, USA, in: *12th International Sedimentology Congress, Abstracts*, pp. 218–219.

Nelson, C. A., 1962, Lower Cambrian–Precambrian succession, White-Inyo Mountains, California, *Geol. Soc. Am. Bull.* **73:**139–144.

Nelson, C. A., 1966a, Geologic map of Waucoba Mountain quadrangle, Inyo and Mono Counties, California, U. S. Geological Survey Geologic Quadrangle Map GQ-528, scale 1:62,500.

Nelson, C. A., 1966b, Geologic map of the Blanco Mountain quadrangle, Inyo and Mono Counties, California, U. S. Geological Survey Geologic Quadrangle Map GQ-529, scale 1:62,500.

Nelson, C. A., 1971, Geologic map of the Waucoba Spring quadrangle, Inyo County, California, Inyo County, California, U. S. Geological Survey Geologic Quadrangle Map, GQ-921, scale 1:62,500.

Nelson, C. A., 1976, Late Precambrian–Early Cambrian stratigraphic and faunal succession of eastern California and the Precambrian–Cambrian boundary in: *Depositional Environments of Lower Paleozoic rocks in the White-Inyo Mountains, Inyo County, California* (J. N. Moore and A. E. Fritsche, eds.), Pacific Section, Society of Economic Paleontologists and Mineralogists Pacific Coast Paleogeography Field Guide 1, pp. 31–42.

Nelson, C. A., and Durham, J. W., 1966, *Guidebook for Field Trip to Pre–Cambrian Succession White-Inyo Mountains, California*, Guide for Geological Society of America Annual Meeting Field Trip.

Okulitch, V., 1954, Archaeocyatha from the Lower Cambrian of Inyo County, California, *J. Paleontol.* **28:**293–296.

Palmer, A. R., 1964, An Unusual Lower Cambrian Trilobite Fauna from Nevada, U. S. Geological Survey Professional Paper 483-F.

Palmer, A. R., 1981, On the correlatability of Grand Cycle tops, in: *Short papers for the Second International Symposium on the Cambrian System* (M. E. Taylor, ed.), U. S. Geological Survey Open-File Report 81-743, p. 193–197.

Pemberton, S. G., and Frey, R. W., 1982, Trace fossil nomenclature and the *Planolites–Paleophycus* dilemma, *J. Paleontol.* **56:**843–881.

Rowell, A. J., 1977, Early Cambrian brachiopods from the southwestern Great Basin of California and Nevada, *J. Paleontol.* **51:**68–85.

Rowland, S. M., 1978, Environmental stratigraphy of the Lower Member of the Poleta Formation (Lower Cambrian), Esmeralda County, Nevada, Ph.D. thesis, University of California, Santa Cruz.

Rowland, S. M., 1981, Were there really framework reefs in the Cambrian?, *Geology* **12:**181–183.

Rowland, S. M., and Carlson, S. J., 1983, *Westagardia gigantea*, a new Lower Cambrian fossil from eastern California, *J. Paleontol.* **57:**1317–1320.

Signor, P. W., and Mount, J. F., 1986a, Lower Cambrian stratigraphic paleontology of the White-Inyo Mountains of eastern California and Esmeralda County, Nevada, in: *Nello Pace Symposium Volume, White Mountain Research Station, August, 1985*, University of California Press, Berkeley, California.

Signor, P. W., and Mount, J. F., 1986b, Position of the Lower Cambrian boundary in the White-Inyo Mountains of California and in Esmeralda County, Nevada, *Newsl. Stratigr.* **16:**9–18.

Signor, P. W., and Mount, J. F., 1989, Paleontology of the Lower Cambrian Waucoban Series in Eastern California and Western Nevada, in: *Late Proterozoic and Cambrian Tectonics, Sedimentation, and Record of Metazoan Radiation in the Western United States* (N. Christie-Blick and M. Levy, eds.), 28th International Geological Congress Field Trip Guidebook T331, American Geophysical Union, pp. 47–53.

Signor, P. W., III, McMenamin, M. A. S., Gevirtzman, D. A., and Mount, J. F., 1983, Two new pre-trilobite faunas from western North America, *Nature* **303:**415–418.

Signor, P. W., Mount, J. F., and Onken, B. R., 1987, A pre-trilobite shelly fauna from eastern California and western Nevada, *J. Paleontol.* **65:**435–448.

Sprinkle, J., 1976, Biostratigraphy and paleoecology of Cambrian echinoderms from the Rocky Mountains, *Brigham Young Univ. Geol. Stud.* **23:**61–73.

Stewart, J. H., 1970, Upper Precambrian and Lower Cambrian strata in the southern Great Basin, California and Nevada, *U. S. Geol. Surv. Spec. Pap.* **620:**1–20.

Taylor, M. E., 1966, Precambrian mollusc-like fossils from Inyo County, California, *Science* **153:**198–201.

Tynan, M. C., 1983, Coral-like microfossils from the Lower Cambrian of California, *J. Paleontol.* **57:**1188–1211.

Walcott, C. D., 1908, Cambrian sections of the cordillera area, *Smithson. Misc. Publ.* **53:**107–130.

Walcott, C. D., 1910, *Olenellus* and other general of the Meronacidae, *Smithson. Misc. Collect.* **53:**229–448.

Walcott, C. D., 1912, *Cambrian Brachiopoda*, U. S. Geological Survey Monograph 51.
Wiggett, G. J., 1973, Occurrence and implications of trace fossils in the Lower Cambrian of the White-Inyo Mountains, *Geol. Soc. Am. Abstr. Progr.* **5**:122.
Yochelson, E. L., Pierce, J. W., and Taylor, M. E., 1970, *Salterella* from the Lower Cambrian of central Nevada, *U. S. Geol. Surv. Prof. Pap.* **643-H**:1–7.

References

Aitken, J. D., 1966, Middle Cambrian to Middle Ordovician cyclic sedimentation, southern Rocky Mountains of Alberta, *Bull. Can. Pet. Geol.* **14**:405–411.
Aitken, J. D., 1978, Revised models for depositional Grand Cycles, Cambrian of the southern Rocky Mountains, Canada, *Bull. Can. Pet. Geol.* **26**:515–542.
Alpert, A. P., 1977, Trace fossils and the basal Cambrian boundary, in: Trace Fossils 2 (T. P. Crimes and J. C. Harper, eds.), *Geol. J. Spec. issue* **9**:1–8.
Bond, G. C., Christie-Blick, N., Kominz, M. A., and Devlin, W. J., 1985, An Early Cambrian rift to post-rift transition in the Cordillera of western North America, *Nature* **316**:742–745.
Bond, G. C., Kominz, M. A., and Grotzinger, J. P., 1988, Cambro-Ordovician eustacy: Evidence from geophysical modelling of subsidence in Cordillera and Appalachian passive margins, in: *New Perspectives in Basin Analysis* (K. L. Kleinspehn and C. Paola, eds.), Springer-Verlag, New York, pp. 129–160.
Bottjer, D., and Jablonski, D., 1988, Paleoenvironmental patterns in the evolution of post-Paleozoic benthic marine invertebrates, *Palaios* **3**:540–560.
Brasier, M. D., 1979, The Cambrian radiation event, in: *The Origin of Major Invertebrate Groups* (M. R. House, ed.), Academic Press, New York, pp. 103–159.
Brasier, M. D., 1982, Sea-level changes, facies changes and the late Precambrian–Cambrian evolutionary explosion, *Precambrian Res.* **17**:105–123.
Brasier, M. D., 1989a, China and the Paleotethyan Belt (India, Pakistan, Kazakhstan, and Mongolia), in: *The Precambrian–Cambrian Boundary* (J. W. Cowie and M. D. Brasier, eds.), Clarendon Press, Oxford, pp. 40–74.
Brasier, M. D., 1989b, Towards a biostratigraphy of the earliest skeletal biotas, in: *The Precambrian–Cambrian Boundary* (J. W. Cowie and M. D. Brasier, eds.), Clarendon Press, Oxford, pp. 117–165.
Carr, T. R., and Kitchell, J. A., 1980, Dynamics of taxonomic diversity, *Paleobiology* **6**:427–443.
Chow, N., and James, N. P., 1987, Cambrian Grand Cycles: A northern Appalachian perspective, *Geol. Soc. Am. Bull.* **93**:735–750.
Christie-Blick, N., Grotzinger, J. P., and von der Borch, C. C., 1988, Sequence stratigraphy in Proterozoic successions, *Geology* **16**:100–104.
Cloud, P. E., 1949, Some problems and patterns of evolution exemplified by fossil invertebrates, *Evolution* **2**:322–350.
Cloud, P. E., and Nelson, C. A., 1966, Phanerozoic-Cryptozoic and related transitions: New evidence, *Science* **154**:766–770.
Crimes, P. T., 1987, Trace fossils and correlation of late Precambrian and early Cambrian strata, *Geol. Mag.* **124**:97–119.
Durham, J. W., 1978, The probable metazoan biota of the Precambrian as indicated by the subsequent record, *Annu. Rev. Earth Planet. Sci.* **6**:21–42.
Fritz, W. H., 1972, Lower Cambrian trilobites from the Sekwi Formationa Type Section, Mackenzie Mountains, Northwestern Canada, *Geol. Surv. Can. Bull.* **212**:1–58.
Fritz, W. H., 1975, Broad correlations of some Lower and Middle Cambrian strata in the North American Cordillera, *Geol. Surv. Can. Pap.* **75-1A**:145–159.
Gangloff, R. A., 1975, The Archaeocyatha of the central and southwestern Great Basin, California and Nevada, Ph.D. dissertation, University of California at Berkeley.
Gangloff, R. A., 1976, Archaeocyatha of eastern California and western Nevada, in: Depositional environments of Lower Paleozoic rocks in the White-Inyo Mountains, Inyo County, California: Pacific Section, Society of Economic Paleontologists and Mineralogists Pacific Coast Paleogeography Field Guide 1 (J. N. Moore and A. E. Fritsche, eds.), pp. 19–30.
Gevirtzman, D. A., and Mount, J. F., 1986, Paleoenvironments of a Tommotian (Earliest Cambrian) Fauna in the Southwestern Great Basin, USA, *J. Sediment. Petrol.* **56**:412–421.

Glaessner, M. F., 1984, *The Dawn of Animal Life*, Cambridge University Press, Cambridge.

Jablonski, D., 1980, Apparent vs real biotic effects of transgressions and regressions, *Paleobiology* **6**:427–443.

Jablonski, D., and Bottjer, D. J., 1983, Soft-bottom epifaunal suspension-feeding assemblages in the Late Cretaceous, in: Biotic interactions in recent and fossil benthic communities (M. J. S. Tevasz and P. L. McCall, eds.), New York, Plenum Press, pp. 747–812.

Jablonski, D., Sepkoski, J. J., Jr. Bottjer, D. J., and Sheehan, P. M., 1983, Onshore–offshore patterns in the evolution of Phanerozoic shelf communities, *Science* **222**:1123–1125.

Lipps, J. H., and Fedonkin, M., 1988, Trace fossils and the Precambrian/Cambrian boundary, *Geol. Soc. Am. Abstr. Prog.* **20**:A256.

Matthews, S. C., and Cowie, J. W., 1979, Early Cambrian transgression, *J. Geol. Soc. Lond.* **136**:133–135.

McKee, E. H., and Gangloff, R. A., 1969, Stratigraphic distribution of archaeocyathids in the Silver Peak Range and the Whie and Inyo Mountains, western Nevada and eastern California, *J. Paleontol.* **43**:716–726.

Missarzhevsky, V. V., and Mambetov, A. J., 1981, Stratigraphy and fauna of Cambrian and Precambrian boundary beds of Maly Karatau, *Trudy Geol. Inst. Akad. Nauka SSSR* **326**:1–90 [in Russian].

Mount, J. F., 1982a, Storm-surge-ebb origin of hummocky cross-stratified units of the Andrews Mountain Member, Campito Formation (Lower Cambrian), White-Inyo Mountains, Eastern California, *J. Sediment. Petrol.* **52**:941–958.

Mount, J. F., 1982b, Earliest Cambrian paleoenvironments of the southern Great Basin, in: *Geology of Selected Areas in the San Bernardino Mountains, Western Mojave Desert and Southern Great Basin, California* (J. D. Cooper, B. W. Troxel, and L. A. Wright, eds.), Geological Society of America Cordilleran Section Meeting Guidebook, pp. 187–193.

Mount, J. F., 1984, Mixing of siliciclastic and carbonate sediments in shallow shelf environments, *Geology* **12**:432–435.

Mount, J. F., and Rowland, S. M., 1981, Grand Cycle A (Lower Cambrian) of the southern Great Basin: A product of differential rates of sea level rise, in: *Short Papers for the Second International Symposium on the Cambrian System* (M. E. Taylor, ed.), U. S. Geological Survey Open-File Report 81-743, pp. 143–146.

Mount, J. F., and Signor, P. W., 1985, Early Cambrian innovation in low-stress environments: Paleoenvironments of Early Cambrian shelly fossils, *Geology* **13**:730–733.

Mount, J. F., and Signor, P. W., 1986, The record of Cambrian Radiation Event in the White-Inyo region, eastern California and western Nevada, in: *Nello Pace Symposium Volume, White Mountain Research Station, August, 1985*.

Mount, J. F., and Signor, P. W., 1989a, Paleoenvironmental context of the metazoan radiation event and its impact on the placement of the Precambrian–Cambrian boundary: examples from the southwestern Great Basin, U.S.A., in: *Late Proterozoic and Cambrian Tectonics, Sedimentation, and Record of Metazoan Radiation in the Western United States* (N. Christie-Blick and M. Levy, eds.), 28th International Geological Congress Field Trip Guidebook T331, American Geophysical Union, pp. 39–46.

Mount, J. F., and Signor, P. W. 1989b, Description of Field Trip Stops, in: *Late Proterozoic and Cambrian Tectonics, Sedimentation, and Record of Metazoan Radiation in the Western United States* (N. Christie-Blick and M. Levy, eds.), 28th International Geological Congress Field Trip Guidebook T331, American Geophysical Union, pp. 88–99.

Mount, J. F., Gevirtzman, D. A., and Signor III, P. W., 1983, Precambrian–Cambrian transition problem in western North America: Part I. Tommotian fauna in the southwestern Great Basin and its implications for the base of the Cambrian System, *Geology* **11**:224–226.

Mount, J. F., Hunt, D. L., Greene, L. R., and Dienger, J., 1991, Depositional systems, biostratigraphy and sequence stratigraphy of Lower Cambrian Grand Cycles, southwestern Great Basin, in: *Paleozoic Paleogeography of the Western United States—II: Pacific Section SEPM* (J. D. Cooper and C. H. Stevens, eds.), **67**:209–229.

Nelson, C. A., 1962, Lower Cambrian–Precambrian succession, White-Inyo Mountains, California, *Geol. Soc. Am. Bull.* **73**:139–144.

Onken, B. R., and Signor, P. W., 1988, Lower Cambrian stratigraphic paleontology of the southwestern Great Basin (White-Inyo Mountains of eastern California and Esmeralda County, Nevada), *Bull. S. Calif. Paleontol. Soc.* **20**:131–150.

Parrish, J. T., Ziegler, A. M., Scotese, C. R., Humphreville, R. G., and Kirschvink, J. L., 1986, Early Cambrian paleogeography, paleoceanography, and phosphorites, in: *Phosphorite Deposits of the World* (J. H. Shergold and P. J. Cook, eds.), Cambridge University Press, Cambridge, pp. 280–294.

Read, J. F., Grotzinger, J. P., Bova, J. A., and Koerschner, W. F., 1986, Models for the generation of carbonate cycles, *Geology* **14**:107–110.

Rowland, S. M., 1984, Were there framework reefs in the Cambrian?, *Geology* **12**:181–183.

Runnegar, B., 1982a, A molecular-clock date for the origin of animal phyla, *Lethia* **15**:199–205.

Runnegar, B., 1982b, The Cambrian explosion: Animals or fossils?, *J. Geol. Soc. Aust.* **29**:395–411.

Scotese, C. R., Bambach, R. K., Barton, C., Van der Voo, R., and Ziegler, 1979, Paleozoic base maps, *J. Geol.* **87**:217–277.

Sepkoski, J. J. Jr., 1978, A kinetic model of Phanerozoic taxonomic diversity. I. Analysis of marine orders, *Paleobiology* **4**:223–251.

Sepkoski, J. J. Jr., 1979, A kinetic model of Phanerozoic taxonomic diversity. II. Early Phanerozoic families and multiple equilibria, *Paleobiology* **5**:221–251.

Sepkoski, J. J. Jr., and Sheehan. P. M., 1983, Diversification, faunal change, and community replacement during the Ordovician radiations, in: Biotic interactions in recent and fossil benthic communities (M. J. S., Tevasz and P. L. McCall, eds.), New York, Plenum Press, pp. 673–717.

Sepkoski, J. J., and Miller, A. I., 1985, Evolutionary faunas and the distribution of Paleozoic benthic communities in space and time, in: *Phanerozoic Diversity Patterns* (J. W. Valentine, eds.), Princeton University Press, Princeton, New Jersey, pp. 153–190.

Signor, P. W., and Mount, J. F., 1986a, Lower Cambrian stratigraphic paleontology of the White-Inyo Mountains of eastern California and Esmeralda County, Nevada, in: *Nello Pace Symposium Volume, White Mountain Research Station, August, 1985*, University of California Press, Berkeley, California.

Signor, P. W., and Mount, J. F., 1986b, Position of the Lower Cambrian boundary in the White-Inyo Mountains of California and in Esmeralda County, Nevada, *Newsl. Stratigr.* **16**:9–18.

Signor, P. W., and Mount, J. F., 1989, Paleontology of the Lower Cambrian Waucoban Series in Eastern California and Western Nevada, in: *Late Proterozoic and Cambrian Tectonics, Sedimentation, and Record of Metazoan Radiation in the Western United States* (N. Christie-Blick and M. Levy, eds.), 28th International Geological Congress Field Trip Guidebook T331, American Geophysical Union, 47–53.

Signor III, P. W., McMenamin, M. A. S., Gevirtzman, D. A., and Mount, J. F., 1983, Two new pre-trilobite faunas from western North America, *Nature* **303**:415–418.

Signor, P. W., Mount, J. F., and Onken, B. R., 1987, A pre-trilobite shelly fauna from eastern California and western Nevada, *J. Paleontol.*

Stanley, S. M., 1976, Fossil data and the Precambrian–Cambrian evolutionary transition, *Am. J. Sci.* **276**:56–76.

Stewart, J. H., 1970, Upper Precambrian and Lower Cambrian strata in the southern Great Basin, California and Nevada, *U. S. Geol. Surv. Spec. Pap.* **620**:1–206.

Stewart, J. H., and Suczeck, 1977, Cambrian and latest Precambrian paleogeography and tectonics in the western United States, in: *Paleozoic Paleogeography of the Western United States, Pacific Coast Paleogeography Symposium 1*, Pacific Section, Society of Economic Paleontologists and Mineralogists, Bakersfield, California, pp. 1–17.

Taylor, M. E., 1966, Precambrian mollusc-like fossils from Inyo County, California, *Science* **153**:198–201.

Theokritoff, G., 1979, Early Cambrian provincialism and biogeographic boundaries in the North Atlantic region, *Lethia* **12**:281–295.

Valentine, J. W., 1973, *Evolutionary Paleoecology of the Marine Biosphere*, Prentice Hall, Engelwood Cliffs, New Jersey.

Voronin, Yu. I., Voronova, L. G., Grigorieva, N. V., Drosdova, N. A., Zhegallo, E. A., Zhuravlev, A. Yu., Ragozina, A. L., Rozanov, A. Yu., Sayutina, T. A., Sysoiev, V. A., and Fonin, V. D., 1982, The Precambrian–Cambrian boundary in the geosynclinal areas (the reference section of Salany-Gol, MPR), *Trans. Joint Soviet–Mongolian Paleontol. Expedition* **18**. [in Russian].

Walcott, C. D., 1908, Cambrian sections of the cordillera area, *Smithson. Misc. Publ.* **53**:107–130.

Walcott, C. D., 1912, *Cambrian Brachiopoda*, U. S. Geological Survey Monograph 51.

Chapter 3

Biological and Biogeochemical Preludes to the Ediacaran Radiation

ANDREW H. KNOLL

1. Introduction

Evolution is sometimes envisioned as a dynamic process played out on an inert or passive planetary platform; however, if the geological record contains a message for evolutionary biology, other than the fact of the fossil record itself, it is that the Earth's surface is in a continual state of change. On various time scales, climates fluctuate, reservoirs and fluxes of biogeochemical cycles vary, sea levels rise and fall, and oceanic water masses form and decay, while continents grow, split apart, drift, and recombine. At the same time, organisms that compete with, prey upon, provide food for, or live in symbiotic association with any given taxon of interest radiate, undergo morphological and physiological change, and become extinct. This is the context of evolution, and paleontological efforts to understand evolutionary events of the past require that we attempt to root those events in the context of ongoing biological and environmental change.

ANDREW H. KNOLL • Botanical Museum, Harvard University, Cambridge, Massachusetts 02138.

Origin and Early Evolution of the Metazoa, edited by Jere H. Lipps and Philip W. Signor. Plenum Press, New York, 1992.

The Ediacaran fauna and other early megascopic animals did not radiate onto an empty or static Earth. In this chapter, I will consider the nature of the biological world into which large animals were introduced, and will further discuss emerging geological data which indicate that the Late Proterozoic era was a period of marked tectonic, climatic, and biogeochemical change. These considerations together constitute the contextual framework within which the pattern of early animal evolution must be evaluated.

2. The Late Riphean Biological World

Precambrian paleontologists love to publish estimates for the antiquity of major biological groups. Unfortunately, all such estimates are seriously hampered by the familiar difficulty of distinguishing between evidence of absence and absence of evidence. During the past decade, the earliest reliable records of both unicellular protists and metaphytes (megascopic, morphologically complex algae) have moved backward into ever earlier epochs of the Proterozoic. At the same time, the Ediacaran floor to the megascopic animal record has increasingly come to be viewed as firm and abrupt.

2.1. Prokaryotes

For more than two decades, paleontologists have recognized that many Proterozoic microfossils resemble living cyanobacteria. Despite the unambiguous nature of this observation, paleobiological interpretation has been clouded by two persistent fears concerning the relationship between preserved morphology and physiology. First, bacteria of widely differing metabolism show morphological convergence. The filamentous trichomes of oscillatorian cyanobacteria are strikingly similar to those of the sulfur-oxidizing bacterium *Beggiatoa,* although comparisons of 16S rRNA sequences show the latter to cluster with the purple bacteria rather than the blue-greens (Woese, 1987). Simple cyanobacterial rods and cocci have numerous analogues among metabolically disparate Eubacteria. Even if a fossil can be interpreted as cyanobacterial with confidence, blue-green evolution nevertheless might have proceeded according to what Schopf (1972) termed (in a now dated metaphor) a "Volkswagen syndrome—little or no evolution of external form concealing marked changes of internal machinery."

Systematic interpretations of Proterozoic prokaryotes depend on three distinct lines of inquiry. In addition to morphological characterization (size–frequency distribution, shape, development, cell division cycles), taphonomic and paleoecological data must be considered (e.g., Knoll, 1985a). An illuminating example is provided by *Polybessurus bipartitus,* a distinctive fossil known from silicified tidal flat carbonates in Upper Proterozoic successions of East Greenland, Spitsbergen, Australia, and Somerset Island, Arctic Canada (Green *et al.*, 1987). *Polybessurus bipartitus* individuals consist of ellipsoidal cells surrounded by multiple envelopes produced asymmetrically to form lamellated stalks that elevated cells above their substrate (Figs. 1A and 1B). Reproduction is inferred to have been by multiple fission, with the subsequent release of baeocytes. Although *P. bipartitus* fossils may occur as solitary individuals within microbial mat as-

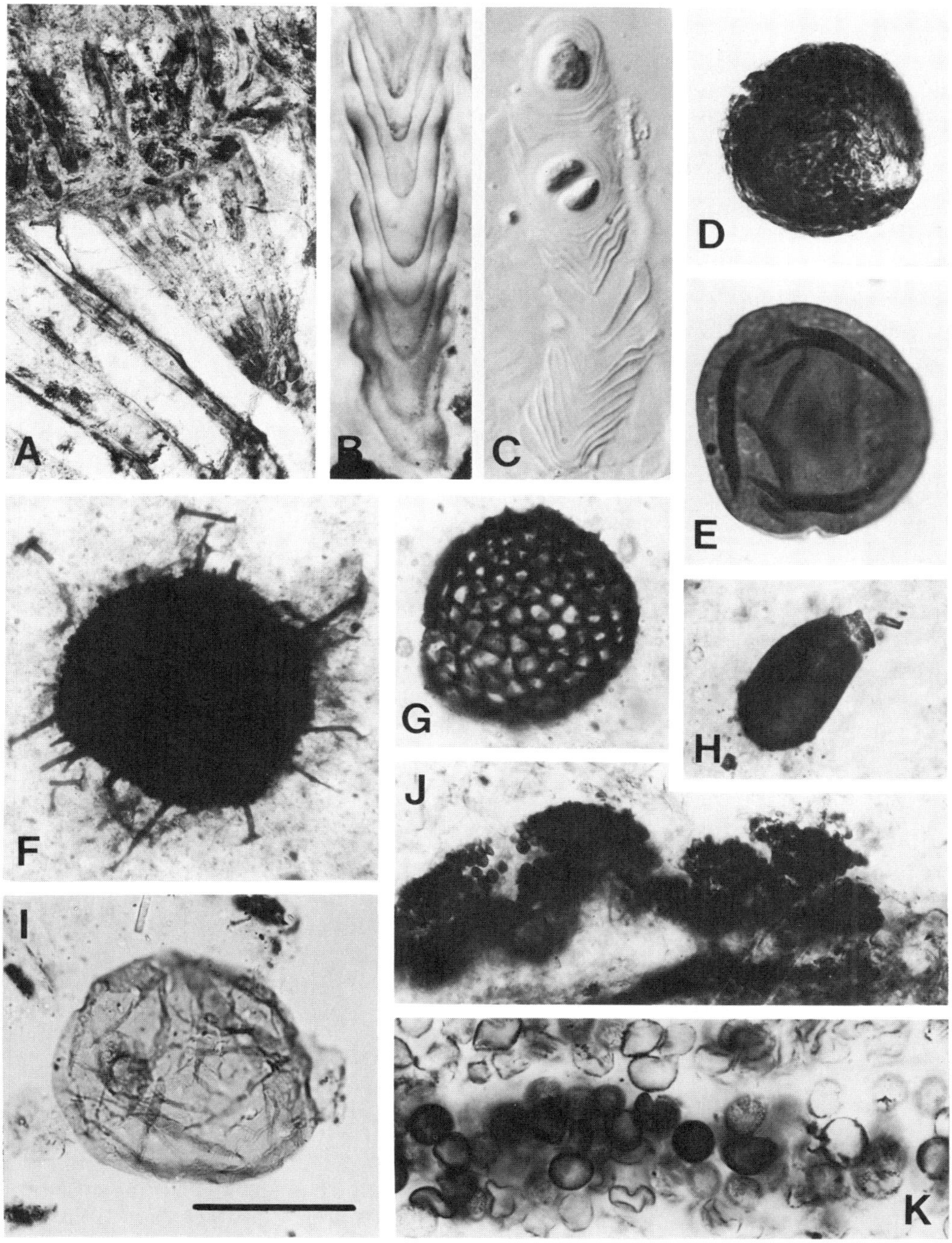

Figure 1. Proterozoic microfossils. (A,B) *Polybessurus bipartitus*, a morphologically distinctive cyanobacterium in silicified peritidal carbonates from the Upper Proterozoic Eleonore Bay Group, East Greenland. (C) Living cyanobacterium from Bahama Banks environment similar to that inferred for *P. bipartitus*. (D,E) Conventional acritarchs from shales of the Upper Proterozoic Visingsö Beds, Sweden, and the Chuar Group, Arizona, respectively. (F,G) Morphologically complex acritarchs from shales of the Upper Proterozoic Svanbergfjellet Formation, Spitsbergen. (H) *Melanocyrillium*, a vase-shaped protist from carbonates of the Upper Proterozoic Elbobreen Formation, Spitsbergen. (I) Leiospherid acritarch from shales of the Middle Proterozoic Mainoru Formation, Roper Group, Australia. (J,K) Silicified lamina-forming protistan fossils in stratiform stromatolites of the Upper Proterozoic Eleonore Bay Group, East Greenland. The bar in part I represents (A) 150 μm; (B) 60 μm; (C,E,G) 20μm; (D,I,K) 40 μm; (F,H) 80 μm; and (J) 200 μm. [Part D courtesy of G. Vidal.]

semblages, they often form dense clusters oriented perpendicular to bedding—indicating localized crust formation in Proterozoic peritidal environments. Physically comparable tidal flat environments in the present-day Bahama Banks contain populations of crust-forming cyanobacteria that are morphologically, developmentally, and behaviorally quite similar to *P. bipartitus* (Fig. 1C). Thus, *P. bipartitus* fossils not only "look like blue-greens," they are morphologically and ecologically comparable to specific extant cyanobacterial populations that live in specifically homologous environments (Green *et al.*, 1987). The implication is that these Proterozoic populations were physiologically as well as morphologically comparable to their modern counterparts.

Polybessurus bipartitus is not an isolated example of a Late Proterozoic population that can sustain detailed, specific comparison to living blue-greens found in similar environments. In fact, such comparisons are more the rule than the exception, suggesting strongly that by Late Proterozoic times cyanobacteria were, in general, physiologically as well as morphologically modern (e.g., Golubic and Hofmann, 1976; Knoll, 1985a; Green *et al.*, 1988). In all probability, most cyanobacteria evolved a modern aspect well before the Late Proterozoic, although how much earlier is uncertain. The partial loss of paleoenvironmental control makes convergence and Schopf's "Volkswagen syndrome" potentially serious problems in the interpretation of Archean and some Early Proterozoic microfossils.

More broadly, geochemistry, microbial physiology, and molecular phylogeny collectively suggest that Eubacteria and Archaebacteria in general are morphologically and physiologically conservative organisms that diversified early in Earth history (Fox *et al.*, 1980; Knoll, 1985b; Woese, 1987). With the important exception of those bacteria that depend on the metabolic products of metazoans and land plants for survival, the prokaryotic flora extant at the dawn of metazoan evolution probably was essentially modern and then, as now, these bacteria held the key to efficient biogeochemical cycling. Bacteria may, in fact, have been instrumental in the evolution of macrophagous animals. Enteric bacteria, living symbiotically within animal hosts, greatly facilitate the catabolic breakdown of ingested food.

2.2. Unicellular Protists

The most abundant fossils in Phanerozoic rocks are not invertebrates; they are the mineralized tests and organic cysts of single-celled protists. Phanerozoic protozoans and algae show patterns of diversification and extinction broadly comparable to those of plants and animals (Tappan and Loeblich, 1972, 1973), and Proterozoic protistan fossils might also exhibit similarly dynamic patterns of evolutionary turnover. This prediction appears to be confirmed by the rapidly improving record of Late Proterozoic acritarchs. Acritarchs are microscopic, organic-walled vesicles of uncertain systematic position—although most are reasonably interpreted in functional terms as the reproductive cysts of planktonic algae. Several Phanerozoic genera formerly placed in this group are now recognized as prasinophyte green algae or dinoflagellates (e.g., Tappan, 1980), and these algal classes, as well as other, possibly extinct groups, likely contributed to the Proterozoic acritarch record.

The pioneering work of Timofeev (1959) showed that Upper Proterozoic strata contain abundant acritarchs of modest morphological complexity and diversity (Figs. 1D and 1E). Assemblages of these routinely encountered microfossils change through time, permitting biostratigraphic subdivision of Upper Proterozoic successions (Jankauskas, 1982; Vidal and Knoll, 1983). Species richness increased throughout the Late Riphean interval (950–680 My ago), only to decline markedly in the Vendian; a pronounced Early Cambrian radiation paralleled that of the metazoans (Volkova *et al.*, 1983, 1985; Vidal and Knoll, 1983).

In recent years, this standard Late Proterozoic acritarch record has been augmented by the discovery of morphologically complex acritarchs in cherts (Knoll, 1984; Yin, 1985; Awramik *et al.*, 1985) and fine-grained shales (Timofeev *et al.*, 1976; Butterfield *et al.*, 1988; Zang, 1988) deposited in normal marine shelf environments. For example, laminated shales in the ca. 700- to 800-My-old Svanbergfjellet Formation, Spitsbergen, contain more than a dozen spinose, process-bearing, or otherwise ornamented acritarch taxa (Butterfield *et al.*, 1988) (Figs. 1F and 1G). Although attaining a level of structural complexity generally associated with Phanerozoic microfossils, these forms, like similar remains in Vendian rocks of China and Australia, are distinct from younger fossils in their commonly large (100–600 μm) size and in the details of their morphology. That is, they appear to be distinctly Proterozoic in aspect. The full paleobiological significance of complex Proterozoic acritarchs is not yet known, owing to their recent discovery and uncertainties surrounding the taphonomic or other environmental circumstances of their distribution; however, the implications are clear. Late Proterozoic acritarchs are significantly more diverse and complex than hitherto appreciated.

Several additional lines of evidence hint at the abundance and diversity of Late Proterozoic protists. Vase-shaped fossils up to 250 μm long and displaying a clear apical collar are tremendously abundant in certain Upper Proterozoic limestones and shales (Bloeser, 1985; Knoll and Vidal, 1980) [occurrences reviewed in Hofmann (1987)] (Fig. 1H). Marked taphonomic differences between these fossils and acritarchs suggest original differences in wall chemistry. Some authors interpret vase-shaped fossils as algal cysts (Bloeser, 1985), but there is a reasonable probability that they are the cysts of heterotrophic protists, or protozoans (e.g., Knoll and Calder, 1983). Some of the small (10–20 μm), morphologically simple spheroids that are nearly ubiquitous in Late Proterozoic microbial mat assemblages may also be the cysts of unicellular protists (Margulis *et al.*, 1983). Especially compelling in this regard are populations of 9- to 16-μm spheroids that form discrete surfaces in tidal flat stratiform stromatolites from the Upper Riphean Eleonore Bay Group, central East Greenland (Green *et al.*, 1989) (Figs. 1J and 1K). These otherwise featureless microfossils commonly contain median splits that strongly support their interpretation as protistan cysts. The green alga *Dunaliella* and the amoeboflagellate protozoan *Paratetramitus* provide structural analogues in present-day mat communities. Molecular geochemistry provides further evidence for a Late Proterozoic protistan radiation; for example, steranes, biomarkers considered diagnostic for eukaryotes, occur in abundance in Upper Proterozoic deposits (Summons *et al.*, 1988a, b).

How far backward can one trace the geological record of eukaryotes? The

1400- to 1500-My-old Roper Group, northern Australia, contains both morphological and molecular evidence of nucleated cells. Peat *et al.* (1978) reported a modest diversity of relatively large spheromorphic microfossils from shales of the McMinn Formation, the uppermost unit of sequence. Comparable fossils are now known to occur throughout the 2000-m-thick group (A. H. Knoll, R. E. Summons, and M. R. Walter, unpublished observations) (Fig. 1I). Roper fossils include populations that closely resemble younger spheromorphic acritarchs of the genus *Leiosphaeridia*, i.e., unornamented, robust-walled vesicles having a tight unimodal size–frequency distribution in the range of 30–60 μm. Other acritarchs are much larger and may display median splits (Peat *et al.*, 1978). None of the populations exhibit sculptured or process-bearing walls, consequently a cyanobacterial interpretation for these vesicles is difficult to rule out unambiguously. On the other hand, most Roper acritarch populations show no positive evidence suggesting a cyanobacterial origin—no evidence of cell division or aggregation and no broad irregular shape variation or size–frequency distribution of the types to be expected if the Roper vesicles were the extracellular envelopes of chroococcalean blue-greens. Locally high concentrations of steranes in Roper bitumens provide independent evidence for the presence of eukaryotes (Jackson *et al.*, 1986; Summons *et al.*, 1988b).

Roper acritarch assemblages are similar to other mid-Proterozoic microfossils reported from North America (Horodyski, 1980), the Soviet Union (Keller and Jankauskas, 1982), and China (Zhang, 1986). All consist exclusively of simple, mostly leiospherid, acritarchs, along with associated prokaryotic filaments. The oldest reported assemblage of this type comes from the Chuanlinggou Formation of China (Zhang, 1986), dated by a whole-rock U–Pb isochron and a Pb–Pb model age on shale of 1922 and 1910 My ago, respectively, as well as by K–Ar dates on phlogopite in intrusive dykes of 1817 and 1875 My ago. Zhang (1986) has questioned the validity of these radiometric determinations; nonetheless, the Chuanlinggou acritarchs are not likely to be younger than Middle Proterozoic in age, and possibly they are Early Proterozoic. Steranes from bitumens in the ca. 1690–My-old McArthur Group, Australia, also provide a faint trail of eukaryotic biology leading backward into latest Early Proterozoic times (Jackson *et al.*, 1986; Summons *et al.*, 1988b).

What, then, can be reasonably inferred about the antiquity of single-celled eukaryotes? Cyst-forming algae existed at least 1400–1500 My ago and underwent a marked morphological diversification during the Late Proterozoic era. The logic of serial endosymbiosis requires that mitochondria-bearing heterotrophic protists predate mitochondria- and plastid-bearing algae, but by how long is uncertain. Morphological paleontology is unlikely to answer this question, but biomarker geochemistry holds exciting promise for deeper penetration into eukaryotic history. The 16S-like rRNAs in living eukaryotes display a sequence variation that matches or exceeds those found for Eubacteria and Archaebacteria, suggesting an early origin of the eukaryotic cytosol (Sogin *et al.*, 1986). While groups such as the euglenids and kinetoplastids (and quite possibly other, as yet unanalyzed groups) appear to have diverged early; the majority of commonly encountered protozoan groups and the ancestors of the fungi, animals, and plants appear to have differentiated during a later radiation. Despite the paucity of the fossil record, one may conclude that Ediacaran metazoans evolved into a world in which both photosynthetic and heterotrophic single-celled protists were diverse, ecologically important components of ecosystems.

2.3. Metaphytes

Morphologically complex, megascopic algae are well represented in Phanerozoic rocks, largely as the calcereous remains of red and green algae. Although not unknown (e.g., Parker and Dawson, 1965; Fry, 1983), the remains of unmineralized metaphytes are rare components of the Phanerozoic fossil record. The evident taphonomic bias against unmineralized seaweeds might be expected to extend backward into the Proterozoic, and this indeed appears to be the case. Nonetheless, carbonaceous compressions of early metaphytes have been reported from Proterozoic rocks, and these are being augmented by the discovery of cellularly preserved metaphyte fossils.

Megascopic carbonaceous compressions have been reported from a number of Proterozoic localities [reviewed by Hofmann (1985a)]. Rightly described as the "orphans of paleontology" (Hofmann, 1985a), most of these remains defy phylogenetic interpretation. Many may be ripped up and redeposited fragments of microbial mats (Horodyski, 1980), but some display a morphologic regularity and determination that precludes such an interpretation. Perhaps best known are the ribbonlike vendotaenids that are widely distributed in Vendian shales (Gnilovskaya, 1979), but these are neither the oldest nor the most complex fossils of this type. *Tawuia dalensis,* first described from the 900- to 750-My-old Little Dal Group, northwestern Canada, is an originally sausage-shaped fossil 2–8 mm wide and 10 to more than 150 mm long (Hofmann and Aitken, 1979; Hofmann, 1985b) (Fig. 2D). Specimens of varying size demonstrate a strong allometric relationship between length and width. The phylogenetic relationships of *T. dalensis* remain uncertain, but it clearly represents a relatively advanced grade of multicellular organization. Another megascopic fossil known from the Little Dal Group, as well as from Upper Proterozoic deposits in China, is *Longfengshania stipitata,* centimeter-scale oblong compressions distinguished by distinct stipes (Du, 1982; Hofmann 1985b) (Fig. 2C). Like *T. dalensis, L. stipitata* is problematic, but almost certainly metaphytic.

Most older compressions are more difficult to interpret. Among the most interesting possible metaphytes are carbonaceous structures collected from the ca. 1400-My-old Belt Group, Montana, by Walcott (1899) and reinterpreted by Walter *et al.* (1976). Walter and colleagues regarded certain of these structures as thalloid red, green, or brown algae. Horodyski (1986), basing his opinions on field as well as museum study, has suggested alternate interpretations for many of these structures, but certain morphologically regular, ribbon-shaped compressions are almost certainly the remains of early seaweeds. The recent discovery of abundant carbonaceous ribbons in dolomitic marls of the ca. 1400-My-old Gaoyuzhuang Formation, China, provides independent evidence in favor of Middle Proterozoic multicellularity (Du *et al.*, 1986). The morphological regularity of these large (1–2 mm wide and up to 50 mm long), sinuous to coiled structures is difficult to reconcile with an abiological or mat fragment origin. The general idea that multicellular eukaryotes existed during the Middle Proterozoic era is also supported by

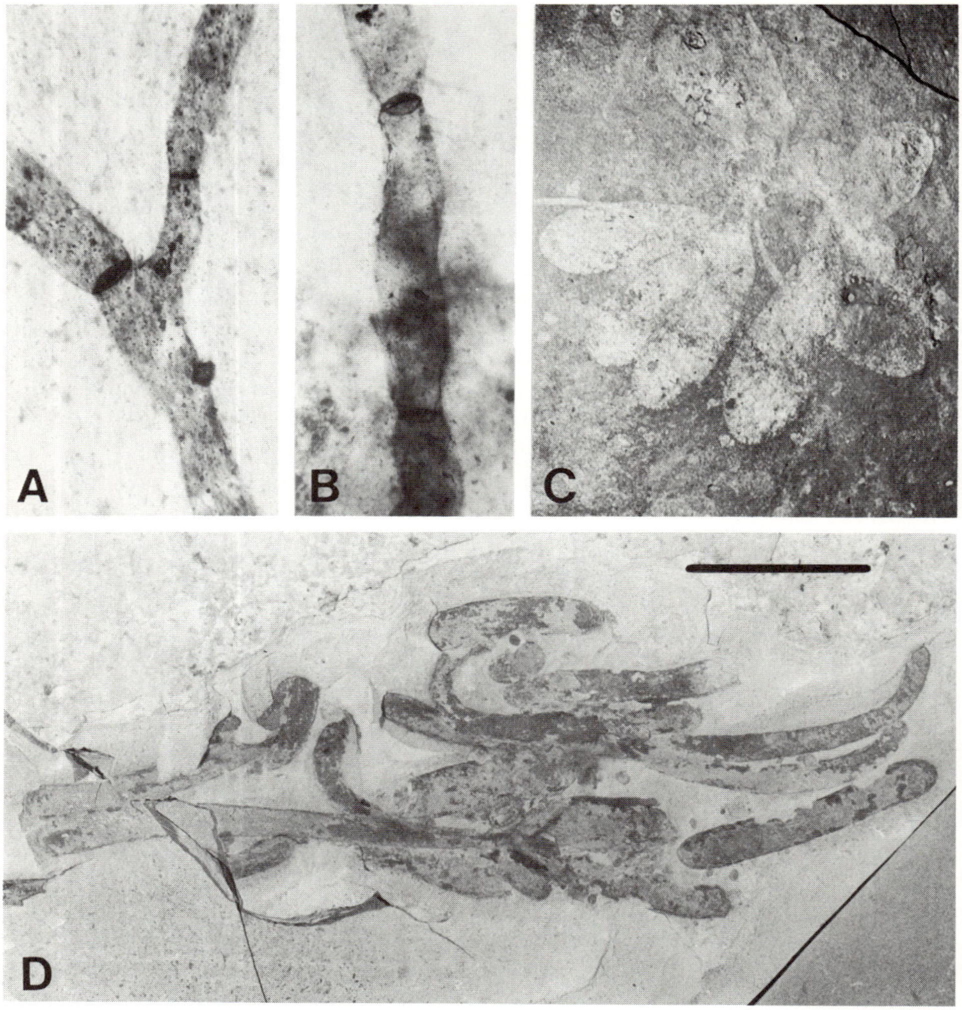

Figure 2. Pre-Ediacaran metaphytes. (A,B) Cellularly preserved cladophoralean green algae from the Upper Proterozoic Svanbergfjellet Formation, Spitsbergen. (C) *Longfengshania stipitata* from the Upper Proterozoic Little Dal Formation, Canada. (D) *Tawuia dalensis,* also from the Little Dal Formation. The bar in part D represents (A) 80 μm; (B) 120 μm; (C) 15 mm; and (D) 20 mm. [Parts C and D courtesy of H. Hofmann.]

unusual trace fossils recently discovered by Grey and Williams (1990; see also Horodyski, 1982) in sandstones of the 1100- to 1300-My-old Manganese Group, Western Australia. These markings closely resemble traces left by species of the segmented brown alga *Hormosira* when stranded along shorelines (K. Grey, personal demonstration); a reasonable abiological alternative has yet to be articulated.

Cellularly preserved remains offer better prospects for phylogenetic interpretation. Recent discoveries in shales and cherts from Spitsbergen (Butterfield *et al.*, 1988) and Somerset Island, Arctic Canada (Butterfield *et al.*, 1990), document the presence of possible cladophoralean (Figs. 2A and 2B) and other green metaphytes, as well as bangialean red algae, in Late Riphean strata.

Whatever the taphonomic and systematic difficulties of tracing early metaphyte evolution, clearly morphologically complex, megascopic algae have a significant Proterozoic fossil record. Red, brown, and several lines of green algae independently evolved metaphytic grades of organization, and likely most, if not all, of these groups attained this status well before the Ediacaran radiation. Indeed, seaweeds may well have appeared relatively soon after the advent of eukaryotic photoautotrophy.

2.4. Pre-Ediacaran Metazoans

Unambiguous metazoan remains are common in marine sediments from Ediacaran times onward; both trace and body fossils tend to occur in more or less dense assemblages that are conspicuously visible in bedding plane or cross-sectional exposure. Pre-Ediacaran strata are starkly different. Upper Riphean and older quartz arenites, shales, and carbonates are routinely devoid of megascopic animal remains or traces. Millimeter-scale laminations in tidal flat to basinal sediments are often preserved undisturbed.

In fact, direct evidence of significantly pre-Ediacaran metazoans is scarce and arguable (Cloud, 1968a; Cloud *et al.*, 1980; Runnegar, 1982a; Glaessner, 1984). Reports of older trace or body fossils appear from time to time, and all deserve to be evaluated on their own merits. But while I keep an open mind, I conclude that, to date, no reported animal structures can be accepted unequivocally as both biogenic and pre-Vendian.* Whatever purported pre-Ediacaran metazoan fossils may be, clearly the dramatic diversification of large animal and trace fossils near the close of the long Proterozoic Eon is not a taphonomic or stratigraphic artifact. It records a genuine radiation.

Like the direct fossil evidence, indirect geological evidence for pre-Ediacaran animals is debatable. Awramik (1971) and Walter and Heys (1985) interpreted secular trends in stromatolite abundance and diversity in terms of early metazoan evolution. Uncontroversially, both interpreted a terminal Proterozoic decline in stromatolite importance to grazing, burrowing, and competition for space by radiating metazoans, although the steepness of the reported Vendian dropoff is surely exacerbated by that interval's brevity and relative paucity of carbonates (Walter and Heys, 1985). Walter and Heys (1985), in particular, interpreted frequency polygon graphs of their data as indicating an overall decay in stromatolite abundance and diversity beginning 700–800 My ago. Unfortunately, the data themselve do not permit this conclusion. Plotted properly as a histogram, they indicate only a significant difference between the Late Riphean and Vendian.

Walter and Heys (1985) further documented a decrease in the abundance of conoidal stromatolites beginning about 1000 My ago, and they suggested that this

*While this chapter was in manuscript form, Hofmann, Narbonne, and Aitken (1990, *Geology* **18**:1199–1202) described possible metazoans from rocks that underlie Vendian tillites in northwestern Canada.

reflects the initial radiation of small metazoans in quiet subtidal environments. This is indeed a possibility; others are that radiating seaweeds coopted available space in such environments, that the entry of eukaryotic algae into mat communities decreased the ability of those communities to build and maintain mats at high angles to the substrate, or that globally statistical differences in environmental representation or sea water chemistry existed that influence comparisons between Middle Riphean and younger carbonates. At present, opinion in this matter is little constrained by evidence.

Runnegar (1982a, 1984) has argued the case for pre-Ediacaran metazoans from a different perspective. Applying molecular clock methodology to sequence divergence in globin genes (and, with less confidence, collagen genes, cytochrome c, and 5S rRNA), he estimated the time of divergence for metazoan phyla to be 800–1000 My ago, significantly before the Ediacaran radiation. The extrapolation of molecular rate data required to make such an estimate has been questioned (e.g., Britten, 1986; Field *et al.*, 1988). Also, Wilson *et al.* (1987), while advocating the use of molecular clocks in evolutionary studies, conceded that the mutations that keep "molecular time" occur at rates more variable than radioactive decay. Thus, time estimates for ancient divergences are subject to significant error. In the case of the globin data, the actual time of metazoan phylum diversification may be even earlier than estimated or later enough to bring this datum in line with the Ediacaran radiation. Nonetheless, 16S-like rRNA data indicating that the protistan ancestors of the fungi, plants, and animals all diverged within a short time interval (Sogin *et al.*, 1986) lend some support to Runnegan's broad estimate in that we *know* that green algae were well established by the Late Riphean.

Logically, there must have been a period of metazoan "prehistory" prior to the Ediacaran radiation—how long is unknown. Molecular estimates of 800–1000 My ago for the origin of Metazoa can be reconciled with the lack of evidence in the Late Riphean geological record if the earliest animals were small and effectively unfossilizable. Tiny, architecturally simple animals such as meiofaunal nematodes may well represent the structural grade of early metazoans. Such an idea is attractive in light of the scenario for latest Proterozoic environmental change presented in the following section; however, direct geological confirmation or refutation will necessarily come from careful petrographic analyses of fabric in Upper Proterozoic shales and carbonates (Awramik *et al.*, 1985).

2.5. Summary of the Late Proterozoic Biota

Clearly, the biosphere into which the Ediacaran fauna radiated was not simply a monotonous expanse of cyanobacterial mats. Microbial mats did, in fact, cover extensive areas of lake bottoms, moist land surfaces, and sea floors, but these communities were themselves complex, containing both autotrophic and heterotrophic eukaryotes as well as cyanobacteria and a host of other, physiologically diverse Eubacteria and Archaebacteria. Metaphytes formed algal lawns in some quiet subtidal environments (Butterfield *et al.*, 1988), and the plankton must have included protistan herbivores and predators, as well as phytoplankton and bacterial decomposers. Archaebacteria and Eubacteria had achieved essentially modern levels of diversity and sophistication. Unicellular protists and metaphytes must have been widely distributed and at least moderately diverse, even though few forms are preserved in the fossil record. Simple metazoans may well have been present.

Ecological complexity did not originate with large animals. Multilevel food webs involving protists and, perhaps, small metazoans as herbivores and predators existed well before the Ediacaran radiation. Save for the biogenic silica cycle, all major geochemical cycles had long been in place. Like ecological complexity, rapid rates of morphological evolution are often assumed to have begun with the rise of the Metazoa, but again, the tempo of morphological evolution had already begun to increase with the radiations and continuing turnover of unicellular protists and metaphytes. Ediacaran metazoans radiated into a taxonomically diverse, ecologically complex, and evolutionarily dynamic world. This fact and the observation that plankton and metaphytes continued to increase in diversity in concert with animals (Knoll, 1987) provide the biological context within which early metazoan evolution must be understood.

3. Biogeochemical Change on the Late Proterozoic Earth

Single-celled eukaryotes existed long before the Ediacaran radiation. Seaweeds may themselves antedate large animals by 650 My or more—a period equal to or greater than the entire recorded history of the metazoans. If the admittedly insecure estimates for the origin of small, architecturally simple animals are correct, then one-third or more of metazoan history elapsed before the Ediacaran and Cambrian radiations. Why did large metazoans appear so late? This question is central to paleontology and evolutionary biology.

Though impossible to dismiss entirely, developmental answers are discouraged by the observation that photosynthetic protists evolved complex multicellularity early in several independent clades. Further, even simple metazoans such as mesozoans, platyhelminths, or nematodes have sophisticated developmental controls (Kenyon, 1988).

The alternative is an extrinsic solution. Perhaps some environmental barrier to the evolution of large animals was removed just prior to the Ediacaran radiation. What kind of geologically plausible environment could support a biota of physiologically varied Eubacteria and Archaebacteria, unicellular eukaryotes, and megascopic algae, but not large animals? The obvious answer is one containing enough free oxygen to support respiration in single-celled heterotrophs (ca. 1% present-day atmospheric levels, PAL) but not enough to support large animals [minimally estimated at ca. 6–10% PAL (Runnegar, 1982b)].

3.1. Oxygen and Metazoan Evolution

The idea that a latest Proterozoic increase in pO_2 made the evolution of large animals possible is not new. Nursall (1959) proposed it and it has been variously championed by Cloud (1968b, 1976), Towe (1970, 1981), Raff and Raff (1970), and Runnegar (1982a, 1982b), among others. The most widely cited exposition of an "oxygen control" hypothesis is that of Berkner and Marshall (1965). These authors suggested that pO_2 first rose above 1% PAL near the end of the Proterozoic Eon, making possible aerobic metabolism and, significantly, providing an ozone shield sufficient to absorb essentially all UV radiation in the lethal band below 290 nm.

Since the articulation of the Berkner–Marshall hypothesis, several generations of atmospheric models and numerous discussions of oxygen and aerobic respiration have been published. On the basis of recent models relating pO_2 to UV absorption by ozone (Blake and Carver, 1977; Levine *et al.*, 1980; Kasting *et al.*, 1985), Berkner and Marshall's (1965) estimate of 1% PAL pO_2 for essentially complete protection against short-wavelength UV radiation is reasonable. As Kasting (1987) put it, "most of the harmful radiation would have been absorbed at O_2 levels above 10^{-2} PAL. . . . Since the atmospheric O_2 abundance should already have exceeded this value by the end of Stage II (ca. 1600 My ago), it seems likely that the Stage III (<1600 Ma) Earth would have been well protected against solar UV." In a review, Chapman and Schopf (1983) similarly concluded that bacterial aerobiosis could have evolved at atmospheric oxygen levels of about 0.5% PAL, with the oxygen-requiring biochemistry of unicellular eukaryotes becoming possible at slightly higher partial pressures. The unequivocal presence of eukaryotic microfossils and steranes (derived from sterols biosynthesized by an O_2-requiring pathway) in 1400- to 1700-My-old rocks indicates that the critical pO_2 was exceeded by this time.

Thus, insofar as metazoan origins are concerned, the problem with the Berkner–Marshall hypothesis is not that the attainment of 1% PAL atmospheric pO_2 was biologically or environmentally insignificant; it is that this threshold must have been crossed 900 My or more before the Ediacaran radiation. This, however, does not mean that atmospheric oxygen levels were irrelevant to the evolution of large animals. Runnegar (1982b) estimated that *Dickinsonia*, a large, flat invertebrate found in several Ediacara-type assemblages, needed a minimum of 6–10% PAL pO_2 to survive. This estimate is based on taxon-independent biophysical requirements for the diffusion of oxygen across multiple cell layers and its movement throughout the body (see also Raff and Raff, 1970; Cloud, 1976). It assumes that Ediacaran animals had well-developed circulatory systems. In the absence of such organographic complexity, Ediacaran animals would have required pO_2 nearly as high as today's. Synthesis of the structurally important protein collagen (Towe, 1970), exercise metabolism, and skeletonization (Rhoads and Morse, 1971) also require oxygen levels well above 1% PAL. Therefore, the hypothesis of a Late Proterozoic increase in atmospheric oxygen still fits the fossil record nicely—if the change in pO_2 was not from less than to greater than 1% PAL, but rather from, say, 1–10% PAL to much higher values (Runnegar, 1982a).

In this regard, I return to the question of pre-Ediacaran animals. As Towe (1981) and Runnegar (1982a, 1984) have pointed out, very small, architecturally simple, and essentially unfossilizable animals have oxygen requirements that differ little from those of unicells. Indeed, some present-day meiofaunal and parasitic animals live under essentially anaerobic conditions. This has prompted speculation that metazoans are primitively anaerobic (Fenchel and Reidl, 1970; Boaden, 1975); however, such organisms still require aerobically synthesized sterols, linking them inexorably, if indirectly, to oxygen-containing environments. The important point here is that the earliest metazoans may well have been small, simple organisms analogous to (although not necessarily homologous with) extant meiofauna such as platyhelminths and nematodes. These animals could proliferate under the low oxygen conditions hypothesized for the Middle and most of the Late Proterozoic eras. The radiation of large metazoans, however, would be made possible by a latest Proterozoic increase in pO_2.

3.2. The Geological Record

The foregoing ideas are appealing, but without independent geological evidence for Late Proterozoic environmental change, they remain speculative. Establishing that a physiologically significant threshold was crossed near the end of the Proterozoic is not easy. Geochemical data suggest that atmospheric oxygen tensions of at least 1% PAL existed much earlier (Holland, 1984), and charcoal in Carboniferous rocks indicates that by this time oxygen concentrations had risen to 60% PAL or more (Cope and Chaloner, 1985); however, there are as yet no geochemical means of documenting intermediate oxygen levels. In their absence, one must minimally make the case that pre-Ediacaran pO_2 was low and that it increased during the latest Proterozoic.

The steady-state pO_2 likely to have been generated by abiological processes is extremely low (Walker, 1977; Canuto *et al.*, 1982, 1983; Kasting, 1985, 1987). Therefore, questions of atmospheric oxygen growth quickly turn to a consideration of photosynthesis. For organisms utilizing Photosystem II—cyanobacteria (usually) and eukaryotes harboring plastids descended from cyanobacterial endosymbionts—the net photosynthetic equation is

$$6CO_2 + 6H_2O \rightarrow C_6H_{12}O_6 + 6O_2$$

The photosynthetic breakdown of water to generate protons and the byproduct O_2 will not necessarily result in increased atmospheric oxygen, because the organic matter and oxygen produced via photosynthesis can be completely respired back to CO_2 and H_2O. The potential for pO_2 increase arises when photosynthetically produced organic matter is buried, leaving the environment enriched in O_2. Of course, this oxygen can be consumed in the oxidation of old organic matter, ferrous iron, reduced sulfur, or other materials. Thus, pO_2 will increase only when the oxidizing potential generated by the burial of photosynthetically produced organic matter exceeds oxygen consumption by weathering and the oxidation of cations in solution.

3.2.1. The Late Proterozoic Carbon Isotope Record

Because of the relationship between oxygen and organic carbon burial, stable carbon isotopes offer a geochemical perspective on the question of paleoatmospheric change. Leaving aside, for the moment, diagenetic or "vital" effects, the ratio of ^{13}C and ^{12}C in sedimentary carbonates and organic matter depends primarily on two things: (1) isotopic fractionation during photosynthetic carbon fixation, which enriches organic matter in ^{12}C, and (2) the relative fluxes of carbonate and organic carbon into sediments. This relationship is well illustrated in Fig. 3 [reproduced from Hayes (1983)]. The ordinate in this diagram indicates

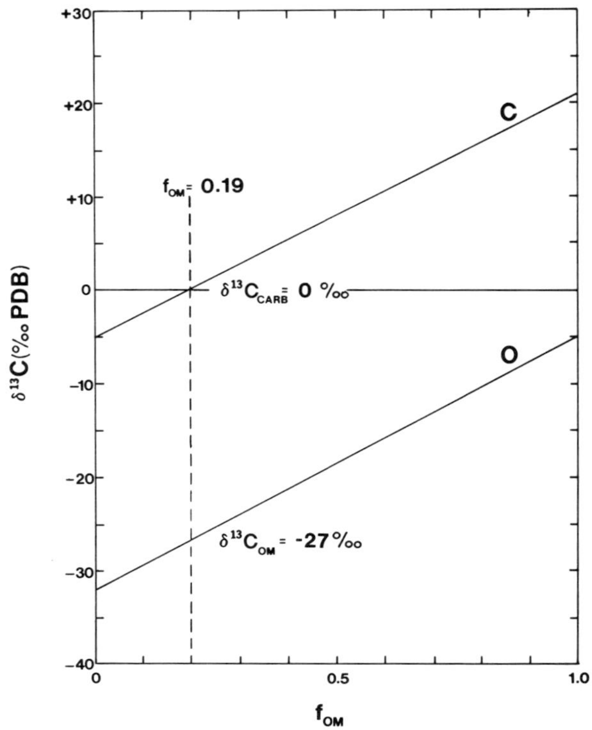

Figure 3. Diagram illustrating the relationship between $\delta^{13}C$ (ordinate, ‰ PDB) and the relative fluxes of carbonate and organic carbon into the sedimentary reservoir (abscissa, f_{OM} is the proportion of carbon buried as organic matter). Diagonal lines *C* and *O* indicate the expected change in carbonate carbon and organic carbon isotopic ratios, respectively, with increasing f_{OM}. The vertical dashed line indicates mean Phanerozoic $\delta^{13}C$ and f_{OM}. [Diagram redrawn from Hayes (1983), with permission.]

$^{13}C/^{12}C$, expressed according to convention as $\delta^{13}C$ (the deviation of the sample from an arbitrarily chosen standard—belemnite carbonate from the Cretaceous Pee Dee Formation, South Carolina):

$$\delta^{13}C = \frac{^{13}C/^{12}C \text{ sample}}{^{13}C/^{12}C \text{ standard}} - 1 \times 10^3 \text{ ‰ PDB}$$

The abscissa indicates the proportion of carbon buried as organic matter. Measurements of mantle carbon indicate that source carbon enters the system at about −5‰ PDB. If all carbon were buried as carbonate, the resulting limestones would have an isotopic signature identical to the input value. Similarly, if all carbon left the system as organic matter, the $\delta^{13}C$ of organic matter would also be −5‰ PDB, regardless of fractionation potential. For intermediate burial proportions, the isotopic signatures for carbonate and organic matter lie along the diagonal lines *C* (carbonate) and *O* (organic matter). The lines are displaced relative to one another because of photosynthetic isotope fractionation; they are parallel because of the requirement for isotopic mass balance between incoming and buried carbon. Thus, carbonates and organic matter newly buried in the present-day oceans have mean isotopic signatures of about 0‰ PDB and −22‰ PDB, respectively, reflecting bulk photosynthetic fractionation ($\Delta\delta$) of roughly 22‰ PDB and an organic matter:carbonate burial ratio of about 1:4. As changes in this burial ratio are thought to depend primarily on variations in the rate of organic matter burial (e.g., Holser, 1984), both carbonate and organic carbon $\delta^{13}C$ values should become higher or lower, respectively, as rates of organic carbon burial increase or decrease.

The carbonate isotopic record thus provides insight into Late Proterozoic environmental change. Broad surveys of carbon isotopes in Archean and Proterozoic rocks (Schidlowski *et al.*, 1975, 1983; Eichmann and Schidlowski, 1975) demonstrate that photosynthesis and the burial of photosynthetically produced organic matter are ancient processes and that isotopic signatures have varied little through time. However, rates of organic carbon burial have not been constant throughout Earth history. Detailed analyses of Phanerozoic carbonates show that $\delta^{13}C$ varies stratigraphically between −2 and +4‰ PDB (Holser, 1984), with short intervals of even heavier carbonate (Holser et al., 1986). Although Proterozoic carbonates should exhibit isotopic variation on similar time scales, broad survey data are not appropriate for addressing this issue. Isolated samples of the type used in published surveys are little constrained in terms of depositional environments or diagenetic history, and they can be correlated only within broad limits.

An alternative is to establish stratigraphic patterns of $\delta^{13}C$ variation within individual sequences, where relative age, depositional facies, and diagenetic alteration are controlled. Figure 4 illustrates the pattern of $\delta^{13}C$ variation within the Upper Proterozoic to Lower Paleozoic Lomfjorden and Hinlopenstretet supergroups of Spitsbergen and correlative successions from Nordaustlandet and East Greenland (Knoll *et al.*, 1986). The three sections represent different portions of what was originally a single sedimentary basin, now tectonically dissected; distances among sample localities at their times of deposition were minimally 650 km.

Patterns of isotopic variation are quite similar among the three sections. In general, the Late Proterozoic carbonates are isotopically heavy ($\delta^{13}C$ = +4 to +7‰ PDB), although this pattern is punctuated by intervals of lighter carbonate (−2 to 0‰ PDB). The $\delta^{13}C$ values for organic carbon covary smoothly with those for carbonates, with a mean $\Delta\delta$ of 28.5‰.

Strong ^{13}C enrichment is unusual in Phanerozoic carbonates. When encountered, it is often ascribed to diagenetic alteration or deposition in a closed basin, although clearly some isotopically heavy carbonates reflect the globally enhanced burial of organic carbon (Holser *et al.*, 1986). The smooth covariance of carbonate and organic carbon isotopic ratios in the Spitsbergen and East Greenland sequences suggests that the stratigraphic curves are not a strong function of diagenesis [see discussion in Knoll *et al.* (1986)]. Combined petrographic–isotopic (microdrilling) analyses of limestones and dolomites also indicate that measured $\delta^{13}C$ values for these carbonates are close to depositional signatures (Fairchild and Spiro, 1987; Kaufman *et al.*, 1991). Sedimentary features that correlate with closed-basin ^{13}C enrichment are lacking—the sequence is virtually free of evaporites, the kerogen content of shales and carbonates is low, and evidence of reduced isotopic fractionation is absent. The $\delta^{13}C$ values are independent of facies within the sequence. At the same time, $\delta^{13}C$ values for individual lithologies are higher than those for similar facies in older or younger sequences.

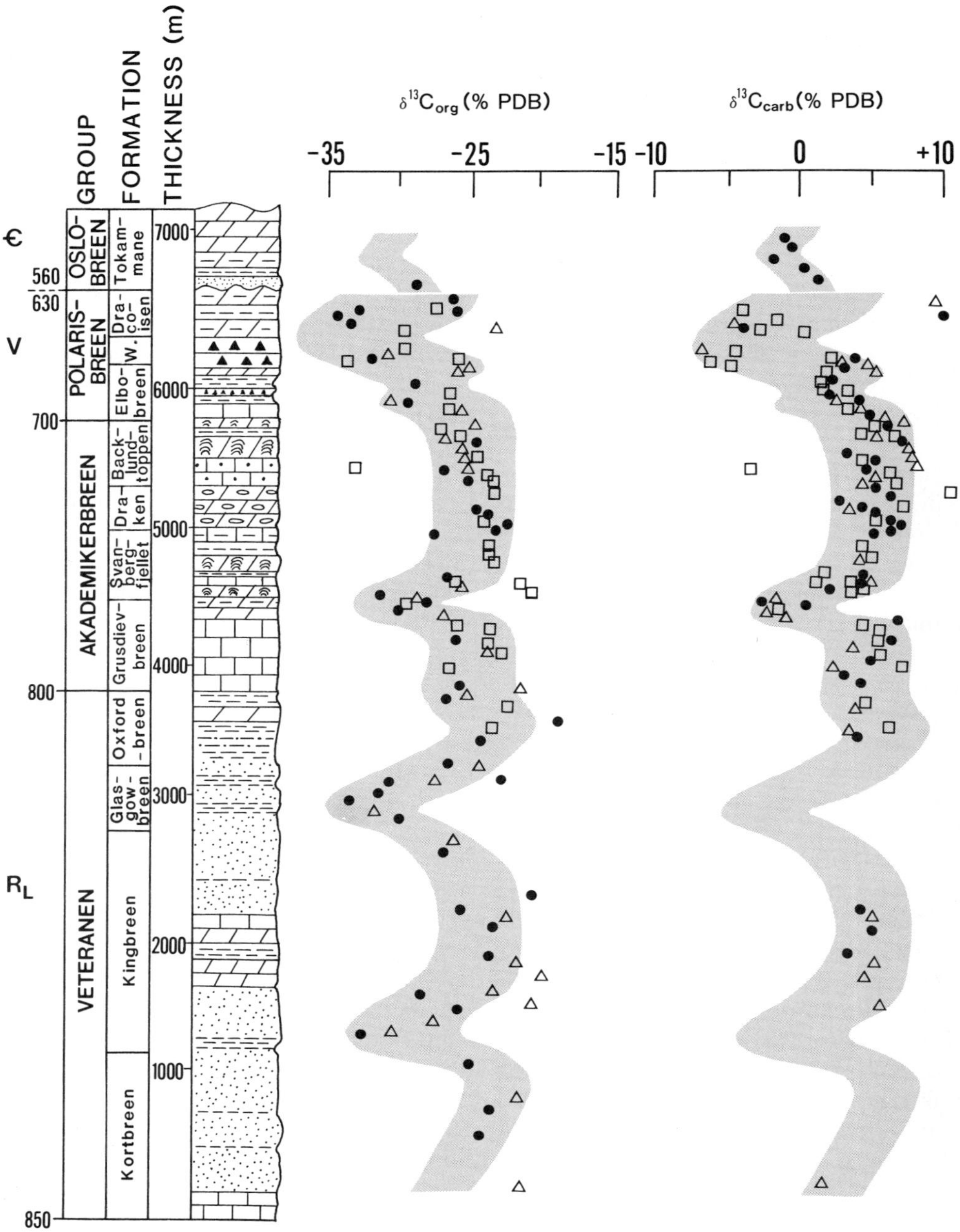

Figure 4. Secular variation of carbonate carbon and organic carbon $\delta^{13}C$ in Upper Proterozoic sequences of northeastern Spitsbergen (circles), Nordaustlandet (triangles), and central East Greenland (squares). The three sequences represent different parts of what was originally a single depositional basin. Correlations among sections are lithostratigraphic and were drawn independently of isotopic data. Curves were drawn using both carbonate and organic carbon ($\delta^{13}C$ + 28.5‰ PDB, the mean difference in $\delta^{13}C$ between carbonate and organic carbon pairs). [Data from Knoll *et al.* (1986).]

The foregoing suggests that the carbon isotopic ratios of Spitsbergen and East Greenland samples reflect, to a first approximation, secular variations in the isotopic composition of Late Proterozoic oceans. The true test of a global signal, however, is the comparison of coeval sequences from various parts of the world. Survey data are consistent with the hypothesis of global ^{13}C enrichment in Upper Proterozoic carbonates and kerogens. Isotopically heavy carbonates are not characteristic of either Cambrian or Middle Proterozoic successions (Schidlowski *et al.*, 1975, 1983; Holser, 1984; Knoll *et al.*, 1986), but isolated samples from five continents indicate that ^{13}C enrichment comparable to that seen in Spitsbergen and East Greenland is common in Upper Proterozoic carbonates. For example, Schidlowski and colleagues (Schidlowski *et al.*, 1975; Banerjee *et al.*, 1986; Sathyanarayan *et al.*, 1987) have published analyses of 43 Middle and Upper Proterozoic carbonate samples from India. The 30 Middle Proterozoic samples have $\delta^{13}C$ values ranging from −3.6 to +1.5‰ PDB ($\bar{x}$ = 0.0‰, S.D. = 1.1‰). Of 13 Upper Proterozoic samples, 12 have values that range from +2.8 to +5.0‰ PDB ($\bar{x}$ = +3.6‰, S.D. = 0.5‰), while the 13th has $\delta^{13}C$ = −0.6‰ PDB. Recently, Strauss (1988) has reported that well-preserved (H/C > 0.2) organic carbon in Upper Proterozoic sedimentary rocks is significantly enriched in ^{13}C relative to comparably well-reserved kerogens in either older Proterozoic or Cambrian samples—an important corroboration of carbonate survey data.

Stratigraphic analyses of $\delta^{13}C$ variation in single basins are few, but increasing. Particularly striking are comparisons of Late Vendian (Ediacaran) sequences from Siberia (Magaritz *et al.*, 1986), China (Lambert *et al.*, 1987, and unpublished data; Wang *et al.*, 1987), Namibia (Kaufman *et al.*, 1991), and the Lesser Himalayas (Aharon *et al.*, 1987). [Possibly coeval carbonates from the Anti-Atlas Mountains, Morocco (Tucker, 1986), are not included here because of uncertainties surrounding their age and the interpretation of extremely heavy carbonates near the base of the sequence (Conway Morris, 1987; Fairchild and Spiro, 1987).] All of the above sequences begin with ^{13}C-depleted carbonates ($\delta^{13}C$ = −4 to −2‰ PDB), increase stratigraphically to a $\delta^{13}C$ peak of +2.5 to +4‰ PDB, and then decrease to moderately light (ca. −1‰ PDB) rocks in the Lower Cambrian. In the Spitsbergen sequence, isotopically light carbonates just above Vendian tillites are overlain by lacustrine dolomites (Fairchild and Hambrey, 1984) and, above a disconformity, typically Cambrian carbonates (Knoll *et al.*, 1968; Fairchild and Spiro, 1987). The "Ediacaran loop" seen elsewhere is not apparent in Spitsbergen—which is actually comforting insofar as acritarch biostratigraphy suggests that the Ediacaran interval is not represented by marine rocks in northeastern Spitsbergen (Knoll and Swett, 1987). Analyses of Upper Riphean sequences are few, but carbonates from the Otavi Group, Namibia, show a stratigraphic pattern of ^{13}C enrichment and depletion generally consistent with that observed in Spitsbergen (Kaufman *et al.*, 1988).

The observed stratigraphic pattern of ^{13}C-enriched Upper Proterozoic carbonates and kerogens separated by isotopically "normal" rocks thus appears to be global in distribution. If isotopically heavy carbonates reflect deposition in basins restricted or closed with respect to CO_2 (cf. Schidlowski *et al.*, 1976; Sathyanarayan *et al.*, 1987), then such basins must have been uniquely widespread during the Late Proterozoic era. The alternative is that the isotopic variation documents periods of globally increased rates of organic carbon deposition punctuated by intervals of moderated accumulation. Given the tectonic setting of many

Upper Proterozoic sequences (discussed below), basin restriction indeed may contribute to anomalously high $\delta^{13}C$ values in some carbonates. In Spitsbergen, for example, some of the ^{13}C enrichment in isotopically heavy samples could reflect basinal conditions; however, local enrichments of as much as 3–4‰ would still suggest that world oceans were enriched in ^{13}C by up to 4‰. Therefore, there appears to be a real underlying pattern of secular variation that may be exacerbated but not obliterated by basinal effects. While acknowledging the need for more and better data, I cautiously accept the second of the two alternative interpretations. During much of the Late Riphean and Vendian intervals, rates of organic carbon burial were as high as or even higher than their Phanerozoic maxima. Four or five times, burial slowed to rates near the Phanerozoic minima, only to increase again—all during the final 300 My of the Proterozoic Eon. A generalized secular curve for Late Proterozoic $\delta^{13}C$ is presented in Fig. 4.

3.2.2. Late Proterozoic Tectonics and Climate

What could account for the unusual variation in rates of organic carbon burial implied by the $\delta^{13}C$ data? In the present-day oceans, high rates of organic carbon burial are not a strong function of surface productivity (Demaison and Moore, 1980; Emerson *et al.*, 1985; Tyson, 1987). Rather, the burial of organic matter is facilitated by rapid clastic sedimentation and combinations of bathymetry and circulation that result in the regional expansion of oxygen-depleted bottom waters (Demaison and Moore, 1980). Phanerozoic episodes of enhanced organic carbon burial in marine sediments suggest similar controls. About 70% of all known petroleum reserves originated in Late Jurassic to Middle Cretaceous source beds (Tissot and Welte, 1984) deposited during a time of widespread continental breakup (with the resultant formation of long, narrow basins), high sea level, and low latitudinal thermal gradients (Schlanger and Jenkyns, 1976; Scholle and Arthur, 1980; de Graciansky *et al.*, 1984).

Data on paleomagnetics, sedimentology/tectonics, and basic dyke swarms indicate that an earlier supercontinent broke apart near the end of the Proterozoic (e.g., Piper, 1976, 1982; Morel and Irving, 1983; Bond *et al.*, 1984; papers in Halls and Fahrig, 1987). Lindsay *et al.* (1987) have observed that in Australia, crustal extension occurred in two distinct phases. At a time 900–800 My ago, a series of extensional basins was initiated over wide areas of the continent, accumulating up to several kilometers of Late Riphean to Vendian fluviatile and shallow marine sediments. Extension was renewed approximately 600 My ago, and at about this time the supercontinent was sundered and ocean basins began to grow. Similar sedimentary/tectonic patterns characterize other continents (e.g., Harland and Gayer, 1972; Fairchild and Hambrey, 1984; Makarevich, 1985; Shpount and Oleinikov, 1987; Bond *et al.*, 1984, and references therein). The Late Proterozoic apparently was a time of unusual if not unique formation of rapidly subsiding extensional basins flooded by marine waters. This development culminated in latest Proterozoic to Early Cambrian continental breakup and a consequent major transgression (Matthews and Cowie, 1979; Bond *et al.*, 1984).

The last 300 My of the Proterozoic was also a time of strong climatic fluctuation, with as many as five distinct ice ages alternating with relatively ice-free periods (Hambrey and Harland, 1985). Following the model of Cretaceous oceans, during Late Proterozoic warm periods, thermohaline circulation probably was characterized by warm, saline bottom waters having a limited capacity to carry oxygen. Given the tectonic/bathymetric disposition of the Earth (and the likelihood that atmospheric pO_2 was much lower than today), during such intervals organic carbon burial would have proceeded at high rates in rapidly subsiding epicratonic basins, as well as in the oceans (see also Walter, 1982). Significant Late Proterozoic oil and gas fields occur in Siberia, Oman, and, perhaps, China, and oil shows have been reported elsewhere in the Soviet Union and in Australia (Postnikova, 1977; Murray *et al.*, 1980; Lagutenkova and Chepikova, 1982; Walter, 1982; Tang and Zhan, 1984; Al-Marjeby and Nash, 1986; Powell, 1987). During glacial epochs, basin drainage occasioned by lowered sea level and vigorous, cold (hence, relatively oxygen-rich) bottom waters would have inhibited the burial of organic matter.

The biogeochemical fluctuation implied by the Late Proterozoic carbon isotope record thus can be explained attractively in terms of tectonic and climatic features known to have characterized the Late Riphean and Vendian Earth. Indeed, possibly high rates of organic carbon burial could have contributed to the initiation of ice ages by drawing down atmospheric CO_2 and decreasing its greenhouse capacity (Walker and Knoll, 1986, and unpublished data), engaging the Earth in an autocyclic climatic fluctuation broken at the close of the era by opening oceans, increased pO_2, and bioturbation.

Direct stratigraphic evidence relates $\delta^{13}C$ fluctuations to climatic change. In Spitsbergen, East Greenland, and China, one horizon of isotopically light carbonates and organic matter occurs in stratigraphic association with Vendian (Varangian) glaciogenic beds. In Namibia, pre-Vendian tillites are also directly associated with isotopically light carbon, and isolated samples of tillite-capping dolomites from Australia show ^{13}C depletion as well (Williams, 1979; Knoll *et al.*, 1986; Wang *et al.*, 1987; Kaufman *et al.*, 1991). As with virtually all other discussions of Late Proterozoic phenomena, more data are needed to clarify relationships, but insofar as tectonics, climate, and biogeochemistry are closely related in the Phanerozoic Eon, they also probably interacted in producing the dramatic environmental fluctuations of the Late Proterozoic era.

3.2.3. Sulfur, Iron, Phosphorites, and Hydrocarbons

Stratigraphically, ^{13}C-enriched carbonates and kerogen occur just where they should if an increase in atmospheric oxygen triggered the Ediacaran radiation. But, as noted earlier, enhanced burial of organic matter need not result in increased pO_2. Is there any *other* geochemical evidence for Late Proterozoic atmospheric change? Conclusive data are hard to come by, but several lines of evidence intriguingly suggest that the answer is "yes."

3.2.3a. The Sulfur Isotope Record. The isotopic record of sedimentary sulfur yields biogeochemical information analogous to that of carbon. During dissimilatory sulfate reduction by sulfate-reducing bacteria, kinetic isotope effects enrich the product H_2S in the lighter isotope, ^{32}S, relative to ^{34}S. Isotopic signatures in sediments primarily reflect this fractionation and the relative fluxes of oxidized (sulfate) and reduced (sulfide) species into and out of the sedimentary reservoir [reviewed by Schidlowski *et al.*, (1983)]. The convention for reporting sulfur isotopic data is analogous to that for carbon:

$$\delta^{34}S = \frac{^{34}S/^{32}S \text{ sample}}{^{34}S/^{32}S \text{ standard}} - 1 \times 10^3 \text{ ‰ CDT}$$

where the standard is troilite from the Canyon Diablo meteorite. The Phanerozoic isotopic curves of carbon and sulfur covary on time scales of 10 My or more, indicating the close redox coupling of these biogeochemical cycles during this eon (Veizer *et al.*, 1980; Holser, 1984). Proterozoic data are few (Fig. 5), and their

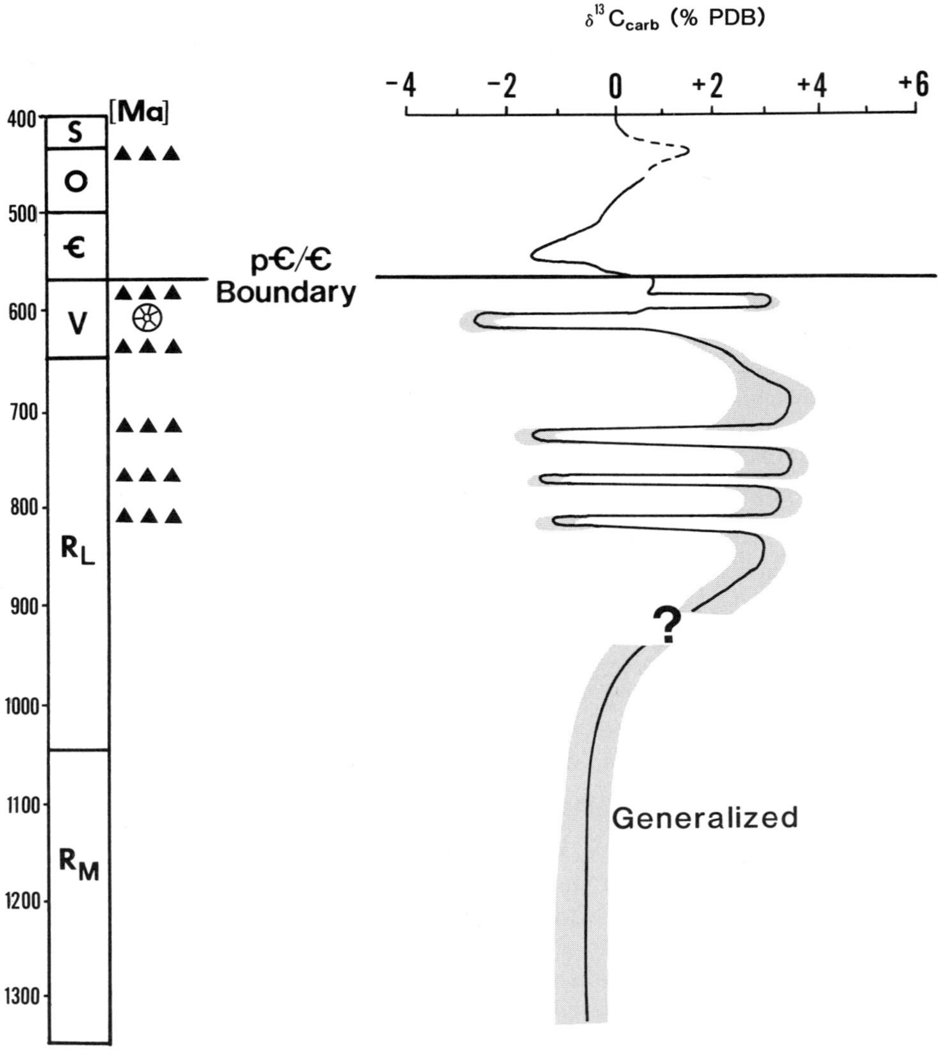

Figure 5. Generalized interpretation of carbonate carbon $\delta^{13}C$ variation through Late Proterozoic and Early Paleozoic time. The data base for sequences older than ca. 900 My is small. S, Silurian; O, Ordovician, €, Cambrian; V, Vendian; R_L, Late Riphean; and R_M, Middle Riphean. Triangles indicate glacial epochs [following Hambrey and Harland (1985), but positioned to reflect the interpreted relationship between ice ages and secular variation in Late Proterozoic carbon isotopes]. The medusoid indicates the stratigraphic level of Ediacaran faunas.

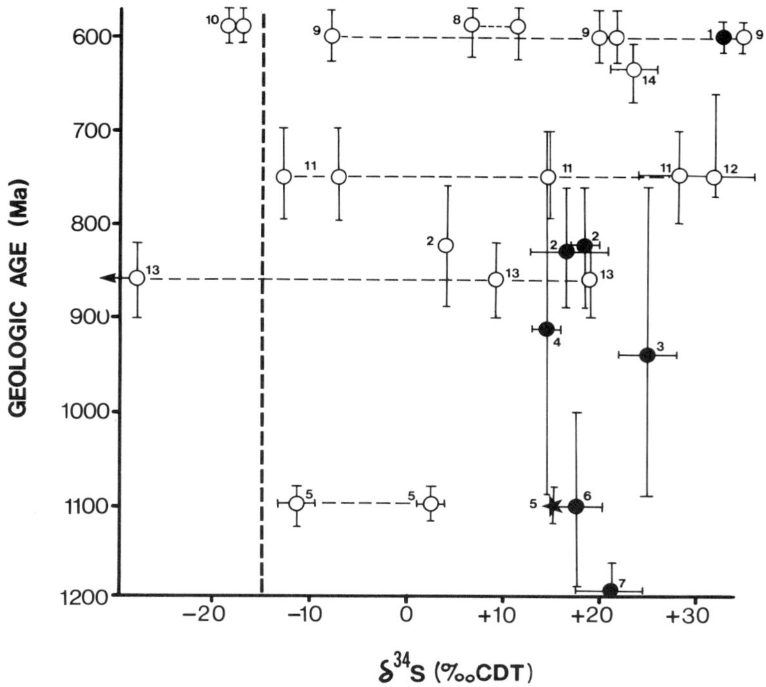

Figure 6. The $\delta^{34}S$ values for Upper Proterozoic sulfate (closed circles) and sulfide (open circles) sulfur. Circles are placed according to estimated age and mean $\delta^{34}S$; vertical lines indicate limits of uncertainty in age estimation; solid horizontal lines indicate one standard deviation about the mean $\delta^{34}S$; dashed horizontal lines connect scattered single $\delta^{34}S$ values for individual formations. The star indicates a reconstituted sulfate value, based on sulfide data. Stratigraphic units are denoted by numbers: (1) Motskaya Formation and underlying, beds, Siberia; (2) Bitter Springs Formation, Australia; (3) Narssarssuk Formation, Greenland; (4) Shaler Group, Canada; (5) Nonesuch Formation, Michigan; (6) Roan Group, Zambia; (7) Grenville metasediments, Canada–United States; (8) Dengying Formation, China; (9) Valdai "Series," East European Platform; (10) Chapel Island Formation, Canada; (11) Lomfjorden Supergroup, Spitsbergen; (12) Tapley Hill Formation, Australia; (13) Chuar Group, Arizona; (14) Doushantuo Formation, China. [Sources of data, Lambert and Donnelly (1990) and references therein, with the exception of Narssarssuk values kindly furnished by E. Perry.]

interpretation must therefore be tentative, but Lambert and Donnelly (1990) have pointed out that if existing measurements are at all representative of the Late Proterozoic Earth, the stratigraphic covariance of carbon and sulfur isotopes characteristic of the Phanerozoic may not hold for earlier periods. A correlation equation fit by Holser (1984) to Phanerozoic isotopic data predicts that Late Proterozoic carbonate $\delta^{13}C$ values of +3 to +4‰ PDB should be accompanied by sulfate $\delta^{34}S$ values of +8 to +11.5‰ CDT, but pre-Ediacaran gypsum and anhydrite have actual $\delta^{34}S$ values of $+20 \pm 5$‰ CDT (Fig. 6). Conversely, the equation predicts $\delta^{13}C$ values of +1 to +2‰ PDB for the $\delta^{34}S$ values measured.

Lambert and Donnelly (1990) also point out that many $\delta^{34}S$ values for Late Proterozoic pyrite are far too high to permit isotopic mass balance with buried sulfate sulfur, given input fluxes and biological fractionation likely to have occurred at that time. Isotopically heavy and variable sulfides are common in younger sediments deposited under brackish to shallow shelf conditions (Schwarcz and Burnie, 1973; Coveney and Shaffer, 1987). Therefore, the suggestion is that isotopically light sulfide sulfur must have been deposited in settings poorly represented by the Proterozoic rock record—continental slopes and ocean bottoms come readily to mind.

Either one of the data sets systematically misrepresents the isotopic composition of the Late Proterozoic ocean (a possibility alluded to earlier for carbon), or there was another major element in the redox balance of Late Proterozoic oceans—one that disappeared just about the time of the Ediacaran radiation. The obvious possibility is ferrous iron in the deep ocean (Derry and Jacobsen, 1988).

3.2.3b. Iron. Upper Proterozoic iron formations are well known (e.g., Young, 1976). Although commonly dismissed as "anomalous," they have the unsettling (for this interpretation) feature of being "anomalous" in significant quantities on six continents. Stratigraphically, Upper Proterozoic iron formations generally occur in association with tillites, and the scenario presented earlier for Late Proterozoic organic carbon burial potentially explains this association as well. Widespread rifting provided a source of ferrous iron which could be stored in deep anoxic ocean waters. The $^{143}Nd/^{144}Nd$ and $^{87}Sr/^{86}Sr$ ratios in Upper Proterozoic carbonates, iron formation, and phosphorites suggest that during the period 900–650 My ago, hydrothermal fluxes were relatively important in determining ocean water chemistry (Veizer *et al.*, 1983; Shaw and Wasserburg, 1985; Derry and Jacobsen, 1988; Derry *et al.*, 1989). With the onset of glaciation, invigorated deep circulation would have transported iron-rich waters onto shallow shelves where contact with oxygen would result in the precipitation of oxide facies iron formation.

Thus, the latest Proterozoic coupling of the carbon and sulfur isotopic curves and the concomitant cessation of iron formation and deposition provide incompletely documented and circumstantial, but tantalizing evidence that pO_2 did indeed increase at this time. The sulfur isotopic mass balance problem noted above is also consistent with this hypothesis—the "missing" isotopically light pyrite may well have been deposited in anoxic slope and ocean basin environments rich in organic matter, ferrous iron, and sulfate-reducing bacteria.

3.2.3c. Molecular Organic Geochemistry. Biomarker molecules in sedimentary organic matter provide another incompletely understood, but intriguing perspective on latest Proterozoic environmental change. Middle and Late Proterozoic bitumens have biomarker signatures that differ in important ways from those of their Phanerozoic counterparts (Summons, 1987; Summons *et al.*, 1988a, b). Of particular significance are the mid-chain monomethyl alkanes that are regular features of Proterozoic organic matter, but occur in much lower relative abundance in bitumens younger than Early Cambrian. According to Summons (1987), "methyl-branching and unsaturation in bacterial fatty acids appear to be mutually exclusive means of controlling membrane fluidity. . . . Mid-chain methylation and wide chain length variation in ester-linked fatty acids may be vestigial strategies for for controling membrane fluidity originating prior to the development of an oxic biosphere and evolution of the now dominant oxygen-dependent desaturation pathways in the biosynthesis of lipids."

Stratigraphic changes in the molecular geochemistry of preserved organic matter may provide an independent approach to the question of latest Proterozoic oxygen levels.

3.2.3d. A Problem and Its Potential Solution. At this point, the astute reader will have noted a significant stratigraphic problem. Anomalously high rates of organic carbon burial commenced at least 800–850 My ago, and, as noted by Knoll *et al.* (1986), the amount of oxygen produced in association with this burial must have been extremely high, far more than enough to bring pO_2 to modern levels. Yet, independent proxies for pO_2 change suggest that atmospheric oxygen concentrations increased rapidly during a short period immediately prior to the Ediacaran radiation, only 600 My ago. How were oxygen levels controlled for the 200 My leading up to this event, and why did they cease functioning when they evidently did? Holland (1989) has suggested that excess production of reduced materials from volcanoes and oceanic ridges could account for the observed carbon isotopic signal. Oxygen produced by the burial of photosynthetically produced organic matter would have been consumed by the oxidation of methane, carbon monoxide, sulfide, and/or metal species. A solution of this type is almost mandated by the isotopic data, but independent, direct evidence is hard to come by. Certainly, the distribution of late Proterozoic iron formations and the anomalously "mantlelike" strontium isotopic ratios noted in Section 3.2.3b would be consistent with, if not proof of, such a scenario. In this regard, it is interesting to note that the base of the Ediacaran radiation sits stratigraphically at or just above a rapid shift in the secular curve of seawater $^{87}Sr/^{86}Sr$ from unusually low to unusually high values.

3.2.4. A Word about Phosphorites

"Perhaps the greatest global phosphogenic episode in geological history" occurred during the Vendian and, especially, Cambrian periods (Notholt and Sheldon, 1986). Cook and Shergold (1984) attributed this event to the turnover and upwelling of previously stratified, anoxic deep waters—a consequence of Vendian glaciation or, more likely, continental breakup and the establishment of new ocean circulation patterns. Parrish *et al.* (1986) cited the additional importance of large-scale Cambrian marine transgression in promoting phosphorite deposition.

It is interesting to consider this event briefly in the context of the larger pattern of Late Proterozoic environmental change under discussion in this chapter. Sedimentary phosphorites occur as scattered deposits in earlier Proterozoic successions, often (but not always) in association with iron formation; however, pre-Vendian accumulations are small compared to those deposited during the Vendian–Cambrian and later in the Phanerozoic Eon. Certainly the establishment of vast new areas of continental shelf and rising sea level were important to Cambrian phosphogenesis, just as similar phenomena were to younger phosphogenic events (Arthur and Jenkyns, 1981). Other factors, however, may contribute to the general enrichment of Phanerozoic sediments in phosphorus relative to those of the Proterozoic. Animal evolution has sometimes been considered as a consequence of Vendian-Cambrian phosphogenesis (e.g., Cook and Shergold, 1984), but the radiation of large animals may itself have facilitated phosphorite deposition by packaging phosphorus (in skeletons and fecal pellets) for efficient sedimentation (Knoll *et al.*, 1986; see also Porter and Robbins, 1981).

If, as I have argued, the deep oceans were permanently swept clean of ferrous iron only in latest Proterozoic times, then another factor is important in the temporal distribution of phosphorites. Ferric iron, precipitated as oxyhydroxides from sea water, scavenges phosphate (Berner, 1973; Froelich *et al.*, 1982; H. D. Holland, personal communication). Thus, the latest Proterozoic elimination of ferrous iron from deep oceans would not only remove an important sink for oxygen, it would remove a major phosphorus sink as well, greatly increasing the probability of phosphorite deposition in the shallow marine environments best represented in ancient sedimentary successions. A corollary, of course, is that more phosphate would be available for recycling. Insofar as more than 90% of the phosphate introduced into the marine photic zone is generated by the recycling of deep water rather than continental weathering (Lerman *et al.*, 1975; Froelich *et al.*, 1982), this could have led to a true increase in the fertility of the oceans near the end of the Proterozic.

4. A Hypothesis Relating Late Proterozoic Biological and Environmental Evolution

Paleontological evidence is beginning to clarify the patterns of Late Proterozoic biological evolution that culminated in the initial radiation of large animals. At the same time, accumulating geological data are deepening our appreciation of this era as a period of major tectonic, biogeochemical, climatic, and atmospheric change. Much remains to be learned about the Late Proterozoic Earth, but it is no longer a *Terra incognita*. Enough data have accumulated to permit the articulation of a hypothesis concerning the relationship between the biological and environmental changes that took place at the end of the long Proterozoic Eon:

Near the end of the Proterozoic Eon, massive rifting and extensional basin formation heralded the breakup of one or a small number of supercontinents. High rates of organic carbon burial associated with this event generated significant amounts of oxygen. Most of this oxygen must have been consumed by the oxidation of sulfides, ferrous iron, and reduced volcanic and hydrothermal gases, but by the middle of the Vendian period, geochemical barriers to oxygen increase had been removed, and pO_2 rose to the point where the shallow oceans could support the physiology of megascopic animals.

Note the form of this hypothesis, as well as its content. A *tectonic* event caused adjustments in the fluxes and reservoirs of *biogeochemical cycles*, precipitating *environmental change* that made possible the *evolution* of a new grade of biological organization. That is, physical and biological Earth history are linked by the one system they share—biogeochemical cycles.

The view advocated here does not require that a Late Proterozoic increase in pO_2 was in and of itself *sufficient* to initiate the Ediacaran revolution. It does, however, hold that the attainment of critical oxygen levels was *necessary* for the evolution of large animals and that, given the stratigraphic congruence of evidence for biological and environmental change, other required steps were either already in place or easily evolved.

This hypothesis accounts for a wide array of geological and paleobiological data. Nonetheless, it is hardly proven. At present many gaps require investigation.

The ideas presented here do focus thinking on the needs for further research, and, of some importance, make specific predictions about the data that will emerge from continuing investigations. Insofar as it predicts the content of observations not yet made, this hypothesis can be falsified in just the way that hypotheses in chemistry and physics are tested by laboratory experiments or astronomical observations. An incomplete sampling of research needs is as follows:

1. There is a need for detailed patterns of stratigraphic $\delta^{13}C$ variation in Middle and Upper Proterozoic basins—the prediction is that a fluctuating pattern of ^{13}C enrichment will characterize Late Proterozoic basins in general, but will not be characteristic of earlier successions.

2. Much improved $\delta^{34}S$ data are needed to complement those for carbon—when known in more detail, will the pre-Vendian sulfur isotope curve really prove unrelated to that of carbon?

3. Isotopic data are needed for carbonates and sulfides deposited in open ocean sediments—the hypothesis predicts that these rocks will contain ^{13}C-enriched carbonate and kerogen, as well as isotopically light sulfides. In practice, this entails the sampling of Upper Proterozoic sediments that are intercalated among or cap ophiolites.

4. Sr and Nd isotopic ratios and rare earth element contents of Upper Proterozoic chemical sediments need further investigation. Will $^{143}Nd/^{144}Nd$ anomalies confirm that hydrothermal activity was relatively important during the Late Proterozic era? Will cerium concentrations confirm the anoxic nature of pre-Ediacaran deep water (cf. Holser *et al.*, 1986)?

5. Insofar as the Late Proterozoic Earth bears comparison with the Jurassic–Cretaceous world, will improved geochemical observations indicate a Mesozoic increase in atmospheric oxygen?

6. Can experimental microbiology shed light on the evolution of lipid biosynthesis, especially in oxygen-depleted environments?

7. Can improved paleontological data refine the presently murky picture of pre-Ediacaran protistan and metaphyte radiations? One might predict, in particular, that careful study will show that green, brown, and red algal seaweeds all diversified long before the Ediacaran radiation.

8. Perhaps most important, we know very little about biological or environmental conditions during the interval preceding the Late Proterozoic era. Will studies of Middle Proterozoic sequences show that Late Proterozoic environmental change was unique or that it was a repeat of earlier events? For example, possibly deep oceans were oxygenated and atmospheric pO_2 relatively high during the Middle Proterozoic era, and the conditions hypothesized for the Late Proterozoic constituted a return to low oxygen conditions comparable to those of the Early Proterozoic Earth.

The specific hypothesis presented in this paper may or may not survive the next decade of research, but the data will not go away. Whether or not the Late Proterozoic Earth evolved just as I have suggested, clearly the Phanerozoic is not the Proterozoic plus large animals. The Late Proterozoic era was a time of dramatic biological, tectonic, climatic, and biogeochemical change that ushered in a new and fundamentally different world. The unique biological and environmental history of this epoch must be taken into account in any attempt to understand early metazoan evolution.

ACKNOWLEDGMENTS. My thinking about Late Proterozoic paleobiology and geology has benefitted from conversations and collaboration with many individuals, among them John Hayes, Keene Swett, Malcolm Walter, Roger Summons, Ian Lambert, Stein Jacobsen, Louis Derry, Jay Kaufman, Roger Buick, Julian Green, Stephen Grant, Nicholas Butterfield, H. D. Holland, M. A. Fedonkin, William Holser, J. J. Sepkoski, and W. B. Harland. Hayes, Holland, Sepkoski, Fedonkin, Walter, and Summons commented on an earlier draft of this paper. Hans Hofmann, Gonzalo Vidal, Nicholas Butterfield, and Julian Green provided some of the photographs used in the figures. Ian Lambert, Eugene Perry, and Roger Summons generously shared unpublished manuscripts and data. I thank them all and note my special appreciation for Preston Cloud, who was a model for interdisciplinary thinking about Earth history. I acknowledge with gratitude fellowship support from the John Simon Guggenheim Foundation. Research leading to this paper was supported by grants from the NSF and NASA.

References

Aharon, P., Schidlowski, M., and Singh, I. B., 1987, Chronostratigraphic markers in end-Precambrian carbon isotope record of the Lesser Himalaya, *Nature* **327:**699–702.

Al-Marjeby, A., and Nash, D., 1986, A summary of the geology and oil habitat of the Eastern Flank Hydrocarbon Province of South Oman, *Mar. Pet. Geol.* **3:**306–314.

Arthur, M. A., and Jenkyns, H. C., 1981, Phosphorites and paleoceanography, *Oceanol. Acta Spec. Publ.* **1981:**83–96.

Awramik, S. M., 1971, Precambrian columnar stromatolite diversity: Reflection of metazoan appearance, *Science* **174:**825–827.

Awramik, S. M., McMennamin, D. S., Yin, C., Zhao, Z., Ding, Q., and Zhang, S., 1985, Prokaryotic and eukaryotic microfossils from a Proterozoic/Phanerozoic transition in China, *Nature* **315:**655–658.

Banerjee, D. M., Schidlowski, M., and Arneth, J. D., 1986, Genesis of Upper Proterozoic–Cambrian phosphorite deposits of India: Isotopic inferences from carbonate fluorapatite, carbonate and organic carbon, *Precambrian Res.* **33:**239–253.

Berkner, L. V., and Marshall, L. C., 1965, On the origin and rise of oxygen concentration in the Earth's atmosphere, *J. Atmos. Sci.* **22:**225–261.

Berner, R. A., 1973, Phosphate removal from sea-water by absorption on volcanogenic ferric oxides, *Earth Planet. Sci. Lett.* **18:**77–86.

Blake, A. J., and Carver, J. H., 1977, The evolutionary role of atmospheric ozone, *J. Atmos. Sci.* **34:**720–728.

Bloeser, B., 1985, *Melanocyrillium*, a new genus of structurally complex Late Proterozoic microfossils from the Kwagunt Formation (Chuar Group), Grand Canyon, Arizona, *J. Paleontol.* **59:**741–765.

Boaden, P. J. S., 1975, Anaerobiosis, meiofauna, and early metazoan evolution, *Zool. Scripta* **4:**21–34.

Bond, G. C., Nickeson, P. A., and Kominz, M. A., 1984, Breakup of a supercontinent between 625 Ma and 555 Ma: New evidence and implications for continental histories, *Earth Planet. Sci. Lett.* **70:**325–345.

Britten, R. J., 1986, Rates of DNA sequence evolution differ between taxonomic groups, *Science* **231:**1393–1398.

Butterfield, N. J., Knoll, A. H., and Swett, K., 1988, Exceptional preservation of fossils in Upper Proterozoic shale, *Nature* **334:**424–427.

Butterfield, N. J., Knoll, A. H., and Swett, K., 1990, A bangiophyte red alga from the Proterozoic of arctic Canada, *Science* **250:**104–107.

Canuto, V. M., Levine, J. S., Augustsson, T. R., and Imhoff, C. L., 1982, UV radiation from the young Sun and oxygen and ozone levels in the prebiological palaeoatmosphere, *Nature* **296:**816–820.

Canuto, V. M., Levine, J. S., Augustsson, T. R., Imhoff, C. L., and Giampapa, M. S., 1983, The young Sun and the atmosphere and photochemistry of the early Earth, *Nature* **305:**281–286.

Chapman, D. J., and Schopf, J. W., 1983, Biological and biochemical effects of the development of an aerobic environment, in: *Earth's Earliest Biosphere: Its Origin and Evolution* (J. W. Schopf, ed.), Princeton University Press, Princeton, New Jersey, pp. 302–320.

Cloud, P., 1968a, Pre-metazoan evolution and the origins of the metazoa, in: *Evolution and Environment* (T. Drake, ed.), Yale University Press, New Haven, Connecticut, pp. 1–72.

Cloud, P. E., 1968b, Atmospheric and hydrospheric evolution on the primitive earth, *Science* **160:**729–736.

Cloud, P., 1976, Beginnings of biospheric evolution and their biogeochemical consequences, *Paleobiology* **2:**351–387.

Cloud, P., Gustafson, L. B., and Watson, J. A. L., 1980, The work of living social insects and the age of the oldest known metazoa, *Science* **210:**1013–1015.

Conway Morris, S., 1987, The search for the Precambrian–Cambrian boundary, *Am. Sci.* **75:**157–167.

Cook, P. J., and Shergold, J. H., 1984, Phosphorus, phosphorites, and skeletal evolution at the Precambrian–Cambrian boundary, *Nature* **308:**231–236.

Cope, M. J., and Chaloner, W. G., 1985, Wildfire: An interaction of biological and physical processes, in: *Geological Factors and the Evolution of Plants* (B. H. Tiffney, ed.), Yale University Press, New Haven, Connecticut, pp. 257–277.

Coveney, R. M., Jr., and Shaffer, N. R., 1988, Sulfur-isotope variations in Pennsylvanian shales of the midwestern United States, *Geology* **16:**18–21.

De Graciansky, P. C., Deroo, G., Herbin, J. P., Montadert, L., Müller, C., Schaaf, A, and Sigal, J., 1984, Ocean-wide stagnation episode in the late Cretaceous, *Nature* **308:**346–349.

Demaison, G. J., and Moore, G. T., 1980, Anoxic environments and oil source bed genesis, *Am. Assoc. Pet. Geol. Bull.* **64:**1179–1209.

Derry, L., and Jacobsen, S., 1988, The Nd and Sr isotopic evolution of Proterozoic seawater, *Geophys. Res. Lett.* **15:**397–400.

Derry, L., Keto, L. S., Jacobsen, S., Knoll, A. H., and Swett, K., 1989, Sr isotopic variations of Upper Proterozoic carbonates from East Greenland and Svalbard, *Geochim. Cosmochim. Acta*, **53:**2331–2339.

Du, R., 1982, The discovery of the fossils such as *Chuaria* in the Qingbaikou System in northwestern Hebei and their significance, *Geol. Rev.* **28:**1–7 [in Chinese, with English abstract].

Du, R., Tian, L., and Li, H., 1986, Discovery of megafossils in the Gaoyuzhuang Formation of the Changchengian System, Jixian, *Acta Geol. Sinica* **1986:**115–120 [in Chinese, with English abstract].

Eichmann, R., and Schidlowski, M., 1975, Isotopic fractionation between coexisting organic carbon–carbonate carbon pairs in Precambrian sediments, *Geochim. Cosmochim. Acta* **39:**585–595.

Emerson, S., Fischer, K., Reimers, C., and Heggie, D., 1985, Organic carbon dynamics and preservation in deep-sea sediments, *Deep Sea Res.* **32:**1–21.

Fairchild, I. J., and Hambrey, M. J., 1984, The Vendian of NE Spitsbergen: Petrogenesis of a dolomite–tillite association, *Precambrian Res.* **26:**111–167.

Fairchild, I. J., and Spiro, B., 1987, Petrological and isotopic implications of some contrasting Late Precambrian carbonates, NE Spitsbergen, *Sedimentology* **34:**973–989.

Fenchel, T., and Riedl, R. J., 1970, The sulfide system: A new biotic community underneath the oxidized layer of marine sand bottoms, *Mar. Biol.***7:**225–268.

Field, K. G., Olsen, G. J., Lane, D. J., Giavannoni, S. J., Ghiselin, M. T., Raff, E. C., Pace, N. R., and Raff, R. A., 1988, Molecular phylogeny of the animal kingdom, *Science* **239:**748–753.

Fox, G. E., *et al.*, 1980, The phylogeny of prokaryotes, *Science* **209:**457–463.

Froelich, P. N., Bender, M. L., Luedtke, N. A., Heath, G. R., and DeVries, T., 1982, The marine phosphorus cycle, *Am. J. Sci.* **282:**474–511.

Fry, W., 1983, An algal flora from the Upper Ordovician of the Lake Winnipeg region, Manitoba, Canada, *Rev. Paleobot. Palynol.* **39:**313–341.

Glaessner, M., 1984, *The Dawn of Animal Life: A Biohistorical Study*, Cambridge University Press, Cambridge.

Gnilovskaya, M. B., 1979, The Venian Metaphyta, *Bull. Cent. Rech. Explor.-Prod. Elf-Aquitaine* **3:**611–618.

Golubic, S., and Hofmann, H. J., 1976, Comparison of modern and mid-Precambrian Entophysalidaceae (Cyanophyta) in stromatolitic algal mats: Cell division and degradation, *J. Paleontol.* **50:**1074–1082.

Green, J. W., Knoll, A. H., Golubic, S., and Swett, K., 1987, Paleobiology of distinctive benthic microfossils from the Upper Proterozoic Limestone–Dolomite "Series", East Greenland, *Am. J. Bot.* **74:**928–940.

Green, J. W., Knoll, A. H., and Swett, K. S., 1988a, Microfossils in oolites and pisolites from the Upper Proterozoic Eleonore Bay Group, central East Greenland, *J. Paleontol.* **62:**835–852.

Green, J. W., Knoll, A. H., and Swett, K., 1989, Microfossils in silicified stromatolitic carbonates from the Upper Proterozoic Limestone–Dolomite "Series", central East Greenland, *Geol. Mag.* **126:**567–585.

Grey, K., and Williams, I. R., 1990, Problematic bedding plane markings from the Middle Proterozoic Manganese Group, Bangemall Basin, Western Australia, *Precambrian Res.* **46:**307–327.

Halls, H. C., and Fahrig, W. F. (eds.), 1987, Mafic Duke Swarms, *Geol. Assoc. Can. Spec. Pap.* **34.**

Hambrey, M. J., and Harland, W. B., 1985, The Late Proterozoic glacial era, *Palaeogeogr. Palaeoclimatol. Palaeoecol.* **51:** 255–272.

Harland, W. B., and Gayer, R. A., 1972, The Arctic Caledonides and earlier oceans, *Geol. Mag.* **109:**289–384.

Hayes, J. M., 1983, Geochemical evidence bearing on the origin of aerobiosis, a speculative hypothesis, in: *Earth's Earliest Biosphere: Its Origin and Evolution* (J. W. Schopf, ed.), Princeton University Press, Princeton, New Jersey, pp. 291–301.

Hofmann, H. J., 1985a, Precambrian carbonaceous megafossils, in: *Paleoalgology: Contemporary Research and Applications* (D. F. Toomey and M. H. Nitecki, eds.), Springer-Verlag, Berlin, pp. 20–33.

Hofmann, H. J., 1985b, The mid-Proterozoic Little Dal macrobiota, Mackenzie Mountains, north-west Canada, *Paleontology* **28:**331–354.

Hofmann, H. J., 1987, Precambrian biostratigraphy, *Geosci. Can.* **14:**135–154.

Hofmann, H. J., and Aitken, J. D., 1979, Precambrian biota from the Little Dal Group, Mackenzie Mountains, northwestern Canada, *Can. J. Earth Sci.* **16:**150–166.

Holland, H. D., 1984, *The Chemical Evolution of the Atmosphere and Oceans*, Princeton University Press, Princeton, New Jersey.

Holland, H. D., 1989, Volcanic gases and isotopic record of carbon and sulfur in sedimentary rocks, in: *Abstracts. 28th International Geological Congress, Washington, D.C.*, Vol. 2, p. 66.

Holser, W. T., 1984, Gradual and abrupt shifts in ocean chemistry during Phanerozoic time, in: *Patterns of Change in Earth Evolution* (H. D. Holland and A. F. Trendall, eds.), Springer-Verlag, Berlin, pp. 123–144.

Holser, W. T., Magaritz, M., and Wright, J., 1986, Chemical and isotopic variations in the world ocean during the Phanerozoic, in: *Global Bio-Events* (O. Walliser, ed.), Springer-Verlag, Berlin, pp. 63–74.

Horodyski, R. J., 1980, Middle Proterozoic shale-facies microbiota from the lower Belt Supergroup, Little Belt Mountains, Montana, *J. Paleontol.* **54:**649–663.

Horodyski, R. J., 1982, Problematic bedding-plane markings from the Middle Proterozoic Appekunny Argillite, Belt Supergroup, northwestern Montana, *J. Paleontol.* **56:**882–889.

Horodyski, R. J., 1986, Walcott's *Helminthoidichnites* assemblage from the Middle Proterozoic Belt Supergroup, *Geol. Soc. Am. Abstr. Progr.* **18:**640.

Jackson, M. J., Powell, T. G., Summons, R. E., and Sweet, I. P., 1986, Hydrocarbon shows and petroleum source rocks in sediments as old as 1.7×10^9 years, *Nature* **322:**727–729.

Jankauskas, T. V., 1982, Microfossils from the Riphean of the Southern Urals, in: *Stratotype of the Riphean: Paleontology and Paleomagnetics* (B. M. Keller, ed.), Nauka, Moscow, pp. 84–120.

Kasting, J. F., 1985, Photochemical consequences of enhanced CO_2 levels in Earth's early atmosphere, in: *The Carbon Cycle and Atmospheric CO_2: Natural Variation Archean to Present* (E. T. Sundquist and W. S. Broecker, eds.), American Geophysical Union, Washington, D.C., pp. 612–622.

Kasting, J. F., 1987, Theoretical constraints on oxygen and carbon dioxide concentrations in the Precambrian atmosphere, *Precambrian Res.* **34:**205–229.

Kasting, J. F., Holland, H. D., and Pinto, J. P., 1985, Oxidant abundances in rainwater and the evolution of atmospheric oxygen, *J. Geophys. Res.* **90:**10497–10510.

Kaufman, A. J., Hayes, J. M., Knoll, A. H., and Germs, J. G. B., 1991, Secular variation of stable carbon isotope ratios in the Upper Proterozoic Otavi and Nama groups, Namibia/South West Africa, *Precambrian Res.* **49:**301–327.

Keller, B. M., and Jankauskas, T. V., 1982, Microfossils in the Riphean stratotype section in the Southern Urals, *Int. Geol. Rev.* **24:**925–933.

Kenyon, C., 1988, The nematode *Caenorhabditis elegans*, *Science* **240:**1448–1453.

Knoll, A. H., 1984, Microbiotas of the Late Precambrian Hunnberg Formation, Nordaustlandet, Svalbard, *J. Paleontol.* **58:**131–162.

Knoll, A. H., 1985a, The distribution and evolution of microbial life in the late Proterozoic era, *Annu. Rev. Microbiol.* **39:**391–417.

Knoll, A. H., 1985b, Patterns of evolution in the Archean and Proterozoic eons, *Paleobiology* **11:**53–64.

Knoll, A. H., 1987, Protists and Phanerozoic evolution in the oceans, in: *Fossil Prokaryotes and Protists: Notes for a Short Course* (J. Lipps and T. Broadhead, eds.), University of Tennessee Press, Knoxville, Tennessee, pp. 248–264.

Knoll, A. H., and Calder, S., 1983, Microbiotas of the late Precambrian Ryssö Formation, Nordaustlandet, Svalbard, *Palaeontology* **26:**467–496.

Knoll, A. H., and Swett, K., 1987, Micropaleontology across the Precambrian–Cambrian boundary in Spitsbergen, *J. Paleontol.* **61:**898–926.

Knoll, A. H., and Vidal, G., 1980, Late Proterozoic vase-shaped microfossils from the Visingsö Beds, Sweden, *Geol. Fören. Stockh. Förh.* **102:**207–211.

Knoll, A. H., Hayes, J. M., Kaufman, A. J., Swett, K., and Lambert, I. B., 1986, Secular variation in carbon isotope ratios from Upper Proterozoic successions of Svalbard and East Greenland, *Nature* **321:**832–838.

Knoll, A. H., Swett, K., and Burkhardt, E., 1989, The paleoenvironmental distribution of microfossils and stromatolites in the Upper Proterozoic Backlundtoppen Formation, Spitsbergen, *J. Paleontol.* **63:**129–145.

Lagutenkova, N. S., and Chepikova, I. K., 1982, *Upper Proterozoic Rocks of the Volga-Ural Region and Their Petroleum Potential,* Nauka, Moscow [in Russian].

Lambert, I. B., and Donnelly, T. H., 1990, The paleoenvironmental significance of trends in sulfur isotope compositions in the Precambrian: A critical review, in: Stable Isotopes and Fluid Processes in Mineralization (H. K. Herbert and S. E. Ho, eds.), *Spec. Publ. 23 Geol. Soc. Austr.*:260–263.

Lambert, I. B., Walter, M. R., Zang, W., Lu, S., and Ma, G., 1987, Paleoenvironment and carbon isotope stratigraphy of Upper Proterozoic carbonates of the Yangtze Platform, *Nature* **325:**140–142.

Lerman, A., Mackenzie, F. T., and Garrels, R. M., 1975, Modeling of geochemical cycles: Phosphorus as an example, *Geol. Soc. Am. Mem.* **142:**205–218.

Levine, J. S., Boughner, R. E., and Smith, K. A., 1980, Ozone, ultraviolet flux, and the temperature of the paleoatmosphere, *Origins Life* **10:**199–213.

Lindsay, J. F., Korsch, R. J., and Wilford, J. R., 1987, Timing the breakup of a Proterozoic supercontinent: Evidence from Australian intracontinental basins, *Geology* **15:**1061–1064.

Magaritz, M., Holser, W. T., and Kirschvink, J. L., 1986, Carbon-isotope events across the Precambrian/Cambrian boundary on the Siberian Platform, *Nature* **320:**258–259.

Makarevich, V. N., 1985, *Geologic Framework and Evolution of Aulacogens and Their Petroleum Potential,* VNIIOENG, Moscow [in Russian].

Margulis, L., Grosovsky, B. D. D., Stolz, J. F., Gong-Collins, E. J., Lenk, S., Read, D., and Lopez-Cortes, A., 1983, Distinctive microbial structures and the pre-Phanerozoic fossil record, *Precambrian Res.* **20:**443–477.

Matthews, S. C., and Cowie, J. W., 1979, Early Cambrian transgressions, *J. Geol. Soc. Lond.* **136:**133–135.

Morel, P., and Irving, E., 1983, Tentative paleocontinental maps for the Early Phanerozoic and Proterozoic, *J. Geol.* **5:**535–561.

Murray, G. E., Kaczor, N. J., and McArthur, R. E., 1980, Indigenous Precambrian petroleum revisited, *Am. Assoc. Pet. Geol. Bull.* **64:**1681–1700.

Notholt, A. J. G., and Sheldon, R. P., 1986, Proterozoic and Cambrian phosphorites—Regional review: World resources, in: *Phosphate Deposits of the World,* Vol. 1, *Proterozoic and Cambrian Phosphorites* (P. J. Cook and J. H. Shergold, eds.), Cambridge University Press, Cambridge, pp. 9–19.

Nursall, J. R., 1959, Oxygen as a prerequisite to the origin of the metazoa, *Nature* **183:**1170–1172.

Parker, B. C., and Dawson, E. Y., 1965, Non-calcareous marine algae from California Miocene deposits, *Nova Hedwig.* **10:**273–295.

Parrish, J. T., Ziegler, A. M., Scotese, C. R., Humphreville, R. G., and Kirschvink, J. L., 1986, Proterozoic and Cambrian phosphorites—Specialist studies: Early Cambrian paleogeography, paleoceanography, and phosphorites, in: *Phosphate Deposits of the World,* Vol. 1, *Proterozoic and Cambrian Phosphorites* (P. J. Cook and J. H. Shergold, eds.), Cambridge University Press, Cambridge, pp. 280–294.

Peat, C. J., Muir, M. D., Plumb, K. A., McKirdy, D. M., and Norvick, M. S., 1978, Proterozoic microfossils from the Roper Group, Northern Territory, Australia, *Bur. Miner. Resour. J. Austr. Geol. Geophys.* **3**:1–17.

Piper, J. D. A., 1976, Paleomagnetic evidence for a Late Proterozoic supercontinent, *Philos. Trans. R. Soc. Lond.* **280A**:469–490.

Piper, J. D. A., 1982, The Precambrian paleomagnetic record: The case for the Proterozoic supercontinent, *Earth Planet, Sci. Lett.* **59**:61–89.

Porter, K. G., and Robbins, E. I., 1981, Zooplankton fecal pellets link fossil fuel and phosphate deposits, *Science* **212**:931–933.

Postnikova, I. E., 1977, *Upper Precambrian Rocks of the Russian Plate and Their Petroleum Potential*, Nauka, Moscow [in Russian].

Powell, T., 1987, Petroleum geology of the Siberian Platform, *Bull. Miner. Resour. Res. Newsl.* **6**:11.

Raff, R. A., and Raff, E. C., 1970, Respiratory mechanisms and the metazoan fossil record, *Nature* **228**:1003–1004.

Rhoads, D. C., and Morse, J. W., 1971, Evolutionary and ecological significance of oxygen-deficient marine basins, *Lethaia* **4**:413–428.

Runnegar, B, 1982a, The Cambrian explosion: Animals or fossils?, *J. Geol. Soc. Aust.* **29**:395–411.

Runnegar, B., 1982b, Oxygen requirements, biology and phylogenetic significance of the late Precambrian worm *Dickinsonia*, and the evolution of the burrowing habit, *Alcheringa* **6**:223–239.

Runnegar, B., 1984, Molecular palaeontology, *Palaeontology* **29**:1–24.

Sathyanarayan, S., Arneth, J. D., and Schidlowski, M., 1987, Stable isotope geochemistry of sedimentary carbonates rom the Proterozoic Kaladgi, Badami, and Bhima groups, Karnataka, India, *Precambrian Res.* **37**:147–156.

Schidlowski, M., Eichmann, R., and Junge, C. E., 1975, Precambrian sedimentary carbonates: Carbon and oxygen isotope geochemistry and implications for the terrestrial oxygen budget, *Precambrian Res.* **2**:1–69.

Schidlowski, M., Eichmann, R., and Junge, C. E., 1976, Carbon isotope geochemistry of the Precambrian Lomagundi carbonate province, Rhodesia, *Geochim. Cosmochim. Acta* **40**:449–455.

Schidlowski, M., Hayes, J. M., and Kaplan, I. R., 1983, Ancient biochemistries, in: *Earth's Earliest Biosphere: Its Origin and Evolution* (J. W. Schopf, ed.), Princeton University Press, Princeton, New Jersey, pp. 149–186.

Schlanger, S. O., and Jenkyns, H. C., 1976, Cretaceous oceanic anoxic events: Causes and consequences, *Geol. Mijnbouw* **55**:179–184.

Scholle, P. A., and Arthur, M. A., 1980, Carbon isotope fluctuations in Createaceous pelagic limestones: Potential stratigraphic and petroleum exploration tool, *Am. Assoc. Pet. Geol. Bull.* **64**:67–87.

Schopf, J. W., 1972, Evolutionary significance of the Bitter Springs (late Precambrian) microflora, in: *Proceedings 24th International Geological Congress*, Section 1, pp. 68–77.

Schwarcz, H. P., and Burnie, S. W., 1973, Influence of sedimentary environments on sulfur isotope ratios in clastic rocks: A review, *Mineral. Deposita* **8**:264–277.

Shaw, H. F., and Wasserburg, G. J., 1985, Sm–Nd in marine carbonates and phosphates: Implications for Nd isotopes in seawater and crustal ages, *Geochim. Cosmochim. Acta* **49**:503–518.

Shpount, B. R., and Oleinikov, B. V., 1987, A comparison of mafic swarms from the Siberian and Russian platforms, in: Mafic Duke Swarms (H. C. Halls and W. F. Fahrig, eds.), *Geol. Assoc. Can. Spec. Paper* **34**:379–383.

Sogin, M. L., Elwood, H. L., and Gunderson, J. H., 1986, Evolutionary diversity of eukaryotic small-subunit rRNA genes, *Proc. Natl. Acad. Sci. USA* **83**:1383–1387.

Strauss, H., 1988, Proterozoic organic carbon—Preservation and record, in: *Abstracts Symposium, The Proterozoic Biosphere: An Interdisciplinary Study*, Los Angeles, California, p. 31.

Summons, R. E., 1987, Branched alkanes from ancient and modern sediments: Isomer discrimination by GC/MS with multiple reaction monitoring, *Org. Geochem.* **11**:281–289.

Summons, R. E., Powell, T. G., and Boreham, C. J., 1988a, Petroleum geology and geochemistry of the Middle Proterozoic McArthur Basin, northern Australia, *Geochim. Cosmochim. Acta* **52**:1747–1763.

Summons, R. E., Volkman, J. K., and Boreham, C. J., 1988b, Dinosterane and other 4-methyl steranes of dinoflagellate origin from sediments and petroleum, Geochim. Cosmochim. Acta **51**:3075–3082.

Tang, Z., and Zhan, S. (eds.), 1984, *Forming Conditions of Sinian Gas Pool in Sichuan Basin*, Beijing Petroleum Geology Symposium, Special Publication, Beijing.

Tappan, H., 1980, *The Paleobiology of Plant Protists*, Freeman, San Francisco.

Tappan, H., and Loeblich, A. R., Jr., 1972, Fluctuating rates of protistan evolution, diversification, and extinction, in: *Proceedings. 24th International Geological Congress*, Section 7, pp. 205–213.

Tappan, H., and Loeblich, A. R., Jr., 1973, Evolution of the oceanic plankton, *Earth Sci. Rev.* **9:**207–240.

Timofeev, B. V., 1959, The oldest flora of Pre-Baltica and its stratigraphic significance, *Trudy VNIGRI* **129:**1–320 [in Russian].

Timofeev, B. V., German, T. N., and Mikhailova, N., 1976, *Plant Microfossils of the Precambrian, Cambrian, and Ordovician,* Nauka, Institute of Precambrian Geology and Geochronology, Leningrad [in Russian].

Tissot, B. P., and Welte, D. H., 1984, *Petroleum Formation and Occurrence,* Springer-Verlag, Berlin.

Towe, K. M., 1970, Oxygen–collagen priority and the early metazoan fossil record, *Proc. Natl. Acad. Sci. USA* **65:**781–788.

Towe, K. M., 1981, Biochemical keys to the emergence of complex life, in: *Life in the Universe* (J. Billingham, ed.), MIT Press, Cambridge, Massachusetts, pp. 297–306.

Tucker, M. E., 1986, Carbon isotope excursions in Precambrian/Cambrian boundary beds, Morocco, *Nature* **319:**48–50.

Tyson, R. V., 1987, The genesis and palynofacies characteristics of marine petroleum source rocks, in: Marine Petroleum Source Rocks (J. Brooks and A. J. Fleet, eds.), *Spec. Publ. Geol. Soc. Lond.* **26:**47–67.

Veizer, J., Holser, W. T., and Wilgus, C. K., 1980, Correlation of $^{13}C/^{12}C$ and $^{34}S/^{32}S$ secular variations, *Geochim. Cosmochim. Acta* **44:**579–587.

Veizer, J., Compston, W., Clauer, N., and Schidlowski, M., 1983, $^{87}Sr/^{86}Sr$ in Late Proterozoic carbonates: Evidence for a "mantle event" at 900 Ma ago, *Geochim. Cosmochim. Acta* **40:**905–914.

Vidal, G., and Knoll, A. H., 1983, Proterozoic plankton, *Geol. Soc. Am. Mem.* **161:**265–277.

Volkova, N., Kirjanov, V. V., Piskun, L. V., Paskeviciene, L. T., and Jankauskas, T. V., 1983, Plant microfossils, in: *Upper Precambrian and Cambrian Paleontology of the East-European Platform* (A. Urbanek and A. Y. Rozanov, eds.), Wydawnnictwa Geologzne, Warsaw, pp. 7–46.

Volkova, N., 1985, Acritarchs and other plant microfossils from the East European Platform, in: *Vendian System,* Vol. 1: Paleontology (B. V. Sokolov and A. B. Ivanovsky, eds.), Nauka, Moscow, pp. 130–139 [in Russian].

Walcott, C. D., 1899, Pre-Cambrian fossiliferous formations, *Geol. Soc. Am. Bull.* **10:**199–244.

Walker, J. C. G., 1977, *Evolution of the Atmosphere,* Macmillan, New York.

Walker, J. C. G., and Knoll, A. H., 1986, Ice ages caused by biological innovation, *EOS* **67:**868 (Abstract).

Walter, M. R., 1982, Proterozoic and Cambrian petroleum, *Annu. Rep. Baas Becking Geobiol. Lab.* **1981:**45–46.

Walter, M. R., and Heys, G. R., 1985, Links between the rise of the Metazoa and the decline of stromatolites, *Precambrian Res.* **29:**149–174.

Walter, M. R., Oehler, J. H., and Oehler, D. Z., 1976, Megascopic algae 1300 million years old from the Belt Supergroup, Montana: A reinterpretation of Walcott's *Helminthoidichnites, J. Paleontol.* **50:**872–881.

Wang, F., Liu, R., and Zhao, D., 1987, Precambrian stromatolite oxygen isotopes from Sichuan-Yunnan area and its significance, *Bull. Chengdu Inst. Geol. Min. Res.* **8:**61–68.

Williams, G. W., 1979, Sedimentology, stable-isotope geochemistry, and palaeoenvironments of dolostones capping late Precambrian glacial sequences in Australia, *J. Geol. Soc. Aust.* **26:**377–386.

Wilson, A. C., Ochman, H., and Prager, E. M., 1987, Molecular time scale for evolution, *Trends Genet.* **3:**241–247.

Woese, C. R., 1987, Bacterial evolution, *Microbiol. Rev.* **51:**221–271.

Yin, L., 1985, Microfossils of the Duoshantuo Formation in the Yangtze Gorge District, Western Hubei, *Palaeontol. Cath.* **2:**229–249.

Young, G. M., 1976, Iron-formation and glaciogenic rocks of the Rapitan Group, Northwest Territories, *Precambrian Res.* **3:**137–158.

Zang, W., and Walter, M. R., 1988, Plankton of the latest Proterozoic: An Ediacaran age flora from the Amadeus Basin, central Australia, *Nature* **337:**632–645.

Zhang, Z., 1986, Clastic facies microfossils from the Chuanlinggou Formation (1800 Ma) near Jixian, North China, *J. Micropalaeontol.* **5:**9–16.

Addendum

This chapter was written in 1988. During the time that it has awaited publication, much has been learned about the Late Proterozoic Earth. Patterns that were hazy four years ago have become clearer, confirming the picture of the Late Proterozoic era as an interval of pronounced biological, biogeochemical, climatic, and tectonic change. That these changes were coupled is supported by mounting stratigraphic data as well as by tectonic and biogeochemical models. A growing array of carbon isotopic data confirms the marked ^{13}C enrichment of many carbonates and organic materials aged 850 to 600 My, as well as the occurrence of short-lived excursions to lower $\delta^{13}C$ immediately following ice ages. The Sr isotopic curve for Late Proterozoic sea water has also become clearer, and hydrothermal activity is now seen as a key to understanding the overall pattern of Late Proterozoic Earth history. Important new hypotheses about Late Proterozoic plate movements may help link the Sr record to fundamental tectonic events. And the Proterozoic paleontological record grows steadily richer. If 1988 brought a nascent sense that the Ediacaran radiation was rooted in a greater pattern of biological and physical change, then 1992 begins with a well-supported conviction that the early evolution of animals must be understood in the context of the tectonic and biogeochemical changes that mark the Late Proterozoic era.

III

Vendian Faunas of the World

Chapter 4

Vendian Faunas and the Early Evolution of Metazoa

M. A. FEDONKIN

1. Introduction

Until recently, paleontological data were almost completely ignored by zoologists studying the origin of major groups of invertebrates and the early phylogeny of the Metazoa, chiefly because Charles Lyell had long ago emphasized the incompleteness of the geological and paleontological record. Charles Darwin adopted the same view and discussed the absence of ancestors of the Cambrian fauna. Only in the last few decades has paleontological information become available that bears on this problem, yet it, too, has been ignored in favor of new methods in the studies of living systems, in particular, from such fields as histology, genetics, biochemistry, and molecular biology. Along with the classical approaches, i.e., comparative anatomy and embryology, neontology dominated the study, not only of the systematics of Recent organisms, but also of the origin and phylogeny of Metazoa.

The discovery of the oldest Vendian (Late Precambrian) faunas in the 1930s, the application of new methods of preparation and study, the recognition of Lagerstätten (extraordinary fossil biotas), and the study of so-called problematic fossils have contributed increasingly to research on the origin and early evolution of Metazoa. The term "problematic fossils" reflects one phase of paleontology when metazoan systematics was fixed at the level of higher-rank taxa. All fossils

M. A. FEDONKIN • Paleontological Institute of the Russian Academy of Sciences, Moscow, B-321, Russia.

Origin and Early Evolution of the Metazoa, edited by Jere H. Lipps and Philip W. Signor. Plenum Press, New York, 1992.

that did not fit into this system were termed problematic, and thus they were commonly excluded from consideration in the historical development of animals. But this material is of primary importance. Problematic fossils have been analyzed in hundreds of articles and dozens of monographs published in the last few years.

The discoveries of unusual organisms which do not fit into existing metazoan systematics and which cannot be placed in the phylogeny of multicellular animals show that in the early evolution of many groups (and the phylogenetic stem of Metazoa as a whole) the diversity of body plans and grades of organization were higher than estimated from recent organisms. In fact these observations were forecast by some zoologists, and explained by Mamkaev's concept of archaic diversity (Mamkaev, 1985, 1986, 1987).

In this chapter, I examine the contribution Vendian fauna makes to the theory of the origin of major groups of invertebrates. In order to evaluate this contribution, I review the available data on the age, paleogeographic, and stratigraphic distribution of the oldest animals, the composition of Precambrian fossil associations, as well as their taphonomy, paleoecology, and systematics.

2. Time–Space Distribution of Precambrian Metazoans

Although prokaryotic fossils are present in rocks 3500 million years (My) old, the remains of multicellular animals appear much later in the fossil record. In fact, trustworthy remains of Metazoa in deposits older than 650 My are unknown.

Almost all articles describing trace and body fossils of metazoans from older rocks deal with objects for which an alternative explanation is possible. These objects are usually abiogenic, namely cracks of drying sediment or syneresis cracks, pseudomorphs after some minerals, various sedimentary structures (current marks, tool marks, turbidite current structures, ripple marks, load casts, water or gas escape structures, etc.), imprints of ice crystals, soft clay pebbles, concretions and nodules, or Liesegang rings that form concentric structures resistant to weathering.

Nonmetazoan biogenic structures sometimes are interpreted as remains of animals—for example, fragments of megascopic algae, bacterial mats, or sapropel organic films, and even stromatolites and microphytolites. In some cases, a very old age for rocks containing true metazoan fossils is based on false initial information.

However, large filamental and segmented microorganisms quite comparable in size to recent meiofauna were described from the 890 to 1000-My-old Lakhanda Formation (Hermann, 1981a) and from the 740 to 850-My-old Miroyedikha Formation (Hermann, 1981b; Veiss, 1984). Some of those organisms, like *Cephalonyx* (Veiss, 1984), show signs of cephalization.

Enigmatic bedding plane markings resembling strings of beads are known from Appekunny Argillite, some 1300 My old, in Montana (Horodyski, 1982). These fossils resemble some forms of Phanerozoic ichnofossils, in particular, the fecal pellets *Neonereites*. Small vase-shaped microfossils first described from the Chuar Group, 750 ± 100 My old (Bloeser *et al.*, 1977) and then discovered elsewhere at eight localities in the world approximately at the same stratigraphic level, have been compared with chitinozoans, which are interpreted by some specialists as eggs or egg cases of early metazoans.

The first definite fossils of multicellular animals appear stratigraphically above the Upper Precambrian tillites that represent the most extensive glaciation in the Earth's history. The Laplandian (Varangian) glaciation took place approximately 650 My ago. The stratigraphic interval between its tillites below and the base of the Cambrian above corresponds more or less to the Terminal Precambrian System. Concepts of this stratigraphic system, known in different places in the world as the Sinian, Vendian, Ediacaran, or Ediacarian according to various authors, are now actively developing. I prefer the name "Vendian" suggested by Sokolov (1952), as the most widely used name for the Terminal Precambrian System in the international geological literature.

The type area of the Vendian is the East European Platform. The lower boundary of the system corresponds to the base of the Laplandian tillites. The thickness of the Vendian deposits in its type area ranges from 200 to 2000 m. The upper boundary is placed between the Rovno Regional Stage (below) and the Lontova Regional Stage, corresponding to the Tommotian Stage of the Lower Cambrian. Some authors consider the Lontova to be pre-Tommotian age, an opinion based on comparative analysis of the stratigraphic distribution of particular acritarch taxa. The age of the upper boundary of the Vendian is about 560 My. Thus the Vendian period embraced an interval of about 80–90 My.

Peculiar megascopic animals that did not have mineral skeletal elements inhabited the Vendian oceans. This fauna, known as the Ediacara fauna, after the name of a fossil locality in South Australia, was discovered by Sprigg (1947, 1949) and became famous after a series of excellent papers by M. F. Glaessner and M. Wade (see references herein, and Jenkins, this volume, Chapter 5). I use the name "Ediacaran fauna" for Precambrian soft-bodied metazoans that are close or identical with respect to taxonomy and grade of organization to the late Precambrian invertebrates discovered at Ediacara, South Australia. The Ediacaran fauna is known now from Africa, Australia, North America, Europe, and Asia. The number of new localities is increasing.

The first specimens of Precambrian metazoans were collected in the Nama Group (Namibia, southwestern Africa) in 1908–1914, and the collection was described later (Gürich, 1930, 1933; Richter, 1955). Further investigations led to the discovery of a large diversity of organisms (Pflug, 1966, 1970a, b, 1971, 1972a, b, 1973; Germs, 1968, 1972a–d, 1973a, b; Crimes and Germs, 1982). Petalonamae, an enigmatic group of soft-bodied metazoans of unknown systematic position, dominate the fossil assemblage from the Nama Group and the oldest known skeletal fossil, *Cloudina*, interpreted as annelid worms, algae, or foraminifera (Germs, 1972d; Glaessner, 1984), or a filter-feeding metazoan of at least cnidarian grade of organization (Grant, 1990) occur there. Rare and problematic medusoids are reported from the lower part of the Precambrian Adoudou Series in Antiatlas, Marocco (Houzay, 1979).

One of the most representative localities of the Ediacaran fauna is in South Australia, where more than 30 species of soft-bodied Metazoa were discovered in the Flinders ranges near Ediacara in the Rawnsley Quartzites of the Wilpena Group (Sprigg, 1947, 1949; Glaessner, 1958, 1959, 1969, 1971, 1976, 1979, 1980, 1984; Glaessner and Wade, 1966, 1971; Wade, 1971, 1972a, b; Glaessner and Walter, 1975; Jenkins, 1985; Jenkins and Gehling, 1978; Gehling, 1984, 1987, 1988). Smaller localities of the Precambrian fauna are known from central Australia (Glaessner, 1969, 1984; Wade, 1969; Glaessner and Walter, 1975).

In western Europe the Precambrian fauna is known from the Charnian Subgroup in the region of Charnwood Forest near Leicester, England (Ford, 1958, 1962, 1963, 1980; Boynton, 1978; Boynton and Ford, 1979). A small locality was also discovered in South Wales (Cope, 1977, 1983). Imprints of medusoids were described from the lower part of the Dividal Group in Norway (Kulling, 1972; Føyn and Glaessner, 1979) and from the San Vito Series of Sardinia (Debrenne and Naud, 1981). A medusoid assemblage has been discovered recently in the Late Proterozoic of northeastern Finnmark (Farmer *et al.*, 1991).

The most diverse fossil metazoan assemblage occurs in the White Sea region in the northern European part of Russia (Keller *et al.*, 1974; Keller and Fedonkin, 1977; Fedonkin, 1977, 1981, 1983a–c, 1984, 1985a–e, 1986, 1987). About 50 species of body fossils and 20 forms of trace fossils have been described from this region. The majority of the fossils were found in fresh unaltered clayey siltstone and sandstone of the Ust-Pinega Formation assigned to the Redkino Regional Stage of the Vendian. Fossils are more abundant in the White Sea region because of a greater diversity of facies than at other localities. These facies reflect different environments that were favorable both for life and subsequent rapid burial and fossilization of dead animals and the preservation of fossils.

The most complete sequence of the Vendian in the European part of the USSR is exposed in the valley of Dniester, Podolia, Ukraine. The first Ediacaran fossils were described from the clastic rocks of the Bernashev Beds of the Yaryshev Formation (Zaika-Novatskii *et al.*, 1968; Zaika-Novatskii and Palij, 1968; Sokolov, 1972). Later, trace and body fossils were found through a wider stratigraphic range (Palij, 1976; Palij *et al.*, 1979; Gureev, 1983, 1984, 1985, 1988). The most diverse Ediacaran assemblage occurs at the base of the transgressive series of the lower Vendian marine sediments (Fedonkin, 1981, 1983b; Velikanov and Gureev, 1984). This assemblage has ten species of invertebrates in common with the type Ediacaran assemblage of South Australia. Stratigraphically, it is about 600 m below the base of the Cambrian.

Several Ediacaran fauna localities have been discovered in the Chernokamenka Formation of the Sylvitsa Series in the Middle Urals (Becker, 1977, 1990; Fedonkin and Krivosheyev, in press) and recently in the Vendian of the South Urals (Becker and Kishka, 1987).

In the Asian part of the USSR, the Precambrian fauna is known mainly in Siberia from the Yudoma complex, which is the Siberian equivalent of the Vendian. Separate discoveries were made in the lower part of the Upper Yudoma Formation in the Yudoma-Maya region and in the lower part of the Kurtun Formation along the Maly Anai River (Sokolov, 1975, 1976). The most representative Vendian fauna was discovered in the Khatyspyt Formation of the Khorbusuonka Series in the middle stream of the Khorbusuonka River on the Olenek Uplift (Sokolov and Fedonkin, 1984; Fedonkin, 1985c, 1990a, b; Vodanjuk, 1989). This locality is unique because Ediacaran metazoans occur in thin, laminated bituminous limestones while all other occurrences worldwide of Precambrian metazoans are in clastic facies.

In China, Precambrian metazoans were found in many places, in particular from the Vu-Khanshan Series on the Liao-Dun Peninsula (Xing and Liu, 1979; Chen, 1991), from the Mashan Group in eastern Heilongjiang Province (Liu, 1981), and from the eastern Yangtze Gorge (Sun, 1986a).

Rather peculiar associations of primitive Coelenterata are known through 2000 m of the Precambrian Conception Group on the Avalon Peninsula, southeastern Newfoundland (Anderson and Misra, 1968; Misra, 1969; Anderson and Conway Morris, 1982). Remnants of Precambrian Metazoa occur in northwestern Canada and in British Columbia (Hofmann, 1981; Hofmann *et al.*, 1983, 1985; Narbonne and Hofmann, 1987; Narbonne and Aitken, 1990). Recently, Pteridiniidae were first found in the United States in North Carolina (Gibson *et al.*, 1984) and new Ediacaran fossils in southern Nevada (Horodyski, 1991).

The chief localities of the Vendian fauna, including the most important, are in lamellar terrigenous, more rarely carbonate and volcanoclastic, deposits in relatively shallow water environments of vast epiplatform seas or at the edges of platforms. Vendian Metazoa also occur in turbiditic deposits, including those deposited in depressions similar to geosynclines. The wide geographic distribution of remains of soft-bodied multicellular animals, their ecological uniformity, and the presence of common forms in widely separated localities, even in mobilistic paleogeographic reconstructions, show that Vendian Metazoa represent a single global fauna and a special stage in the evolution of multicellular animals. The wide distribution and abundance of the fossils in Vendian deposits indicate that they can be used as guide fossils for biostratigraphic correlation as well as for solving questions about the paleogeography of marine deposits of the late Precambrian.

A certain regularity is observed in the stratigraphic distribution of the Ediacaran fauna. The oldest invertebrate fossils are most characteristic of the stratigraphic interval between the Laplandian (Varangian) tillites (or a gap in the geological record relates to a eustatic lowering of sea level during this glacial period) and recognized by the sudden appearance of the Lower Cambrian Tommotian Stage skeletal fossils. This interval is characterized by a quantitative and qualitative nonuniformity in the distribution of soft-bodied metazoan fossils. The zone of their abundant occurrence is confined to the lower half of the Vendian, corresponding to the Redkino Regional Stage (the lower half of the Valdai Series) in the European part of the USSR. A major part of the typical Ediacaran forms disappear in the upper half of the Vendian, corresponding to the Kotlin and Rovno Regional Stages (upper half of the Valdai Series and the basal part of the Baltic Series), which may represent a mass extinction. Numerous ichnofossils and rare medusoids continue into this stratigraphic interval. The quantity, diversity, and size of trace fossils increase at the levels of the Rovno Horizon of the Russian Platform and the Nemakit-Daldyn Horizon of the Siberian Platform, marking the end of the Vendian in those regions. Imprints of soft-bodied invertebrates are relatively rare.

A similar sequence of body and trace fossils is observed in other complete Precambrian sections in the world. The abundance and morphological diversity of Metazoa in rocks stratigraphically separated from the Varangian tillites by a short interval indicate a high probability of finding Metazoa remains in pretillite deposits. This possibility is supported by the recent discovery of medusiform fossils from intertillite beds in northwestern Canada (Hofmann, Narbonne, and Aitken, 1990).

Metazoan trace and body fossils have been described in many papers from rocks older than 620–640 my, although most of them are actually inorganic structures of diverse nature (pseudofossils), dubiofossils, organic films, parts of bacteria mats, or algae. In other cases, true fossils are described, but in these cases Vendian or younger fossils were thought to be much older (Sun, 1986b; Sun, Wang, and Zhou, 1986; Vidal and Moszydlowska, 1987).

All Proterozoic problematic objects that might be metazoan remains or traces should be studied because indirect evidence indicates that the phylogenetic origin of invertebrates extends into the pre-Vendian geological record. This indirect evidence includes: (1) the high taxonomic differentiation of the Vendian animals; (2) the presence of advanced metazoan phyla in the Ediacaran fauna; and (3) molecular clock data.

3. Preservation and Interpretation of Vendian Soft-Bodied Metazoans

The Vendian period was unique in many respects, including taphonomy. In the later paleontological record, the preservation of soft-bodied organisms is a rare, exceptional phenomenon, but in the Vendian it is typical. The causes of this difference should be sought in abiotic factors of the environment rather than in the peculiar features of the Vendian organisms.

Among abiotic factors promoting the burial and fossilization of invertebrates are gigantic tidal flats that served as traps for animals during tides as well as the vast shallow water zones with abundant benthic inhabitants where an active hydrodynamic regime and an intensive erosion of the land, not protected by vegetation, promoted a rapid burial of dead or even living organisms, especially during storms.

Filter-feeders are well known for cleaning marine water. Active filtration is a perfect mechanisms for extracting suspended matter from the water and it is also important because filter-feeders "pack" nondigested remains of food into fecal pellets, which sink to the sea floor much faster than fine particles in suspension.

In fossil associations of Ediacaran fauna, no remains are known of active filter-feeders similar to sponges, archaeocyaths, or brachiopods, the first known radiation of which took place in the Early Cambrian.

Vendian taphonomy can be inferred from this paleontological record. In the absence of active filter-feeders, turbidity led to a good lighting and warming of only the surface layer of water, which was enriched in oxygen as a result of the photosynthetic activity of planktonic algae and cyanobacteria. The water below was cold and poorly oxidized. The mixing of water masses was very slow because of stratification as the result of a sharp difference in the temperature of the upper and lower parts of the water. The bottom water in relatively deep places did not favor life, and that decreased the activity of organisms that could decompose buried organic matter. The presence of a large quantity of clayey, argillaceous particles in the water and sediment decreased the permeability of the sediment to gases and fluids and increased its cementing properties, the rate of early diagenesis, and lithification.

The most important biotic events that promoted the preservation of Vendian invertebrates were probably the following: a high benthic population density in zones with active hydrodynamic conditions; a considerable quantity of attached forms that had structures capable of preservation; a low activity level of scavengers and vagile predators; and little processing of sediment by deposit feeders and other bottom animals. These factors allowed the preservation of undisturbed structures in soft sediment. Weak biological processing of the sediment may have allowed cyanobacterial mats to develop on the sediment surface, which might have stabilized the sediment and prevented the destruction of biogenic structures in the littoral shallow water (Fig. 35).

Moreover, bacterial mats probably hindered aeration and the escape of decay products from the sediment. Products from the biotic decomposition of the organic matter of buried animals might form an environment in sediment where the early diagenetic lithification was more rapid due to specific physical and chemical processes. Apparently only early and rapid diagenesis of sediment preserved the Ediacaran animals before decomposition. Traces of decomposition are typically not observed on Precambrian fossils. I assume a temporary mineralization around a buried and decayed body that, however, did not lead to the formation of concretions similar to well-known ones from the Pennsylvanian Mason Creek (Nitecki, 1979). When the process of decomposition and homogenization of the geochemical environment in the sediment was completed, this mineralization, which was probably dispersed, disappeared. This phenomenon could promote the very good preservation of Vendian soft-bodied organisms in sediments of different facies, especially in relatively coarse-grained clastic sediments like the Rawnsley Quartzites of South Australia (Glaessner, 1984) or the Lomosovo Beds of the Mogilev Formation in Podolia, Ukraine (Fedonkin, 1983b, 1987), rocks that seem to be the least favorable for the fossilization of soft-bodied animals, especially medusoids.

The more or less adequate fossil record of soft-bodied metazoans apparently ends long before the beginning of the Cambrian period. In all complete sections of the Vendian, in particular on the Russian Platform (Fedonkin, 1987) or on the Avalon Peninsula, Newfoundland (Anderson and Conway Morris, 1982), abundant Ediacaran fossils are confined to the lower half of the Terminal Precambrian System, and they are usually separated from the base of the Cambrian by thick deposits ranging from several hundred meters on platforms to several kilometers in miogeosynclines or in other basins with high sedimentation rates.

Although the upper half of the Vendian has abundant vendotaenian algae, acritarchs, small trace fossils of benthic invertebrates, in particular *Palaeopasichnus delicatus* Palij and *Harlaniella podolica* Sokolov, and even rare imprints of medusoids, a rich and diverse Ediacaran fauna has not been found. The disappearance of the Ediacaran fauna long before the beginning of the Paleozoic could be related to the extinction of some primitive groups and to a sharp decrease in body sizes of other more advanced groups, which later led to a rapid radiation of the Cambrian skeletal fauna. Some of the first Cambrian skeletal invertebrates are very small.

The absence of a fossil record of most of the soft-bodied animals in the second half or at the end of the Vendian may have resulted from a change in general taphonomic processes; in particular, an increased activity of destructive animals, and replacement of benthic cyanobacterial communities by eukaryotic algae and benthic invertebrates. Possibly this absence coincides with the appearance of meiofauna, although no direct evidence supports this supposition. But the meiofauna could have played a decisive role in changing the density of the population of decomposers, such as fungi or bacteria, and thus increased the efficiency of their activity. Of particular importance would be the movement of meiofauna in the sediment, which would improve aeration and promote the transportation of decay products to the surface. In the absence of cyanobacterial mats this process could be more effective and lead to a rapid decomposition of buried animals while sediment diagenesis was slower (Fedonkin, 1987).

At the end of the Vendian and the beginning of the Cambrian, the colonization of the marine sediment by megasopic vagile invertebrates sharply increased, as indicated by a high diversity of ichnofossils and by the degree of the biological processing of sediment (Fedonkin, 1977, 1981).

The preservation of Ediacaran invertebrates as imprints or molds has a certain peculiarity that is not characteristic of skeletal fauna of later periods. In the majority of cases only one side of the animal or its part, especially of large forms, is observable. The correlation of this side with an opposite one is not always possible. Different sides of animals as well as different degrees of decomposition could be described as different taxa, though evident signs of decay are very rarely observed in Ediacaran fossils.

Small organisms with a thin and soft body could be preserved as imprints that show details of the inner anatomy; for example, canals of the gastrovascular system in medusoids such as *Albumares brunsae* Fedonkin (Fig. 21). The preservation of such structures may result from a higher protein content, but the mechanical resistance could be provided also by the presence in the canals of gases and fluids formed as a result of the decomposition of the body and which maintained the turgor of these canals until they were imprinted in sediment. These circumstances probably allowed the formation of complicated molds (Fig. 10), which show two or more microbedding planes with topographically different systems of organs or tissues (Wade, 1968).

4. Systematics of Vendian Invertebrates

The systematic position of Vendian animals is controversial. Many structures in the imprints of these fossils are interpreted in different ways. These difficulties result from the huge evolutionary distance separating the Vendian fauna from recent organisms. The ancestors of the Vendian fauna are unknown, and Paleozoic and younger descendants are not evident. In initial studies of Precambrian animals, the morphology and the systematic position of invertebrates was interpreted differently. However, for a long time paleontologists thought that Precambrian organisms belonged to the same phyla, classes, and even orders as known from the later paleontological or modern records.

For example, Glaessner (1984) places Precambrian animals in the following taxa: Phylum Coelenterata (classes Hydrozoa, Scyphozoa, and Conulata, medusae of uncertain affinity, and problematic Petalonamae); Phylum Annelida (Class Polychaeta); Phylum Arthropoda (Superclass Trilobitomorpha or Chelicerata of uncertain class and Superclass Crustacea; Class Branchiopoda); Phylum Pogonophora, Phylum Echiurida, as well as some forms of uncertain systematic position. A possible representative of the echinoderms was described recently from the Ediacara assemblage by Gehling (1987). However, the classification of the Precambrian animals within the framework of later invertebrate systematics produces many contradictions. Therefore, other approaches and principles of classification have been developed recently.

A completely different interpretation of Ediacaran fossils has been developed by the Germans. H. D. Pflug, in particular, proposed the high-rank taxon Petalonama, now widely accepted by paleontologists (Pflug, 1970a, b; 1972a, b; 1973). His phylogenetic models may be unacceptable, but his morphological interpretation of the enigmatic leaflike and bag-shaped fossils certainly merits consideration. Seilacher (1984, 1989) offered a new functional, morphological, and taxonomic interpretation of the Precambrian megascopic fossils. Having noted that the Vendian fauna cannot be easily related phylogenetically with later invertebrates and stressing the fact of unusual preservation of Ediacaran fossils, Seilacher inferred that Precambrian organisms do not have Recent analogs and are therefore unique. He believed that they had an extensive body surface resulting in a complicated relief and they had low body volume because they were relatively flat. The high surface–volume ratio of the body allowed the absorption of oxygen and organic matter dissolved in water by diffusion through the body wall. Thus, a mouth, digestive tract, or respiratory organs were not necessary.

All the characters of Ediacaran fossils used by Seilacher to demonstrate their unusual and nonmetazoan organization can be used equally to support the opposite interpretation of those fossils as known metazoans. The unusually abundant preservation of soft-bodied invertebrates at that time may have resulted from biotic and abiotic factors of the environment rather than in the uniqueness of the Vendian organisms. The study of the taphonomic peculiarities of the Phanerozoic soft-bodied invertebrates that are not so rare in the fossil record can provide a key to the problem of the extremely abundant remains of nonskeletonized animals in the Precambrian.

The comparative body plan analysis of the Vendian soft-bodied metazoans carried out by Fedonkin (1983a, 1984, 1985a, b, e, 1987) has led to a different system for the Precambrian animal world: Phylum Coelenterata (classes Cyclozoa, Inordozoa, Trilobozoa, and ?Scyphozoa); Phylum Proarticulata (classes Dipleurozoa and Vendiamorpha); Phylum Arthropoda (classes Paratrilobita and Crustacea). Probable representatives of the phyla Plathelminthes, Annelida, and Mollusca can be tentatively identified by rare problematic body imprints and numerous and diverse trace fossils.

The recent tendency of Proterozoic paleozoologists is to consider the Vendian soft-bodied metazoans as representatives of short-lived groups that went extinct well before the Cambrian period. This can be supported partially by paleontological data. Nevertheless, the apparent absence of wide phylogenetic connections between the faunas of these two periods may reflect a purely taphonomic difference. Thus, when comparing the animals of the Vendian with those of the Cambrian, different categories of fossils are compared: soft body imprints and skeletal remains.

The metazoan fossil record of the Late Proterozoic may not be representative of the total biota. The general impression that the diversity of the Vendian Period is much smaller than that of the Cambrian may well be a preservational artifact.

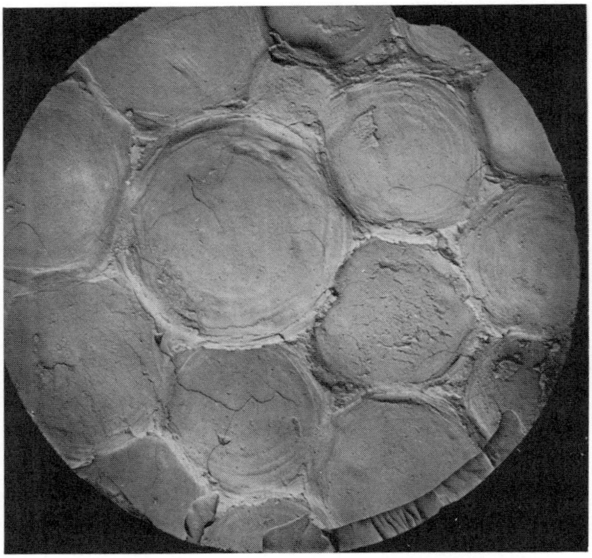

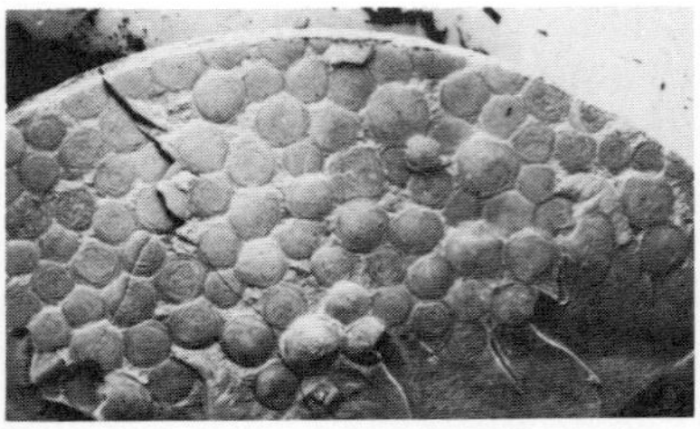

Figure 1. Left: *Beltanelloides sorichevae* Sokolov, ×0.85, White Sea region, USSR; above: *Beltanelloides sorichevae* Sokolov, ×1, White Sea region, USSR.

4.1. Coelenterata: Solitary Forms

Among Precambrian macroscopic metazoans the simplest morphologically but most numerous are coelenterates with a monaxonic heteropolar symmetry of infinitely high order (Figs. 1–14). Representatives of this group have common peculiarities, such as type of symmetry, concentric body plan resulting from rhythmic growth, the absence of well-developed radial elements in the wide gastral cavity, the absence of radially-arranged branching canals of the gastrovascular system, the presence of different modes of asexual reproduction, par-

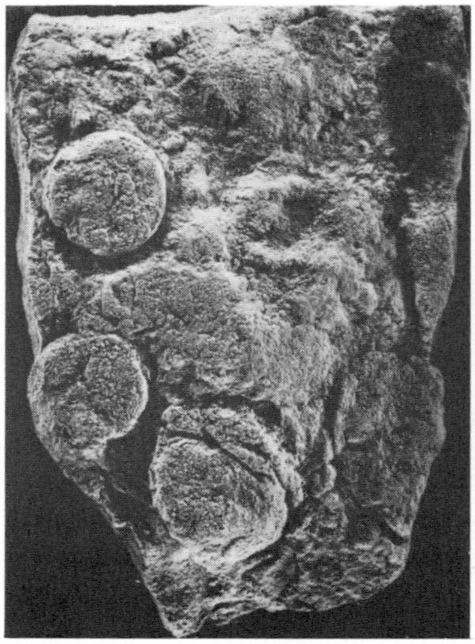

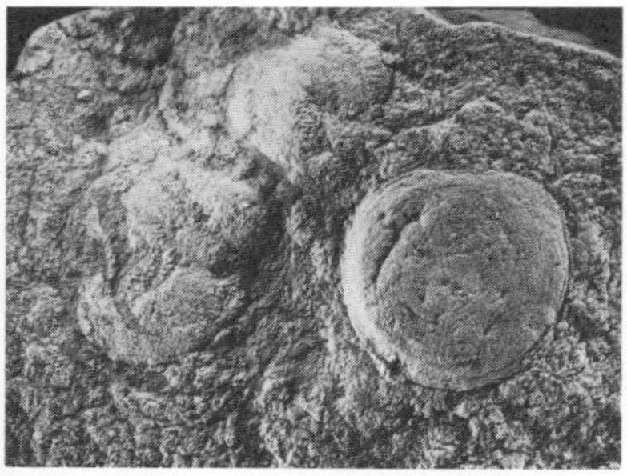

Figure 2. Left: *Nemiana simplex* Palij, ×1, White Sea region, USSR; above: *Nemiana simplex* Palij, ×0.5, White Sea region, USSR.

Figure 3. Left: *Paliella patelliformis* Fedonkin, ×0.5, White Sea region, USSR; right: *Paliella patelliformis* Fedonkin, ×0.7, Olenyok Upland, north of Siberian Platform, USSR.

ticularly simple longitudinal fission, and a predominantly sedentary mode of life. All these are indicative of the primitive and archaic character of the group as well as of the monophyletic origin of its representatives. This group of the Vendian Radialia is considered as a special Class Cyclozoa in the Coelenterata (Fedonkin, 1983a, 1984, 1987).

The combination of concentric and radial types of the symmetry in the body plan of some Precambrian medusoids, for instance, in *Eoporpita* (Fig. 14), may indicate a phylogenetic relationship of the Cyclozoa with the more advanced radially-arranged Coelenterata. Among the cyclozoans, sedentary forms dominate while pleistonic and planktonic animals make up a smaller part. This proportion of ecological characters in the most primitive Precambrian Coelenterata may be

Figure 4. *Paliella patelliformis* Fedonkin, ×0.9, Dniester River Valley, Podolia, USSR.

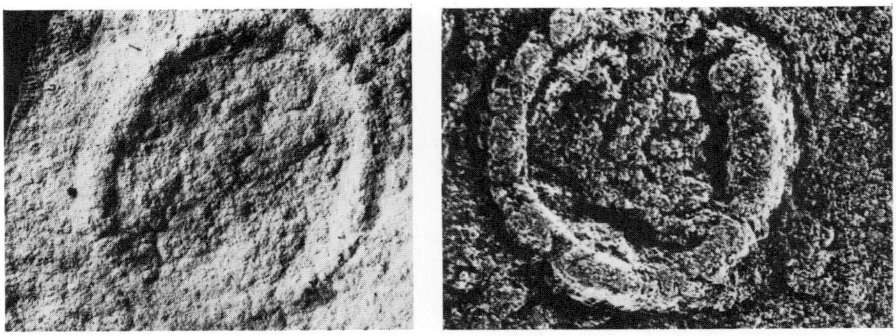

Figure 5. Left: *Nimbia occlusa* Fedonkin, ×1, White Sea region, USSR; right: *Nimbia occlusa* Fedonkin, ×2, White Sea region, USSR.

evidence for the primary nature of the polypoid stage in the evolution of the metagenesis of Cnidaria. A varied sea floor environment with greater number of habitats than in the pelagic zone might have promoted a diversity of forms.

Medusae with monaxonic heteropolar symmetry, a radial body plan, and an axis of uncertain order comprise a large diverse group among the Vendian radially-arranged coelenterates (Figs. 15–20). Radial symmetry of uncertain order in Vendian forms embraces the whole organism as well as different systems of organs, e.g., marginal tentacles, the channels of the gastrovascular system, gonads, and gastral cavity.

The main characteristic of this group is an increase in the order of radial symmetry during ontogeny, i.e., new antimeres were added while the organism

Figure 6. *Nimbia dniesteri* Fedonkin, ×0.9, Dniester River Valley, Podolia, USSR.

Figure 7. *Cyclomedusa plana* Glaessner and Wade, aboral side, fragment of umbrella, ×0.5, White Sea region, USSR.

grew, but no order or any succession in their formation has been observed. The number of antimeres does not correspond to any particular number. The forms with such a type of symmetry and the primitive growth pattern were united in the class Inordozoa (Fedonkin, 1984, 1985b, c, e, 1986, 1987).

Despite the primitive mode of the formation of antimeres, the Vendian Inordozoa are of a higher organization than the Cyclozoa. Inordozoans commonly have a well-arranged system of tentacles, various and complicated systems of

Figure 8. *Cyclomedusa plana* Glaessner and Wade, oral side, ×1, White Sea region, USSR.

Figure 9. *Protodipleurosoma rugulosum* Fedonkin [the two structures at the edge of the umbrella (right and lower) can be interpreted as the budding daughter individuals], ×0.5, White Sea region, USSR.

gastrovascular canals that branch several times toward the periphery and may form a network (plexus). A reproductive system (gonads) in many forms of the group indicates the prevalence of sexual reproduction in their development. On the contrary, examples of asexual reproduction in this group are rare. The Inordozoa are closer to later cnidarians in level of organization than they are to the Class Cyclozoa.

Figure 10. *Ediacaria* sp. (a few bedding planes preserve the features of an internal and external morphology of the organism, and a discoidal gastral cavity is preserved as a mold), ×1, Dniester River Valley, Podolia, USSR.

Figure 11. Left: *Ediacaria flindersi* Sprigg (oral side, proximal ends of the marginal tentacles are situated slightly inward of the edge of the umbrella), ×0.8, Dniester River Valley, Podolia, USSR; right: *Ediacaria flindersi* Sprigg (aboral side), ×0.5, White Sea region, USSR.

Coelenterates with a consistent radial symmetry (Figs. 21–24) are not as numerous among the Vendia Radialia as the representatives of Cyclozoa and Inordozoa. Possibly the regulation of radial symmetry resulting from the stabilization of the number of antimeres was a relatively late phenomenon representing a specific and main evolutionary path of Coelenterata. Vendian medusae with an axis of consistent radial symmetry are more similar to living forms of Cnidaria than they are to Precambrian contemporaries. Nevertheless, this group has characters not typical of the coelenterates or of recent metazoans in general.

One such feature is a threefold symmetry. This symmetry occurs in recent cnidarians as an exception or as a teratism and in other groups of Metazoa it is known as a secondary phenomenon, e.g., in nematodes and gastrotrichs. But this symmetry was widespread in the Vendian medusae, e.g., *Skinnera, Tribrachidium, Albumares, and Anfesta* (Glaessner and Wade, 1966; Wade, 1969; Keller and Fedonkin, 1977; Fedonkin, 1981, 1983a).

A unique threefold radial symmetry makes the group sufficiently homogeneous in a phylogenetic sense to recognize it as a taxon of high rank: a special

Figure 12. *Cyclomedusa* sp. (aboral side of two individuals soon after fission), ×0.9, Dniester River Valley, Podolia, USSR.

Figure 13. *Ovatoscutum concentricum* Glaessner and Wade (numerous folds indicate that the animal had a thin and elastic body), ×1, White Sea region, USSR.

class, Trilobozoa (Coelenterata). This class might include also a group of oldest skeletal organisms known as Augustiochreidae (Valkov and Sysoiev, 1970) or Anabaritidae (Missarzhevski, 1974; Glaessner, 1979), which appeared at the very end of the Vendian and became widespread at the beginning of the Early Cambrian. The relation of the organisms united in the subclass Augustiomedusae (Valkov, 1982) with the Class Scyphozoa is not evident, as a three-ray symmetry is not characteristic of this class of cnidarians. A fourfold symmetry prevails in Scyphozoa, so the Class Trilobozoa is reasonably recognized as a specific class of Coelenterata similar to the Class Scyphozoa in organization, but having a different symmetry. It is as homogeneous as scyphozoans with respect to symmetry. Recently an attempt has been made to consider Trilobozoa as a separate phylum

Figure 14. *Eoporpita medusa* Wade (oral side, radially oriented tentacles arranged into concentric series), ×1, White Sea region, USSR.

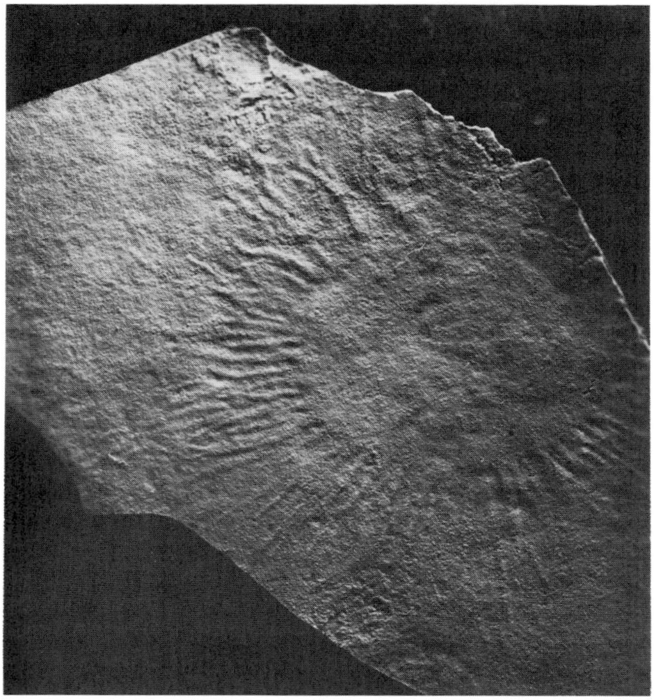

Figure 15. *Hiemalora stellaris* Fedonkin, ×0.7, Olenyok Upland, north of Siberian Platform, USSR.

Figure 16. *Hiemalora stellaris* Fedonkin, ×0.9, White Sea region, USSR.

Figure 17. *Hiemalora stellaris* Fedonkin (small pustules scattered over the umbrella can be interpreted as numerous reproductive organs or as small juvenile medusae in the aboral reproductive bag), ×1.2, White Sea region, USSR.

(Runnegar, 1989) but without an indication of where it should be placed, e.g., among Diploblastica or among Triploblastica.

The forms with the fourfold symmetry comprise an essential part of the Precambrian Radialia with the stable order of the radial symmetry, e.g., *Conomedusites, Kimberella, Ichnusina, Persimedusites,* and *Staurinidia* (Glaessner and Wade, 1966; Wade, 1969, 1972a; Hahn and Pflug, 1980; Debrenne and Naud, 1981; Fedonkin, 1983a). Their organization as well as their typical scyphozoan symmetry indicate that these small jellyfishes belong to the Class Scyphozoa. Indirect evidence for such a supposition is *Conomedusites,* the only form with a

Figure 18. Left: *Irridinitus multiradiatus* Fedonkin, ×2, Dniester River Valley, Podolia, USSR; right: *Bonata septata* Fedonkin, ×1, White Sea region, USSR.

Figure 19. *Evmiaksia aksionovi* Fedonkin, ×1, White Sea region, USSR.

theca among the fourfold coelenterates (Fig. 24). This genus was placed into the Class Conulata (Glaessner, 1979), and originally the subclass Conulata was placed in Class Scyphozoa (Moore and Harrington, 1956).

Although the forms with the stable order of the radial symmetry are relatively rare among the Vendian Radialia, they show considerable diversity in their body plan. They are characterized by a varied gastrovascular system (single and paired, branching and nonbranching radial channels). In some forms an umbrella is divided into narrow branching lobes, inside of which may be channels (*Anfesta*, Fig. 21). Such organization is possibly the initial one of the forms with thin branching channels of the gastrovascular system.

Unlike the forms of Cyclozoa and Inordozoa, the Vendian coelenterates with a stable order of symmetry do not show any sign of asexual reproduction. They have

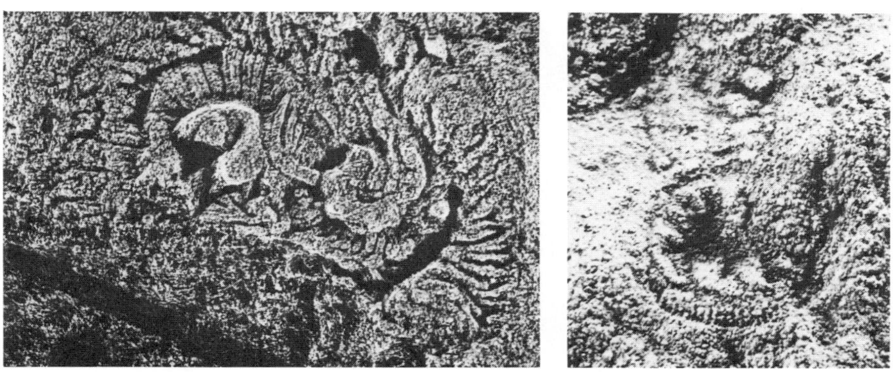

Figure 20. Left: *Pomoria corolliformis* Fedonkin (oral side of two individuals demonstrates the marginal and the circum-oral sets of the tentacles), ×2, White Sea region, USSR; right: *Armillifera parva* Fedonkin, ×2, White Sea region, USSR.

Figure 21. Left: *Albumares brunsae* Fedonkin, ×5, White Sea region, USSR; right: *Anfesta stankovskii* Fedonkin, ×1, White Sea region, USSR.

a diversity of reproductive organs both with regard to shape and in position that reflects the higher organization.

4.2. Colonial Forms: Petalonamae

A large part of the Precambrian fauna consists of colonial forms that show a wide morphological diversity, a high integration of colonies, and an absence of definite signs of functional polymorphism of zooids. There are double pinnate-shaped, fan-shaped, and comb-shaped colonies. Dominant are double pinnate-shaped colonies (Figs. 27 and 28–30) that were attached to loose sediment with a

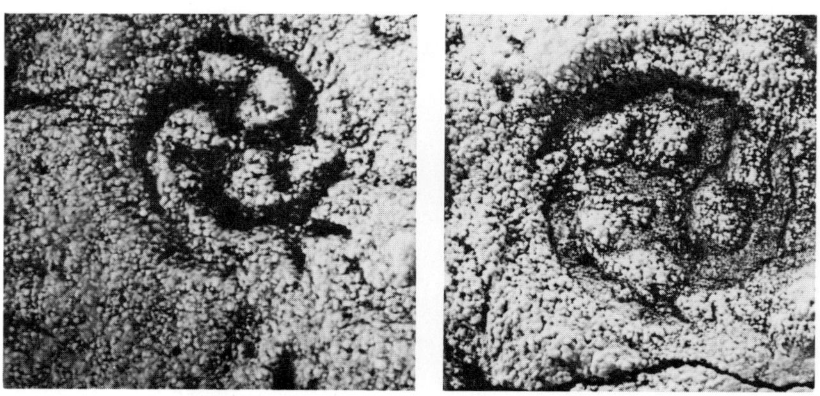

Figure 22. Left: *Tribrachidium heraldicum* Glaessner, ×4, White Sea region, USSR; right: *Tribrachidium heraldicum* Glaessner, ×4, White Sea region, USSR.

Figure 23. Left: *Tribrachidium heraldicum* Glaessner, ×2.5, White Sea region, USSR; right: *Tribrachidium heraldicum* Glaessner (the rarely preserved oral surface of this sedentary organism demonstrates radially arranged fine feeding grooves that fall into three commalike collecting depressions), ×4, White Sea region, USSR.

basal part of the main stem of the colony being buried (attaching disc, egg-shaped bulges, etc.), but pelagic forms also occur. Double pinnate-shaped colonies seem to be similar to pennatularias only in appearance, as solitary coral polyps have not been discovered in the Precambrian. On the other hand, these frondlike forms cannot be assigned to any other phylum but the Coelenterata. A more precise systematic position of the Vendian colonial forms cannot be determined because of an absence of data about the structure of separate zooids or any hard parts (spicules, sclerites, etc.). Among colonial organisms are forms with a monopodial growth, such as *Charniodiscus, Vaizitsinia,* and *Rangea,* and with simpodial growth, such as *Charnia* and *Glaessnerina.*

Glaessner (1979) considered these forms as problematic Coelenterata, uniting then with other leaf-shaped, ribbed organisms of unclear systematic position into the group Petalonamae, separated as a phylum (Pflug, 1970a, b). Colonial Coelenterata should be separated from Petalonamae, which are enigmatic organisms in

Figure 24. *Conomedusites lobatus* Glaessner and Wade, ×0.6, Dniester River Valley, Podolia, USSR.

Figure 25. Left: *Pteridinium nenoxa* Keller, ×0.8, White Sea region, USSR; above: *Pteridinium nenoxa* Keller, ×0.7, White Sea region, USSR.

Figure 26. *Pteridinium nenoxa* Keller [organism preserved with its axis oriented vertically in the bed of the siltstone (life position?)], ×0.7, White Sea region, USSR.

Figure 27. Left: *Ramellina pennata* Fedonkin, ×2, White Sea region, USSR; right: *Inkrylovia lata* Fedonkin, ×0.7, White Sea region, USSR.

Figure 28. *Charnia masoni* Ford, ×1, White Sea region, USSR.

many respects (Figs. 25–27, 31, and 32). This group includes *Pteridinium*, *Ernietta*, *Baikalina*, *Namalia*, *Nasepia*, *Inkrylovia*, *Valdainia*, and *Podolimirus*. A group of Petalonamae is represented now by collections of fossils of unclear systematic position, though many of them have some common features, in particular a leaf-shaped or bag-shaped body, a primary "segmentation" or ribbed surface without any traces of the secondary division of these segments. With current knowledge of these organisms, which are sufficiently different in morphology, uniting them into taxa of supergeneric ranks is difficult. Some workers propose to assign Petalonamae to Coelenterata of unclear systematic position (Glaessner and Walter, 1975; Glaessner, 1984). Recently, Seilacher (1984) suggested that some Petalonamae were segmented organisms with a structure similar to an air mattress that permitted a high inner pressure of the body fluid (hydrostatic skeleton). These animals utilized dissolved organic matter absorbed from water through the body integuments (Seilacher, 1984; Gould, 1984). The Petalonamae could also be related to the Tunicata (Sokolov, 1976; Chistiakov *et al.*, 1984). Additional collections and detailed studies of fossils will probably reveal the nature of these organ-

Figure 29. *Charnia masoni* Ford, ×0.6, Olenyok Upland, north of Siberian Platform, USSR.

isms. Nevertheless, there is no reason to consider petalonamas anything other than Diploblastica.

4.3. Solitary Bilateral Organisms

An overwhelming majority of known Vendian bilateral metazoans are represented by flat-bodied forms. Various segmented animals dominate, including polymeric and oligomeric creatures with signs of homonomous and heteronomous metamerism as well as with a characteristic imperfect segmentation and symmetry of sliding reflection. Very few forms have been described as bilateral nonsegmented metazoans, e.g., *Protechiurus* (Glaessner, 1979), *Platypholinia,* and *Vladimissa* (Fedonkin, 1985a–c), but these fossils are represented by or are unique specimens and poor morphology makes their interpretation difficult (Figs. 33–34).

Most Vendian Bilateria are characterized by peculiarities that are not observed in many multicellular animals of later epochs. The most important peculiarity is the plane of symmetry of sliding reflection, which is characteristic of

Figure 30. *Charnia masoni* Ford (juvenile colony), ×1, Olenyok Upland, north of Siberian Platform, USSR.

very primitive Bilateria, in particular of polymeric forms with homonomous metamerism (Family Dickinsoniidae) as well as of oligomeric Bilateria with evident signs of cephalization (Family Vendomiidae).

Thin sheetlike forms of the Family Dickinsoniidae have the lowest organization among segmented forms (Figs. 35–37). They were first considered as Coelenterata of the Class Dipleurozoa (Harrington and Moore, 1956b), then as annelids (Glaessner and Wade, 1966; Wade, 1972b; Runnegar, 1982), and finally as flat worms (Fedonkin, 1981). The analysis of symmetry of representatives of this family and the comparison of their body plan with the architecture of Coelenterata, Turbelliaria, and Annelida indicate that the Dickinsoniidae are an independent branch of the oldest Bilateria (Fedonkin, 1983a). The theoretical possibility of deriving the symmetry of Dickinsoniidae from that of the most archaic Radialia with an axis of rotation of infinitely high order and concentric organization (Class Cyclozoa, e.g., *Kullingia–Ovatoscutum–Chondroplon–Dickinsonia*) or with an axis of variable order and radial organization (i.e., *Irridinites–Dickinsonia*) indicates that this group may be extremely primitive (Fig. 44). The Dickinsoniidae might represent one of the branches of Bilateria that separated from Radialia earlier than others did. Whether this attempt was successful or not is difficult to judge. Because other, more highly organized Bilateria (which might have the phylogenetic connections to their Paleozoic counterparts) existed in the Vendian

Figure 31. *Inkrylovia lata* Fedonkin, ×1, White Sea region, USSR.

besides the Dickinsoniidae, the Dickinsoniidae went extinct in the Vendian, as neither they nor similar forms are found in the Late Vendian or the Cambrian. Thus the Dickinsoniidae comprise the Class Dipleurozoa, and are placed not within the Coelenterata, but rather within primitive Bilateria of the Phylum Proarticulata (Fedonkin, 1983a).

Animals with flat oval bodies and large cephalic areas with an axial lobe and segments behind it are placed in an independent group of Vendian Bilateria, the Family Vendomiidae. The segments of the left and right sides are placed opposite each other or they alternate. The group represented by *Praecambridium, Vendia, Vendomia,* and *Onega* (Keller and Fedonkin, 1977) tends to develop a large cephalic area, which seems to be a progressive character (Figs. 37–39). But the

Figure 32. *Archangelia valdaica* Fedonkin, ×1, White Sea region, USSR.

presence of a plane of symmetry of sliding reflection within the area of segments, as in *Vendia* (Keller, 1969), and the flat oval body without lateral appendages may indicate that the group is an extremely archaic, special branch in the ancestry of true Articulata, which are considered a special class of the Phylum Proarticulata (Fedonkin, 1983a).

Praecambridium may not be a true segmented animal, although it was compared with the arthropods and annelids (Glaessner, 1984; Birket-Smith, 1981b). Among Vendomiidae, only *Vendomia* can be attributed to normal segmented animals and even compared to trilobite larvae. Interpretation of *Praecambridium*,

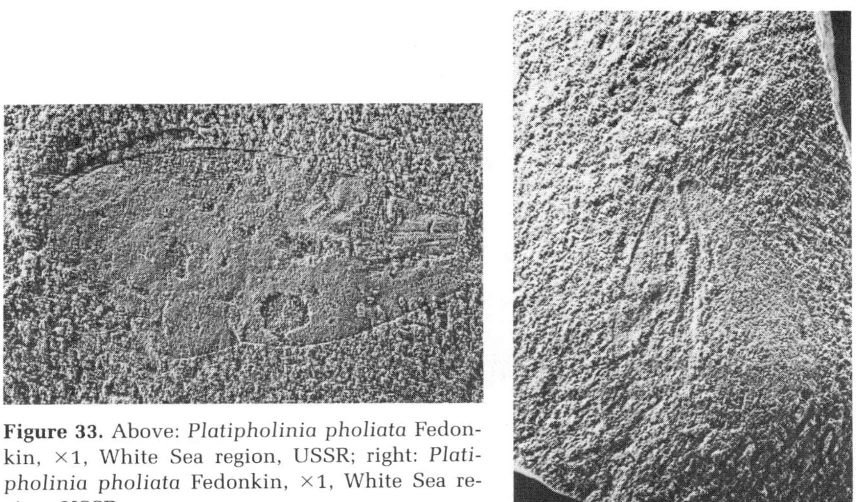

Figure 33. Above: *Platipholinia pholiata* Fedonkin, ×1, White Sea region, USSR; right: *Platipholinia pholiata* Fedonkin, ×1, White Sea region, USSR.

Figure 34. *Vladimissa missarzhevskii* Fedonkin, ×1, White Sea region, USSR.

Vendomia, and *Onega* as larvae cannot be supported yet by transitional fossils between them and any known large Precambrian animals. Thus, small bilateral vendomiids seem to be adult forms.

Some Vendian Bilateria, including the vendomiids, are difficult to place in recognized Metazoan higher taxa. Representatives of these groups are commonly characterized by an unusual complex of features, each of which, taken individually, is typical of widely separated phyla of later epochs.

A well-known example of such a combination of different phyletic characters is *Spriggina* with its horseshoe-shaped head that resembles the head shield of trilobites and body with lateral appendages resembling some annelids (Glaessner, 1984). *Spriggina* has recently been reinterpreted (Seilacher, 1989) as an organism that is certainly unrelated to any segmented invertebrate, based on the discovery of alternation of the left and the right semisegments (the absence of true segmenta-

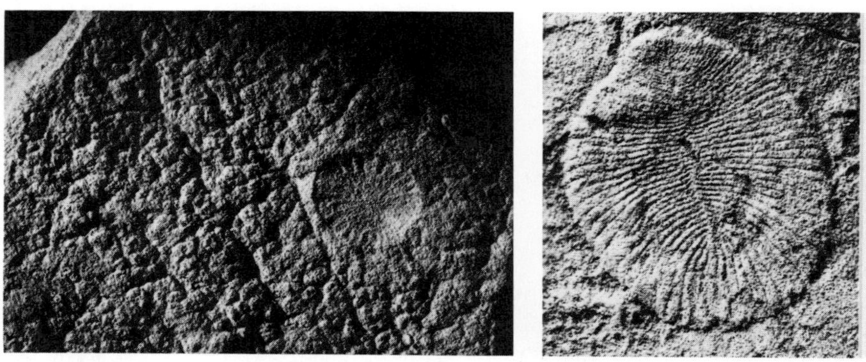

Figure 35. Left: *Dickinsonia costata* Sprigg, (juvenile individual; wrinkled surface of the bedding plane probably reflects the texture of the microbial mat that covered the surface of the sediment soon after the moment of burial of the animal), ×1, White Sea region, USSR; right: *Dickinsonia costata* Sprigg, ×1, White Sea region, USSR.

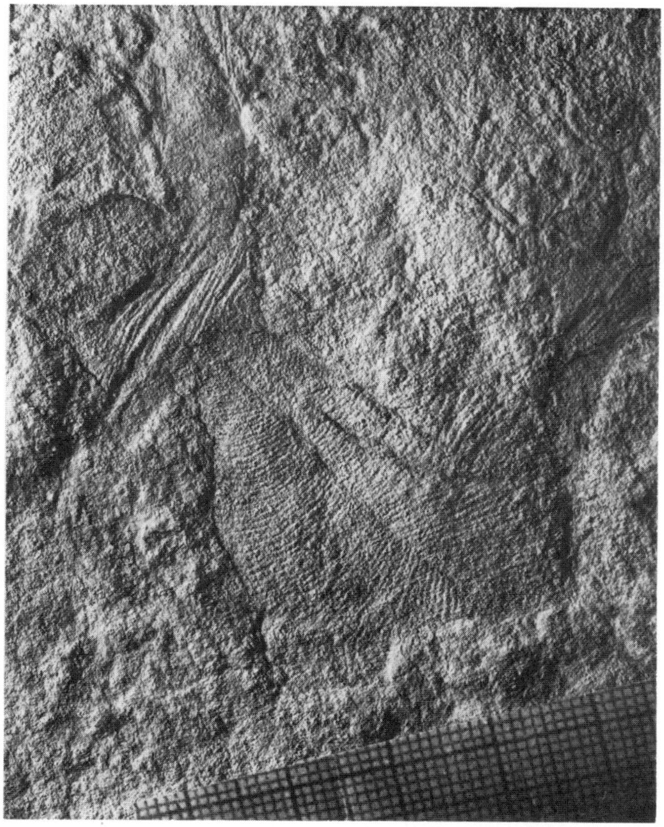

Figure 36. *Dickinsonia tenuis* Glaessner and Wade, ×1.4, Dniester River Valley, Podolia, USSR.

tion). Furthermore, the numerous lateral ridges in the elongate leaf-shaped fossil are not actually the appendages, but rather the lateral relief of the body.

True segmented animals resembling or belonging to the arthropods did live in the Vendian oceans, and some can be compared with Paleozoic counterparts. For example, *Parvancorina* Glaessner had a shieldlike, rather soft carapace with a faint marginal rim and elevated anterolateral and median smooth dorsal ridges. Approximately five pairs of stout anterior appendages are followed by up to 20 pairs of posterior filiform appendages. The similarity of *Parvancorina* to the Paleozoic arthropods of the *Marellomorpha* group may indicate that it is close to the ancestors of the Crustacea (Glaessner, 1979).

Two other animals resemble some crustaceans, but may not belong to this class of arthropods (Fedonkin, 1985b, c, 1990a, b). *Bomakellia* is rather large with a semilunar head, elongated body, and lateral appendages (Fig. 41). Its dorsal side has paired hillocks. The lateral flat appendages have a feather-shaped relief and ovate outline resembling epipodites and pre-epipodites of some crustaceans, although the same is true for the gills of some marine annelids. *Bomakellia's* large head resembles trilobite head shields, but other characters make the comparison unlikely. The body is not subdivided into segmented axial rings. The appendages

Figure 37. Left: *Paleoplatoda segmentata* Fedonkin, ×1, White Sea region, USSR; right: *Vendia sokolovi* Keller, ×4.5, Yarensk borehole, depth 1552 m.

Figure 38. *Onega stepanovi* Fedonkin, ×6, White Sea region, USSR (latex cast).

Figure 39. Left: *Vendomia menneri* Keller, ×5, White Sea region, USSR; right: *Pseudovendia* sp., ×1.5, White Sea region, USSR.

are much larger posteriorly. High relief of the fossil, particularly in its axial part, as well as the presence of a narrow ridge that outlines the mold along the distal ends of appendages, could be interpreted as a thin carapace that covered the body of *Bomakellia*. No evidence exists for feet, except for a few linear structures along the axis of the body at the posterior, slightly broken, end. The position of those linear structures in the host rock corresponds to the ventral side of this animal; nevertheless, these structures are not positively imprints of the feet.

Another form, described as *Mialsemia* (Fig. 40), seems to have a body plan related to that of *Bomakellia*, although it differs in proportions and some details. A smooth cigar-shaped body bears lateral, flat, featherlike appendages covered by a wide carapace. The relatively small semilunar head is poorly preserved and partially distorted. Both forms are known from single specimens in the Ust-Pinega Formation (Redkino Horizon) of the Vendian along the White Sea coast, northern European part of the USSR.

Figure 40. *Mialsemia semichatovi* Fedonkin, ×1, White Sea region, USSR.

Figure 41. *Bomakellia kelleri* Fedonkin, ×0.7, White Sea region, USSR.

The body plan of these animals does not match any existing invertebrate group. Recently I suggested placing both forms in a special class, the Paratrilobita (Fedonkin, 1985b, c, 1987, 1990a, b).

5. Early History of Invertebrates Based on the Vendian Fossil Record

Although data on the Vendian fauna are certainly incomplete, new, and controversial, they cannot be ignored in the discussion of the origin and early history of metazoans. Here I touch on some aspects of the problem—the initial morphological evolution of coelenterates and the origin of metagenesis in this group, and the origin of metamerism.

Although direct paleontological data on the evolution of coelenterates in the Precambrian are lacking, comparative morphology of the Vendian solitary Radialia can be used to infer their phylogeny. Comparison of the Vendian coelenterates and their Recent counterparts shows that very important changes took place with respect to body plan evolution over 600 million years. One of the three chief body plans in Vendian solitary coelenterates completely dominates the modern coelenterates—a radial body plan with a stable order of the symmetry.

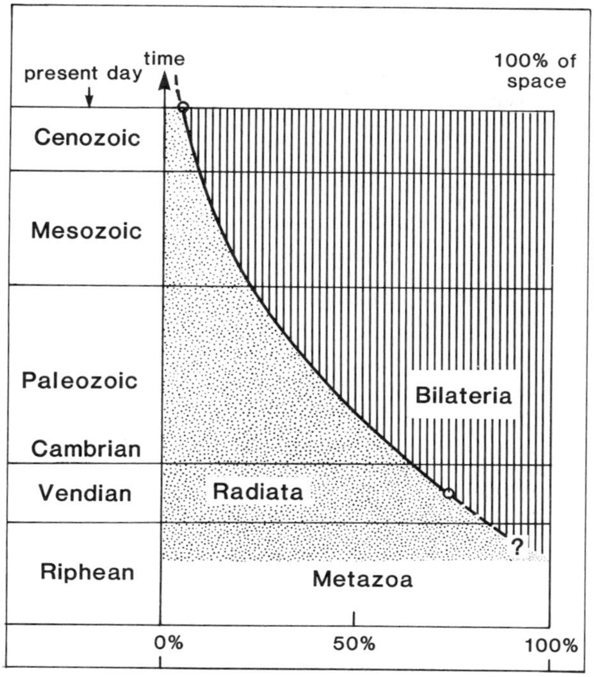

Figure 42. Comparison of the ratio of the number of species in Radiata to that of Bilateria in the present day and in the Vendian fossil record; this supports the idea of the primary nature of the coelenterate-grade animals (Diploblastica) in the metazoan phylogeny. [From Fedonkin (1985e).]

A large number of fundamental characters, such as body plan, that reflect not only symmetry but also the mode of growth and other formative processes during ontogeny, reflects the phenomenon known as archaic diversity (Mamkayev, 1968, 1979, 1987). Archaic diversity can be recognized at different taxonomic levels as well as with regard to different classes of characters of the organism at the stage of initial radiation of some particular group. Diversity at the level of the body plan may indicate that the Vendian coelenterates are very close to their common ancestor both in phylogeny and chronology.

Body plan analysis of the Vendian Radialia reveals that organization and morphology became more complex, new physiological systems appeared and improved, and the order of radial symmetry decreased (Fedonkin, 1984, 1985b, e, 1987). These trends can be observed in forms from Cyclozoa through Inordozoa to Trilobozoa. In other words, in parallel with the change in order of symmetry from an infinitely high order of radial symmetry through an uncertain order to the constant stable order of radial symmetry, tentacles, gastrovascular systems, and gonads also appeared (Fig. 43), supporting Beklemishev's (1970) conclusion on the basis of a comparative anatomical analysis of Recent invertebrates that there is certain correlation of the impoverishment of the body symmetry with the growth of the general complexity of the organisms.

If the simple *Nemiana*-like polyps represent direct descendants of the oldest and the most primitive Radialia, then the initial form of all coelenterates was a spherical or loaf-shaped polyp that did not have tentacles, had an undifferentiated

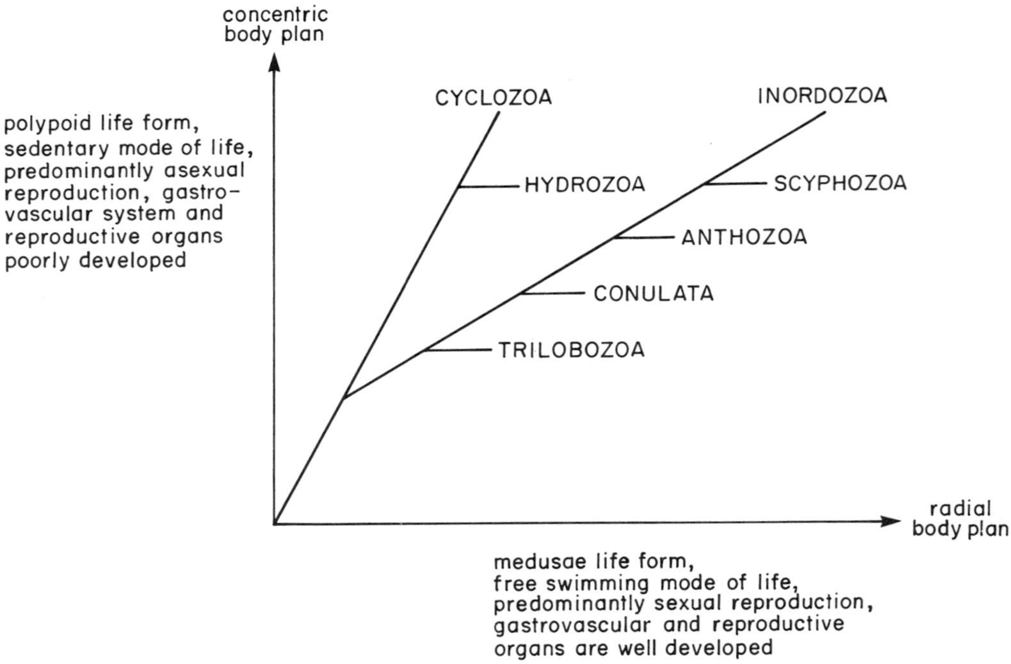

Figure 43. Model for the early evolution of coelenterates in morphological space between the two main morphological trends.

gastral cavity, and had a well-developed mesoglea that bore a supporting function. The further evolution of coelenterates could follow two principal directions—a concentric flattening and a radial development of structures.

The flattening of the gastral cavity from spherical to discoidal, observable in some Vendian sedentary medusoids, resulted in a concentric body plan. Possibly some forms, e.g., *Cyclomedusa* and *Kullingia,* had concentric septa in the gastral cavity that promoted the enlargement of its surface. A wide disc-shaped stomach, the edges of which could penetrate into the peripheral part of the body, allowed a normal lateral distribution of food, and provided the possibility for the existence of a medusa life form at the rather early and primitive organization of the Cyclozoa. The cyclozoans are the oldest class among known coelenterates. This hypothesis is supported by the fact that this group shows the highest morphological diversity, predominantly sessile mode of life, and a lack of any structures that could be interpreted as reproductive organs, although evidence of different kinds of asexual reproduction is present.

The concentric organization of the cyclozoan body could well be the result of rather primitive, periodic additive growth. But later on, the concentric body plan allowed the development of the patelloid, conical and medusalike forms of coelenterates with (1) polymerization that gave polycycloidal forms, (2) oligomerization that gave oligocycloidal forms, and (3) stabilization of the number of functional differentiated concentric zones. Together with these trends of morphological evolution in the most primitive phylogenetic branch of coelenterates,

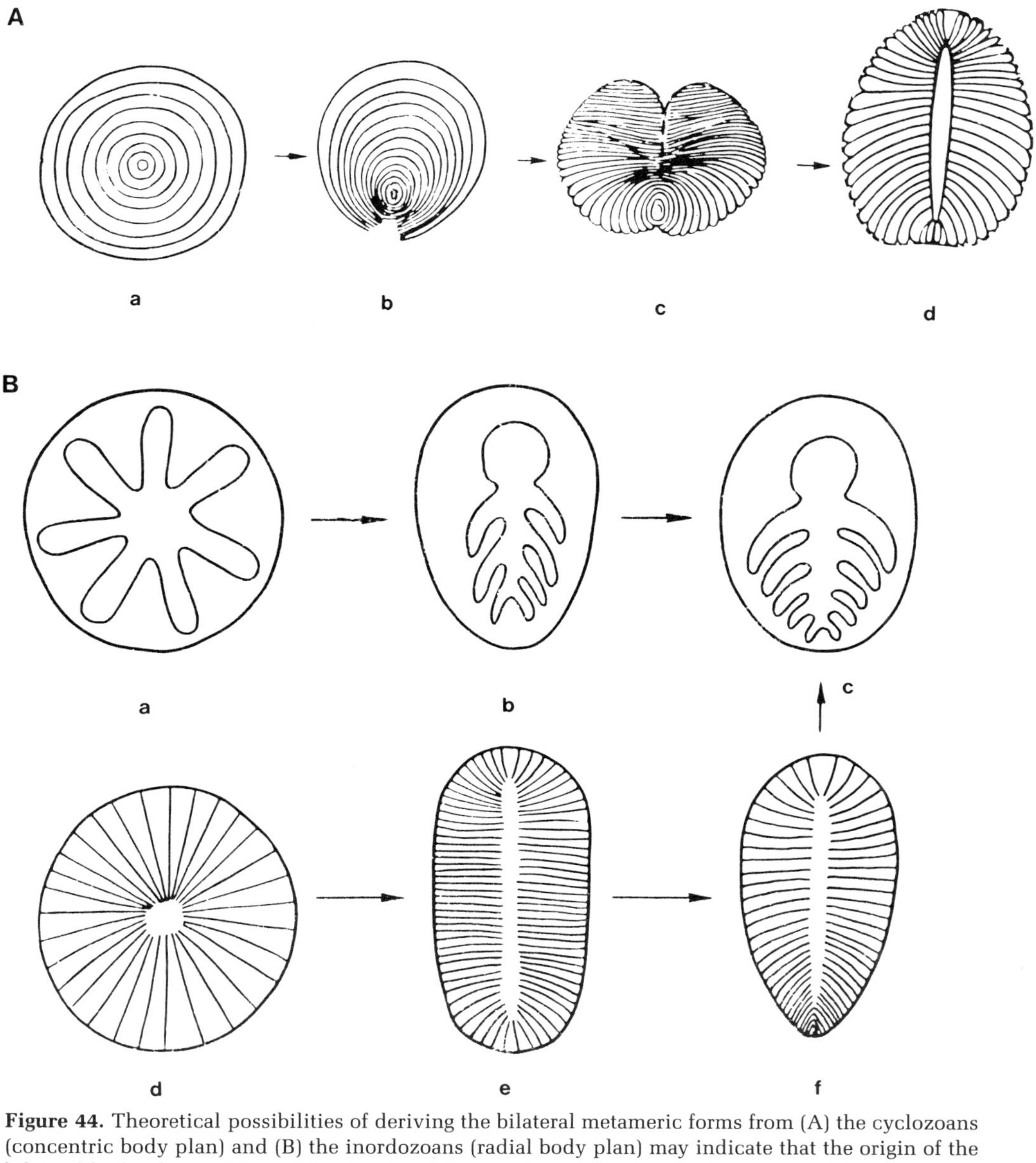

Figure 44. Theoretical possibilities of deriving the bilateral metameric forms from (A) the cyclozoans (concentric body plan) and (B) the inordozoans (radial body plan) may indicate that the origin of the bilateral body plan and the origin of metamerism could have been related processes at the very early stages of metazoan evolution. [Modified from Fedonkin (1985a).] A: a, *Kullingia*; b, *Ovatoscutum*; c, *Chondroplon*; d, *Dickinsonia*; B: a, *Bonata*; b, *Vendia*; c, *Praecambridium*; d, *Irridinitus*; e, intermediate form; f, *Dickinsonia*.

the polycycloidal body plan combined with bilateral symmetry, and the oligocycloidal body plan combined with the radial organization. Combinations of concentric and radial symmetries in some Vendian coelenterates can be considered as evidence for a phylogenetic link between Cyclozoa and Inordozoa and, on the other hand, it shows a conventional character that spans the two classes.

The second direction of early morphological evolution of coelenterates was variation on radial organization. The most important step here may have been the formation of radial relief in the internal surface of mesogloea along with the flattening of the body and the gastral cavity. Radial furrows might have permitted the development of wide discoidal stomachs and prevented a complete closure of the peripheral outer zone. Further development of a medusoid life form may have resulted in the raising of the oral and aboral sides of the enthoderma at the peripheral zone of the umbrella, leaving the radial channels empty. The Vendian fossil record thus suggests a possible origin for the gastrovascular system.

The uncertain order of the radial symmetry typical of the Vendian Inordozoa is very rare in living cnidarians and is considered a primitive archaic feature. This type of symmetry prevailed in many groups of the Vendian coelenterata of relatively high organization. The radial symmetry of uncertain order may represent an important and, possibly, a necessary stage in the early evolution of many phyletic lines of Coelenterata on the way to the Phanerozoic cnidarians with their stable body symmetries.

Many structures and the organ systems characteristic of recent Cnidaria seem to have been formed in the Precambrian Radialia at the stage of radial symmetry of uncertain order, e.g., tentacles, gastrovascular channels, and gonads.

And again, comparative body plan analysis of the Vendian coelenterates shows how those systems of organs could have evolved. In simplest forms with a nondifferentiated umbrella, as in *Hiemalora* (Fig. 17), reproductive organs may be irregularly arranged. With the appearance of concentric organization of the umbrella, the gonads could be arranged in one of those circular zones, as in *Armillifera* (Fig. 20), *Evmiaksia* (Fig. 19), or *Elasenia,* but, like other organs in all representatives of Inordozoa, the number of gonads increased while the organism was growing. An adult definitive stage may not have been fixed by the number of antimeres or the order of the symmetry. The next step in the evolutionary development of the reproductive system was the stabilization of the number of gonads that was connected with the general oligomerization trends and stabilization of the order of the symmetry reflected in other physiological systems.

I stress that the difference in the symmetries between known classes of the Vendian coelenterates reflects, primarily, the difference in the mode of growth. This fundamental character, along with others, puts all classes equally distant from each other in the phylogenetic tree of early coelenterates.

The Vendian fossil record sheds light on the problem of the origin of metagenesis in cnidarians. The majority of zoologists assume that the polyp life form is the primary and most ancient evolutionary stage, while others believe that a medusa life form is primary and the polyp is a secondarily simplified form of the life cycle.

The arguments of those who think that in the life cycle of Coelenterata the generation of medusoids appeared later than polyps seem to be better substantiated, and the Vendian material supports it. Many primitive Cyclozoa were sedentary but demonstrated a characteristic medusoid form and a thick mesoglea usually typical of medusae. These low conical polyps can be interpreted as sedentary medusoids or as an intermediate or transitional stage from polyp to medusa. Some medusa-shaped polyps might have changed their mode of life and become pelagic and thus solved the problem of dispersal.

Assuming an older pre-Vendian origin of many groups of Coelenterata, the regression associated with the Varangian (Laplandian) glaciation may have influenced the elimination of attached polyps and favored the development of metagenetic and purely medusoid forms. The most important step in this model is the initial position of Cyclozoa, in particular medusoid-shaped sedentary polyps from which a true medusa life form could evolve with all the progressive consequences of this innovation.

Analysis of the body plan of the Vendian Radialia, made in short here, indicates that the early evolution of Coelenterata in general proceeded from the concentric to radial body play and from an uncertain order of radial symmetry to a stable one. This evolutionary path was marked by those classes of animals that became extinct in the second half of the Vendian or in the Early Paleozoic.

Concluding the discussion on some aspects of the early evolution of coelenterates in the light of the Vendian fossil record, I note one important fact: all the localities of Precambrian metazoans containing the remains of the Vendian fauna demonstrate a strong domination by the solitary and colonial forms of coelenterates. This observation can be interpreted to favor an old idea on the primary nature of radial symmetry in the ontogeny and phylogeny of Eumetazoa (Fig. 42). Thus, early in metazoan history Radialia may have dominated numerically, and Diploblastica had more numerous phyla than did Cnidaria and Ctenophora, which survived into the recent.

Of those Vendian metazoans that are interpreted as Bilateria or Triploblastica, only a very few doubtful forms can be considered nonsegmented animals, while the large majority of solitary bilateral organisms demonstrate some sort of metamerism, although rather peculiar in some cases. This fact indicates that the origin of Bilateria (Triploblastica) and the origin of metamerism might be related evolutionary processes. Body plan analysis of the Vendian metazoans carried out in detail earlier (Fedonkin, 1983a–c, 1984, 1985a, b, 1987, 1990a) gives new insight to the protaxony–plagioaxony problem of comparative anatomy. The Vendian metazoan record can be interpreted in terms of a plagiaxony model by taking into account the following observations: (1) the definite domination of segmented forms over nonsegmented ones; (2) the presence of primitive organisms with a large number of metamers and with preserved relics of radial arrangement at their juvenile stages; and (3) the presence of polymerous and oligomerous animals with alternating semisegments that reflect a rather peculiar and primitive mode of growth. The symmetry of sliding reflection of the Vendian Proarticulata seems a natural, and even necessary, stage in the phylogenetic process of the formation of true metamerism, if articulated forms are derived from such simple coelenterates as the Vendian Inordozoa (Fig. 44). This evolutionary process could be traced and indicated by those unusual animals such as *Vendia* or *Dickinsonia* that are not yet true segmented animals according to recent understanding, e.g., not Articulata.

One of the problems of the Vendian fauna is an absence of evident ancestors of the important Cambrian invertebrate groups, including Archaeocyatha, Mollusca, Brachiopoda and Echinodermata, all of which appear early in the Lower Cambrian as well-differentiated phyletic lines. They must have had some period of Precambrian evolution. But a low species diversity and a prevalence of monotypic genera may indicate a relatively short time interval between the rise of these groups and the appearance of their ability to form a skeleton.

Skeletalization developed gradually during the Vendian. During the first half of the period, annelidlike animals with chitinoid, comblike jaws appeared. These have been named *Redkinia spinosa* Sokolov. Chitinoid tubes of sabelliditids appear at the same level as well as the calcareous tubular fossil *Cloudina* Germs. The very end of the Vendian saw a wide distribution of tubular shells, sclerites, and conodontlike forms. Small sizes of the oldest shelly fossils and wide geographic distribution could indicate that the Precambrian ancestors had small body sizes and a planktonic mode of life.

Possibly in parallel with the extinction of some groups of fauna, there was a decrease in body sizes for other groups in the second half of the Vendian. This could explain the sharp impoverishment, if not a gap, in the fossil record of invertebrates of the Late Vendian. But the decrease in body size could have led to the oligomerization in many primitive polymerous forms. And this in its turn could have resulted in a certain increase in the level of organization and/or even in the specialization of some forms. The appearance of some groups of meiofauna might be connected with this event.

A progressive colonization of the sea floor and initiation of burrowing and mud-eating played an important role in the evolution of the animal world. A critical time in this history was at the end of the Vendian period when biological processing of sediments increased greatly and many new groups of invertebrates began to inhabit the sea floor. The body size of the in fauna, represented mainly by soft-bodied animals, also increased at this time. All these phenomena, as well as the formation of skeletons in other groups, were possibly adaptive and reflect the increasing predation of vagile animals. Burrowing and the formation of skeletons had extremely important biological and evolutionary consequences that are not yet entirely understood by paleontologists or zoologists.

References

Anderson, M. M., and Conway Morris, S., 1982, A review, with description of four unusual forms of the soft-bodied fauna of the Conception and St. John's Groups (Late Precambrian), Avalon Peninsula, Newfoundland, in: *Proceedings Third North American Paleontological Convention,* (B. Mamet and M. J. Copeland, eds.), Vol. 1, pp. 1–8.

Anderson, M. M., and Misra, S. B., 1968, Fossils found in the Precambrian Conception Group in Southeastern Newfoundland, *Nature* **220**:680–681.

Becker, Yu. R., 1977, First paleontological finds in the Riphean of the Urals, *Izv. Akad. Nauk SSSR Ser. Geol.* **3**:90–100 [in Russian].

Becker, Yu. R., and Kishka, N. V., 1987, First Vendian biota of the South Urals, in: *Abstracts. 3rd All-Union Symposium on Paleontology of the Precambrian and Early Cambrian,* Karelian Branch, USSR Academy of Sciences, Petrozavodsk, USSR, pp. 8–9 [in Russian].

Becker, Yu. R., 1990, Vendian Metazoa from the Urals, in: *The Vendian System,* Vol. 1, *Paleontology* (B. S. Sokolov and A. B. Iwanowski, eds.), Springer-Verlag, New York, pp. 121–131.

Beklemishev, V. N., 1970, *Principles of Comparative Anatomy of Invertebrates,* Vol. 1, Oliver and Boyd Edinburgh.

Birket-Smith, S. J. R., 1981a, A reconstruction of the Precambrian *Spriggina*, *Zool. Jahrb. Anat.* **105**:237–258.

Birket-Smith, S. J. R., 1981b, Is *Praecambridium* a juvenile *Spriggina*?, *Zool. Jahrb. Anat.* **106**:233–235.

Bloeser, B., Schopf, J. W., Horodyski, R. J., and Breed, W. J., 1977, Chitinozoans from the Late Precambrian Chuar Group of the Grand Canyon, Arizona, *Science* **195**:676–679.

Boynton, H. E., 1978, Fossils from the Pre-Cambrian of Charnwood Forest, Leicestershire, *Mercian Geol.* **6**(4):291–296.

Boynton, H. E., and Ford, T. D., 1979, *Pseudovendia charnwoodensis*—A new Precambrian arthropod from Charnwood Forest, Leicestershire, *Mercian Geol.* **7**:175–177.

Chen Menge, 1991, Discussion on the stratigraphic significance of macrofossils from the late Precambrian sequence in southern Liaoning Province, *Scientia Geologica Sinica*, **4(2)**:120–128.

Chistiakov, V. G., Kalmykova, N. A., Nessov, L. A., and Syslov, G. A., 1984, On the presence of Vendian deposits in the middle stream of Onega River and on the possible existence of tunicates (Tunicata: Chordata) in the Precambrian, *Vestnik Leningrads. Gos. Univ.* **6**:11–19 [in Russian].

Cope, J. C. W., 1977, An Ediacara fauna from South Wales, *Nature* **268**:624.

Cope, J. C. W., 1983, Precambrian fossils of the Carmarthen area, Dyfed, *Nature Wales* **1**(2):11–16. 16.

Crimes, T. P., and Germs, G. J. B., 1982, Trace fossils from the Nama Group (Precambrian–Cambrian) of Southwest Africa (Namibia), *J. Paleontol.* **56**:890–907.

Debrenne, F., and Naud, G., 1981, Méduses et traces fossils supposées précambriennes dans la formation de San Vito, Sarrabus, Sud-Est de la Sardaigne, *Bull. Soc. Geol. France 7* **23**(1):23–31.

Farmer, J. D., Vidal, G., Strauss, H., Moszydlowska, M., Ahlberg, P., Siedlecka, A., 1991, Ediacaran medusoids from the Innerelv Member (late Proterozoic), Tanafjorden area of N. E. Finnmark. *Geol. Soc. Am., Abstracts* **23**(5):A97.

Fedonkin, M. A., 1977, Precambrian–Cambrian ichnocoenoses of the East-European platform, in: Trace Fossils 2 (T. P. Crimes and J. C. Harper, eds.), *Geol. J. Spec. issue.* **9**:183–194.

Fedonkin, M. A., 1981, *White Sea Biota of the Vendian (Precambrian Non-skeletal Fauna of the Russian Platform North)*, Nauka, Moscow [in Russian].

Fedonkin, M. A., 1983a, The organic world of the Vendian, *Itogi Nauki Tekh. Ser. Strat. Paleontol.* **12**:1–127 [in Russian].

Fedonkin, M. A., 1983b, Non-skeletal fauna of Podolia (Dniester River valley), in: *The Vendian of the Ukraine* (V. A. Velikanov, E. A. Assejeva, and M. A. Fedonkin), Naukova Dumka, Kiev, pp. 128–139 [in Russian].

Fedonkin, M. A., 1983c, Promorphology of Vendian Radialia as a key to understanding the early evolution of Coelenterata, in: *Fourth International Symposium on Fossil Cnidaria. Abstracts*, Washington, D.C., p. 6.

Fedonkin, M. A., 1984, Promorphology of the Vendian Radialia, in: *Stratigraphy and Paleontology of the Earliest Phanerozoic*, Nauka, Moscow, pp. 30–58 [in Russian].

Fedonkin, M. A., 1985a, Promorphology of the Vendian Bilateria and the problem of the origin of metamerism of Articulata, in: *Problematics of the Late Precambrian and Paleozoic*, Nauka, Moscow, pp. 79–92 [in Russian].

Fedonkin, M. A., 1985b, Non-skeletal fauna of the Vendian: Promorphological analysis, in: *The Vendian System*, Vol. 1, *Paleontology* (B. S. Sokolov and A. B. Iwanowski, eds.), Nauka, Moscow, pp. 10–60 [in Russian].

Fedonkin, M. A., 1985c, Systematic description of the Vendian Metazoa, in: *The Vendian System*, Vol. 1, *Paleontology* (B. S. Sokolov and A. B. Iwanowski, eds.), Nauka, Moscow, pp. 70–106 [in Russian].

Fedonkin, M. A., 1985d, Paleoichnology of the Vendian Metazoa, in: *The Vendian System*, Vol. 1, *Paleontology* (B. S. Sokolov and A. B. Iwanowski, eds.), Nauka, Moscow, pp. 112–117 [in Russian].

Fedonkin, M. A., 1985e, Precambrian metazoans: The problems of preservation, systematics and evolution, *Philos. Trans. R. Soc. Lond. B* **311**:27–45.

Fedonkin, M. A., 1986, Precambrian problematic animals: Their body plan and phylogeny, in: *Problematic Fossil Taxa* (A. Hoffman and M. H. Nitecki, eds.), Oxford University Press, Oxford, pp. 59–67.

Fedonkin, M. A., 1987, Non-skeletal fauna of the Vendian and its place in metazoan evolution, *Trudy Paleontol. Inst. Akad. Nauk SSSR* **226**:1–176 [in Russian].

Fedonkin, M. A., 1990a, Non-skeletal fauna of the Vendian: Promorphological Analysis, in: *The Vend-*

ian System, Vol. 1, *Paleontology* (B. S. Sokolov and A. B. Iwanowski, eds.), Springer-Verlag, New York, pp. 7–70.

Fedonkin, M. A., 1990b, Systematic description of the Vendian Metazoa, *The Vendian System,* Vol. 1, *Paleontology* (B. S. Sokolov and A. B. Iwanowski, eds.), Springer-Verlag, New York, pp. 71–120.

Fedonkin, M. A., and Krivosheyev, V. I., in press, Ediacaran fauna in the Vendian of the Central Urals, in: *Proceedings Fifth North American paleontological Convention,* Chicago.

Ford, T. D., 1958, Pre-Cambrian fossils from Charnwood Forest, *Proc. Yorkshire Geol. Soc.* **31:**211–217.

Ford, T. D., 1962, The oldest fossils, *New Sci.* **15:**191–194.

Ford, T. D., 1963, The Pre-Cambrian fossils of Charnwood Forest, *Trans. Leicestershire Lit. Philos. Soc.* **57:**57–62.

Ford, T. D., 1980, The Ediacaran fossils of Charnwood Forest, Leicestershire, *Proc. Geol. Assoc.* **91:**81–83.

Føyn, S., and Glaessner, M. F., 1979, *Platysolenites,* other animals, and the Precambrian–Cambrian transition in Norway, *Norsk Geol. Tidsskrift* **59:**25–46.

Gehling, J. C., 1984, New Ediacaran fossils from the Flinders Ranges, South Australia, in: *Abstracts. 27th International Geological Congress, Moscow,* Nauka, Moscow, Vol. 1, p. 272.

Gehling, J. G., 1987, Earliest known echinoderm—A new Ediacaran fossil from the Pound Subgroup of South Australia, *Alcheringa* **11:**337–345.

Gehling, J. G., 1988, A cnidarian of actinian-grade from the Ediacaran Pound Subgroup, South Australia, *Alcheringa* **12:**299–314.

Germs, G. J. B., 1968, Discovery of a new fossil in the Nama System, South West Africa, *Nature* **219:**53–54.

Germs, G. J. B., 1972a, Trace fossils from the Nama System, South West Africa, *J. Paleontol.* **46:**864–870.

Germs, G. J. B., 1972b, Thin concentric structures of biologic origin from the Nama System, South West Africa, *Bull. Geol. Soc. Am.* **83:**463–466.

Germs, G. J. B., 1972c, The stratigraphy and paleontology of the Lower Nama Group, South West Africa, *Univ. Cape Town Dept. Geol. Precambrian Res. Unit Bull.* **12:**1–250.

Germs, G. J. B., 1972d, New shelly fossils from Nama Group, South West Africa, *Am. J. Sci.* **272:**752–761.

Germs, G. J. B., 1973a, A reinterpretation of *Rangea schneiderhoehni* and the discovery of a related new fossil from the Nama Group, South West Africa, *Lethaia* **6:**1–10.

Germs, G. J. B., 1973b, Possible sprigginid worm and a new trace fossils from the Nama Group, South West Africa, *Geology* **1:**69–70.

Gibson, G. G., Teeter, S. A., and Fedonkin, M. A., 1984, Ediacarian fossils from the Carolina slate belt, Stanly County, North Carolina, *Geology* **12:**387–390.

Glaessner, M. F., 1958, New fossils from the base of the Cambrian in South Australia, *Trans. R. Soc. S. Aust.* **81:**185–188.

Glaessner, M. F., 1959, The oldest fossil faunas of South Australia. *Geol. Rundsch.* **47:**522–531.

Glaessner, M. F., 1969, Trace fossils from the Precambrian and basal Cambrian, *Lethaia* **2:**369–393.

Glaessner, M. F., 1971, The genus *Conomedusites* Glaessner et Wade and the diversification of the Cnidaria, *Palaeontol. Z.* **45:**7–17.

Glaessner, M. F., 1976, A new genus of Late Precambrian polychaete worms from South Australia, *Trans. R. Soc. S. Aust.* **100:**169–170.

Glaessner, M. F., 1979, Precambrian, in: *Treatise on Invertebrate Paleontology,* Part A, *Introduction. Fossilization (Taphonomy) Biogeography* (R. A. Robison and C. Teichert, eds.), Geological Society of America and University of Kansas Press, Lawrence, Kansas, pp. A79–A118.

Glaessner, M. F., 1980, *Parvancorina*—An arthropod from the Late Precambrian (Ediacarian) of South Australia, *Ann. Naturhist. Mus. Wien* **83:**83–90.

Glaessner, M. F., 1984, *The Dawn of Animal Life: A Biohistorical Study,* Cambridge University Press, Cambridge.

Glaessner, M. F., and Wade, M. 1966, The Late Precambrian fossils from Ediacara, South Australia, *Palaeontology* **9:**599–628.

Glaessner, M. F., and Wade, M., 1971, *Praecambridium*—A primitive arthropoda, *Lethaia* **4:**71–77.

Glaessner, M. F., and Walter, M. R., 1975, New Precambrian fossils from the Arumbera Sandstone, Northern Territory, Australia, *Alcheringa* **1:**59–69.

Gould, S. J., 1984, The Ediacaran experiment, *Nat. Hist.* **2:**14–23.

Grant, S. W. F., 1990, Shell structure and distribution of *Cloudina*, a potential index fossil for the terminal Proterozoic, *Am. J. Sci.* **290-A:**261–294.

Gureev, Yu. A., 1983, On a new form of trace fossils from the Lower Baltic deposits in Podolia, Dniester Valley, *Paleontol. Sb.* **20:**70–73 [in Russian].

Gureev, Yu. A., 1984, Bioglyphs of Paleozoic kind in the Vendian of Podolia and their stratigraphic potential, *Doklady UkrSSR Akad. Nauk Ser. B Geol. Khim. Biol. Nauki* **4:**5–8.

Gureev, Yu. A., 1985, Vendiata—Primitive Precambrian Radialia, in: *Problematics of the Late Precambrian and Paleozoic*, Nauka, Moscow, pp. 92–103.

Gureev, Yu. A., 1988, Non-skeletal fauna, in: *Biostratigraphy and Paleogeographic Reconstructions of the Precambrian of the Ukraine*, Naukova Dumka, Kiev, pp. 65–80 [in Russian].

Gürich, G., 1930, Die bisland ältesten Spuren von Organismen in Südafrika, in: *International Geological Congress, Pretoria (1929)*, Vol. 15, pp. 670–680.

Gürich, G., 1933, Die Kuibis-Fossilen der Nama Formation von Südwestafrica, *Paleontol. Z.* **15:**137–154.

Hahn, G., and Pflug, H. D., 1980, Ein neuer Medusen-Fund aus dem Jung-Präkambrium von Zentral-Iran, *Senckenb. Lethaea* **60:**449–461.

Harrington, H. J., and Moore, R. C., 1956a, Scyphomedusae, in: *Treatise in Invertebrate Paleontology*, Part F, *Coelenterata* (R. C. Moore, ed.), Geological Society of America, New York, and University of Kansas Press, Lawrence, Kansas, p. F38.

Harrington, H. J., and Moore, R. C., 1956b, Dipleurozoa, in: *Treatise on Invertebrate Palaeontology*, Part F, *Coelenterata* (R. C. Moore, ed.), Geological Society of America, New York, and University of Kansas Press, Lawrence, Kansas, pp. F24–F27.

Hermann, T. N., 1981a, Filamentous microorganisms from Lakhanda Formation, Maya River, *Paleontol. Zh.* **2:**126–131 [in Russian].

Hermann, T. N., 1981b, Finding of filamentous algae in Miroyedikha Formation, Upper Precambrian, *Paleontol. Zh.* **4:**118–121 [in Russian].

Hofmann, H. J., 1981, First record of a Late Proterozoic faunal assemblage in the North American Cordillera, *Lethaia* **14:**303–310.

Hofmann, H. J., Fritz, W. H., and Narbonne, G. M., 1983, Ediacaran (Precambrian) fossils from the Wernecke Mountains, Northwestern Canada, *Science* **221:**455–457.

Hofmann, H. J., Mountjoy, E. W., and Teitz, M. W., 1985, Ediacaran fossils from the Miette Group, Rocky Mountains, British Columbia, Canada, *Geology* **13:**819–821.

Hofmann, H. J., Narbonne, G. M., and Aitken, J. D., 1990, Ediacaran remains from intertillite beds in northwestern Canada, *Geology* **18:**1199–1202.

Horodyski, R. J., 1982, Problematic bedding-plane markings from the Middle Proterozoic Appekunny Argillite, Belt Supergroup, Northwestern Montana, *J. Paleontol.* **56:**882–889.

Horodyski, R. J., 1991, Late Proterozoic megafossils from southern Nevada, *Geol. Soc. Am., Abstracts* **23**(5):A163.

Houzay, J. P., 1979, Empreintes attribuables à des méduses dans la série de base de l'Adoudounien (Precambrian terminal de l'Anti-Atlas, Maroc), *Géol. Med.* **6:**379–384.

Jenkins, R. J. F., 1985, The enigmatic Ediacaran (Late Precambrian) genus *Rangea* and related forms, *Paleobiology* **11:**336–355.

Jenkins, R. J. F., and Gehling, J. C., 1978, A review of the frond-like fossils of the Ediacara assemblage, *Rec. S. Aust. Mus.* **17:**347–359.

Keller, B. M., 1969, Imprint of an unknown animal from the Valday Series of the Russian Platform, in: Tommotian Stage and the Cambrian lower Boundary Problem (A. Y. Rosanov, ed.), *Trudy Geol. Inst.* **206:**175–176 [in Russian].

Keller, B. M., and Fedonkin, M. A., 1976, New findings of fossils in the Valdai Series (Precambrian) on the Syuzma River, *Izv. Akad. Nauk SSSR Ser. Geol.* **3:**38–44.

Keller, B. M., and Fedonkin, M. A., 1977, New organic fossil finds in the Precambrian Valday Series along the Syuz'ma River, *Int. Geol. Rev.* **19:**924–930.

Keller, B. M., Menner, V. V., Stepanov, V. A., and Chumakov, N. M., 1974, New finds of Metazoa in the Vendomian of the Russian Platform, *Izv. Akad. Nauk SSSR Ser. Geol.* **12:**130–134 [in Russian].

Kulling, O., 1972, The Swedish Caledonides, in: *Scandinavian Caledonides* (T. Strand and O. Kulling, eds.), Wiley, London, pp. 147–302.

Liu Xiaoliang, 1981, Metazoa fossils from Mashan Group near Jixi, Heilongjang, *Bull. Chin. Acad. Geol. Sci.* **3:**71–83.

Mamkaev, J. V., 1985, The principle of morphological radiation and its significance for evolutionary morphology, in: *Evolution and Morphogenesis*, Praha, Academia, pp. 421–427.

Mamkaev, J. V., 1986, Initial morphological diversity as a criterion in deciphering turbellarian phylogeny, *Hydrobiologia* **132:**31–33.

Mamkaev, Yu. V., 1987, Ciliated worms and the methodological principles of evolutionary morphology, in: Morphology of Ciliated Worms, Pogonophorans and Ascidians, *Trudy Zool. Inst. SSSR Akad. Nauk* **167:**4–33 [in Russian].

Misra, S. B., 1969, Late Precambrian (?) fossils from southeastern Newfoundland, *Geol. Soc. Am. Bull.* **80:**2133–2140.

Missarzhevski, V. V., 1974, New data on the earliest fossils of the Early Cambrian of the Siberian Platform, in: *Biostratigraphy and Paleontology of the Lower Cambrian of Europe and Northern Asia*, Nauka, Moscow, pp. 179–189 [in Russian].

Moore, R. C., and Harrington, H. J., 1956, Conulata, in: *Treatise on Invertebrate Paleontology*, Part F, *Coelenterata* (R. C. Moore, ed.), Geological Society of America and University of Kansas Press, Lawrence, Kansas, pp. F54–F56.

Narbonne, G. M., and Aitken, J. D., 1990, Ediacaran fossils from the Sekwi Brook area, Mackenzie Mountains, Northwestern Canada, *Paleontology* **33** (Part 4): 945–980.

Narbonne, G. M., and Hofmann, H. J., 1987, Ediacaran biota of the Wernecke Mountains, Yukon, Canada, *Paleontology* **30:**647–676.

Nitecki, M. H. (ed.), 1979, *Mason Creek Fossils*, Academic Press, New York.

Palij, V. M., 1976, Remains of non-skeletal fauna and traces of life activity from deposits of the Upper Precambrian and Lower Cambrian of Podolia, in: *Paleontology and Stratigraphy of Upper Precambrian and Lower Paleozoic in south-west East-European Platform*, Naukova Dumka, Kiev, pp. 63–77 [in Russian].

Palij, V. M., Posti, E., and Fedonkin, M. A., 1979, Soft-bodied Metazoa and fossil traces of animals in the Vendian and Early Cambrian, in: *Paleontology of Upper Precambrian and Cambrian Deposits of the East-European Platform*, Nauka, Moscow, pp. 49–82 [in Russian].

Pflug, H. D., 1966, Neue Fossilreste aus den Nama-Schichten in Südwest-Afrika, *Palaeontol. Z.* **40:**14–25.

Pflug, H. D., 1970a, Zur Fauna der Nama-Schichten in Südwest-Afrika. I. Pteridinia, Bau und systematische Zugehörigkeit, *Palaentogr. Abt. A.* **134:**226–262.

Pflug, H. D., 1970b, Zur fauna der Nama-Schichten in Südwest-Africa. II. Rangeidae. Bau und systematische Zugehörigkeit, *Palaeontogr. Abt. A* **135:**198–231.

Pflug, H. D., 1971, Neue Fossilfunde im Jung-Präkambrium und ihre Aussagen zur Entstehung der höheren Tiere, *Geol. Rundsch.* **60:**1340–1350.

Pflug, H. D., 1972a, Systematik der Jung-Präkambrischen Petaloname Pflug 1970, *Palaeontol. Z.* **46:**56–67.

Pflug, H. D., 1972b, Zur Fauna der Nama-Schichten in Südwest-Afrika. III Erniettmorpha, Bau und Systematik, *Palaeontogr. Abt. A* **139:**134–170.

Pflug, H. D., 1973, Zur Fauna der Nama-Schichten in Südwest-Afrika. IV. Mikroskopische Anatomie der Petalo-Organismen, *Paleontogr. Abt. A* **144:**166–202.

Richter, R., 1955, Die ältesten Fossien Süd-Africas, *Senckenb. Lethaea* **36:**243–89.

Runnegar, B., 1982, Oxygen requirements, biology and phylogenetic significance of the late Precambrian worm *Dickinsonia* and the evolution of the burrowing habit, *Alcheringa* **6:**223–239.

Runnegar, B., 1989, Proterozoic evolution of Metazoa, in: *Abstracts. 28th International Geological Congress, Washington, D.C.* Vol. 2, pp. 733–735.

Seilacher, A., 1984, Late Precambrian and Early Cambrian Metazoa: Preservational or real extinctions?, in: *Patterns of Change in Earth Evolution* (H. D. Holland and A. F. Trendall, eds.), Springer-Verlag, Berlin, pp. 159–168.

Seilacher, A., 1989, Organismic construction in the Proterozoic biosphere, *Lethaia* **22:**229–239.

Sokolov, B. S., 1952, On the age of the old sedimentary cover of the Russian Platform, *Izv. Akad. Nauk SSSR Ser. Geol.* **5:**21–31 [in Russian].

Sokolov, B. S., 1972, The Vendian stage in Earth history, in: *Paleontology, 24th Session. Reports of Soviet Geologists*, Nauka, Moscow, pp. 114–123 [in Russian].

Sokolov, B. S., 1975, On paleontological finds in the pre-Usol deposits at Irkutsk Amphitheatre, *Trudy Inst. Geol. Geofiz. Sibirsk. Otd. SSSR Akad. Nauk* **232:**112–117 [in Russian].

Sokolov, B. S., 1976, Precambrian Metazoa and the Vendian–Cambrian boundary, *Paleontol. Zh.* **1:**3–18 [in Russian].

Sokolov, B. S., and Fedonkin, M. A., 1984, The Vendian as the Terminal System of the Precambrian, *Episodes* **7**:12–19.

Sprigg, R. C., 1947, Early Cambrian (?) jellyfishes from the Flinders Ranges, South Australia, *Trans. R. Soc. S. Aust.* **71**:212–224.

Sprigg, R. C., 1949, Early Cambrian "jellyfishes" of Ediacara, South Australia and Mount John, Kimberley District, Western Australia, *Trans. R. Soc. S. Aust.* **73**:72–99.

Sun Weiguo, 1986a, Late Precambrian pennatulids (sea pens) from the eastern Yangtze Gorge, China: *Paracharnia gen. nov., Precambrian Research,* **31**:361–375.

Sun Weiguo, 1986b, Are there Pre-Ediacarien metazoans? *Precambrian Research* **41**:409.

Sun Weiguo, Wang Guixiang, and Zhoe Benhe, 1986, Macroscopic worm-like body fossils from the Upper Precambrian (900–700 Ma), Huainan district, Anhui, China and their stratigraphic and evolutionary significance, *Precambrian Research* **31**:377–403.

Valkov, A. K., 1982, *Biostratigraphy of the Lower Cambrian of the East Siberian Platform (Uchur-Maya Region),* Nauka, Moscow [in Russian].

Valkov, A. K., and Sysoiev, V. A., 1970, Cambrian augustiochreids of Siberia, in: *Stratigraphy and Paleontology of Proterozoic and Cambrian of the East Siberian Platform,* Yakutsk Publishing House, Yakutsk, pp. 94–100 [in Russian].

Veiss, A. F., 1984, Microfossils from Upper Riphean of Turukhansk Region, *Paleontol. Zh.* **2**:102–108 [in Russian].

Velikanov, V. A., and Gureev, Yu. A., 1983, New finds of non-skeletal Metazoa and bioglyphs in Mogilyov Formation of the Vendian, Podolia, *Doklady Akad. Nauk SSSR* **272**:427–429 [in Russian].

Vidal, G., and Moszydlowska, M., 1987, Further reflections on metazoan evolution, *Precambrian Research* **36**:345–348.

Vodanjuk, S. A., 1989, Remains of non-skeletal metazoans from Khatyspyt Formation of Olehyok Uplift, in: Late Precambrian and early Paleozoic of Siberia. *Trudy Inst. Geol. Geofiz., Sibirsk. Otd. SSSR Akad. Nauk,* Novosibirsk, pp. 61–74 [in Russian].

Wade, M., 1968, Preservation of soft-bodied animals in the Precambrian sandstones in Ediacara, South Australia, *Lethaia* **1**:238–267.

Wade, M., 1969, Medusae from Uppermost Precambrian or Cambrian sandstones, Central Australia, *Palaeontology* **12**:351–365.

Wade, M., 1971, Bilaterial Precambrian Chondrophores from the Ediacara Fauna, South Australia, *Proc. R. Soc. Victoria* **84**:183–188.

Wade, M., 1972a, Hydrozoa and Scyphozoa and other medusoids from the Ediacara fauna South Australia, *Palaeontology* **15**:197–225.

Wade, M., 1972b, *Dickinsonia:* Polychaete worms from the late Precambrian Ediacara fauna, South Australia, *Mem. Queensl. Mus.* **16**:171–190.

Wang Hongzhen and Lin Baoyu, 1983, Fossil Cnidaria from the Sinian System of China, in: *Fourth International Symposium on Fossil Cnidaria,* Washington, D. C., p. 38.

Xing, Y., and Liu, G., 1979, Coelenterate fossils from the Sinian System of Southern Liaoning and its stratigraphic significance, *Acta Geol. Sinica* **53**:167–172.

Zaika-Novatski, V. S., and Palij, V. M., 1968, New data concerning the problematic imprints from the Vendian deposits of Podolia, *Paleontol. Sb. Lvov Univ.* 5 **1**:130–135 [in Russian].

Zaika-Novatski, V. S., Velikanov, V. A., and Koval, A. P., 1968, First representative of Ediacaran fauna in the Vendian of the Russian Platform (Upper Precambrian), *Paleontol. Zh.* **2**:269–270 [in Russian].

Chapter 5

Functional and Ecological Aspects of Ediacaran Assemblages

RICHARD J. F. JENKINS

1. Introduction: Discovery and Significance of Ediacaran Remains

Reference to fossil imprints of soft-bodied Ediacaran metazoans made by Hill and Bonney (1877, p. 757) recorded two of "those curious arrangements of concentric rings which have been supposed to be organisms" present on one of the bedding faces of the North Quarry, Woodhouse Eaves in Charnwood, Forest, Leicestershire, England (see Ford, 1958, 1963); the markings were dismissed as being "accidental . . . (and) inorganic." Early this century, P. Range and H. Schneiderhöhn collected fossil remains of equivalent age at Kuibis Farm in South West Africa (Namibia), and the organic nature of this material was confirmed by Gürich (1929, 1933). The history of discovery of such fossils during the mid part of the century (Sprigg, 1947, 1949; Ford, 1958, 1963, 1968, 1979a,b, 1981; Anderson and Misra, 1968) and the subsequent finding of similar materials widely sited about the globe are well known (e.g., Glaessner, 1984; Hofmann, 1987).

RICHARD J. F. JENKINS • Department of Geology and Geophysics, University of Adelaide, Adelaide, South Australia 5000.

Origin and Early Evolution of the Metazoa, edited by Jere H. Lipps and Philip W. Signor. Plenum Press, New York, 1992.

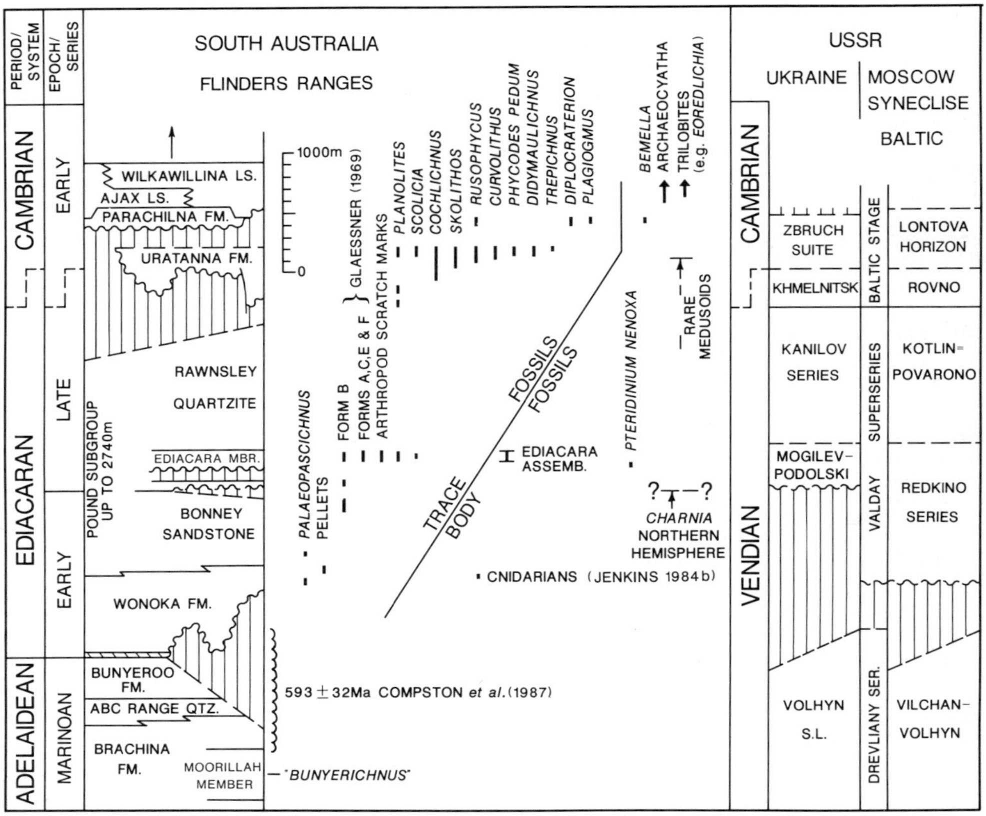

Figure 1. Correlation between the terminal Precambrian and Early Cambrian of the Flinders Ranges, South Australia, and equivalent successions on the Russian Platform, based on Ediacaran faunal elements, occurrences of distinctive trace fossils, and early skeletonized remains.

Soviet authors and others (this volume) assign soft-bodied assemblages of the late Precambrian to the Vendian, a time–rock division based on strata spread widely over the Russian Platform (e.g., Sokolov and Fedonkin, 1984). Although Sokolov (1952) nominated the Vendian as comprising the Valday Superseries (Fig. 1), the interval subsequently proved to be commonly fossiliferous, and current usage of the term includes underlying unfossiliferous tillitic deposits and volcanogenic intervals confined to narrow troughs or aulacogens (Makhnach *et al.*, 1986; Yakobson, 1987). The tillitic deposits are correlated with the Laplandian or Varangian Glaciation of Scandinavia. Volcanics associated with an upper Laplandian tillite in the Middle Ural Mountains have given a Rb–Sr isochron age of 621 ± 9 million years (My) and pebbles from related subvolcanic granitic intrusives occur at the base of overlying Valdaian rocks (Krasnobaev and Semikhatov, 1986).

Comparable fossil remains in the Flinders Ranges occur widely in a narrow interval (Ediacara Member) of the Rawnsely Quartzite, Pound Subgroup (Fig. 1), which succeeds a great thickness of mainly clastic sediments long recognized as the Adelaide Series (David, 1922; Howchin, 1925, p. 37) or Adelaide (Adelaidean)

System (Mawson and Sprigg, 1950). The Pound Subgroup and underlying Wonoka Formation together comprise a cycle of sedimentation in the Flinders Ranges unknown in the Mount Lofty Ranges of the Adelaide region, the type area of the Adelaidean, and have been nominated the stratotype for an "Ediacaran" System (Jenkins, 1981). This division is potentially distinguished biostratigraphically by purported indications of unskeletonized animal life (Jenkins, 1984a, 1985); it overlaps the Valdaian. I am not persuaded that there are any indications of metazoans within Adelaidean sediments (e.g., Jenkins *et al.*, 1981; Jenkins, 1986) and for this reason do not favor the downwardly extended Ediacarian System of Cloud and Glaessner (1982).

The widespread discoveries of fossil remains of Ediacaran aspect have generated continued excitement in respect of their possible evolutionary significance, and in terms of recent surrealistic interpretations of their morphology and inferred interrelationships (Pflug, 1970a,b, 1972a,b; Fedonkin 1985a,b; Gould, 1984, 1985; Seilacher, 1984, 1985, 1989; McMenamin, 1986; and Bergstrom, 1989, 1990), a reexamination of the actual fossil material is timely. This study attempts to appraise the geological setting and paleoecology of such remains at four major localities.

A growing body of opinion links the prolific radiation of Ediacaran life to the advent of an oxygenated atmosphere (Runnegar, 1982b, 1991; Glaessner, 1984; Jenkins, 1991; Knoll, 1991), and there is presently considerable research on isotopic data bearing on this question (e.g., Knoll *et al.*, 1986). While a gradual increase in free oxygen as a result of processes of photosynthesis has commonly been considered (Runnegar, 1982b), modeling of oceanic advective diffusion processes suggests that aeration of surface waters of the sea is enhanced and extends to relatively greater depth at times of glaciation (Wilde, 1987). Sinking of oxygen-rich, cold, surface waters at high latitudes oxygenates the ocean deeps (Deacon, 1984). Possibly the degree of aeration of the oceans at the close of the Precambrian fluctuated in response to the unprecedented ca. 100-My-long cyclical global refrigeration that occurred at this time, and the radiation of higher life forms was triggered when concentrations of free oxygen oscillated through some optimal level (Knoll *et al.*, 1986; Knoll, 1991; Jenkins, 1991; Runnegar, 1991).

Repositories for specimens referred to in this chapter are prefixed Pf. for the collection of H. D. Pflug, Giessen, West Germany, A.U. for the paleontogical collection of Adelaide University, South Australia, and S.A.M. for the South Australian Museum, Adelaide, South Australia.

2. Avalon Peninsula, Newfoundland

2.1. Geological Setting

Spectacular Ediacaran fossil remains on the Avalon Peninsula, Newfoundland (Anderson and Misra, 1968; Misra, 1969, 1971; Anderson, 1976, 1978; Anderson and Conway Morris, 1982) occur high in the Conception Group and in parts of the St. John's Group (Hofmann and King, 1979; Williams and King, 1979; Landing *et al.*, 1988; Conway Morris, 1989a) (Fig. 2). Sites of fossil finds are shown on a more recent geological map (Department of Mines and Energy, Newfoundland, 1985).

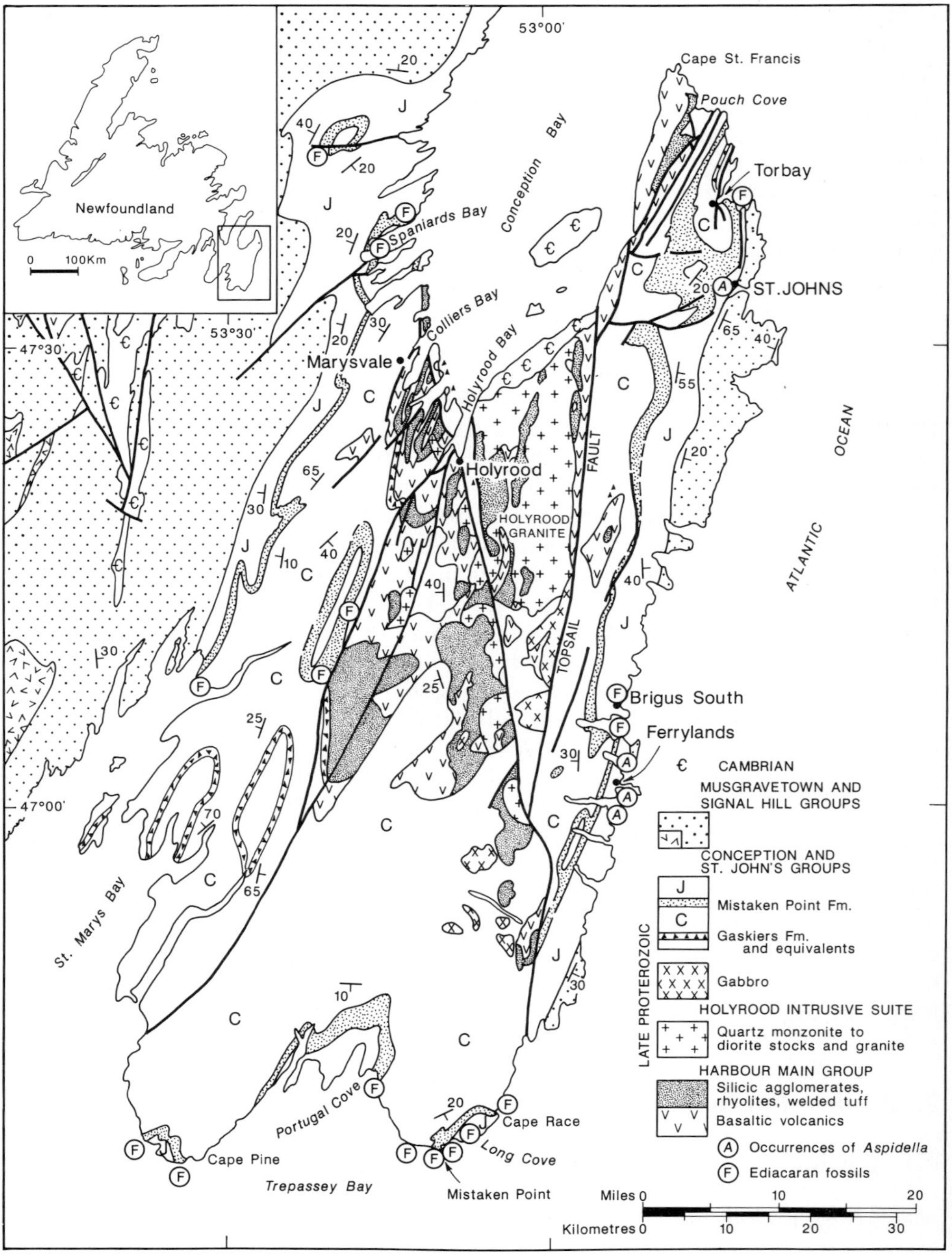

Figure 2. Simplified geological map of southeastern Newfoundland showing major stratal divisions and intrusives, and occurrences of Ediacaran fossil remains. [Based on Department of Mines and Energy, Newfoundland (1985).]

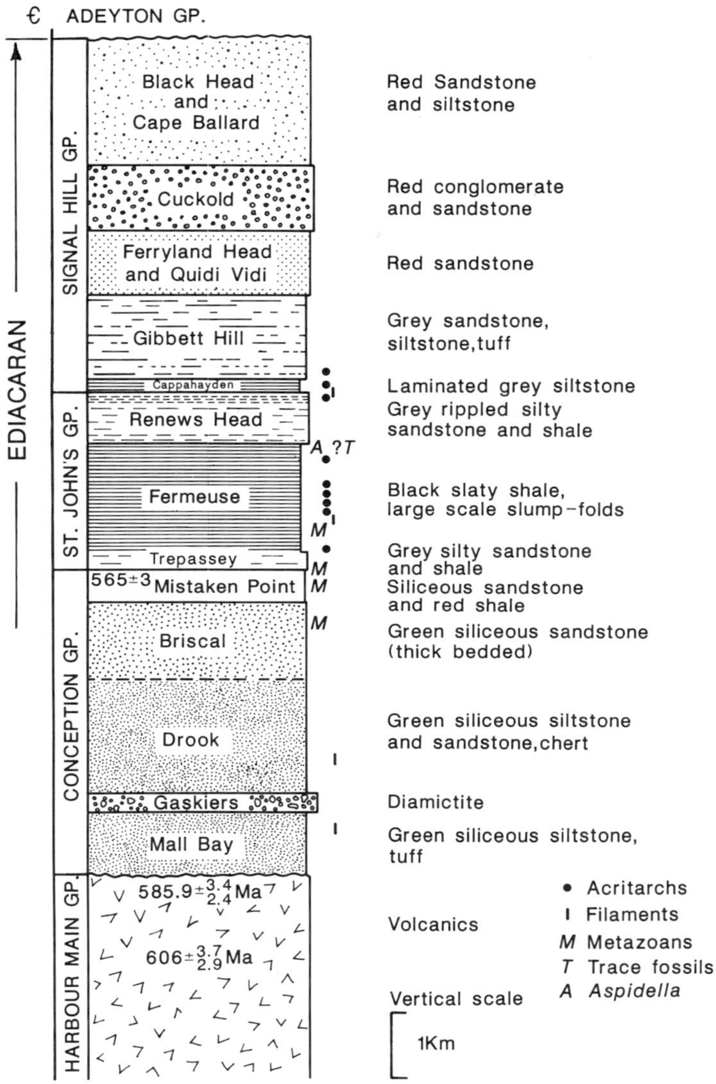

Figure 3. Late Proterozoic stratigraphic succession for the Avalon Peninsula, Newfoundland, and levels of fossil finds. [After Hofmann and King (1979).]

Centered near Holyrood, Conception Bay, is a complex of volcanic rocks and intercalated sediments termed the Harbour Main Group (Rose, 1952; McCartney, 1967). The volcanic rocks are bimodal, rhylolitic to andesitic with agglomerates, tuffs, and numerous ignimbrites, or basaltic in composition (McCartney, 1967; O'Brien *et al.*, 1983). Krogh *et al.* (1988) obtained a U–Pb zircon age of 620.5 (+2.1, −1.8) My for the Holyrood Granite, a large subvolcanic intrusive body of coarse-grained granite lying to the east of Holyrood. At the eastern margin of the inlier of volcanics, sediments resembling the Conception Group are intercalated with thin intervals (2–20 m) of altered mafic rocks that possibly represent basaltic flows. Most contacts are sheared. North, between Pouch Cove and Cape St. Francis, pillow lavas are intercalated with sediments similar to the Conception Group (e.g., Papezik, 1982). A rhyolite dyke near Cape St. Francis provided the youngest reliable age obtained for the Harbour Main Group, 585.9 (+3.4, −2.4) My (Krogh *et al.*, 1988).

In the Holyrood–Marysvale region and to the south, aerial basaltic flows and an underlying ignimbrite dated at 606 (+3.7, −2.9) My (Krogh *et al.*, 1988) were overstepped unconformably by marine sediments of the Conception Group somewhat before and perhaps up to about the time of deposition of glacigenic mixtites equivalent to the Gaskiers Formation.

The ca. 5000-m-thick Conception Group and lower part of the succeeding St. John's Group (ca. 2000 m overall) mainly consist of turbidites (Anderson, 1978; O'Brien *et al.*, 1983; Gardiner and Hiscott, 1984; Landing *et al.*, 1988; Conway Morris, 1989a). Megascopic fossil remains occur in the Briscal Formation and overlying Mistaken Point Formation of the upper Conception Group, and the Trepassey Formation and succeeding Fermeuse Formation of the St. John's Group (Fig. 3). A tuff band in the fossiliferous part of the Mistaken Point Formation at Mistaken Point has given a U–Pb zircon age of 565 ± 3 My (Benus, in 1988). The site of deposition was possibly a major ensialic rift (Strong *et al.*, 1978; O'Brien *et al.*, 1983). Hence the turbidites seem unlikely to have formed at truly oceanic depths, but perhaps accumulated in a setting such as a continental terrace (e.g., Benus, 1988).

Mainly medium-bedded turbidites in the Mistaken Point Formation are gray to maroon and relatively fine-grained, with cycles T_{abcde_3} or T_{bcde_3}. Whitish intervals at the T_{de} levels may indicate a tuffaceous component, and the formation tends to become more siliceous up section. The well-known fossil site at Mistaken Point comprises thin to thick (10–80 cm), $T_{cde_2e_3f}$ and T_{bcde_3f} turbidites showing cycles of thickening and thinning over a vertical distance of 6–8 m, and in terms of studies such as those of Pickering (1983, 1985), probably are outer fan deposits of a lobe fringe setting. The cyclical ridging on the tops of some beds is tectonic in origin.

The more sandy, gray, T_{bcde} turbidites of the Trepassey Formation at Mistaken Point are thin- to medium-bedded and show an overall regularity in thickness. At Long Beach, thin, $T_{cde_2e_3f}$ and $T_{bcde_{1-3}f}$ turbidites are rather siliceous. Lobe fringe and fan fringe deposits may be represented.

At Cape Race, slumped deposits in the Fermeuse Formation resemble the "fan lateral margin deposits" illustrated by Pickering (1983, Fig. 6). Higher parts of the stratigraphic succession are progradational and generally indicate shallowing upward (Williams and King, 1979; O'Brien *et al.* 1983).

2.2. Fossil Remains

Fossil remains in the Briscal, Mistaken Point, and Trepassey Formations occur at the interface between tops of turbidites and covering layers of volcanic ash varying from a few millimeters to 0.5 m thick. The ash comprises a fraction consisting of euhedral crystals (mainly feldspar) of submillimeter to millimeter dimensions and phyllosilicates altered from glass shards that are sometimes indicated by relict outlines. Evidently the tuffs were generated by nearby large phreatomagmatic explosions that dispersed steam-buoyed ash in an apron over the sea. The crystals settled fastest through the water. As the density of unskeletonized marine creatures is almost the same as seawater, organisms dispersed in the water column could not have settled faster than the crystals; hence, the fossil assemblages must comprise benthonic taxa and represent life communities (Anderson, 1978). The organisms preserved are apparently not metaphytes, as they presumably lived below the photic zone, and for the same reason cannot have been nurtured by contained photosynthetic symbionts as inferred by Seilacher (1984, 1989); their widespread occurrence and presence on numerous bedding planes would seem to obviate the suggestion of McMenamin (1987) and McMenamin and McMenamin (1990) that their distribution may be related to hydrothermal vents and a supposed chemotrophic metabolism. The corollary is that they were heterotrophic and presumably metazoans. Bottom waters of the ocean must have been sufficiently well oxygenated to maintain heterotrophic metabolism in reasonably large animals (up to ca. 50 cm in major dimensions), and some form of adequate suspended nutrient was necessary to support the dense and diverse communities preserved (Fig. 4).

Fossilization seems to have involved soft, fleshy parts of the organisms being pushed down into the underlying muddy substrate, and tough or resilient elements such as stalks, stolons, and "branch" axes supporting the covering tuff, with mud from below filling the molds as decomposition proceeded (Anderson, 1978). The crystal layer of the tuffs is sufficiently coarse to composite-mold the imprints, obscuring detail, or to persist in adhering to molds during exposure and weathering, obscuring aspects of the morphology, a problem limiting interpretation.

Except for the recognition of several taxa occurring on other continents, none of the remains in the turbidite assemblages has been named formally. The assemblages are said to include some 20 genera (Anderson, 1978) and at least 30 species (Anderson and Conway Morris, 1982). A few specimens are on display in the Royal Ontario Museum, Toronto. Five main groups of body fossils comprise the assemblages, ranking in approximate decreasing order of abundance as set out below; this listing is not comprehensive.

1. "Spindle-shaped forms" in the upper parts of the Mistaken Point Formation.
2. Complexely branched forms including the "bushlike form," probably the "pectinate forms," and animals with numerous fronds joined by branching or zigzag connections (Misra, 1969, pl. 3B, pl. 4—form above scale container); upper parts of the Mistaken Point Formation, the Trespassey Formation, and the lower Fermeuse Formation.
3. Singly frondose forms with stalks expanded basally or terminating in a disc; *Charnia masoni* Ford, 1958, and about four species that may be loosely grouped in *Charniodiscus* Ford, 1958; the upper part of the Mistaken Point Formation, the Trepassey Formation, and the lower Fermeuse Formation.
4. Discoidal organisms either with radial "lobes" or irregular "lobes"; the Briscal, Mistaken Point, and Trepassey Formations.

Figure 4. Reconstruction of life forms of the upper part of the Mistaken Point Formation, Avalon Peninsula, portrayed approximately one third natural size: (A) *Charnia masoni* Ford, 1958; (B) spindle-shaped fossils; (C) *Charniodiscus concentricus* Ford, 1958; (D) lobate discoidal remains; (E) bushlike fossils; (F) branched frondose fossil with fronds showing coarse secondary structure somewhat resembling that in *Glaessnerina* Germs, 1983; (G) *Charniodiscus* Ford, 1958, sp.; (H) pectinate organism reconstructed as bearing leaf-shaped fronds; (I) dichotomously branched frondose organism; $T_{bcde_2e_3f}$ Bouma divisions of turbidite substrate.

Figure 4. (*Continued*)

5. Discoidal organisms with strongly developed annulations (Anderson, 1978, Fig. 3); ? mid-Fermeuse Formation.

Different kinds of organisms occur in varying abundance on individual fossiliferous bedding planes; thus, for example, bushlike forms are common on some surfaces in the upper Mistaken Point Formation and other complexely branched animals are plentiful on different surfaces in the same unit.

Spindle-shaped forms (Anderson and Conway Morris, 1982, pp. 3–7) comprise perhaps 75% overall of the individual fossil remains in the upper Mistaken Point Formation. Their branched structure is indicated by external molds in underlying sediment. The organism is remarkable in that it shows no obvious polarity; that is, both ends of the tapering "spindle" seem the same and are not identifiably proximal or distal. The relative spacing of the groups of "branches" in specimens in parallel orientations is interpreted by Anderson and Conway Morris (1982) as indicating that the form was contractile; the orientation relative to penetrative cleavage has also led to distortion. All specimens show similar overall characteristics; as the form is not found attached to a stem and there is no anchoring device to hold the creature erect, it probably actually lived lying on one side, the aspect that is invariably preserved. The "branches" made hollow impressions in the substrate. Thus the part of the "branches" actually molded was convex, the "reverse" side in terms of my analysis (Jenkins, 1985) of the comparable organism *Rangea* Gürich, 1929, from the Ediacaran of Namibia. The opposite side or "obverse" side that faced directly toward the water column is suggested to have supported the structures involved in feeding/respiration. Overlap of the "branches" and their differing curvature suggest that they had some freedom of movement relative to each other, and photographs of specimens show that individual "branches" were capable of expanding to form small fronds. Groups of these contractile fronds were arranged in clusters that approximately alternated on either side of the "midline."

The bifurcating sets of fronds may have formed hollow arrangements, the reverse sides of which touched down on the substrate and supported the whole, horizontally lying colony. Pairs of recurved fronds also provided support. The fronds show alternate, chevron alignments of millimeter "secondary" subdivisions, which are rarely seen to be divided into fine, serial, tertiary divisions 0.3–0.5 mm wide; these fine divisions curve toward the "obverse" cavity and may represent "polyp" anthosteles. This animal was probably rather tough or muscular and exquisite molds are commonly preserved. It evidently grew from the midpoint of an axial stolon, which budded new clusters of fronds as it propagated toward either pointed tip.

While the bushlike form (ca. 10 cm in height) is described as having a more or less globular "body" with "its outer surface deeply and sharply corrugated or plicated meridionally" (Anderson and Conway Morris, 1982, p. 3), I consider that its geometry more likely reflects overlapping of rather broad "fronds" arranged in a conical circlet. The fronds were evidently divided into serial, curved "branches" (upper middle plate 1, Fig. 3, of Anderson and Conway Morris, 1982) that probably formed an alternating arrangement. There is evidence of a finer subdivision within these secondary branches. Certainly, the form had a holdfast in the substrate during life, but specimens may have been displaced as they were laid flat just before or during burial. This animal resembles an unnamed organism from Charnwood Forest.

The Mistaken Point assemblage includes a variety of animals comprising relatively large fronds (ca. 15 cm long or larger) joined in series to robust stolons. A specimen of the "pectinate" organism in the Royal Ontario Museum shows that the serial branches from the stolon pass into the axes of poorly preserved fronds with rather widely spaced "branches." The main stolon is oddly bent where its two main branches emerge from a stout stem. In other forms, fronds are joined to stolons in an alternate, biserial manner; in one rare form the symmetry suggests that the zigzag stolon propagated in two directions from a median point. Fronds sometimes seem to represent an overlapping jumble arising from irregular stolons, surely an indication of accidental flattening of several series on top of one another. Another small, rare taxon shows a bifurcating pattern of stolons extending from a suboval holdfast.

The initial stem or stolon of these forms evidently represents the primary polyp or oozoid, or its first budded daughters, and provided support for the arrangement of fronds projecting up into the water column. Seemingly, the stems were tough or rather resistant, indicating the presence of mesoglea and/or strong musculation if these organisms were cnidarians. Commonly, the individual fronds are poorly preserved, perhaps because their integument was thin, or damage occurred as they were laid over at the time of burial. The best-preserved examples show wide, curved lateral "branches" with robust secondary divisions much as in *Glaessnerina* Germs, 1973. Their affinities may be anthozoan.

Charnia is notable for the preservation of the flattened outline of its bulbous holdfast in numerous specimens, the identifiable examples of which all show the same general characteristics of the frond. This poses an intriguing question as to whether this organism is preserved in one particular orientation, or if, as seems intrinsically more likely, both sides of its frond reflected fundamentally similar constructional architecture. Remains of several forms broadly referable to *Charniodiscus* occur at similar stratigraphic levels. Their structure is less clearly seen than in material from the Flinders Ranges. One with a large holdfast disc is probably identifiable as *Charniodiscus concentricus* Ford, 1958. All of the organisms so far described can probably be characterized as suspension feeders.

The discoidal organisms pose major problems for interpretation because the characteristics evident are limited to one side of the disc and apparently relate to the external surface. None apparently shows a mouth or any certain feeding device. Among the more common forms are discoidal remains up to over 40 cm in diameter preserved as casts of the upward-facing surface and showing 5–10 broad radiating lobes and swales between. Presumably parts of the structure differed in their degree of resilience. Still other discoidal creatures show irregular lobes. Even more bizarre are irregular subovoid "discs" up to ca. 50 cm in longest dimension, and showing casts of numerous rounded protruberances arranged irregularly and projecting as much as 1 cm up into the covering tuff; these and other irregularly lobate discoidal forms first appear in the Briscal Formation. Seemingly, no fine structure is apparent. Still another kind of discoidal organism up to ca. 24 cm in diameter comprises depressed molds showing 16–20 rounded peripheral lobes and subsidiary "blisters" and a central zone with some irregular complexity (Misra, 1969, pl. 4,A; Misra, 1971, Fig. 2A). The relationships of all these remains is intrinsically indeterminate, but possibly those showing relatively irregular lobes or protruberances were near sponges (though lacking obvious oscula), and those with more regular radial lobation were possibly polypoid and likely cnidarians.

The star-shaped form from the Trepassey Formation described by Anderson and Conway Morris (1982, pp. 7–8, Fig. 4; 1989a, Fig. 2.4c) is indicated by unusual dark material in the sediment, lending emphasis to the possibility that it is a trace fossil.

The discoidal, annulate organism found as deep impressions in the plane of hiatus between successive turbidites in the mid Fermeuse Formation is another enigma. The deep impressions may imply that it was partly buried during life, and the annulation possibly resulted from the impaction of a ringed, cone-shaped morphology of polyplike aspect. In this respect it is similar to the Australian form *Cyclomedusa davidi* Sprigg, 1947, which also probably lived attached to the substrate (see below, Section 5.2), but the Newfoundland taxon is distinctive in its strongly developed and rather more regular annulation.

The widely occurring, supposed pseudofossil *Aspidella terranovica* Billings, 1872, present in intercalated thin sandstones and shales or ripple-bedded, flaggy sandstones near the top of the Fermeuse Formation, also shows annulation and occasional radial structures similar to *C. davidi*, and a small central plug of sand (Conway Morris, 1989a) resembles the fill of attachment stalks of such sedentary, discoidal forms in the Flinders Ranges. The flaggy sandstone facies in which *A. terranovica* may occur bears an evident similarity to certain of the fossiliferous Ediacaran strata in the Flinders Ranges. In view of these observations, I believe that *A. terranovica* is possibly organic (Jenkins, 1989), and that the numerous larger discoidal forms associated with it in the Ferrylands area may be equivalents of the "medusiform" remains of South Australia; some show an outer zone of radiating "filaments" resembling tentacles. Rare vermiform, tubular fills seem to be trace fossils.

3. Charnwood Forest, Leicestershire, England

3.1. Geological Setting

The mildly metamorphosed and well cleaved, ca. 3500-m-thick sedimentary sequence of the Charnian Supergroup is subdivided in upward succession into the Blackbrook, Maplewell, and Brand Groups (Ford, 1979a) and crops out as a large anticline plunging toward the southeast (Watts, 1947; Mosely and Ford, 1985) (Fig. 5). The Blackbrook Group comprises chiefly fine- to coarse-grained tuffs and tuffaceous pelites and greywackes; the Maplewell Group includes coarse-grained tuffs, volcanic breccia and agglomerates, slumped greywackes, and tuffaceous pelites; the Brand Group contains a tuffaceous conglomerate and quartz arenite, but is generally pelitic. At least locally, the Brand Group is disconformable above the Maplewell Group (Mosely and Ford, 1985, p. 13). Several volcanic centers occur in the western limb of the regional anticline, but cross-faulting poses difficulties in correlating the eruptive complexes with other parts of the sequence; the abundance of tuffs diminishes in higher parts of the Maplewell Group.

A conglomerate near High Tor Farm, about 1 km southeast of a porphyry in the Whitwick Complex, includes highly angular, subrounded, and rounded clasts of a variety of siliceous volcanic rocks and may represent a slide of volcanic rubble with some stream influence.

3.2. Fossil Remains

Fossil remains present in each of the three identified sedimentary groups attest to their general marine nature and common slump deposits in the Black-

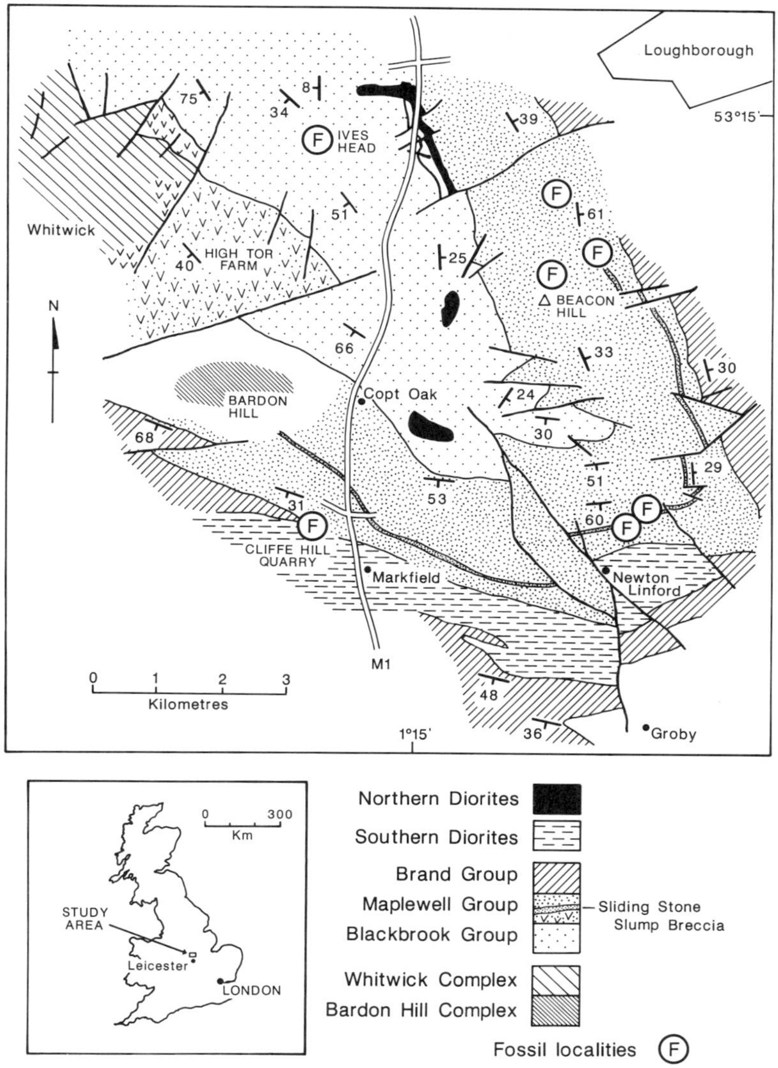

Figure 5. Simplified geological map of the Charnian Complex, Leicestershire, England, and localities of Ediacaran fossil remains. [After Mosely and Ford (1985).]

brook and Maplewell Groups indicate slopes. Current rippling occurs in parts of the main fossiliferous interval of the Maplewell Group, but wave oscillation ripples seem to be absent, suggesting deposition below storm wave base.

Fossil remains in the Old John Member at the top of the Beacon Hill Formation and in the lower part of the Hallgate Member of the Bradgate Formation, Maplewell Group, are well known (Ford, 1958, 1963, 1968, 1979a, b, 1981; Boynton, 1978; Boynton and Ford, 1979; Jenkins, 1985). These comprise various discoidal forms, the frondose *Charnia masoni* and *Charniodiscus concentricus*, a comparatively large bushlike form with multiple fronds, and a putative arthropod, *Pseudovendia charnwoodensis* Boynton and Ford, 1979. The preservation of the latter renders it indeterminate. The frondose forms are evidently confined to the lower part of the Hallgate Member. Discoidal remains up to 12 cm in diameter and with their surfaces showing casts or molds of irregular structures have been located by Helen Boynton (1978, and personal communication 1981) in tuffaceous beds of the Lubcloud Greywackes Member, Ives Head Formation, Blackbrook Group, at Ives Head Hill. These resemble the discoidal forms with irregular lobes in the Briscal Formation, southern Avalon Peninsula.

Charnia is a relatively elongate, somewhat ovate, stalked frond with oblique primary divisions ("branches") arranged in an alternate chevron pattern and distinctly sigmoidal in shape, tapered in both directions. Curiously, the secondary divisions of the branches are narrow near the midline of the frond, thence widen, and narrow again toward the margins. The secondary divisions themselves are marked by 5–7 oblique tertiary grooves slightly angled relative to the midline of the frond (Fedonkin, 1983, pl. 98, Fig. 37; Jenkins, 1985, Fig. 7b). Fossils of *Charnia* in Newfoundland show the strong stalk and bulbous holdfast. An incomplete frond in outcrop on Memorial Crag, Bradgate Park, is over 58 cm long and in excess of 22 cm broad. The tertiary structures on the branches are interpreted by Glaessner (1984, p. 87) as the actual molds of polyp anthosteles, and, while this seems possible, the molds are not so perfectly preserved as to make this notion unequivocal. *Charnia* evidently had just the inflated, foliate "pneu-structure" and "quilting" that Seilacher (1984, 1985, 1989) postulates is indicative of a new phylum, but as modern pennatulaceans largely support themselves hydrostatically by distending their body spaces with water (Brafield, 1969, pl. 1a, b), this trait is in essence one of their diagnostic attributes.

Ediacaran sediments of generally shallow-water, shelf/basinal settings in northern parts of the USSR (Fedonkin, 1985b) include examples of *Charnia*, some of which apparently show a three-dimensional form of preservation (Fedonkin, 1985a, pl. 13, Fig. 3). Fedonkin interprets this material as indicating that both sides of the frond were similar, but the illustrated specimen does not show if only one frond is truly involved, or if several fronds are juxtaposed with a thin layer of sediment between. Other frondlike forms to which it may be allied are known to comprise multiple fronds (e.g., *Rangea* Gürich, 1929). The uniform mode of preservation of most specimens of *Charnia* suggests that it was foliate or flat-sided, with the frond "leaf"-like in general configuration.

Charniodiscus concentricus is perhaps best interpreted in respect of its well-preserved allies from the Flinders Ranges (Fig. 6). The curvature of its "branches" is a specific characteristic seen also in Newfoundland material. Discoidal markings with a cast of a stem emanating near the center are probable holdfasts of *Charniodiscus* from which the frond was either torn away or projected through the sediment that buried the holdfast and was not preserved.

The several large bushlike organisms present on Memorial Crag (Ford, 1979b, pp. 13, 14; Ford, 1981; Jenkins, 1985, Figs. 6, 7D1-2) consisted of many overlapping foliate elements. In the specimen I previously described (Jenkins, 1985), the foliate elements seem to extend from a central boss in a crudely radial pattern; the boss probably represents a holdfast, perhaps with an infilling of sand that acted as ballast rather in the manner described (Seilacher, 1983, 1985) in some sand- or mud-sticking Paleozoic coelenterates. Two other specimens do not show bosses. One individual is 38 cm in its longest diameter; another dubiously related form is ca. 47 cm in longest dimension.

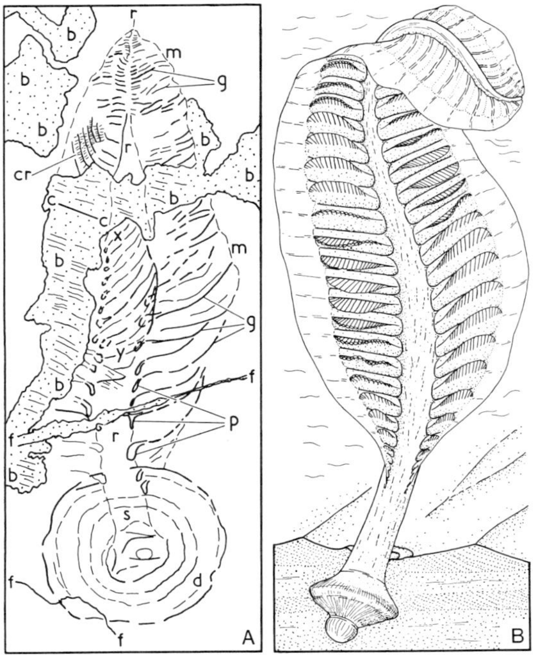

Figure 6. (A) Outline drawing of the holotype of *Charniodiscus concentricus* Ford, 1958, from the Charnian compared with (B) a reconstruction of *Charniodiscus oppositus* Jenkins and Gehling, 1978, from the Flinders Ranges; both ca. one fifth natural size. f, fractures in rock surface; b, projecting irregularities in matrix; d, disc; s, stalk; r, rhachis of frond; m, outline of frond; g, grooves between branches of frond; p, pockets formed by these grooves; cr, faint cross-structure on branches; x–y, overfolded portion of frond. [After Jenkins and Gehling (1978).]

Small areas of preserved detail show that that the major foliate elements in these animals consisted of chevron arrangements of "small fronds" or "branches" varying from ca. 2 to ca. 6 cm long. The small fronds evidently had a plastic or "spongy" consistency and may be preserved in either a distended or a "squashed," foreshortened manner. Each small frond itself comprised an alternate, chevron array of perhaps 7–21 curved "secondary branches." These varied from millimeter dimensions up to possibly 2.5 cm long and were further subdivided into "third-order" structures consisting of sharply separated, subcylindrical elements 0.5–4 mm long and 0.1–0.7 mm in width. The latter were probably the "polyps" of the animal and there is evidence that they also formed a chevron arrangement. The major foliate elements surely projected up into the water column and the "polyps" probably faced into the hollow formed by the encircling cone of fronds. Sufficient details are preserved to indicate that the number of "polyps" in the colony exceeded 170,000, and was very likely about 500,000.

Discoidal organisms present in the Maplewell Group comprise rather small, flat ovoids with a slightly raised rim and various circular markings with a raised central zone, or forms with two inner grooves, one outlining a median boss; rare discs 7.5–22 cm in longest diameter show numerous annulations. The abundance of the simple discoidal forms with few rings contrasts markedly with the prevalence of lobate circular forms in the turbidite settings of the Avalon Peninsula, but compares closely with the common occurrence of nondescript circular bodies in the Ediacara assemblage of the Flinders Ranges. While these remains are medusiform in general form, their simple geometry precludes further analysis. Discoidal remains with numerous annulations (Ford, 1968, pl. 1, Fig. c) may represent floats of chondrophores. Discoidal imprints from probably the lower Brand Group resemble *Ediacaria* Sprigg, 1947.

4. Namibia, Southwestern Africa

4.1. Geological Setting

The Nama Group comprises platform sediments of Late Precambrian to Cambrian age cropping out in southern Namibia and south of the Orange River (Germs, 1972a, 1974; Kröner *et al.*, 1980; Tankard *et al.*, 1982, pp. 288–299; Germs *et al.*, 1986). The main region of deposition is subdivided by a basement ridge at Osis Farm. Trace fossils in the sediments north of the Osis Ridge (Crimes and Germs, 1982) suggest that the greater part of the sequence is of Cambrian age though this evidence has been critically questioned (Grant, 1990); Ahrendt *et al.*, (1978) obtained a K–Ar dating of 530 ± 10 My related to folding and synkinematic metamorphism of these sediments. More recent K-Ar work confirms this result (Horstmann *et al.*, 1990) or may indicate metamorphic alteration of micas beginning a little earlier (Stanistreet, Kukla, and Henry, 1991). South of the Osis Ridge, indications of soft-bodied organisms and sparse trace fossils of Ediacaran aspect are present in the Kuibis Subgroup and older parts of the overlying Schwarzrand Subgroup (Germs, 1972a, c, 1973a, b; Jenkins, 1985). Based on organic-walled microfossils obtained from bores, Germs *et al.* (1986) claim an "unequivocal assignment of latest Proterozoic (Valdaian) age" for the Kuibis Subgroup and parts of the Schwarzrand Subgroup on both sides of the Osis Ridge (cf. Grant, 1990). In the southern subbasin, trace fossils above the local disconformity at the base of the Huns Member of the Totem Formation, Schwarzrand Subgroup, seem of equivocal late Valdaian/Early Cambrian kind and the succeeding Nomtsas Formation may be Cambrian.

Within the Dabis Formation of the Kuibis Subgroup distal alluvial-plain deposition is suggested to grade up into sandstones that originated as tidal-channel fills and fan deltas along a low-energy shoreline (Tankard *et al.*, 1982). The fan deltas are indicated by large, gradationally-based sigmoidal foresets in lower parts of the fossiliferous Kliphoek Member. Overlying shales may have been deposited offshore. Mudcracks, shale clasts, herringbone cross-stratification, and gypsum in sandstones and siltstones at the top of the Kliphoek Member indicate tidal deposition.

The quartzite of the Niederhagen Member of the Nudaus Formation, lower Schwarzrand Subgroup, includes mudcracks; salt casts occur at a subjacent level (Germs, 1972a, 1974). The Nudaus Formation chiefly consists of shale and minor intercalated sandstones with limestones prevalent toward the southwest. Beds of fine sandstone present at the top of the Kliphoek Member and in the upper "shale" of the Nudaus Formation and parts of the succeeding quartzitic Nasep

Formation show prominent flute casts on soles and sometimes include small-scale crossbedding and slumps, consistent with deposition from density currents (Germs, 1972a, p. 47). Localized glaciation occurred in the region of the Klein Karas Mountains during the time of deposition of the Nudaus Formation (Germs, 1974; Tankard *et al.*, 1982).

Quartz arenites, limestones, and shales of the Nasep Formation represent near-shore marine depositional environments and the arenites were evidently subjected to strong unidirectional tidal currents (Germs, 1972a, 1974).

4.2. Fossil Remains

Rangea schneiderhoehni Gürich, 1929, is recorded from coarse orthoquartzites in the Kliphoek Member of the Dabis Formation together with *Ernietta plateauensis* Pflug, 1966 (= *Namibia villiersiensis* Germs, 1968, ? = *Kuibisia glabra* Hahn and Pflug, 1985); *Pteridinium simplex* Gürich, 1933, is present in the transitional zone of quartzite to shale in the Kliphoek Member, where it is evidently in direct association with *Rangea*, and also occurs with *R. schneiderhoehni* in the Niederhagen Member, Nudaus Formation (Germs, 1972a, 1973a, 1974). The now lost holotype of the otherwise unknown, problematic *Orthogonium parallelum* Gürich, 1930, came from the Kliphoek Member, as did the purported alcyonacean *Ausia fenestrata* Hahn and Pflug, 1985, casts of *Nemiana simplex* Palij, 1976 [pseudo-*Bergaueria* of Germs (1972a pp. 118–119, pl. 29, Figs. 2–3], and suggested sprigginid (Germs, 1973b) and echiurid worms (Glaessner, 1979a). The supposed worm remains are each represented by unique specimens and will not be further discussed. Other irregular, small discoidal structures (Germs, 1972a, p. 207, pl. 26, Fig. 3; Crimes and Germs, 1982, pl. 1, Fig. 1) from upper parts of the Kuibis Subgroup provide little paleoecological information. *E. plateauensis* occurs together with "*Namibia*" in the topmost beds of the Kliphoek Member. Of the 28 species within 13 genera, four families, and two orders that Pflug (1972a) collectively referred to as members of the "Erniettamorpha," all seem referable to *E. plateauensis*.

Nasepia altae Germs, 1972, a probable specimen of *Pteridinium*, and the lost holotype of *Paramedusium africanum* Gürich, 1930, were collected from the top of the Nasep Formation. Casts of annulate discoidal organisms resembling *Cyclomedusa davidi* occur in limestones of the Huns Member of the Totem Formation.

Calcified multilayered tubular shells of *Cloudina* Germs, 1972b (see Glaessner, 1976a; Conway Morris, Mattes, and Chen, 1990) occur in carbonate facies of the Kuibis and Schwarzrand Subgroups on either side of the Osis Ridge. "Worm" tubes of agglutinated sediment described as *Archaeichnium haughtoni* Glaessner, 1963 (see Glaessner, 1978) are from the top of the Nasep Formation. These tubular forms fall outside the scope of the present study.

A particular aspect of the Nama assemblages is that many of the remains of the soft-bodied forms evidence three-dimensional shapes. Two main kinds of preservation are usual. In the more common mode (H. D. Pflug, personal communication, 1977), frondose organisms and medusiform creatures are preserved lying parallel or subparallel to bedding planes. In the second mode of preservation, the

Figure 7. View of lower side of an impure sandstone bed in which a group of *Ernietta plateauensis* Pflug, 1966, are preserved in three-dimensional manner, Kliphoek Member of Dabis Formation, Nama Group, Plateau Farm west of Schakalskuppe, Namibia. Scale divisions in centimeters.

fossil specimens comprise curved or rounded forms that project through relatively homogeneous, medium-grained, feldspathic sandstone (quartzite) in a rather random way. The long fronds of *Pteridinium* are commonly twisted, buckled, and overfolded (Glaessner, 1979b, Fig. 11, 1a-b), suggesting that they were enveloped by fluidized sand. Specimens of *Rangea* are preserved with several adjacent fronds separated by intervening pods of sediment. *Ernietta* may occur in groups or clusters with bases of their vase- or sack-shaped bodies flattened at bedding planes (Fig. 7). These remains were probably buried by density currents generated as a result of wave movement and rips at times of storms. That latent molds made by sand grains pressing against cadavers could survive decomposition and collapse of body parts and subsequent diagenetic compaction seems improbable. Presumably some process led to fixing of the molds prior to final decay of the organic tissues, perhaps by the early precipitation of a film of mineral material in the interstices of the matrix.

Silicon may complex with chelating catechols to become 700 times more soluble than in water alone and subsequent changes in Eh and pH cause its rapid reprecipitation as an elastic, acqueous gel (Jones and Fitzgerald, 1986) that might well serve to bind sediment grains. Catechols are known to be released during anaerobic decay of algae. If no significant interstitial biota had evolved during the

Ediacaran, perhaps sediments tended to be unusually charged with sapropels and decay of cadavers may have served to localize rapid silica dissolution and re-precipitation over a matter of days.

Rangea schneiderhoehni seems mainly to have been preserved in storm-deposited sandstones. The numerous studies concerning its morphology or possible relationships (see Jenkins, 1985) show that it consisted of overlapping or conjoined fronds [petaloids in Pflug's (1970b) terminology] closely juxtaposed about a large central polyp or oozoid (Fig. 8). There is limited evidence that the oozoid extended into a short stalk that expanded into a relatively large bulb. No evidence of musculation is associated with the bulb, which is sediment-packed and includes a few small shale chips in the one reasonably well-preserved example—possibly it formed a weighted ballast. At least four fronds and perhaps more seem to have been arranged back-to-back around the central polyp and were probably connected to it basally by stolons. Elements comprising the fronds show primary, secondary, and tertiary patterns of sequential branching (Pflug, 1970b; Jenkins, 1985).

The configuration of *Rangea* is convergent with that of the spindle-shaped and bushlike forms from the Avalon Peninsula and the bushlike forms from Charnwood Forest. There is also a similarity with respect to the fine subdivision of the secondary divisions of the fronds in *Charnia*. All of these organisms may be considered members of the order Rangeomorpha Pflug, 1972a, and evidently bore myriads of tiny polyps that derived their food from suspended organic matter. The creatures were sufficiently large and robust as to require some form of stiffening of body tissues, presumably a connective mesogloea, in addition to the hydrostatic inflation of the colony. They were certainly of cnidarian grade, and in view of the absence of mesoglea in hydrozoans (Hyman, 1940, pp. 538–539), may be considered either ancestral to or close analogues of the Anthozoa. Several specimens of *Rangea*-like forms have recently been found in the Flinders Ranges (e.g., Gheling, 1991, Fig. 1).

The occurrence of *Rangea* and *Pteridinium simplex* in direct association suggests an overlap of the environments they inhabited. The main parts of the integument of *Pteridinium* tended to buckle or form rounded curves rather than crease or fold sharply, indicating it was rather tough and perhaps rubbery or leathery in consistency (Jenkins, 1986), and the extremely elongate, fusiform shape seems ideally suited to life in streaming currents, perhaps in rip channels. This is consistent with the common current alignment of specimens (e.g., Richter, 1955, pl. 4, Fig. 5a). Numerous examples preserved in three-dimensional form show that the "frond" consisted of three vanes or fusiform, half-leaf-shaped, foliate structures joined at a common (central) axis (Richter, 1955; Glaessner, 1963; Glaessner and Wade, 1966; Pflug, 1970a; Germs, 1972a) (Fig. 9). The curvature of ribs and grooves on the vanes indicates a polarity, but neither end of the frond has been seen in material available for study. The damaged ends possibly resulted from individuals being ripped from anchorage and tumbled about prior to burial.

The significance of the ribbing is revealed by specimens in which chambers or spaces within the medial parts of the vanes are at least partly infilled by sediment and preserved as internal molds. Thus, rounded ridges on a surface representing the exterior of a vane correspond to the site of an underlying, elongate, narrow chamber or tube, while the sharply incised grooves between (or

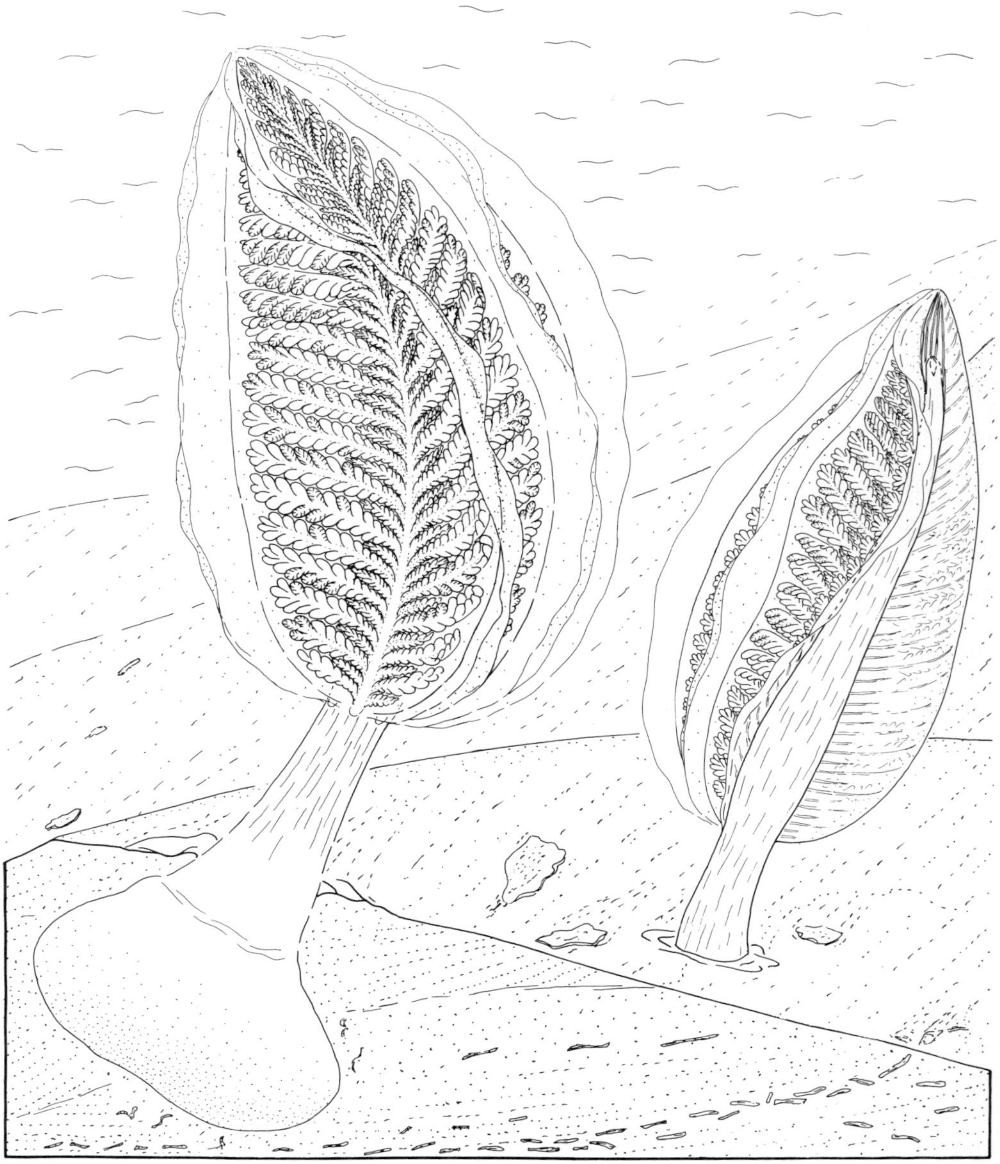

Figure 8. Reconstruction of specimens of *Rangea schneiderhoehni* Gürich, 1929, shown in inferred life position anchored in current-bedded sand, ca. 0.52 times natural size. The individual on the left is shown in "ventral" view, that on the right in a "dorsal" aspect with the tapering contractile stem or oozoid largely evident. At least four fronds, and possibly as many as six or more, were clustered together back to back about the oozoid. [From Jenkins (1985).]

tracelike lines on the lateral parts of the vane, Fig. 9) mark the place of insertion of the membranes or septa which divide each chamber from the next. There is evidence of at least two series of overlapping chambers forming each vane. Two vanes seem to be identical in structure and the chambers of the one are offset relative to those of the other with the blind ends of the chambers fitting together to

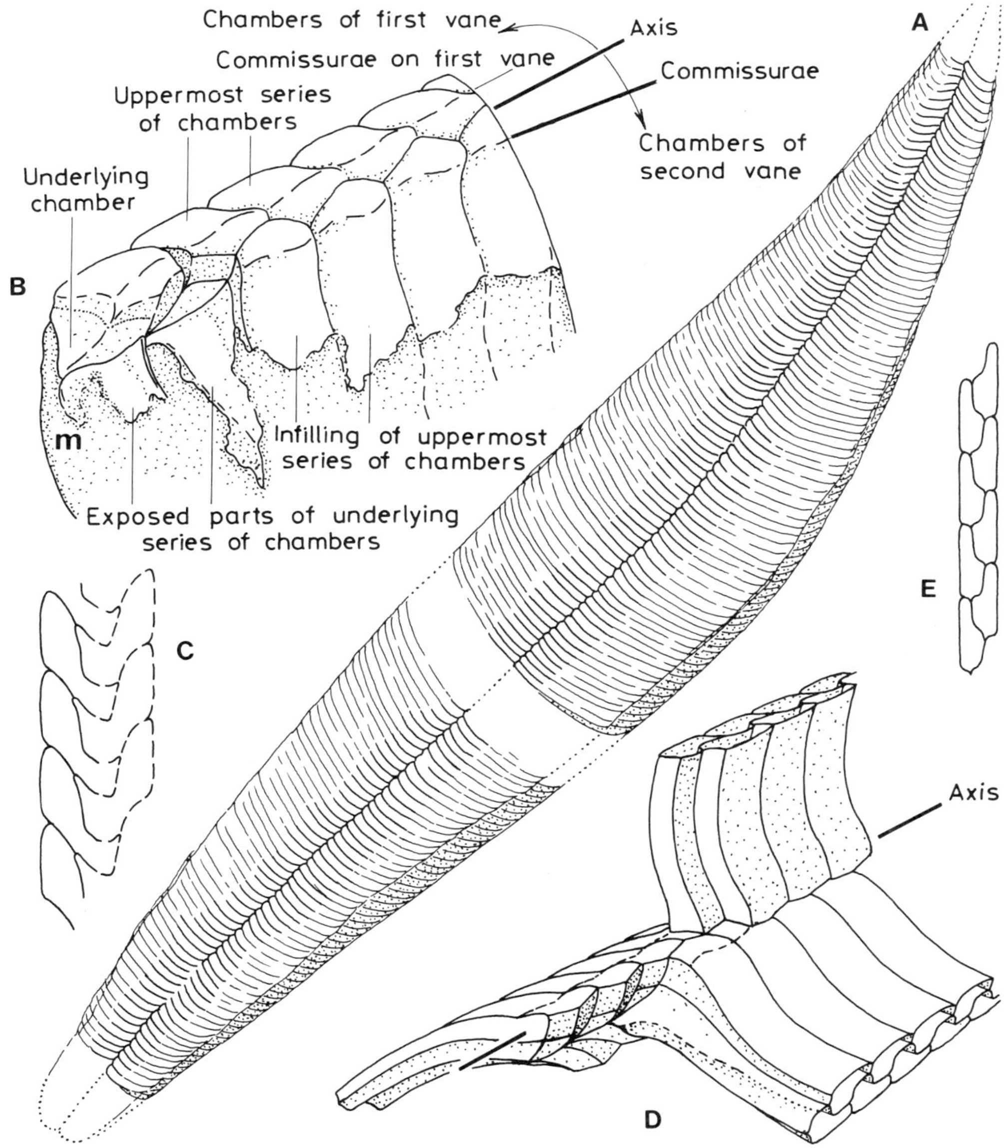

Figure 9. The frondose organism *Pteridinium simplex* Gürich, 1933, a representative of the Phylum Petalonamae occurring in the Nama Group. (A) general form of frond, ca. 0.225 times natural size; (B) three-dimensionally preserved speciman Pf. No. 399 from the Kliphoek Member of the Dabis Formation, Plateau Farm, Namibia, suggesting the sectional form of the chambers near the axis (C) and the way in which the vanes join axially (D); (E) section of chambers near edge of vane; m, matrix. Diagrams B–E are ca. 2.25 times natural size.

make a zigzag axial trace which seems to represent a thin septum. The third vane evidently attached to the axial region of the other two vanes along the zone of commissurae. Presumably the species grew by adding further chambers or tubes at one end of the frond.

Pflug (1970a,b, 1972a,b) and others have repeatedly suggested that

Pteridinium resembles and is allied to the engimatic sack-shaped remains of *Ernietta plateauensis*. Both possess a zigzag suture from which extends a subparallel ribbing. However, *Ernietta* is subcylindrical or polyplike in form. As in *Pteridinium*, the surficial ribbing corresponds to underlying tubular structures or narrow chambers which are also preserved as internal molds by infilling sediment. Up to three sets or layers of tubes are evident in *Ernietta*. More remarkably, there seems to be an approximately similar number of chambers in each set (ca. 64) regardless of the size of the organism. Some specimens show a transverse groove that possibly marks a zone of "budding" of some secondary part of the structure. The tissues of *Ernietta* appear to have been flexible and soft, with rare specimens showing fine markings or creases transverse to the more conspicuous ribbing. The mature *Ernietta* organism was gregarious in habit, with numerous individuals living close together and evidently attached to the sea bottom (Fig. 7).

Ernietta may be interpreted as an organism of approximately coelenterate grade (Glaessner and Walter, 1975). However, its detailed morphology seemingly shows no close resemblance to any known present-day coelenterate. The precise arrangement and number of tubes in *Ernietta* have important implications concerning the mode of growth of the form, indicating that the basal part of the sack was complete or entire prior to upward growth of the side walls. The juvenile presumably existed as a more or less discoidal, leaf-shaped form, with up to 30 or so lateral "ribs" on either side. It may have been either pelagic or attached. When the organism attained the number of ribs of a mature individual it began to develope into the "polyp" shape with longitudinal and transverse increase in all membranes. I consider that *Pteridinium* and *Ernietta* are not coelenterates, but validly represent the new Phylum Petalonamae Pflug, 1970a.

5. Ediacara, South Australia

5.1. Geological Setting

The Rawnsley Quartzite was also deposited in a shallow, epicontinental basin, and shows evidence of tidal sedimentation (Jenkins *et al.*, 1983). There are divergent opinions concerning inferred environments of deposition of the fossiliferous Ediacara Member (Preiss, 1987). Jenkins *et al.* (1983) maintained ideas of Sprigg (1947) and Glaessner (1961) that fossilization of the soft-bodied fauna was principally linked to diurnal movement of sands during tidal deposition. Extending some of the notions of Goldring and Curnow (1967), and Wade's (1968) conclusion that the fossils represent remains interred by sands laid down in some depth of water, Gehling (1982, 1983, 1987, 1988, 1991) considers that deposition occurred in an outer shelf setting below fair-weather wave base, with burial of the organisms by storm-surge sands (cf. Mount, 1989, 1991). A. Seilacher (Yale, Tübingen, personal communication 1985) also adheres to the idea of storm-related sedimentation, but holds that the presence of ubiquitous wave oscillation and interference ripples in the fossiliferous beds is characteristic of a shoreface. Ideas of storm deposition are opposite to Wade's (1968) conclusion that fossilization occurred in regions of lessened hydrodynamic energy in the lee of rising diapirs that impeded the predominant wave fetch.

Recent investigations disclose large polygonal desiccation cracks in highly fossiliferous parts of the section at Ediacara Range and in facies D of Jenkins *et al.* (1983) in part of the Heysen Range. This seems to potentially substantiate the suggestion of deposition in tidal regimes subject to emergence, and the intense concentric crinkling of some of the remains resulted from drying.

Groupings of like organisms in the fossil assemblages may indicate preservation of virtually intact segments of the community, and delicate epichnial trace fossils are common. Hence, fossilization initiated by an energetic process such as a storm seems inconceivable. Laminar streaming at the first stage of tidal creep is a possible way of gently transporting sand grains to initiate burial of organisms stuck down or clinging to the gluey surface of drying algal films. On the other hand, the more complete benthic communities that are represented must indeed have inhabited subtidal settings (cf. Nedin and Jenkins, 1991). A listing of taxa occurring in the Ediacara assemblage is presented in Table I.

5.2. Fossil Remains

A recently discovered "nested" set of large, muddy-sandstone flags located by N. Lemon in the central western Flinders Ranges includes counterpart surfaces marked by parallel-crested ripples and preserves hyporelief casts and negative epirelief molds of ten large individuals of *Ediacaria flindersi* (15–23 cm diameter) as well as several smaller indeterminate imprints. A thin, shaley layer separating the fossiliferous surfaces is intruded by sinusoidal and polygonal, sand-filled mudcracks and a second shaley interlayer ca. 5 cm stratigraphically above displays another set of mudcracks. The organic remains show saucer-shaped configurations with relief of up to 15 mm; a central cylindrical element extends down into the material of the underbed. The central structure is generally filled with course sand (now sandstone), but in a specimen in which this material has broken away, the "stalk" seems to terminate at a mud clast about 0.8 cm deeper in the bed than the adjacent central regions of the disc. Evidently the organisms were preserved in life position attached to mudflakes or to some layer in the substrate and they had colonized a shallow pool with a sandy bottom prone to rippling by currents. Seemingly, the pool became desiccated on at least two occasions. The mode of preservation is closely comparable with the example of *Ediacaria* illustrated by Wade (1968, Fig. 12A–C) or the "*Cyclomedusa davidi*" shown in Wade (1968, Fig. 13A, B).

The creatures were apparently bowl-shaped, with a short "stalk" forming an attachment, and on flattening during burial and compaction, the "sides" of the bowl were creased into a variable number of rather irregular annulations. A central zone of strong relief in the casts is circumscribed by a marked groove or fold where it meets a flat peripheral flange at about two thirds the radius of the whole; this probably represents an abrupt change from a relatively thick integument to a thinner membrane.

One of the specimens shows a thin layer of sediment 1–2 mm thick broken away from part of the epirelief mold to reveal a parting formed by a membrane beneath; this underlying parting shows an annulation not seen in the covering one. Yet another membrane is molded over part of the sediment filling the central

TABLE I. Composition of the Ediacara Assemblage of the Flinders Ranges, South Australia

Taxa recognized herein	Likely synonyms
Coelenterata	
Phylum Cnidaria	
Class Hydrozoa	
Order Hydroida, Suborder Chondrophorina	
Family Chondrophoridae	
Chondroplon bilobatum Wade, 1971	
Ovatoscutum concentricum Glaessner and Wade, 1966	
?Family Porpitidae	
Eoporpita medusa Wade, 1972	
Suborder uncertain	
Family Conchopeltidae	
Conomedusites lobatus Glaessner and Wade, 1966	
Class Cubozoa	
?Order Carybdeida	
Kimberella quadrata (Glaessner and Wade, 1966)	
?Class Scyphozoa	
Family uncertain	
Rugoconites enigmaticus Glaessner and Wade, 1966	? = *Lorenzenites rarus* Glaessner and Wade, 1966
Wadea tenuirugosus (Wade, 1972a) gen. nov.	
Sedentary medusoids or polypoids	
Ediacaria flindersi Sprigg, 1947	= *Beltanella gilesi* Sprigg, 1947
Cyclomedusa davidi Sprigg, 1947	= *Cyclomedusa plana* Glaessner and Wade, 1966
	= *Cyclomedusa radiata* Sprigg, 1949
Inaria karli Gehling, 1988	= *Spriggia annulata* (Sprigg, 1949)
Medusinites asteroids Sprigg, 1949	= *Spriggia wadea* Sun, 1986a
Hiemalora stellaris (Fedonkin, 1980)	
Mawsonites spriggi Glaessner and Wade, 1966	
Mawsonites randellensis Sun, 1986	
Colonial Cnidaria, possibly related to Pennatulacea	
Charniodiscus arboreus (Glaessner, 1959)	
Charniodiscus longus (Glaessner and Wade, 1966)	
Charniodiscus oppositus Jenkins and Gehling, 1978	
Glaessnerina grandis Glaessner and Wade, 1966	

Order Rangeomorpha
Rangea ? schneiderhoehni Gürich, 1929
Phylum Petalonamae
Family Pteridiniidae
Pteridinium nenoxa Keller, 1974
? *Pteridinium simplex* Gürich, 1933
Phyllozoon hanseni Jenkins and Gehling, 1978
Phylum uncertain
Brachina delicata Wade, 1972
Nemiana simplex Palij, 1976

Coelomata
Phylum Annelida
Class Polychaeta
Family Dickinsoniidae
Dickinsonia costata Sprigg, 1947 = *Dickinsonia elongata* Glaessner and Wade, 1966
Dickinsonia lissa Wade, 1972 = *Dickinsonia brachina* Wade, 1966
Dickinsonia rex nom. nov. [for *Dickinsonia elongata* Glaessner and Wade, 1966 (*sensu lato*)]
Dickinsonia tenuis Glaessner and Wade, 1966
Family Sprigginidae
Spriggina floundersi Glaessner, 1958
Articulate of uncertain placement
Marywadea ovata (Glaessner, 1976)
Phylum Arthropoda
Family Parvancorinidae
Parvancorina minchami Glaessner, 1959
Family Vendomiidae
Praecambridium sigillum Glaessner and Wade, 1966
?Family Trilobita
Unnamed genus and species
Phylum Tribrachidia nov.
Family Tribrachididae nov.
Tribrachidium heraldicum Glaessner, 1959
Phylum Echinodermata
Class Edrioasteroidea
Arkarua adami Gehling, 1987

"stalk" and an equivalent parting extends over a sector of the inner disc. The partings and the thin layers of sediment enveloped between indicate that the organism is preserved in a three-dimensional manner, with the "stalk" and a presumed large "gastric cavity" partly infilled by detritus, and perhaps another slice of the fill within a collapsed "stomodaeum." In other material, the mouth of *Ediacaria* may be represented by a circle with radial puckers and sometimes a creased center. Overfolding usually obscures the periphery of the organism.

Glaessner and Wade (1966) considered *Beltanella gilesi* synonymous with *E. flindersi* and this was formalized by Wade (1972a). However, the round "gonads" evident on the holotype of *B. gilesi* are not accidental as suggested by Glaessner and Wade (1966), because in addition to the three circular imprints clearly marked, two more are reasonably well evident and a further two are faintly marked, a total of seven of the likely eight present. Several other specimens show similar round structures in equivalent positions. In the holotype, weak radial traces possibly representing septa extend between the presumed gonads; specimen A.U. F17328 suggests an octamerous arrangement of the septa, and in certain lighting a zone showing poorly preserved tentacles is seen to rim the gastric space. An annulate zone of numerous overlapping tentacles is well displayed in A.U. F17327.

The synonomy of *Beltanella* and *Ediacaria* may be reasonably founded. Evidently *Beltanella* shows composite molds of the gonads and/or septa on a surface indicating the adoral tissues of the gastric space (Fig. 10A). The arrangement of gonads and septa somewhat resembles that in the modern scyphozoan order Coronata (Russell, 1970, pp. 2–5); perhaps *Ediacaria* was an evolutionary intermediary between a polyp and a medusa. Indeterminate specimens that might be loosely compared with *Ediacaria* reach about 1 m in diameter.

Seilacher (1984, 1985) is incorrect in considering *Mawsonites spriggi* to be a trace fossil consisting of a "central vertical burrow surrounded by several tiers of sharply separated radial lobes." Even cursory study of the holotype (e.g., Glaessner and Wade, 1966, pl. 99, Fig. 1) and other specimens shows the strong lobation is an integral part of a continuous surface representing a cast (? or mold) of the animal's body. The expression of the lobation differs between specimens and in its least development constitutes separate pustules (parts of A.U. F16719). The wrinkling of the center indicates that this was peaked or conical.

In view of the variability of the genus, the distinction between *Mawsonites randellensis* Sun, 1986c, and the type species may be regarded with caution. Sun's (1986c) reconstruction of *M. randellensis* as a likely coronate medusa is speculative, as none of the concentric grooves can be regarded as especially significant in respect of the others. A large new specimen (Fig. 11) shows a groove possibly marking the edge of some gastric element, and circular grooves toward the periphery, but these characteristics do not prove it to be a medusoid, and as the specimen shows moderate relief, it could well have been attached to the substrate.

Sprigg's (1947, pp. 220–221) description of *Cyclomedusa davidi* from a single rather poorly preserved specimen follows his nomination of *Ediacaria flindersi* (Sprigg, 1947, pp. 215–218) in the same publication. The well-preserved discoidal form Sprigg (1949, p. 86) designated as *Tateana inflata* has page priority over *Cyclomedusa radiata,* which likely represents the same species.

A thin sandstone bed exposed over an area of perhaps 5 × 3 m at a site in the western Flinders Ranges has sagged downward over underlying shales and sandstones fretted by salt weathering. Overturning of various flags broken from this bed revealed abundant medusiform fossils whose characteristics show a graded series varying from *Cyclomedusa plana* to *C. davidi sensu stricto* and *Spriggia annulata,* and through morphologies resembling *E. flindersi.* Examples of *Medusinites asteroides* are present as well. The bed shows a rippled top and presumably the sands that formed it were rapidly tracted over a muddy expanse on which numerous of these discoidal creatures were situated. The large number of individuals of rather uniform morphology (*C. plana*) suggests the sampling of part of a "bloom" or "swarm," and while perhaps more than one form is present, certainly one taxon predominates, and in terms of the overlap in size of all morphologies represented, the five nominate taxa may reflect differences in preservation of only one species. The question of a possible taxonomic relationship with *E. flindersi* is not resolved because none of the specimens correspond closely with the holotype of the latter. Sun (1986a) argued for the unity of *C. davidi* and *T. inflata,* and new material indicates an intergrading series between *C. plana* and *T. inflata* (= *C. radiata*), substantiating the synonomy of all the local forms of *Cyclomedusa.*

The overlapping excentric rugae characterizing some specimens assigned to *Cyclomedusa* led Wade (1972a) to deduce that the creature did not swim like a medusoid, but that an overfolded conical base possibly formed attachment to the substrate. Specimens commonly show a prominent central boss (e.g., Wade, 1968, Fig. 13A, B) that parallels the "stalk" in the new discoveries of *Ediacaria.* The annulation of rugae may be largely an artifact of flattening with burial. A regular radial arrangement of resistive fibers or ribs extended through the integument of the disc. Certain circumstances linked to the taphonomy of the remains, perhaps partial decomposition of the integument of dead individuals, seems to have accentuated the ribbing so that it imprinted into the covering sands. Failure to appreciate the extent of the ribs has led to incorrect identification of tentacles (Jenkins, 1984b, Fig. 5). Rare specimens in which the sand fill of a large internal space has partly broken away (Jenkins, 1984b, Fig. 6; Sun, 1986a, Figs. 3B and 5) show a membrane marked by fine radial striae or pleating. Stems and partial fronds attached to or associated with specimens in the type series of *Cyclomedusa* suggest that discoidal remains of this kind may commonly represent the anchoring devices of sea-pen-like organisms (Jenkins, 1989).

Rugoconites enigmaticus and *Rugoconites tenuirugosus* both show reticulate patterns of grooves that bifurcate toward the margin of the disc. Both may show portions of a smooth annulus around the disc suggesting a marginal flange. I am of the opinion that two distinct genera are represented and that either *R. engimaticus* was rather variable or that it includes two species.

In *R. enigmaticus sensu stricto* the reticulate grooves form a relatively coarse pattern and are chiefly subradial in disposition. Hollow molds of resistive elements may be present at the center of the disc, which commonly seems to have collapsed downward about these structures. *R. engimaticus sensu lato* (Wade, 1972a, pl. 43, Fig. 3) shows an even coarser and more polygonal pattern of grooves which branch rather abruptly near the periphery. Fine striae extending across the marginal flange and beyond have been interpreted as molds of tentacles. A Seilacher (personal communication 1985) disagrees that tentacles are represented,

Figure 10. Reconstruction of subtidal biota represented in the Ediacara assemblage of the Flinders Ranges. Organisms are shown at approximately true to .18 times natural size. (A) *Edicaria flindersi* (Sprigg, 1947), with sectioned individual portraying stomodaeum, mesenteries, and gonads; (B) *Tribrachidium heraldicum* (Glaessner, 1959); (C) *Parvancorina minchami* (Glaessner, 1959); (D) *Praecambridium sigillium* (Glaessner and Wade, 1966); (E) burrowing organisms with three sclerotized elements on anterior; (F) *Dickinsonia costata* (Sprigg, 1947), juveniles and mature examples; (G) scratch marks made by feeding arthropods; (H) soft-bodied trilobite; (I) dead and torn specimens of *Cyclomedusa* (Sprigg, 1947) partly exhumed from the substrate; (J) *Kimberella quadrata* (Glaessner and Wade, 1966); (K) *Charniodiscus longus* (Glaessner and Wade, 1966); (L) *Rugoconites enigmaticus* (Glaessner and Wade, 1966); (M) *Nemiana simplex* (Palij, 1976); (N) *Charniodiscus oppositus* (Jenkins and Gehling, 1978); (O) *Ovatoscutum concentricum* (Glaessner and Wade, 1966); (P) *Eoporpita medusa* (Wade, 1972); (Q) *Medusinites asteroides* (Sprigg, 1949), an anemone-like creature.

Figure 10. (*Continued*)

arguing that they would be too fragile to remain in a subparallel array. However, if the creature was stranded, a final side-slip or the last swirl of water could have distended the tentacles on the sector of the periphery where they were evident, and may also explain the folding of the flange on an opposite part of the margin.

R. enigmaticus commonly occurs as a negative hyporelief and new specimens show marked collapse craters in the covering sands, suggesting a thick mesoglea. Three "gonads" are indicated in rare specimens, but usually the number is unclear; the quadrate center of one remains and an approximately quadrate pattern in the bifurcation of the radial grooves in others may reflect a near relationship with scyphozoans (Jenkins, 1984b).

Figure 11. *Mawsonites randellensis* Sun, 1986c, from upper part of Ediacara Member, Rawnsley Quartzite, at Bunyeroo Gorge, western Flinders Ranges; hyporelief cast photographed from a plaster copy. Scale divisions in centimeters.

Wadea gen. nov., erected to include *Rugoconites tenuirugosus*, is distinguished from *Rugoconites* by the finer pattern and more radial disposition of the grooves. The holotype, A.U. F17461, shows the possible mold of a round mouth enclosed by a second circle suggested to indicate the stomach (Wade, 1972a, pl. 43, Fig. 5); stomach pouches and robust gonads seem not to have been present. Its shallow impression probably reflects a much thinner bell than in *Rugoconites*. Wade's (1972a) indication that foam may fill the gastrovascular system of stranded modern medusoids, or that early decomposition of food materials may distend the canals with gas, provides an actualistic explanation of why circulatory networks have acted as resistive elements during the process of molding of *Rugoconites* and *Wadea*.

New finds have been made of material resembling the form *Eoporpita medusa* (Wade, 1972a), which is related to the porpitid chondrophores (e.g., Jenkins, 1984b, Fig. 3). One new specimen rather clearly preserves the central "polyp," which was about seven times the diameter of the tentacle-like structures ringing it. The majority of specimens show the "polypoid" aspect facing downward, the likely position of grounding if such float-buoyed creatures were stranded on tidal flats. The common preservation of such remains in a three-dimensional manner, showing perhaps 9–14 mm of relief, with fine sand intimately filling the coelenteric cavities of verrucae or tentacles and thus casting their form, however, lends credence to the possible interpretation of some as actinians (e.g., Jenkins, 1989).

Fedonkin (1985a, b) considered *Eoporpita* relatively primitive because, with the exception of the central individual, the "polyps" do not seem to show the polymorphism that tends to be present in modern chondrophores. *Hiemalora stellaris* has recently been located in an offshore depositional setting in the Flinders Ranges and was most likely a kind of actinian.

The recent claim that *Chondroplon bilobata* is a distorted specimen of a large *Dickinsonia* Sprigg, 1947, rather than the float of a bilateral chondrophore (Hofmann, 1988) is unlikely, as it is clearly chambered in section, a characteristic never observed in *Dickinsonia*. *Ovatoscutum* also shows the geometry of a bilateral chondrophore. It was possibly peaked in shape, somewhat as in the modern *Velella* Lamarck, 1801, and on flattening with burial, part of the float usually split along the midline to form a V-shaped reentrant.

Evidence concerning the morphology of *Kimberella quadrata* was reviewed by Jenkins (1984b). The general shape of the elongate bell of this taxon is well established, and transversely furrowed sectors suggest strong swimming muscles. The puckered gonads may be exquisitely preserved and one individual (Jenkins, 1984b, Fig. 9) shows a groove and deep pit at the expected position of a rhopalium. The supposed number of four gonads is not seen (McMenamin, 1988; McMenamin and McMenamin, 1990), but the two or (?) three evident seem consistent with the likely vagaries of compaction attending preservation (Jenkins, 1989). These animals were directly ancestral to the chirodropid sea wasps and were fast-swimming pelagic predators like their modern descendants.

Occasional specimens of *Charniodiscus* show that a disc with a robust stalk emerging from it centrally joined the axial stem of the frond. Parts of the disc and stalk are sometimes preserved with sand packing internal spaces, indicating that in life they were largely hollow or a very large canal ran the length of the stalk. Stalks sometimes extend from a round plug of sand that presumably filled a crater formed where a buried disc had collapsed within the substrate. Radial lines suggesting a strong musculature mark the outer flange of the disc of *C. oppositus* (Fig. 7). The musculature of the disc of *C. arboreus* seems to have been principally concentric.

The discs were anchoring devices that could be worked down into the substrate where their inflation by fluid held the animals in place; their apparent flat shape resulted from sediment compaction and they were presumably subspherical in life. The frond projected into the water column where its polyps could feed; the holotype specimen of *C. oppositus* is remarkably preserved (Jenkins and Gehling, 1978, Fig. 6) perhaps as a result of early silicification, and suggests that the foliate base of the frond was inflated with fluid like a ribbed air-mattress and was an organ of hydrostatic support. Other specimens indicate needlelike spicules embedded in the integument of the stem and in the "branches" of the frond and between the polyp anthosteles (the tough "cups" into which the tentaculate parts of the polyps could withdraw) on the flaplike "polyp leaves." No trace of the possible mineral material of the spicules remains. The fluid-inflated basal lamella of the frond evidently tended to deflate on the death of the animal, leaving the stiffened "branches" to show through like the ribs of a mummified carcass.

The majority of specimens of *Charniodiscus* possibly represent storm-damaged remains torn free of their anchorages. *C. arboreus* may have inhabited tide pools and probably grew about 40 cm tall. *C. oppositus* is mainly found broken from its anchor and the large size of some specimens in an indication that it lived offshore. The supportive stalk alone is up to 30 cm long and the frond grew probably in excess of 1 m tall and at least 28 cm wide. Such a "giant" located by J. G. Gehling and his students at Ediacara Range shows the spicular supports of individual polyp leaves separated by as much as 8–9 mm; there is an indication that the polyps were only 1.5–2 mm in width and grouped in clusters somewhat as in the modern *Pteroeides* Herklots, 1858. Organisms comparable to these forms of *Charniodiscus* occur in the Mid-Cambrian Burgess Shale fauna (Conway Morris, 1989b, Fig. 5A).

C. longus also reached large size, and was very elongate, one incomplete frond extending over 39 cm in length (Sun, 1986b, Fig. 3). A number of flags broken from one bed show a remarkable association of individuals preserved on the upper surface of a sandy layer and aligned by the current (Wade, 1968) away from round, conical craters that evidently formed above the anchoring discs. The current distorted the polyp leaves and "branches" into deep pockets that became packed with sand; numerous specimens at varied stages of growth overlap. The long, tapering frond of *C. longus* was evidently an adaptation to allow it to stream out in currents, and clusters of these animals probably inhabited channels.

Extremely rare, *Glaessnerina* Germs, 1983 is in enigmatic genus, and all known specimens from the Flinders Ranges are incomplete. One specimen of *G. grandis* seems to show fleshy branches slightly separated. The "divisions" of the branches may represent large polyp anthosteles; the widest of these are near the axis of the frond and they become progressively narrower toward the tips of the branches as in a modern sea-pen such as *Pennatula* Linné, 1758. The polyps needed access to free water and probably projected away from the colony in life; compression on burial could account for their zigzag overlap along the midline of the frond.

The second species belonging to *Glaessnerina* had relatively narrow, numerous polyp anthosteles; one small specimen shows a cast of well-preserved anthosteles in three-dimensional relief.

The finding of specimens resembling *Rangea schneiderhoehni* in the Ediacara Member (cf. Gehling, 1991) is an important indication that the Ediacaran rocks of the Nama Group are similar in age to the Rawnsley Quartzite. *Pteridinium nenoxa* occurs in thick-bedded quartzites below the Ediacara Member *sensu stricto* at the south end of Ediacara Range. It also is important biostratigraphically, as this taxon is present in northern Russia (e.g., Fedonkin, 1981; Sokolov and Fedonkin, 1984) and North Carolina (Gibson *et al.*, 1984). The beds in which it was found at Ediacara Range are unlayered internally and may have formed as a result of fluidized sands settling offshore following storms. Recent exploration of these beds revealed poorly preserved, large organic remains resembling *Pteridinium simplex*.

Phyllozoon hanseni is similar to *Pteridinium*, but its frond seems to have been planar, bilateral rather than comprising three vanes. The frond had a membraneous integument stiffened by resistive elements that formed the zigzag axis and lateral ribs. These were sufficiently strong to form counterpart molds where separate specimens overlap. Numerous fronds may occur on a single bedding plane. The manner in which *Phyllozoon* lived is unknown, but the form is rare at Ediacara Range, where emergent tidal facies are represented, but locally common elsewhere. Hence it may have favored waters a little offshore, and was presumably benthonic, rarely subject to stranding; individuals perhaps had a weak attachment to stolons and tended to grow in clusters.

While the possible annelid relationships of *Dickinsonia* have commonly been disavowed (Termier and Termier, 1968; Conway Morris, 1981; Fedonkin, 1981, 1985a,b; Seilacher, 1984, 1985, 1989; Gould, 1984, 1985), none of the objections raised against this classification are substantive. One objection leveled is that the apparent segmentation may alternate along the midline of the animal. However, the elytral ridges of modern errant polychaetes are bedded in muscular tissues of the body wall and may have deeply penetrating roots to which flexural muscles are attached (cf. Graff, 1888). In life, the elytral ridges are motile, and may be flexed to and fro by these muscles, thus probably playing a role in respiration (Dam, 1940; Borradaile *et al.*, 1963, p. 276 and Fig. 181). The notopodial–elytral ridges probably were similarly flexural in *Dickinsonia*, as their broadened distal ends are flattened in opposite directions in the clearly molded large individual of *D. costata* figured by Wade (1972b, pl. 5, Fig. 2). If, during morbidity prior to death, the elytral ridges came to lie in opposite directions and the worm underwent contraction, their inner ends could be drawn into an alternate arrangement.

Dickinsonia closely resembles the modern worm *Spinther* Johnston, 1845, which is parasitic on sponges. Indeed, the two genera differ provably in only two major aspects; *Dickinsonia* commonly had more segments, but lacked the massive, deep-rooted hooks that the modern animal uses to attach to its host. I agree with Wade (1972b) that the apparent similarity between the two is remarkable; the likelihood of iterative evolution occurring over the long time separation involved (Runnegar, 1982a, 1991) is a question difficult to judge.

A well-preserved dorsal aspect of the anterior end of a large *D. costata* from northern Edicara Range shows a flattened area of the axis where Wade (1972b) suggested the placement of the prostomium, and this organ seems to be indicated itself by a small, faint, raised circle (Fig. 12). Present also in the specimen just mentioned and overlapping the presumed prostomium is a pair of faint divergent grooves corresponding in position to the hind margins of the pharyngeal sack and esophagus in *Spinther* (cf. Graff, 1888, Figs. 7, 8, and pl. 1, Figs. 6, 7). The length of these grooves suggests that the fused anterior pair of segments (Glaessner and Wade, 1966) and as many as seven pairs of segments behind were "cephalized" about the esophagus; alternatively and perhaps about equally as likely, the grooves could reflect a composite mold of the partly everted pharynx.

The strong medial rib commonly present in finely segmented dickinsoniids

Figure 12. Well-preserved anterior end of *Dickinsonia costata* from Ediacara Member of Rawnsley Quartzite, northern Ediacara Range; (A) enlargement showing margins of ?pharangeal sack (arrowed) and esophagus (os); (B) esophageal region shown in part A differently illuminated to disclose prostomial spot, which is arrowed and partly outlined. Scale divisions in centimeters; parts A and B are on the same scale.

and specimens that likely have undergone some degree of contraction probably represents a detritus-packed intestine (Runnegar, 1982a, 1991). In *Spinther* an intriguing double intestine is present; a rather broad lower intestine with mainly segmental, paired diverticulae extends from the postpharyngeal region to the anus, and joining this lower intestine near the rear is an upper blind intestine, also with cecae, and commonly containing elongate masses of fecal material. Food may be secondarily digested in the blind intestine by activity of protozoa or bacteria. The dorsal ridge in the fossil worms sometimes pinches and swells along the body suggesting uneven packing by feces. Well-preserved specimens may show a slightly elevated (in latex molds corresponding to casts) zone along the midline; this tends to taper toward the pharynx and constricts near the presumed anus. Such a tract is rather clearly evident in the counterpart of the large specimen illustrated by Wade (1972b, pl. 7, Fig. 2) and is up to about ⅛ the width of the worm; likely, this represents the ventral intestine.

Of several specimens cited as showing gastric cecae (Glaessner and Wade, 1966; Wade, 1972b), one seems to be a poorly preserved worm that shrunk greatly after making a mark in the sediment (S.A.M. P14434). Sediment-infilled, positive hyporeliefs appear to represent cecae in the specimen illustrated by Glaessner and Wade (1966, pl. 101, Fig. 4), and Peter Haines has recently found another individual with some cecae evident. The specimen illustrated by Glaessner and Wade is strongly deformed, with the lower surface principally represented, and the notopodial–elytral ridges of the sheared dorsal surface mainly impressed faintly as composite molds; the worm seems to have had about 40 segments. The casts of the cecae form an approximately symmetrical distribution about a smooth central zone presumed to match the placement of the ventral intestine; the more anterior cecae curve rearward relative to the displaced segmentation of the dorsal integument; the more rearward cecae are approximately intersegmental (parallel to the segmentation). Cecae of about the first third of the worm are branched; some 16 cecal branches, variably complete, are evident on the left side of the specimen and about 11 are on the right. Branching cecae may occur in *Spinther* (e.g., Graff, 1888, Fig. 8, and pl. 1, Figs. 6, 7), where their transgression across segmental boundries is possible because of the filamentous and incomplete nature of the intersegmental spetal muscles. Manton (1967) concluded that the poor development of septal muscles precluded *Spinther* from being very active.

Specimens of *D. costata* commonly show double outlines indicating that they had undergone contraction (e.g., Wade, 1972b, pl. 5, Fig. 1) (Fig. 13) and hence the animals had muscular body walls. The longitudinal–elliptical muscles are shown in numerous examples by ridges shaped like magnetic lines of force between poles at the front and rear of the animal. In extreme instances of contraction the width of the worm is reduced by as much as one half, and assuming that the coelome was fluid-filled, the animal presumably increased some four times in thickness (Runnegar, 1982a, 1991). This is a further indication that the intersegmental muscles were rather tenuous or filamentous.

Though *Dickinsonia* is commonly assumed to be a flat animal, this may be partly an artifact of its common mode of preservation sandwiched between resistive sand layers. Some 5% of known individuals of *D. costata* are preserved in a deformed, asymmetrical manner as though a rather underfilled sausage of integument about the coelome had rolled sideways. *D. costata* grew up to ca. 115 segments and were ca. 23 cm wide and ca. 27 cm long.

A large species with many segments has been identified as *Dickinsonia elongata* Glaessner and Wade, 1966. Unfortunately, the "juvenile" individual designated as the holotype cannot be differentiated from contracted large specimens of *D. costata* and hence the name "*elongata*" is a junior synonym of this. The large worms included in "*elongata*" certainly belong to a validly distinct taxon. A partial specimen discovered by Bruce Runnegar is more than 69 cm long and this together with other incomplete examples suggests that these worms reached a length of ca. 1 m. The replacement name of *Dickinsonia rex* nom. nov. nominated for this prince of worms is based on the specimen illustrated by Wade (1972b, pl. 7, Fig. 2) as the holotype. It is distinguished from *D. costata* by its elongate shape and large number of segments. A beautifully preserved large specimen (42 cm long, Fig. 14) discovered by Victor Gostin shows a weakly contracted rear end (more than 91 segments) shaped like a beaver's tail and 0.46 the width of the anterior body (ca. 220 segments). The narrowly segmented, nominate taxon *Dickinsonia brachina* Wade, 1972b, likely represents expanded examples of the wrinkly segmented, highly contracted worm diagnosed as *Dickinsonia lissa* Wade, 1972b. *D. lissa* has page priority.

At least the smaller specimens of *D. costata* seem to have inhabited the extreme shallows, their flat shape perhaps favoring life in tide pools. Specimens of *D. rex* in the same beds are commonly overfolded or incomplete, suggesting current transport of cadavers prior to burial. Large *D. costata* and complete spec-

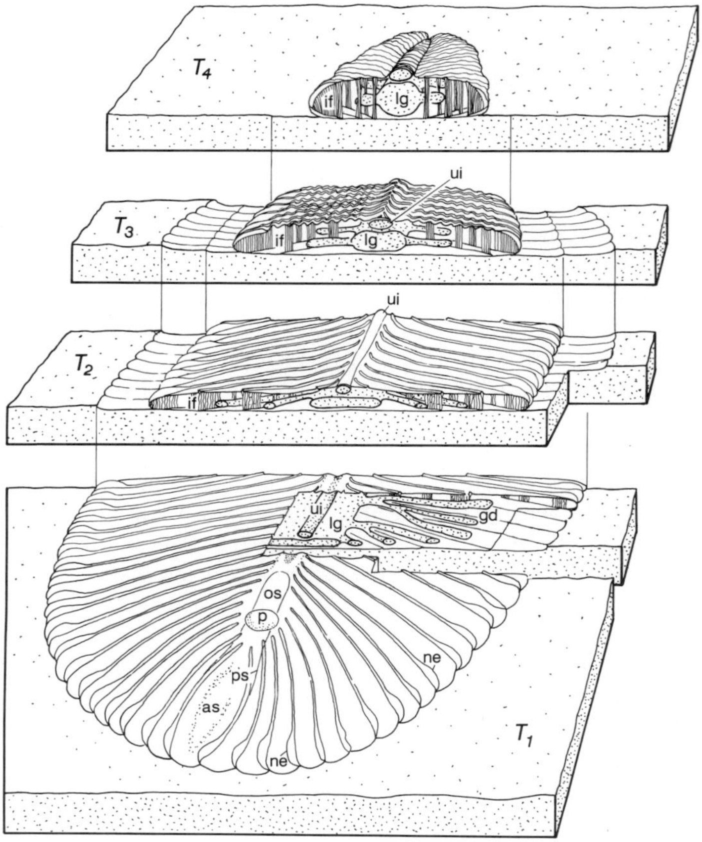

Figure 13. Taphonomic changes attending the preservation of *Dickinsonia costata*. T_1 represents the live animal with flexible notopodial–elytral ridges (ne), fused anterior segments (as), margins of pharangeal sack (ps), and esophagus (os), prostomium (p), inferred lower gut (lg), and gastric diverticulae (gd), hypothetical upper intestine (ui), and postulated interseptal filaments (if). T_2, during early morbidity, the stranded animal undergoes contraction, leaving a mark on the substrate. T_3, during late morbidity, the animal contracts further, with gut contents inflating intestines, contracted longitudinal muscles wrinkling the upper body wall, and resistive ends of elytral ridges pulled into an alternate arrangement (zigzag). T_4, stage of extreme contraction on or after death with stretched interseptal filaments further wrinkling the surface, and feces-packed upper intestine becoming prominent; such remains have usually been moved around by currents. Covering sands may mold and preserve the animal at any stage. About one half natural size.

imens of *D. rex* tend to occur in thick or massive sand beds [facies 'D' of Jenkins *et al.* (1983)] together with fronds of *Charniodiscus* and sometimes *Phyllozoon*. This setting may have been one of tide pools and sand waves in the lower part of the tidal range. *D. tenuis* and *D. lissa* are uncommon, but cosmopolitan.

In view of the weak septal musculation inferred, dickinsoniids were probably not highly active worms, but crept slowly about by expanding and contracting areas of the body in a somewhat ameboid fashion, browsing on algal films and decaying saprophels at the sediment–water interface. This leisurely way of life of such meaty "pancakes" surely reflects an absence of predators.

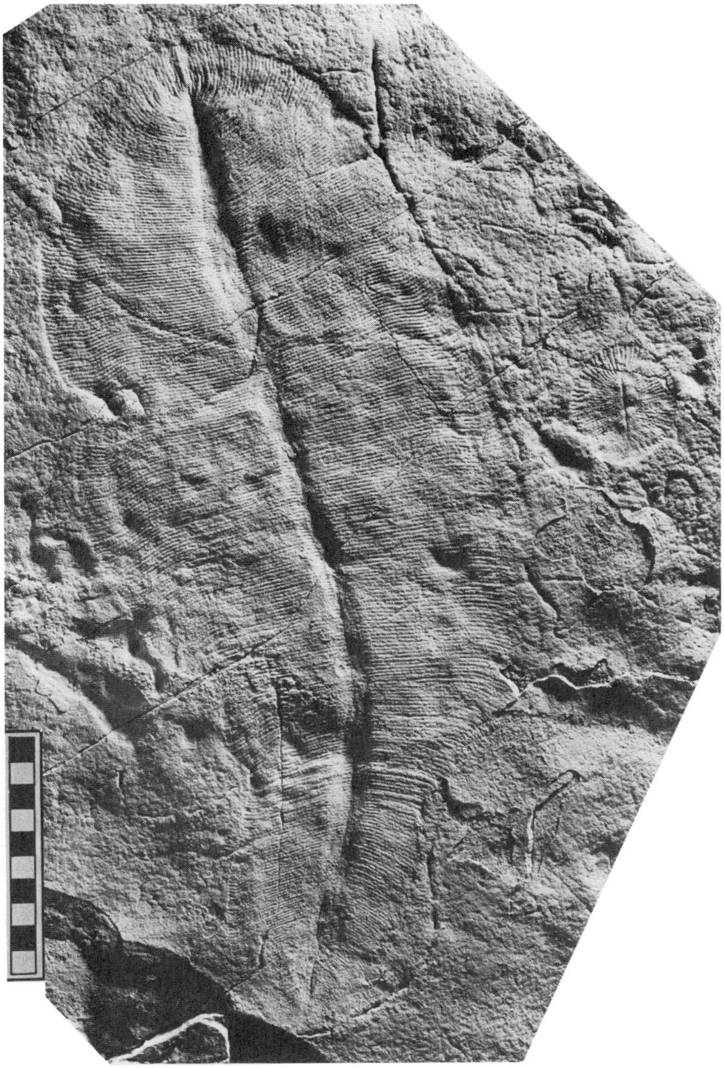

Figure 14. *Dickinsonia rex* nom. nov. (large individual) from upper part of Ediacara Member of Rawnsley Quartzite, western Flinders Ranges, and showing well-defined outline of ?lower intestinal tract and tail shaped like a beaver's paddle; a specimen of *Dickinsonia costata* is situated to the right. Scale in centimeters.

With their large crescentic "prostomium" and tapered, multisegmented bodies, *Spriggina* and *Marywadea* resemble the outline of trilobites (Runnegar, 1982b; Conway Morris, 1985) and are classified by some as arthropods (Birket-Smith, 1981; Runnegar, 1991). True limbs are never apparent and the complex interpretation of Birket-Smith (1981) is overly influenced by grain boundaries in the course matrix.

Well-preserved specimens of *Spriggina floundersi* show imprints of setae extending from the notopodial inflections of the segments. Suggestions that the lateral processes were too closely oppressed to function as parapodia do not take into account that the animals underwent a degree of compression on burial and the body extensions probably rotated to some extent and tended to broaden. A pair of molds of structures within the "peristomium" may represent sclerotized jaws. The animal had a relatively tough or resilient body as shown by the concave molds left in the covering sands. Its interpretation as an errant annelid still seems plausible.

A relatively large and beautifully preserved specimen (Glaessner, 1976b) referred to *Marywadea* shows a broad anterior end with rather clearly defined branching gastric diverticulae extending along either side of the axis (Gehling, 1991, Fig. 1). This characteristic certainly resembles the inferred placement of genal cecae in trilobites, but ramifying gastric cecae are known in living polychaetes such as the Spintheridae and Myzostomia.

The cecae of the form just described are similar to those of one of the arthropods in the assemblage, *Praecambridium sigillum*. The affinities of this animal are well argued by Glaessner and Wade (1971) and Glaessner (1984). It is evidently related to the trilobitoid *Vendia sokolovi* Keller, 1969, and the meraspidlike *Vendomia menneri* Keller, 1976, from the Vendian of northern Russia. The asymmetry of *Vendia* noted by Bergstrom (1989) does not preclude its arthropod affinities, because if it had a thin cuticle, distortion could have occurred on burial.

The kite- or shield-shaped *Parvancorina minchami*, which varies from 3 to 35 mm long, is also usually considered as an arthropod (e.g. Glaessner, 1980, 1984; Gehling, 1991). This animal is reasonably abundant at many sites in the Flinders Ranges. J. G. Gehling has discovered specimens buried on top surfaces of thick mud layers by sands that settled from density currents. These remains show an unusually tall, three-dimensional shape suggesting that the more common mode of preservation where remains are associated with thin mud films sandwiched between sand layers is attended by significant flattening (Gehling, 1987, 1988). Rubber molds of the less-deformed specimens show the central element high-standing and distinctively humped at about two thirds of its length from the anterior. The integument of the "shield" (? carapace) seems commonly to have collapsed downward, presumably because the space beneath remained empty of sediment as the animal was interred. Some small specimens are arched in shape. *Parvancorina* possibly may be compared with the arthropod *Burgessia bella* Walcott, 1912, from the Burgess Shale of British Columbia. The axial ridge and anchor "tines" of *Parvancorina* may be analogous to the midgut and two large diverging gastric diverticulae present in *Burgessia*. Seemingly, the anterior of *Parvancorina* became concertinaed with compaction on burial, obscuring the origins of the backwardly splayed cephalic limbs. About ten pairs of limbs seem to have been present on the hind axis (eight pairs in *Burgessia*) and complex overprinting in one individual (Glaessner, 1980, pl. 2, Fig. 4) suggests the possibility that the limbs were biramous, perhaps with gill branches that functioned as oxygen exchange organs.

That relatively large arthropods were denizens of the Ediacaran biota is now reasonably well established. Gehling has found a specimen of *Diplichnites* Dawson, 1873, indistinguishable from examples in the local Early Cambrian. Numerous other bedding surfaces show either molds of scratch marks (lower surfaces of quartzite flags) (Gehling, 1991, Fig. 3) or the actual scratches (tops of flags) made by arthropods raking the surface for detritus. The scratches are paired, indicating a double claw on each telopodite, and occur in fan-shaped arrays up to 7 cm wide and 14 cm long. From the numbers of scratches associated in each array, the animal is estimated to have had about 12–13 pairs of limbs, and from the dimensions of the feeding marks, the creature was as long as ca. 4 cm and had elongate legs.

A half dozen or so specimens of the possible originator of the raking marks have been discovered recently by Gehling and myself. This animal is a "soft-bodied trilobite" with the characteristic tripartite divisions of the body. A semicircular cephalon with a well-marked border, conspicuous recurved eye ridges, and median glabella is followed by a broad prothorax of 13 segments showing well-defined pleural grooves, and a tapered opisthothorax of approximately eight narrow segments; the ovate pygidium is too small to show any detail in the single specimen in which it is preserved. Transversely, the animal had a raised median axis and wide, flat pleural fields (Fig. 15). Distortion of the specimens indicates a degree of elasticity in the integument. Its observed characteristics occur in a variety of Paleozoic trilobites, but it can probably be considered to differ from all known forms at the ordinal level.

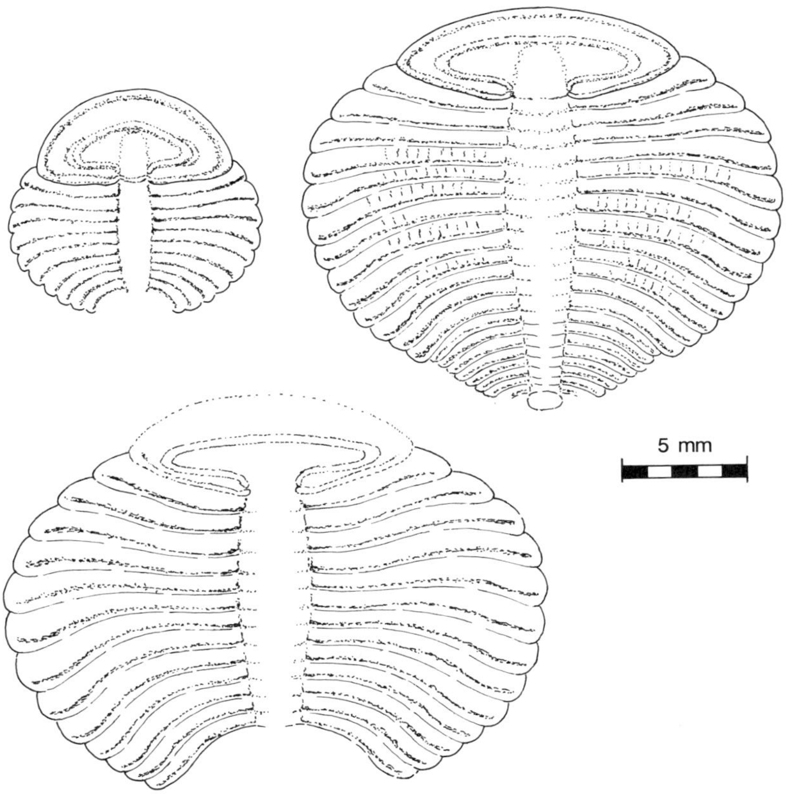

Figure 15. Reconstructions of several instars of a recently discovered "soft-bodied trilobite" in Ediacara Member of Rawnsley Quartzite, central Flinders Ranges.

Tribrachidium heraldicum, currently being redescribed, seems to have consisted of a sacklike body with a relatively resilient dorsal disc bearing three crescentic arms that were probably disposed about a central mouth. Between the arms were three resilient organs interpreted as possible gonads. The outer margin of each arm bore a row of stout filaments which may have been arranged in a double series with every second one at a higher level than its neighbor, much as the filaments are positioned on the lophophore of brachiopods. Exceptionally well-preserved specimens seem to show a layer of tissue or a "caul" extending over the arm-bearing disc; this membrane formed three crescentic winglike elements supported by radiating "ribs" ["bristles" of Glaessner and Wade (1966); radial elements in Fig. 3b of Sokolov and Fedonkin (1984) and Figures 2 and 3 of Gehling (1991)] that terminated in recurved whiskers. The animal probably lived with its body shallowly buried in the upper substrate, and the winglike apparati of the caul protected the underlying gonads and lophophores but admitted currents of water through the screen of marginal whiskers. Glaessner (1960, 1984) identified *Tribrachidium* as belonging to a new phylum and I am in agreement that it represents a new division approaching the organization of the Echinodermata. *Arkarua adami* is convincing as an ancestral echinoderm resembling Cambrian edrioasteroids. Evident sutures in its outer disc suggest that it possessed at least a partial skeleton of mineralized plates (Gehling, 1987, p. 344).

6. Conclusions

Fossil assemblages of soft-bodied, megascopic remains of Ediacaran aspect now reported from all major continents except Antarctica (e.g., Hofmann, 1987) include occurrences of the same species at various localities, suggesting near equivalence in age. One such assemblage at Mistaken Point is dated directly at ca. 565 My. Other workers have commonly linked the finds to a relatively wide age spectrum, recently even adducing the occurrence of these kinds of fossils in sequential stratotectonic cycles (Harris and Glover, 1988). The localities cited in this latter study are quite widely separated (ca. 145 km) in a complex tectonic province.

Frondose taxa predominating in distal turbidites of the Avalon Peninsula are buried beneath the crystal fraction of covering tuffs, indicating that the remains constitute benthonic life assemblages, and it is likely that the organisms represented were heterotrophs. Forms comprising numerous complexly conjoined small fronds are most common, and the evolutionary grade of these and other remains with both multiple and single larger fronds is at least comparable with that of living anthozoans. The possible relationships of associated discoidal remains are problematic. Elements of the Newfoundland assemblages also lived in the volcanic shelf setting represented by the Charnian Complex. Rare, closely annulate, discoidal forms present at the latter occurrence could be remains of chondrophores.

Rangea and enigmatic members of the Phylum Petalonamae occurring in the Nama Group seem to have been denizens of shelf and tidal settings. Remains that were enveloped by storm-generated density currents remarkably preserve three-dimensional aspects of their original morphology and early diagenetic silicification may have acted as a fixative.

New finds in tidal or subtidal sediments in the Flinders Ranges highlight the variable apparent morphology of the discoidal "medusiform" creatures depending on taphonomic processes attending burial. Their diversity is surely less than previously supposed and instances of three-dimensional modes of preservation documenting attachments to the substrate show that mainly benthonic forms are represented. Variable preservation of tentacles or polypoids, gonads, and mesenteries, or chambered floats suggests affinities or parallelism with living cnidarians.

Arthropod trace fossils and body imprints of a generalized "soft-bodied trilobite" underscore evolutionary grades attained during the Late Ediacaran. *Tribrachidium* evidences a unique phyletic experiment and an early edrioasteroid already shows a pattern of sutures suggesting it had a mineralized skeleton.

Four alone of the 30 or so reported world Ediacaran fossil sites indicate that at this time heterotrophic invertebrates had colonized the sea from relatively deep rifts to tidal shallows, and evidently the waters were oxygenated. The sheer profusion of fossil remains in appropriate preservational settings suggests an abundance of food probably supplied by a rich phytoplankton. The appearance of animals as advanced as trilobites and echinoderms portends the explosive radiation characterizing the Cambrian.

ACKNOWLEDGMENTS. Dr. M. M. Anderson of Memorial University, St. John's, Newfoundland, kindly demonstrated local Ediacaran fossil sites. Gratitude is expressed to R. Barrett for photography, S. Proferes for assistance with drafting, and S. Tsemitsidis for typing. This research was supported by Study Leave Grants from the University of Adelaide and funding from Esso Australia Limited.

References

Ahrendt, H., Hunziker, J. C., and Weber, K., 1978, Age and degree of metamorphism and time of nappe emplacement along the southern margin of the Damara orogen/Namibia (SW-Africa), *Geol. Rundsch.* **67**(2):719–742.

Anderson, M. M., 1976, Fossil Metazoa of the Late-Precambrian Avalon fauna, southeastern Newfoundland, *Geol. Soc. Am. Abstr. Progr.* **8**(6):754.

Anderson, M. M., 1978, Ediacara fauna, in: *McGraw-Hill Yearbook of Science and Technology*, McGraw-Hill, New York, pp. 146–149.

Anderson, M. M., and Conway Morris, S., 1982, A review, with descriptions of four unusual forms, of the soft-bodied fauna of the Conception and St. John's Groups (late-Precambrian), Avalon Peninsula, Newfoundland, in: *Proceedings. Third North American Palaeontological Convention* (B. Mamet and M. J. Copeland, eds.), Vol. 1, pp. 1–8.

Anderson, M. M., and Misra, S. B., 1968, Fossils found in the Precambrian Conception Group in Southeastern Newfoundland, *Nature* **220**(5168):680–681.

Benus, A. P., 1988, Sedimentological context of a deep-water Ediacaran fauna (Mistaken Point Formation, Avalon Zone, eastern Newfoundland), *Bull. N.Y. St. Mus.* **463**:8–9.

Bergstrom, J., 1989, The origin of animal phyla and the new phylum Procoelomata, *Lethaia* **22**(3):259–269.

Bergstrom, J., 1990, Precambrian trace fossils and the rise of bilaterian animals, *Ichnos* **1**(1):3–13.

Birket-Smith, S. J. R., 1981, A reconstruction of the Precambrian *Spriggina*, *Zool. Jahrb. Anat.* **105**(2):237–258.

Borradaile, L. A., Potts, F. A., Eatham, L. E. S., and Saunders, J. T., 1963, *The Invertebrata* (rev. by G. A. Keskut), 4th ed., Cambridge University Press, Cambridge.

Boynton, H. E., 1978, Fossils from the Pre-Cambrian of Charnwood Forest, Leicestershire, *Mercian Geol.* **6**(4):291–296.

Boynton, H. E., and Ford, T. D., 1979, *Pseudovendia charnwoodensis*—A new Precambrian arthropod from Charnwood Forest, Leicestershire, *Mercian Geol.* **7**(2):175–177.

Brafield, A. E., 1969, Water movements in the pennatulid coelenterate *Pteroides griesum*, *J. Zool. Lond.* **158**(3):317–325.

Cloud, P., and Glaessner, M. F., 1982, The Ediacarian Period and System: Metazoa inherit the Earth, *Science* **217**(4562):783–792.

Compston, W., Williams, I. S., Jenkins, R. J. F., Gostin, V. A., and Haines, P. W., 1987, Zircon age evidence for the Late Precambrian Acraman ejecta blanket, *Aust. J. Earth Sci.* **34**(4):435–445.

Conway Morris, S., 1981, Parasites and the fossil record, *Parasitology* **82**(3):489–509.

Conway Morris, S., 1985, The Ediacaran biota and early metazoan evolution, *Geol. Mag.* **122**(1):77–81.

Conway Morris, S., 1989a. South-eastern Newfoundland and adjacent areas (Avalon Zone), in: *The Precambrian–Cambrian Boundary* (J. W. Cowie and M. D. Brasier, eds), Clarendon Press, Oxford, pp. 7–39.

Conway Morris, S., 1989b. The Burgess Shale fauna and the Cambrian explosion, *Science* **246**(4928):339–346.

Conway Morris, S., Mattes, B. W., and Chen Menge, 1990, The early skeletal organism *Cloudina*: New occurrences from Oman and possibly China, *Am. J. Sci.* **290-A**:245–260.

Crimes, T. P., and Germs, G. J. B., 1982, Trace fossils from the Nama Group (Precambrian–Cambrian) of southwest Africa (Namibia), *J. Paleontol.* **56**(4):890–907.

Dam, L. van, 1940, On the mechanism of ventilation in *Aphrodite aculeata*, *J. Exp. Biol.* **17**(1):1–7.

David, T. W. E., 1922, Occurrence of remains of small Crustacea in the Proterozoic (?) or Lower Cambrian (?) rocks of Reynella, near Adelaide, *Trans. R. Soc. S. Aust.* **46**:6–8.

Deacon, G. E. R., 1984, Oxygen in Antarctic water, *Deep-Sea Res.* **31**(11):1369–1371.

Department of Mines and Energy, Newfoundland, 1985, Mineral Occurrence Map Avalon Peninsula Newfoundland, Map 85–62, Mineral Development Division, Department of Mines and Energy, St. John's, Newfoundland.

Fedonkin, M. A., 1980, New Precambrian Coelenterata in the north of the Russian Platform, *Paleont. Zh.* **1980**(2):7–15. [in Russian].

Fedonkin, M. A., 1981, *White Sea Biota of the Vendian (Precambrian Non-skeletal Fauna of the Russian Platform North)*, Nauka, Moscow [in Russian].

Fedonkin, M. A., 1983, The organic world of the Vendian, *Itogi Nauki Tekh. Ser. Strat. Paleontol.* **12**:1–127 [in Russian].

Fedonkin, M. A., 1985a, Non-skeletal Vendian fauna: Promorphological analysis, in: *The Vendian System*, Vol. 1, *Paleontology* (B. S. Sokolov and A. B. Ivanovskii, eds.), Nauka, Moscow, pp. 10–69 [in Russian].

Fedonkin, M. A., 1985b, Precambrian metazoans: The problems of preservation, systematics and evolution, *Philos. Trans. R. Soc. Lond. B* **311**(1148):27–45.

Ford, T. D., 1958, Precambrian fossils from Charnwood Forest, *Proc. Yorkshire Geol. Soc.* **31**(3):211–217.

Ford, T. D., 1963, The Pre-Cambrian fossils of Charnwood Forest, *Trans. Leicester Lit. Philos. Soc.* **57**:57–62.

Ford, T. D., 1968, B. The Precambrian palaeontology of Charnwood Forest, in: *The Geology of the East Midlands* (P. C. Sylvester Bradley and T. D. Ford, eds.), Leicester University Press, Leicester, England, pp. 12–14.

Ford, T. D., 1979a, The history of the study of the Precambrian rocks at Charnwood Forest, England, in: History of Concepts in Precambrian Geology (W. O. Kupsch and W. A. S. Sarjeant eds.), *Geol. Assoc. Can. Spec. Pap.* **19**:65–80.

Ford, T. D., 1979b, Precambrian fossils and the origin of the Phanerozoic phyla, in: *The Origin of Major Invertebrate Groups* (M. R. House, ed.), Academic Press, New York, pp. 7–21.

Ford, T. D., 1981, The Ediacaran fossils of Charnwood Forest, Leicestershire, *Proc. Geol. Assoc. Lond.* **91**(142):81–83.

Gardiner, S., and Hiscott, R. N., 1984, Sedimentology and basin analysis of the Precambrian Conception Group, Avalon Zone, Newfoundland, *Geol. Surv. Can. Pap.* **84-1A**:213–216.

Gehling, J. G., 1982, The sedimentology and stratigraphy of the Precambrian Pound Subgroup, Central Flinders Ranges, South Australia, M.Sc. thesis, University of Adelaide, South Australia.

Gehling, J. G., 1983, The Ediacara Member: A shallowing upward submarine fan sequence, within the Pound Subgroup, *Geol. Soc. Aust. Abstr.* **10**:52–54.

Gehling, J. G., 1987, Earliest-known echinoderm—A new Ediacaran fossil from the Pound Subgroup of South Australia, *Alcheringa* **11**(4):337–345.

Gehling, J. G., 1988, A cnidarian of actinian-grade from Ediacaran Pound Subgroup, South Australia. *Alcheringa* **12**(4):299–314.

Gehling, J. G., 1991, The case for Ediacaran fossil roots to the metazoan tree, in: *The World of Martin F. Glaessner* (B. P. Radhakrishna, ed.), *Geol. Soc. India Mem.* **20**:181–223, Geological Society of India, Bangalore.

Germs, G. J. B., 1972a, The stratigraphy and paleontology of the lower Nama Group, South West Africa, *Univ. Cape Town Dept. Geol. Precambrian Res. Unit Bull.* **12**:1–250.

Germs, G. J. B., 1972b, New Shelly fossils from the Nama Group, South West Africa, *Am. J. Sci.* **272**(8):752–761.

Germs, G. J. B., 1972c, Trace fossils from the Nama Group, South West Africa, *J. Paleontol.* **46**(6):864–870.

Germs, G. J. B., 1973a, A reinterpretation of *Rangea schneiderhoehni* and the discovery of a related new fossil from the Nama Group, South West Africa, *Lethaia* **6**(1):1–10.

Germs, G. J. B., 1973b, Possible sprigginid worm and a new trace fossil from the Nama Group, South West Africa, *Geology* **1**(2):69–70.

Germs, G. J. B., 1974, The Nama Group in South West Africa and its relationship to the Pan-African Geosyncline, *J. Geol.* **82**(3):301–317.

Germs, G. J. B., Knoll, A. H., and Vidal, G., 1986, Latest Proterozoic microfossils from the Nama Group, Namibia (South West Africa), *Precambrian Res.* **32**(1):45–63.

Gibson, G. G., Teeter, S. A., and Fedonkin, M. A., 1984, Ediacarian fossils from the Carolina slate belt, Stanly County, North Carolina, *Geology* **12**(7):387–390.

Glaessner, M. F., 1959, Fauna, in: The geology and Late Precambrian fauna of the Ediacara Fossil Reserve, *Rec. S. Aust. Mus.* **13**(3):377–401.

Glaessner, M. F., 1960, Precambrian fossils from South Australia, in: *Reports. 21st International Geological Congress*, Part 22, pp. 59–64.

Glaessner, M. F., 1961, Pre-Cambrian animals, *Sci. Am.* **204**(3):72–78.

Glaessner, M. F., 1963, Zur Kenntnis der Nama-Fossilien Sudwest-Afrikas, *Ann. Naturhist. Mus. Wien* **66**:113–120.

Glaessner, M. F., 1969, Trace fossils from the Precambrian and basal Cambrian, *Lethaia* **2**(4):369–393.

Glaessner, M. F., 1976a, Early Phanerozoic annelid worms and their geological and biological significance, *J. Geol. Soc. Lond.* **132**(3):259–275.

Glaessner, M. F., 1976b, A new genus of Late Precambrian polychaete worms from South Australia, *Trans. R. Soc. S. Aust.* **100**(3):169–170.

Glaessner, M. F., 1978, Re-examination of *Archaeichnium*, a fossil from the Nama Group, *Ann. S. Afr. Mus.* **74**(13):335–342.

Glaessner, M. F., 1979a, An echiurid worm from the Late Precambrian, *Lethaia* **12**(2):121–124.

Glaessner, M. F., 1979b, Precambrian, in: *Treatise on Invertebrate Paleontology*, Part A, *Introduction. Fossilization (Taphonomy) Biogeography and Biostratigraphy* (R. A. Robinson and C. Teichert, eds.), Geological Society of America, and University of Kansas Press, Lawrence, Kansas, pp. A79–A118.

Glaessner, M. F., 1980, *Parvancorina*—An arthropod from the Late Precambrian (Ediacarian) of South Australia, *Ann. Naturhist. Mus. Wien* **83**:83–90.

Glaessner, M. F., 1984, *The Dawn of Animal Life: A Biohistorical Study*, Cambridge University Press, Cambridge.

Glaessner, M. F., and Wade, M., 1966, The Late Precambrian fossils from Ediacara, South Australia, *Palaeontology* **9**(4):599–628.

Glaessner, M. F., and Wade, M., 1971, *Praecambridium*—A primitive arthropod, *Lethaia* **4**(1):71–77.

Glaessner, M. F., and Walter, M. R., 1975, New Precambrian fossils from the Arumbera Sandstone, Northern Territory, Australia, *Alcheringa* **1**(1):59–69.

Goldring, R., and Curnow, C. N., 1967, The stratigraphy and facies of the Late Precambrian at Ediacara, South Australia, *J. Geol. Soc. Aust.* **14**(2):195–214.

Gould, S. J., 1984, The Ediacaran experiment, *Nat. Hist.* **93**(3):14–23.

Gould, S. J., 1985, Treasures in a taxonomic wastebasket, *Nat. Hist.* **94**(12):22–33.

Graff, L. von, 1888, Die Annelidengattung *Spinther*, *Z. Wiss. Zool.* **46**:1–66.

Grant, S. W. F., 1990, Shell structure and distribution of *Cloudina*, a potential index fossil for the terminal Proterozoic. *Am. J. Sci.* **290-A**:261–294.

Gürich, G., 1929, Die altesten fossilien Sudafrikas, *Z. Prakt. Geol.* **37**(6):85.

Gürich, G., 1933, Die Kuibis-Fossilien der Nama-Formation von Sudwestafrika, *Palaeontol. Z.* **15**(2–3):137–154.

Harris, C. W. and Glover III, L., 1988, The regional extent of the ca. 600 Ma Virgilina deformation: Implications for stratigraphic correlation in the Carolina terrane, *Geol. Soc. Am. Bull.* **100**(2):299–310.

Hill, E., and Bonney, T. G., 1877, The Precarboniferous rocks of Charnwood Forest, *Q. J. Geol. Soc. Lond.* **33**:754–789.

Hofmann, H. J., 1987, Precambrian biostratigraphy, *Geosci. Can.* **14**(3):135–154.

Hofmann, H. J., 1988, An alternative interpretation of the Ediacaran (Precambrian) chondrophore *Chondroplon* Wade, *Alcheringa* **12**(4):315–318.

Hofmann, H. J., and King, A. F., 1979, Late Precambrian microfossils, southeastern Newfoundland, *Geol. Surv. Can. Curr. Res. B Pap.* **79-1B**:83–98.

Horstmann, V. E., Ahrendt, H., Clauer, N., and Porada, H., 1990, The metamorphic history of the Damara Orogen based on K/Ar data of detrital white micas from the Nama Group, Namibia, *Precambrian Res.* **48**(1/2):41–61.

Howchin, W., 1925, *The Building of Australia and the Succession of Life: With Special Reference to South Australia*, Part 1, Government Printer, Adelaide.

Hyman, L. H., 1940, *The Invertebrates: Protozoa through Ctenophora*, McGraw-Hill, New York.

Jenkins, R. J. F., 1981, The concept of an 'Ediacaran Period' and its stratigraphic significance in Australia, *Trans. R. Soc. S. Aust.* **105**(4):179–194.

Jenkins, R. J. F., 1984a, Ediacaran events: Boundary relationships and correlation of key section, especially in 'Armorica', *Geol. Mag.* **121**(6):635–643.

Jenkins, R. J. F., 1984b, Interpreting the oldest fossil cnidarians, *Palaeontogr.* **54**:95–104.

Jenkins, R. J. F., 1985, The enigmatic Ediacaran (late Precambrian) genus *Rangea* and related forms, *Paleobiology* **11**(3):336–355.

Jenkins, R. J. F., 1986, Are enigmatic markings in Adelaidean of Flinders Ranges fossil ice-tracks?, *Nature* **323**(6087):472.

Jenkins, R. J. F., 1989, The "supposed terminal Precambrian extinction event" in relation to the Cnidaria, *Mem. Ass. Australas Palaeontols* **8**:307–317.

Jenkins, R. J. F., 1991, The early environment, in: *Animal Life Without Oxygen* (C. Bryant, ed.), Chapman and Hall, London, pp. 38–64.

Jenkins, R. J. F., and Gehling, J. G., 1978, A review of the frond-like fossils of the Ediacara assemblage, *Rec. S. Aust. Mus.* **17**(23):347–359.

Jenkins, R. J. F., Plummer, P. S., and Moriarty, K. C., 1981, Late Precambrian pseudofossils from the Flinders Ranges, South Australia, *Trans. R. Soc. S. Aust.* **105**(2):67–83.

Jenkins, R. J. F., Ford, C. H., and Gehling, J. G., 1983, The Ediacara Member of the Rawnsley Quartzite: The context of the Ediacara assemblage (late Precambrian, Flinders Ranges), *J. Geol. Soc. Aust.* **30**(1):101–119.

Jones, J. B., and Fitzgerald, M. J., 1986, Silica-rich layering at Blanche Point, South Australia, *Aust. J. Earth Sci.* **33**(4):529–551.

Knoll, A. H., Hayes, J. M., Kaufman, A. J., Swett, K., and Lambert, I. B., 1986, Secular variation in carbon isotope ratios from Upper Proterozoic successions of Svalbard and East Greenland, *Nature* **321**(6073):832–838.

Krasnobaev, A. A., and Semikhatov, M. A., 1986, Geochronological scale of the Upper Proterozoic (Riphean and Vendian) of the USSR: The present state, in: *Metody isotopnoy geologii i geokhronologicheskaya shkala*, Nauka, Moscow, pp. 159–183 [in Russian].

Krogh, T. E., Strong, D. F., and Papezik, V. S., 1983, Precise U–Pb ages of zircons from volcanic and plutonic units in the Avalon Peninsula, *Geol. Soc. Am. Prog. Abstr.* **15**(3):135.

Krogh, T. E., Strong, D. F., O'Brien, S. J., and Papezik, V. S., 1988, Precise U-Pb zircon dates from the Avalon Terrane in Newfoundland, *Can. J. Earth Sci.* **25**(3):442–453.

Kröner, A., McWilliams, M. O., Germs, G. J. B., Reid, A. B., and Schalk, K. E. L., 1980, Paleomagnetism of late Precambrian to early Paleozoic mixtite-bearing formations in Namibia (South West Africa): The Nama Group and Blaubeker Formation, *Am. J. Sci.* **280**(9):942–968.

Landing, E., Narbonne, G. M., Myrow, P., Benus, A. P., and Anderson, M. M., 1988, Faunas and depositional environments of the Upper Precambrian through Lower Cambrian, southeastern Newfoundland, *Bull. N.Y. St. Mus.* **463**:18–52.

Makhnach, A. S., Veretennikov, N. V., and Shkurator, V. I., 1986, Vendian rocks of the western part of the East European Platform: Stratotype range, boundaries and principles of their establishment, *Geol. Mag.* **123**(4):349–356.

Manton, S. M., 1967, The polychaete *Spinther* and the origin of the Athropoda, *J. Nat. Hist.* **1**(1):1–22.

Mawson, D., and Sprigg, R. C., 1950, Subdivision of the Adelaide System, *Aust. J. Sci.* **13**(1):69–72.

McCartney, W. D., 1967, Whitbourne map-area, Newfoundland, *Geol. Surv. Can. Mem.* **341**:1–135.

McMenamin, M. A. S., 1986, The garden of Ediacara, *Palaios* **1**(2):178–182.

McMenamin, M. A. S., 1987, The emergence of animals, *Sci. Am.* **256**(4):84–92.

McMenamin, M. A. S. and McMenamin, D. L. S., 1990, *The Emergence of Animals: The Cambrian Breakthrough*, Columbia University Press, New York.

Misra, S. B., 1969, Late Precambrian (?) fossils from Southeastern Newfoundland, *Geol. Soc. Am. Bull.* **80**(11):2133–2139.

Misra, S. B., 1971, Stratigraphy and depositional history of late Precambrian coelenterate-bering rocks, southeastern Newfoundland, *Geol. Soc. Am. Bull.* **82**(4):979–987.

Mosely, J., and Ford, T. D., 1985, A stratigraphic revision of the Late Precambrian rocks of Charnwood Forest, Leicestershire, *Mercian Geol.* **10**(1):1–18.

Mount, J. F., 1989, Re-evaluation of unconformities separating the "Ediacaran" and Cambrian Systems, South Australia, *Palaios* **4**(4):366–373.

Mount, J. F., 1991, Re-evaluation of unconformities separating the "Ediacaran" and Cambrian Systems, South Australia: Reply, *Palaios* **6**(1):105–107.

Nedin, C., and Jenkins, R. J. F., 1991. Evaluation of unconformities separating the "Ediacaran" and Cambrian Systems, South Australia: Comment, *Palaios* **6**(1):102–105.

O'Brien, S. J., Wardle, R. J., and King, A. F., 1983, The Avalon Zone: A Pan-African terrane in the Appalachian orogen of Canada, *Geol. J.* **18**(3):195–222.

Papezik, V. S., 1982, Stop 0–4: Harbour Main Volcanics, in: *Field Guide for Avalon and Meguma Zones* (A. F. King, ed.), the Caledonide Orogen IGCP Project 27, NATO Advanced Study Institute Atlantic Canada, Memorial University, St. Johns, Newfoundland, pp. 42–43.

Pflug, H. D., 1970a, Zur Fauna der Nama-Schichten in Sudwest-Afrika. 1. Pteridinia, Bau und systematische Zugerhorigkeit, *Palaeontogr. Abt. A.* **134**(4–6):226–262.

Pflug, H. D., 1970b, Zur Fauna der Nama-Schichten in Sudwest-Africa. II. Rangidae, bau und systematische Zugehorigkeit, *Palaeontogr. Abt. A.* **135**(3–6):198–231.

Pflug, H. D., 1972a, Zur Fauna der Nama-Schichten in Sudwest-Afrika. 111. Erniettomorpha, beu und Systematik, *Palaeontologr. Abt. A.* **139**(4–6):134–170.

Pflug, H. D., 1972b, Systematik der jungprakambrischen Petalonamae Pflug 1970, *Palaeontol. Z.* **46**(1/2):56–67.

Pickering, K. T., 1983, Transitional submarine fan deposits from the late Precambrian Kongsfjord formation submarine fan, N. E. Finnmark, N. Norway, *Sedimentology* **30**(2):181–199.

Pickering, K. T., 1985, Kongsfjord turbidite system, Norway, in: *Submarine Fans and Related Turbidite Systems* (A. H. Bouma, W. R. Normark, and N. E. Barnes, ed.), Springer-Verlag, New York, pp. 237–244.

Preiss, W. V. (Compiler), 1987, The Adelaide Geosyncline—Late Proterozoic stratigraphy, sedimentation, palaeontology and tectonics, *Bull. Geol. Surv. S. Aust.* **53**:1–438.

Richter, R., 1955, Die altesten Fossilien Sud-Afrikas, *Senckenb. Lethaea* **36**(3/4):243–289.

Rose, E. R., 1952, Torbay map-area, Newfoundland, *Geol. Surv. Can. Mem.* **265**:1–64.

Runnegar, B., 1982a, Oxygen requirements, biology and phylogenetic significance of the late Precambrian worm *Dickinsonia*, and the evolution of the burrowing habit, *Alcheringa* **6**(3):223–239.

Runnegar, B., 1982b, The Cambrian explosion: Animals or fossils?, *J. Geol. Soc. Aust.* **29**(4):395–411.

Runnegar, B., 1991, Oxygen and the early evolution of the Metazoa, in: *Animal Life Without Oxygen* (C. Bryant, ed.), Chapman and Hall, London, pp. 65–87.

Russell, F. S., 1970. *The medusae of the British Isles II. Pelagic Scyphozoa with a supplement to the first volume on Hydromedusae*, Cambridge University Press, Cambridge.

Seilacher, A., 1983, Paleozoic sandstones in southern Jordan: Trace fossils depositional environments and biogeography, in: *Geology of Jordan* (A. M. Abed and H. M. Khaled, eds.), Jordan Geologists Association, pp. 209–222.

Seilacher, A., 1984, Late Precambrian and early Cambrian Metazoa: Preservational or real extinctions?, in: *Patterns of Change in Earth Evolution* (H. D. Holland and A. F. Trendall eds.), Springer-Verlag, Berlin, pp. 159–168.

Seilacher, A., 1985, Discussion of Precambrian metazoans, *Philos. Trans. R. Soc. Lond. B* **311**(1148):47–48.

Seilacher, A., 1989, Vendozoa: Organismic construction in the Proterozoic biosphere, *Lethaia* **22**(3):229–239.

Sokolov, B. S., 1952, On the age of the ancient sedimentary cover of the Russian Platform, *Izv. Akad. Nauk SSSR Ser. Geol.* **1952**(5):21–31 [in Russian].

Sokolov, B. S., and Fedonkin, M. A., 1984, The Vendian as the terminal system of the Precambrian, *Episodes* **7**(1):12–19.

Sprigg, R. C., 1947, Early Cambrian (?) jellyfishes from the Flinders Ranges, South Australia, *Trans. R. Soc. S. Aust.* **71**(2):212–224.

Sprigg, R. C., 1949, Early Cambrian "Jellyfishes" of Ediacara, South Australia and Mount John, Kimberley District, Western Australia, *Trans. Roy. Soc. S. Aust.* **73**(1):72–99.

Stanistreet, I. G., Kukla, P. A., and Henry, G. 1991, Sedimentary basinal responses to a Late Precambrian Wilson Cycle: The Damara Orogen and Nama Foreland, Namibia, *J. African Earth Sci.* **13**(1):141–156.

Strong, D. F., O'Brien, S. J., Taylor, S. W., Strong, P. G., and Wilton, D. H., 1978, Aborted Proterozoic rifting in eastern Newfoundland, *Can. J. Earth Sci.* **15**(1):117–131.

Sun, W. G., 1986a, Precambrian medusoids: The *Cyclomedusa* plexus and *Cyclomedusa*-like pseudofossils, *Precambrian Res.* **31**(4):325–360.

Sun, W. G., 1986b, Late Precambrian pennatulids (sea pens) from the eastern Yangtze Gorge, China: *Paracharnia* gen. nov. *Precambrian Res.* **31**(4):361–375.

Sun, W. G., 1986c, Late Precambrian scyphozoan medusa *Mawsonites randellensis* sp. nov. and its significance in the Ediacara metazoan assemblage, South Australia, *Alcheringa* **10**(3):169–181.

Tankard, A. J., Jackson, M. P. A., Eriksson, K. A., Hobday, D. K., Hunter, D. R., and Minter, W. E. L., 1982, *Crustal Evolution of Southern Africa: 3.8 Billion Years of Earth History,* Springer-Verlag, New York.

Termier, H., and Termier, G., 1968, *Évolution et biocinèse. Les invertébrés dans l'histoire du monde vivant,* Masson, Paris.

Wade, M., 1968, Preservation of soft-bodied animals in Precambrian sandstones at Ediacara, South Australia, *Lethaia* **1**(3):238–267.

Wade, M., 1972a, Hydrozoa and Scyphozoa and other medusoids from the Precambrian Ediacara fauna, South Australia, *Palaeontology* **15**(2):197–225.

Wade, M., 1972b, *Dickinsonia:* Polychaete worms from the late Precambrian Ediacara fauna, South Australia, *Mem. Queensl. Mus.* **16**(2):171–190.

Watts, W. W., 1947, *Geology of the Ancient Rocks of Charnwood Forest,* Leicester Literary and Philosophical Society, Leicester, England.

Wilde, P., 1987, Model of progressive ventilation of the Late Precambrian–Early Paleozoic ocean, *Am. J. Sci.* **287**(5):442–459.

Williams, H., and King, A. F., 1979. Trepassey map-area, Newfoundland, *Geol. Surv. Can. Mem.* **389**:1–24.

Yakobson, K. E., 1987, Vendian strata in their type area, *Geol. Mag.* **124**(1):73–78.

Chapter 6

The Record of Trace Fossils across the Proterozoic–Cambrian Boundary

THOMAS PETER CRIMES

1. Introduction

Numerous studies have shown that there is a dramatic increase in the abundance and diversity of trace fossils at about the base of the Cambrian (Seilacher, 1956; Daily, 1972; Crimes, 1975, 1987; Alpert, 1977; Fedonkin, 1980a,b; Brasier, 1982; Urbanek and Rozanov, 1983; Crimes and Anderson, 1985; Fritz and Crimes, 1985; Nowlan *et al.*, 1985; Paczesna, 1986; Narbonne *et al.*, 1987). Accordingly, the possibility of using trace fossils to define or correlate the base of the Cambrian System has been widely discussed (Daily 1972; Alpert, 1977; Fedonkin, 1980a; Crimes and Anderson, 1985; Crimes, 1987; Narbonne *et al.*, 1987).

THOMAS PETER CRIMES • Department of Earth Sciences, University of Liverpool, Liverpool L69 BBX, England.

Origin and Early Evolution of the Metazoa, edited by Jere H. Lipps and Philip W. Signor. Plenum Press, New York, 1992.

Trace fossils, being markings left by animals moving on or within the sediment during or shortly after deposition, have a form which is therefore a function of (1) the size and shape of the animal and (2) its behavior. In some cases, the trace fossil closely reflects the shape of the animal which produced it. This is true of the trilobite resting traces (*Rusophycus*) and is beautifully illustrated in material from the Upper Cambrian of the Holy Cross Mountains, Poland, where the precise form of the undersurface of the trilobite can be seen (Radwanski and Roniewicz, 1963; Orlowski *et al.*, 1970). Similarly, the resting traces of echinoids show distinctive features of their producers (Smith and Crimes, 1983). In examples of this type, the evolutionary development of the animals would be reflected in the trace fossils and any increase in animal diversity would lead to an increase in trace fossil diversity. Most trace fossils, however, do not clearly reflect the form of their producers and are more a response to behavior. Any worm or wormlike animal can make a wide variety of trace fossils, dependent on the path of its trail or burrow. A simple nonbranching straight burrow would classify as *Planolites*, whereas a similar but tightly meandering burrow might fall within *Helminthoida*. Increase in diversity of trace fossils might, at least in part, reflect evolution of behavior patterns. Such an increase cannot therefore be taken to imply a directly equivalent increase in animal diversity. The evidence for increase in trace fossil and animal diversity needs to be considered quite separately.

The purpose of this chapter is first to identify and discuss the trace fossil diversity increase in selected Proterozoic–Early Cambrian sections, worldwide, and second to analyze this increase in terms of the evolution of behavior patterns.

2. The Worldwide Trace Fossil Diversity Increase during the Late Proterozoic and Phanerozoic

An increase in trace fossil diversity from Precambrian to Cambrian rocks was first suggested by Seilacher (1956). He drew particularly from his own observations in the Grand Canyon, Arizona, and Salt Range, Pakistan, and concluded that Precambrian traces were rare, but by the Cambrian, traces were much more abundant and showed more complex behavior patterns. He saw this as proof of an "explosive evolution" at the beginning of Cambrian time and suggested that where other fossils are not preserved, trace fossils may be used to determine the lower boundary of the Cambrian System.

Changes in trace fossil diversity through Phanerozoic time were considered by Crimes (1974) and Seilacher (1974). Figure 1 has been constructed from the data contained in Table 2 of Crimes (1974). It shows a low diversity in the Precambrian, with only four ichnogenera recognized, but a rapid increase in the Cambrian to a total of 33. There is a further small increase in the Ordovocian, but thereafter the overall diversity remains approximately constant. The Precambrian includes only inchnogenera of shallow-water or facies-crossing types. The diversification in the Cambrian is also largely the result of an increase in the number of these same types. The number of shallow-water inchnogenera then remains approximately constant or decreases slightly from the Cambrian throughout Phanerozoic time. The diversity of deep-water trace fossils is low in the Cambrian, increases in the Ordovician, and then remains constant until the Cretaceous, when there is a significant increase.

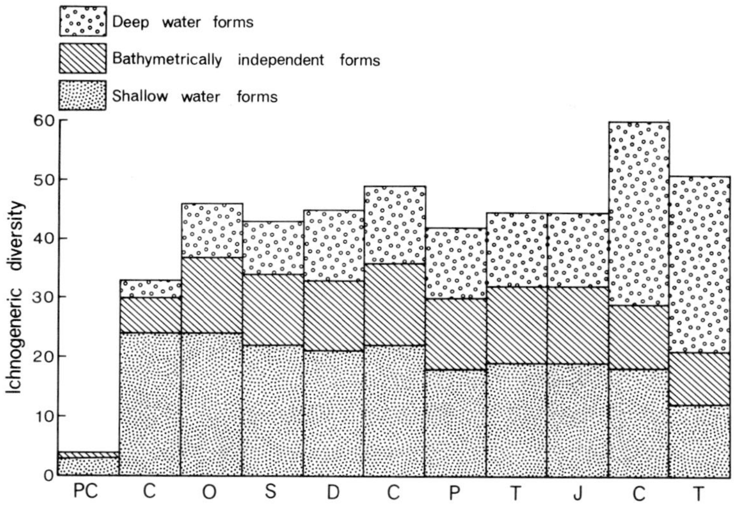

Figure 1. Ichnogeneric diversity from the Precambrian to the Tertiary. [Data from Crimes (1974, Table 2).]

Life apparently first developed in shallow-water environments and only gradually was there any significant colonization of the deep oceans. Crimes (1974, p. 329) suggested that the initial colonization of the deep oceans may have been stimulated by high dispersal pressures in the shallow-water niches. He pointed out, however, that trace fossil abundance and diversity indicated that such pressures reached a high level as early as Lower Cambrian times, yet significant colonization of the deep oceans was apparently delayed until the Ordovician, when dispersal pressures did not appear to have been much higher. This delay may have been the result of low oxygen concentration in the early deep seas or it might reflect inadequate supplies of organic detritus within the muds of the early deep ocean floor.

The deep-sea ichnospectrum is dominated by tightly meandering, spiral, and patterned traces, many of which indicate a carefully programmed behavioral response. In the past, these responses, and therefore the resultant traces, were assumed to have evolved in the deep sea. *Nereites* and *Helminthoida* were analyzed behaviorally by Seilacher (1967) and Raup and Seilacher (1969) and interpreted in terms of evolutionary optimization of available food (Papentin, 1973; Seilacher, 1974). Seilacher (1974) suggested that complex meanders did not appear before the Late Mesozoic and he showed examples, such as *Nereites*, with only simple meanders in the Cambrian, tighter meanders in the Devonian, and more regular and tighter ones by the Cretaceous. There is, however, a paradox in the argument for such evolutionary development in the deep seas. Why would such a closely programmed response develop in the deep sea, where suitable food-rich niches extend over very large areas? A "worm," for example, could probably acquire adequate food by traveling in a straight line all its life.

The clue to this dilemma may, however, come from work on shallow-water late Precambrian and Early Cambrian successions. Crimes and Anderson (1985) reported that some trace fossils of deep-water type occur in a shallow-water

Precambrian–Cambrian succession in the Burin Peninsula, Newfoundland, and this has been confirmed by subsequent studies (Narbonne *et al.*, 1987). The peritidal and shelf deposits of the Chapel Island Formation have examples of the "deep-water" ichnogenera *Helminthoida crassa* Shafhäutl, *Helminthopsis*, *Hormosiroidea*, and *Taphrhelminthopsis*, while the sand-bar and channel-fill deposits of the overlying Random Formation have *Paleodictyon* and *Squamodictyon*. *Helminthoida crassa*, *Paleodictyon*, and *Squamodictyon* have never been reported from post-Cambrian shallow-water sequences anywhere in the world, but are common in later deep-water flysch successions. This indicates that these behavioral patterns evolved in shallow water in late Precambrian to Early Cambrian times and only later migrated to the deep oceans. Pickerill (1980, p. 1268) also notes a significant Ordovician radiation of behavioral diversity in the deep sea. Apparently during the Early Cambrian, all available shallow-water niches were occupied and quiet-water muddy areas provided a suitable site for animals to meander or tunnel close to the surface. The areal extent of quiet-water muddy niches is relatively low in shallow-water environments. This would therefore provide a stimulus to animals to graze their habitat efficiently and hence produce very regular meanders. This response is shown by the grazing trail illustrated by Glaessner (1969, Fig. 5c,d) from the late Precambrian Pound Quartzite of South Australia or the examples of *Helminthoida crassa* and *H. miocenica* Sacco figured by Crimes and Anderson (1985, Fig. 7) from the Early Cambrian Chapel Island Formation of Newfoundland. As animal diversity increased rapidly during the Cambrian and dispersal pressures built up, animals behaving in this way and adapted to living on or within mud would be advantageously placed to migrate to deeper-water environments where extensive areas of muddy substrate exist. Therefore not only did these behavior patterns start in shallow water, but also their evolutionary development was far more advanced in these environments by Early Cambrian times than was envisaged by Seilacher (1974). Contrary to his model, examples of *Helminthoida crassa* from Mesozoic–Tertiary flysch successions show little advance over their counterparts in the Early Cambrian of Newfoundland.

3. The Diversity Increase in the Late Proterozoic and Early Cambrian

Recently, there has been much work on the trace fossils of Proterozoic–Cambrian boundary sequences and this is summarized in Crimes (1987).

In North America, there have been investigations in the White-Inyo Mountains, California (Alpert 1975, 1976, 1977), Esmeralda County, Nevada (Mount *et al.*, 1983), the Rocky Mountains, Canada (Young, 1972), Cassiar Mountains, British Columbia, Canada (Fritz and Crimes, 1985), Mackenzie Mountains, Yukon and North West Territory, Canada (Fritz, 1980), and Newfoundland, Canada (Crimes and Anderson, 1985; Narbonne *et al.*, 1987). Investigations were undertaken in East Greenland by Cowie and Spencer (1970) and in the English Midlands by Brasier *et al.* (1978) and Brasier and Hewitt (1979). Sections in north Spain were considered by Crimes *et al.* (1977), while stratigraphically equivalent rocks in the south were studied by Brasier *et al.* (1979), Perejon (1982), and Fedonkin *et al.*

(1983). Sequences in Finnmark, Norway, were described by Banks (1970) and in Poland by Fedonkin (1977) and Paczesna (1986). Extensive work has been undertaken in the USSR, principally in the White Sea and Baltic regions (Fedonkin 1976, 1977, 1979, 1980a; Urbanek and Rozanov, 1983). In India, there have been studies by Bhargava and Srikantia (1982), Shah and Sudan (1983), and Raina *et al.* (1983).

Sections in China, particularly that at Meishucun, near Kunming have been described, together with their trace fossils, by Jiang *et al.* (1982) and Crimes and Jiang (1986).

In the southern hemisphere, there has been work in Argentina (Aceñolaza, 1978; Aceñolaza and Toselli, 1981; Aceñolaza and Durand, 1983), Namibia (Germs, 1972; Crimes and Germs, 1982), and Australia (Glaessner, 1969, 1984; Daily 1972, 1973; Webby, 1970, 1984).

These studies have all demonstrated an increase in trace fossil abundance and diversity at about the Proterozoic–Cambrian boundary. Attempts to relate more precisely this diversity increase to the boundary are frustrated because no type section for the strata on either side of the boundary has been agreed and there are great problems in paleontological correlation at this level. The most widely used stratigraphic subdivisions are those employed in the USSR. In this scheme, the highest Proterozoic rocks are included within the Vendian, while the lowest Cambrian strata are referred to the Tommotian and the first trilobite-bearing rocks to the Atdabanian. For convenience in this chapter, Vendian strata will be considered as Proterozoic, Tommotian strata will be treated as transitional, with Atdabanian strata referred to as Early Cambrian. Even within the USSR, there are unresolved correlation problems. Traditionally, the Rovno Horizon, which is the lowest member of the Baltic Series, has been considered to be Tommotian and this view will be maintained here, but recently, without detailed discussion, Sokolov and Fedonkin (1984) place the Rovno Horizon as the terminal unit of the Vendian in the Russian Platform.

Correlations between sections in many parts of the world using trace fossils and supported by other criteria (e.g., small shelly fossils) have been tentatively made by Crimes (1987). Using these correlations, the following list of trace fossil occurrences can be compiled for worldwide strata of the Proterozoic–Cambrian boundary interval. In drawing up this list, trace fossils which have been recorded from a lower unit and also from a much higher level are assumed to be present in the intervening period. For example, *Conichnus* has been reported from the Lower Tommotian (Narbonne *et al.* 1987) and is also known from the Ordovician (Häntzschel, 1975, p. W52), and is therefore assumed to exist in the intervening period.

Vendian
Arenicolites, Bergaueria, Bilinichnus, Brooksella, Buthotrephis, Cochlichnus, Curvolithus, Didymaulichnus, Gordia, Harlaniella, Intrites, Nenoxites, Neonereites, Nereites, Palaeopascichnus, Planolites, Scolicia, Skolithos, Torrowangea, Vendichnus, Vimenites.

Lower Tommotian
Arenicolites, Bergaueria, Bilinichnus, Brooksella, Buthotrephis, Chondrites, Cochlichnus, Conichnus, Curvolithus, Didymaulichnus, Gordia, Gyrolithes,

Monomorphichnus, Muensteria, Neonereites, Nereites, Phycodes, Planolites, Protopaleodictyon, Scolicia, Skolithos, Teichichnus, Torrawangea, Treptichnus.

Upper Tommotian
Arenicolites, Astropolichnus, Belorhaphe, Bergaueria, Brooksella, Buthotrephis, Chondrites, Cochlichnus, Conichnus, Cosmorhaphe, Curvolithus, Didymaulichnus, Diplichnites, Diplocraterion, Gordia, Gyrolithes, Helminthoida, Helminthopsis, Hermosiroidea, Mammilichnis, Monomorphichnus, Muensteria, Neonereites, Nereites, Phycodes, Plagiogmus, Planolites, Protopaleodictyon, Rhizocorallium, Rusophycus, Scolicia, Skolithos, Taphrhelminthopsis, Teichichnus, Torrowangea, Treptichnus.

Atdabanian
Archaeonassa, Arenicolites, Arthrophycus, Asteriacites, Astropolichnus, Belorhaphe, Bergaueria, Brooksella, Buthotrephis, Chondrites, Cochlichnus, Conichnus, Cosmorhaphe, Cruziana, Curvolithus, Dactyloidites, Dictyodora, Didymaulichnus, Diplichnites, Dimorphichnus, Diplocraterion, Dolopichnus, Gordia, Gyrolithes, Halopoa, Helminthoida, Helminthopsis, Hormosiroidea, Isopodichnus, Laevicyclus, Mammilichnis, Monocraterion, Monomorphichnus, Muensteria, Neonereites, Nereites, Paleodictyon, Phycodes, Planolites, Protichnites, Protopaleodictyon, Rhizocorallium, Rosselia, Rusophycus, Scolicia, Skolithos, Squamodictyon, Syringomorpha, Taphrhelminthopsis, Teichichnus, Torrowangea, Treptichnus, Zoophycos.

No doubt as correlations are refined, there will be changes in this list, but utilizing the data, Fig. 2 shows the increasing ichnogeneric diversity across the boundary interval. Six ichnogenera are restricted to the Vendian and most of these (e.g., *Nenoxites, Vimenites*) are rather different from the longer-ranging Paleozoic types.

Each of the other horizons include only two ichnogenera which do not range up into younger strata. There is a steady increase in diversity from the Vendian

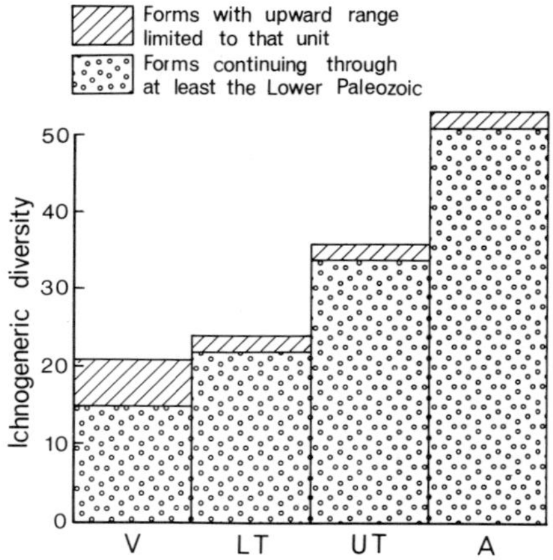

Figure 2. Ichnogeneric diversity across the Precambrian-Cambrian boundary interval; V, Vendian; LT, Lower Tommotian; UT, Upper Tommotian; A, Atdabanian.

through to the Atdabanian, without any clear "explosive evolution." The diversity in the Atdabanian (53 ichnogenera) is higher than that recorded in Fig. 1, which was based on older and more restricted data. Notwithstanding recent detailed investigations of the ichnofauna of younger Phanerozoic strata, later diversity increases in shallow-water facies are unlikely to match that seen in Proterozoic–Cambrian boundary sequences.

Details of this diversity increase will now be considered by reference to four localities where measured sections are available and an extensive ichnofauna has been recorded.

3.1. White-Inyo Mountains, California

One of the earliest well-documented examples showing a diversity increase at this level was in the White-Inyo Mountains, California, described by Alpert (1976, 1977).

The lowest units, the Wyman Formation and Reed Dolomite, yielded only the simple unbranching tubular burrows classified as *Planolites*.

The overlying Deep Spring formation had a more diverse ichnofauna consisting of *Diplichnites, Monocraterion, Monomorphichnus, Phycodes, Planolites, Rusophycus, Scolicia,* and *Skolithos*.

The lower member of the Campito Formation had additionally *Archaeonassa, Belorhaphe, Cochlichnus,* and *Helminthopsis*, while the upper member also had *Arthrophycus, Bergaueria, Cruziana, Teichichnus,* and *Zoophycos*.

The overlying Harkless Formation also yielded *Dolopichnus, Laevicyclus,* and *Psammichnites*.

This increase in trace fossil diversity takes place through a thickness of 3000 m of strata and is plotted graphically in Fig. 3 with the tentative correlations with the Russian sequence discussed in Crimes (1987) also shown.

The problem with this section is the extensive development of limestones and dolomites, two rock types which rarely yield trace fossils, in the Reed Dolomite and lower part of the Deep Spring Formation. The main diversity increase therefore appears to occur rather late in strata suggested to be of upper Tommotian and Atdabanian age, but this may well only be a reflection of the lack of suitable facies at lower levels in the section.

3.2. Cassiar Mountains, British Columbia, Canada

Recent work in the Cassiar Mountains by Fritz and Crimes (1985) included a measured section and trace fossil descriptions from the Stelkuz and Boya Formations.

The lowest investigated part of the Stelkuz Formation yielded only *Planolites* and *Gordia*, but *Chondrites* and *Helminthopsis* were found 37 m higher. *Didymaulichnus, Neonereites, Skolithos,* and *Taphrhelminthopsis* occur about 200 m above in strata suggested to be at about the Vendian–Tommotian boundary. A disconformity occurs about 100 m higher, followed by strata-bearing *Diplocraterion, Skolithos,* and *Teichichnus*, and then *Plagiogmus*. Approximately

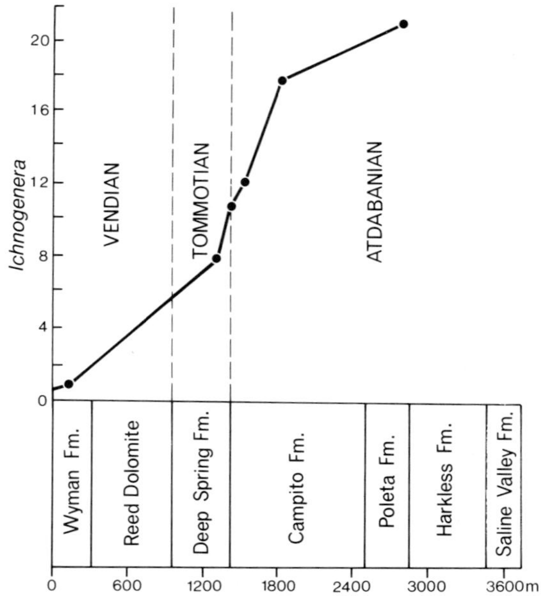

Figure 3. Ichnogeneric diversity across the Precambrian–Cambrian boundary interval in the White-Inyo Mountains, California. [Data from Alpert (1976, 1977).]

150 m above, in the Boya Formation, there is *Monomorphichnus*, and 125 m higher, *Phycodes* and *Treptichnus* have been found. Finally, *Cruziana* and *Rusophycus* occur near the top of the Boya Formation in rocks suggested to be of Atdabanian age and almost 1000 m above the first trace fossil recorded from the underlying Skelkuz Formation.

Ichnogeneric diversity is plotted graphically in Fig. 4 against the thickness of strata above the base of the measured section. A steady rise in diversity can be seen, but the position of the Vendian–Tommotian boundary in this section is rather uncertain and may be placed too high in the figure (Crimes, 1987, p. 107).

3.3. Burin Peninsula, Newfoundland, Canada

This area has yielded the most extensive suite of trace fossils of any Proterozoic–Cambrian boundary sequence (Fig. 5). Details have been given in Crimes and Anderson (1985) and Narbonne *et al.* (1987). The succession consists of the Chapel Island Formation, approximately 1000 m thick and divided into five members, and the overlying Random Formation, about 175 m thick.

Member 1 of the Chapel Island Formation is considered to be of Vendian age and has *Buthotrephis, Gordia, Harlaniella, Planolites*, and *Torrowangea*. There is a dramatic increase in diversity in Member 2A, which has additionally *Arenicolites, Bergaueria, Cochlichnus, Conichnus, Curvolithus, Didymaulichnus, Dimorphichnus, Gyrolithes, Helminthopsis, Monomorphichnus, Neonereites, Nereites, Palaeopascichnus, Palaeophycus, Phycodes, Protopaleodic-*

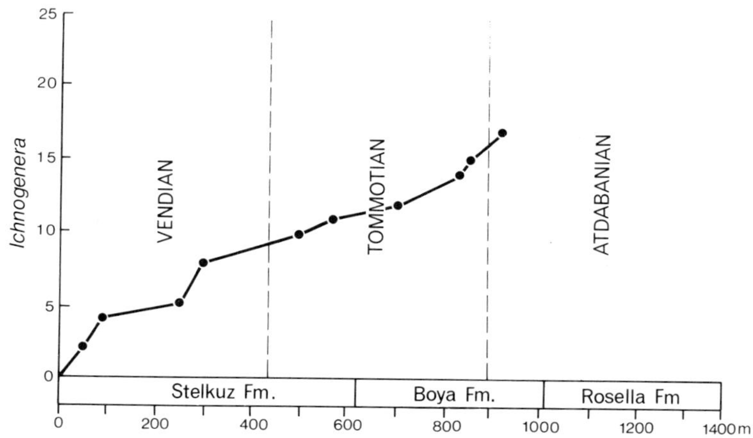

Figure 4. Ichnogeneric diversity across the Precambrian–Cambrian boundary interval in the Cassiar Mountains, British Columbia. [Data from Fritz and Crimes (1985).]

tyon, *Rusophycus*, *Skolithos*, *Taphrhelminthopsis*, and *Treptichnus*. The Vendian–Tommotian boundary is either at, or a few meters above, the base of Member 2A (Crimes and Anderson, 1985; Crimes, 1987; Narbonne *et al.* 1987). The increase in diversity therefore occurs at about this boundary.

Diversity continues to increase slowly through the section with the addition of *Helminthoida*, *Hormosiroidea*, and *Teichichnus* in Member 2B, *Astropolichnus*, *Diplocraterion*, and *Muensteria* in Member 3, and, finally, *Cruziana*, *Paleodictyon*, *Scolicia*, and *Squamodicyton* in the Random Formation. This is summarized in Fig. 5, where the diversity in each unit is plotted at the top of the

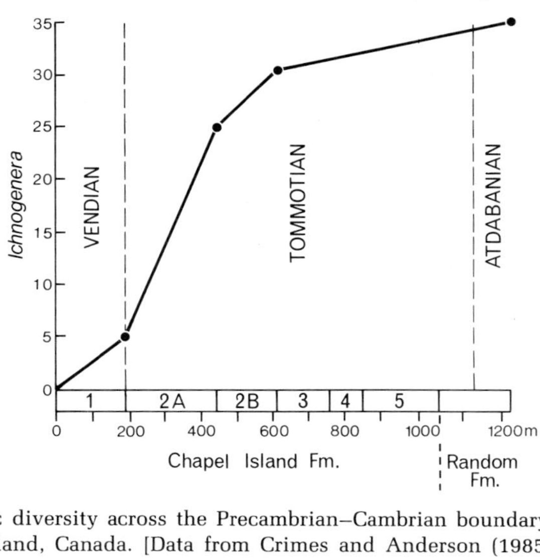

Figure 5. Ichnogeneric diversity across the Precambrian–Cambrian boundary interval in the Burin Peninsula, Newfoundland, Canada. [Data from Crimes and Anderson (1985) and Narbonne *et. al* (1987).]

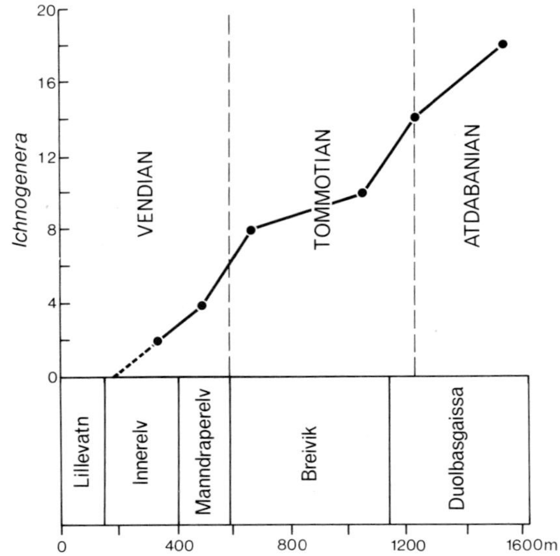

Figure 6. Ichnogeneric diversity across the Precambrian–Cambrian boundary interval in the Tanafjord district, Finnmark, north Norway. [Data from Banks (1970).]

unit because the data in Crimes and Anderson (1985) are only recorded on a member by member basis. The figure shows the sharp rise in diversity near the base of Member 2A. A preliminary facies analysis by Crimes and Anderson (1985, pp. 313–314) and a more detailed analysis reported in Narbonne *et al.* (1987) show that this diversity increase does not precisely coincide with any significant lithologic or known environmental change.

3.4. Tanafjord, Finnmark, Norway

This excellently exposed section is known through the work of Banks (1970).

The oldest trace fossils (*Arenicolites* and *Skolithos*) occur 200 m above the Vendian tillite in the middle of the Innerelv Member. The overlying Manndraperelv Member also has *Curvolithos* and *Planolites*. The Vendian–Tommotian boundary is probably at the junction between the Manndraperelv Member and overlying Lower Breivik Member, which has yielded *Cochlichnus, Phycodes, Treptichnus,* and *Rusophycus*. Further increases in diversity are reflected in the incoming of *Gyrolithes* and *Teichichnus* in the Upper Breivik Member; *Cruziana, Dimorphichnus, Diplocraterion,* and *Plagiogmus* in the Lower Duolbasgaissa Member; and finally, *Diplichnites, Rhizocorallium, Syringomorpha,* and radial burrows in the Upper Duolbasgaissa Member.

The diversity increase is illustrated in Fig. 6 and the tentatively inferred correlations with the Russian sequences are shown. The 1600-m-thick section appears to reflect a steady increase in diversity from the Vendian to the Atdabanian.

3.5. Steady or Explosive Evolution?

The foregoing sections provide conflicting evidence regarding the hypothesis of "explosive evolution" first suggested by Seilacher (1956) from his analysis of trace fossil diversity across the Precambrian–Cambrian boundary interval. It would be unwise to read too much into the results reported here, since the correlations are far from certain and the trace fossil collections in many sections are not yet very extensive. However, the picture apparent from the worldwide review in Fig. 2 is one of a steady increase in diversity from the Vendian to the Atdabanian. The same appears to be true from detailed work in the sections in the White-Inyo Mountains (Fig. 3), Cassiar Mountains (Fig. 4), and Finnmark (Fig. 6). In all these cases, the plot of diversity against thickness of strata above an arbitrary base approximates to a straight line. There is, of course, no reason to suppose that rate of sedimentation was constant in any of the sequences nor that all facies represented would support exactly the same diversity of trace. However, all three sequences are over 1000 m thick and composed predominantly of shallow-water clastic sediments. Therefore nothing suggests that rates of sedimentation or potential to support and preserve a wide diversity of trace fossils changed dramatically during deposition of the sequence. The conclusion based on these sections and the general worldwide analysis would therefore be that there may well have been a steady rise in trace fossil diversity across the Precambrian–Cambrian boundary.

The evidence from the Burin Peninsula (Fig. 5) is, however, somewhat different, with a rapid increase in diversity in Member 2A of the Chapel Island Formation. The diversity rises from 5 ichnogenera to 25 in less than 200 m of strata. Indeed, as the detailed work of Narbonne *et al.* (1987, Fig. 5) has shown, much of this increase in diversity occurs in the lowest 5 m of Member 2. No dramatic change in facies occurs at this point, and Narbonne *et al.* (1987, p. 1290) conclude that sedimentation rate was relatively uniform and essentially continuous across the Precambrian–Cambrian boundary. However, they do record phosphatic shale clasts at the contact between Members 1 and 2 (Narbonne *et al.* 1987, p. 1290) and this might indicate at least a slight pause in sedimentation.

The dramatic increase in diversity occurring at about the base of Member 2A is well down the succession. According to the small shelly fossils, the Tommotian–Atdabanian boundary is probably over 700 m above, in the Bonavista Formation (Bengtson and Fletcher, 1983; Narbonne *et al.*, 1987). The diversity increase is therefore likely to be at or below the base of the Tommotian. The Newfoundland section may therefore be out of step with the other areas in that there appears to be an earlier and more dramatic increase in diversity. In view of the broad similarity in the lithologic successions in all these sections, the influence of a factor other than substrate may have been important. One possibility would be for the climate in Newfoundland to have been more favorable. Recently, attempts have been made to reconstruct the late Precambrian and Early Cambrian continents (Donovan, 1987; Piper, 1987). Figure 7 shows the postulated situations of the continents in the Vendian (625 My ago) and at about the base of the Tommotian (575 My ago). Newfoundland moved from a latitude of 30° at 625 My ago to 25° at 575 My ago. Conversely, the sections in western Canada, including the Cassiar Mountains, moved from approximately the equator at 625 My ago to 60° at 575 My ago. Finnmark is suggested to have been at 40° at 625 My ago and 20° at 575 My ago.

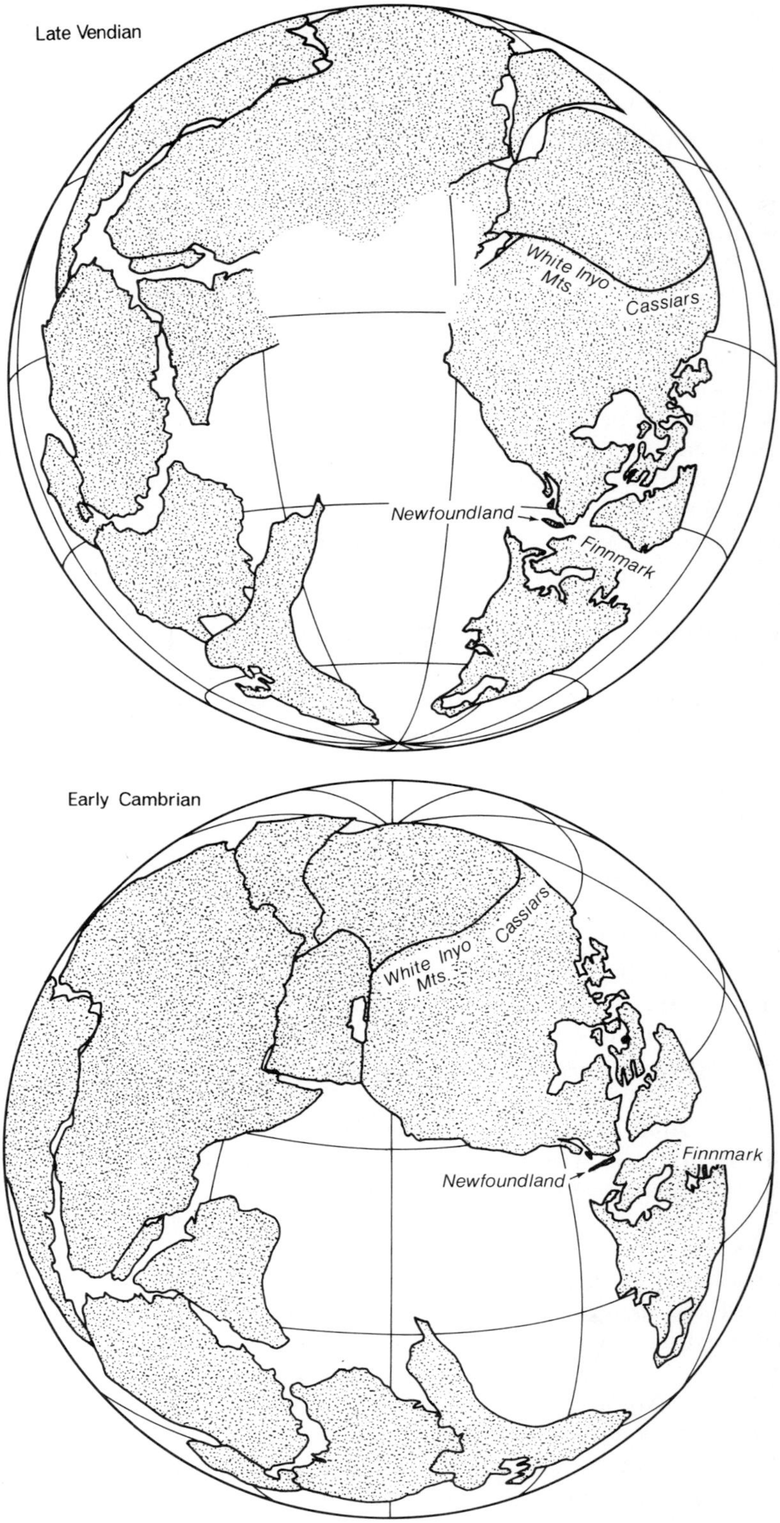

Figure 7. The postulated situations of the continents in the Vendian (625 My ago) and at about the base of the Tommotian (575 My ago). [From Piper (1987).]

The position of California is very conjectural, but it may have been 10° at 625 and then 60° at 575 My ago. These results would therefore place Newfoundland in the tropics at about the time when the dramatic increase in diversity took place. Animal abundance and diversity is highest in tropical latitudes (Fischer, 1960; Valentine, 1973). At this time, shortly after the Vendian worldwide glaciation, temperature gradients between the equator and poles would be high. The position of the Cassiar Mountains at the equator in Vendian times, if correct, could have provided appropriate conditions for a greater diversity increase there than in the other sections. To some extent this appears to be the case, with eight ichnogenera recorded, but the position of the Vendian–Tommotian boundary in this section is far from reliably established and any conclusions are speculative. It is, however, at least consistent with the available evidence to suggest that the most rapid diversity increase took place low down in the Cambrian succession in Newfoundland at a time when it may have been in low tropical latitudes. The animals producing these traces then migrated to the other areas during the remainder of the Tommotian and early Atdabanian.

4. Evolution of Trace Fossils across the Proterozoic–Cambrian Boundary

4.1. Introduction

The evolution of behavioral diversity in the oceans was first discussed by Crimes (1974) and Seilacher (1974). They showed independently that there was a rapid rise in trace fossil diversity at about the Proterozoic–Cambrian boundary, but neither analyzed the evolution of trace fossil types which contributed to this diversity increase.

Seilacher (1977) discussed the evolution of trace fossil communities through Phanerozoic time and inferred that gradually improving foraging programs and sensory monitoring led to more complex and perfect trace fossil search patterns. This improvement was claimed to take place over a long period so that, for example, *Nereites* showed imperfect, relatively open meanders in the Cambrian, closer, more carefully directed meanders in the Devonian, and tight meanders in the Cretaceous. Similar changes are deduced for *Scolicia* (Seilacher, 1977, Fig. 6). Fedonkin (1978, Fig. 2) showed an evolutionary progression from sinusoidal burrows (*Cochlichnus*) in the Cambrian, to relatively open meanders (*Helminthopsis*) in the Ordovician, to tighter meanders (*Parahelminthoida*) in the Carboniferous, to tight, nearly perfectly directed meanders (*Helminthoida*) by the Cretaceous. The concept that these burrow patterns developed gradually over hundreds of millions of years during the Phanerozoic was, however, contradicted by the discovery of tight, carefully directed meanders (*Helminthoida crassa*) in rocks of probable lower Tommotian age in Newfoundland (Crimes and Anderson, 1985).

Fedonkin (1978, 1980a, b) examined the evolution of behavior patterns across the Precambrian–Cambrian boundary interval. The Vendian has only simple feeding traces, including the open meandering *Nenoxites*, the tight meandering *Palaeopascichnus*, and the corkscrewing *Harlaniella*, while more complex branching burrows (*Chondrites, Phycodes, Treptichnus*), spiraling traces (*Gyrolithes*), and spreite burrows (*Teichichnus*) occur in the Tommotian. A complex evolutionary sequence of trace fossils was suggested (Fedonkin, 1978, Fig. 3). Similar changes were also inferred by Paczesna (1986) from her study of Upper Vendian and Lower Cambrian strata of southeastern Poland. She claimed that Vendian traces are simple and consist mainly of primitive feeding burrows (e.g., *Torrowangea*) with a few crawling traces (e.g., *Gordia*) and grazing traces (*Palaeopascichnus*). All were considered to be of one trophic type, deposit feeders, grazing on or within the sediment. Vertical traces were completely absent. The Lower Cambrian strata showed a rapid increase in ethological diversity with dwelling structures (e.g., *Bergaueria, Monocraterion*), feeding traces (e.g., *Treptichnus, Phycodes*), and crawling traces (e.g., *Bilinichnus*) all common. Two tropic groups were represented: suspension feeders and deposit feeders.

4.2. Evolutionary Development of the Main Phanerozoic Trace Fossil Lineages

Consideration of the incoming of different trace fossils in worldwide sequences allows a provisional picture of the evolutionary development of the main Phanerozoic trace fossil lineages (Fig. 8).

Simple unbranching sediment-filled burrows or possibly, in some cases, surface trails are generally included in the ichnogenus *Planolites*. They appear first in many sections worldwide and typically the earliest are small, with diameters of only a few millimeters. They are recorded widely in Vendian strata and increase in size significantly up into the Tommotian. Examples of this type of trace from the late Precambrian Pound Quartzite (Glaessner, 1969) are, however, relatively large. Pelleted traces such as *Neonereites* and *Torrowangea* also appear well below the top of the Vendian, although commonly a little above the simpler *Planolites*. Bilobed burrows such as *Didymaulichnus* often occur next and are followed by simple vertical burrows (*Skolithos*). Although these appear in the Vendian, deep forms, extending through more than a few centimeters, are only found higher up. The earliest coelenterate resting traces (*Intrites*) are also encountered in the Vendian, with the larger *Bergaueria* possibly not appearing until the Tommotian. Simple arthropod scratch marks (*Monomorphichnus*) probably exist in the uppermost part of the Vendian (Jenkins *et al.*, 1983), but are more common in younger rocks. More complex arthropod traces, such as *Rusophycus* and *Cruziana*, only appear in Tommotian to Atdabanian strata. Radiating traces (e.g., *Asterichnus*) come in at about the base of the Tommotian, at a similar horizon to the first development of true branching burrows (e.g., *Palaeophycus*). The much more complex spreite burrows are probably the last to evolve and include *Teichichnus*.

Each of these lineages shows the development of more complex forms as an evolutionary progression and this is considered below.

4.3. Evolutionary Development within Some of the Main Lineages

4.3.1. Simple Sediment-Filled Burrows

The development of these burrows is illustrated in Fig. 9. The very simple *Planolites* is a more or less straight burrow which ranges from at least the early

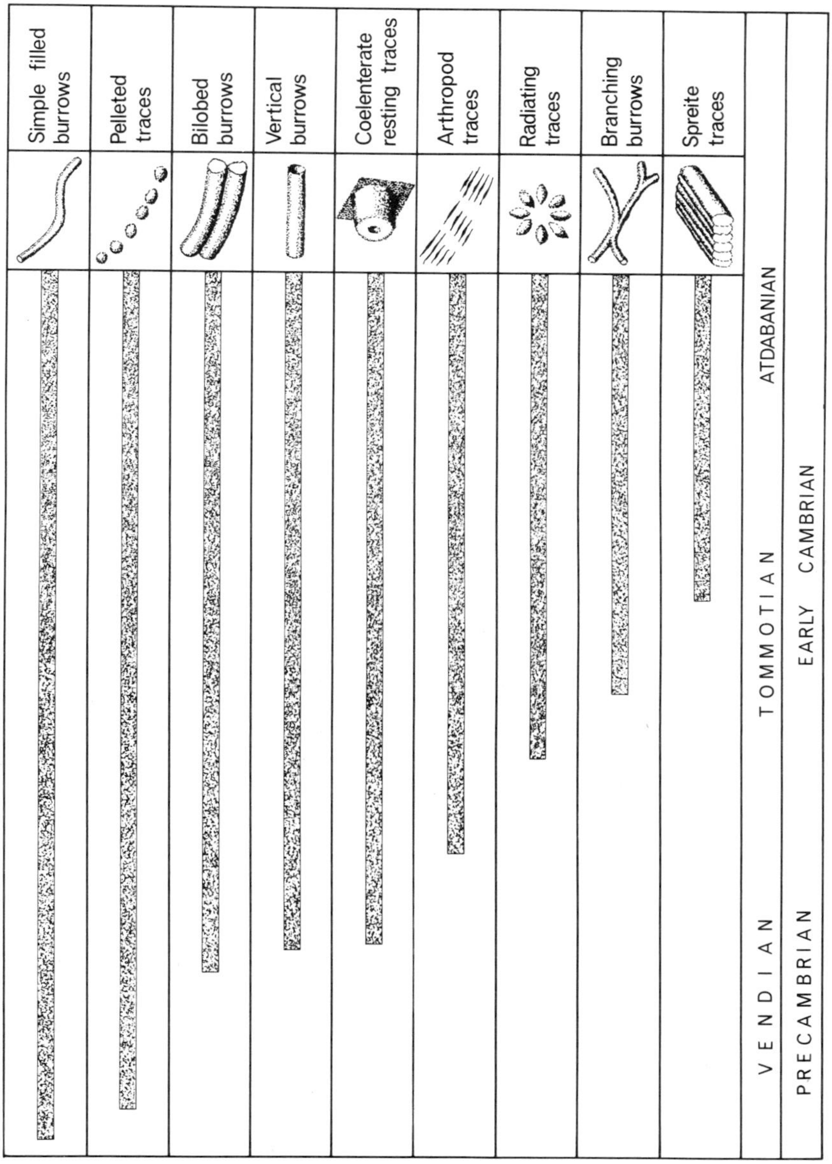

Figure 8. Evolutionary development of the main Phanerozoic trace fossil lineages across the Precambrian–Cambrian boundary interval.

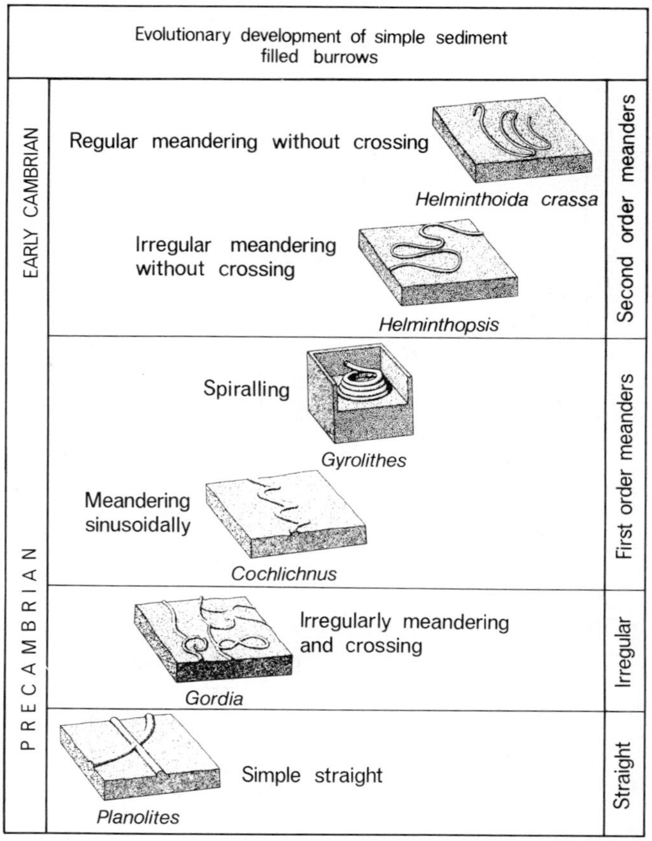

Figure 9. Evolutionary development of simple sediment-filled burrows across the Precambrian–Cambrian boundary interval.

Vendian and possibly earlier. Irregularly meandering burrows with crossings (e.g., *Gordia*) also occur in the Vendian and represent slightly more complex behavior. *Cochlichnus* shows regular sinusoidal first-order meanders and typically occurs first at about the Vendian–Tommotian boundary, followed by the tighter, vertically spiraling, first-order meanders of *Gyrolithes*. *Helminthopsis,* which consists of irregular but noncrossing first- and second-order meanders, normally occurs first well up in the Tommotian and is followed by the similar but more regular *Helminthoida crassa*. This evolutionary progression is well seen in the Burin Peninsula, Newfoundland (Crimes and Anderson, 1985; Narbonne *et al.* 1987), where *Planolites* occurs lowest, in the transitional beds between the Rencontre and Chapel Island Formation, *Gordia* is in Member 1 of the Chapel Island Formation, with *Cochlichnus* at the base of Member 2 and *Gyrolithes* slightly higher. Regular *Helminthopsis* and *Helminthoida crassa* are present in Member 3. The evolutionary progression therefore seems to be from more or less straight burrows, through irregular meanders with crossings, to regular first-order meanders, and then to second-order meanders, which represent the highest behavioral development.

4.3.2. Branching Burrows

The inferred path by which branching burrows evolved is shown in Fig. 10. Starting from simple nonbranching burrows (*Planolites*), there was development in the late Vendian of irregularly branching burrows usually classified as *Buthotrephis* or *Palaeophycus*. In the earliest Tommotian strata are found regular branching burrow systems, including *Phycodes pedum* Seilacher, *Phycodes coronatum* Crimes and Anderson, *Treptichnus*, and *Chrondites*. At about this time, the first irregular burrow networks (*Protopaleodictyon*) occur and these develop into more regular part networks (*Belorhaphe*). More complex regular networks (*Paleodictyon, Squamodictyon*) have so far only been found near the top of the Tommotian. The Vendian therefore only has irregularly branching burrows, but at about the base of the Tommotian regular branching burrow systems developed, followed by irregular and then regular networks. The branching burrow systems probably represent feeding burrows, but the later regular networks might indicate the maintenance of open tunnel systems (cf. Seilacher, 1977).

4.3.3. Complex Burrows

There are two forms of more complex burrows which are well represented in Proterozoic–Cambrian sequences (Fig. 11).

Sediment-stuffed burrows appear in the Vendian with *Neonereites*, which consist of either a single line of pellets (*N. uniserialis* Seilacher) or a double line (*N. biserialis* Seilacher). There are also lobate stuffed burrows (*Nereites*) in the Vendian Pound Quartzite of South Australia (Glaessner, 1969, p. 380, Fig. 5b) and the probably Vendian Nadaus Formation of Namibia (Crimes and Germs, 1982, pl. 2, Fig. 8). Although a burrow origin for these traces is most likely, possibly some could be trials formed at the sediment–water interface. The upper part of the Vendian has *Harlaniella* and this seems to be a stuffed corkscrewlike burrow. At about the Vendian–Tommotian boundary, the first bilobed stuffed burrows of *Scolicia* type appear. The irregularly meandering, bilobed, stuffed burrow *Taphrhelminthopsis circularis* Crimes normally only occurs in the upper part of the Tommotian, followed by the very complexely stuffed burrow *Plagiogmus*. Both of these burrows appear to have a range restricted to the Tommotian to Lower Atdabanian.

Spreite burrows are, in general, absent until the Tommotian. The simple vertical wall-like *Teichichnus* is normally first to appear. *Diplocraterion* is more complex, with its vertical U-tube with spreite preserved between the limbs of the U. It normally occurs slightly later than *Teichichnus*, although abberant and very small forms of *Diplocraterion* have been recorded from the lower part of the Schwarzrand Supergroup in Namibia which might have an upper Vendian age (Crimes and Germs, 1982). *Rhizocorallium* is similar to *Diplocraterion*, but with the U-tube horizontal. It is encountered first in younger rocks and has been reported from the Neobolus Beds in Pakistan (Seilacher, 1955) and the upper member of the Duolbasgaissa Formation in Finnmark, Norway (Banks, 1970), both above the appearance of the first trilobite and in rocks of Atdabanian age. More complex spreite structures such as *Dictyodora* and *Zoophycos* also appear well above the first trilobites. *Dictyodora* has been reported from the Neobulus Beds in Pakistan (Seilacher, 1955) and *Zoophycos* from the middle of the Campito Forma-

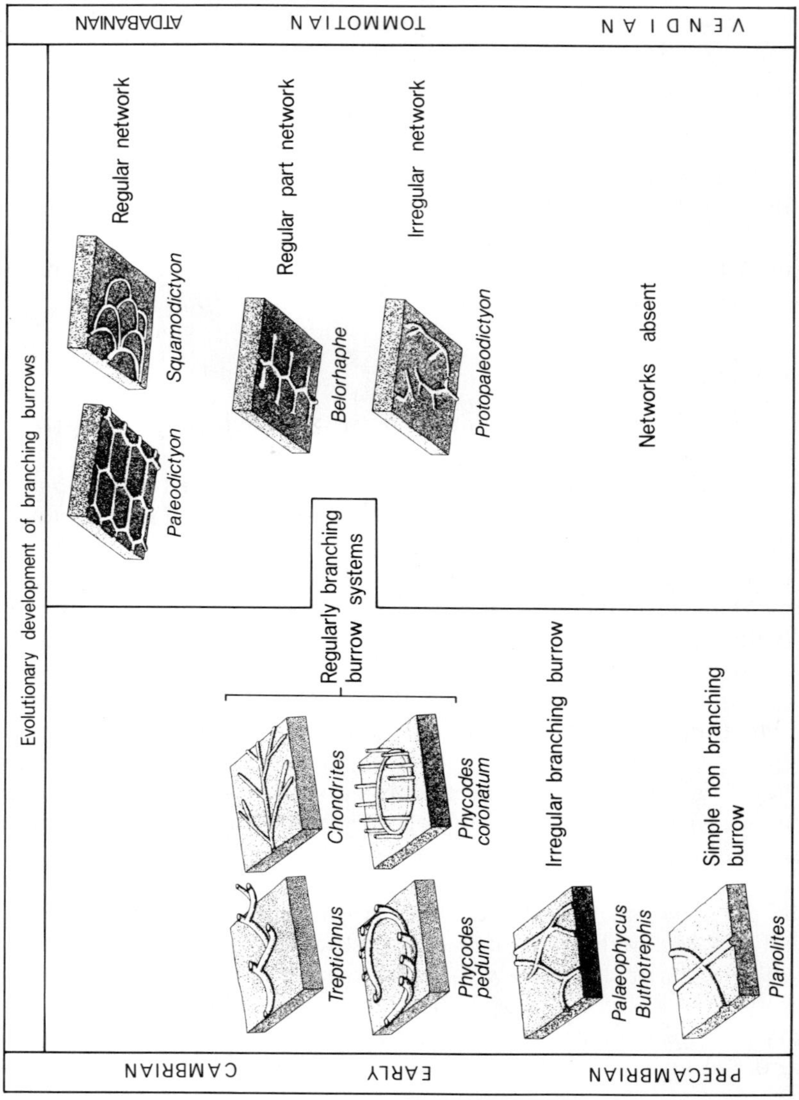

Figure 10. Evolutionary development of branching burrows across the Precambrian–Cambrian boundary interval.

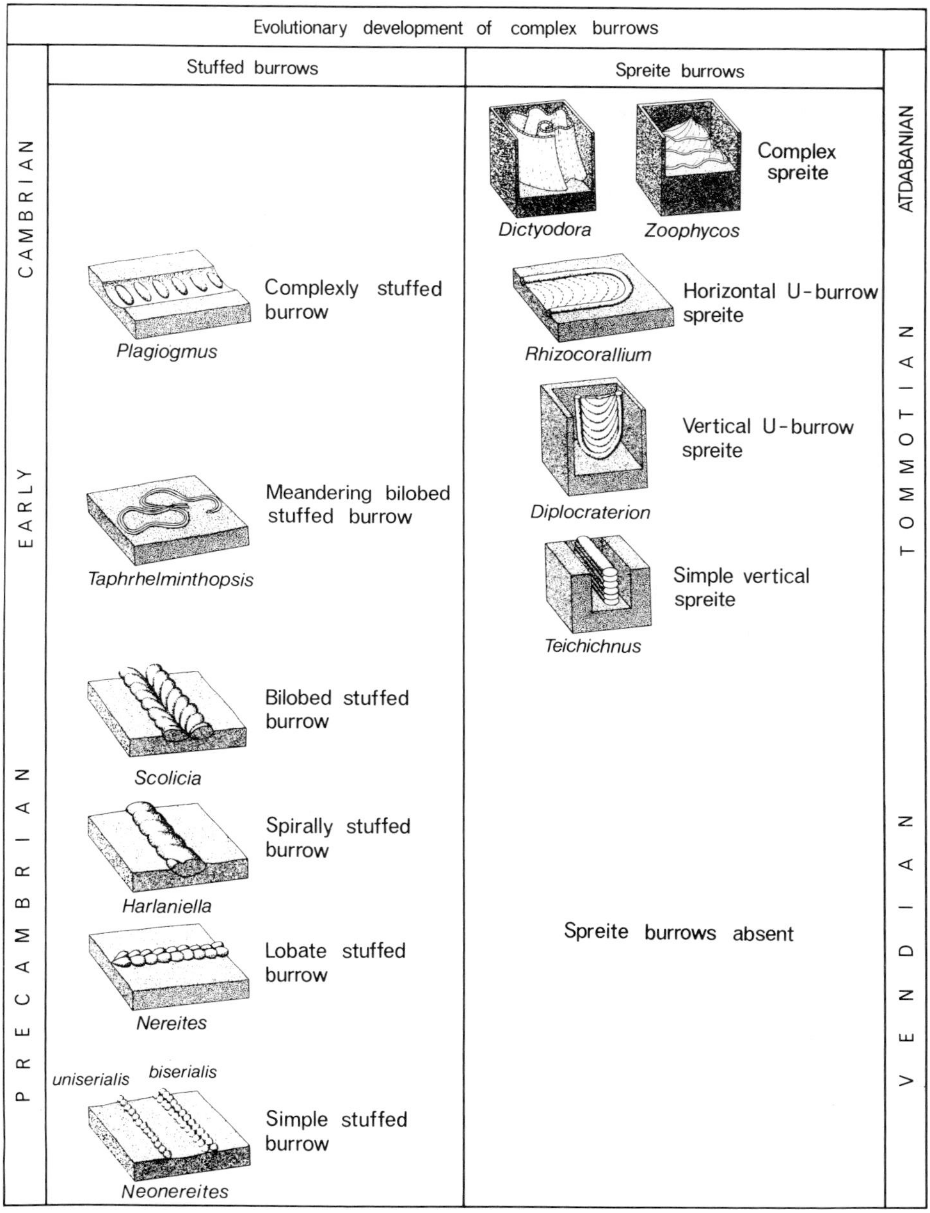

Figure 11. Evolutionary development of stuffed and spreite burrows across the Precambrian–Cambrian boundary interval.

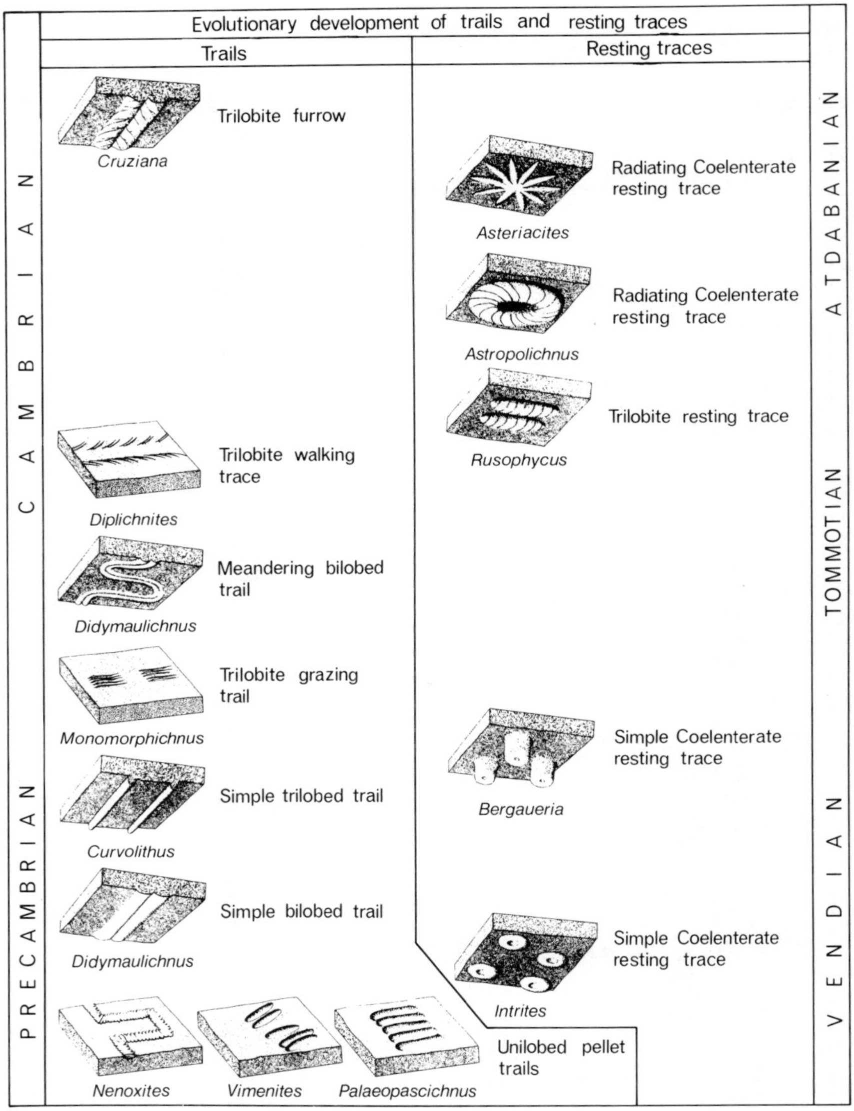

Figure 12. Evolutionary development of trails and resting traces across the Precambrian–Cambrian boundary interval.

tion of the White-Inyo Mountains, California, about 100 m above the first occurrence of the trilobite *Fallotaspis*.

4.3.4. Trails

Trails and furrows formed at the sediment–water interface (and illustrated in Fig. 12) are not always possible to distinguish from horizontal burrows which are mostly produced along sand–mud interfaces. The traces discussed in this section are, however, believed to have been made mostly or entirely at the sediment–water interface.

Trails of this type seem to have been a major part of the ichnospectrum in Vendian times, and *Nenoxites, Vimenites,* and *Palaeopascichnus* (Fig. 12) are restricted to rocks of this age (Sokolov and Ivanovskii, 1985). Simple forms of *Didymaulichnus* normally appear higher, near the top of the Vendian, as in the White Sea region (M. A. Fedonkin, personal communication 1981), Cassiar Mountains (Fritz and Crimes, 1985), and Rocky Mountains (Young, 1972), while the more specialized *Didymaulichnus miettensis* Young, with its lateral bevels, probably first occurs slightly higher, at about the Vendian–Tommotian boundary and dies out well below the first trilobite (Crimes, 1987, p. 102).

Curvolithus, which is simple trail with two sharp lobes and an intervening ribbonlike lobe, seems to occur immediately above the base of the Tommotian in Newfoundland (Crimes and Anderson, 1985) and also in the Mandraperelv Member in Finnmark, Norway (Banks, 1970, pl. 2d) considered to be at about the Vendian–Tommotian boundary (Crimes, 1987, Fig. 13).

The first trace attributed to arthropods, possibly forerunners of the trilobites, has been recorded as the grazing trail *Monomorphichnus* from the late Vendian Rawnsley Quartzite of South Australia by Jenkins *et al.* (1983) and I have seen similar specimens in the Pound Quartzite.

Regularly meandering forms of *Didymaulichnus* have been found by B. S. Sokolov (personal communication 1985) in the lower Tommotian Kessinsa Formation of Yakutia, USSR.

Trilobite walking trails (*Diplichnites*) normally appear slightly higher, usually in the upper Tommotian (Crimes, 1987). The trilobite furrow *Cruziana* is commonly only recorded not far below the lowest trilobite in rocks of Lower Atdabanian age (Crimes, 1987).

4.3.5. Resting Traces

The earliest resting traces of Metazoa (Fig. 12) are probably the small (1–5 mm diameter) sac-like bodies with a minute well-defined circular central depression classified as *Intrites.* They have been recorded from the upper Vendian Redkino Horizon in the White Sea region of the USSR (Fedonkin, 1980a, b). Small *Bergaueria* recorded from the Kuibis Subgroup of Namibia by Crimes and Germs (1982) can also be ascribed to this ichnogenus.

These forms probably represent coelenterate resting traces, as does *Bergaueria,* which generally appears in the Tommotian (Crimes, 1987), although it has been recorded as low as the Vendian in the USSR (Sokolov and Ivanovskii, 1985). The first trilobite resting traces (*Rusophycus*) do not appear before the Upper Tommotian Lontova Horizon in the USSR (Sokolov and Ivanovskii, 1985). *Astropolichnus* is a radiating coelenterate trace with a central axial cylinder. It is a more complex than *Bergaueria* and comes in later, being most commonly encountered at about the Tommotian–Atdabanian boundary, although it may be somewhat lower in Newfoundland, where it has been recorded in Member 3 of the Chapel Island Formation, which might be of Lower Tommotian age (Crimes and Anderson, 1985). Other radiating traces normally appear higher, including *Asteriacites.* This has been recorded from the Harkless Formation of the White Inyo

Mountains (Alpert, 1976), a unit well above the first trilobite occurrence and probably of Late Atdabanian or younger age. It has, however, also been found in Unit 3 of the Zhongyicun Member at Meishucun, China (Crimes and Jiang, 1986) in rocks variously described as Vendian or Atdabanian.

5. Ethological Changes across the Proterozoic–Cambrian Boundary

Assigning trace fossils to one of Seilacher's (1953) ethological groups is by no means easy and this is particularly true of some of the rare, and in some cases poorly described or figured, Proterozoic or Early Cambrian ichnogenera. Some forms, such as *Didymaulichnus*, while generally considered to be crawling trails, might be burrows, while others, such as *Nereites*, might be a feeding burrow or alternatively a grazing burrow, depending on how well-ordered any meanders were. Notwithstanding these difficulties, Proterozoic and Early Cambrian traces have been assigned to ethological groups as follows:

Feeding traces: *Archaeonassa, Arthrophycus, Belorhaphe, Brooksella, Buthotrephis, Chondrites, Cochlichnus, Dictyodora, Gyrolithes, Halopoa, Harlaniella, Hormosiroidea, Muensteria, Neonereites, Nereites, Palaeopascichnus, Phycodes, Plagiogmus, Protopaleodictyon, Teichichnus, Torrowangea, Treptichnus, Zoophycos.*

Dwelling traces: *Arenicolites, Diplocraterion, Dolopichnus, Laevicyclus, Rhizocorallium, Rosselia, Skolithos.*

Resting traces: *Asteriacites, Astropolichnus, Bergaueria, Conichnus, Dactyloidites, Intrites, Isopodichnus, Mammillichnis, Rusophycus, Vendichnus.*

Crawling traces: *Bilinichnus, Cruziana, Curvolithus, Didymaulichnus, Diplichnites, Protichnites, Scolicia.*

Grazing traces: *Cosmorhaphe, Dimorphichnus, Gordia, Helminthoida, Helminthopsis, Monomorphichnus, Nenoxites, Paleodictyon, Squamodictyon, Syringomorpha, Taphrhelminthopsis.*

Thus, taking this population overall, there are 41% feeding traces, 12% dwelling traces, 17% resting traces, 12% crawling traces, and 19% grazing traces. However, more interesting to consider are the gradual changes in the ethological composition of the ichnofauna in passing from the Vendian to the Atdabanian. Taking the known trace fossil occurrences at the different stratigraphic levels as recorded in Section 3, the ethological changes can be determined (Table I).

TABLE I. Distribution of Ethological Groups

Ethological Group	Vendian	Lower Tommotian	Upper Tommotian	Atdabanian
Feeding	52%	57%	47%	40%
Dwelling	10%	9%	11%	15%
Resting	14%	9%	14%	15%
Crawling	14%	17%	11%	11%
Grazing	10%	9%	17%	19%

Feeding traces dominated the ichnospectrum in the Vendian and lower Tommotian, but then suffered a steady decline. This is a relatively simple type of activity not generally involving complex behavioral responses. Its dominance of the Vendian ichnospectrum is therefore to be expected. Its relative decline is a consequence of the gradual development of more highly evolved and complex behavior patterns, so that from the lower Tommotian to the Atdabanian there is a significant increase in dwelling, resting, and grazing traces. The ethological changes therefore reflect the evolutionary development of the Metazoa and their behavior patterns.

6. Conclusions

The main conclusions to be drawn from this account of trace fossils across the Proterozoic–Phanerozoic boundary interval are as follows.

1. In shallow-water niches, there was a steady increase in trace fossil abundance and diversity starting in the late Vendian and reaching very high levels by the early Atdabanian.

2. There was little further increase in shallow-water trace fossil diversity throughout the Phanerozoic.

3. The diversity increase may possibly have been slightly earlier and more rapid in Newfoundland, where paleomagnetic evidence suggests tropical latitudes which would provide excellent conditions for evolution and high population density.

4. Deep-water Phanerozoic trace fossils evolved in shallow water in Proterozoic to Early Cambrian seas and rapidly achieved a high level of sophistication, including carefully controlled behavioral responses. The efficiency of the basic behavior patterns improved little after the Early Cambrian.

5. The main trace fossil lineages evolved during the Vendian and Tommotian, with the more primitive lines (e.g., simple unbranched burrows, pelleted traces, bilobed burrows) appearing in the Vendian, but more advanced forms (e.g., branching or spreite burrows) not occurring until the Tommotian.

6. Each lineage shows an evolutionary progression with increasing complexity and more advanced behavioral programming. Most lineages achieved a high degree of behavioral perfection by Atdabanian times.

7. Despite the early and successful evolution of behavior patterns in shallow-water niches and the development of high dispersal pressures, trace-making animals did not significantly colonize deep-sea environments until Ordovician times.

Acknowledgment. I thank J. Lynch for redrawing the line diagrams.

References

Aceñolaza, F. G., 1978, El Paleozoico inferior de Argentina segun sus trazas fosiles, *Ameghiniana* **15**:15–64.

Aceñolaza, F. G., and Durand, F., 1983, Trazas fósiles del basamento cristalino del noroeste Argentino, *Bol. Asoc. Geol. Cordoba* **2**:45–55.

Aceñolaza, F. G., and Toselli, A. J., 1981, *Geologia del noroeste Argentino,* Universidad Nacional de Tucuman, Publication 1287.

Alpert, S. P., 1975, *Planolites* and *Skolithos* from the Upper Precambrian–Lower Cambrian, White Inyo Mountains, California, *J. Paleontol.* **49:**508–521.

Alpert, S. P., 1976, Trilobite and star-like trace fossils from the White Inyo Mountains, California, *J. Paleontol.* **50:**226–239.

Alpert, S. P., 1977, Trace fossils and the basal Cambrian boundary, in: Trace Fossils 2 (T. P. Crimes and J. C. Harper, eds.), *Geol. J. Spec. Issue* **9:**1–8.

Banks, N. L., 1970, Trace fossils from the Late Precambrian and Lower Cambrian of Finnmark, Norway, in: Trace Fossils (T. P. Crimes and J. C. Harper, eds.), *Geol. J. Spec. Issue* **3:**19–34.

Bengtson, S., and Fletcher, T. P., 1983, The oldest sequence of skeletal fossils in the Lower Cambrian of southeastern Newfoundland, *Can. J. Earth Sci.* **20:**525–536.

Bhargava, O. N., and Srikantia, S. V., 1982, *Taphrohelminthopsis circularissic* from ?Cambrian sediments of southeast Kashmir Valley, *J. Geol. Soc. India* **23:**406–407.

Brasier, M. D., 1982, Sea-level changes, facies changes and the Late Precambrian–Early Cambrian evolutionary explosion, *Precambrian Research* **17:**105–123.

Brasier, M. D., and Hewitt, R. A., 1979, Environmental setting of fossiliferous rocks from the uppermost Proterozoic–Lower Cambrian of central England, *Palaeogeogr. Palaeoclimatol. Palaeoecol.* **27:**35–57.

Brasier, M. D., Hewitt, R. A., and Brasier, C. J., 1978, On the Late Precambrian–Early Cambrian Hartshill Formation of Warwickshire, *Geol. Mag.* **115:**21–36.

Brasier, M. D., Perejon, A., and San Jose, M. A., 1979, Discovery of an important fossiliferous Precambrian–Cambrian sequence in Spain, *Estud. Geol. Inst. Inv. Geol. "Lucas Mallada"* **35:**379–383.

Cowie, J. W., and Spencer, A. M., 1970, Trace fossils from the Late Precambrian/Lower Cambrian of East Greenland, in: Trace Fossils (T. P. Crimes and J. C. Harper, eds.), *Geol. J. Spec. Issue* **3:**91–100.

Crimes, T. P., 1974, Colonisation of the early ocean floor, *Nature* **248:**328–330.

Crimes, T. P., 1975, The stratigraphical significance of trace fossils, in: *The Study of Trace Fossils* (R. W. Frey, ed.), Springer-Verlag, New York, pp. 109–130.

Crimes, T. P., 1987, Trace fossils and correlation of Late Precambrian and Early Cambrian strata, *Geol. Mag.* **124:**97–119.

Crimes, T. P., and Anderson, M. M., 1985, Trace fossils from Late Precambrian–Early Cambrian strata of southeastern Newfoundland (Canada): Temporal and environmental implications, *J. Paleontol.* **50:**310–343.

Crimes, T. P., and Germs, G. J. B., 1982, Trace fossils from the Nama Group (Precambrian–Cambrian) of southwest Africa (Namibia), *J. Paleontol.* **56:**890–907.

Crimes, T. P., and Jiang Zhiwen, 1986, Trace fossils from the Precambrian–Cambrian boundary candidate at Meishucun, Jinning, Yunnan, China, *Geol. Mag.* **123:**641–649.

Crimes, T. P., Legg, I., Marcos, A., and Arboleya, M., 1977, ?Late Precambrian–low Lower Cambrian trace fossils from Spain, in: Trace Fossils 2 (T. P. Crimes and J. C. Harper, eds.), *Geol. J. Spec. Issue* **9:**91–138.

Daily, B., 1972, The base of the Cambrian and the first Cambrian faunas, Centre Precambrian Res. Univ. Adelaide S. Aust. Spec. Pap. **1:**13–27.

Daily, B., 1973, Discovery and significance of basal Cambrian Uratanna Formation, Mt. Scott Range, Flinders Ranges, South Australia, *Search* **4:**202–205.

Donovan, S. K., 1987, The fit of the continents in the late Precambrian, *Nature* **327:**139–141.

Fedonkin, M. A., 1976, Traces of multicellular animals from the Valdai Series, *Izv. Akad. Nauk SSSR Ser. Geol.* **4:**129–132 [in Russian].

Fedonkin, M. A., 1977, Precambrian–Cambrian ichnocoenoses of the east European platform, in: Trace fossils 2 (T. P. Crimes and J. C. Harper, eds.), *Geol. J. Spec. Issue* **9:**183–194.

Fedonkin, M. A., 1978, Ancient trace fossils and the ways of behavioural evolution of mud eaters, *Palaeontol. J.* **12:**106–112.

Fedonkin, M. A., 1979, Paleoichnology of Precambrian and Early Cambrian, in: *Paleontology of Precambrian and Early Cambrian* (B. S. Sokolov, ed.), USSR Academy of Sciences, Leningrad, pp. 183–192 [in Russian].

Fedonkin, M. A., 1980a, Early stages of evolution of Metazoa on the basis of palaeoichnological data, *Izv. Akad. Nauk SSSR Ser. geol.* **2:**226–233 [in Russian].

Fedonkin, M. A., 1980b, Fossil traces of Precambrian Metazoa, *Izv. Akad. Nauk SSSR Ser. Geol.* **1:**39–46 [in Russian].

Fedonkin, M. A., Linan, E., and Perejon, A., 1983, Icnofosiles de las rocas precambrico–cambricas de la Sierra de Cordoba, Espana, *Bol. R. Soc. Espanola Hist. Nat. (Geol.)* **81**:125–138.

Fischer, A. G., 1960, Latitudinal variations in organic diversity, *Evolution* **14**:64–81.

Fritz, W. H., 1980, International Precambrian–Cambrian boundary working groups 1979 field study to Mackenzie Mountains, Northwest Territories, Canada, *Curr. Res. A Geol. Surv. Can. Pap.* **80-1A**:41–45.

Fritz, W. H., and Crimes, T. P., 1985, Lithology, trace fossils, and correlation of Precambrian–Cambrian boundary beds, Cassiar Mountains, north-central British Columbia, Canada, *Geol. Surv. Can. Pap.* **83-13**:1–24.

Germs, G. J. B., 1972, Trace fossils from the Nama Group, south-west Africa, *J. Paleontol.* **46**:864–870.

Glaessner, M. F., 1969, Trace fossils from the Precambrian and basal Cambrian, *Lethaia* **2**:369–393.

Glaessner, M. F., 1984, *The Dawn of Animal Life,* Cambridge University Press, Cambridge.

Häntzschel, W., 1975, Trace fossils and problematica, in: *Treatise on Invertebrate Paleontology: Part W, Miscellanea, Supplement 1* (C. Teichert, ed.), Geological Society of America and University of Kansas Press, Boulder, Colorado, and Laurence, Kansas.

Jenkins, R. J. F., Ford, C. H., and Gehling, J. G., 1983, The Ediacara Member of the Rawnsley Quartzite: The context of the Ediacara assemblage (Late Precambrian, Flinders Ranges), *J. Geol. Soc. Aust.* **30**:101–119.

Jiang Zhiwen, Luo Huilin, and Zang Shishan, 1982, Trace fossils of the Meishucun Stage (Lowermost Cambrian) from the Meishucun section in China, *Geol. Rev.* **28**:7–13.

Mount, J. F., Gevirtzman, D. A., and Signor, P. W., 1983, Precambrian–Cambrian transition problem in western North America: Part I, Tommotian fauna in the southwestern Great Basin and its implications for the base of the Cambrian System, *Geology* **11**:224–226.

Narbonne, G. M., Myrow, P. M., Landing, E., and Anderson, M. M., 1987, A candidate stratotype for the Precambrian–Cambrian boundary, Fortune Head, Burin Peninsula, southeastern Newfoundland, *Can. J. Earth Sci.* **34**:1277–1293.

Nowlan, G. S., Narbonne, G. M., and Fritz, W. H., 1985, Small shelly fossils and trace fossils near the Precambrian–Cambrian boundary in the Yukon Territory, Canada, *Lethaia* **18**:233–256.

Orlowski, S., Radwanski, A., and Roniewicz, P., 1970, The trilobite ichnocoenoses in the Cambrian sequence of the Holy Cross Mountains, in: Trace Fossils (T. P. Crimes and J. C. Harper, eds.) *Geol. J. Spec. Issue* **3**:345–360.

Paczesna, J., 1986, Upper Vendian and Lower Cambrian ichnocoenoses of Lublin region, *Bull. Inst. Geologicznego* **355**:31–47.

Papentin, F., 1973, A Darwinian evolutionary system, *J. Theor. Biol.* **39**:397–405.

Perejon, A., 1982, Problematica paleontological del limite Precambrico–Cambrico en Espana, *Real Acad. Cienc. Exactas Fis. Nat.* **1982**:95–109.

Pickerill, R., 1980, Phanerozoic flysch trace fossil diversity—Observations based on an Ordovician flysch ichnofauna from the Aroostook–Metapedia Carbonate belt of northern New Brunswick, *Can. J. Earth Sci.* **17**:1259–1270.

Piper, J. D. A., 1987, *Palaeomagnetism and the Continental Crust,* Wiley, New York.

Radwanski, A., and Roniewicz, P., 1963, Upper Cambrian trilobite ichnocoenosis from Wielka Wisniowka (Holy Cross Mountains, Poland), *Acta Palaeontol. Polon.* **8**:259–280.

Raina, B. K., Kumar, G., Bhargava, O. N., and Sharma, V. P., 1983, ?Precambrian–low Lower Cambrian ichofossils from the Lolab Valley, Kashmir, Himalaya, India, *J. Palaeontol. Soc. India* **28**:91–94.

Raup, D. M., and Seilacher, A., 1969, Fossil foraging behavior: Computer simulations, *Science* **166**:994–995.

Seilacher, A., 1953, Studien zur palichnologie. I Uber die methoden der palichnologie, *Neues Jahrb. Geol. Palaeontol. Abh.* **96**:421–452.

Seilacher, A., 1955, Spuren und facies in unterkambrium, in: Beitrage zur kenntnis des kambriums in der Salt Range (Pakistan) (O. H. Schindewolf and A. Seilacher eds.), *Akad. Wiss. Lit. Mainz Math.-Nat. Kl. Abh.* **10**:11–143.

Seilacher, A., 1956, Der Beginn des kambriums als biologische Wende, *Neues Jahrb. Geol. Palaeontol. Abh.* **103**:155-180.

Seilacher, A., 1967, Vorzeitliche Maanderspuren, in: *Die Strassen der Tiere* (H. Hediger ed.), Vieweg, Braunschweig, pp. 294–306.

Seilacher, A., 1974, Flysch trace fossils: Evolution of behavioural diversity in the deep-sea, *Neues Jahrb. Geol. Palaeontol.* **4**:233–245.

Seilacher, A., 1977, Evolution of trace fossil communities, in: *Patterns of Evolution As Indicated by the Fossil Record* (A. Hallam, ed.), Elsevier, Amsterdam, pp. 357–376.

Shah, S. K., and Sudan, C. S., 1983, Trace fossils from the Cambrian of Kashmir and their stratigraphic significance, *J. Geol. Soc. India* **24**:194–202.

Smith, A. B., and Crimes, T. P., 1983, Trace fossils formed by heart urchins—a study of *Scolicia* and related traces, *Lethaia* **16**:79–92.

Sokolov, B. S., and Fedonkin, M. A., 1984, The Vendian as the terminal system of the Precambrian, *Episodes* **7**:12–19.

Sokolov, B. S., and Ivanovskii, A. B., 1985, *The Vendian System*, Vol. 1, Academy of Sciences of the USSR, Moscow [in Russian].

Urbanek, A., and Rozanov, A. Yu., 1983, *Upper Precambrian and Cambrian Palaeontology of the East-European platform*, Wydawynictwa Geologiczna, Warsaw.

Valentine, J. W., 1973, *Evolutionary Paleoecology of the Marine Biosphere*, Prentice Hall, Englewood Cliffs, New Jersey.

Webby, B. D., 1970, Late Precambrian trace fossils from New South Wales, *Lethaia* **3**:79–109.

Webby, B. D., 1984, Precambrian–Cambrian trace fossils from western New South Wales, *Aust. J. Earth Sci.* **31**:427–437.

Young, F. G., 1972, Early Cambrian and older trace fossils from the southern Cordillera of Canada, *Can. J. Earth Sci.* **9**:1–17.

IV

Radiations of the Cambrian Faunas

Chapter 7

The Lower Cambrian Fossil Record of the Soviet Union

ALEXEI YU. ROZANOV and ANDREY YU. ZHURAVLEV

1. Introduction

Lower Cambrian deposits are found throughout the Soviet Union and Mongolia (Fig. 1). They occur over the entire Siberian Platform, much of the East European Platform, and widely in the fold belts bordering the southern Siberian Platform (the Altay-Sayan Fold Belt, Mongolia–Okhotsk Fold Belt), Kazakhstan, Central Asia, and the Urals. In most cases, they are conformably underlain by Vendian rocks. The most important sections containing Early Cambrian fossils are in Sibe-

ALEXEI YU. ROZANOV and ANDREY YU. ZHURAVLEV • Paleontological Institute of the Russian Academy of Sciences, Moscow, B-321, Russia.

Origin and Early Evolution of the Metazoa, edited by Jere H. Lipps and Philip W. Signor. Plenum Press, New York, 1992.

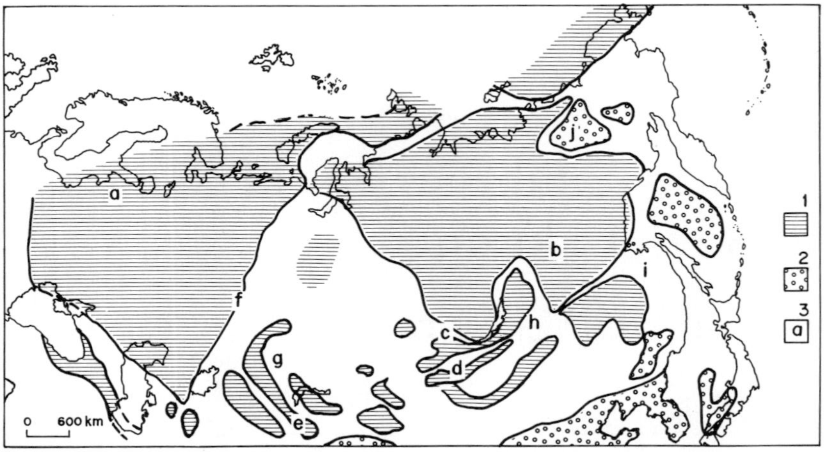

Figure 1. Tectonic scheme of the USSR and nearby territories [modified from Yanshin *et al.* (1984)]. (1, 2) Ancient cratons including framing mobile belts: (1) Laurasian affinity; (2) "Pacific" affinity. (3) The main regions of Cambrian sedimentation: (a) Baltic Syneclise of the East European Platform, (b) Lena–Aldan area of the Siberian Platform, (c) Altay-Sayan Fold Belt, (d) western Mongolia, (e) Maly Karatau (south Kazakhstan), (f) South Urals, (g) central Kazakhstan, (h) Transbaikalian, (i) Far East, (j) Kolyma Basin.

ria, the East European Platform and Kazakhstan in the USSR, and in western Mongolia. The key region for the Cambrian stages adopted in the USSR is the Siberian Platform.

2. Geographic Setting and Stratigraphy

2.1. The Siberian Platform

Three large Lower Cambrian facies (Fig. 2) are present on the Siberian Platform (Khomentovsky and Repina, 1965; Savitsky *et al.*, 1972; Jegorova *et al.*, 1976; Rozanov and Sokolov, 1984). The Turukhansk-Irkutsk-Olekma (Western) facies is characterized chiefly by lagoonal deposits, most of which are represented by dolomites, except in the Irkutsk region, where there are both dolomites and salts. The Yudoma-Olenek (Eastern) facies is typically off-shelf with bituminous limestones, black shales, and cherts, and, particularly in the Botomian, with bituminous carbonate shales and cherts with phosphate nodules. The first Cambrian bituminous rocks on the Siberian Platform and elsewhere occur at the end of the Atdabanian and beginning of the Botomian. The Anabar-Sinsk (Transitional) facies, located between the other two, is characterized by patch reefs and bioherms. The three facies cut across the Siberian Platform from northeast to southwest.

Rocks of the Western Region contain few fossils except for fairly rich faunas (commonly, trilobites) in rocks deposited at maximum transgressions. These faunas allow correlation of the western type sections with those of other regions (Khomentovsky and Repina, 1965; Krasnov *et al.*, 1983). In the Eastern Region, the upper part of the Lower Cambrian is characterized by trilobites, and the lower half

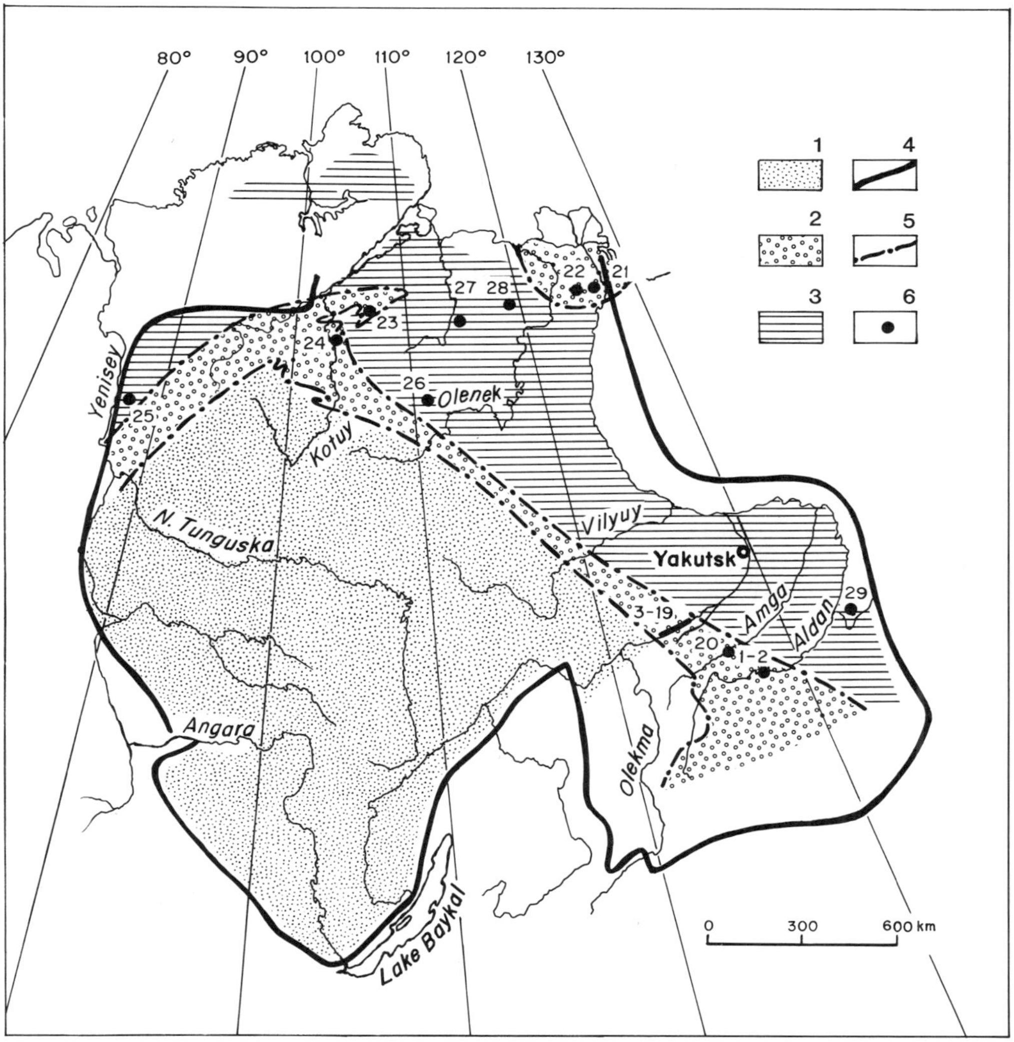

Figure 2. The Lower Cambrian facies regions of the Siberian Platform [modified from Savitsky and Astashkin (1979)]. (1) Turukhansk-Irkutsk-Olekma (Western) Facies Region (lagoonal deposits). (2) Anabar-Sinsk (Transitional) Facies Region (reef complex deposits). (3) Yudoma-Olenek (Eastern) Facies Region (bituminous off-shelf deposits), (4) Siberian Platform boundary, (5) reef belt boundary, (6) section numbers: 1–2 and 3–19, Aldan River and middle Lena River; 20, Amga River; 21, 22, Khara-Ulakh Mountains; 23, Khorbusuonka River; 24, Fomich River, Kotuy River, near the mouth of the Kotuykan River; 25, Sukharikha River; 26, Low Arga-Sala and Olenek Rivers; 27, Nekkekit River; 28, Bol'shaya Kuonamka River; 29, Yudoma River.

of the Lower Cambrian, especially the Tommotian, is represented, as in the Transitional Region, by red limestones. The Transitional Region is characterized by archaeocyanthan–algal bioherms, except in the east, where archaeocyaths are rare and shallow-water carbonate rocks typical of barrier reef environments are present (limestone, oolites, bioclastic rock, bioturbated carbonates, secondary dolomites). The distribution of these deposits and their relationship with the Western and Eastern Region deposits are exposed along the middle Lena River (Fig. 3). The

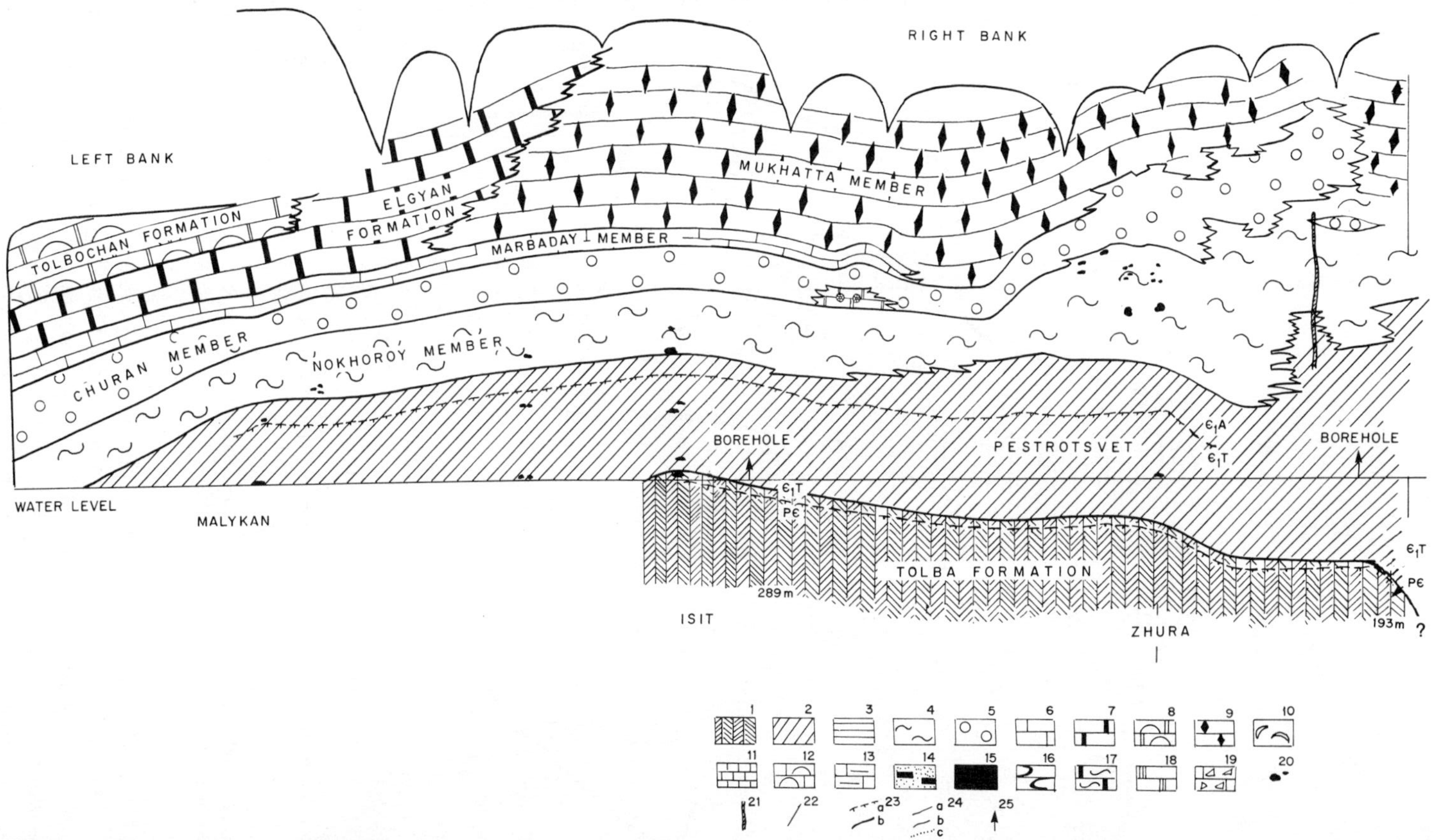

Figure 3. Geologic profile section of the middle Lena River area from the Tolba River to Elanskoe Village [modified from Astashkin *et al.* (1984)] (1) Tolba Formation: gray patch-laminated dolostone; (2) Pestrotsvet Formation: red and gray plane-parallel-laminated argillaceous limestone with dolomite bands, red and gray wavy and nodular bed limestone, red and yellow cross-parting dolostone; (4) Nokhoroy Member: gray wavy bed limestone and argillaceous dolostone, gray bedded limestone, yellow-gray bedded dolostone, and yellow-gray wavy bed dolostone; (5) Churan Member: light yellow oolitic and apoolitic clastic dolostone, interbedding of oolitic and fine-grained dolostones; (6) Marbaday Member: white calcareous sandstone

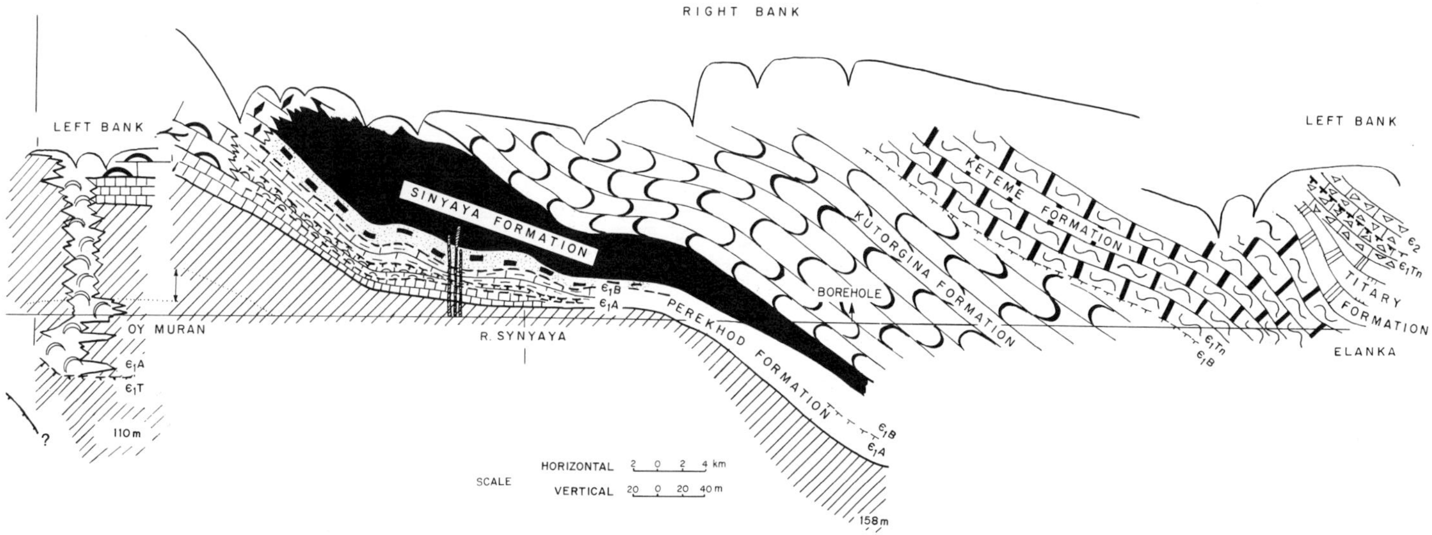

and algal detrital limestone, yellow and gray plane-parallel-laminated thin-parting dolostone; (7) El'gyan Formation: yellow-gray, massive, cavernous, weakly bituminous dolostone, interbedding of yellow-gray dolostone and gray-brown limestone; (8) Tolbachan Formation: interbedding of massive stromatolite and thin, plane-parallel-laminated dolostones, argillaceous dolostone, and dolomitic marl; (9) Mukhatta Member: coarse-grained massive dolostone with fine-grained, thin-laminated dolostone bands, yellow, coarse-grained, massive, cavernous dolostone with dolomitic sandstone and grainstone bands; (10) Oymuran Reef Massif: algal biohermal and biostromal limestone and dolostone, clastic and taphostromal limestone, dolomitic and calcareous archaeocyathan–algal bioherms; (11–14) Perekhod Formation: (11) Member I, greenish-gray and red bedded argillaceous limestone; (12) Member II, light-colored laminated limestone; (13) Member III, light-colored red and green argillaceous limestone; (14) Member IV, gray, dark-gray, and black, thin parallel-laminated limestone; (15) Sinsk Formation: dark gray and black, thin parallel-laminated bituminous limestone; (16) Kutorgina Formation: gray-brown bituminous limestone and dolostone; (17) Keteme Formation: white wavy bed limestone with dolomite bands and subunits; (18) Titary Formation: yellow, coarse-grained, cavernous, thick-bedded dolostone; (19) Elanskoe Formation: light-colored clastic limestone with dolomite bands and shell hash; (20) archaeocyanthan–algal and algal bioherm; (21) dolerite dike; (22) fault; (23) boundaries: (a) stage, (b) Oymuran Massif; (24) boundaries: (a) formation, (b) member, (c) marker band; (25) borehole. Stage index: $Є_1T$, Tommotian; $Є_1A$, Atdabanian; $Є_1B$, Botomian; $Є_1T_n$, Toyonian.

Lower Cambrian zones of the Transitional and Eastern Facies Regions are reliably correlated (Fig. 4).

Placement of the lower and upper boundaries of the Lower Cambrian varies. The boundary is usually put between the Lower and Middle Cambrian at the base of the *Schistocephalus* Zone, although it is also placed for convenience at the base of the *Anabaraspis splendens* Zone. The type sections of the Precambrian–Cambrian boundary in the southern part of the Siberian Platform (Aldan–Lena Rivers area) and the Anabar Massif and Olenek Uplift in the north of the Platform have been correlated differently (Fig. 5). The middle Aldan River section is the type section of the lower boundary of the Cambrian (Rozanov *et al.*, 1969; Cowie and Rozanov, 1983; Rozanov, 1984a; Rozanov and Sokolov, 1984) and the stratotype of the Tommotian Stage. The lower boundary lies at the base of the *Nochoroicyathus* (=*Aldanocyathus*) *sunnaginicus* Zone, which Fedorov (1982) divided into a less diverse lower part and a richer upper part. This boundary in the sections of the Anabar Massif, Olenek Uplift, and the Uchur-Maya Area is not easily defined. Zones with abundant skeletal fossils older than the Tommotian have been recognized in those sections (Missarzhevsky, 1982, 1983, 1989; Valkov, 1982, 1987). The problem is unresolved, but here we accept the correlation of the lower boundary of the *sunnaginicus* Zone with the beds underlying the Koril member, and equivalence of the Manykay Stage proposed by Missarzhevsky with the Nemakit-Daldyn Stage. The disagreement results from facies differences between small shelly fossil assemblages, because archaeocyaths are absent in the northern sections, and because the zones have been defined differently.

Early Cambrian Assemblage or Oppelian Zones were defined by Rozanov and Missarzhevsky (1966; Rozanov *et al.*, 1969) on the Siberian Platform in the Tommotian Stage using species ranges rather than the appearance of new taxa. However, Early Cambrian taxa gradually appear at the beginning of the Cambrian and continue for a long time and are not so useful in defining assemblage zones. The zones of the Siberian Platform were based chiefly on evolutionary changes in archaeocyaths and on occurrences of small shelly fossils, as were the zones of western Mongolia (Voronin *et al.*, 1982; Esakova, 1990; Zhegallo, 1990). Later schemes were based on various fossil groups (Missarzhevsky, 1982, 1983, 1989; Valkov, 1982, 1987; Sysoiev, 1968, 1972; Pelman, 1977, 1985). Missarzhevsky defined zones on different groups of small shelly fossils, but he ignored type sections for zones and boundaries. Therefore, his zonal boundaries change as correlations do, and the fossil composition of his zones varies accordingly.

At present, the most useful zones and precise boundaries are based on separate fossil groups. The zonal sequence, however, must be established in type sections where boundaries between successive zones can be determined. Thus, on the Siberian Platform, we use only those zonal subdivisions where type sections are designated. On the Siberian Platform, sections in the Transitional Region where facies vary greatly are most useful for correlation (Rozanov and Sokolov, 1984) because of favorable conditions for maximal development and diversification of many groups. This region was the center of origin of some groups (for example, archaeocyaths), and hence, the subdivisions of the Siberian Platform are recognizable elsewhere in the world (Fig. 6).

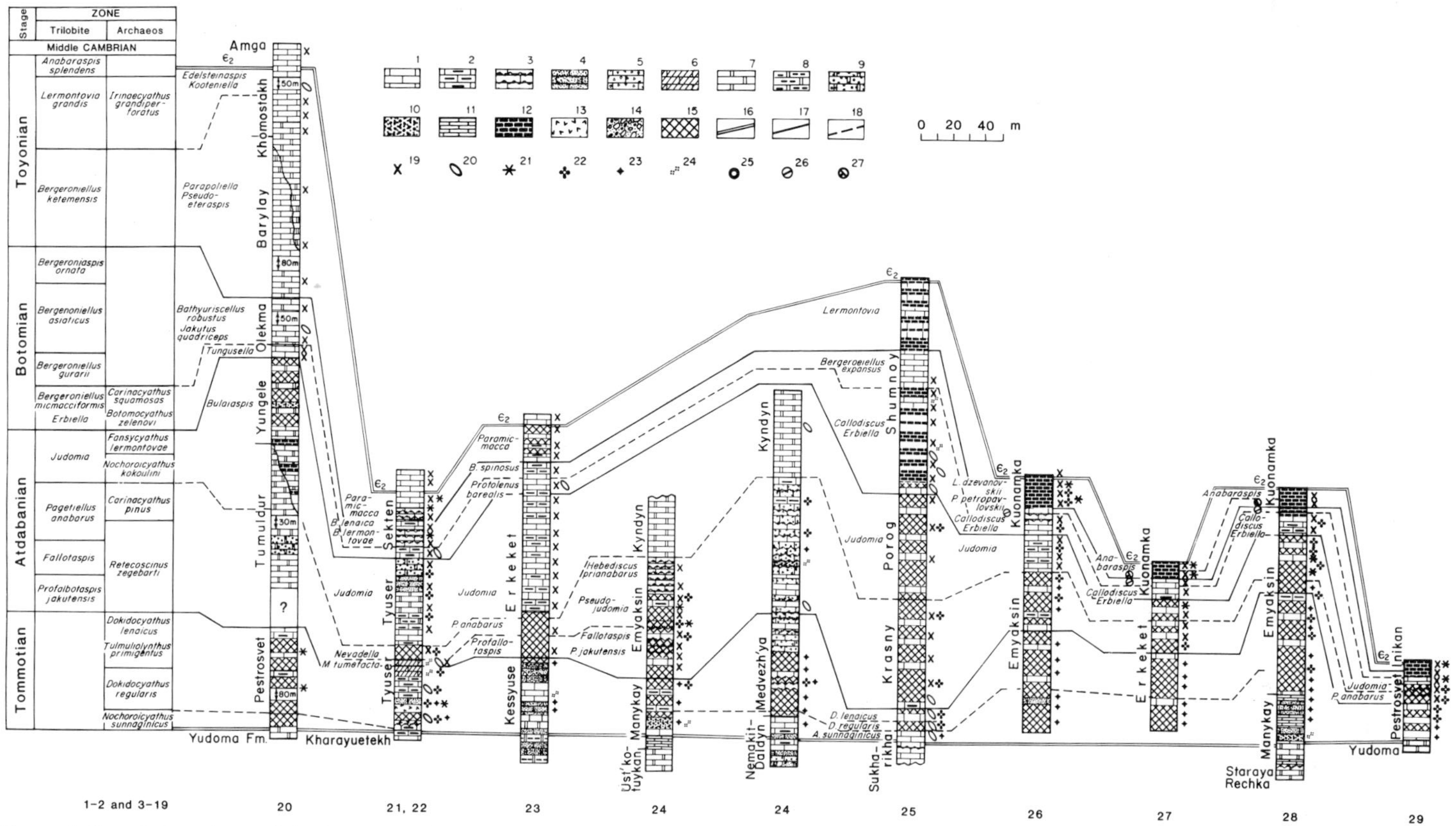

Figure 4. Correlation chart of the Lower Cambrian sections of the Siberian Platform [modified from Rozanov and Sokolov (1984)]. (1) limestone; (2–6) limestones: (2) argillaceous clayey, (3) wavy bed, (4) arenaceous, (6) algal; (7) dolomitic; (8–10) dolostones; (8) argillaceous, (9) oolitic, (10) algal; (11) sandstone; (12) calcareous-argillaceous bituminous shale and chert; (13) extrusive dike; (14) congolmerate; (15) variegated (mainly red) carbonate-argillaceous rock; (16–18) boundaries: (16) series, (17) stages, (18) zones; (19–23) faunal assemblages: (19) trilobites, (20) archaeocyaths; (21) brachiopods, (22) hyoliths, (23) shelly fossils and molluscoids; (24) calcareous algae; (25) bed with *Calodiscus–Erbiella*; (26) bed with *Bergeroniellus expansus*; (27) bed with *Lermontovia dzevanovskii–Paramicmacca petropavlovskii*. Section numbers are the same as in Fig. 2.

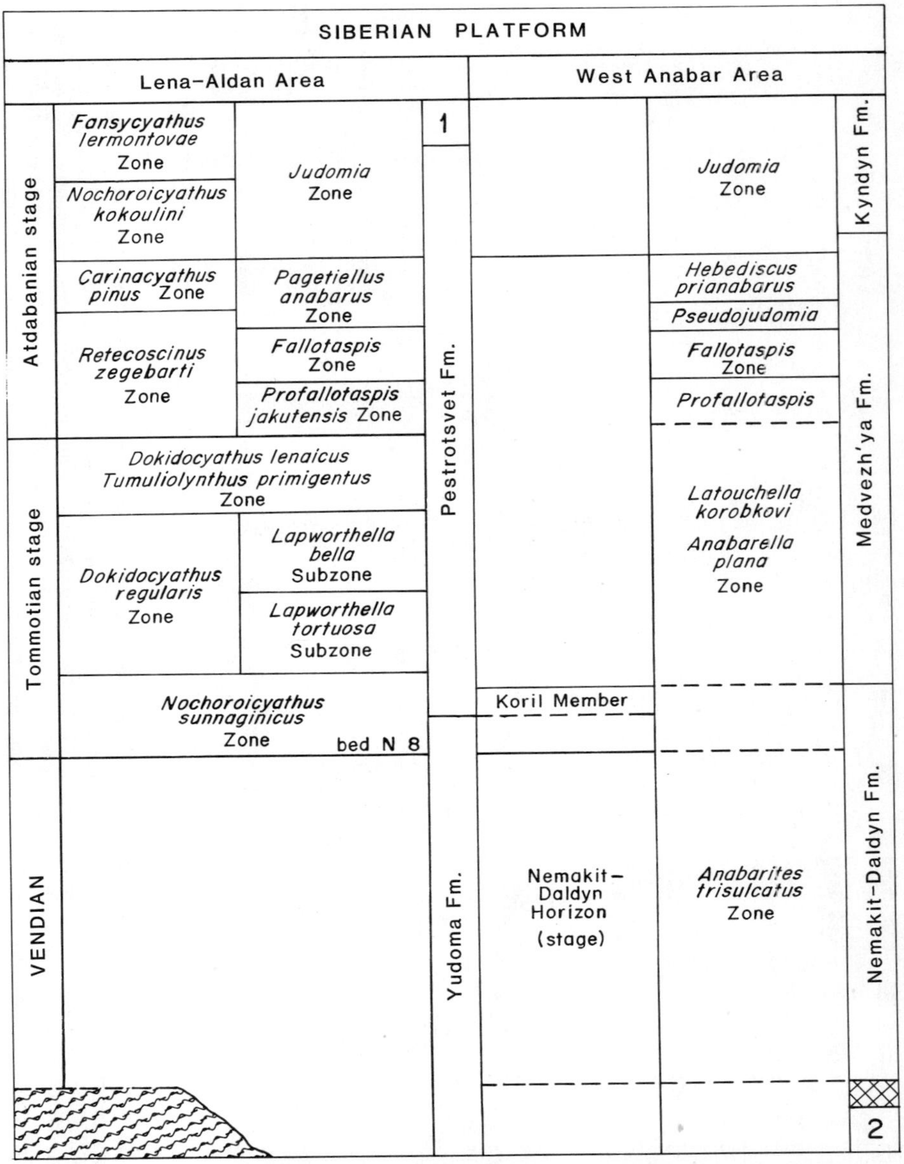

Figure 5. Correlation chart of the Upper Precambrian and Lower Cambrian boundary sections of the Lena–Aldan and West Anabar areas of the Siberian Platform. (1) Perehod Formation, (2) Staraya Rechka Formation.

2.2. The East European Platform

The Lower Cambrian deposits of the East European Platform are siliciclastic. The Lontova Horizon (Regional Stage) is correlated approximately with the Tommotian Stage. The Lontova rocks are the last deposits in an old Vendian basin (Fig. 7). At the outset of the Dominopol' (=Talsy) time (the Atdabanian), a sharp struc-

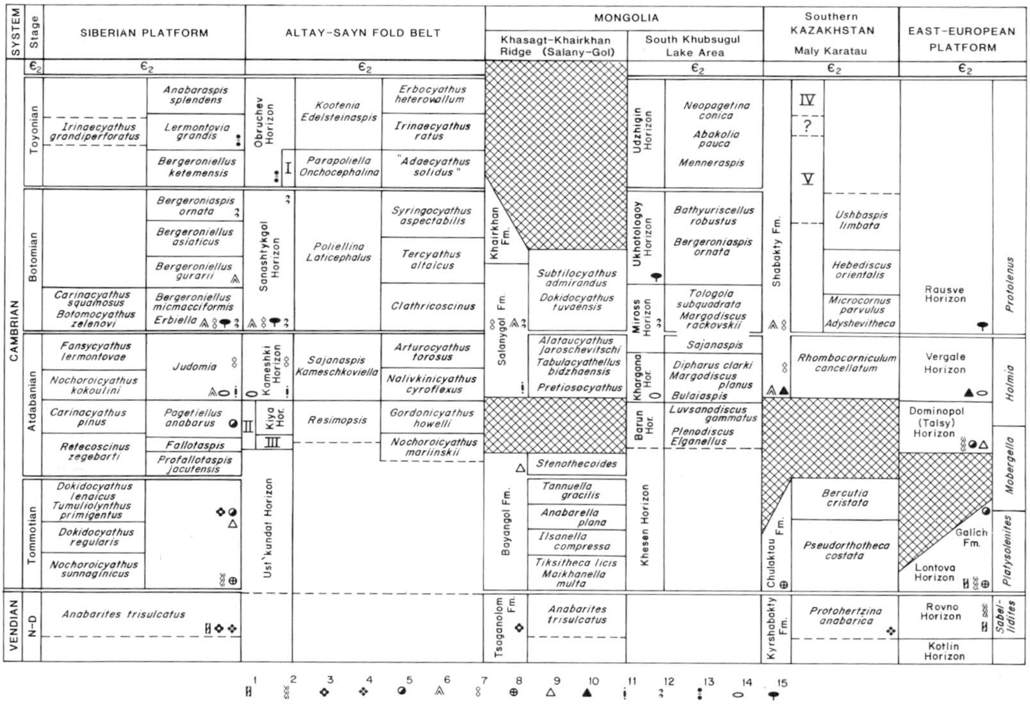

Figure 6. Correlation chart of the reference Lower Cambrian sections of the USSR and Mongolia. (I) Solontsy Subhorizon, (II) Bazaikha Superhorizon, (III) Natalievka Horizon, (IV) *Probowmania asiatica* Zone, (V) *Redlichia chinensis–Kootenia gimmelfarbi* Zone. (1) Sabelliditids; (2) *Platysolenites;* (3) *Anabarites trisulcatus;* (4) *Protohertzina anabarica;* (5) *Mobergella;* (6) *Rhombocorniculum cancellatum;* (7) *Microdictyon;* (8–10) acritarch assemblages: (8) Lontova, (9) Dominopol' (Talsy), (10) Vergale; (11–13) archaeocyathan assemblages: (11) upper Atdabanian, (12) Botomian, (13) Toyonian; (14, 15) trilobites: (14) Nevadiidae, Holmiidae, Dolerolenidae, (15) Protoleninae, Bergeroniellinae.

tural change of the entire East European Platform took place. Most Early Cambrian sedimentation was in the western East European Platform (Poland, western Ukraine, Byelorussia, and the southern Baltic coast) (Fig. 8) (Rozanov and Łydka, 1987). The following Lower Cambrian horizons (regional stages) are recognized: Lontova, Dominolpol', Vergale, and Rausve (Fig. 6) (Keller and Rozanov, 1979; Brangulis *et al.*, 1981; Mens *et al.*, 1990).

The most complete sections of the Lontova Horizon are in boreholes in the Moscow Syneclise, the Ukraine, Eastern Poland, and the southern Baltic coast, with isolated outcrops in Estonia and near St. Petersburg. It consists chiefly of the famous "blue clays" with interbeds of siltstones and sandstones and a coarse-grained coastal facies in some regions. Abundant acritarchs and *Platysolenites*, rare sabelliditids, single *Aldanella, Anabarella,* poorly preserved hyoliths, and trace fossils occur in the clayey beds (Urbanek and Rozanov, 1983; Rozanov and Sokolov, 1984; Mens *et al.*, 1987, 1990). The Lontova Horizon is usually about 100 m thick, but in some regions, such as the central part of the Moscow Syneclise, it reaches 300 m. The upper part of the section has more varied facies including bioturbated rocks, marker bands of ferrous oolites, *Skolithos* sandstones, and other types. As in the Lontova Horizon, abundant acritarchs occur throughout the

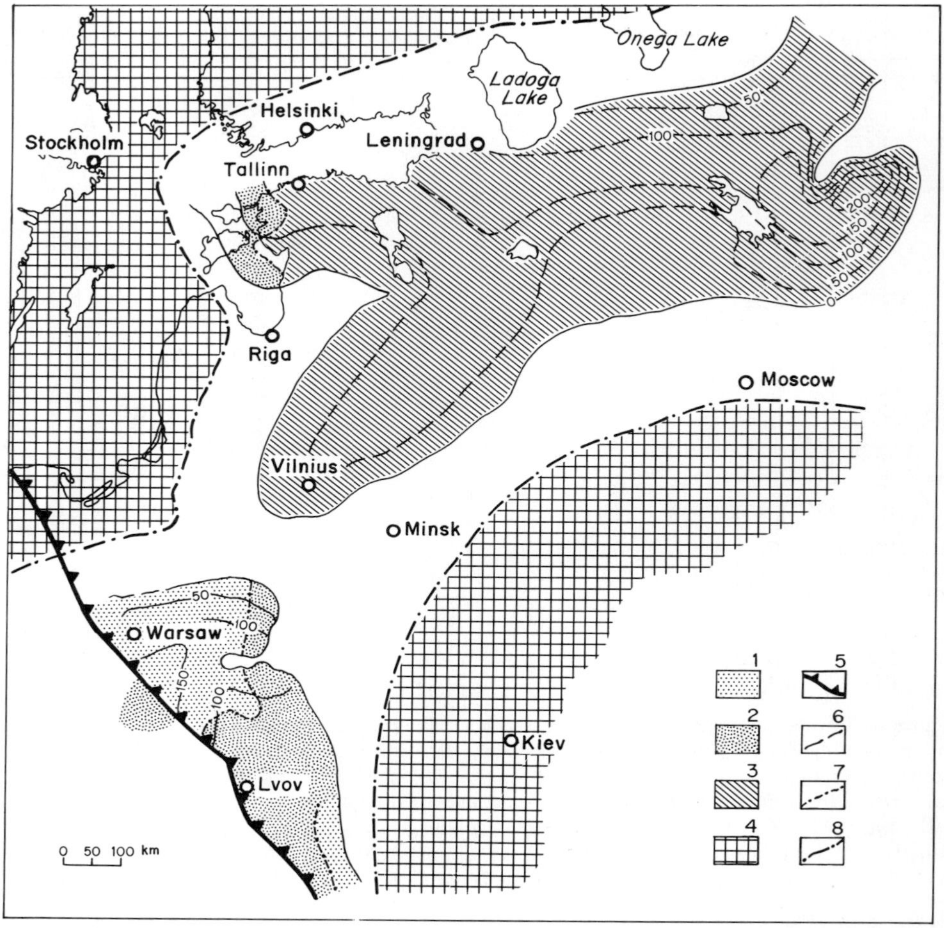

Figure 7. Lithopaleogeographic map of the western East European Platform, Lower Cambrian, Lontova Horizon [modified from Rozanov and Łydka (1987)]. (1) Mainly sandy rocks, (2) mainly silty rocks, (3) mainly argillaceous rocks, (4) crystalline rocks, (5) the Teisseyre–Tornquist Zone, (6) established and inferred isopachs, (7) boundaries of lithofacies zones, (8) inferred limits of sedimentation areas.

entire section. *Vorborthella* and inarticulate brachiopods are present, together with rare trilobites and other fossils.

The East European Platform, as compared with other regions, has a remarkable lateral continuity of facies. In spite of this, local formations are recognized in each administrative region (Mens *et al.*, 1987, 1990). The correlation of sections in the East European Platform with those of the Siberian Platform thus varies according to different authors. The Upper Vendian Rovno Horizon is correlated with the Nemakit-Daldyn Horizon based on sabelliditids and similar acritarch assemblages. The Lontova acritarch assemblage also occurs at the top of the Nemakit-Daldyn Formation (Kirjanov, 1987), and *Aldanella* and *Anabarella* are present in the Lontova Horizon and at the top of the section in the Moscow Syneclise (Glebovo Formation). *Mobergella* occurs in the Lontova Horizon, indicating that it is Tommotian. Some workers (for example, Khomentovsky, 1974; Moczydłowska

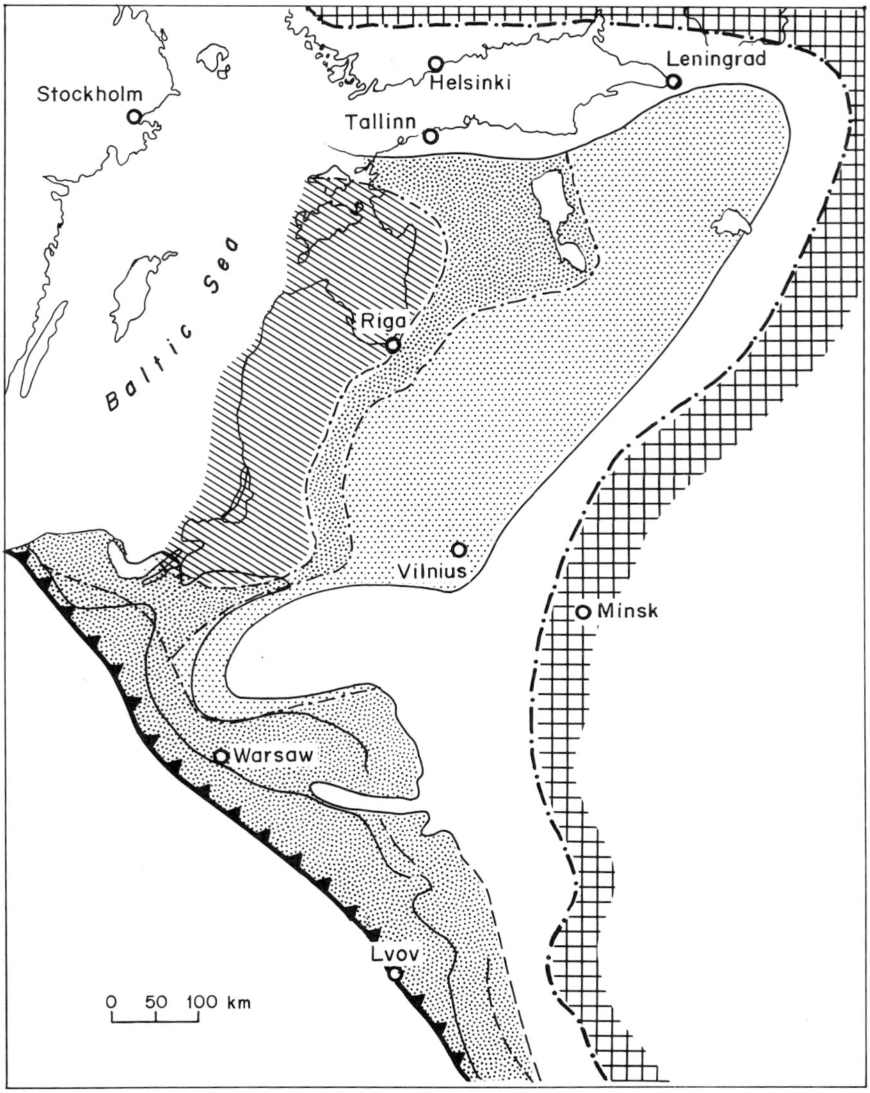

Figure 8. Lithopaleogeographic map of the western East European Platform. Lower Cambrian, Vergale, and Rausve Horizons [modified from Rozanov and Lydka (1987)]. For key, see Fig. 7.

and Vidal, 1988) believe that Tommotian sediments are absent on the East European Platform but that they would occur between the Lontova and Dominopol' deposits. Archaic olenellids occur in the Dominopol' and Vergale Horizons, indicating a correlation with the Atdabanian Stage of the Siberian Platform. Also, *Strenuaeva*, which is characteristic of northern Europe, is found in Siberia, *Mobergella* occurs in the Dominopol' Horizon, and the appearance of protolenid trilobites in the base of the Rausve Horizon indicates a correlation with the Botomian.

2.3. The Altay-Sayan Fold Belt

The Altay-Sayan Fold Belt borders the southwestern part of the Siberian Platform. This region is important in determining the stratigraphic and geographic distribution of Lower Cambrian fossils. Here Repina *et al.* (1964) and Konjuschkov (1972) proposed stage subdivisions of the Lower Cambrian.

Within the Altay-Sayan Fold Belt are provinces differing sharply in structure and facies, which has caused disagreement in the interpretation of the succession of rock sequences and faunal assemblages. Some of these problems have been clarified by data from the Cambrian sections on the Siberian Platform.

Although different provincial successions were identified within the Altay-Sayan Fold Belt, the horizons (regional stages) defined are recognized throughout the entire Fold Belt: Ust'kundat, Bazaikha, Kameshki, Sanashtykgol, and Obruchev. The Bazaikha Horizon, considered to be a superhorizon, is subdivided into the Natal'evka and Kiya Horizons. The Solontsy Subhorizon lies in the lower part of the Obruchev Horizon (Fig. 6). The subdivision of the Lower Cambrian is best in the carbonate sequences of Bazaikha and Kiya Rivers and Batenevsky Ridge (Kuznetsky Alatau).

The Lower Cambrian of the Altay-Sayan Fold Belt is characterized by the richest assemblages of algae, archaeocyaths, and trilobites (Repina *et al.*, 1964; Zhuravleva *et al.*, 1967; Osadchaja *et al.*, 1979). Other groups are sporadically distributed and occur (especially small shelly fossils) in terrigenous-carbonate facies. The sections of the Altay-Sayan Fold Belt and the Siberian Platform are easily correlated (Fig. 6) because trilobites and archaeocyaths are so abundant. However, the lower boundary of the Botomian Stage is unclear, the lower part of the Sanashtykgol Horizon may be Atdabanian in age, and the placement of the Botomian–Toyonian Stage boundary is also under discussion. The correlation of Precambrian–Cambrian boundary is also difficult. These deposits are well-developed in the Altay-Sayan Fold Belt, but they, with the exception of a few localities, lack fossils other than algae and sponge spicules. Tommotian archaeocyaths are absent in the Altay-Sayan Fold Belt. Studies of the Precambrian–Cambrian boundary sequences in the southern Siberian folded regions did not yield substantial information on the distribution of fossils. In Mongolia, however, fossils provide the necessary data.

2.4. Mongolia

Because Mongolia lies among the Siberian Platform, the Altay-Sayan Fold Belt, and China, the fossil assemblages resemble those of the other regions. Trilobites and archaeocyaths are most similar to those of the Altay–Sayan Fold Belt, leading to the initial use for Mongolia of the stratigraphic nomenclature of the other regions. In the last decade, assemblages of mollusks, hyoliths, brachiopods, and small shelly fossils have been found that are most similar to those of the Siberian Platform and China (Korobov and Missarzhevsky, 1977; Drozdova *et al.*, 1981; Voronin *et al.*, 1982).

The most completely studied sections are in two regions: northern Mongolia in the Khubsugul Lake area and western Mongolia in the Dzabkhan structural-facies province. In northern Mongolia, Korobov (1980, 1989) reported associations of trilobite species and genera typical both of the Siberian Platform and the Altay-Sayan Fold Belt. These trilobites allow recognition of the Atdabanian, Botomian, and Toyonian Stages. The phosphate-bearing rocks, which Luvsandanzan and Rozanov (1984) assigned to the Tommotian, of the lower part of the section contain very rare fossils. Recently, the acritarch *Tasmanites bobrowskii* Warynska, the problematic algae *Spirellus columnaris* Jiang, and *Archaeooides*-like spheres were found in the phosphorite beds of the Khubsugul Lake Area (Zhegallo and Ragozina, 1990).

In spite of rather rich assemblages of trilobites and Altay-Sayan and Siberian index fossils, the correlation of the sections in this region with those of other regions remains uncertain because the age of the Miross Horizon, defined by Korobov (1980) (Fig. 6), is questionable. Both typical Atdabanian *Judomia, Bulaiaspis,* and other fossils and typical Sanashtykgolian *Shivelicus, Serrodiscus, Tuvanella,* and other trilobite genera occur in the Miross Horizon.

Precambrian–Cambrian sections in the Dzabkhan province contain the richest assemblages of small shelly fossils (Voronin *et al.*, 1982). Here, in the geosyncline region, the faunal assemblages are similar to those of the Tommotian of the Siberian Platform and the Meishucunian of China. The rocks are characterized by rhythmic carbonate and terrigenous interbeds with bands of phosphorites. Chiefly the carbonate beds (usually several dozens of centimeters to 1 m thick) contain small shelly fossils. In some cases, the carbonate beds are comprised only of algae. Beds of phosphorites consist of shells and thus they appear to be the first microshell-hash type ore in the Earth history (Rozanov and Zhegallo, 1989). A similar situation occurs in China. In spite of these similarities, Precambrian–Cambrian boundary sections in Siberia and China still cannot be correlated.

2.5. Maly Karatau

The Maly Karatau region contains important Lower Cambrian fossils. Karatau is the northeastern spur of the Tien Shan Mountains and was well studied because of its large phosphorite deposits. But only in the past 10–15 years have Mambetov and Repina (1979) and Missarzhevsky and Mambetov (1981) described typical Siberian Platform fossils: *Protohertzina anabarica* Missarzhevsky, *Tiksitheca licis* Missarzhevsky, *Rhombocorniculum cancellatum* (Cobbold), and others. Thus, Maly Karatau can be correlated, with assurance, with the Siberian Platform (Fig. 6). A series of forms described originally from China was also recorded here, including cambroclaves which in fact are the same as the Chinese zhijinitids (Bengston *et al.*, 1990). They occur in the Atdabanian of Maly Karatau but in the Meishucunian in China. Of no less interest is the occurrence of the trilobite *Redlichia* in this region.

3. The Fossil Record

Here we consider all the fossil groups found in the Lower Cambrian of the USSR and Mongolia, even if they are represented only by one or two specimens, except for those with obscure stratigraphic and geographic data and that are not in collections. We examined the majority of the fossils discussed in this chapter ourselves. The acritarch and trace fossil data are summarized from the literature. Archaeocyaths are described by Debrenne in this volume, Chapter 13; we discuss them only briefly. Most of the fossils and their geographic distribution in the Lower Cambrian sections of the Anabar–Sinsk and Yudoma–Olenek Facies Regions of the Siberian Platform, western Mongolia (Dzabkhan or Tsaganolom structural province) and the East European Platform are indicated in Figs. 9–12. The new genera and species of Missarzhevsky (1989) are difficult to consider because of inadequate description, but most of them are probably synonyms of known taxa.

3.1. Skeletal Fossils

3.1.1. Possible Protists

Not long ago protists were considered to be present in the upper Precambrian–Lower Cambrian. Reitlinger (1948; Bykova and Reitlinger, 1959) described in thin sections *Obruchevella, Glomovertella, Cavifera,* and *Syniella* as foraminifera with complex tests from the Sinsk Formation (Botomian Stage) of the middle Lena River. Vologdin (1958) found *Obruchevella* and named seven new genera in the Lower Cambrian of Tuva, and put them in the Order Reitlingerellida. All of these are now recognized as oblique thin sections of the calcareous alga *Obruchevella delicata* Reitlinger; it most probably belongs among the Oscillatoriales. Pieces of similar etched specimens from the Middle Cambrian of Sardinia were also referred to the foraminifera by Cherchi and Schroeder (1985).

Other Siberian "foraminifers" assigned to *Jaraktina* Galperova are found in the Mota Group (Vendian–Lower Cambrian?) of the Irkutsk Amphitheater [*J. complanata* Galperova in Akulcheva *et al.* (1981)] and in the upper part of the Perekhod Formation (Botomian stage) of the middle Lena River (Vassiljeva, 1983). "*J.*" *lenensis and "J." impr처pera* were assigned to the same genus because they both have a globose shape, small size (0.35–1.2 mm in diameter), and no indications of a foramen or pores. The main features of foraminifera, however, are not present in *Jaraktina. J. complanata* is hollow, thin-walled, smooth, and somewhat irregular in shape. We examined this material and concluded that it probably represents the remains of oncolites. Thick-walled "*J.*" *lenensis* and "*J.*" *improspera* (Figs. 13a and 13b) look very much like the Chinese problematic fossils *Archaeooides* Qian and *Olivooides* Qian. *Archaeooides* also is known from the same locality as "*J.*" *lenensis* (Sokolov and Zhuravleva, 1983).

A hypothetical Early Cambrian agglutinated foraminifera *Platysolenites* Eichwald (Glaessner, 1975) occurs throughout the upper Rovno, Lontova, and Dominopol' Horizons of northern Europe. Brasier (1982) included it with *Bathysiphon* as central in the development of many groups of foraminifera. *Platysolenites* tubes (Fig. 12, panel 19), however, were shown by Rozanov (1979, 1983a) to be a nonagglutinating form and probably consist of primary amorphous silica.

Many of the other "foraminifer" remains from the East European Platform and northern Europe (Brasier, 1982; Makhnach *et al.*, 1985) are evidently

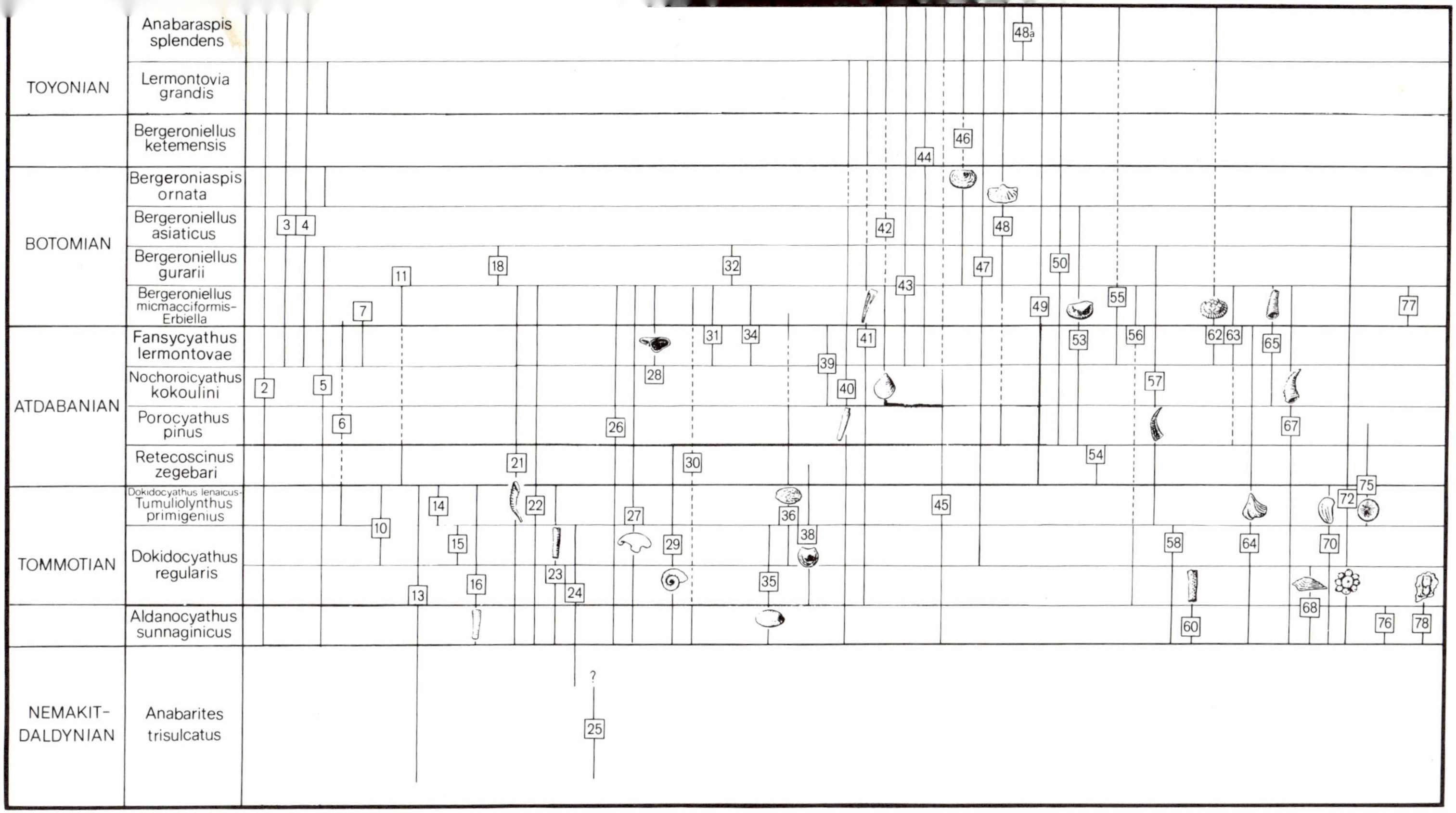

Figure 9a. Stratigraphic distribution of fauna in the Lower Cambrian sections of the Anabar-Sinsk Facies Region of the Siberian Platform. Key to Figs. 9–12: (1) agglutinating foraminifers, (2) hexactinellids, (3) demosponges, (4) calcareous sponges, (5) archaeocyaths, (6) radiocyaths, (7) cribricyaths, (8) branching khasaktiids, (9) *Velumbrella*, (10) *Cysticyathus*, (11) hydroconozoans, (12) *Cambroporella*, (13) *Anabarites*, (14) *Selindeochrea* and similar forms, (15) *Mareochrea* and *similar forms, (16) Tiksitheca*, (17) *Kugdatheca*, (18) palaeoscolecids, (19) *Platysolenites*, (20) *Volborthella*, (21) Hyolithellidae, (22) Torellellidae, (23) coleolids, *(24) Cambrotubulus*, (25) sabelliditids, (26) cap-shaped shells, (27) Coreospiridae, (28) Pelagiellidae, (29) Aldanellidae, (30) Yochelcionellidae, (31) Sinuitidae, (32) Multifariidae, (33) *Rozanoviella*, (34) *Beshtashella*, (35) rostroconchs, (36) pelecypods, (37) *Maikhanella* (=?*Canopoconus*), (38) *Aldanolina*, (39) "*Mirabella*," (40) Orthothecimorpha, (41) Hyolithimorpha, (42) stenothecoids, (43) lingulids, (44) acrotretids, (45) paterinids, (46) kutorginids, (47) obolellids, (48) orthids, (48a) pentamerids, (49) polymeroid trilobites, (50) miomeroid trilobites, (51) *Gdowia* and *Liwia*, (52) *Cassubia*, (53) bradoriids, (54) *Isoxys*, (55) echinoderms, (56) conodontomorphs, (57) *Rhombocorniculum*, (58) *Fomitchella*, (59) *Salanacus*, (60) *Tommotitubulus*, (61) *Mongolitubulus*, (62) *Hadimopanella*, (63) *Microdictyon*, (64) tommotiids, (65) kelanellids, (66) tannuolinids, (67) lapworthellids, *(68) Sunnaginia*, (69) *Stefania*, (70) halkieriids, (71) siphogonuchitids, (72) chancelloriids, (73) *Cloudina?*, (74) *Cambrocassis*, (75) *Mobergella*, (76) *Markuelia*, (77) *Archaeooides*-like, (78) *Tumulduria*.

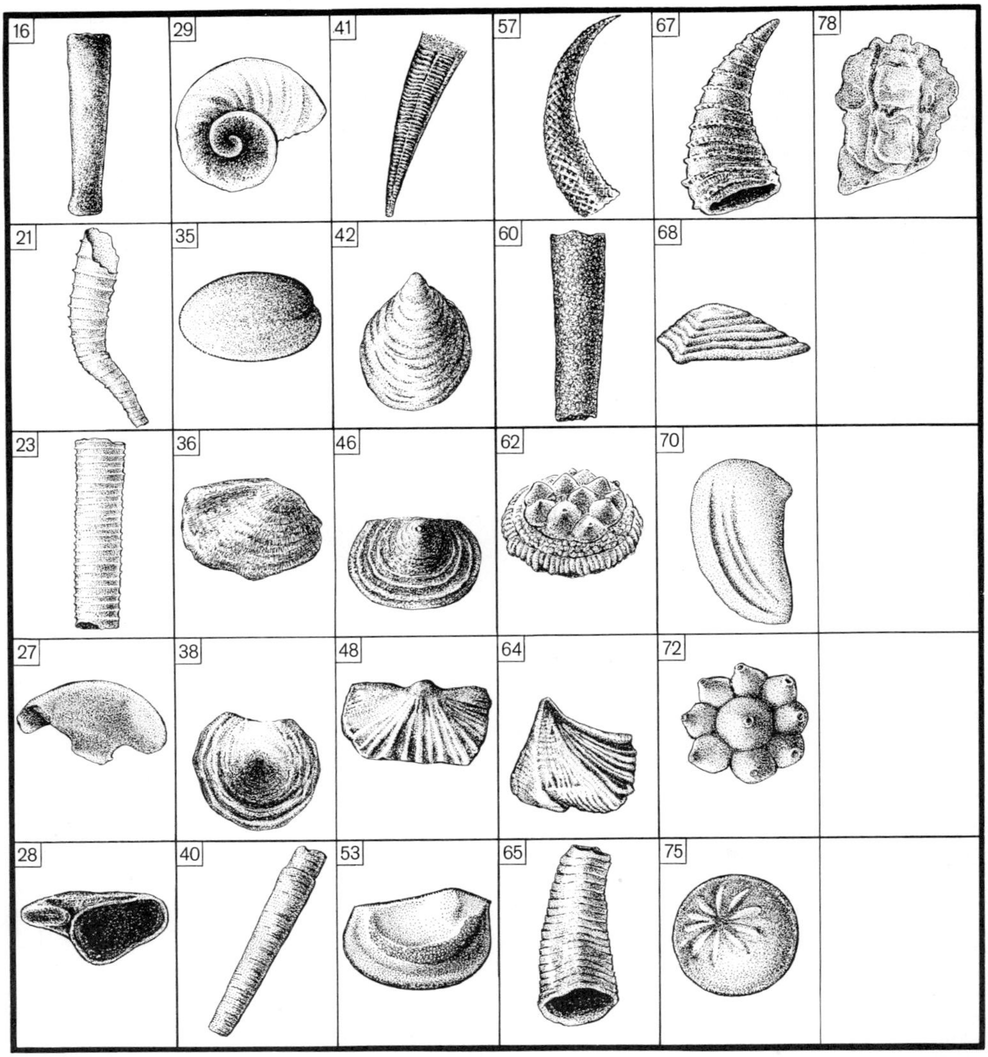

Figure 9b. Illustrations of selected Lower Cambrian fossils of the Anabar-Sinsk Facies Region of the Siberian Platform. See Fig. 9a for key to illustrations.

Platysolenites fragments, including the forms described as *Yanishevskytes petropolitanus* (Yanishevsky) (Urbanek and Rozanov, 1983). "*Leukatiella*" also occurs in the same region (Dominopol' and Vergale Horizons), and is most probably an agglutinating foraminifera (Mens *et al.*, 1987). Missarzhevsky (Missarzhevsky and Mambetov, 1981) described small tubes with transverse bulges as *Karatubulus* (Fig. 13c) from the Atdabanian part of the Shabakty Formation of Maly Karatau. They resemble some tubular foraminifera according to Rozanov (1986), but their composition is phosphatic (?secondary).

Cambrian "radiolaria," including Early Cambrian ones, were studied by

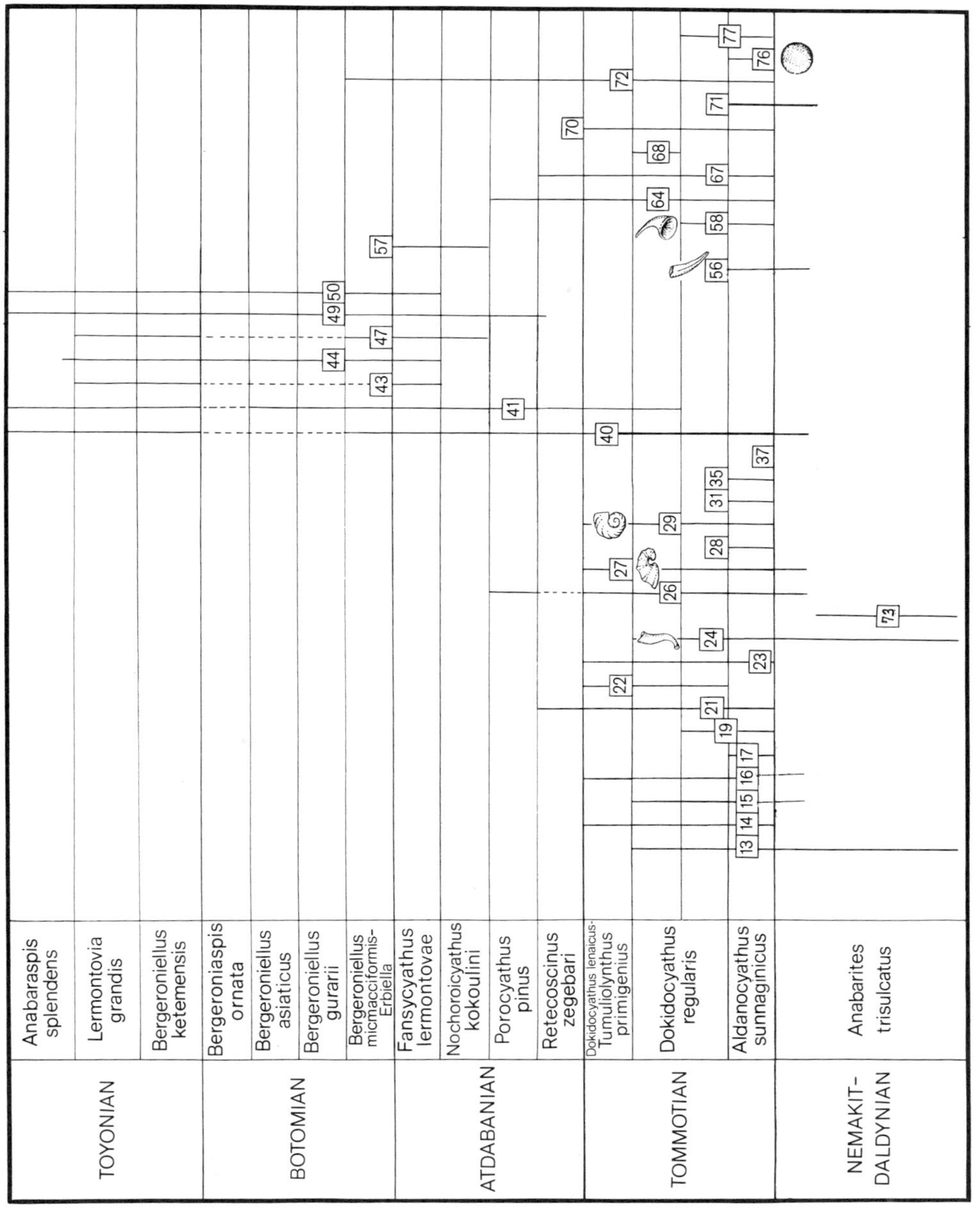

Figure 10a. Stratigraphic distribution of the fauna in the Lower Cambrian sections of the Yudoma-Olenek Facies Region of the Siberian Platform. Key to illustrations in Fig. 9a.

Figure 10b. Illustrations of selected Lower Cambrian fossils of the Yudoma-Olenek Facies Region of the Siberian Platform. See Fig. 9a for key to illustrations.

Nazarov (1973, 1974, 1975) in thin sections. The oldest "radiolaria," *Paleocenosphaera* Nazarov and *Paleoxiphospaera* Nazarov from the Atdabanian Stage of the Batenevsky Ridge, occur in shallow-water limestones with numerous algae and sometimes with archaeocyaths. Unlike true radiolaria, these globose structures have no central cavity; they resemble spicules of demosponges. Rocks

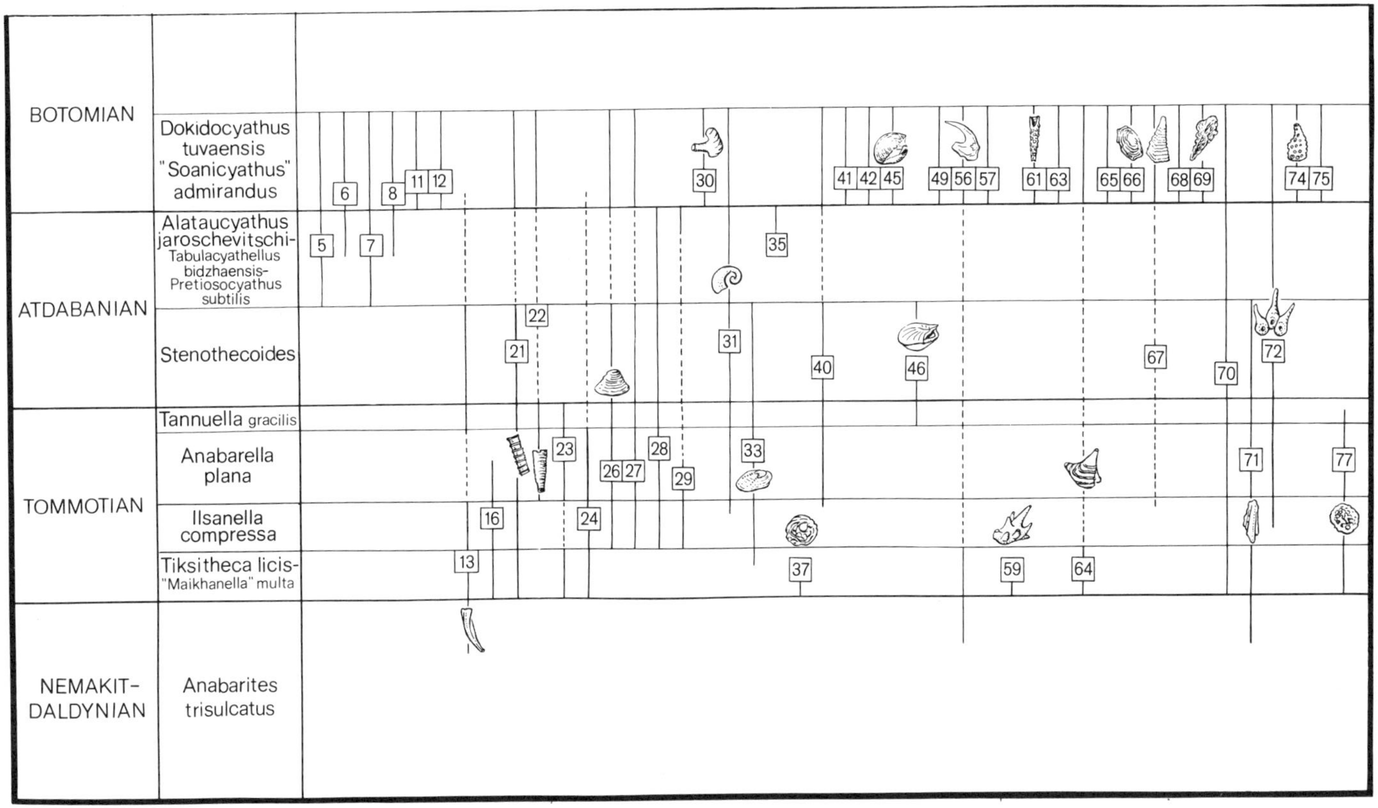

Figure 11a. Stratigraphic distribution of the fauna in the Lower Cambrian sections of western Mongolia (Dzabkhan or Tsaganolom structural province). Key to illustrations in Fig. 9a.

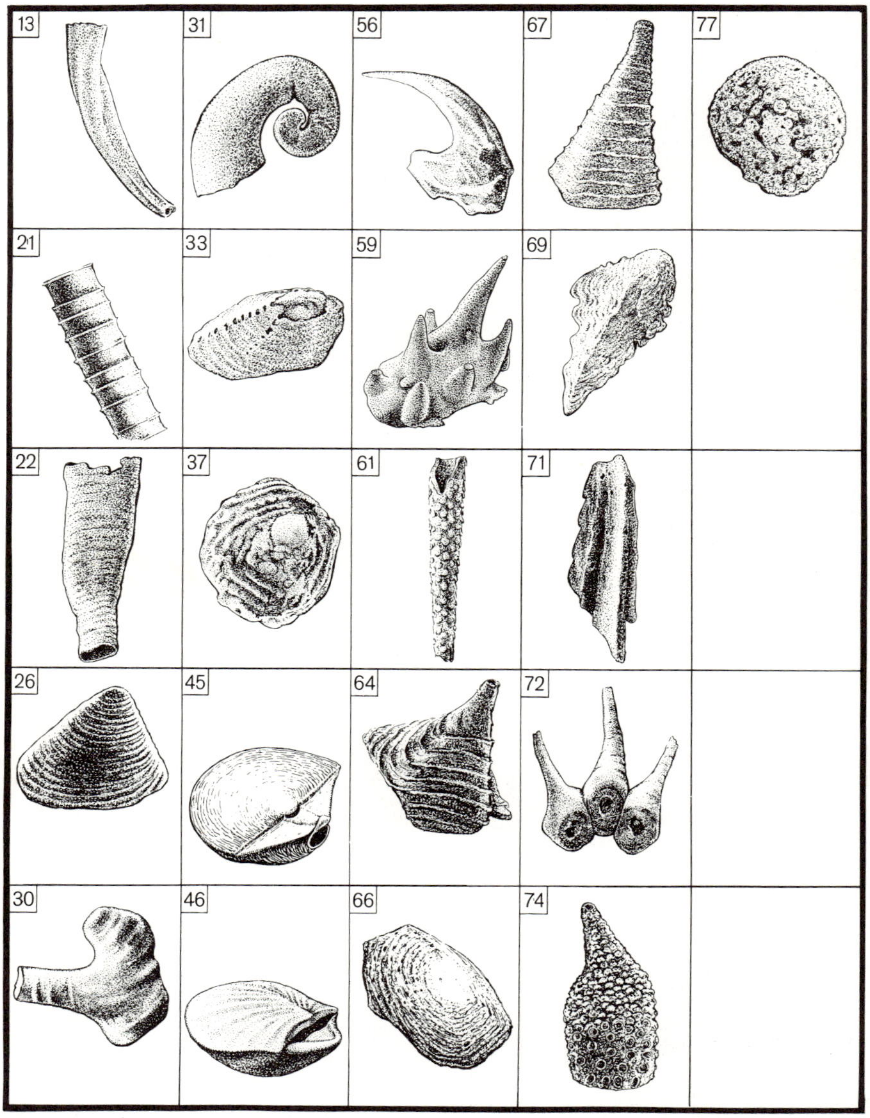

Figure 11b. Illustrations of selected Lower Cambrian fossils of western Mongolia (Dzabkhan or Tsaganolom structural province). See Fig. 9a for key to illustrations.

similar to those from which the "radiolaria" were described contain micrasters. "*Lithapium* Haeckel" described by Nazarov (1973) probably are pentactines of hexactinellids and the problematic *Azyrtalia* apparently are sections of pentactines. Bengtson (1986a) described similar sections in his work on spicules from the Upper Cambrian of Queensland.

Thus, the first probable agglutinating foraminifer remains are not known before the Atdabanian and radiolaria have not yet been found in deposits older than the Middle Cambrian.

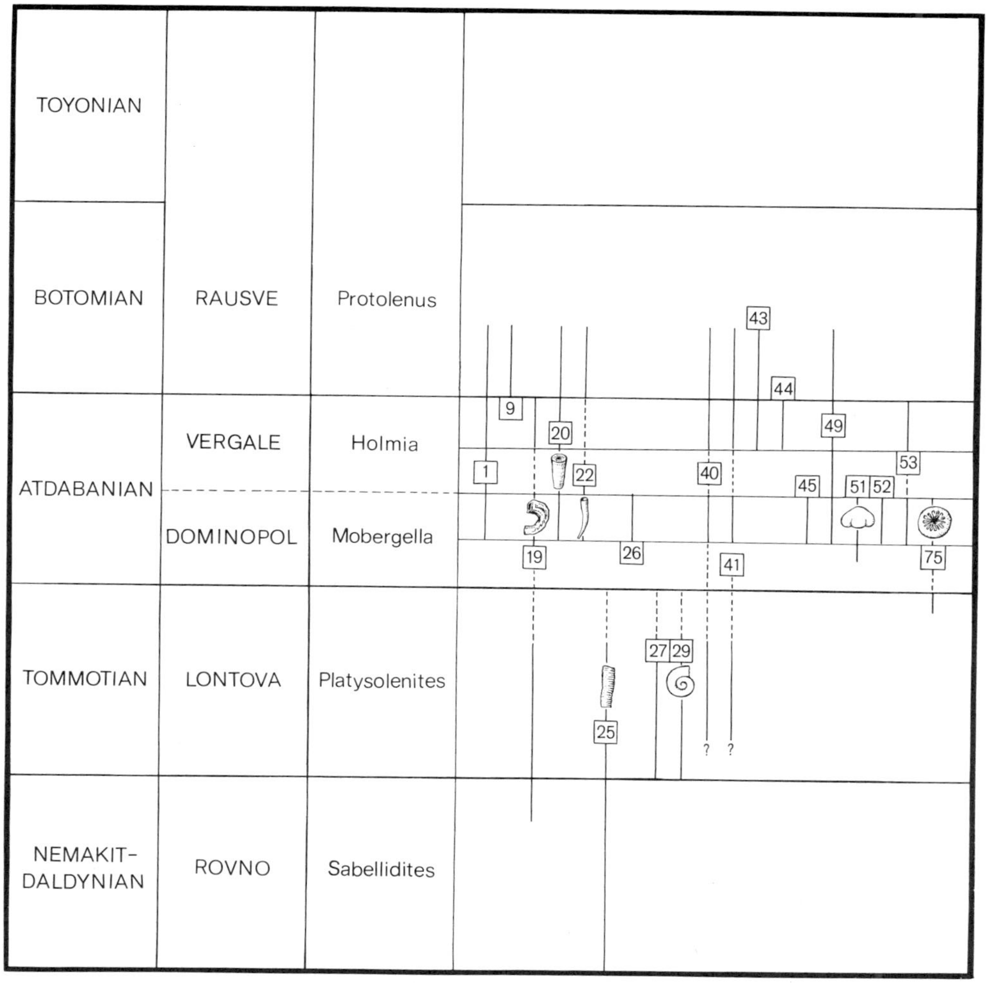

Figure 12a. Stratigraphic distribution of fauna in the Lower Cambrian sections of the East European Platform. Key to illustrations in Fig. 9a.

3.1.2. Sponges and Spongiomorphs

Isolated sponge spicules occur commonly in the Lower Cambrian of Siberia (Tarnovsky and Astashkin, 1973; Zhuravleva and Luchinina, 1980; Sokolov and Zhuravleva, 1983). They are described mostly from thin sections; therefore their affinity is rather difficult to establish.

Spicules belonging to both recent phyla or subphyla (Cellularia and Symplasma) are found in the older part of the Lower Cambrian. On the Siberian Platform isolated stauractines probably belonging to Protospongiidae (Hexacinellida) occur in the lowermost part of the Tommotian (Sokolov and Zhuravleva, 1983) and true hexactines are present in the upper one fourth of the Atdabanian (Shabalov *et al.*, 1987) (Fig. 14a). In the Botomian Stage (Kuonamka Formation), pentactines are found together with hexactines; they form spongolites

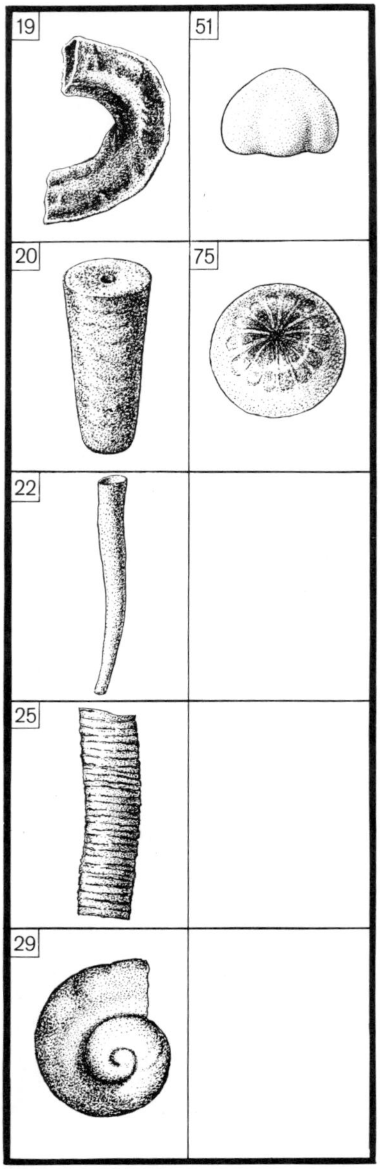

Figure 12b. Illustrations of selected Lower Cambrian fossils of the East European Platform. See Fig. 9a for key to illustrations.

(Fedorov and Pereladov, 1987). Pentactines also known under the name *Sulugurella* Fedorov occurred at the very base of the Tommotian on the Aldan and Dzhanda rivers (Pelman *et al.*, 1990). Among the oldest hexactinellids is *Multivasculatus* (Howell and Van Houten, 1940) from the Middle Cambrian of North America, which are very similar in shape to *Aulophycus* (Fenton and Fenton, 1939) present in the same deposits. Studies of *Aulophycus* from the Lower Cambrian of the Siberian Platform (Astashkin, 1985) show that it is most likely a bioturbation structure. Spicules are found rather seldom in *Aulophycus*, but they

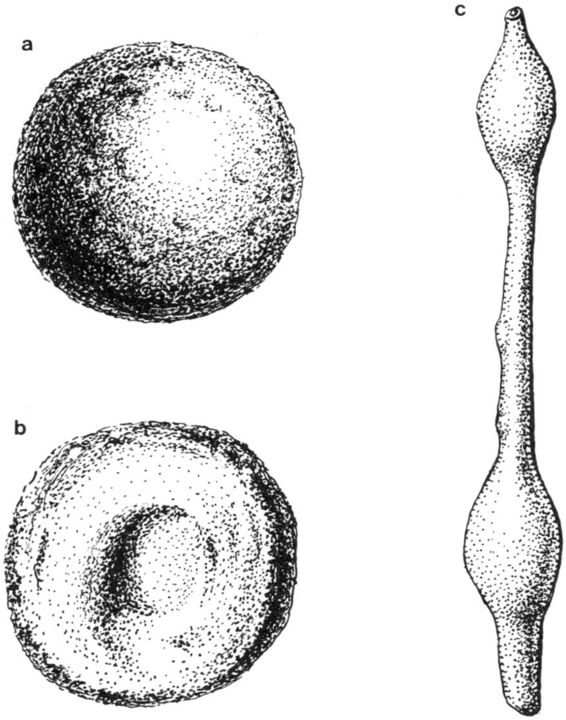

Figure 13. *"Jaraktina"* and *Karatubulus*. (a) *"Jaraktina" lenensis* Vassiljeva, holotype VNIGRI N 732/1, (b) *"Jaraktina" improspera* Vassiljeva, holotype VNIGRI N 732/3, Middle Lena River, Achchagyy-Tuoydakh Creek, Atdabanian stage; (c) *Karatubulus nodosus* Missarzhevsky, holotype GIN N 4296/27, Maly Karatau, Botomian stage. Scale bar 100μm. (VNIGRI: All-Union Research Geological Institute, Ministry of Geology. GIN: Geological Institute of the USSR Academy of Sciences.)

are commonly poorly preserved and occur with a considerable admixture of quartz grains.

True demosponge spicules are present in the upper one fourth of the Atdabanian Stage (Emyaksin Formation). They are described in open nomenclature by Fedorov (Shabanov *et al.*, 1987, forms 1, 5–7) and are represented by stauracts with various additional elements. At the base of the Botomian, asters are probably present. In the Lower Botomian (Sinsk Formation) on the Siberian Platform, a single complete demosponge, consisting of bundles of monaxones, was described by Gorjansky (1977) as *Lenica*, although similar forms were described by Walcott (1920) from the Middle Cambrian of North America as *Hamptonia* (J. K. Rigby, personal communication).

The first calcareous sponge spicules with triradiate symmetry (Fig. 14b) occur in the Atdabanian (Shabanov *et al.*, 1987, forms 2–4, 15). Rigby and Toomey (1978) assign them to Calcarea, while Mostler (1985) puts them in the Heteractinida, and Fedorov (Shabanov *et al.*, 1987) in the Lithistida. The regular triradiate symmetry and original composition, characteristic of Polyactinellidae, suggest that they might belong to the subclass Calcinea. Spicules described by Fedorov as form 4 belong to genera *Phobetractinia* Reif and *Dodecaactinella* Reif, and as form

Figure 14. Sponges and spongiomorphs. (a) Hexactine spicule, PIN N 4274/3; (b) triactine spicule of *Dodecaactinella* sp., PIN N 4274/4; (c) *Cjulankella asimmetrica* Fedorov, PIN N 4274/5, middle Lena River, mouth of the Sinyaya River, Botomiam Stage; (d) archaeocyathan cup *Palaeoconularia prima* Tchudinova, holotype PIN N 1577, Western Sayan, Botomian Stage; (e) a fragment of radiacyathan skeleton *Girphanovella georgensis* (Rozanov), PIN N 3900/34; (f) cribricyathan skeleton, PIN N 4137/17-6, Tuva, Shivelig-Khem River, Botomian Stage. Scale bar 1 mm. (PIN: Paleontological Institute of the USSR Academy of Sciences.)

14 to *Phobetractinia*. According to Bengtson *et al.* (1990), all of them are synonyms of *Dodecaactinella* Reif.

The vast majority of spicules known from the Lower Cambrian of the Siberian Platform are difficult to assign to particular classes of sponges. These include the "poriferid" from the Tommotian Stage of the Aldan River (Matthews and Missarzhevsky, 1975), *Disparella* Fedorov and *Cjulanciella* Fedorov (Fig. 14c) from the Kuonamka Formation (Botomian Stage, Middle Cambrian) (Fedorov and Pereladov, 1987), and *Heterostella* Fedorov, *Inflexiostella* Fedorov, and other species from the Emyaksin Formation (Atdabanian Stage) (Shabanov *et al.*, 1987). The last two genera are placed by Fedorov in the Class (Order) Heteractinida. We consider this order artificial; some fossils assigned to it (hollow chancelloriids) most likely are not sponges (Bengtson and Missarzhevsky, 1981), whereas others (radiocyaths) are not formed by spicules (Debrenne *et al.*, 1970; Zhuravlev, 1986a). *Cjulanciella* and Fedorov's forms 8 and 9 may be dermal spicules of demosponges.

Lenastella Missarzhevsky, a starlike, six-rayed fossil described as a sclerite (Missarzhevsky and Mambetov, 1981), is most likely an artifact. *L. araniformis* Missarzhevsky (type species) from the Atdabanian of the Shabakty Formation of the Maly Karatau are spicules with six rays placed in the same plane, resembling those of the problematic sponge *Eiffelia* Walcott. *L. aculeata* Missarzhevsky and *L. umbonata* Missarzhevsky from the Atdabanian of Kirgiziya are probably spicules of Astraeospongiidae because, in addition to radially divergent rays, they also have a seventh, perpendicular ray. *L. mucronata* Missarzhevsky from the Atdabanian of the Lena River is probably a partially etched radiocyath.

The most recent intensive studies on Recent rigid sponges having a massive calcareous skeleton, with or without spicules (Hartman and Goreau, 1970; Vacelet, 1985), resulted in a new view of many problematic Paleozoic organisms. Increasingly, paleontologists now believe that the typical Early Cambrian archaeocyaths are actually sponges without spicules (Debrenne and Vacelet, 1984; Zhuravlev, 1989; Debrenne *et al.*, 1989). The evolution of archaeocyaths, traced by changes in the cup structure (Rozanov, 1973), was rather dynamic. The principal four stages of archaeocyathan development (oligomerization of body pores; appearance of compensatory skeletal elements; oligomerization and recombination of the compensatory elements; exhaustion of morphological abilities) underlie the four-stage scale of the Lower Cambrian used in the USSR. The oldest archaeocyaths occur at the base of the Tommotian in the Lena–Aldan and Igarka areas of the Anabar-Sinsk Facies Region. Extinction of archaeocyaths at the end of the Early Cambrian was due to a number of causes, chiefly the cessation of development of new skeletal elements or their combinations, and the isolation of basins and reduction of suitable habitats associated with the regression in the Toyonian (the Hawke Bay event) (Zhuravlev, 1986b). Archaeocyaths have been intensively studied in the USSR (e.g., Vologdin, 1932; Zhuravleva, 1960, 1963; Zhuravleva *et al.*, 1964, 1967, 1970; Repina *et al.*, 1964; Rozanov and Missarzhevsky, 1966; Datzenko *et al.*, 1968; Rozanov *et al.*, 1969; Okuneva and Repina, 1973; Beljaeva *et al.*, 1975; Jazmir *et al.*, 1975; Jegorova *et al.*, 1976; Osadchaja *et al.*, 1979; Sokolov and Zhuravleva, 1983). Supposed thalamid sponge *Blastulospongia* Pickett and Jell has been described from the *sunnaginicus* zone of the Aldan River and Olenek Uplift (Khomentovsky *et al.*, 1990), but its poriferan affinity has been recently questioned by Conway Morris and Chen (1990).

Radiocyaths, another exceptional Early Cambrian group, considered for a long time to belong to archaeocyaths (Okulitch, 1955), may also be sponges without spicules. The first radiocyaths (*Gonamispongia* Korshunov) appear in the last one fourth of the Tommotian Stage on the Siberian Platform. Their developmental peak is observed at the end of the Atdabanian and the beginning of the Botomian age, when they become widespread across the Siberian Platform, in Tuva, the

Transbaikalian, Mongolia, Morocco, Australia, and Antarctica, and are represented by the most typical genera of radiocyaths—*Radiocyathus* Bedford and Bedford and *Girphanovella* Zhuravleva (Figs. 14e and 15a). Subspherical skeletons of these organisms are constructed from meromes, each consisting of a proximal and a distal nesaster with a shaft connecting them. Distal nesasters with fused rays form the outer wall of the skeleton, and proximal nesasters form the inner wall. The shape of meromes and their interstitial growth pattern indicate that these organisms belong to the sponges. However, meromes may bifurcate, and their granular microstructure could not be formed even secondarily as a result of diagenesis (Zhuravlev, 1986a). On the other hand, skeleton growth in radiocyaths, because of the interstitial addition of meromes, makes us doubt that they are receptaculitids, which might well be algae (Nitecki, 1967; Beadle, 1988), despite their great similarity in general plan. We do not yet know the exact time of radiocyathan extinction, but they evidently did not persist into the Toyonian.

Khasaktiids (Sayutina, 1980) appear at the end of the Tommotian (the filmy forms *Vittia* Sayutina and *Khasaktia* Sayutina) on the Siberian Platform. These two forms sometimes are difficult to distinguish from holdfasts (epitheca) of archaeocyaths. Other fossils assigned to the Khasaktiidae appear in the early Botomian and are mainly characteristic of fold regions (the South Urals, the Altay-Sayan Fold Belt, Mongolia), but are not known yet in Toyonian deposits. They are dendroid (modular), multilayered, calcareous structures—*Edelsteinia* Vologdin, *Rackovskia* Vologdin, and *Drosdovia* Sayutina (Figs. 15d and 16a). Several synonyms of *Edelsteinia* are recorded: *Tannuolaia* Vologdin, *Sarchoina* Korde, and *Astronia* Korde (Vologdin, 1967; Korde, 1990a). According to Korde (1986), these are Coelenterata, but Sayutina (1980) notes that the similarity of the skeletal structure is evidence that these organisms are ancestors of stromatoporoids. According to Webby (1986), khasaktiids might not be related to stromatoporoids, which attained a similar skeletal structure. Based on their microstructure (Debrenne *et al.*, 1990), we believe they were corallomorphs rather than sponges of the stromatoporoid grade of organization.

Among the most enigmatic Early Cambrian groups are the cribricyaths. They have a calcareous, narrow, conical, horn-shaped cup with a bilaterally symmetrical cross section. The concave side of the horn is usually flattened, and the convex side may bear a longitudinal rib. Two orders of cribricyaths can be distinguished (Jankauskas, 1973). The first, Cribricyathida, has a clathrate outer wall and a subsquare section of the cup (Fig. 15b). The second, Pterocyathida, has an outer wall which is built by transversely bent plates, evidently twisted along the cup axis, forming a ribbonlike spiral. In both of the orders there are one- and two-walled forms. The inner wall is located eccentrically, fusing into the outer wall on its flattened side. Cribricyaths appeared in the first half of the Atdabanian in Kuznetsky Alatau and Eastern Sayan. They were distributed across the Altay-Sayan Fold Belt, Mongolia, the Transbaikalian, the Far East, and the South Urals (Jankauskas, 1972; Okuneva and Repina, 1973; Beljaeva, 1985; Sayutina, 1985a). Practically all cribricyathan remains occur in fold regions where they are found in organic carbonate facies. At present only one species is recognized in the upper part of the Atdabanian from the Siberian Platform (Sundukov and Zhuravlev, 1989). Cribricyaths probably became extinct by the Late Botomian. Zhuravleva and Okuneva (1981) and Beljaeva (1985) believe that cribriciaths were peculiar forms of archaeocyaths, citing archaeocyathan cups with cribriciaths adhered to

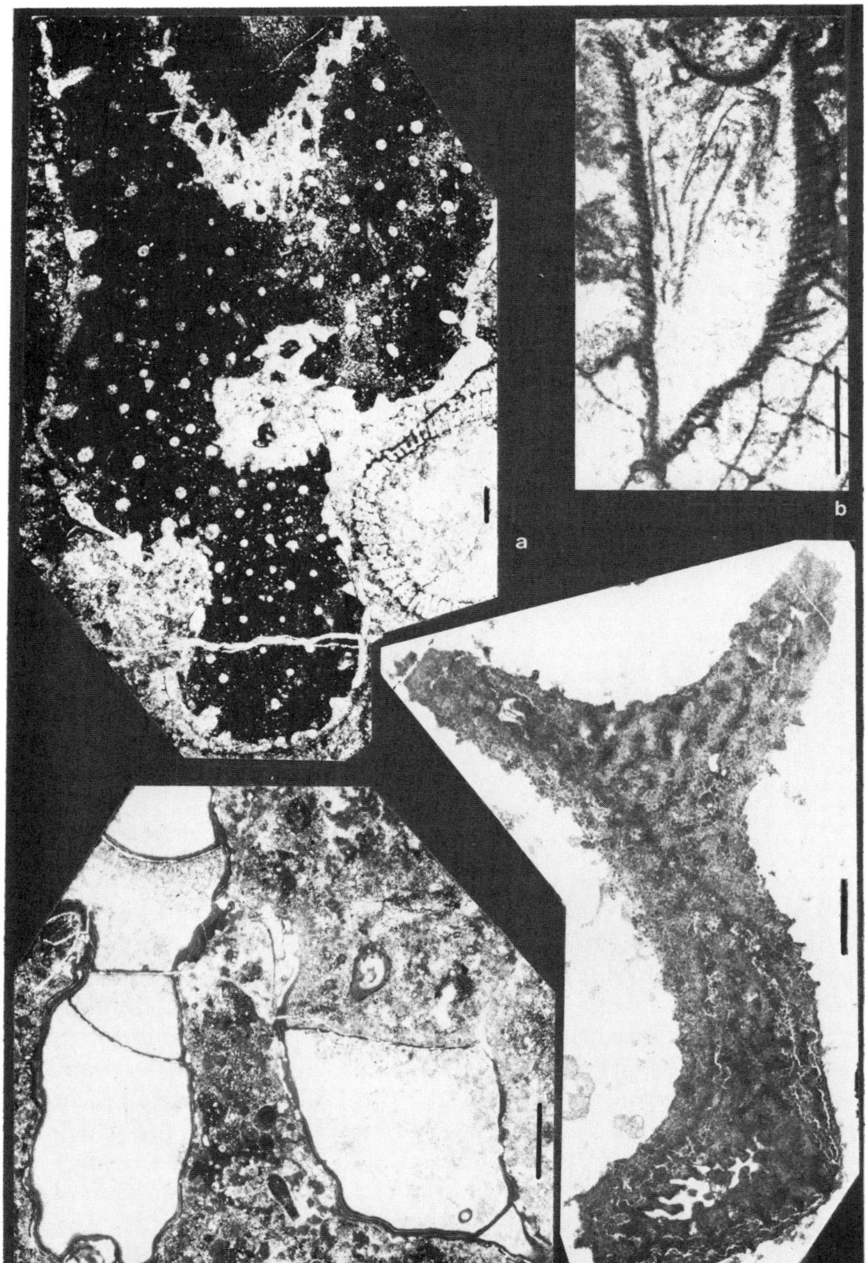

Figure 15. Spongiomorphs and ?coelenterates. (a) Oblique longitudinal section of *Girphanovella georgensis* (Rozanov), PIN N 3302/162b, western Mongolia, Zuune-Arts Mountain, Botomian Stage; (b) oblique longitudinal section of cribricyath *Abicyathus asimmetricus* (Vologdin), anchored on a *Coscinocyathus* sp. cup, PIN N 4274/6, Altay-Sayan Fold Belt, Azyrtal Ridge, Botomian Stage; (c) longitudina section of *Cysticyathus tunicatus* Zhuravleva, PIN N 4274/7, middle Lena River, Tiktirikteekh Creek, Tommotian Stage; (d) longitudinal section of *Edelsteinia vologdini* Sayutina, PIN N 3302/130, western Mongolia, Salany-Gol Creek, Botomian Stage. Scale bar 1 mm.

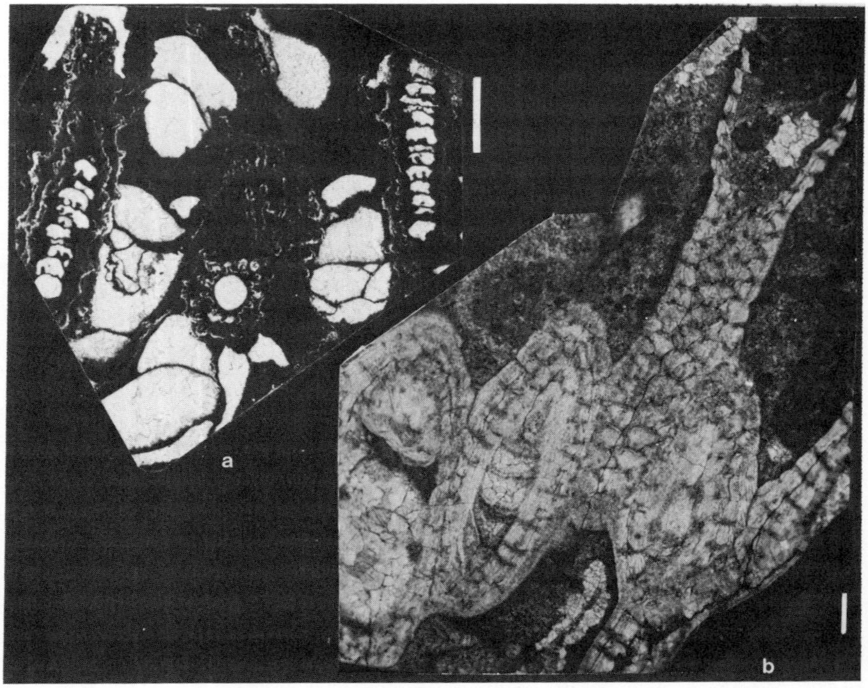

Figure 16. Coelenterates? (a) Longitudinal section of the colony *Drosdovia aenigmatica* Sayutina, holotype PIN N 3302/107, western Montolia, Salany-Gol Creek; (b) oblique longitudinal section of *Cambroporella tuvensis* Korde, PIN N 3175/111, western Mongolia, Seer' Mountains, Botomian Stage. Scale bar 1 mm.

them as evidence. This does not rule out, however, a symbiosis between those organisms. On the contrary, Jankauskas (1969, 1972) considers cribricyaths a separate phylum because the morphology and ontogeny of the cup are distinct from those of archaeocyathan cups.

Glaessner (1976) assigned cribricyaths to the polychaetes, but the peculiar microstructure of their cup (Rozanov and Sayutina, 1982), similar to those of archaeocyaths and certain algae, shows this is wrong. Glaessner's opinion was influenced to a great extent by an apparent similarity of cribricyathan and *Cloudina* cups in thin sections. Germs (1972) first noted this similarity when he described *Cloudina* from the Precambrian Kuibis subgroup in Namibia. On the assumption that these tubes were cribricyaths, Ediacaran-type faunas were assumed to be possibly Tommotian. *Cloudina* was compared with single-walled Pterocyathida. However, *Cloudina* tubes consist of series of multilayered, tubular structures closed at the base (Hanh and Pflug, 1985; Grant, 1990). Its growth appears to have been periodic, whereas the wall of pterocyathids is continuous and with a single twist in the spiral chamber (Fig 14f). The tube surface in *Cloudina* is therefore complicated by numerous wedgelike prominences, but in cribricyaths it is a single spiral rib.

Apart from *Tyrkanispongia* Vologdin and Drosdova and some spicules described in China (Qian and Yin, 1984), spiculate sponges are absent from the

Precambrian (Rigby, 1986). *Tyrkanispongia* includes the Cambrian problematic fossils *Protohertzina, Sachites, Markuelia*, and others. It is not a sponge, and it was placed in the Precambrian by mistake (Vologdin and Drosdova, 1970). Chinese spicules are probably no older than the Tommotian. Thus, spiculate sponges and all sponges with mineralized skeletons appear at the very beginning of the Cambrian and no earlier. At the base of the Tommotian, hexactinellid spicules and archaeocyaths are present, and by the middle Lower Cambrian, if not earlier, these fossils are found together with demosponges, calcareous sponges, and radiocyaths. The three Recent classes of sponges existed by the end of the first half of the Early Cambrian. The oldest spicules have the simplest structure. On the whole, this evidence substantiates Rigby's (1986) hypothesis that sponges evolved from thin-walled to thick-walled forms and developed dermal skeletal elements. However, Rigby's suggestion that at least hexactinellids arose on relatively deep-water shelves and moved to a shallow-water environment only by the end of the Middle Cambrian has no basis. The oldest spicules of all three classes are found in shallow-water deposits, in interbiohermal beds, and even in bioherms, and only in the Botomian are they found in outer shelf or slope facies. The Early Cambrian reefs were formed mainly by filter-feeders (archaeocyaths) and algae. The most common Early Cambrian sponges and spongiomorphs that had rigid skeletons, and evidently cribricyaths similar to them, became extinct by the end of the Early Cambrian or perhaps lost their mineralized skeleton at that time.

3.1.3. Coelenterates?

Few coelenterates occur in the Cambrian in general and in the Lower Cambrian in particular. One of the most likely coelenterates in the USSR, *Tabulaconus* Handfield (Debrenne *et al.*, 1987), was found only in Koryakia (Zhuravlev, 1988). *Cysticyathus* Zhuravleva appears close to *Tabulaconus*. It is a relatively large, irregular to conical, multilayered, nonporous cup (Fig. 15c) found on the Siberian Platform in the upper half of the Tommotian Stage. *Cysticyathus* shows a skeletal microstructure similar to that of the Cambrian corallomorph *Khasaktia* Sayutina, but different from that of the archaeocyaths and *Tabulaconus* (Debrenne *et al.*, 1990).

Other possible coelenterates include the Hydroconozoa (Korde, 1963). They are small, single or modular, conical structures with transverse constrictions and a clear pediclelike part containing an axial canal (Fig. 17). The canal opens into a bowl-like depression in the upper part of the skeleton. In thin sections the skeleton consists of transparent calcareous layers. In the lower part of the skeleton is a disclike bulge. The skeleton's outer surface, known from etched dolomitized specimens, is wrinkled and does not have any elaborate sculpture, as shown by Korde (cited in Drushchits, 1974). The skeletal microstructure of *Hydroconus* is most like that of true corals (Lafuste *et al.*, 1990). Hydroconozoa arose at the beginning of the Atdabanian on the Siberian Platform and disappeared at the end of the Botomian. They also occur in the Altay-Sayan Fold Belt and in Mongolia. Mesozoic structures assigned to Hydroconozoa by Korde (1975) have little in common with this group. *Dasyconus* Korde (Kashina and Sayutina, 1986) is similar to *Hydroconus* in its structure; it also has a multilayered calcareous skeleton, but it could be pierced with pores.

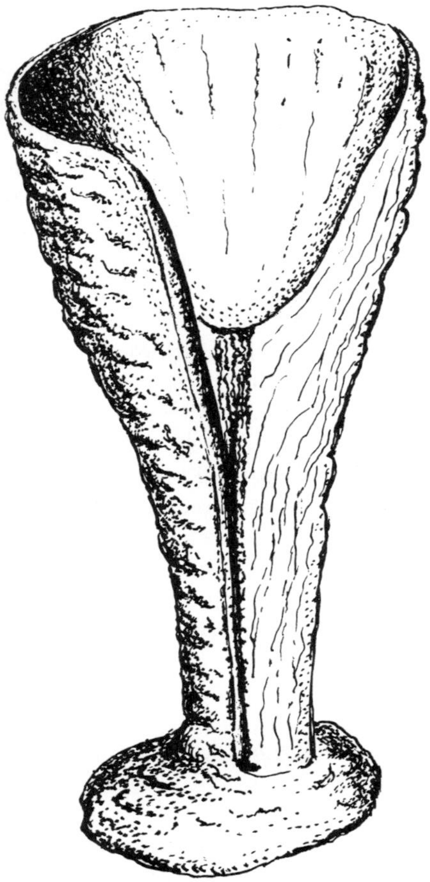

Figure 17. Reconstruction of *Hydroconus mirabilis* Korde. Scale bar 0.5 mm.

Vologdin (1959) described an archaeocyathan–coral community from Tuva, but he mistook a section of the alga *Bija* Vologdin for the coral *Cambrophyllum* Fritz and Howell (Menner *et al.*, 1960). Tchudinova (1959) recorded "conularia" in the Botomian of Western Sayan as a new genus *Palaeoconularia* (Fig. 14d). We reexamined the type material (specimen and thin sections) with the help of I. I. Tchudinova, and we conclude that it is, in fact, an archaeocyathan cup similar to "*Rhabdocyathella*" *baileyi* Vologdin. "*Stromatocerium*" (Khalfina and Yaworsky, 1974), considered to be a stromatoporoid, is a *nomen nudum*, although the illustrated sections from the Toyonian of Kuznetsky Alatau resemble modular Lower Cambrian corals of South Australia (Lafuste *et al.*, 1988).

The anabaritids may be Cambrian coelenterates (Missarzhevsky, 1974; Valkov, 1982; Fedonkin, 1983). Valkov placed a number of late Vendian–Early Cambrian small shelly fossils, including Lobiochreidae, Angustiochreidae, Anabaritidae, Aculeochreidae, *Cambrotubulus* Missarzhevsky, *Kugdatheca* Missarzhevsky, and *Tiksitheca* Missarzhevsky (Fig. 9, panel 16, and Fig. 11, panel 13), in the subclass Angustiomedusa, which he assigned to the class Scyphozoa. Fedonkin assumed that some Vendian structures, such as *Tribrachidium* Glaessner, might belong to the same subclass. These authors noted a similarity between Angustiomedusa and Conularia. Angustiomedusa, however, are triradiate, whereas Conularia and Scyphozoa have a fourfold symmetry. Triradiate symmetry, a conical shape, and small size (3–5 mm long) are common features for all Vendian Angustiomedusa representatives of the "subclass." According to Valkov (1982), this subclass is also characterized by two-layered walls connecting two-walled septa, by chevronlike sculpture, and by a transverse segmentation that has not been observed in any forms assigned to anabaritids. Whether all anabaritid tubes had a closed apex is also unclear. We do not recognize new families and orders among anabaritids, because of existing confusion (Valkov and Sysoiev, 1969, 1970; Missarzhevsky, 1974; Valkov, 1982), but we describe some concrete representatives of this interesting group.

Anabarites tricarinatus Missarzhevsky, *Anabaritellus* Missarzhevsky, and *Selindeochrea* Valkov have a twisted conical tube with three longitudinal prominences distinct in transverse section—the shape of the mold is coincident with the tube's external shape. The prominences change into straight, thin or undulating lobes stretching along the entire tube (Fig. 18a). The width of the lobes is about equal to the diameter of the tube aperture (approximately 1 mm). The outer and inner surfaces of the tube are smooth. The presence of outer lobes suggests that these forms might have an inner skeleton (Abaimova, 1978). Similar fossils occur in Nemakit-Daldyn and the Tommotian in Anabar-Sinsk and in the Yudoma-Olenek Facies Regions of the Siberian Platform.

The triradiate conical tubes (*Mariochrea* Valkov, *Gastreochrea* Valkov, and the Aculeochreidae) are characterized by a distinct transverse separation of the tube into rings. The outer shape of the tube is round and the inner shape is tripartite (Fig. 18b). The separate rings of the tube are trifoliate funnels set one into the other, which easily split apart. Valkov (1982) established that these rings were described as the operculum of *Coleolella billingsi* Missarzhevsky (Rozanov *et al.*, 1969). The lower margin of each funnel bears spinelike prominences seen as three rows of different-shaped furrows on the tube mold. The Aculeochreidae may be molds of *Mariochrea* and *Gastreochrea*. Such structures are found throughout the Tommotian Stage of the Anabar-Sinsk and Yudoma-Olenek Facies Regions of the Siberian Platform.

A third group of triradiate tubes contains the greatest number of forms, including *Anabarites trisulcatus* Missarzhevsky, *Angustiochrea* Valkov and Sysoeiv, and others. These simple tubes have three longitudinal ribs on the inner surface but only smooth growth lines on the outer surface. *A. trisulcatus* appears at the base of the Nemakit-Daldyn Horizon of the Vendian. On the Siberian Platform the last typical anabaritids are known in the *D. lenaicus–T. primigenius* Zone of the Tommotian Stage. They or similar fossils are found not only on the Siberian Platform, but also in Mongolia, Kazakhstan, Central Asia, the Lesser Himalayas, Iran, northwest Canada, Newfoundland, Sweden, south China, and Australia. They occur in the Australia up to the Botomian (Jell *et al.*, 1990).

Among these numerous structures with similar triradiate symmetry may be true coelenterates, for example, the three-lobed *Selindeochrea* and similar forms. Two other groups are sometimes ascribed to the coelenterates: the Cambrian "stromatoporoids" (Yaworsky, 1932; Khalfina and Yaworsky, 1974) and *Cambro-*

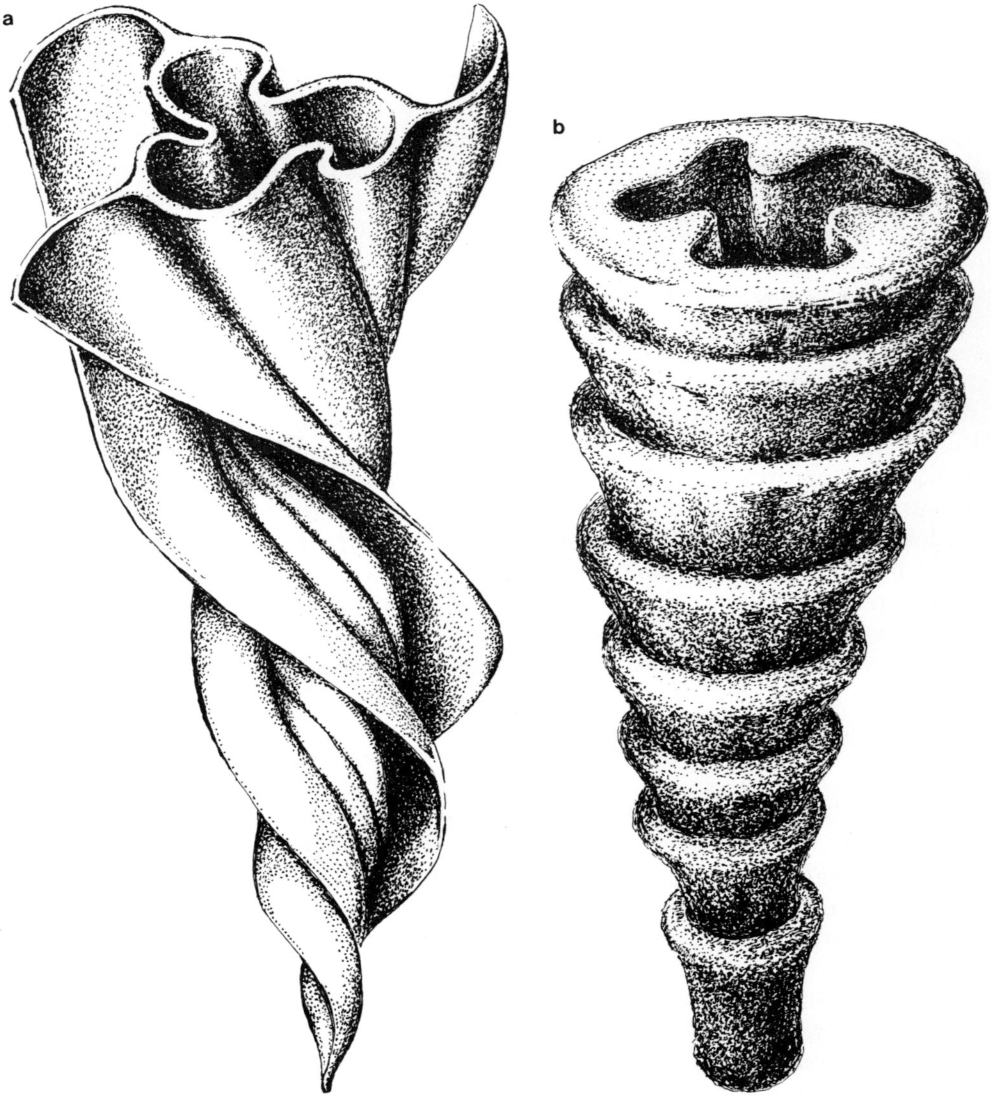

Figure 18. Reconstruction of (a) *Anabarites tricarinatus* Missarzhevsky and (b) *Mariochrea sinuosa* Valkov. Scale bar 100 μm.

porella Korde and similar forms (Korde, 1959; Sayutina, 1985b). The archaeocyathan origin of Cambrian "stromatoporoids" is undoubted. Zhuravlev (Voronova *et al.*, 1987) put them into the order Kazachstanicyathida of irregular archaeocyaths. They have a morphology typical of many irregular archaeocyaths and a cup ontogeny similar to that of the regular archaeocyath order Capsulocyathida. Such ontogeny of the skeleton is not observed in true stromatoporoids. Even if some stromatoporoids are derived from some irregulars (Zhuravlev, 1985), the whole group is, nevertheless, of sponge affinity (Vacelet, 1985; Wood and Reitner, 1986; Boyajian and LaBarbera, 1987).

Cambroporella (Fig. 16b) has a laminated calcareous skeleton with regularly arranged pores; it was compared with porous hydroconozoans by Sayutina (1985b), although it does possess tabulalike structures. However, the typical features of this form indicate it is a dasyclade green algae (Bassoullet *et al.*, 1979; Chuvashov *et al.*, 1987). *Cambroporella* together with *Amgaella* Korde, *Yakutina* Korde (=*Siberiella* Korde), and some similar forms belong to the family Seletonellaceae. This family was typical of the Cambrian (beginning in the Botomian) of the Siberian Platform, the Altay-Sayan Fold Belt, Mongolia, and Kazakhstan. *Cambroporella*, however, possesses tabulalike structures (Fig. 16b).

Two compound cup-like things have been described by Korde (1959) from the Siberian Platform. One of them, *Lenaella* Korde from the upper Atdadanian of the middle Lena River greatly resembles Australian *Tunkia* Bedford and Bedford, but this find has never been mentioned there again. *Coelenteratella* Korde from the base of the Middle Cambrian of the Amga River are *nomen dubium*.

As a whole, the probable Cambrian skeletal coelenterates informally called Corallomorpha by Jell (1984) are chiefly solitary vase-shaped empty cups with a varied and different microstructure (Zhuravlev, 1991). Their diversification, expressed in the appearance of many short-lived groups, the relationship among which is difficult to understand, is similar to the pattern of diversification of otherCambrian groups. Cambrian problematic skeletal and soft-bodied coelenterates are very rare (see Section 3.2), in spite of the widely accepted concept (Fedonkin, 1983; Glaessner, 1984; Sokolov and Ivanovsky, 1985) of the relationship between diverse Vendian "coelenterates" and Phanerozoic representatives of the same phylum.

3.1.4. "Worm Tubes"

Many Cambrian tubicolous fossils may be worm tubes. Four main groups can be distinguished: the siliceous tubes of *Platysolenites* Eichwald and *Volborthella* Schmidt, phosphate tubes of hyolithelminths, calcareous tubes of coleolids and similar structures, and sabelliditids with chitinlike tubes. The siliceous tubes of *Platysolenites antiquissimus* Eichwald and *Volborthella tenuis* Schmidt (Fig. 12, panels 19 and 20) are characteristic of the siliciclastic facies of the East European Platform. They occur in the Rovno (starting in the second half of the *Sabellidites cambriensis* Zone), Lontova, and Dominopol' Horizons. They also occur in Scandinavia (Mens *et al.*, 1987), Wales (Rushton, 1978), England (Brasier, 1986), western North America (Firby-Durham, 1977), and the northern Siberian Platform in Anabar Massif (Missarzhevsky, 1983, 1989), but only in siliciclastic facies and within the interval from the Tommotian to the lower part of the Atdabanian Stage. *Platysolenites* tubes, discussed in Section 3.1.1, may have the narrowest end randomly bent. The fragments of such tubes have been described as *Spirosolenites* Glaessner.

Volborthella appears in the Dominopol' Horizon and is distributed at least in the lowermost Rausve Horizon, but is characteristic of the upper Dominopol' and Vergale Horizons of the East European Platform and of the corresponding level in Spitsbergen and northern Europe (Mens *et al.*, 1987). It occurs in the western part of the United States as *Volborthella* (Lipps and Sylvester, 1968) and *Campitius* (Firby and Durham, 1974) and perhaps in western Canada (Fritz, 1980) within the limits of the "*Fallotaspis*" Zone in all these regions. Lipps and Sylvester (1968) and Glaessner (1976) assigned *Volborthella* to the annelids. *Volborthella* built its tube incrementally with amorphous silica in a layer-by-layer crystallization, which gives the appearance of a divided tube with its parts similar to phragmacone chambers. The thinner layers consist of cristobalite and tridymite, and the thicker ones are composed of quartz (Rozanov, 1983b), contradicting Yochelson's suggestions (1977, 1983) that *Volborthella* is a diagenetically transformed *Salterella* Billings. Therefore the interpretation of tubes from Scotland (Yochelson, 1983) and Spitsbergen (Lauritzen and Yochelson, 1982) as poorly preserved *Salterella* may be questionable. Furthermore, *Volborthella* occurs chiefly in the Atdabanian and *Salterella* in the Toyonian (Yochelson, 1981b; Voronova *et al.* 1987). *Onuphionella* Kirjanov (Krandievsky *et al.*, 1968) is also a "terrigenous" (?agglutinated) tube from the Tommotian of the Baltic Group on the East European Platform. It is a tube with imbricated mica flakes covering its exterior normal to the tube axis.

Phosphate tubes are assigned to the hyolithelminthes, which is divided into the following families: Hyolithellidae (*Hyolithellus* Billings, *Pseudorthotheca* Cobbold) and Torellellidae (*Torellella* Holm, *Rushtonia* Cobbold and Pocock, *Torellelloides* Meshkova, *Koksuja* Missarzhevsky). Rozanov (1986) suggested that the family Mongolitubulidae (*Mongolitubulus* Missarzhevsky, *Tommotitubulus* Fedorov) should also be placed in this order. Both tubes have an elaborate surface ornamentation and they belong among conodontomorphs rather than among hyolithelminths. Hollow tubes of hyolithelminths are characterized by a clear transverse sculpture, and the length of the tube exceeds its diameter. They are often flexed randomly, may have a pedicle on the narrow end, and have a smooth inner surface. Hyolithellidae are round (Fig. 9, panel 21, and Fig. 11, panel 21) and Torellellidae are elliptical in section (Fig. 11, panel 22, and Fig. 12, panel 22). The tubes are phosphatic (the composition of *Koksuja* and *Rustonia* is unknown). According to Missarzhevsky (1984), the phosphatic tube of *Pseudorthotheca costata* Mambetov from the Chulaktau Formation (Tommotian Stage) of Maly Karatau might be a secondary transformed calcareous tube of *Coleolella billingsi* Missarzhevsky. The hyolithelminth tube wall has two platey layers of different orientation (Grigorieva, 1980), an inner layer parallel with the tube surface and an outer one oblique to the inner layer. The two-layered structure, the arbitrary tube flexure, and the lack of regularity in the arrangement of the sculpture are the only characters that indicate a relationship between hyolithelminths and Recent worm tubes.

On the Siberian Platform the first hyolithelminths (*Hyolithellus* and *Torellella*) are found at the base of the Tommotian Stage. Finds of *Hyolithellus* sp. on the Aldan River 21–23.5 m below the top of the Yudoma Formation (Valkov, 1983) is evidently a mistake. Those fossils are more like fragments of *Sinotubulites* Chen, Chen and Qian or *Nevadatubulus* Signor, Mount and Onken, both of which in turn could be synonyms of *Cloudina* Germs (Grant, 1990). (The last one is now found in the upper Vendian of the Yudoma-Maya Area [Fedorov and Zhuravlev, personal communication]). The tube wall of these genera consists of nested funnels (Signor *et al.*, 1987). The upper limit of *Torellella* on the Siberian Platform is poorly known, but it does occur in the lower Botomian. *Hyolithellus* is found up to the Middle Cambrian (Meshkova, 1985). *Torellelloides* and *Rushtonia* have been found only in the upper Tommotian and the lower Atdabanian Stages. On the East European Platform *Torellella* is known in the Dominopol' and Rausve

Horizons. In Kazakhstan and Mongolia, *Hyolithellus* and *Torellella* span the lower Tommotian–Upper Cambrian (Koneva *et al.*, 1990) and *Koksuja* ranges from the upper Atdabanian through the lower Botomian Stages.

The Coleolidae (*Coleolus* Hall, *Coleoloides* Walcott, *Coleolella* Missarzhevsky) have calcareous tubes that are conical, slightly bent at random, and thick-walled with a distinct oblique or transverse sculpture (Fig. 9, panel 23). All three genera occur on the Siberian Platform from the bottom of the Tommotian Stage to the beginning of the Atdabanian Stage. These forms are not known from the Lower Cambrian of the East European Platform or Kazakhstan, but *Coleolella* is found in the Middle Cambrian of Central Asia (Meshkova, 1985). *Cambrotubulus* (Fig. 9, panel 24, and Fig. 10, panel 24) resembles the coleolids more than the anabaritids. Its tubes are irregularly flexed in different planes, are round in section, and have thin growth lines on the outer surface and a smooth inner surface. *Cambrotubulus* is characteristic of the Nemakit-Daldyn Horizon and the Tommotian Stage of the Siberian Platform, Kazakhstan, and Mongolia, but in the latter region it also occurs (redeposited?) in the Botomian.

The last group of tubes, the sabelliditids, are very long, thin-walled, cylindrical, and commonly flattened with a wrinkled surface (Fig. 12, panel 25). The sabelliditid tube microstructure differs (Urbanek and Mierzejewska, 1983; Ivantsov, 1989) from some pogonophorans, to which Sokolov (1965, 1972) believed they belonged. To confirm this speculation, not only pogonophores, but graptolite and pterobranch microstructure should be studied. Sabelliditids are characteristic mainly of the upper Vendian of the East European Platform. They are present in the Redkino Horizon, and three genera (*Sabellidites* Yanishevsky, *Paleolina* Sokolov, and *Sokoloviina* Kirjanov) are found in the upper Rovno Horizon (Mens *et al.*, 1987). Only *Sabellidites cambriensis* Yanishevsky is well represented in the Lontova Horizon; it rarely is found in the Vergale Horizon, and is also known in the Tommotian in the southern Urals. Sabelliditids, represented by *Calyptrina* Sokolov (Nemakit-Daldyn Formation in the western Anabar Massif) and *Paleolina* (Sukharikha Formation in the northern Krasnoyarsk Region and some other localities), are found in the lower Nemakit-Daldyn Horizon on the Siberian Platform.

In the living fauna, about 20 wormlike phyla were formerly classified together, although they are not related. Wormlike fossils are known that cannot be classified with modern worms (Conway Morris and Robison, 1986; Hoffman and Nitecki, 1986). The earliest Cambrian tubes discussed here are seldom compared to any group of wormlike organisms other than polychaetes and pogonophorans. They could, however, belong to animals whose modern descendants look quite different. For example, hyolithelminth tubes might be ancestors of the lophophorates.

3.1.5. Mollusk-like Fossils

The fossil record of mollusk-like shells on the Siberian Platform is generalized from Rozanov *et al.* (1969), Sokolov and Zhuravleva (1983), Rozanov and Sokolov (1984), Bokova (1985, 1990), Khomentovsky and Karlova (1986, 1989), Jermak and Pelman (1986), Missarzhevsky (1989), and Vassiljeva and Rudavskaja (1989).

The oldest mollusk-like fossils occur in the uppermost part of the Nemakit-

Daldyn Horizon (the western Anabar Massif and Olenek Uplift), and are represented by shells of *Purella* Missarzhevsky and *Latouchella* Cobbold (= ?*Oelandiella* Vostokova). *Purella* is bilaterally symmetrical, slightly compressed laterally with a tightly curved apex and keel (Fig. 9, panel 26), and cannot be a mollusk at all. *Latouchella* has an asymmetrical conch with a low spiral and oval aperture. Species of *Stenothecoides* and *Securiconus* recorded by Bokova (1985) are assigned to *Purella* and to *Latouchella*, respectively. Bokova (1985, pl. II, Fig. 7) also noted retiform problematic fossils at this level which may be fragments of *Canopoconus* Jiang or *Maikhanella* H. Zhegallo.

Many more mollusk-like fossils are now known from the Siberian Platform mainly in the Tommotian (Missarzhevsky, 1989; Khomentovsky *et al.*, 1990; Pelman *et al.*, 1990; Vassiljeva, 1990), but some of them are synonyms (e.g., *Tuoraconus* Missarzhevsky = *Lenoconus* Vassiljeva) and may also have relatives among Chinese fauna.

In the *sunnaginicus* Zone of the southern Siberian Platform, molluscoid shells are very diverse. Only on the Aldan River (in the Pestrotsvet Formation), in addition to the forms previously noted from this level, have numerous Helcionellida (*Bemella* Missarzhevsky, *Igorella* Missarzhevsky, *Ilsanella* Missarzhevsky = *Ginella* Missarzhevsky) been found. From the equivalent level on the Gonam River, *Gonamella* Valkov and Karlova, and on the Selinde River, *Salanyella* Missarzhevsky and *Hamusella* Valkov, were described. In addition to *Latouchella*, the Corespiridae are represented by *Pseudoyangtzespira* Bokova and Stenothecidae by *Anabarella* Vostokova (Fig. 9, panel 27, and Fig. 10, panel 27). On the Selinde River, new families appear and among them are asymmetrically tightly-coiled Pelagiellidae (*Nomgoliella* Missarzhevsky) and true gastropods in the Aldanellidae (*Aldanella* Vostokova, *Barskovia* Golubev, and *Paraaldanella* Golubev) (Fig. 9, panel 29, and Fig. 10, panel 29). Shells with a snorkel (*Yochelcionella*) also occur in this zone (Fedorov, 1986) and Sinuitidae (*Khairkhania* Missarzhevsky) have been found on the Selinde River. Rostroconchs *Watsonella* Grabau (= *Heraultipegma* Pojeta and Runnegar) (Fig. 9, panel 35) characterize the *sunnaginicus* Zone. Kerber (1988) and Landing (1989) discussed the synonymy of *Watsonella* and *Heraultipegma*. Both genera, however, were described from fossils with different kinds of preservation, and the Siberian rostroconchs are preserved in yet another way.

In the *regulares* Zone, an assemblage of completely problematic, cap-shaped fossils is known as *Aldanolina* Pelman (Fig. 9, panel 38). They are low, bilaterally symmetrical shells with a round apex corresponding to the limit of the aperture with a widely triangular indenture under the apex. *Aldanolina* perhaps may be a halkieriid shield. In the lower part of the Lena River at Khara-Ulakh, in the equivalent of the *regularis* Zone in the Tyusser Formation, the oldest pelecypods *Fordilla* Barrande (Fig. 9, panel 36) and *Buluniella* Jermak were recently found. Jermak (1986) considers *F. sibirica* Krasilova as a junior synonym of *F. troyensis* Barrande. However, only *F. sibirica*'s hinge teeth have been described (Krasilova, 1977; Jermak, 1986), but the hinge is unknown in *F. troyensis* (Pojeta *et al.*, 1973; Yochelson, 1981a). This is probably because *F. sibirica* is preserved almost always as molds of shells with closed valves and very seldom as separate valves [in Jermak's (1986) material the proportion of molds and valves is 30:2]. *F. troyensis*, on the contrary, usually occurs as separate valves (Yochelson, 1981a). The species also differ in valve ornamentation: in *F. sibirica* the outer surface of the valves has dense, clearly distinct small ribs. If the difference in ornamentation is of no importance, then the difference in the structure of the hinge probably indicates a higher taxonomic level than species. *Tannuella* Missarzhevsky occurs in the lower Atdabanian on the Selinde and Merdvezh'ya Rivers. It is the most symmetrical and largest Helcionellidae with a central apex. The size of the aperture of the shell is up to 3–4 cm. An enigmatic form of "*Mirabella*" Barskova is known from the upper Atdabanian in the Kolyma River basin (Barskova, 1988). These shells have an eccentric curved apex with a wide process under it. Their shape slightly resembles *Mackenziella* H. Zhegallo from the Sekwi Formation (upper part of "*Fallotaspis*" Zone) of western Canada. The latter carries a symmetrical pair of muscle scars of the subapical process (Voronova *et al.*, 1987). In the upper zone of the Atdabanian Stage at Khara-Ulakh (Tyusser Formation) appear true *Pelagiella* Matthew (Fig. 9, panel 28) and Sinuitidae (*Michniakia* Missarzhevsky). *Pelagiella* and *Yochelcionella* are especially characteristic of the entire *lermontovae* Zone of the Atdabanian Stage and of the *squamosus–zelenovi* Zone of the Botomian (Missarzhevsky and Mambetov, 1981; Rozanov and Sokolov, 1984; Aksarina, 1984; Voronova *et al.*, 1986, 1987). Shells of the Multifariidae (*Protowenella* Runnegar and Jell) occur in the Sinsk Formation (*gurarii* Zone of the Botomian Stage).

In Mongolia the succession of molluscoids has been well studied in the Dzabkhan or Tsaganolom structural province (Voronin *et al.*, 1982; Voronova *et al.*, 1986). In beds with *Tiksitheca licis–Maikhanella multa* (the Bayan-gol Formation), *Maikhanella* (Fig. 11, panel 37) and *Rozanoviella* Missarzhevsky (Fig. 11, panel 33) occur. In *Rozanoviella* a series of holes runs along the shell and an axis extends from the apex to the margin and resembles the traces of an overgrown sinus like that in the Bellerophontida. At the next stratigraphic level (beds with *Ilsanella compressa*) conchs are more abundant and represented by Helcionellidae (*Ilsanella, Salanyella*), *Coreospirida* (*Latouchella*), Stenothecidae (*Anabarella*), Pelagiellidae (*Nomgoliella*), Aldanellidae (*Barskovia*), and Sinuitidae (*Khairkhania*) (Fig. 11, panel 31). In the beds with *Tannuella gracilis,* this assemblage is completed with *Bemella* and *Tannuella* (Fig. 11, panel 26). In the Atdabanian part of the Salany-gol Formation *Aldanella, Pelagiella,* and *Watsonella* occur. Numerous Yochelcionellidae (Fig. 11, panel 30) are distributed in the Botomian part of western Mongolian sections.

The molluscan assemblages of the Maly Karatau (Missarzhevsky and Mambetov, 1981) are poorer than those of Siberia and Mongolia. In the Chulaktau Formation (Tommotian Stage, *Pseudorthotheca costata* Zone) only Coreospiridae (*Chabaktiella* Missarzhevsky) are known. In the lower part of the Shabakty Formation (the *Rhombocorniculum cancellatum* Zone), two assemblages are represented. The lower part of the Zone (Atdabanian Stage) is characterized by *Igorella, Yochelcionella,* and *Protowenella* and the upper part (Botomian Stage), in addition to them, by *Purella, Pelagiella,* problematic Onychochilidae (*Beshtashella* Missarzhevsky), and Tryblidiidae (*Aktugaia* Missarzhevsky). On the East European Platform mollusks are unknown except for *Aldanella kunda* (Opik), *A. polonica* Lendzion (Fig. 12, panel 29), and *Anabarella* (Öpik, 1926; Lendzion, 1977; Posti, 1978) in the Lontova Horizon. A possible Lower Cambrian cephalopod, *Vologdinella* Balashova, was recorded from central Kazakhstan (Vologdin, 1931), but the only specimen known could be assigned to the group that includes *Salterella*.

During the last 15 years a large number of papers have been published on the systematics of ancient mollusks and their early evolution (e.g., Runnegar and Jell, 1976; Runnegar and Pojeta, 1974; Rozov, 1976; Pojeta and Runnegar, 1976; Starobogatov, 1977; Yochelson, 1979, 1981a; Pojeta, 1980; Dzik, 1987; Runnegar, 1981, 1983; Stanley, 1982; Yu, 1984). These are based mainly on the morphology of Cambrian mollusks and to a lesser extent on their stratigraphic occurrence. Therefore more precise stratigraphic data may change the conclusions. For example, Yochelson (1981a) wrote that *Fordilla* could not have originated from *Heraultipegma,* because the latter is younger, but later *Heraultiperma* was found at the base of the Tommotian Stage (Rozanov and Sokolov, 1984; Khomentovsky and Karlova, 1986). At present practically all groups of Early Cambrian molluscoids have been found in the *sunnaginicus* Zone of the Tommotian Stage or at least in the overlying *regularis* Zone. Therefore no phylogenetic scheme for the origin of mollusk classes can be based on stratigraphic data. Data on the morphology of Cambrian mollusks in Siberia and neighboring regions are extremely rare; descriptions are based mainly on molds. For example, the Family Yangtzespiriidae was characterized (Yu, 1987) by the presence of a posterior canal on the aperture. This feature is very rarely preserved and so molds might also be assigned to Coreospiridae. In this respect, bivalves are better (Jermak, 1986; Krasilova, 1987). Of all the fossils assigned to monoplacophorans, paired muscle scars are known only in *Aktugaia* (Missarzhevsky, 1976), but this form is possibly *Mobergella sensu lato*. All other conchs and tightly coiled shells can be placed with the same degree of uncertainty either among monoplacophorans or gastropods, or in extinct classes of mollusks or even nonmollusks (*Canopoconus, Rozanoviella, Aldanolina, "Mirabella"*). Gradual morphological transitions observed between different conchs, including *Watsonella,* and tightly coiled shells, and their occurrence at the same level may indicate that Early Cambrian molluscoids are a separate single animal group.

Hyoliths are a diverse and widely distributed group of Early Cambrian organisms, placed either among mollusks (Marek and Yochelson, 1976; Dzik, 1978) or in a separate phylum (Runnegar *et al.*, 1975; Sysoiev, 1976b). In the USSR they occur in all four Lower Cambrian stages on the Siberian Platform (Sokolov and Zhuravleva, 1983; Rozanov and Sokolov, 1984), in the Kolyma Basin (Tkachenko *et al.*, 1987), on the East European Platform (Mens *et al.*, 1987), in the Altay-Sayan Fold Belt (Sysoiev, 1970), in Kazakhstan and Central Asia (Missarzhevsky and Mambetov, 1981), and in the Transbaikalian and Mongolia (Voronin *et al.*, 1982). The stratigraphic succession and morphological changes of hyoliths have been most fully studied on the Siberian Platform (Sysoiev, 1962, 1968, 1972, 1976a, b; Meshkova, 1974a; Valkov, 1975; Sokolov and Zhuravleva, 1983; Rozanov and Sokolov, 1984). In the Anabar-Sinsk Facies Region, the oldest hyoliths are found in the upper Nemakit-Daldyn and *sunnaginicus* Zone. Orthothecida are represented by the Tchuranithecidae (*Turcutheca* Missarzhevsky, *Ovalitheca* Sysoiev, *Ensitheca* Valkov) and Allathecidae (*Allatheca* Missarzhevsky), with Exilithecida including the Exilithecidae (*Exilitheca* Sysoiev = *Egdetheca* Missarzhevsky), Nicathecidae (*Nicatheca* Valkov), and Circothecida (*Spinulitheca* Sysoiev, *Ladatheca* Sysoiev, *Loculitheca* Sysoiev). These numerous forms as well as the problematic

Conotheca Missarzhevsky and *Laratheca* Missarzhevsky belong to Orthothecimorpha. This class consists of hyoliths whose conical valve has an aperture without a lip (Fig. 9, panel 40) and the operculum (the second valve) is flat or a low cone. Apertural appendages (helens) in Orthothecimorpha are either absent or, according to Dzik (1980), not calcified. Apertures in the Orthothecimorpha of the *sunnaginicus* Zone may be round, oval, elliptical, or, more seldom, oval–triangular in section, i.e., the dorsal and ventral side of the shell are weakly expressed. By the end of the *sunnaginicus* Zone, all three orders of Orthothecimorpha were present.

The hyoliths evidently radiated in the *regularis* Zone. The Orthothecimorpha reached their maximum diversity in the entire Early Cambrian. Among the Tchuranithecidae, *Tchuranitheca* Sysoiev and *Uniformitheca* Sysoiev appeared with a prominent ventral apertural edge which is characteristic of this family. The Allathecidae reached their peak (*Eonovitatus* Sysoiev and others). The Isitithecidae arose with a lenticular cross section, and the first hyoliths with round to trapezoidal cross sections appear among Exilithecida and Obliquathecidae. The *Dokidocyathus regularis* Zone is marked by the appearance of the first true Hyolithimorpha with prominent lips and appendages (Fig. 9, panel 41). Two suborders appear at once, including the Aimitidae (*Oxytus* Sysoiev), Crestjahitidae (*Crestjahitus* Sysoiev, *Burithes* Missarzhevsky, *Jakuticornus* Sysoiev), and Notabilitoidei with Notabilitidae (*Notabilitus* Sysoiev, *Oblisicornus* Sysoiev), Dorsojugatidae (*Dorsojugatus* Sysoiev), and Nelegerocornidae (*Microcornus* Mambetov, *Yakutolituus* Missarzhevsky). The simultaneous appearance of such diversity indicates that, in this case, too, the appearance is not of the group itself, but of its skeletonization. Among Orthothecimorpha (*Exilitheca multa* Sysoiev) and especially Hypolithimorpha, forms were already present in the *regularis* Zone well-distinguished dorsal and ventral sides of the shell.

A striking parallel occurs in the morphological changes in Orthothecimorpha and Hyolithimorpha. Thus, in the middle Tommotian both were mainly represented by species with round to triangular and lenticular sections of the shell. At the end of the Tommotian and the beginning of the Atdabanian, the species with trapezoidal sections appear in both groups. The same changes are observed within the orders (Exilithecidae to Obliquathecidae) and families (*Eonovitatus* to *Majatheca* Missarzhevsky, *Isititheca* Sysoiev to *Plicitheca* Sysoiev) and within genus [*Dorsojugatus sedecostatus* (Sysoiev) to *D. multicostatus* (Sysoiev)] in the Atdabanian (Sysoiev, 1972).

The next important stage of hyolithan evolution was among Orthothecimorpha. At the beginning of this stage, Tetrathecidae with a polyhedral cross section of the axially twisted shell, Novitatidae with a concave ventral side of the shell, and *Lenatheca* Missarzhevsky with a heart-shaped aperture appeared. In the second half of the Atdabanian, new families arose among Hyolithimorpha (Angusticornidae and Galicornidae), including *Trapezovitus* Sysoiev and *Grandicornus* Sysoiev, the largest Early Cambrian hyolith.

At the beginning of the Botomian, the Orthothecimorpha and Hyolithimorpha were rather diverse. By the end of the Toyonian, their number and diversity decreased at all levels. This late Early Cambrian decrease in diversity resembles the pattern of archaeocyathan development on the Siberian Platform, probably because hyoliths reach their maximum diversity in interbiohermal facies. Indeed, as soon as archaeocyathan–algal bioherms arose in the Toyonian (*grandis* Zone), hyoliths (*Sokolovitheca* Sysoiev and *Ketemecornus* Sysoiev) immediately appeared. In the Cambrian of Kazakhstan and Mongolia hyoliths are scarce and occur in the carbonate parts of sections. In the siliciclastic facies of the East European Platform, they are rare. The interbiohermal facies, characterized by fine carbonate detritus with considerable argillaceous admixture, were favored by hyoliths because they were detritophagous (Runnegar *et al.*, 1975; Marek and Yochelson, 1976; Dzik, 1980; Sysoiev, 1981, 1984).

Hyoliths are considered in the Matthew–Sysoiev model as an independent branch of lophophorates. According to this model, the helens supported the lophophore. Dzik (1980) assigned hyoliths to mollusks and noted a morphological similarity of their flat valve with gastropod opercula, with the exception of the asymmetry of the latter. However, this element is also morphologically similar to brachiopod valves (e.g., Bassett and Yochelson, 1979). Unlike Sysoiev, we do not think that the septa separated gas camera in the initial part of the hyolithan shell. Perhaps these septa served only to separate a living part of the valve, as in the attached productid *Richthofenia*.

Stenothecoids are another mysterious and exclusively Cambrian group of molluscoid organisms. The oldest stenothecoids are known from the Lower Atdabanian (*Stenothecoides*? in the Kuznetsky Alatau and Mongolia) (Aksarina and Pelman, 1978; Voronin *et al.*, 1982) and actually belong to mollusks. Stenothecoids (*Cambridium* Horny) (Fig. 9, panel 42), which appeared on the Siberian Platform in the second half of the Atdabanian stage (Sokolov and Zhuravleva, 1983), existed before the Early Botomian stage, and the same species also appeared in the Middle Toyonian. However, stenothecoids, such as *Bagenovia* Radugin, *Bagenoviella* Aksarina, *Sulcocarina* Aksarina, *Makarakia* Aksarina, and others, are especially diverse and abundant in the second half of the Atdabanian–Toyonian stages of the fold regions of the Kuznetsky Alatau, Mountain Shoria, Western Sayan, Mountain Altay, Tuva, Central Kazakhstan, and Mongolia (Korobov and Missarzhevsky, 1977; Aksarina and Pelman, 1978; Koneva, 1979; Pelman, 1985). Isolated stenothecoids are found in the Botomian of the Maly Karatau (Missarzhevsky and Mambetov, 1981).

Stenothecoids were considered an independent, extinct mollusk class (Aksarina, 1968; Yochelson, 1969), although Rozov (1984) suggested that they were a separate phylum, the Stenothecata, based on the combination of a bivalved carbonate shell with an incomplete symmetry (the beaks of valves are shifted relative to the longitudinal axis; the shell has unequal valves) of the segmented soft body and the absence of a hinge tooth. In addition, looplike scars on the mold of *Stenothecella* Aksarina are somewhat shifted along the longitudinal axis of the shell, unlike muscle scars. This system of scars closely resembles orthothecid intestinal loops. Considering all this, Rozov considered stenothecates to be closer to brachiopods than to mollusks. But 1 year later, teeth were found in stenothecoid shell of *Cambridium* and *Serioides* Pelman from the Botomian of western Mongolia. Between the tooth area and an apical process is a shallow groove where a ligament may have been. In this case, a lip, under the beak of the dorsal valve of some of the species *Cambridium* and *Katunioides* Aksarina, could serve to strengthen a ligament (Pelman, 1985). Most stenothecoid valves occur in archaeocyathan reefs. Large bivalve, stenothecoid shells with unequal valves most likely belonged to slightly moving organisms.

3.1.6. Arthropods

In the Tommotian stage, no arthropods have been found, except for trace fossils. Identification of *Tumulduria incomperta* Missarzhevsky from the base of the Pestrotsvet Formation of the Aldan River with trilobites (Fedorov *et al.*, 1979) proved false (Bengtson *et al.*, 1987). In the very base of the Atdabanian stage on the Siberian Platform, polymerid trilobites and a problematic bivalve crustacean (*Isoxys* Walcott) are present (Ivantsov, 1990) (Fig. 19c). Both of them occur near the lower boundary of the Atdabanian Stage in one section (Zhurinsky Mys) in the middle Lena River. A little higher, but also in the lower Atdabanian, bradoriid crustaceans occur (Grigorieva *et al.*, 1983). On the East European Platform, in Poland, in the Dominopol' Horizon, another group of problematic arthropods (*Cassubia* Lendzion and *Liwia* Lendzion) is known (Lendzion, 1975; Dzik and Lendzion, 1988). The first of these is put now in the Anomalocaridida (class Dicephalosomita), and the second is assigned to trilobites with nonmineralized

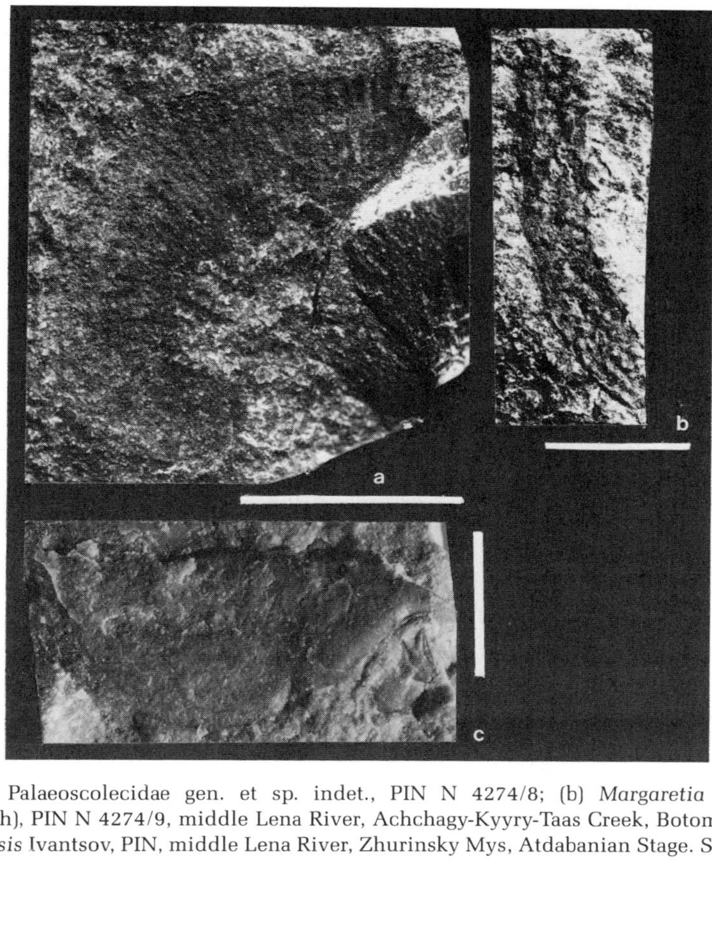

Figure 19. (a) Palaeoscolecidae gen. et sp. indet., PIN N 4274/8; (b) *Margaretia antiquissima* (Krischtofovich), PIN N 4274/9, middle Lena River, Achchagy-Kyyry-Taas Creek, Botomian Stage; (c) *Isoxys zhurensis* Ivantsov, PIN, middle Lena River, Zhurinsky Mys, Atdabanian Stage. Scale bar 1 cm.

dorsal exoskeleton, the Nektaspida. Yanishevsky (1950) described "the oldest trilobite" *Gdowia assatkini* Yanishevsky from deposits probably belonging to the lower Dominopol' Horizon. *Gdowia* is quite possibly a *Liwia* cephalon (Fig. 12, panel 51).

The oldest trilobites (*Profallotaspis* Repina) (Fig. 20a) belong to the family Fallotaspididae of the suborder Olenellina. *Profallotaspis* is an immediate ancestor of the widely distributed genus *Fallotaspis* Hupe (Fig. 20b). Deposits with *Fallotaspis* were supposed to be of the same age (Fritz, 1972) all over the world. Analysis of the fauna accompanying these trilobites has shown that the *Fallotaspis* biozone includes the upper three quarters of the Atdabanian Stage (Repina, 1986; Voronova *et al.*, 1987). The oldest species of this genus occur on the Siberian Platform, and the youngest ones in western North America. The oldest trilobites, Olenellina, had a cephalon with signs of segmentation, a multisegmented thorax, a small pygidium, but no facial sutures, which was an important landmark in the evolution of this group. In Repina's opinion, trilobites with librigenae had great advantages in molting. *Bigotina* Cobbold (Suborder Redlichiina) (Fig. 20f) was probably the oldest trilobite with librigenae, which might have originated directly from *Profallotaspis* (Repina, 1983). Miomerid trilobites appeared somewhat later than polymerids and, quite possibly, independently of them. In the *anabarus* Zone of the Siberian platform, two genera in the family Pagetidae, *Pagetiellus* Lermontova and *Triangullina* Repina, appear at once.

The oldest bradoriids, a bivalve crustacean related to ostracods, are also found in the *anabarus* Zone. In the upper Atdabanian, bradoriids are present on the East European Platform (Melnikova, 1987), in the northeastern Kazakhstan (Melnikova, 1990), and in the lower Botomian in the Transbaikalian (Melnikova, 1988). In the Botomian, the first palaeocopids (*Ushkarella* Koneva) probably occur in Central Kazakhstan (Koneva, 1979). The supposed ostracod *Motina* Galperova from the Mota Group on the Siberian Platform (Akulcheva *et al.*, 1981) is an oncolite envelope.

By the end of the Botomian, trilobite species and probably nontrilobite arthropods (Whittington, 1979; Müller, 1983) dominated the invertebrates.

3.1.7. Brachiopods

The first brachiopods are found in the *sunnaginicus* Zone of the Tommotian stage, in the lower part of the Pestrotsvet Formation on the Aldan River. They are inarticulates with a phosphate shell belonging to the genus *Aldanotreta* Pelman (Pelman, 1977) which represents the Order Paterinida. In the *regularis* Zone on the Siberian Platform, *Nochoroiella* Pelman (Rozanov and Sokolov, 1984) occurs, which Pelman (Grigorieva *et al.*, 1983) put in the Order Obolellida. In the Atdabanian Stage, true articulate brachiopods of the Order Orthida and true inarticulate ones of the Obolellida (Sokolov and Zhuravleva, 1983; Rozanov and Sokolov, 1984) appeared on the Siberian Platform. Orthids (Fig. 9, panel 48) are represented by *Nisusia?* (Ushatinskaya, 1986a) in the *anabarus* Zone. In the Altay-Sayan Fold Belt (the Kuznetsky Alatau), the problematic brachiopod *Kundatella* Aksarina is known at the very base of the Atdabanian Stage (the upper part of the Ust'kundat Horizon) (Aksarina and Pelman, 1978). The oldest Lingulida on the Siberian and East European Platforms evidently occur in the base of the Boto-

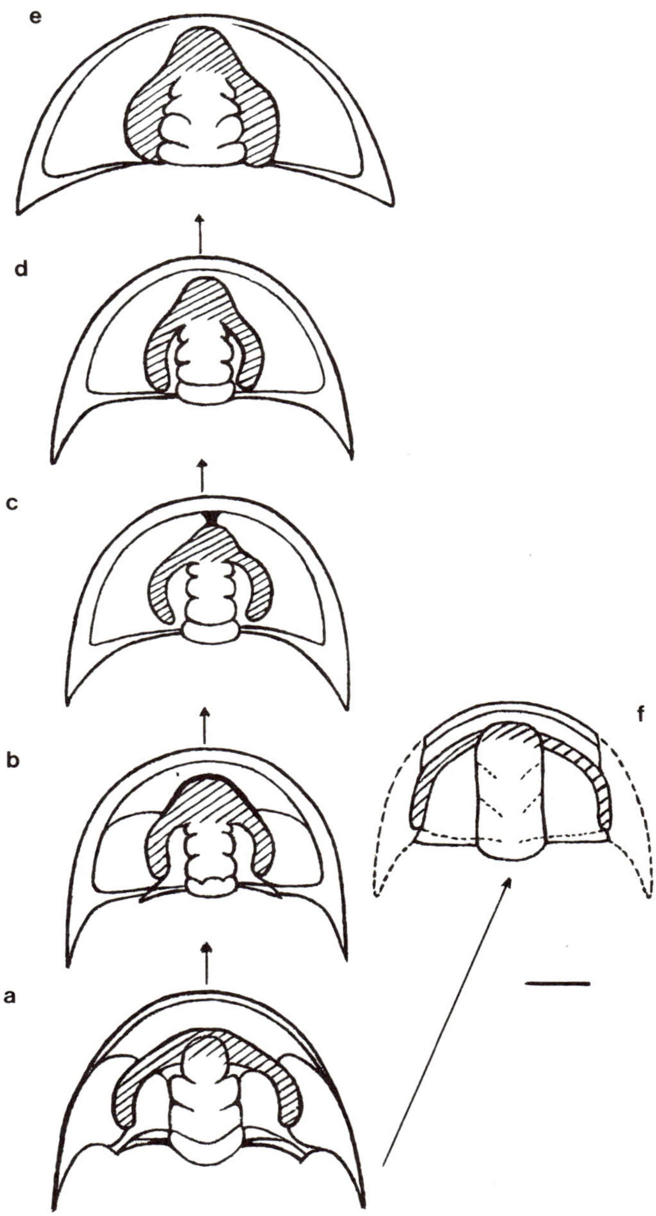

Figure 20. Cephalons of olenellid trilobites in the sequence of their appearance on the Siberian Platform [modified from Repina, (1983)]. (a) *Profallotaspis jakutensis* Repina, (b) *Fallotaspis siberica* Repina, (c) *Nevadella* sp., (d) *Judomia tera* Lazarenko, (e) *Judomiella heba* Lazarenko, (f) *Bigotina* (*Bigotinella*) *malycanica* Suvorova. Antennal segment is lined. Scale bar 2 mm.

mian. The first Actrotretida on the Siberian Platform are represented by the genus *Botsfordia* Matthew in the *lermontovae* Zone.

On the Siberian Platform, in the upper Botomian, Kutorginida (*Kutorgina* Billings) (Fig. 9, panel 46) are known. In Mongolia, the first Kutorginida are known in the upper Tommotian. They are *Khasagtina* Ushatinskaya (Fig. 11, panel 46) from the Bayan-gol Formation (Ushatinskaya, 1987), described earlier as *Kundatella* sp. (Voronin *et al.*, 1982). An Atdabanian age for these sediments cannot be ruled out (Kirschvink *et al.*, 1987). In the uppermost Toyonian, on the southern Siberian Platform (the Chara River), *Tcharella* Andreeva has been found, which possibly represents another order of articulate brachiopods, the Pentamerida (Andreeva, 1987).

Ushatinskaya (1986b) noted that the most distinct differences in the Cambrian brachiopod assemblages are associated with their facies. Usually species with calcareous shells are characteristic of the carbonate facies of the Siberian Platform. Phosphate shells prevail on the East European Platform.

New data on the ontogeny of recent brachiopods and on their shell composition and microstructure have changed opinions about this group (Wright, 1979; Rowell, 1982; Gorjansky and Popov, 1986). Articulation appears to be present in the Obolellida (Gorjansky and Popov, 1986; Ushatinskaya, 1988). Besides, morphological features of the shell of the Botomian paterinids (*Salanygolina* Ushatinskaya) (Fig. 11, panel 45) resemble those of articulates. With an equal degree of probability, these data indicate that obolellids, for example, belong to articulate brachiopods (Gorjansky and Popov, 1986) or that the process of "articulation" developed independently in different branches of brachiopods in the Early Cambrian.

3.1.8. Echinoderm Plates

Complete echinoderm skeletons (Eocrinoidea and Homostelea) are known on the Siberian Platform only from the Middle Cambrian (Jakovlev, 1956; Rozhnov, 1990), but plates which were undoubtedly skeletal components of these organisms occur in the *lermontovae* Zone of the Atdabanian Stage. From the upper part of the Emyaksin Formation, plates were described as problematic *Daniella* Zhuravleva (Shabanov *et al.*, 1987) (Fig. 21a). Echinoderm plates of the Botomian age are found in the eastern Transbaikalian (Figs. 21b and 21c). Plates from these localities are characterized by sutural pores. Such pores are present in plates of various Cambrian echinoderm classes (Sprinkle, 1981), but the morphological types indicate different Eocrinoidea. The oldest finds of echinoderm plates and skeletons in North America, Siberia, and other regions correspond to about the end of the Atdabanian.

3.1.9. Conodontomorphs

Here we consider diverse toothlike fossils attributed to protocondonts, paraconodonts, and euconodonts without regard to their systematic position. The oldest conodontomorph is *Protohertzina* Missarzhevsky (Fig. 10, panel 56, and Fig. 11, panel 56), which occurs at the base of the Nemakit-Daldyn Horizon (Missarzhevsky, 1973). *Protohertzina* was widespread over the entire Siberian Plat-

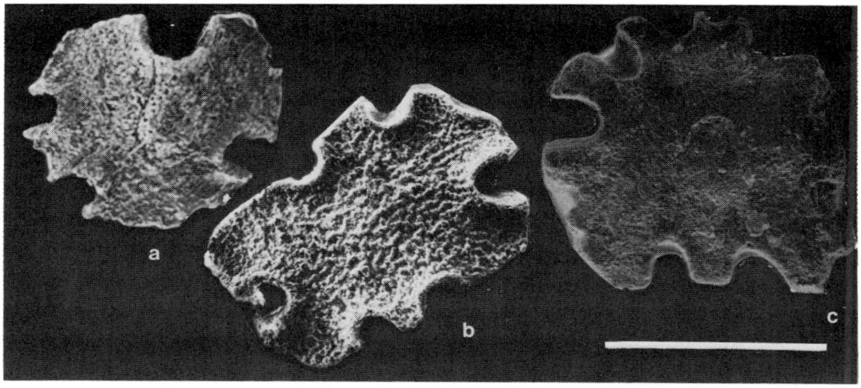

Figure 21. Echinoderm plates. (a) PIN N 4274/10, Siberian Platform, central part, Atdabanian Stage; (b, c) PIN NN 4274/11, 4274/12, eastern Transbaikalian, Botomian Stage. Scale bar 1 mm.

form, and in Kazakhstan, Mongolia, China, the Lesser Himalayas, and western North America (Conway Morris and Fritz, 1980; Missarzhevsky and Mambetov, 1981; Luo *et al.*, 1984; Nowlan *et al.*, 1985; Brasier and Singh, 1987; Hamdi, 1989). In the Maly Karatau, Mountain Shoria, and Mongolia, *Protohertzina cultrata* Missarzhevsky occurs up to the Botomian. *Protohertzina* is a bilaterally symmetrical tooth curved in the plane of symmetry, with a deep basal cavity. The concave side bears a median keel, the convex one is rounded, and the sides are screened off by lateral ribs so that the apertural section is round to pentagonal.

Fomitchella Missarzhevsky (Fig. 10, panel 58) is reported from the Tommotian on the Siberian Platform (Rozanov *et al.*, 1969) and from the same level in Maly Karatau (Missarzhevsky and Mambetov, 1981). It is a small, wide conical tooth with thin-layered walls. The problematic conodontmorph *Kijacus* Missarzhevsky is a spiniform, slightly curved tube with a round cross section. It occurs in the upper Tommotian (the Ust'kundat Horizon) of the Kuznetsky Alatau. Initially, this structure was described as *Hyolithellus kijanicus* Missarzhevsky (Rozanov and Missarzhevsky, 1966). The toothlike structure with elaborate cancellated ornamentation *Rhombocorniculum* Walliser (Fig. 9, panel 57, and Fig. 11, panel 57) (Sokolov and Zhuravleva, 1983; Rozanov and Sokolov, 1984) is widespread over the Siberian Platform, from approximately the base of Atdabanian Stage. In the lower Atdabanian and, probably, the upper Tommotian (*lenaicus–primigenius* Zone), this genus is represented by the species *R. insolutum* Missarzhevsky, but in the middle Atdabanian (*pinus* Zone), by *R. cancellatum* (Cobbold). The latter is present throughout the middle Botomian (*asiaticus* Zone) and occurs in different facies (in shallow-water facies of the Pestrotsvet Formation as well as shelf margin facies of the Sinsk Formation). Most of *R. cancellatum* occur, apparently, within the *kokoulini* Zone of the Atdabanian and the *squamosus–zelenovi* Zone of the Botomian. At the same time, *R. cancellatum* is reported practically everywhere (Landing *et al.*, 1980; Missarzhevsky and Mambetov, 1981; Aksarina, 1984; Brasier, 1986).

From the upper Atdabanian of the Maly Karatau (Shabakty Formation) several species of conodontomorphs were described: *Amphigeisina renae* Mambetov and Missarzhevsky, *Hertzina busulcata?* Müller, and *Oneotodus? triangulus* Mambetov and Missarzhevsky (Missarzhevsky and Mambetov, 1981). *Mongolodus* Missarzhevsky and *Mongoliacus* Missarzhevsky (Missarzhevsky, 1977) were previously reported from the Botomian of Mongolia (the Salany-gol Formation). *Glauderia* Ch. Poulsen and "*Prooneotodus*"? are recognized in the Botomian of the Mountain Shoria (Aksarina, 1984). Most Lower Cambrian conodontomorphs (*Protohertzina*, *Amphigeisina* Bengtson, *Glauderia*, and others) are assigned to the protoconodonts according to the classification of Bengtson (1976, 1983) and Miller (1984). Szaniawski (1982) suggested a chaetognath affinity for protoconodonts. The systematic position of *Rhombocorniculum* is not clear. *Fomitchella*, by Dzik's (1986b) assertion, can even be attributed to ancient chordates. However, Hinz (1987) described conodontlike spines on the plates of *Microdictyon* which evidently have no relation to chordates, hemichordates, or chaetognaths.

Among conodontomorphs are the ornamented tubular "Mongolitubulidae." *Tommotitubulus* (Fig. 9, panel 60) comes from the base of the Pestrotsvet Formation of the Aldan River (Tommotian Stage) (Fedorov, 1986) and *Mongolitubulus* (Fig. 11, panel 61) occurs in the Botomian of Mongolia, Maly Karatau, and Kirgizia (Missarzhevsky and Mambetov, 1981). Upper Cambrian *Mamillodus* Abaimova of the Maly Karatau (Abaimova, 1983) resembles both of these genera. Elaborate ornamentation of these tubes does not permit them to be put in the hyolithelminths. *Rukhinella* Borovko (L. Ye. Popov *et al.*, 1989) from the Upper Cambrian of the Leningrad region could be a derivative of a *Mongolitubulus*-bearing animal with very elaborated ornamentation on the tube. *Koksodus* Missarzhevsky and *Salanacus* N. Grigorieva (Fig. 11, panel 59) from the upper Tommotian of the Maly Karatau and Mongolia are curved plates with several conical teeth (Missarzhevsky and Mambetov, 1981; Voronin *et al.*, 1982). The shape and tooth size of these structures are widely varied, so they are probably the same species.

3.1.10. Small Problematic Fossils

Here we deal with those Lower Cambrian skeletal fossils whose systematic position is obscure. Most of them, such as Tommotiida *sensu lato*, coeloscleritophorans, cambroclaves, utahphosphids, and *Microdictyon*, are sclerites. Sclerites are separate "scales" that articulate in an assemblage (scleritome). New finds (Chen *et al.*, 1989; Hinz *et al.*, 1990; Conway Morris and Peel, 1990) have confirmed the previous sclerital interpretations of many small shelly fossils, but as a whole the affinity of sclerite-bearing organisms is not yet entirely clear. Halkieriids could be mollusks or worms *sensu lato*. Utahphosphids were segmented worms, but not necessarily annelids.

3.1.10a. Tommotiids. One of the most widespread and typical Lower Cambrian sclerite groups are the phosphatic tommotiids. There are several versions of tommotiid systematics: (1) Missarzhevsky and Grigorieva (1981; Sokolov and Zhuravleva, 1983); (2) Landing (1984); and (3) Dzik (1986a). Except Dzik, these authors believe that tommotiids are an independent order-rank group of unclear systematic affinity. Dzik put the Tommotiids as an order together with the Sachitida, Turrelepadida, and others in the Class Machaeridia. He proposed a bilaterally symmetrical metameric initial arrangement of several kinds of sclerites on a wormlike animal. But Dzik's system poses some questions. Because of the speculative bilaterally symmetrical pattern of tommotiid and sachitid sclerite arrangement, he rejects radially symmetrical chancelloriid sclerites from the latter. But chancelloriid growth (Bengtson and Missarzhevsky, 1981) is much closer to sachitids than the latter is to tommotiids or turrelepadids. In addition, among chancelloriids are sclerites such as *Archiasterella* Sdzuy (Fig. 22b) that are bilaterally symmetrical and, hence, questionably belong to sedentary animals such as inferred for chancelloriids. Some morphological resemblance between the sclerites of Turrilepadida, Tommotiida, and Sachitida is explained simply because they belong to bilateral symmetrical organisms. However, some data (Bengtson and Conway Morris, 1984) indicate that sachitid sclerites were secreted and molted. In contrast, tommotiid sclerites (Landing, 1984) apparently grew by accretion around their base during the entire life of a sclerite-bearing animal. The composition of tommotiid sclerites is constant regardless of their occurrence in different facies. Their microstructure, even if secondary, has some analogy with that of phosphatic brachiopods (L. Ye. Popov and Ushatinskaya, 1986). Sachitid sclerite preservation is very different in similar occurrences.

Taking everything into account, we prefer, like Bengtson (1970, 1977, 1985, 1986b), Conway Morris (1982; Bengtson and Conway Morris, 1984), Esakova and Missarzhevsky (Missarzhevsky and Grigorieva, 1981; Sokolov and Zhuravleva, 1983), and Landing (1984) to regard the Tommotiida, Coeloscleritophora (Halkieriidae, etc.), and Machaeridia as independent groups. In the Order Tommotiida we include, as Missarzhevsky and Esakova do, the families Tommotiidae, Lapworthellidae, and Kelanellidae [=Lapworthellidae according to Landing (1984)]. Landing (1984) attributes Tannuolinidae and Sunnaginidae to the Tommotiida. Mitral *Tommotia* Missarzhevsky are dextral and sinistral, pyramidal, conoidal sclerites often bearing four distinct lobes; different ornamentation resembling ridges is typical of both dextral and sinistral sclerites.

The first Tommotiids on the Siberian Platform are recorded from the base of the Tommotian stage (Rozanov *et al.*, 1969; Sokolov and Zhuravleva, 1983; Rozanov and Sokolov, 1984) and are represented by two genera, *Tommotia* and *Camenella* (Fig. 9, panel 64) [= mitral and stellate sclerites of the *Camenella* scleritome according to Bengtson (1970)]. Tommotiid diversity peaked in the Tommotian. They seldom occur in the Atdabanian and are not recorded above its top. Tommotiids are more typical of the Anabar-Sinsk than the Yudoma-Olenek Facies Region. In Mongolia, Tommotiids (*Tommotia plana* (Missarzhevsky), *T. mongolica* N. Grigorieva, *T.* cf *baltica* Bengtson, and *Camenella parilobata* Bengtson (Fig. 11, panel 64) are reported from the Tommotian, Atdabanian, and, probably, Botomian stages of the Khasagt-Khairkhan Ridge (Voronin *et al.*, 1982; Bengtson, 1986b). The Botomian age of the Mongolian tommotiids is dubious because they were derived from redeposited boulders of the Khairkhan Formation. On the Maly Karatau (Missarzhevsky and Mambetov, 1981), tommotiids are represented by the genera *Bercutia* Missarzhevsky and *Geresia* Missarzhevsky in the Tommotian part of the Chulaktau Formation and *Ninella* Missarzhevsky in the Atdabanian (Shabakty Formation). According to Landing (1984), *Bercutia* and *Geresia* are synonyms and belong to Lapworthellidae. *Camenella korolevi* Missarzhevsky (Missarzhevsky and Mambetov, 1981) is also described from the Chulaktau Formation, but Bengtson (1986b) believes that this species sharply differs from the typical *Camenella*. Rare *Camenella* is recorded in Tommotian Stage of the Kolyma Basin (Tkachenko *et al.*, 1987).

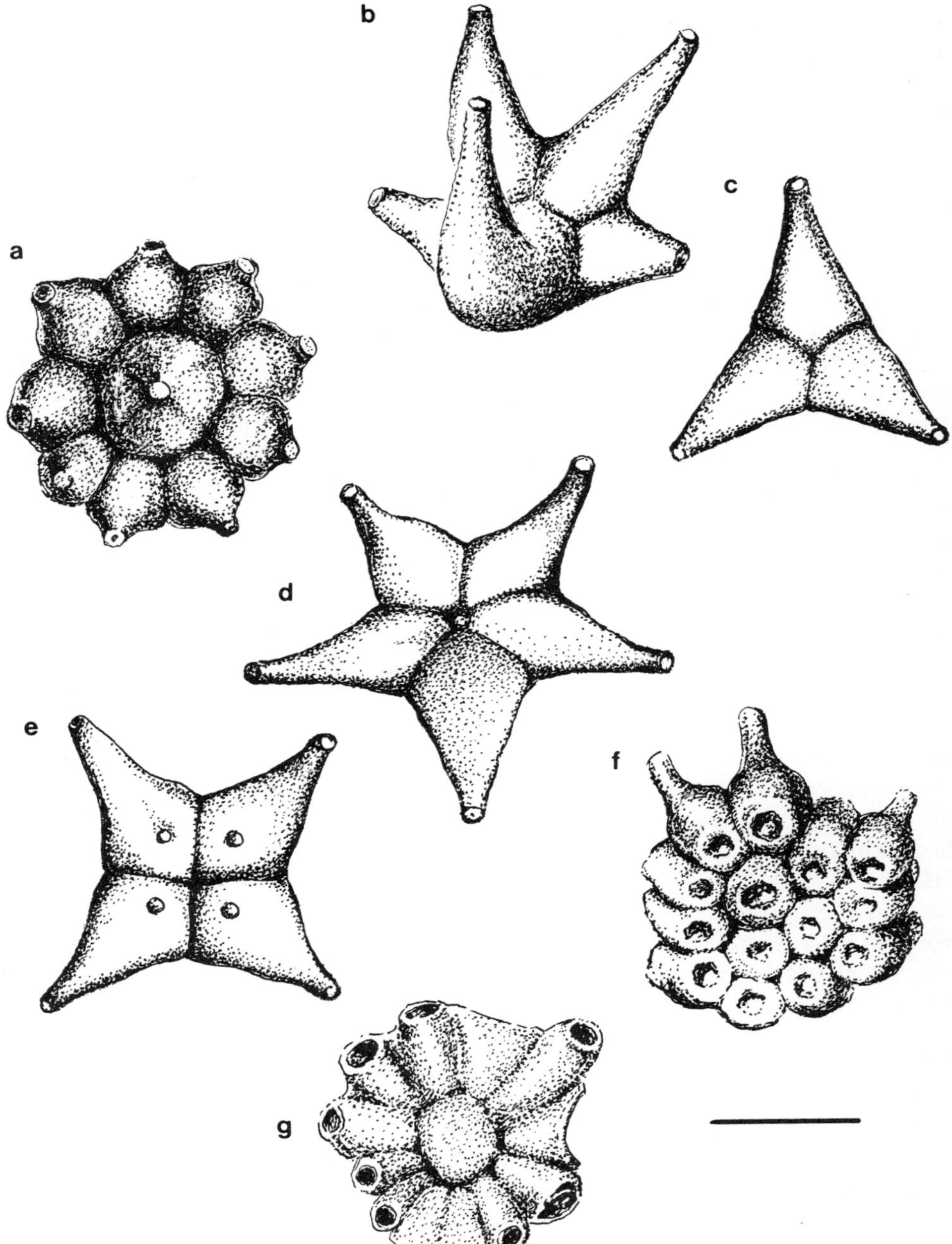

Figure 22. Chancelloriid sclerites. (A) *Chancelloria* Walcott, (b) *Archiasterella* Sdzuy, (c) *Allonnia* Doré and Reid, (d) *Aldania* Vassiljeva, (e) *Onychia* Jiang, (f) *Platyspinites* Vassiljeva, (g) *Ginospina* Missarzhevsky (*Stellaria* Vassiljeva). Scale bar 0.5 mm.

Kelanellids are morphologically closer to tommotiids, but differ by septae in the sclerite cavity (Missarzhevsky and Grigorieva, 1981). On the Siberian Platform the oldest kelanellids, *Lugoviella* Grigorieva (Fig. 9, panel 65), are found in the *squamosus–zelenovi* Zone of the Botomian. Nondefinitive fragments of kelanellids were previously reported from the *kokoulini* Zone of the Atdabanian (Sokolov and Zhuravleva, 1983; Rozanov and Sokolov, 1984). *Lugoviella* occurs in the Botomian in the area between the Dzabkhan and Khungiy Rivers of western Mongolia (Voronova *et al.*, 1986). Kelanellids are more typical of the Altay-Sayan Fold Belt, where they occur from the upper Atdabanian (*Kelanella* Missarzhevsky in Mountain Altay) to the lower Middle Cambrian (*Sonella* Missarzhevsky and Grigorieva) (Rozanov and Missarzhevsky, 1966; Missarzhevsky and Grigorieva, 1981).

A transverse arrangement of ornamentation elements (ridges and tubercules) and a narrow hornlike conical shape with slightly asymmetrical curved top are typical of Lapworthellidae in contrast to the Tommotiidae and Kelanellidae. Like tommotiids, lapworthellids appear on the Siberian Platform at the base of the Tommotian (Rozanov *et al.*, 1969; Fedorov, 1984; Khomentovsky and Karlova, 1986), but occur up to the lower Botomian (Sokolov and Zhuravleva, 1983; Rozanov and Sokolov, 1984) (Fig. 9, panel 67). In Mongolia, on the Khasagt-Khairkhan Ridge, lapworthellids were reported from the middle Tommotian up to the first half of the Botomian (Voronin *et al.*, 1982) (Fig. 11, panel 67). Lapworthellids occur in the Atdabanian of the Mountain Altay (Rozanov and Missarzhevsky, 1966) and the Botomian of the Kolyma Basin (Tkachenko et al., 1987). The sclerites of *Lapworthella* Cobbold are characterized by wide species variability (Matthews, 1973; Conway Morris and Fritz, 1984). But *Lapworthella* is, nevertheless, one of few problematic fossils that can be used to subdivide the Lower Cambrian rather effectively. The *regularis* Zone is, for example, subdivided into the *Lapworthella tortuosa* and *L. bella* Subzones (Rozanov *et al.*, 1969), although at present the first species is also described from the *sunnaginicus* Zone (Fedorov, 1984).

The Tannuolinidae as well as Tommotiidae are represented by two kinds of sclerites (Fonin and Smirnova, 1967). Convex, dextral and sinistral sclerites have the cap shape with a rounded-pentagonal aperture. The upper margin of the sclerite forms a duplicatura over the aperture. Convex sclerites bear ornamentation like coarse growth lines; one side is pseudoporous, the other has true pores. Flat sclerites have a depression along the axis; the top of the sclerite is curved like a beak. In Tuva, Mountain Shoria, and Mongolia, tannuolinids occur in the Botomian (Fonin and Smirnova, 1967; Aksarina, 1984; Voronova *et al.*, 1986) (Fig. 11, panel 66). Curiously, Korde (1990b) described in thin sections of *Tannuolina multifora* seven new species of the new genus *Tuvinia*, which in its turn is assigned to a new class of colonial Cambrian problematic fossils.

At present only *Sunnaginia* Missarzhevsky (Landing, 1984) can be assigned to the Sunnaginidae. These are dextral and sinistral hooflike, irregular, triangular sclerites with thick-layered massive walls ornamented by coarse growth lines. On the Siberian Platform, *Sunnaginia* is recorded only from the *sunnaginicus* and *regularis* Zones of the Tommotian Stage (Sokolov and Zhuravleva, 1983; Rozanov and Sokolov, 1984) (Fig. 9, panel 68). *Eccentrotheca?* sp. reported by Brasier (1986) from the Aldan River is, apparently, a fragment of *S. imbricata* Missarzhevsky. Species of *Sunnaginia* outside the Siberian Platform are known from the younger deposits. *S. acuta* N. Grigorieva occurs in the Botomian of Mongolia (Voronin *et al.*, 1982). In Nova Scotia, Newfoundland, and England (Landing *et al.*, 1980; Bengtson and Fletcher, 1983; Brasier, 1986), different species of *Sunnaginia* occur in the deposits correlated with the Atdabanian and the lower Botomian.

The scalelike asymmetric sclerites *Stefania* N. Grigorieva (Fig. 11, panel 69), known only from the Botomian of western Mongolia (Voronin *et al.*, 1982), resemble *Sunnaginia* and *Eccentrotheca* Landing, Nowlan and Fletcher in the ornamented massive shell and the rounded top. Perhaps these three genera should be joined in one family.

3.1.10b. Coeloscleritophorans. The class Coeloscleritophora (Bengtson and Missarzhevsky, 1981), representing the next large group of sclerites, includes the families Sachitidae, Halkieriidae, Wiwaxiidae, Siphogonuchitidae, and the Chancelloriidae (Bengtson and Conway Morris, 1984; Qian and Bengston, 1989; Bengtson *et al.*, 1990). *Wiwaxia*, however, may be excluded from the Coeloscleritophora according to Butterfield (1990). These sclerites have a relatively thin wall and a large inner cavity, which was open to the outside through a restricted basal foramen. Separate sclerites were fused forming a closely interlocking array over the body. Such fused assemblages are known in *Wiwaxia* Walcott (Conway Morris, 1985), *Chancelloria* Walcott and *Allonnia* Doré (Walcott, 1920; Rigby, 1978), *Dabashanites* Chen (Chen, 1979), *Siphogonuchites* Qian (Voronin *et al.*, 1982), *Drepanochites* Qian and Jiang (Qian and Bengtson, 1989), and now in *Halkieria* (Conway Morris and Peel, 1990). As Esakova (1987) notes, in the Siphogonuchitidae, the scleritomes consist of both sclerites of the same morphological type and those of different types. Sclerites were fused either by imbrication or close fitting of their walls. Missarzhevsky (1989) considered *Maikhanella* to be a complete scleritome of a siphogonuchitid. A more plausible assumption is that *Maikhanella* and similar scaley caps might be the shields of a siphogonuchitid-bearing animal. The Halkieriidae, and most Chancelloriidae consisted of several morphological types of sclerites (Rigby, 1978; Bengtson and Conway Morris, 1984). Halkieriids had several morphotypes of dextral and sinistral sclerites (palmate, siculate, spiniform, cultrate) and anterior and posterior monoplacophoranlike shields. A flattened cross section and a base turned with respect to the sclerite's plane on which a restricted foramen is placed are characteristic of all the sclerite types. The lower surface is smooth, the upper one may bear one or several longitudinal ribs; both surfaces are covered by fine transverse wrinkling. On the Siberian Platform, Halkieriidae (*Halkieria* Poulsen) (Fig. 9, panel 70) now are known only in the Tommotian (Sokolov and Zhuravleva, 1983; Rozanov and Sokolov, 1984) and on the Maly Karatau, in western Mongolia on the Khasagt-Khairkhan Ridge, they occur from the Tommotian to the lower Botomian (Missarzhevsky and Mambetov, 1981; Voronin *et al.*, 1982), and *Halkieria* was noted in the Botomian of the Mountain Shoria (Aksarina, 1984). Earlier Siberian halkieriids were most often assigned to *Sachites* Meshkova (Meshkova, 1969, 1974b; Sokolov and Zhuravleva, 1983).

The Siphogonuchitidae includes oblong sclerites twisted slightly in a spiral, with a polygonal section; keels are developed at the junction of the sclerite's facets. On the Siberian Platform, all the known siphogonuchitids (*Siphogonuchites*) originate from the Yudoma-Olenek Facies Region (Bokova, 1985, 1987; Khomentovsky and Karlova, 1989) and, evidently, from the uppermost Nemakit-Daldyn and the *sunnaginicus* Zone. In western Mongolia, siphogonuchitids (*Lopochites* Qian, *Siphogonuchites* Qian) (Fig. 11, panel 71) are also mainly associated with the Tommotian Stage, and only one species, *L. latazonalis* Qian, is found in the lowermost Atdabanian (Voronin *et al.*, 1982).

Sclerites fused in rosettes are assigned to the Chancelloriidae. A rosette usually consists of four or more radial spines of a narrow, conical shape with a bulge on the proximal end and a central spine perpendicular to them, which may also be of a conical or subspherical shape (Fig. 22). Foramena of all spines are arranged on one basal plane of the rosette. Each spine has its own wall. Sometimes a central spine may be absent, and a rosette consists of only four or three (*Allonnia*) radial spines. Two-spined (*Dimidia* Jiang) and single-spined (*Monospinites* Vassiljeva and Sayutina) chancelloriids were described, but they may be only partial compound rosettes. This is indicated by the presence of facets (areas of interlocking) in such forms. The "Halkeriid" *Manotes* Meshkova (1984) probably is a radial spine of a chancelloriid rosette, and *Rhabdochites* described from the Siberian Platform (Khomentovsky and Karlova, 1986) represents central spines.

A systematic arrangement of chancelloriids, usable in practice, is nonexistent. On one hand, the organism's exoskeleton is known to possibly consist of rosettes of different structure (Rigby, 1978); on the other hand, efforts have been made to create a chancelloriid system (Vassiljeva, 1985; Vassiljeva and Sayutina, 1988) which would take into account only the structure of separated sclerites. Since chancelloriids are of little stratigraphic interest, they are very rarely described, which makes an estimation of their geographic and stratigraphic distribution more difficult. On the Siberian Platform, chancelloriids occur in different facies of the Tommotian–Botomian (Zhuravleva and Korde, 1955; Gorjansky, 1973; Rozanov and Sokolov, 1984; Vassiljeva, 1985) (Fig. 9, panel 72). In Mongolia and on the Maly Karatau, chancelloriids were noted in the Tommotian–Botomian interval (Missarzhevsky and Mambetov, 1981; Voronin *et al.*, 1982) (Fig. 11, panel 72). Chancelloriids have also been found in Tuva, Mountain Altay, Kuznetsky Alatau, Central Asia, and the Far East (Zhuravleva and Korde, 1955; Romanenko, 1968; Zhuravleva *et al.*, 1970).

3.1.10c. Cambroclaves. The Family Zhijinitidae Qian, 1978 (= Cambroclavitidae Mambetov, 1979) should probably be considered sclerites. They are hollow, bilaterally symmetric sclerites consisting of a flat oval plate and a spine perpendicular to the plate in its anterior part. In *Cambroclavus* Mambetov a spine and a plate have separate cavities, and in *Pseudoclavus* Mambetov they are fused together (Fig. 23b). *Zhijinites* is represented in Talassky Ala-Too by *Cambroclavus undulatus* Mambetov (Bengtson *et al.*, 1990). Yu (1984) has found a cambroclavid scleritome, in which every subsequent sclerite overlies the flattened, refined part of the previous plate (Fig. 23a). However, he described this scleritome as a new representative of the polyplacophoran family Yangtzechitonidae. Similar sclerites completely reject the supposed cambroclavid affinity to Acanthocephala (Qian and Yin, 1984). Cambroclavids are known from South Kazakhstan, the Mountain Shoria, and South China. They occur in the Maly Karatau and the Talassky Alatau

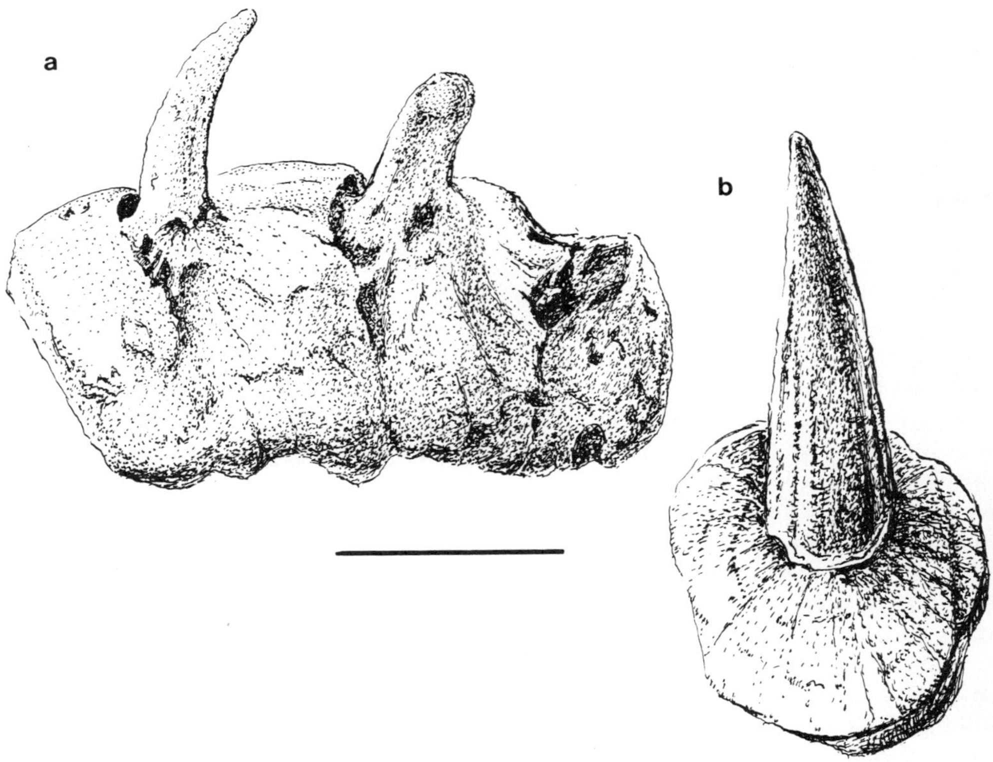

Figure 23. Cambroclaves *Cambroclavus* Mambetov. (a) Scleritome (=*Yantzechiton elongatus* Yu), China, eastern Yunnan, Meishucunian Stage; (b) separate sclerite [=*Zhijinites undulatus* (Mambetov)], Maly Karatau, Atdabanian Stage. Scale bar 0.5 mm.

from the second half of the Atdabanian Stage to the lower Botomian (Mambetov and Repina, 1979; Missarzhevsky and Mambetov, 1981). In the Mountain Shoria, they occur in the Botomian (Aksarina, 1984).

3.1.10d. *Cambrocassis.* *Cambrocassis* Missarzhevsky (Fig. 11, panel 74) may also be a sclerite. It is a conoidal, helmet-shaped, bilaterally symmetrical sclerite; its outer surface is covered with columnlike tubercules. This form occurs in the Botomian of Mongolia, the Batenevsky Ridge, and the Mountain Shoria (Missarzhevsky, 1977; Aksarina, 1984).

3.1.10e. Utahphosphids. Another group of sclerites is united in the Utahphosphidae, including *Utahphospha* Müller and Miller and *Hadimopanella* Gedik (= *Lenargyrion* Bengtson) (Wrona, 1987), or the hadimopanellidae, including *Hadimopanella* and *Kaimenella* Märss (Märss, 1988). *Kaimenella* is a morphologically transitional form between *Hadimopanella* and *Milaculum* Müller. Dzik (1986b) noted the histological similarity of *Hadimopanella* to *Milaculum*. *Utahphospha* is an aggregate which consists of several *Hadimopanella* (Peel and Larsen, 1984; Wrona, 1987). Of all these sclerites, only *Hadimopanella* and, recently, *Utahphospha* have been found in the Lower Cambrian. They are very small (40–450 μm in diameter), buttonlike phosphate sclerites. In the sclerite, one can clearly distinguish caping, basal, and inner layers (Märss, 1988). Dzik (1986b) believes that this sclerite should be assigned to the vertebrates. Surprisingly, the *Hadimopanella*-bearing organism turned out to be a worm belonging to the Palaeoscolecida (Hinz *et al.*, 1990). The most ancient *Hadimopanella*, *H. knappologica* (Bengtson) (Fig. 9, panel 62), is known in the *lermontovae* Zone of the Atdabanian Stage on the Siberian Platform (Bengtson, 1977; Rozanov and Sokolov, 1984). In this region, it also occurs in the lower Botomian and in the Toyonian (Sokolov and Zhuravleva, 1983; Shabanov *et al.*, 1987). In the Botomian of the Maly Karatau, problematic *Resegia* Missarzhevsky (Missarzhevsky and Mambetov, 1981) occur which resemble *Hadimopanella apicata* Wrona, without surface tubercules.

3.1.10f. *Microdictyon*. Evidently, the oldest typical netlike structure, *Microdictyon* Bengtson, Matthews and Missarzhevsky, also occurs in the *lermontovae* Zone on the Siberian Platform. The even older *Microdictyon? tenuiporatum* Bengtson, Matthews and Missarzhevsky occurs in the upper Tommotian, but its assignment to that genus is doubtful. Recently Bengtson *et al.* (1986) reviewed in detail all the materials available on *Microdictyon*. A *Microdictyon*-bearing animal was recently described (Chen *et al.*, 1989), with the plates placed just over a number of limbs on a wormlike body.

3.1.10g. *Mobergella*. Phosphate disc-shaped problematic *Mobergella* Hedström (Fig. 9, panel 75, and Fig. 12, panel 75) with seven pairs of bilaterally symmetrical muscle impressions or ribs on the usually concave side (Bengtson, 1968) are characteristic of the Lower Cambrian. It could be an operculum of some animal living in tubes (Fisher, 1962) or sclerites. On the Siberian Platform, the oldest *Mobergella*, *M. radiolata* Bengtson, occurs from the base of the *lenaicus–primigenius* Zone of the Tommotian Stage into the *pinus* Zone of the Atdabanian (Sokolov and Zhuravleva, 1983; Rozanov and Sokolov, 1984), but it is especially abundant in the *zegebarti* Zone. On the East European Platform, *Mobergella* is characteristic of the Dominopol' Horizon (the lower half of the Atdabanian Stage) (Mens *et al.*, 1987). In this region the oldest *Mobergella* occurs in the Glebovo Member (Lontova Horizon) (Rozanov, 1973; Kirsanov, 1974). True *Mobergella* are known in siliciclastic sediments of the Seleta Synclinorium (central Kazakhstan) (Koneva, 1983), where they evidently also occur in the Botomian. *Mobergella* is recorded in the Botomian of western Mongolia (Voronin *et al.*, 1982), but they may have come from redeposited boulders. "*Mobergella*" *scutata* Missarzhevsky from the Botomian of the Maly Karatau have only four pairs of marks on the shell (Missarzhevsky and Mambetov, 1981), and according to the definition, should be excluded from this genus. The *Mobergella* Range–Zone is probably limited by the interval of the *lenaicus–primigenius* Zone to the *pinus* Zone.

3.1.10h. *Tumulduria*. Problematic phosphatic, bilaterally symmetrical *Tumulduria* Missarzhevsky plates (Fig. 9, panel 78) were recently described in detail by Bengtson *et al.* (1987). The plates are known only from the *sunnaginicus* Zone of the Tommotian Stage on the Aldan River.

3.1.10i. Globose Structures. The last group is made up of problematic spherical fossils. *Markuelia* Valkov, up to 1 mm in diameter (Fig. 10, panel 76), has a smooth outer surface and an inner cavity divided into chambers. *Markuelia* was previously described and referred to by Valkov (1983) and Valkov and Karlova (1984), but Valkov (1986) validated the genus with the first mention of the type species and its holotype. *Markuelia* is found as yet only in the *sunnaginicus* Zone in the southern part of the Siberian Platform. *Markuelia* is possibly a mold of *Archaeooides*, which looks like a flattened hollow ball with papillae on the surface. On the Siberian Platform, *Archaeooides* was found at the base of the Tommotian (Khomentovsky and Karlova, 1986) and in the lower Botomian (Sokolov and Zhuravleva, 1983). In western Mongolia, it occurs on the Khasagt-Khairkhan Ridge (Fig. 11, panel 77) in the lower part of the Bayan-gol Formation (Tommotian Stage) (Voronin *et al.*, 1982). In the Maly Karatau, *Gaparella* Missarzhevsky was described from the Botomian part of the Shabakty Formation; this problematic fossil is very similar to *Archaeooides* and may also be its synonym. Close affinity of these Botomian *Archaeooides*-like structures to *Aetholicopalla* Conway Morris cannot be excluded.

The study of the Lower Cambrian problematic fossils is in the initial stages, and knowledge will increase with every passing year. On the other hand, clearly their great morphological variability has resulted in many more described forms than really exist, because some separate parts of scleritomes have been described as independent species or genera. The most recent example is "colonial" *Tuvina* Korde, which is a thin section of *Tannuolina* only.

3.2. Soft-Bodied Fossils

Barskov and Zhuravlev (1988) recently reexamined museum collections of the meagre soft-bodied Cambrian fossils of the USSR. Many medusoids and worms described from the Siberian Platform (Milashev, 1958; Miroshnikov and Kravtsov, 1960, 1965) are trace fossils. *Dickinsonia*, described by Borovikov (1976) from the Lower Cambrian of Kazakhstan, may be a trace fossil (Fig. 24), as our study of the type material in the Central Geological Museum, Leningrad, has shown. *Eomedusa* Popov (Yu. N. Popov, 1967; Datzenko *et al.*, 1968) from the Upper Cambrian of northern Krasnoyarsk Region belongs unquestionably to a soft-bodied animal. The tripartite structure of the *Eomedusa* umbrella (Fig. 25) closely resembles *Velumbrella* Stasinska from the Botomian of the Holy Cross Mountains (Stasinska, 1960). Both of them evidently belong to chondrophores, as suggested by Scrutton (1979) and Stanley (1986). Barskov and Zhuravlev (1988) described the annelid *Protoscolex? tchopkoensis* Barskov and A. Zhuravlev, belonging to the Class Palaescolecida, from the same region. Another palaeoscolecid of uncertain affinity was found by A. Zhuravlev in the Sinsk Formation (Botomian stage) of the middle Lena River (Fig. 19a). Voropinov (1957) found one arthropod species in abundance in the Verkholensk Formation (of Middle–Upper Cambrian age) of the upper Lena River. We cannot exclude the possibility that these are trace fossils (Fig. 26).

Repina and Okuneva (1969) described several nontrilobite arthropods from the Middle and Upper Cambrian of the Khanka Massif of the Far East. At present, a Lower Cambrian age for these finds cannot be eliminated because a similar fauna is found throughout the Cambrian. Two of these forms—*Maritimella* Repina and Okuneva and *Orientella* Repina and Okuneva—were placed in the Order Nektaspida. Whittington (1977) believed that only *Maritimella* belongs to this order. The two other genera, *Khankaspis* Repina and Okuneva and *Beckwithia* Resser, were placed in the Aglaspida. Because appendages in these forms have not been preserved, their position among arthropods is still under discussion.

Figure 24. A trace fossil described as "*Dickinsonia*" by Borovikov (1976). Type specimen $CNIGR_m$ N 11461, Maly Karatau, Botomian Stage. Scale bar 1 cm. ($CNIGR_m$: Central Scientific-Research Geological-Exploratory Museum, St. Petersburg.)

Two facts draw attention. First, almost all the soft-bodied fossils on the Siberian Platform occur in facies close to those of the outer detrital belt of Laurentia (Astashkin *et al.*, 1984). Thus, along with palaeoscolecids, there are North American forms of the sponge *Hamptonia* and the alga *Margaretia* Walcott (= *Aldanophyton* Krishtofovich) (Fig. 19b) in the Sinsk Formation. Second, the Siberian discovery of a number of North American genera (*Margaretia, Hamptonia*) confirms the thesis of Conway Morris *et al.* (1987) on the conservatism of the fauna of the "Burgess Shale" type.

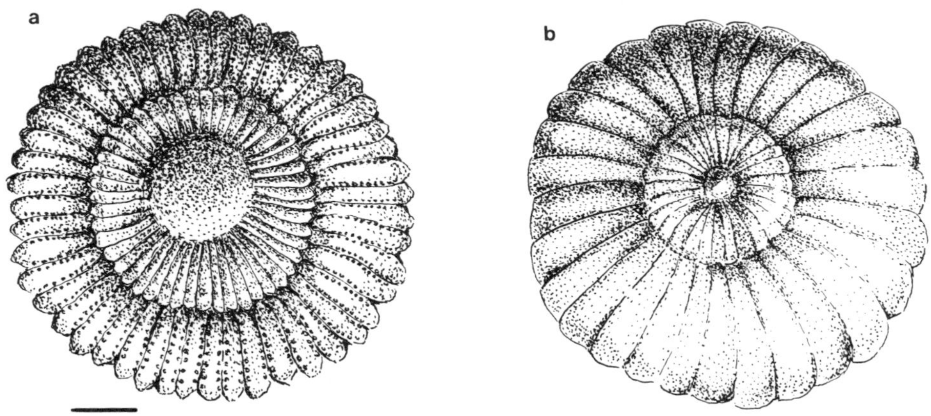

Figure 25. Chondrophorines. (a) *Eomedusa datzenkoi* (Popov), Siberian Platform, north of the Krasnoyarsk Region, Upper Cambrian; (b) *Velumbrella czarnockii* Stasinska, Holy Cross Mountains, Botomian Stage. Scale bar 1 cm.

Figure 26. "Arthropods," PIN N 1143, upper Lena River, Kachug, ? Middle Cambrian. Scale bar 1 cm.

3.3. Algae

The pattern of development of Early Cambrian algae with a calcareous thallus was studied in detail by Luchinina (1975), Voronova and Radionova (1976), Stepanova and Luchinina (1982), Riding and Voronova (1984, 1985), Chuvashov and Riding (1984), Voronova and Drosdova (1986), and many others. In the upper part of the Nemakit-Daldyn Horizon, on the northern Siberian Platform on the Kotuy River, *Renalcis polymorphus* Maslov are known (Luchinina, 1985). From the *sunnaginicus* Zone of the Tommotian Stage on the Siberian Platform, the following forms are known: botryoidal—*Renalcis* Vologdin, *Gemma* Luchinina, *Tarthinia* Drosdova; tubiform—*Girvanella* Nicholson and Etheridge, *Protoortonella* Luchinina, *Batinevia* Korde, *Botomaella* Korde, *Botominella* Reitlinger, *Proaulopora* Krasnopeeva, and *Obruchevella;* and delicate dendritic—*Korilophyton* Voronova and *Angulocelluloria* Vologdin. In the *regularis* Zone, in addition to these forms are the real dendritic algae *Epiphyton* Bornemann, *Chabakovia* Vologdin, *Tubomorphophyton* Korde, and *Gordonophyton* Korde. *Obruchevella* is a Cyanophyta. All the rest of these algae, by their sizes and simple morphology, are rather close to Cyanobacteria, but are characterized by a different wall calcification, and, according to Luchinina (1987), may represent a single group that became extinct by the end of the Mesozoic. This supposition is also substantiated by the ecologic characteristics of such algae and their constant coexistence with various groups of animals. Riding (1982; Chuvashov and Riding, 1984) thinks, however, that the bulk of these algae still were cyanobacteria, and their extinction at the Mesozoic–Cenozoic boundary was caused by a global change in the chemical composition of the ocean water (Mg/Ca ion ratio). The presumed Rhodophyta (*Bija*) is known from the Botomian. On the Siberian Platform, *Bija* is described (Voronova and Radionova, 1976) from the Shumnoy Formation of the Igarka Area and in the Altay-Sayan Fold Belt from the Sanashtykgol Horizon (Vologdin, 1932). Riding and Voronova (1985) suggest that *Bija* may be a synonym of *Solenopora* Dybowski.

Uncalcified algae are known mainly from the Vendian–Cambrian transitional sediments of the East European Platform (Gnilovskaya, 1976). They are probable metaphytes called vendotaenids. Of vendotaenids, only *Tyrasotaenia* Gnilovskaya is known in the Lower Cambrian (Lontova Horizon). This alga has a thin, wrinkled thallus and a homogeneous surface.

The problematic metaphyte *Aldanophyton* (Krischtofovich, 1953), known from the Sinsk (Botomian Stage) and Inikan (Middle Cambrian) Formations of the southern Siberian Platform, is a wide ribbon with a denticulate edge, covered by regularly arranged prominences of irregular rhombic shape (Fig. 19b). All its features indicate that *Aldanophyton* may be a junior synonym of *Margaretia* described by Walcott from the Middle Cambrian of Utah.

Acritarchs, widespread in the Lower Cambrian, are phytoplankton. Volkova (1985) showed several stages in the development of this group in the Early Cambrian, using material from the East European Platform. In the Lontova Horizon, assemblages of simple Vendian forms were joined by discoid acritarchs (*Granomarginata* Naumova and others) and porous prasinophytes (*Tasmanites* Newton). The variety of acritarchs increased sharply in the Dominopol' Horizon. In the Rausve Horizon, acritarchs appeared with long (*Deunffia* Downie) and vellumlike (*Liepaina* Jankauskas and Volkova) processes and other compound forms. Rudavaskaja and Vassiljeva's data (1984, 1985) on the northern Siberian Platform contradict these proposals, because there rare acanthomorphs were found in the Olenek River and Khara-Ulakh sections in the Tommotian Stage, which is at the Lontova level. According to Kirjanov (1987), however, these assemblages, which he considers the Chuskun and the Chekurovka, respectively, are of the same age, with a gap in sedimentation on the East European Platform at the boundary of the Lontova and Dominopol' Stages. In addition, *Comasphaeridium brachyspinosum* (Kirjanov) and *C. strigosum* (Jankauskas), known in the Chekurovka assemblage, are not now placed among true acanthomorphs. Burzin (1987) suggested that the rapid development of spiny acritarchs in the Dominopol' Stage was caused by the evolution of a new size group of planktivorous animals of Recent copepod size. Larvae of trilobites diversifying at that time might be such animals.

3.4. Trace Fossils

Changes in trace fossil assemblages at the Vendian–Cambrian boundary on the East European Platform were considered in detail by Fedonkin (1981; Urbanek and Rozanov, 1983). His views were substantiated by a study on a continuous section of the upper Vendian–Lower Cambrian platform of Poland (Paczesna, 1986). The most significant shifts in the taxonomic, morphological, and ethological diversification of trace fossils are observed closer to the boundary between the Rovno (*Sabellidites cambriensis* Zone) and the Lontova (*Platysolenites antiquissimus* Zone) Horizons than to the boundary between the Vendian and Tommotian Stages. Thus, the diversification of Paleozoic-type trace fossils coincides in time with the Tommotian Stage diversification of small shelly fossils, but did not precede it, as supposed earlier (Sepkoski, 1983).

The Vendian–Cambrian trace fossil record is best studied in siliciclastic facies (Alpert, 1977; Fedonkin, 1981; Fritz and Crimes, 1985; Crimes and Anderson, 1985); carbonate facies have not been studied much. Some data on the trace fossil assemblages in carbonates come from the Siberian Platform (Rozanov and Sokolov, 1984). In the Vendian part of the Yudoma Formation along the Aldan River, *Planolites* sp. and *Cochlichnus* sp. have been found in massive, thickly bedded dolomites. *Planolites beverleyensis* (Billings), *P. striatus* (Hall), *Rhizocorallium jenense* Zenker, and *Chondrites* sp. occur in the Tommotian (*regularis* and *lenaicus–primigenius* Zones) on the Aldan and Lena Rivers in argillaceous thin-bedded, red limestones of the Pestrotsvet Formation. In similar limestones in the Atdabanian Stage, *Plagiogmus* sp. was found. *Dactyloidites* Hall was noted in the Krasny Porog Formation (Atdabanian and Botomian Stages) on the Sukharikha River (Luchinina and Meshkova, 1969). Trace fossils are known in terrigenous beds of the Kessyuse Formation on the northern Siberian platform, in the lower Olenek River (Sokolov and Ivanovsky, 1985). *Gordia* sp., *Phycodes* sp., and *Olenichnus irregularis* Fedonkin occur in the Nemakit–Daldyn Horizon, and *Planolites* sp. and *Didymaulichnus meandriformis* Fedonkin occur in the *sunnaginicus* Zone of the Tommotian Stage.

4. Discussion and Conclusion

The first true fossils with skeletons are found in the Redkino Horizon of the Vendian: sabelliditids and the problematic toothlike fossil *Redkinia*. This Horizon is commonly considered as an age equivalent to the Ediacaran *sensu stricto*. *Cloudina* is apparently found at the same level in Namibia. Both sabelliditids and *Redkinia* have an organic skeleton, whereas *Cloudina* tubes may be organic and mineral (Hanh and Pflug, 1985).

The next step in skeletal mineralization occurs in the Nemakit–Daldyn (Rovno) age, at the time of sabelliditids' peak and the appearance of anabaritids, possible chaetognaths (*Protohertzina*), and problematic tubes. At the end of that age, single caps and phosphate skeletal elements only (protoconodonts, anabaritids) appeared.

One of the most significant steps in skeletal mineralization occurs at the very beginning of the Cambrian in the early Tommotian. Practically simultaneously, hexactinellids, archaeocyaths, inarticulate brachiopods, orthothecimorphs, gastropods (*Aldanella*), rostroconchs, and a number of other molluscoids, tommotiids, siphogonuchitids, chancelloriids, halkieriids, hyolithelminths, coleolids, and other diverse problematic fossils appeared. In many groups (for instance, archaeocyaths, orthothecimorphs, mollusks) scores of forms, which belong to various groups of rather high taxonomic rank, appear suddenly. The early Tommotian is the time of mass appearance of algae with a calcareous thallus. In addition, the first siliceous skeletons (in hexactinellids and *Platysolenites*) appeared in the early Tommotian. Thus, all the common types of skeletal mineralization among metazoans appeared in the Tommotian.

In the early Atdabanian, skeletal mineralization in arthropods started. Polymerid and miomerid trilobites, phyllocarids (*Isoxys*), and bradoriids occur first within the *zegebarti* Zone on the Siberian Platform. At about the same level on the East European Platform, trilobitomorph remains with weakly mineralized exoskeletons (*Gdowia–Liwia*) and anomalocarids (*Cassubia*) are found. In the Early Atdabanian, apart from arthropods, articulate brachiopods and possibly cribricyaths and hydroconozoans acquired skeletons.

Another Early Cambrian stage of skeletal mineralization started at the very end of the Atdabanian and the beginning of the Botomian, when "skeletal" echinoderms and a number of new groups of problematic fossils arose. The next well-known stage of skeletal mineralization occurred later, in the Ordovician (corals, bryozoans, and others). Thus, the ability to make a skeleton by different groups was apparently not a sudden and quick phenomenon, as noted previously, but a number of steps were involved. Within the interval between the middle Vendian and the Ordovician, there were at least six stages of the mass appearance of different groups of skeletal organisms. (According to A. Rozanov, the Riphean stage may also be added: *Udokania* Leites and camasiids might be examples of the first attempts at skeleton building.)

Different skeletal types appeared in approximately the following sequence: organic—in the middle Vendian; phosphate—at the end of the Vendian; and siliceous and carbonate—at the Tommotian. The first agglutinated shells also appeared at the beginning of the Cambrian (Atdabanian Stage) (Fig. 27). In addition, many groups of animals were represented by very different forms at the same time as skeletal mineralization was occurring. In other words, the first skeletal representatives of many phyla already belong to several orders or even classes. Sponges, mollusks, arthropods, and echinoderms are especially obvious examples. Probably, the appearance of mineralized skeletons caused a burst of diversification of these groups. Considering the stage-by-stage nature of skeletal development in various groups and the use of different materials for building skeletons, we note a remarkable coincidence in the time and space distribution of maximum phosphate storage and the development of small shelly fossils with phosphate skeletons. They are localized within the Asian territory and limited mainly to the Tommotian.

The problems of paleogeography and paleobiogeography have recently been

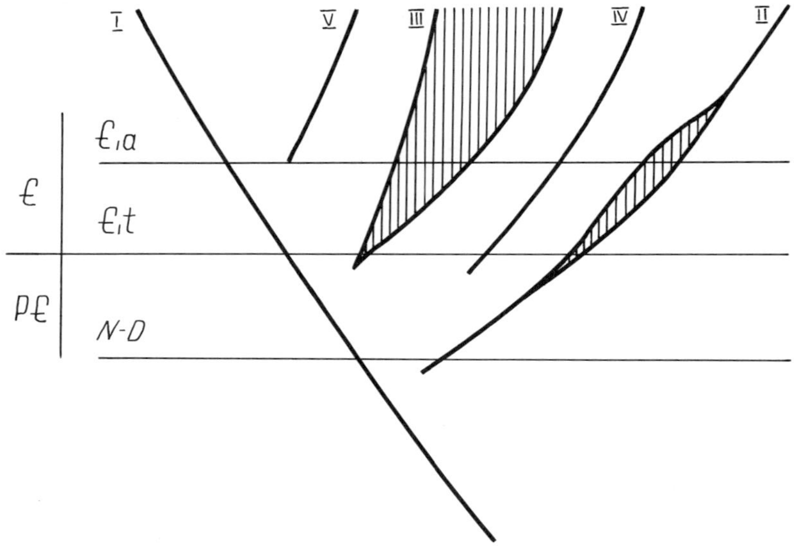

Figure 27. Appearance of various skeletal types. (I) organic, (II) phosphate, (III) calcareous, (IV) siliceous, (V) agglutinated. N-D, Nemakit-Daldyn Horizon; $Є_1$t, Tommotian Stage; $Є_1$a, Atdabanian Stage.

thoroughly considered (Rozanov, 1984b, 1985; Zhuravlev, 1986b). Three main provinces may be distinguished for the Tommotian age: (1) Laurentia and Baltia; (2) Siberia; (3) Gondwana. The main differences among these three provinces lie not so much in faunal assemblages as in types of sedimentation. Terrigenous deposits are typical of Laurentia and Baltia, and were caused by great land masses on these continents, and not the position of the continents in high latitudes. In Siberia at the Anabar Massif where Tommotian terrigenous deposits occur, *Platysolenites* and the typical northern European assemblage of acritarchs (Missarzhevsky, 1983; Rudavskaja and Vassiljeva, 1985) are present. Gondwana is characterized by a predominance of dolomite and phosphatic deposits, and Siberia by limestones. Again, in spite of some difference in faunal assemblages caused by the prevalence of certain facies, the boundaries of these provinces are not very distinct. Mixed Siberian–Gondwana assemblages are spread not only in Mongolia and Kazakhstan, but on the Siberian Platform as well. In its Yudoma-Olenek Facies Region, which is connected with Anabar-Sinsk, many "Chinese" fossils were found recently, such as *Blastulospongia, Canopoconus, Obtusoconus, Lophotheca, Siphogonuchites,* and others (e.g., Bokova, 1985; Khomentovsky and Karlova, 1986; Khomentovsky *et al.*, 1990; Pelman *et al.*, 1990). More careful comparison of Siberian and Chinese fauna, including the so-called endemics, is necessary to carry out in the nearest future. For example, good preservation of *Tumulduria* figures by Khomentovsky *et al.* (1990, Figs. 3–7) and redescription of some Chinese paracarinachitids by Conway Morris and Chen (1991, Fig. 12e,f) pose a question on their possible affinity.

The Siberian Platform was the center of origin of many groups (Rozanov, 1980), and had, of course, its own features in the Tommotian (archaeocyaths, radiocyaths, abundant hyoliths, hyolithelminths, tommotiids) because it had the most favorable environments as compared with other continents. The Siberian Platform was right on the paleoequator (Kirschvink and Rozanov, 1984) and had a huge shallow-water epicontinental basin without large land masses but with contrasting conditions of salinity in different areas.

In the Atdabanian, paleobiogeographic subdivisions at high levels are still difficult to distinguish because, by the middle of the Atdabanian, such groups as archaeocyaths, trilobites, acritarchs, calcareous algae, and others were widely distributed. Practically no difference exists between archaeocyanthan assemblages of Siberia, Australia, China, southern Europe, and northern Africa (Rozanov, 1980; Zhuravlev, 1986b), between acritarchs of Laurentia, Greenland, Baltia, and Mongolia (Downie, 1982; Voronin *et al.*, 1982, Vidal and Knoll, 1983; Moczydłowska and Vidal, 1986), or between calcareous algae of the Siberian Platform, the Altay-Sayan Fold Belt, Mongolia, southern Europe, and northern Africa (Voronova and Drosdova, 1986). *Rhombocorniculum, Microdictyon* and many molluscoids were also cosmopolitan at that time.

Provincialism of Cambrian faunas began to increase only in the Botomian (Rozanov, 1984b; Repina, 1985; Zhuravlev, 1986b). But the cause of that provincialism cannot be a climatic zoning. The Cambrian thermal regime (Chumakov, 1984), confirmed by data on glauconite minerals, was similar to that of the Jurassic (Nikolaeva, 1981). According to these data, the temperature of seawater in Baltia was about 28°C, and on the Siberian Platform and its Anabar-Sinsk Facies Region, about 30–38°C. All the principal epicontinental basins were in low latitudes, as confirmed by paleomagnetic data obtained on fossiliferous sections in Australia, China, Mexico, and Mongolia, and on the East European and Siberian Platforms (Kirschvink, 1978; Khramov and Rodionov, 1980; Barr and Kirschvink, 1983; Kirschvink and Rozanov, 1984; Zhang and Zhang, 1984; Kirschvink *et al.*, 1987). Therefore, the chief reason for increasing provincialism is, in our opinion, the fragmentation of the Paleopangea supercontinent (Kirschvink *et al.*, 1984). Increasing provincialism caused further diversification of Cambrian organisms (Conway Morris, 1987; Zhuravlev, 1987).

An unexpected conclusion can be derived from Early Cambrian biogeography. No matter how continents are arranged based on verified faunal, paleomagnetic, and lithologic data, they cannot be placed in such a way that all basins with similar faunas are geographically related as well. The size of the present day globe is too large to allow a biogeographically-correct fit.

ACKNOWLEDGMENTS. We are very grateful to Drs. Jere H. Lipps and Philip W. Signor for inviting us to prepare this chapter. We thank Dr. V. A. Astashkin, M. B. Burzin, Dr. F. Debrenne, A. B. Fedorov, Dr. N. V. Esakova, A. Yu. Ivantsov, Dr. V. A. Luchinina, Dr. L. M. Melnikova, Dr. T. A. Sayutina, Dr. V. A. Sysoiev, Dr. A. I. Varlamov, Dr. G. T. Ushatinskaya, A. V. Zakharov, and E. A. Zhegallo for significant advice and useful comments on the contents of the chapter and for materials provided for our use. We express our thanks to L. N. Boldyreva, N. I. Krasnova, and S. I. Solovjeva for their hard work in translating the manuscript, T. I. Gerasimova for the drawings, and N. V. Chernova and M. N. Zhuravleva for technical assistance. Dr. Jere H. Lipps provided considerable editorial assistance, and Phyllis Thompson drafted our charts and maps for publication. None of these people are responsible for the content of this chapter, however.

References

Abaimova, G. P., 1978, Anabaritids—The oldest fossils with calcareous skeletons, in: New Material on the Stratigraphy and Paleontology of Siberia (S. V. Sukhov, ed.), *Trudy Sibirsk. Nauchno-Issled. Inst. Geol. Geofiz. Mineral.* **260:**77–83 [in Russian].

Abaimova, G. P., 1983, Conodonts and other Middle-Upper Cambrian remains of the Maly Karatau, in: *The Lower Paleozoic Stratigraphy and Paleontology of Kazakhstan* (M. K. Apollonov, S. M. Bandaletov, and N. K. Ivshin, eds.), Nauka Kazakh SSR, Alma-Ata, USSR, pp. 96–104 [in Russian].

Aksarina, N. A., 1968, Probivalvia—A new class of ancient molluscs, in: *New Data on Geology and Natural Resources of Western Siberia,* Tomsk, USSR, Vol. 3, pp. 77–86 [in Russian].

Aksarina, N. A., 1984, About the additional paleontologic characteristics of the Adiak Formation in Kamzas Village area (Mountain Shoria), in: *Nature and Economics of the Kuzbass, Theses of the Forthcoming Conference,* Novokuznetsk, USSR, pp. 77–86 [in Russian].

Aksarina, N. P., and Pelman, Yu. L., 1979, Cambrian Brachiopods and Bivalve Molluscs of Siberia, *Trudy Inst. Geol. Geofiz. Sibirsk. Otd. Akad. Nauk SSSR* 362 [in Russian].

Akulcheva, Z. A., Galperova, E. M., Drobkova, E. L., Lysova, L. A., Titorenko, T. N., Trestchetenkova, A. A., and Fajzulina, A. H., 1981, The Mota deposits and their analogs in the Irkutsk Amphitheater, in: Precambrian and Cambrian Boundary Deposits of the Siberian Platform (Biostratigraphy, Paleontology, Environments) (N. P. Meshkova and I. V. Nikolaeva, eds.), *Trudy Inst. Geol. Geofiz. Sibirsk. Otd. Akad. Nauk SSSR* **475:**65–139 [in Russian].

Alpert, S. P., 1977, Trace fossils and the basal Cambrian boundary, *Geol. J., Spec. Issue* **9:**1–8.

Andreeva, O. N., 1987, Cambrian articulate brachiopods, *Paleontol. Zh.* **1987**(4)**:**31–40 [in Russian].

Astashkin, V. A., 1985, Problematic rock-forming organisms in the Lower Cambrian sediments of the Siberian Platform, in: Problematics of the Late Precambrian and Paleozoic (B. S. Sokolov and I. T. Zhuravleva, eds.), *Trudy Inst. Geol. Geofiz. Sibirsk. Otd. Akad. Nauk SSSR* **632:**144–149 [in Russian].

Astashkin, V. A., Varlamov, A. I., Gubina, N. K., Ekhanin, A. E., Pereladov, V. S., Romenko, V. I., Sukhov, S. S., Umperovich, N. V., Fedorov, A. B., Fedyanin, A. P., Shishkin, B. B., and Khobnya, E. I., 1984, *Geology and Oil Pool Perspectives of the Reef System of the Cambrian of the Siberian Platform,* Nedra, Moscow [in Russian].

Barr, T. D., and Kirschvink, J. L., 1983, The paleoposition of North America in the Early Paleozoic: New data from the Caborca sequence in Sonora, Mexico, *EOS Trans. Am. Geophys. Union* **64:**689–690.

Barskov, I. S., and Zhuravlev A. Yu., 1988, Soft-bodied organisms from the Cambrian of the Siberian Platform, *Paleontol. Zh.* **1988**(1)**:**3–9 [in Russian].

Barskova, M. I., 1988, New molluscs from the Lower Cambrian strata of the Kolyma Uplift, *Paleontol. Zh.* **1988**(1)**:**101–105 [in Russian].

Bassett, M. G., and Yochelson, E. L., 1979, *Crania sedgwici* revised as a hyolith operculum, *Lethaia* **12:**307–312.

Bassoullet, J. P., Bernier, P., Deloffre, R., Genot, P., Jaffrezo, M., and Vachard, D., 1979, Essae de classification des Dasycladales en tribus (Attempt to classify Dasycladales in tribes), *Bull. Centr. Rech. Explor.-Prod. Elf-Aquitaine* **3:**429–442.

Beadle, S. C., 1988, Dasylcads, cyclocrinitids and receptaculitids: Comparative morphology and paleoecology, *Lethaia* **21:**1–12.

Beljaeva, G. V., 1985, More about cribricyaths, in: Problematics of the Late Precambrian and Paleozoic (B. S. Sokolov and I. T. Zhuravleva, eds.), *Trudy Inst. Geol. Geofiz. Sibirsk. Otd. Akad. Nauk SSSR* **632:**33–38 [in Russian].

Beljaeva, G. V., Luchinina, V. A., Nazarov, B. B., Repina, L. N., and Sobolev, L. P., 1975, Cambrian fauna and flora of the Dzhagdy Ridge (Far East), *Trudy Inst. Geol. Geofiz. Sibirsk. Otd. Akad. Nauk SSSR* **226** [in Russian].

Bengtson, S., 1968, The problematic genus *Mobergella* from the Lower Cambrian of the Baltic area, *Lethaia* **1:**325–351.

Bengtson, S., 1970, The Lower Cambrian fossil *Tommotia, Lethaia* **3:**363–392.

Bengtson, S., 1976, The structure of some Middle Cambrian conodonts, and the early evolution of conodont structure and function, *Lethaia* **9:**185–206.

Bengtson, S., 1977, Early Cambrian button-shaped phosphatic microfossils from the Siberian Platform, *Palaeontology* **20:**751–762.

Bengtson, S., 1983, The early history of Conodonta, *Fossils Strata* **15:**5–19.

Bengtson, S., 1985, Redescription of the Lower Cambrian *Halkieria obliqua* Poulsen, *Geol. Fören. Stockh. Förh.* **107**:101–106.

Bengtson, S., 1986a, Siliceous microfossils from the Upper Cambrian of Queensland, *Alcheringa* **10**:195–216.

Bengtson, S., 1986b, A new Mongolian species of the Lower Cambrian genus *Camenella* and the problems of the scleritome-based taxonomy of the Tommotiidae, *Palaeontol. Z.* **60**:45–55.

Bengtson, S., and Conway Morris, S., 1984, A comparative study of Lower Cambrian *Halkieria* and Middle Cambrian *Wiwaxia*, *Lethaia* **17**:307–329.

Bengtson, S., and Fletcher, T. P., 1983, The oldest sequence of skeletal fossils in the Lower Cambrian of southwestern Newfoundland, *Can. J. Earth Sci.* **20**:525–536.

Bengtson, S., and Missarzhevsky, V. V., 1981, Coeloscleritophora—A major group of enigmatic Cambrian metazoans, in: *Short Papers for the Second International Symposium on the Cambrian System 1981* (M. E. Taylor, ed.), U. S. Geological Survey Open-File Report 81-743, pp. 19–21.

Bengtson, S., Matthews, S. C., and Missarzhevsky, V., 1986, The Cambrian netlike fossil *Microdictyon*, in: *Problematic Fossil Taxa* (A. Hoffman and M. H. Nitecki, eds.), Clarendon Press, Oxford, pp. 97–115.

Bengtson, S., Fedorov, A. B., Missarzhevsky, V. V., Rozanov, A. Yu., Zhegallo, E. A., and Zhuravlev, A. Yu., 1987, *Tummulduria incomperta* and the case for Tommotian trilobites, *Lethaia* **20**:361–370.

Bengtson, S., Conway Morris, S., Cooper, B., Jell, P. A., and Runnegar, B. N., 1990, Early Cambrian fossils from South Australia, *Mem. Assoc. Aust. Palaeontol.* **9**:1–364.

Bokova, A. R., 1985, The ancient assemblage of Cambrian organisms of West Prianabarie, in: *Late Precambrian and Early Paleozoic Stratigraphy of Siberia (Vendian and Riphean)* (V. V. Knomentovsky, ed.), Institute of Geology and Geophysics, Siberian Branch, USSR Academy of Sciences, Novosibirsk, USSR, pp. 13–28 [in Russian].

Bokova, A. R., 1987, Shelly fossil assemblages of the Kessyuse Formation (Olenek Uplift), in: *Abstracts of Reports, 3rd Symposium of Paleontology of the Precambrian and the Early Cambrian, 1987*, Karelian Branch, USSR Academy of Sciences, Petrozavodsk, USSR, pp. 9–11 [in Russian].

Bokova, A. R., 1990, New Lower Cambrian gastropods of the Siberian Platform, *Paleontol. Zh.* **1990**(2):123–126 [in Russian].

Borovikov, L. I., 1976, First find of fossil remains of *Dickinsonia* in the Lower Cambrian sediments on the territory of the USSR, *Doklady Akad. Nauk SSSR* **231**(5):1182–1184.

Boyajian, G. E., and LaBarbera, M., 1987, Biomechanical analysis of passive flow of stromatoporoids—Morphologic, paleoecologic, and systematic implications, *Lethaia* **20**:223–229.

Brangulis, A. P., Kirjanov, V. V., Mens, K. A., and Rozanov, A. Yu., 1981, Vendian and Cambrian paleogeography of the East-European Platform, in: *Short Papers for the Second International Symposium on the Cambrian System 1981* (M. E. Taylor, ed.), U. S. Geological Survey Open-File Report 81-743, pp. 26–28.

Brasier, M. D., 1982, Foraminiferid architectural history: A review using the Min LOC and PI Methods, *J. Micropalaeontol.* **1**:95–105.

Brasier, M. D., and Singh, P., 1987, Microfossils and Precambrian–Cambrian boundary stratigraphy at Maldeota, Lesser Himalaya, *Geol. Mag.* **124**:323–345.

Burzin, M. B., 1987, Defense strategy from predation, pelagial trophic and gradational structure, and the stages of acritarch morphologic evolution in the Late Precambrian and Early Cambrian, in: *Abstracts of Reports, 3rd All-Union Symposium on Paleontology of the Precambrian and Early Cambrian, 1987*, Karelian Branch, USSR Academy of Sciences, Petrozavodsk, USSR, pp. 12–14 [in Russian].

Butterfield, N. J., 1990, A reassessment of the enigmatic Burgess Shale fossil *Wiwaxia corrugata* (Matthew) and its relationship to the polychaete *Canadia spinosa* Walcott, *Paleobiology* **16**:287–303.

Bykova, E. V., and Reitlinger, E. A., 1959, Order Astrorhizida, in: *Osnovy Paleontologii: Obshchaya Chast, Prosteyshie* (D. M. Rauzer-Chernoussova and A. V. Fursenko, eds.), USSR Academy of Sciences, Moscow, pp. 168–177 [in Russian].

Chen Junyuan, Hou Xianguang, and Erdtmann, B. D., 1989, New soft-bodied fossil fauna near the base of the Cambrian System at Chengjiang, Eastern Yunnan, China, in: *Developments in Geosciences. Contributions to the 28th International Geological Congress, 1989, in Washington, D. C.*, Academia Sinica, Science Press, Beijing, pp. 265–278.

Chen Menge, 1979, Some skeletal fossils from the phosphatic sequence, early Lower Cambrian, South China, *Scientia Geol. Sinica* **4**:187–189 [in Chinese with English abstract].

Cherchi, A., and Schroeder, R., 1985, Middle Cambrian foraminifera and other microfossils from SW Sardinia, *Boll. Soc. Paleontol. Ital.* **23**:149–160.

Chumakov, N. M., 1984, The main glacial events of the past and their geological significance, *Izv. Akad. Nauk SSSR Ser. Geol.* **1984**(7):35–53 [in Russian].

Chuvashov, B., and Riding, R., 1984, Principal marine calcareous algae, *Paleontology* **27**:487–500.

Chuvashov, B. I., Luchinina, V. A., Shuysky, V. P., Shaikin, I. M., Berchenko, O. I., Istchenko, A. A., Saltovskaja, V. D., and Schirschova, D. I., 1987, Fossil Calcareous Algae (Morphology, Systematics and Methods of Study), *Trudy Inst. Geol. Geofiz. Sibirsk. Otd. Akad. Nauk SSSR* **674** [in Russian].

Conway Morris, S., 1982, *Wiwaxia corruga* (Matthew), a problematical Middle Cambrian animal from the Burgess Shale of British Columbia, in: *Proceedings. Third North American Paleontological Convention, Montreal*, Vol. 3, pp. 93–98.

Conway Morris, S., 1985, The Middle Cambrian metazoan *Wiwaxia corrugata* (Matthew) from the Burgess Shale and *Ogygopsis* Shale, British Columbia, Canada, *Philos. Trans. R. Soc. Lond. B* **307**:507–586.

Conway Morris, S., 1987, The search for the Precambrian–Cambrian boundary, *Sci. Am.* **75**:156–167.

Conway Morris, S., and Chen Menge, 1990, *Blastulospongia polytreta* n. sp., an enigmatic organism from the Lower Cambrian of Hubei, China, *J. Paleontol.* **64**:26–30.

Conway Morris, S., and Chen Menge, 1991, Cambroclaves and paracarinachitids, early skeletal problematics from the Lower Cambrian of South China, *Palaeontology* **34**:357–397.

Conway Morris., S., and Fritz, W. A., 1980, Shelly microfossils near the Precambrian–Cambrian boundary, Mackenzie Montains, northwestern Canada, *Nature* **286**:381–384.

Conway Morris, S., and Fritz, W. A., 1984, *Lapworthella filigrana* n. sp. (incertae sedis) from the Lower Cambrian of the Cassiar Mountains, northern British Columbia, Canada, with comments on possible levels of competition in the early Cambrian, *Paläontologische Zeitschrift* **58**:197–209.

Conway Morris, S., and Peel, J. S., 1990, Articulated halkieriids from the Lower Cambrian of north Greenland, *Nature* **345**(6278):802–805.

Conway Morris, S., and Robison, R. A., 1986, Middle Cambrian priapulids and other soft-bodied fossils from Utah and Spain, *Univ. Kans. Paleontol. Contrib.* **117**:1–22.

Conway Morris, S., Peel, J. S., Higgins, A. K., Soper, N. J., and Davis, N. C., 1987, A Burgess Shale-like fauna from the Lower Cambrian of North Greenland, *Nature* **326**:181–183.

Cowie, J. W., and Rozanov, A. Yu., 1983, Precambrian–Cambrian boundary candidate, Aldan River, Yakutia, USSR, *Geol. Mag.* **120**:129–139.

Crimes, T. P., and Anderson, M. M., 1985, Trace fossils from Late Precambrian–Early Cambrian strata of Southeastern Newfoundland (Canada): Temporal and environmental implications, *J. Paleontol.* **59**:310–343.

Datzenko, V. A., Zhuravleva, I. T., Lazarenko, N. P., Popov, Yu. N., and Tchernysheva, N. E., 1968, Biostratigraphy and fauna of Cambrian deposits of the North-West Siberian Platform, *Trudy Nauchno-Issled. Inst. Geol. Arkt.* **155** [in Russian].

Debrenne, F., and Vacelet, J., 1984, Archaeocyatha: Is the sponge model consistent with their structural organization?, *Palaeontogr. Am.* **54**:358–369.

Debrenne, F., Termier, H., and Termier, G., 1970. Radiocyatha. Une nouvelle classe d'organismes primitifs du Cambrien inférieur, *Bull. Soc. Géol. Fr. 7* **13**:439–444.

Debrenne, F. M., Gangloff, R. A., and Lafuste, J. G., 1987, *Tabulaconus* Handfield: Microstructure and its implication in the taxonomy of primitive corals, *J. Paleontol.* **61**:1–9.

Debrenne, F., Zhuravlev, A. Yu., and Rozanov, A. Yu., 1989, Regular Archaeocyaths, *Trudy Paleontol. Inst. Akad. Nauk SSSR* **223** [in Russian].

Debrenne, F., Lafuste, J., and Zhuravlev, A., 1990, Coralomorphes et spongiomorphes à l'aube du Cambrien, *Bull. Mus. Natn Hist. Nat.*, Paris, **12**(1):17–39.

Downie, C., 1982, Lower Cambrian acritarchs from Scotland, Norway, Greenland and Canada, *Trans. R. Soc. Edinb. Earth Sci.* **72**:257–285.

Drozdova, N. A., Fonin, V. D., Grigorjeva, N. V., Ragozina, A. L., Rozanov, A. Yu., Saytina, T. A., Voronova, L. G., Voronin, Yu. I., and Zhegallo, E. A., 1981, Precambrian–Cambrian reference section in Mongolia, in: *Papers for the Second International Symposium on the Cambrian System 1981* (M. E. Taylor, ed.), U. S. Geological Survey Open-File Report 81-743, pp. 76–77.

Drushchits, V. V., 1974, *Paleontology of Invertebrates*, Moscow State University Publishing House, Moscow [in Russian].

Dzik, J., 1978, Larval development of hyoliths, *Lethaia* **11**:293–299.

Dzik, J., 1980, Ontogeny of *Bacrtotheca* and related hyoliths, *Geol. Fören. Stockh. Förh.* **102:**223–233.

Dzik, J., 1986, Turrilepadida and other Machaeridia, in: *Problematic Fossil Taxa* (A. Hoffman and M. H. Nitecki, eds.), Clarendon Press, Oxford, pp. 116–134.

Dzik, J., 1987, Origin of Cephalopoda, *Acta Palaeontol. Polon.* **26:**161–191.

Dzik, J., and Lendzion, K., 1988, The oldest arthropods of the East European Platform, *Lethaia* **21:**29–38.

Esakova, N. V., 1987, Siphogonuchitids, in: *Abstracts of Reports. 3rd All-Union Symposium on Paleontology of the Precambrian and Early Cambrian*, 1987, Karelian Branch, USSR Academy of Sciences, Petrozavodsk, pp. 37–38 [in Russian].

Esakova, N. V., 1990, Small shelly fossils in Precambrian–Cambrian boundary deposits from western Mongolia, in: *Third International Symposium on the Cambrian System, Abstracts* (L. N. Repina and A. J. Zhuravlev, eds.), Subcommission on Cambrian Stratigraphy, Commission on Stratigraphy, International Union of Geological Sciences, Akad. Nauk SSSR, Novosibirsk, p. 93.

Fedonkin, M. A., 1981, White Sea Biota of the Vendian (Precambrian Non-skeletal Fauna of the Russian Platform North), *Trudy Geol. Inst. Akad. Nauk SSSR* **342** [in Russian].

Fedonkin, M. A., 1983, Organic world of the Vendian, *Itogi Nauki i Tekhniki, ser. stratigrafiya, Paleontologiya* **12** [in Russian].

Fedorov, A. B., 1982, Biostratigraphy and facies of the oldest Lower Cambrian horizon of the Middle Aldan River, in: *Stratigraphy and Facies of the Siberian Sedimentary Basins* (V. A. Astashkin, ed.), Siberian Scientific Research Institute of Geology, Geophysics and Mineral Resources, Novosibirsk, USSR, pp. 18–25 [in Russian].

Fedorov, A. B., 1984, New representatives of the shelly fossils in the Precambrian–Cambrian stratotype sections of the Siberian Platform (Aldan, Kotuy Rivers), in: *New Species of Ancient Invertebrates and Plants of the Siberian Oil-Gas Fields* (S. P. Bulynnikova and I. G. Klimova, eds.), Siberian Scientific Research Institute of Geology, Geophysics and Mineral Resources, Novosibirsk, USSR, pp. 5–9 [in Russian].

Fedorov, A. B., 1986, New tubicolous problematics from the stratotype of the Tommotian Stage, *Paleontol. Zh.* **1986**(3)**:**110–112 [in Russian].

Fedorov, A. B., and Pereladov, V. S., 1987, Siliceous sponge spicules from the Kuonamka Formation of the North-East Siberian Platform, in: *New Species of Ancient Plants and Invertebrates of the Phanerozoic of Siberia* (S. P. Bulynnikova and I. G. Klimova, eds.), Siberian Scientific Research Institute of Geology, Geophysics and Mineral Resources, Novosibirsk, USSR, pp. 36–46 [in Russian].

Fedorov, A. B., Jegorova, L. I., and Savitsky, V. E., 1979, The first find of ancient trilobites in the lower part of the stratotype of the Lower Cambrian Tommotian Stage (Aldan River), *Doklady Akad. Nauk SSSR* **249:**1188–1190 [in Russian].

Fenton, C., and Fenton, M., 1939, Pre-cambrian and Paleozoic algae, *Bull. Geol. Soc. Am.* **50:**89–126.

Firby, J. B., and Durham, J. W., 1974, Molluscan radula from earliest Cambrian, *J. Paleontol.* **48:**1109–1119.

Firby-Durham, J., 1977, *Platysolenites* from the Lower Cambrian of California, *Izv. Akad. Nauk SSSR Ser. Geol.* **1977**(9)**:**146–149 [in Russian].

Fisher, D. W., 1962, Small conoidal shells of uncertain affinities, in: *Treatise on Invertebrate Paleontology*, Part W, *Miscellanea* (R. C. Moore, ed.), Geological Society of America and University of Kansas Press, Lawrence, Kansas, pp. W98–W143.

Fonin, V. D., and Smirnova, T. N., 1967, A new group of problematic Early Cambrian organisms, and some methods of their preparation, *Paleontol. Zh.* **1967**(2)**:**15–27 [in Russian].

Fritz, W. H., 1972, Lower Cambrian trilobites from the Sekwi Formation type section, Mackenzie Mountains, northwestern Canada, *Geol. Surv. Can. Bull.* **212:**1–90.

Fritz, W. H., 1980, International Precambrian–Cambrian Boundary Working Group's field study to Mackenzie Mountains, Northwest Territories, Canada, *Geol. Surv. Can. Pap.* **80-1A:**41–45.

Fritz, W. H., and Crimes, T. P., 1985, Lithology, trace fossils, and correlation of Precambrian–Cambrian boundary beds, Cassiar Mountains, North-Central British Columbia, *Geol. Surv. Can. Pap.* **83-13:**1–24.

Germs, G. J., 1972, New shelly fossils from Nama Group, South West Africa, *Am. J. Sci.* **272:**752–761.

Glaessner, M. F., 1975, The oldest Foraminifera, *Bull. Bur. Miner. Resour. Geol. Geophys. Aust.* **192:**61–65.

Glaessner, M. F., 1976, Early Phanerozoic annelid worms and their geological and biological significance, *J. Geol. Soc. Lond.* **132:**259–275.

Glaessner, M. F., 1984, *Dawn of Animal Life: A Biohistorical Study,* Cambridge University Press, Cambridge.

Gnilovskaya, M. B., 1976, Vendian Metaphyta, in: *Abstracts of Papers for All-Union Symposium, Paleontology of the Precambrian and Early Cambrian, Novosibirsk, USSR,* Institute of Geology and Geophysics, Siberian Branch, USSR Academy of Sciences, Novosibirsk, pp. 19–21 [in Russia].

Gorjansky, V. Yu., 1973, On the necessity of excluding *Chancelloria* Walcott from phylum Porifera, in: Problems of Lower Cambrian Paleontology and Biostratigraphy of Siberia and the Far East (I. T. Zhuravleva, ed.), *Trudy Inst. Geol. Geofiz. Sibirsk. Otd. Akad. Nauk SSSR* **49:**39–44 [in Russian].

Gorjansky, V. Yu., 1977, The first find of sponge remains in the Lower Cambrian of East Siberia, *Ezhegodnik Vses. Paleontol. Obshch.* **20:**274–278 [in Russian].

Gorjansky, V. Yu., and Popov, L. Ye., 1986, On the origin and systematic position of the calcareous shelled inarticulate brachiopods, *Lethaia* **19:**233–240.

Grant, S. W. F., 1990, Shell structure and distribution of *Cloudina,* a potential index fossil for the Terminal Proterozoic, *Am. J. Sci.* **290A:**261–294.

Grigorieva, N. V., 1980, On the question of hyolithelminth microstructure research, in: *26th International Geological Congress, Paleontological and Stratigraphic Reports of Soviet Geologists,* Nauka, Moscow, pp. 49–55 [in Russian with English abstract].

Grigorieva, N. V., Melnikova, L. M., and Pelman, Yu. L., 1983, Brachiopods, ostracods (bradoriids) and problematica from the stratotype area of the Lower Cambrian stages, *Paleontol. Zh.* **1983**(3)**:**54–58 [in Russian].

Hamdi, B., 1989, Stratigraphy and palaeontology of the Late Precambrian to Early Cambrian in the Alborz Mountains, Northern Iran, *Geol. Surv. Iran Rep.* **59:**1–41.

Hanh, G., and Pflug, H. D., 1985, Die Cloudinidae n. fam., Kalk–Rohren aus dem Vendium und Unter-Cambrium, *Senkenberg. Lethaea* **65:**413–431.

Hartman, W. D., and Goreau T. F., 1970, Jamaican coralline sponges: Their morphology, ecology and fossil relatives, *Zool. Soc. Lond. Symp.* **25:**205–243.

Hinz, I., 1987, The Lower Cambrian microfauna of Comley and Rushton, Shropshire, England, *Palaeontogr. Abt. A* **198:**41–100.

Hinz, I., Kraft, P., Mergl, M., and Müller, K. J., 1990, The problematic *Hadimopanella, Kaimenella, Milaculum* and *Utahphospha* identified as sclerites of Palaeoscolecida, *Lethaia* **23:**217–221.

Hoffman, A., and Nitecki, M. H. (eds.), 1986, *Problematic Fossil Taxa,* Clarendon Press, Oxford.

Howell, B. F., and Van Houten, F. B., 1940, A new sponge from the Cambrian of Wyoming, *Bull. Wagner Free Inst. Sci.* **15:**1–8.

Ivantsov, A. Yu., 1989, Ultrastructure of *Sabellidites cambriensis* Jan. tubes (pogonophores?) of the Vendian, *Byull. Mosk. Obshch. Ispyt. Prir.* **64**(5)**:**129 [in Russian].

Ivantsov, A. Yu., 1990, First finds of phyllocarids in the Lower Cambrian of Yakutia, *Paleontol. Zh.* **1990**(2)**:**130–132.

Jakovlev, N. N., 1956, The first find of a crinoid in the Cambrian of the USSR, *Doklady Akad. Nauk SSSR* **108:**726–727 [in Russian].

Jankauskas, T. V., 1969, Pterocyathids of the Lower Cambrian of Krasnoyarsk Ridge (Eastern Sayan), in: *Lower Cambrian Biostratigraphy and Paleontology of Siberia and the Far East* (I. T. Zhuravleva, ed.), Nauka, Moscow, pp. 114–157 [in Russian].

Jankauskas, T. V., 1972, Cribicyaths of the Siberian Lower Cambrian, in: *Problems of the Lower Cambrian Biostratigraphy and Paleontology of Siberia* (I. T. Zhuravleva, ed.), Nauka, Moscow, pp. 161–183 [in Russian].

Jankauskas, T. V., 1973, The study of Cambrian Cribricyathi in the USSR, in: Problems of the Lower Cambrian Paleontology and Biostratigraphy of Siberia and the Far East (I. T. Zhuravleva, ed.), *Trudy Inst. Geol. Geofiz. Sibirsk. Otd. Akad. Nauk SSSR* **49:**45–53 Nauka, [in Russian].

Jazmir, M. M., Dalmatov, B. A., and Jazmir, I. K., 1975, *Atlas of the Palaeozoic and Mesozoic Fauna and Flora of Buryatian ASSR: Palaeozoic,* Nedra, Moscow [in Russian].

Jegorova, L. I., Shabanov, Yu. Ya., Rozanov, A. Yu., Savitsky, V. E., Tchernyshova, N. E., and Shishkin, B. B., 1976, Elanka and Kuonamka Facies-stratotypes of the Lower Boundary of the Middle Cambrian in Siberia, *Trudy Sibirsk. Nauchno-Issled. Inst. Geol. Geofiz. Mineral. Syr.* **211** [in Russian].

Jell, J. S., 1984, Cambrian Cnidarians with mineralized skeletons, *Paleontogr. Am.* **54:**105–109.

Jell, P. A., Gravestock, D. I., and Zhuravlev, A. Yu., 1990, Yorke Peninsula (South Australia) Lower Cambrian section: A key to China-Siberia correlation, in: *Third International Symposium on the*

Cambrian System, Abstracts (L. N. Repina and A. J. Zhuravlev, eds.), Subcommission on Cambrian Stratigraphy, Commission on Stratigraphy, International Union of Geological Sciences, Akad. Nauk SSSR, Novosibirsk, p. 117.

Jermak, V. V., 1986, The Early Cambrian fordillids (Bivalvia) from the Northern Siberian Platform, in: Cambrian Biostratigraphy and Paleontology of North Asia (I. T. Zhuravleva, ed.), *Trudy Inst. Geol. Geofiz. Sibirsk. Otd. Akad. Nauk SSSR* **669:**183–188, Nauka, Moscow [in Russian].

Jermak, V. V., and Pelman, Yu. L., 1986, Some Cambrian Mollusks and brachiopods of the northern Kharaulakh, in: Cambrian Biostratigraphy and Paleontology of North Asia (I. T. Zhuravleva, ed.), *Trudy Inst. Geol. Geofiz. Sibirsk. Otd. Akad. Nauk SSSR* **669:**188–200 [in Russian].

Kashina, L. N., and Sayutina, T. A., 1986, New find of Hydroconozoa in the Lower Cambrian of the Eastern Sayan, in: Cambrian Biostratigraphy and Paleontology of North Asia (I. T. Zhuravleva, ed.), *Trudy Inst. Geol. Geofiz. Sibirsk. Otd. Akad. Nauk SSSR* **669:**209–212 [in Russian].

Keller, B. M., and Rozanov, A. Yu. (eds.), 1979, *Upper Precambrian and Cambrian Stratigraphy of the Western Part of the East-European Platform*, Nauka, Moscow [in Russian].

Kerber, M., 1988, Mikrofossilien aus unterkambrischen Gesteinen der Montagne Noire, Frankreich, *Palaeontogr. Abt. A* **202:**127–203.

Khalfina, V. K., and Yaworsky, V. I., 1974, On the evolution of stromatoporoids, in: *Ancient Cnidaria* (B. S. Sokolov, ed.), Nauka, Novosibirsk, USSR, Vol. 1, pp. 23–37 [in Russian].

Khomentovsky, V. V., 1974, The principles of the determination of the Vendian as a Paleozoic System, in: Contributions to Stratigraphy (A. L. Yanshin, ed.), *Trudy Inst. Geol. Geofiz. Sibirsk. Otd. Akad. Nauk SSSR* **276:**33–70 [in Russian].

Khomentovsky, V. V., and Karlova, G. A., 1986, On the Pestrotsvet Formation lower boundary in the Aldan River Basin, in: *Late Precambrian and Early Paleozoic of Siberia. The Siberian Platform and Outer Zone of Altay-Sayan Fold Belt* (V. V. Khomentovsky, ed.), Institute of Geology and Geophysics, Siberian Branch, USSR Academy of Sciences, Novosibirsk, USSR, pp. 3–22 [in Russian].

Khomentovsky, V. V., and Karlova, G. A., 1989, Vendian–Cambrian strata of the Dzhanda River and their analogies in the reference sections of Eastern Siberia, in: *Late Precambrian and Early Paleozoic of Siberia. Actual Problems of Stratigraphy* (V. V. Khomentovsky and Yu. K. Sovetov, eds.), Institute of Geology and Geophysics, Siberian Branch, USSR Academy of Sciences, Novosibirsk, SSSR, pp. 23–61 [in Russian].

Khomentovsky, V. V., and Repina, L. N., 1965, *Lower Cambrian of the Stratotype Section of Siberia*, Nauka, Moscow [in Russian].

Khomentovsky, V. V., Valkov, A. K., and Karlova, G. A., 1990, New data on the biostratigraphy of transitional Vendian-Cambrian strata in the middle Aldan River basin, in: *Late Precambrian and Early Paleozoic of Siberia (Problems of the Regional Stratigraphy)* (V. V. Khomentovsky and A. S. Gibsher, eds.), Institute of Geology and Geophysics, Siberian Branch, USSR Academy of Sciences, Novosibirsk, pp. 3–57 [in Russian].

Khramov, A. N., and Rodionov, V. P., 1980, Paleomagnetism and horizontal movement of the Russian and the Siberian Platforms, in: *26th International Geological Congress, Reports of Soviet Geologists and Geophysicists*, Nauka, Moscow, pp. 72–79 [in Russian with English abstract].

Kirjanov, V. V., 1987, The succession of acritarch assemblages in Precambrian–Cambrian boundary strata of the East-European and Siberian Platforms, in: *Theses of Reports, 3rd All-Union Symposium on Paleontology of Precambrian and Early Cambrian, 1987*, Karelian Branch, USSR Academy of Sciences, Petrozavodsk, USSR, pp. 44–45 [in Russian].

Kirsanov, V. V., 1974, Questions of the stratigraphy of the Vendian and Cambrian boundary beds in the central areas of the East-European Platform, in: *Lower Cambrian Biostratigraphy and Paleontology of Europe and Northern Asia* (I. T. Zhuravleva and A. Yu. Rozanov, eds.), Nauka, Moscow, pp. 5–21 [in Russian].

Kirschvink, J. L., 1978, The Precambrian–Cambrian boundary problem: Magnetostratigraphy of the Amadeus Basin, Central Australia, *Geol. Mag.* **115:**139–150.

Kirschvink, J. L., and Rozanov, A. Yu., 1984, Magnetostratigraphy of lower Cambrian strata from the Siberian Platform: A paleomagnetic pole and a preliminary polarity time scale, *Geol. Mag.* **121:**189–203.

Kirschvink, J. L., Barr, T. D., Zhuravlev, A. Yu., and Rozanov, A. Yu., 1984, The destruction of Paleopangea (?) in the Early Cambrian, in: *Abstracts. 27th International Geological Congress Moscow 1984*, Nauka, Moscow, Volume 4, p. 113.

Kirschvink, J. L., Budney, C. J., and Zhuravlev, A. Yu., 1987, Magnetostratigraphy of the Precambrian–Cambrian reference section near Salany-Gol, Western Mongolia: Comparison with the Siberian Platform, *Geol. Soc. Am. Abstr. Progr.* **19:**728.

Koneva, S. P., 1979, *Stenothecoids and Inarticulate Brachiopods of the Lower and Lower Middle Cambrian of the Central Kazakhstan,* Nauka, Kazakh SSR Publishing House, Alma-Ata, USSR [in Russian].

Koneva, S. P., 1983, *Mobergella* from the Lower Cambrian of Seleta Synclinorium, in: *The Lower Paleozoic Stratigraphy and Paleontology of Kazakhstan* (M. K. Apollonov, S. M. Bandaletov, and N. K. Ivshin, eds.), Nauka, Kazakh SSR Publishing House, Alma-Ata, USSR, pp. 110–112 [in Russian].

Koneva, S. P., Popov, L. E., Ushatinskaya, G. T., and Esakova, N. V., 1990, Inarticulate brachiopods (acrotretids) and microproblematics from the Upper Cambrian of northeastern Kazakhstan, in: Cambrian Biostratigraphy and Paleontology of Northern Asia (L. N. Repina, ed.), *Trudy Inst. Geol. Geofiz. Sibirsk. Otd. Akad. Nauk SSSR* **765:**158–170 [in Russian].

Konjuschkov, K. N., 1972, The problem of the stage subdivision of the Lower Cambrian, in: *Problems of the Lower Cambrian Biostratigraphy and Paleontology of Siberia* (I. T. Zhuravleva, ed.), Nauka, Moscow, pp. 7–14 [in Russian].

Korde, K. B., 1959, Problematic remains from Cambrian strata of the South-East Siberian Platform, *Doklady Akad. Nauk SSSR* **125:**625–627 [in Russian].

Korde, K. B., 1963, Hydroconozoa—A new class of coelenterate animals, *Paleontol. Zh.* **1963**(2)**:**20–25 [in Russian].

Korde, K. B., 1975, Hydroconozoa from Cretaceous and Paleocene deposits of Crimea, in: *Evolution and Change of Organic Kingdom at the Mesozoic–Caenozoic Boundary, New Data on the Evolution of Fauna* (V. N. Shimansky and A. N. Soloviev, eds.). Nauka, Moscow, pp. 32–38 [in Russian].

Korde, K. B., 1986, About ancestors of the orders Scyphozoa and Edelsteiniida, in: *Phanerozoic Reefs and Corals of the USSR, Transactions 5th All-Union Symposium on Corals and Reefs, Dushanbe, 1983* (B. S. Sokolov, ed.), Nauka, Moscow, pp. 20–23 [in Russian].

Korde, K. B., 1990a, Edelstéiniida Korde-former problematics, in: Fossil Problematics of the USSR (B. S. Sokolov and I. T. Zhuravleva, eds.), *Trudy Inst. Geol. Geofiz. Sibirsk Otd. Akad. Nauk SSSR* **783:**50–69 [in Russian].

Korde, K. B., 1990b, Tuvinoidea—A New group of Cambrian organisms with an unknown systematic position, in: Fossil problematics of the USSR (B. S. Sokolov and I. T. Zhuravleva, eds.), *Trudy Inst. Geol. Geofiz. Sibirsk. Otd. Akad. Nauk SSSR* **783:**129–146 [in Russian].

Korobov, M. N., 1980, Biostratigraphy and miomeric trilobites of the Lower Cambrian of Mongolia, in: Biostratigraphy of the Lower Cambrian and the Carboniferous of Mongolia (V. V. Menner and S. V. Meien, eds.), *Trans. Joint Soviet–Mongolian Geol. Expedition* **26:**5–108 Nauka, [in Russian].

Korobov, M. N., 1989, Lower Cambrian biostratigraphy and polymeroid trilobites of Mongolia, *Trans. Soviet–Mongolian Geol. Expedition* **46** [in Russian].

Korobov, M. N., and Missarzhevsky, V. V., 1977, On the Cambrian and Precambrian boundary beds of West Mongolia, in: *Paleozoic Invertebrates of Mongolia* (Yu. I. Voronin, ed.), *Trans. Joint Soviet-Mongolian Paleontol. Expedition* **1977:**7–9 [in Russian].

Krandievsky, V. S., Ishchenko, T. A., and Kirjanov, V. V., 1968, *Lower Palaeozoic Palaeontology and Stratigraphy of Volyn'-Podolia,* Naukova Dumka, Kiev [in Russian].

Krasilova, I. N., 1977, Fordillids (Bivalvia) from the Lower Cambrian of the Siberian Platform, *Paleontol. Zh.* **1977**(2)**:**42–48 [in Russian].

Krasilova, I. N., 1987, Earlier representatives of bivalved molluscs. *Paleontol. Zh.* **1987**(4)**:**24–30 [in Russian].

Krasnov, V. I., Savitsky, V. E., Tesakov, Yu. I., and Khomentovsky, V. V. (eds.), 1983. *Decrees of the All-Union Stratigraphical Meeting on the Precambrian, Paleozoic and Quaternary Systems of Middle Siberia, Novosibirsk, 1979,* Part 1 (*Upper Proterozoic and Lower Paleozoic*), Cartographic Publishing House of All-Union Geological Institute, Leningrad, USSR [in Russian].

Krischtofovich, A. N., 1953, The discovery of lycopsid-like plant in the Cambrian of East Siberia, *Doklady Akad. Nauk SSSR* **91:**1377–1379 [in Russian].

Lafuste, J., Jell, J. S., Gandin, A., Gravestock, D., and Debrenne, F., 1988, New tabuloid form from the Lower Cambrian of South Australia, in: *Fifth International Symposium Fossil Cnidaria, Brisbane, Australia, Programme and Handbook,* p. 48.

Lafuste, J., Debrenne, F., and Zhuravlev, A., 1990, Les fuscinules, type nouveau de biocristaux dans le squelette d'*Hydroconus* Korde, 1963, Corallomorphe du Cambrien inferieur, *C. R. Acad. Sci. Paris II* **310:**1553–1559.

Landing, E., 1984, Skeleton of lapworthellids and the suprageneric classification of tommotiids (Early and Middle Cambrian phosphatic problematica), *J. Paleontol.* **58:**1380–1398.

Landing, E., 1989, Paleoecology and distribution of the Early Cambrian rostroconch *Watsonella crosbyi* Grabau, *J. Paleontol.* **63:**566–573.

Landing, E., Nowlan, G. S., and Fletcher, T. P., 1980, A microfauna associated with Early Cambrina trilobites of the *Callavia* Zone, northern Antigonish Highlands, Nova Scotia, *Can. J. Earth Sci.* **17**(3):400–418.

Lauritzen, Q., and Yochelson, E. L., 1982, *Salterella rugosa* (Early Cambrian: Agmata) on Nordaustlandet and Spitsbergen, Svalbard, *Polar Res.* **1982**(1):5–10.

Lendzion, K., 1975, Fauna of the *Mobergella* zone in the Polish Lower Cambrian *Kwartalnik Geologiczny* **19**(2):237–242.

Lendzion, K., 1977, First gastropod fauna from the Klimontovian stage (Lower Cambrian) of the South-Eastern Poland, *Kwartalnik Geologiczny* **21**(2):239–244.

Lipps, J. H., and Sylvester, A. G., 1968, The enigmatic Cambrian fossil *Volborthella* and its occurrence in California, *J. Paleontol.* **42:**329–336.

Luchinina, V. A., 1975, Paleoalgologic characteristics of the Early Cambrian of the Siberian Platform, *Trudy Inst. Geol. Geofiz. Sibirsk. Otd. Akad. Nauk SSSR* **216** [in Russian].

Luchinina, V. A., 1985, Algal buildups of the Early Paleozoic on the North Siberian Platform, in: Environment and Life in the Geological Past, Paleobasins and their Inhabitants (O. A. Betekhtina and I. T. Zhuravleva, eds.), *Trudy Inst. Geol. Geofiz. Sibirsk. Otd. Akad. Nauk SSSR* **628:**45–50 [in Russian].

Luchinina, V. A., 1987, Calcareous algae of the Siberian Platform in the transitional Precambrian–Cambrian strata, in: *Theses of Reports, 3rd All-Union Symposium on Paleontology of the Precambrian and the Early Cambrian, 1987,* Karelian Branch, USSR Academy of Sciences, Petrozavodsk, USSR, pp. 60–62 [in Russian].

Luchinina, V. A., and Meshkova, N. P., 1969, The first find of *Dactyloidites asteroides* in the Lower Cambrian of Siberia, in: *Lower Cambrian Biostratigraphy and Paleontology of Siberia and the Far East* (I. T. Zhuravleva, ed.), Nauka, Moscow, pp. 180–181 [in Russian].

Luo Huilin, Jiang Zhiwen, Wu Xiche, Song Xueliang, Ouyang Lin, Xing Yusheng, Liu Guizhi, Zhang Shishan, Tao Yonghe, and Liu Mingquan, 1984, *The Sinian–Cambrian Boundary Stratotype Section at Meishucun, Jinning, Yunnan, China,* People's Publishing House, Kunming, Yunnan, China.

Luvsandanzan, B., and Rozanov, A. Yu., 1984, The age of the ancient phosphates of Asia, *Doklady Akad. Nauk SSSR* **277:**164–167 [in Russian].

Makhnach, A. S., Shkuratov, V. I., Zinovenko, G. S., and Piskun, L. V., 1985, *The Cambrian of Byelorussia,* Nauka i Tekhnika, Minsk, USSR [in Russian].

Mambetov, A. M., and Repina, L. N., 1979, The Lower Cambrian of the Talassky Ala-Too and its correlation with the sections of Maly Karatau and the Siberian Platform, in: Lower Cambrian Biostratigraphy and Paleontology of Siberia (I. T. Zhuravleva and N. P. Meshkova, eds.), *Trudy Inst. Geol. Geofiz. Sibirsk. Otd. Akad. Nauk SSSR* **406:**99–138 [in Russian].

Marek, L., and Yochelson, E. L., 1976, Aspects of the biology of Hyolitha (Mollusca), *Lethaia* **9:**65–82.

Märss, T., 1988, Early Palaeozoic hadimopanellids of Estonia and Kirgizia (USSR), *Eesti NSV Tead. Akad. Toim. Geol.* **37**(1):10–17.

Matthews, S. C., 1973, Lapworthellids from the Lower Cambrian *Strenuella* Limestone at Comley, Shropshire, *Palaeontology* **16:**139–148.

Matthews, S. C., and Missarzhevsky, V. V., 1975, Small shelly fossils of late Precambrian and early Cambrian age: A review of recent work, *J. Geol. Soc. Lond.* **131:**289–304.

Melnikova, L. M., 1987, Bradoriids from the Tiskre Formation (Lower Cambrian) of Estonia, *Paleontol. Zh.* **1987**(1):128–131 [in Russian].

Melnikova, L. M., 1988, Some bradoriids (Crustacea) from the Botomian Stage of the eastern Transbaikalian *Paleontol. Zh.* **1988**(1):114–117 [in Russian].

Melnikova, L. M., 1990, Early and Late Cambrian Bradoriida (ostracods) of northeastern Kazakhstan, in: *Cambrian Biostratigraphy and Palaeontology of Northern Asia* (L. N. Repina, ed.), *Trudy Inst. Geol. Geofiz. Sibirsk Otd. Akad. Nauk SSSR* **632:**127–133 [in Russian].

Menner, V. V., Pokrovskaia, N. V., and Rozanov, A. Yu., 1960, The "Upper-Cambrian" archaeocyathan–coral coenose in the Tannu-Ola Ridge (Tuva), *Izv. Akad. Nauk SSSR Ser. Geol.* **1960**(7):99–100 [in Russian].

Mens, K., Bergstrom, J., and Lendzion, K., 1987, *The Cambrian System on the East European Platform (Correlation Chart and Explanatory Notes)*, Valgus Publishers, Tallin, USSR [in Russian with English chart].

Mens, K., Bergstrom, J., and Lendzion, K., 1990, *The Cambrian System on the East European Platform*, International Union Geol. Sci. Publication 25.

Meshkova, N. P., 1969, On the question of the paleontological characteristics of the Lower Cambrian deposits of the Siberian Platform, in: *Lower Cambrian Biostratigraphy and Paleontology of Siberia and the Far East* (I. T. Zhuravleva, ed.), Nauka, Moscow, pp. 158–174 [in Russian].

Meshkova, N. P., 1974a, Hyoliths of the Lower Cambrian of the Siberian Platform, *Trudy Inst. Geol. Geofiz. Sibirsk. Otd. Akad. Nauk SSSR* **97** [in Russian].

Meshkova, N. P., 1974b, Cambroscleritida incertae sedis—A new order of Cambrian fossils, in: *Lower Cambrian Biostratigraphy and Paleontology of Europe and Northern Asia* (I. T. Zhuravleva and A. Yu. Rozanov, eds.), Nauka, Moscow, pp. 190–193 [in Russian].

Meshkova, N. P., 1984, New oldest wiwaxiids of Siberia, in: Problematics of the Paleozoic and the Mesozoic (B. S. Sokolov, ed.), *Trudy Inst. Geol. Geofiz. Sibirsk. Otd. Akad. Nauk SSSR* **597:**133–136 [in Russian].

Meshkova, N. P., 1985, New tubicolous problematics from the Middle Cambrian of Siberia and Central Asia, *Trudy Inst. Geol. Geofiz, Siberian Otd.* **632:**127–133 [in Russian].

Milashev, V. A., 1958, Discovery of an Upper Cambrian medusoid in the Middle Olenek area of the Siberian Platform, in: *Collective Book on Paleontology and Biostratigraphy,* Scientific Research Institute on the Geology of the Arctic, Leningrad, Vol. 7, pp. 19–21 [in Russian].

Miller, J. F., 1984, Cambrian and earliest Ordovician conodont evolution, biofacies, and provincialism, *Geol. Soc. Am. Spec. Pap.* **196:**43–68.

Miroshnikov, L. D., and Kravtsov, A. G., 1960, Rare paleontological remains and trace fossils in the Late Cambrian strata of the North-West Siberian Platform, *Trudy Nauchno-Issled. Inst. Geol. Arkt.* **3:**28–41 [in Russian].

Miroshnikov, L. D., and Kravtsov, A. G., 1965, Late Cambrian scyphomedusoids of the Siberian Platform, *Ezhegodik Vses. Paleontol. Obshch.* **17:**46–66 [in Russian].

Missarzhevsky, V. V., 1973, Conodont-like organisms from boundary layers of the Cambrian and the Precambrian in the Siberian Platform and Kazakhstan, in: Problems of the Lower Cambrian Paleontology and Biostratigraphy of Siberia and the Far East (I. T. Zhuravleva, ed.), *Trudy Inst. Geol. Geofiz. Sibirsk. Otd. Akad. Nauk SSSR* **49:**53–57 [in Russian].

Missarzhevsky, V. V., 1974, New data on Early Cambrian ancient fossils of the Siberian Platform, in: *Lower Cambrian Biostratigraphy and Paleontology of Europe and Northern Asia* (I. T. Zhuravleva and A. Yu. Rozanov, eds.), Nauka, Moscow, pp. 179–189 [in Russian].

Missarzhevsky, V. V., 1976, New data on Early Cambrian monoplacophorans, *Paleontol. Zh.* **1976**(2):129–131 [in Russian].

Missarzhevsky, V. V., 1977, Conodonts(?) and Cambrian phosphatic problematics from Mongolia and Siberia, in: Paleozoic Invertebrates of Mongolia (Yu. I. Voronin, ed.), *Trans. Joint Soviet–Mongolian Paleontol. Expedition* **1977:**10–19 [in Russian].

Missarzhevsky, V. V., 1982, Subdivision and correlation of the Precambrian and Cambrian boundary beds using some groups of the oldest skeletal organisms, *Byull. Mosk. Obshch. Ispyt. Prir. Otd. Geol.* **57**(5):52–67 [in Russian].

Missarzhevsky, V. V., 1983, Stratigraphy of the oldest Phanerozoic rocks of the Anabar Massif, *Sovetsk. Geol.* **1983**(9):62–73 [in Russian].

Missarzhevsky, V. V., 1984, Maly Karatau, in: Rozanov, A. Yu., and Sokolov, B. S., (eds.), Lower Cambrian stage subdivision. Stratigraphy (A. Yu. Rozanov and B. S. Sokolov, eds.). Akad. Nauk SSSR, Nauka, Moscow, pp. 123–127 [in Russian].

Missarzhevsky, V. V., 1989, Oldest skeletal fossils and stratigraphy of the Precambrian and Cambrian boundary beds, *Trudy Geol. Inst. Akad. Nauk SSSR* **443:**1–237 [in Russian].

Missarzhevsky, V. V., and Grigorieva, N. V., 1981, New representatives of the order Tommotiida, *Paleontol. Zh.* **1981**(4):91–97 (in Russian).

Missarzhevsky, V. V., and Mambetov, A. M., 1981, Stratigraphy and fauna of Cambrian and Precambrian boundary beds of Maly Karatau, *Trudy Geol. Inst. Akad. Nauk SSSR* **326:**1–92 [in Russian].

Moczydłowska, M., and Vidal, G., 1986, Lower Cambrian acritarch zonation in southern Scandinavia and southeastern Poland, *Geol. Fören. Stockh. Förh.* **108:**201–223.

Moczydłowska, M., and Vidal, G., 1988, How old is the Tommotian?, *Geology* **16:**166–168.
Mostler, H., 1985, Neue heteractinide Spongien (Calcispongea) aus dem Unter- und Mittelkambrium Sudwestsardiniens, *Bur. Nat. Med. Verein Innsbruck* **72:**7–32.
Müller, K. J., 1983, Crustacea with preserved soft parts from the Upper Cambrian of Sweden, *Lethaia* **16:**93–109.
Nazarov, B. B., 1973, Radiolarians from lower horizons of the Cambrian of the Batenevsky Ridge, in: Problems of the Lower Cambrian Paleontology and Biostratigraphy of Siberia and the Far East (I. T. Zhuravleva, ed.), *Trudy Inst. Geol. Geofiz. Sibirsk. Otd. Akad. Nauk SSSR* **49:**5–13 [in Russian].
Nazarov, B. B., 1974, Problematic siliceous structures from Lower Paleozoic of Kazakhstan, in: *Lower Cambrian Biostratigraphy and Paleontology of Europe and Northern Asia* (I. T. Zhuravleva and A. Yu. Rozanov, eds.), Nauka, Moscow, pp. 110–112 [in Russian].
Nazarov, B. B., 1975, Lower-Middle Paleozoic radiolarians of Kazakhstan, *Trudy Geol. Inst. Akad. Nauk SSSR* **275:**1–203 [in Russian].
Nikolaeva, I. V., 1981, Facies zoning of chemical composition of glauconite minerals and factors affecting them, in: Mineralogy and Geochemistry of Glauconite (I. V. Nikolaeva and D. K. Arkhipenko, eds.), *Trudy Inst. Geol. Geofiz. Sibirsk. Otd. Akad. Nauk SSSR* **515:**4–41 [in Russian].
Nitecki, M. H., 1967, Systematic position of receptaculitids, *Abstracts Geological Society America Annual Meeting,* pp. 165–167.
Nowlan, G. S., Narbonne, G. M., and Fritz, W. H., 1985, Small shelly fossils and trace fossils near the Precambrian–Cambrian boundary in the Yukon Territory, Canada, *Lethaia* **18:**233–256.
Okulitch, V. J., 1955, Archaeocyatha, in: *Treatise on Invertebrate Paleontology,* Part E, *Archaeocyatha and Porifera* (R. C. Moore, ed.), Geological Society of America and University of Kansas Press, Lawrence, Kansas, pp. E1–E20.
Okuneva, O. G., and Repina, L. N., 1973, Cambrian biostratigraphy and fauna of Primorie, *Trudy Inst. Geol. Geofiz. Sibirsk. Akad. Nauk SSSR* **37** [in Russian].
Öpik, A., 1926, Uber den estländischen blauen Ton, *Tartu Ülikooli Geol. Inst. Toim.* **6:**39–47.
Osadchaja, D. V., Kashina, L. N., Zhuravleva, I. T., Borodina, N. P., and Bojarinov, A. S., 1979, *Stratigraphy and Archaeocyaths from the Lower Cambrian of Altay-Sayan Fold Belt,* Nauka, Moscow [in Russian].
Paczesna, J., 1986, Upper Vendian and Lower Cambrian ichnocoenoses of Lublin region, *Biul. Inst. Geologicznego* **358:**31–47.
Peel, J. S., and Larsen, N. H., 1984, *Hadimopanella apicata* from the Lower Cambrian of western North Greenland, *Rapp. Grønlands Geol. Unders.* **121:**89–96.
Pelman, Yu. L., 1977, Early and Middle Cambrian inarticulate brachiopods from the Siberian Platform, *Trudy Inst. Geol. Geofiz. Sibirsk. Otd. Akad. Nauk SSSR* **316** [in Russian].
Pelman, Yu. L., 1985, New stenothecoids from the Lower Cambrian of West Mongolia, in: Problematics of the Late Precambrian and Paleozoic (B. S. Sokolov and I. T. Zhuravleva, eds.), *Trudy Inst. Geol. Geofiz. Sibirsk. Otd. Akad. Nauk SSSR* **632:**103–114 [in Russian].
Pelman, Yu. L., Jermak, V. V., Fedorov, A. B., Luchinina, V. A., Zhuravleva, I. T., Repina, L. N., Bondarev, V. I., and Borodaevskaya, Z. V., 1990, New data on the Upper Precambrian and Lower Cambrian stratigraphy and paleontology of the Dzhanda River (right tributary of the Aldan River), in: *Cambrian Biostratigraphy and Paleontology of Northern Asia* (L. N. Repina, ed.), *Trudy Inst. Geol Geofiz. Sibirsk. Otd. Akad. Nauk SSSR* **765:**3–93 [in Russian].
Pojeta, J., 1980, Molluscan phylogeny, *Tulane Stud. Geol. Paleontol.* **16:**55–80.
Pojeta, J., and Runnegar, B., 1976, The paleontology of rostroconch mollusks and the early history of the phylum Mollusca, *U. S. Geol. Surv. Prof. Pap.* **968:**1–88.
Pojeta, J., Runnegar, B., and Kriz, J., 1973, *Fordilla troyensis* Barrande: The oldest known pelecypod, *Science* **180:**866–868.
Popov, L. Ye., and Ushatinskaya, G. T., 1986, On secondary alteration of the microstructure in calcium-phosphatic inarticulate brachiopods, *Izv. Akad. Nauk SSSR Ser. Geol.* **1986**(10)**:**135–137 [in Russian].
Popov, L. Ye., Khazanovich, K. K., Borovko, N. G., Sergeeva, S. P., and Sobolevskaya, R. F., 1989, Reference sections and stratigraphy of Cambrian–Ordovician Obolus Formation on the Northeastern Russian Platform, *Trudy Mezhvedomstvennogo Stratigraficheskogo Komiteta* **18** [in Russian].
Popov, Yu. N., 1967, A new Cambrian scyphozoan, *Paleontol. Zh.* **1967**(2)**:**122–123 [in Russian].

Posti, E., 1978, New finds of platysolenitids and gastropods from the Lontova Horizon of Estonia, *Eesti NSV Tead. Akad. Toim. Geol.* **27**(3):103–107 [in Russian with English abstract].

Qian Yi and Bengtson, S., 1989, Paleontology and biostratigraphy of the Early Cambrian Meishucunian Stage in Yunnan Province, South China, *Fossils Strata* **24**:1–156.

Qian Yi and Yin Gong-zheng, 1984, Zhijinitidae and its stratigraphical significance, *Acta Palaeontol. Sinica* **23**:215–223 [in Chinese with English abstract].

Reitlinger, E. A., 1948, Cambrian foraminifers of Yakutia, *Byull. Mosk. Obshch. Ispyt. Prir. Otd. Geol.* **23**(2):77–81 [in Russian].

Repina, L. N., 1983, On the problem of the systematics in the trilobite suborder Olenellina, in: Morphology and Systematics of Phanerozoic Invertebrates (A. S. Dagis and V. N. Dubatolov, eds.), *Trudy Inst. Geol. Geofiz. Sibirsk. Otd. Akad. Nauk SSSR* **538**:114–127 [in Russian].

Repina, L. N., 1985, Early Cambrian seas of the world and paleobiogeographic subdivisions according to trilobites, in: Environment and Life in the Geological Past, Paleobasins and Their Inhabitants (O. A. Betekhtina and I. T. Zhuravleva, eds.), *Trudy Inst. Geol. Geofiz. Sibirsk. Otd. Akad. Nauk SSSR* **628**:5–17 [in Russian].

Repina, L. N., 1986, Global correlation of the Lower Cambrian stage subdivisions according to trilobites, in: Cambrian Biostratigraphy and Paleontology of Northern Asia (I. T. Zhuravleva, ed.), *Trudy Inst. Geol. Geofiz. Sibirsk. Otd. Akad. Nauk SSSR* **669**:4–25 [in Russian].

Repina, L. N., and Okuneva, O. G., 1969, Cambrian arthropods of Primorie, *Paleontol. Zh.* **1969**(1):106–114 [in Russian].

Repina, L. N., and Zhuravlev, A. J. (eds.), 1990, *Abstracts of the Third International Symposium on the Cambrian System, 1–9 August 1990, Novosibirsk, USSR*, Institute of Geology and Geophysics, Siberian Division, USSR Academy of Sciences, Novosibirsk, USSR [in Russian and English].

Repina, L. N., Khomentovsky, V. V., Zhuravleva, I. T., and Rozanov, A. Yu., 1964, *Lower Cambrian Biostratigraphy of the Sayan-Altay Fold Belt*, Nauka, Moscow [in Russian].

Riding, R., 1982, Cyanophyte calcification and changes in ocean chemistry, *Nature* **299**:814–815.

Riding, R., and Voronova, L., 1984, Assemblages of calcareous algae near the Precambrian/Cambrian boundary in Siberia and Mongolia, *Geol. Mag.* **121**:205–210.

Riding, R., and Voronova, L., 1985, Morphological groups and series in Cambrian calcareous algae, in: *Paleoalgology: Contemporary Research and Applications* (D. F. Toomey and M. H. Nitecki, eds.), Springer-Verlag, Berlin, pp. 56–78.

Rigby, J. K., 1978, Porifera of the Middle Cambrian Wheeler Shale from the Wheeler Amphitheater, House Range, in Western Utah, *J. Paleontol.* **52**:1325–1345.

Rigby, J. K., 1986, Sponges of the Burgess Shale (Middle Cambrian) British Columbia, *Palaeontogr. Can.* **2**:1–155.

Rigby, J. K., and Toomey, D. F., 1978, A distinctive sponge spicule assemblage from organic buildups in the Lower Ordovician of Southern Oklahoma, *J. Paleontol.* **52**:501–506.

Romanenko, E. V., 1968, Cambrian sponges of the order Heteractinellida of Altay, *Paleontol. Zh.* **1968**(2):134–136 [in Russian].

Rowell, A. J., 1982, The monophyletic origin of the Brachiopoda, *Lethaia* **15**:299–307.

Rozanov, A. Yu., 1973, Regularities in the morphological evolution of archaeocyaths and the issues of stage subdivision of the Lower Cambrian, *Trudy Geol. Inst. Akad. Nauk SSSR* **241** [in Russian].

Rozanov, A. Yu., 1979, Some problems in study of ancient skeletal organisms, *Byull. Mosk. Obshch. Ispyt. Prir. Otd. Geol.* **54**(3):62–69 [in Russian].

Rozanov, A. Yu., 1980, Centers of origin of Cambrian faunas, in: *26th International Geological Congress. Paleontological and Stratigraphic Reports of Soviet Geologists*, Nauka, Moscow, pp. 30–34 [in Russian with English abstract].

Rozanov, A. Yu., 1983a, *Platysolenites*, in: *Upper Precambrian and Cambrian Palaeontology of the East-European Platform* (A. Urbanek and A. Yu. Rozanov, eds.), Wydawnictwa Geologiczhe, Warsaw, pp. 94–100.

Rozanov, A. Yu., 1983b, *Volborthella*, in: *Upper Precambrian and Cambrian Palaeontology of the East-European Platform* (A. Urbanek and A. Yu. Rozanov, eds.), Wydawnictwa Geologiczne, Warsaw, pp. 111–113.

Rozanov, A. Yu., 1984a, The Precambrian–Cambrian Boundary in Siberia, *Episodes* **7**(1):20–24.

Rozanov, A. Yu., 1984b, Some aspects of studies on bio- and paleogeography of the Cambrian, in: *Proceedings of the 27th International Geological Congress, Moscow 1984*, Volume 2: *Palaeontology*, VNU Science Press, Utrecht, pp. 143–157.

Rozanov, A. Yu., 1985, Paleogeography and paleobiogeography of the Early Cambrian, *Vestnik Akad. Nauk SSSR* **1985**(11):78–92 [in Russian].
Rozanov, A. Yu., 1986, Problematica of the Early Cambrian, in: *Problematic Fossil Taxa* (A. Hoffman and M. H. Nitecki, eds.), Clarendon Press, Oxford, pp. 87–96.
Rozanov, A. Yu., and Łydka, K. (eds.), 1987, *Palaeogeography and Lithology of the Vendian and Cambrian of the Western East-European Platform*, Wydawnictwa Geologiczne, Warsaw.
Rozanov, A. Yu., and Missarzhevsky, V. V., 1966, Biostratigraphy and fauna of the Lower Cambrian Horizons, *Trudy Geol. Inst. Akad. Nauk SSSR* **148** [in Russian].
Rozanov, A. Yu., and Sayutina, T. A., 1982, Microstructures of certain algae, archaeocyathids and cribricyathids, in: *Proceedings. Third North American Paleontological Convention, Montreal*, Vol. 2, pp. 453–455.
Rozanov, A. Yu., and Sokolov, B. S., eds., 1984, Lower Cambrian stage subdivision. Stratigraphy. Akad. Nauk SSSR, Nauka, Moscow [in Russian].
Rozanov, A. Yu., and Zhegallo, E. A., 1989, On the problem of the genesis of the ancient phosphorites from Asia, *Litologiya Poleznye Iskopaemye* **1989**(3):67–82 [in Russian].
Rozanov, A. Yu., Missarzhevsky, V. V., Volkova, N. A., Voronova, L. G., Krylov, I. N., Keller, B. M., Korolyuk, I. K., Lendzion, K., Michniak, R., Pykhova, N. G., and Sidorov, A. D., 1969, The Tommotian Stage and the Cambrian Lower Boundary problem, *Trudy Geol. Inst. Akad. Nauk SSSR* **206** Moscow [in Russian] [English translation, M. E. Raaben (ed.), Amerind Publishing Co., New Delhi, India (1981)].
Rozhnov, S. V., 1990, Cambrian echinoderms on the territory of the USSR, in: *Third International Symposium on the Cambrian System, Abstracts* (L. N. Repina and A. J. Zhuravlev, eds.), Subcommission on Cambrian Stratigraphy, Commission on Stratigraphy, International Union of Geological Sciences, Akad. Nauk SSSR, Novosibirsk, p. 153.
Rozov, S. N., 1976, About classification of monoplacophorans, in: New Directions of Research in Paleontology (A. N. Oleynikov, ed.), *Trudy Sessii Vses. Paleontol. Obshch.* **15:**91–102 [in Russian].
Rozov, S. N., 1984, Morphology, terminology and systematic affinity of stenothecoids, in: Problematics of the Paleozoic and the Mesozoic (B. S. Sokolov, ed.), *Trudy Inst. Geol. Geofiz. Sibirsk. Otd. Akad. Nauk SSSR* **597:**117–133 [in Russian].
Rudavskaja, V. A., and Vassiljeva, N. I., 1984, First finds of Lukati acritarchs in the Lower Cambrian of Chekurovka section of Eastern Siberia, *Doklady Akad. Nauk SSSR* **279:**1454–1456 [in Russian].
Rudavskaja, V. A., and Vassiljeva, N. I., 1985, Acritarchs and skeletal problematics on the boundaries of the Vendian, Tommotian and Atdabanian Stages, in: *Late Precambrian and Early Paleozoic of the Siberian Platform*, All-Union Oil Scientific-Research Geological-Exploratory Institute, Leningrad, pp. 51–57 [in Russian].
Runnegar, B., 1981, Muscle scars, shell form and torsion in Cambrian and Ordovician univalved molluscs, *Lethaia* **14:**311–322.
Runnegar, B., 1983, Molluscan phylogeny revised, *Mem. Assoc. Aust. Palaeontol.* **1:**121–144.
Runnegar, B., and Jell, P. A., 1976, Australian Middle Cambrian molluscs and their bearing on early molluscan evolution, *Alcheringa* **1:**109–138.
Runnegar, B., and Pojeta, J., Jr., 1974, Molluscan phylogeny: The paleontological viewpoint, *Science* **186:**311–317.
Runnegar, B., Pojeta, J., Morris, N. J., Taylor, J. D., Taylor, M. E., and McClung, G., 1975, Biology of the Hyolitha, *Lethaia* **61:**181–191.
Rushton, A. W. A., 1978, Description of the macrofossils from the Dolwen Formation, *Bull. Geol. Surv. Great Britain* **61:**46–48.
Savitsky, V. E., and Astashkin, V. A., 1979, The significance and development of reef building in the Cambrian history of the Siberian Platform, in: *Geology of Reef Systems of the Cambrian of Western Yakutia*, Siberian Scientific Research Institute of Geology, Geophysics and Mineral Resources, Novosibirsk, USSR, pp. 5–18 [in Russian].
Savitsky, V. E., Evtushenko, V. M., Jegorova, L. I., Kontorovich, A. E., and Shabanov, Yu. Ya., 1972, Cambrian of the Siberian Platform (The Yudoma-Olenek Type of Section. The Kuonamka Complex of Deposits), *Trudy Sibirsk. Nauchno-Issled. Inst. Geol. Geofiz. Mineral. Syr.* **130** [in Russian].
Sayutina, T. A., 1980, Early Cambrian family Khasaktiidae fam. nov.—Possible stromatoporoids, *Paleontol. Zh.* **1980**(4):13–28 [in Russian].
Sayutina, T. A., 1985a, Cribricyaths of the Lower Cambrian in Mongolia, in: Problematics of the Late

Precambrian and Paleozoic (B. S. Sokolov and I. T. Zhuravleva, eds.), *Trudy Inst. Geol. Geofiz. Sibirsk. Otd. Akad. Nauk SSSR* **632:**38–44 [in Russian].

Sayutina, T. A., 1985b, On the revision of the genus *Yakovlevites* Korde, 1979, in: Problematics of the Late Precambrian and Paleozoic (B. S. Sokolov and I. T. Zhuravleva, eds.), *Trudy Inst. Geol. Geofiz. Sibirsk. Otd. Akad. Nauk SSSR* **632:**70–74 [in Russian].

Scrutton, C. T., 1979, Early fossil cnidarians, in: *The Origin of Major Invertebrate groups* (M. R. House, ed.), Academic Press, New York, pp. 161–207.

Sepkoski, J. J., Jr., 1983, Precambrian–Cambrian boundary: The spike is driven and the monolith crumbles, *Paleobiology* **9:**199–206.

Shabanov, Yu. Ya., Astashkin, V. A., Pegel, T. V., Jegorova, L. I., Zhuravleva, I. T., Pelman, Yu. L., Sundukov, V. M., Stepanova, M. V., Sukhov, S. S., Fedorov, A. B., Shishkin, B. B., Vaganova, N. V., Jermak, V. I., Ryabukha, K. V., Yadrenkina, A. G., Abaimova, G. P., Lopushinsakaya, T. V., Sychev, O. V., and Moskalenko, T. A., 1987, *Lower Paleozoic of the Southwestern Slope of the Anabar Antecline (According to Boring Data),* Nauka, Novosibirsk, USSR [in Russian].

Signor, P. W., Mount, J. F., and Onken, B. B., 1987, A pre-trilobite shelly fauna from the White-Inyo region of eastern California and western Nevada, *J. Paleontol.* **61:**425–438.

Sokolov, B. S., 1965, Ancient deposits of the Early Cambrian and sabelliditids, in: *Theses of Reports of All-Union Symposium on Precambrian and Cambrian Paleontology,* Institute of Geology and Geophysics, Siberian Branch, USSR Academy of Sciences, Novosibirsk, USSR, pp. 78–91 [in Russian].

Sokolov, B. S., 1972, Vendian and Early Cambrian Sabelliditida (Pogonophora) of the USSR, in: *International Geological Congress, 23rd Session, Czechoslovakia 1968, Evolution, Ostracoda, Paleoecology and Paleobiogeography, Other Subjects,* Wydawnictwa Geologiczne, Warsaw, pp. 79–86.

Sokolov, B. S., and Ivanovsky, A. B. (eds.), 1985, *The Vendian System (Historical-Geological and Paleontological Grounding 1: Paleontology),* Nauka, Moscow [in Russian].

Sokolov, B. S., and Zhuravleva, I. T. (eds), 1983, *Lower Cambrian Stage Subdivision of Siberia, Atlas of Fossils,* Nauka, Moscow [in Russian].

Sprinkle, J., 1981, Diversity and evolutionary patterns of Cambrian echinoderms, in: *Short Papers for the Second International Symposium on the Cambrian System 1981* (M. E. Taylor, ed.), U. S. Geological Survey Open-File Report 81-743, pp. 219–221.

Stanley, G. D., Jr., 1982, Paleozoic chondrophores (medusoid hydrozoans) and their implications for the problematical mollusc-like fossils, in: *Proceedings. Third North American Paleontological Convention, Montreal,* Vol. 3, pp. 504–504.

Stanley, G. D., Jr., 1986, Chondrophorine hydrozoans as problematic fossils, in: *Problematic Fossil Taxa* (A. Hoffman and M. H. Nitecki, eds.), Clarendon Press, Oxford, pp. 68–86.

Starobogatov, Ya. I., 1977, Systematic affinity of concardiids and the system of Paleozoic Septibranchia (Bivalvia), *Byull. Mosk. Obshch. Ispyt. Prir. Otd. Geol.* **52**(4):125–139 [in Russian].

Stasinska, A., 1960, *Velumbrella czarnockii* n. gen., n. sp.—Meduse du Cambrien Inférieur des Monts de Sainte-Croiz, *Acta Paleontol. Polon.* **5:**337–348.

Stepanova, M. V., and Luchinina, V. A., 1982, The Lower Cambrian boundary according to calcareous algae, in: *Boundaries of the Main Subdivisions of the Phanerozoic of Siberia* (V. I. Krasnov, ed.), Siberian Scientific Research Institute of Geology, Geophysics and Mineral Resources, Novosibirsk, USSR, pp. 39–46 [in Russian].

Sundukov, V. M., and Fedorov, A. B., 1986, Paleontological characteristic and age of beds with algal-archaeocyathan bioherms from the Medvezh'ya River, in: *Cambrian Biostratigraphy and Paleontology of North Asia* (I. T. Zhuravleva, ed.), *Trudy Inst. geol. Geofiz. Sibirsk. Otd. Akad. Nauk SSSR* **669:**108–119 [in Russian].

Sundukov, V. M., and Zhuravlev, A. Yu., 1989, First finds of cribricyaths in the Lower Cambrian of the Siberian Platform, *Paleontol. Zh.* **1989**(3):101–102 [in Russian].

Sysoiev, V. A., 1962, *Cambrian Hyoliths of Northern Slope of the Aldan Shield,* USSR Academy of Sciences, Moscow [in Russian].

Sysoiev, V. A., 1968, *Stratigraphy and Hyoliths of the Oldest Strata of the Lower Cambrian of the Siberian Platform,* Yakutsk Publishing House, Yakutsk, USSR [in Russian].

Sysoiev, V. A., 1970, On a new Lower Cambrian hyolith family, in: *Stratigraphy and Paleontology of the Proterozoic and Cambrian of the East Siberian Platform* (A. K. Bobrov, ed.), Yakutsk Publishing House, Yakutsk, USSR, pp. 109–115 [in Russian].

Sysoiev, V. A., 1972, *Biostratigraphy and Orthothecimorph Hyoliths of the Lower Cambrian of the Siberian Platform*, Nauka, Moscow [in Russian].

Sysoiev, V. A., 1976a, Stages and certain regularities of Early Cambrian hyolith evolution, in: New Directions of Research in Paleontology (A. N. Oleynikov, ed.), *Trudy Sessii Vses. Paleontol. Obshch.* **15**:110–118 [in Russian].

Sysoiev, V. A., 1976b, About the systematics and systematic affinity of hyoliths, in: *The Chief Problems of Animal Systematics* (V. N. Shimansky, ed.), Paleontological Institute of the Academy of Sciences, Moscow, pp. 28–34 [in Russian].

Sysoiev, V. A., 1981, From the history of hyolith studies to the discussion of the hyolith systematic affinity, in: Problematics of the Phanerozoic (B. S. Sokolov, ed.), *Trudy Inst. Geol. Geofiz. Sibirski. Otd. Akad. Nauk SSSR* **481**:85–92 [in Russian].

Sysoiev, V. A., 1984, Morphology and systematic affinity of hyoliths, *Paleontol. Zh.* **1984**(3):3–14 [in Russian].

Szaniawski, H., 1982, Chaetognath grasping spines recognized among Cambrian protocondonts, *J. Paleontol.* **56**:806–810.

Tarnovsky, S. L., and Astashkin, V. A., 1973, On sponges in the Precambrian and new data on ancient stratigraphy of Bellyk Belogorie of Eastern Sayan, in: Materials on Stratigraphy and Paleontology of Siberia (S. V. Sukhov, ed.), *Trudy Sibirsk. Nauchno-Issled. Inst. Geol. Geofiz. Mineral. Syr.* **173**:68–73 [in Russian].

Tchudinova, I. I., 1959, The discovery of conulariids in the Lower Cambrian of Western Sayan, *Paleontol. Zh.* **1959**(2):53–55 [in Russian].

Tkachenko, V. I., Ushatinskaya, G. T., Zhuravlev, A. Yu., and Repina, L. N., 1987, Cambrian sediments of the Kolyma Uplift, *Izv. Akad. Nauk SSSR Ser. Geol.* **1987**(8):55–62 [in Russian].

Urbanek, A., and Mierzejewska, G., 1983, The fine structure of zooidal tubes in Sabelliditida and Pogonophora, in: *Upper Precambrian and Cambrian Palaeontology of the East-European Platform* (A. Urbanek and A. Yu Rozanov, eds.), Wydawnictwa Geologiczne, Warsaw, pp. 100–111.

Urbanek, A., and Rozanov, A. Yu. (eds.), 1983, *Upper Precambrian and Cambrian Palaeontology of the East-European Platform*, Wydownictwa Geologiczne, Warsaw.

Ushatinskaya, G. T., 1986a, The discovery of the oldest articulate brachiopod, *Paleontol. Zh.* **1986**(4):102–103 [in Russian].

Ushatinskaya, G. T., 1986b, The Cambrian brachiopods (Review of localities and some principles of geographical distribution), in: Problems of Paleobiogeography of Asia, (A. Yu. Rozanov, ed.), *Trans. Soviet–Mongolian Paleontol. Expedition* **29**:8–34 [in Russian].

Ushatinskaya, G. T., 1987, Unusual inarticulate brachiopods from the Lower Cambrian of Mongolia, *Paleontol. Zh.* **1987**(2):62–68 [in Russian].

Ushatinskaya, G. T., 1988, Obolellids (brachiopods) with articulated hinged valves from the Lower Cambrian of the Transbaikalian, *Paleontol. Zh.* **1988**(1):34–39 [in Russian].

Vacelet, J., 1985, Coralline sponges and the evolution of Porifera, in: *The Origins and Relationships of Lower Invertebrates* (S. Conway Morris, J. D. George, R. Gibson, and H. M. Platt, eds.), Clarendon Press, Oxford, pp. 1–13.

Valkov, A. K., 1975, *Biostratigraphy and Hyoliths of the Northeastern Siberian Platform*, Nauka Publishing House, Moscow [in Russian].

Valkov, A. K., 1982, *Lower Cambrian Biostratigraphy of the East Siberian Platform (Uchur-Maya Area)*, Nauka, Moscow [in Russian].

Valkov, A. K., 1983, Distribution of the oldest shelly fossils and the correlation of the Cambrian lower boundary in the southeastern part of the Siberian Platform, in: *Late Precambrian and Early Paleozoic of Siberia, Vendian Deposits* (V. V. Khomentovsky, ed.), Institute of Geology and Geophysics, Siberian Branch, USSR Academy of Sciences, Novosibirsk, USSR, pp. 37–48 [in Russian].

Valkov, A. K., 1987, *Lower Cambrian Biostratigraphy of the East Siberian Platform, Yudoma-Olenek Region*, Inst. Geol. Akad. Nauk CCCP, Siberian Branch, Moscow [in Russian].

Valkov, A. K., and Karlova, G. A., 1984, Fauna from the transitional Vendian–Cambrian strata in the low Gonam River, in: *Late Precambrian and Early Paleozoic Stratigraphy, Middle Siberia* (V. V. Khomentovsky, ed.), Institute of Geology and Geophysics, Siberian Branch, USSR Academy of Sciences, Novosibirsk, USSR, pp. 12–41 [in Russian].

Valkov, A. K., and Sysoiev, V. A., 1969, On a new group of animals (angustiochreids) from the oldest

Cambrian strata of Siberia, *Theses of the 15th Session of the All-Union Paleontological Society*, All Union Paleontological Society, Leningrad, pp. 15–16 [in Russian].

Valkov, A. K., and Sysoiev, V. A., 1970, Angustiochreids of the Cambrian of Siberia, in: *Stratigraphy and Paleontology of the Proterozoic and Cambrian of the East Siberian Platform* (A. K. Bobrov, ed.), Yakutsk Publishing House, Yakutsk, USSR, pp. 94–100 [in Russian].

Vassiljeva, N. I., 1983, On the find of problematic microorganisms in the Aldan Stage strata of the middle Lena River, in: *The Significance of Microfauna in Studying Deposits of Continents and Seas*, All-Union Oil Scientific Research Geological-Exploratory Institute, Leningrad, pp. 123–126 [in Russian].

Vassiljeva, N. I., 1985, On the systematics of the order Chancelloriida Walcott, 1920 (incertae sedis) from the Lower Cambrian of the eastern part of the Siberian Platform, in: Problematics of the Late Precambrian and Paleozoic (B. S. Sokolov and I. T. Zhuravleva, eds.), *Trudy Inst. Geol. Geofiz. Sibirsk. Otd. Akad. Nauk SSSR*, **632**:115–126, Nauka Publishing House, Moscow [in Russian].

Vassiljeva, N. I., 1990, New Early Cambrian gastropods of the Siberian Platform and problems of their systematics, in: *Microfauna of USSR (Problems of the Systematics and Biostratigraphy)* (A. I. Nikolaev, ed.), All-Union Oil Scientific-Research Geological-Exploratory Institute, Leningrad, pp. 4–21 [in Russian].

Vassiljeva, N. I., and Rudavskaja, V. A., 1989, Regularities in the distribution of fauna and phytoplanktonic associations from the Vendian and Lower Cambrian boundary beds on the Siberian Platform, in: *Methodic Aspects of Stratigraphic Researches in Oil and Gas Bearing Basins* (M. S. Mesezhnikov and S. A. Chirva, eds.), VNIGRI Press, Leningrad, pp. 69–79 [in Russian].

Vassiljeva, N. I., and Sayutina, T. A., 1988, Morphological diversity of chancelloriid sclerites, in: Cambrian of Siberia and Middle Asia (I. T. Zhuravleva and L. N. Repina, eds.), *Trudy Inst. Geol. Geofiz. Sibirsk. Otd. Akad. Nauk SSSR* **720**:190–198 [in Russian].

Vidal, G., and Knoll, A. H., 1983, Proterozoic plankton, *Geol. Soc. Am. Mem.* **161**:265–277.

Volkova, N. A., 1985, The main acritarch types of the Late Vendian and the Cambrian of the East-European Platform, in: *Abstracts of Reports of the 5th All-Union Palynological Conference, Saratov, 1985, Part 1*, Geological Institute of the USSR Academy of Sciences, Moscow, pp. 50–51 [in Russian].

Vologdin, A. G., 1931, On some fossils from the Palaeozoic of the Chingiz Ridge, *Ezhegodnik Russk. Paleontol. Obshch.* **9**:131–142 [in Russian and English].

Vologdin, A. G., 1932, *Archaeocyaths of Siberia, Issue 2, Fauna of the Cambrian Limestones of Altay*, State Scientific-Technical Geological-Exploratory Publishing House, Moscow [in Russian].

Vologdin, A. G., 1958, Lower Cambrian foraminifers of Tuva, *Doklady Akad. Nauk SSSR* **120**:405–408 [in Russian].

Vologdin, A. G., 1959, An Upper Cambrian archaeocyathan–coral coenose in the Tannu-Ola Ridge, Tuva, *Doklady Akad. Nauk SSSR* **129**:670–673 [in Russian].

Vologdin, A. G., 1967, On the remains of the Siphonales from the Lower Cambrian of the Tannu-Ola Ridge (Tuva), *Doklady Akad. Nauk SSSR* **174**:952–955 [in Russian].

Vologdin, A. G., and Drosdova, N. A., 1970, A new find of ancient fauna, *Doklady Akad. Nauk SSSR* **190**:195–197 [in Russian].

Voronin, Yu. I., Voronova, L. G., Grigorieva, N. V., Drosdova, N. A., Zhegallo, E. A., Zhuravlev, A. Yu., Ragozina, A. L., Rozanov, A. Yu., Sayutina, T. A., Sysoiev, V. A., and Fonin, V. D., 1982, *The Precambrian–Cambrian boundary in the geosynclinal areas (the reference section of Salany-Gol, MPR), Trans. Joint Soviet–Mongolian Paleontol. Expedition* **18** [in Russian].

Voronova, L. G., and Drosdova, N. A., 1986, On the problem of Early Cambrian biogeography (according to the studies of algae, in: Problems of paleogeography of Asia (A. Yu. Rozanov, ed.), *Trans. Joint Soviet–Mongolian Paleontol. Expedition* **29**:34–55 [in Russian].

Voronova, L. G., and Radionova, E. P., 1976, Paleozoic algae and microphytolites, *Trudy Geol. Inst. Akad. Nauk SSSR* **294** [in Russian].

Voronova, L. G., Voronin, Yu. I., Drosdova, N. A., Esakova, N. G., Zhegallo, E. A., Zhuravlev, A. Yu., Luchinina, V. A., Meshkova, N. P., Ragozina, A. L., Sayutina, T. A., and Fonin, V. D., 1986, Organic remains in the Lower Cambrian Deposits between the Dzabkhan and Khungui Rivers (Mongolia), in: Cambrian Biostratigraphy and Paleontology of North Asia (I. T. Zhuravleva, ed.), *Trudy Inst. Geol. Geofiz. Sibirsk. Otd. Akad. Nauk SSSR* **669**:163–168 [in Russian].

Voronova, L. G., Drosdova, N. A., Esakova, N. V., Zheghallo, E. A., Zhuravlev, A. Yu., Rozanov, A. Yu.,

Sayutina, T. A., and Ushatinskaya, G. T., 1987, Lower Cambrian fossils of the Mackenzie Mountains (Canada), *Trudy Paleontol. Inst. Akad. Nauk SSSR* **224** [in Russian].

Voropinov, V. S., 1957, First finds of the fauna in the Verkholensk Formation, *Doklady Akad. Nauk SSSR* **114:**1291–1293 [in Russian].

Walcott, C. D., 1920, Cambrian geology and paleontology. IV. Middle Cambrian Spongiae, *Smithson. Misc. Collect.* **67**(6):261–364.

Webby, B. D., 1986, Early stromatoporoids, in: *Problematic Fossil Taxa* (A. Hoffman and M. H. Nitecki, eds.), Clarendon Press, Oxford, pp. 148–166.

Whittington, H. B., 1977, The Middle Cambrian trilobite *Naraoia*, Burgess Shale, British Columbia, *Philos. Trans. R. Soc. Lond. B* **280:**409–443.

Whittington, H. B., 1979, Early arthropods, their appendages and relationships, in: *The Origin of the Major Invertebrate Groups* (M. R. House, ed.), Academic Press, New York, pp. 253–268.

Wood, R. A., and Reitner, J., 1986, Poriferan affinity of Mesozoic stromatoporoids, *Paleontology* **29:**469–473.

Wright, A. D., 1979, Brachiopod radiation, in: *The Origin of Major Invertebrate Groups* (M. R. House, ed.), Academic Press, New York, pp. 235–252.

Wrona, R., 1987, Cambrian microfossil *Hadimopanella* Gedik from glacial erratics in West Antarctica, *Paleontol. Polon.* **49:**37–48.

Yanishevsky, M. E., 1950, The earliest trilobite from Lower Cambrian blue clay—*Gdowia assathini* gen. et sp. nov., in: *Questions of Paleontology*, Leningrad State University Publishing House, Leningrad, Vol. 1, pp. 32–40 [in Russian].

Yanshin, A. L., Khain, V. E., and Gatinsky, Yu. G., 1984, The main problems of Asian tectonics, in: *Tectonics of Asia, 27th International Geological Congress, Colloquium Reports* (A. L. Yanshin, ed.), Nauka, Moscow, Vol. 5, pp. 3–10 [in Russian].

Yaworsky, V. I., 1932, Ein Stromatoporenfund im Kambrium, *Zentralbl. Min. Geol. Paleont. B* **12:**613–616.

Yochelson, E. L., 1969, Stenothecoida, a proposed new class of Cambrian Mollusca, *Lethaia* **2:**49–62.

Yochelson, E. L., 1977, Agmata, a proposed extinct phylum of Early Cambrian age, *J. Paleontol.* **51:**437–454.

Yochelson, E. L., 1979, Early radiation of mollusca and mollusc-like groups, in: *The Origin of Major Invertebrate Groups* (M. R. House, ed.), Academic Press, New York, pp. 323–358.

Yochelson, E. L., 1981a, *Fordilla troyensis* Barrande: "The oldest known pelecypod" may not be a pelecypod, *J. Paleontol.* **55:**113–125.

Yochelson, E. L., 1981b, A survey of *Salterella* (phylum Agmata), in: *Short Papers for the Second International Symposium on the Cambrian System 1981* (M. E. Taylor, ed.), U. S. Geological Survey Open-File Report 81-743, pp. 244–248.

Yochelson, E. L., 1983, *Salterella* (Early Cambrian Agmata) from the Scottish Highlands, *Palaeontology* **26:**253–260.

Yu Wen, 1984, Early Cambrian molluscan faunas of Meishucun Stage with special reference to Precambrian–Cambrian boundary, in: *Developments in Geosciences, Contributions to the 27th International Geological Congress, 1984, Moscow*, Academia Sinica, Science Press, Beijing, pp. 21–35.

Yu Wen, 1987, Yangtze micromolluscan fauna in Yangtze Region of China with notes on Precambrian-Cambrian boundary, in: *Stratigraphy and Palaeontology of Systemic Boundaries in China. Precambrian-Cambrian Boundary*, Vol. 1, Nanjing University Publishing House, Nanjing, China, pp. 19–344.

Zhang Huimin and Zhang Wenzhi, 1984, Middle and Upper Proterozoic magnetostratigraphy and tectonic evolution in Eastern China, in: *Scientific Papers on Geology for International Exchange, Series 2, Prepared for the 27th International Geological Congress*, pp. 151–162 [in Chinese with English abstract].

Zhegallo, E. A., 1990, Lower Cambrian mollusks and their significance for the zonal stratigraphy of western Mongolia, in: *Third International Symposium on the Cambrian System, Novosibirsk, USSR Abstracts*, Institute of Geology and Geophysics, Siberian Branch, USSR Academy of Sciences, Novosibirsk, p. 33 [in Russian].

Zhegallo, E. A., and Ragozina, A. L., 1990, Microfossils of Lower Cambrian phosphorite-bearing deposits from the Khubsugul Lake area in Mongolia, in: *Third International Symposium on the Cambrian System, Abstracts* (L. N. Repina and A. J. Zhuravlev, eds.), Subcommission on Cambrian

Stratigraphy, Commission on Stratigraphy, International Union of Geological Sciences, Akad. Nauk SSSR, Novosibirsk, p. 176.

Zhuravlev, A. Yu., 1985, Recent archaeocyaths?, in: Problematics of the Late Precambrian and Paleozoic (B. S. Sokolov and I. T. Zhuravleva, eds.) *Trudy Inst. Geol. Geofiz. Sibirsk. Otd. Akad. Nauk SSSR* **632:**24–33 [in Russian].

Zhuravlev, A. Yu., 1986a, Radiocyathids, in: *Problematic Fossil Taxa* (A. Hoffman and M. H. Nitecki, eds.), Clarendon Press, Oxford, pp. 35–44.

Zhuravlev, A. Yu., 1986b, Evolution of archaeocyaths and palaeobiogeography of the Early Cambrian, *Geol. Mag.* **123:**377–385.

Zhuravlev, A. Yu., 1987, Global paleogreographic changes at the Vendian–Cambrian boundary, in: *Abstracts of Reports, 3rd All-Union Symposium of Paleontology of the Precambrian and the Early Cambrian, 1987,* Karelian Branch, USSR Academy of Sciences, Petrozavodsk, USSR, pp. 38–39 [in Russian].

Zhuravlev, A. Yu., 1988, Archaeocyaths from the Lower Cambrian of the extreme Far East of the USSR, in: *Cambrian of Siberia and Middle Asia* (I. T. Zhuravleva and L. N. Repina, eds.), Nauka, Moscow, pp. 97–110 [in Russian].

Zhuravlev, A. Yu., 1989, Poriferan aspects of archaeocyathan skeletal function, *Mem. Assoc. Aust. Palaeontol.* **8:**387–399.

Zhuravlev, A. Yu., 1991, Early Cambrian microstructural diversification of Cnidarians, in: *Sixth International Symposium on Fossil Cnidaria, Münster, Westphalia, Federal Republic of Germany, Abstracts,* pp. 108–109.

Zhuravleva, I. T., 1960, *Archaeocyaths of the Siberian Platform,* USSR Academy of Sciences, Moscow [in Russian].

Zhuravleva, I. T., 1963, *Archaeocyaths of Siberia: One-walled Archaeocyaths.* USSR Academy of Sciences, Moscow [in Russian].

Zhuravleva, I. T., and Korde, K. B., 1955, The discovery of *Chancelloria* Walcott sponge in Lower Cambrian deposits of Siberia, *Doklady Akad. Nauk SSSR* **104:**474–477 [in Russian].

Zhuravleva, I. T., and Luchinina, V. A., 1980, Paleontological characteristics of the Chingin and the Lower Monok Formations of the northern slope of West Sayan, in: *Cambrian of the Altay-Sayan Fold Belt* (I. T. Zhuravleva, ed.), Nauka, Moscow, pp. 161–172 [in Russian].

Zhuravleva, I. T., and Okuneva, O. G., 1981, On the nature of cribricyaths, in: Problematics of the Phanerozoic (B. S. Sokolov, ed.), *Trudy Inst. Geol. Geofiz. Sibirsk. Otd. Akad. Nauk SSSR* **481:**23–30 [in Russian].

Zhuravleva, I. T., Konjuschkov, K. N., and Rozanov, A. Yu., 1964, *Archaeocyaths of Siberia: Two-Walled Archaeocyaths,* Nauka, Moscow [in Russian].

Zhuravleva, I. T., Zadorozhnaya, N. M., Osadchaja, D. V., Pokrovskaja, N. V., Rodionova, N. M., and Fonin, V. D., 1967, *Lower Cambrian Fauna of Tuva (Shivelig-Khem Key-Section),* Nauka, Moscow [in Russian].

Zhuravleva, I. T., Repina, L. N., Yaskovich, B. V., Hajrullina, T. I., Poniklenko, I. A., and Luchinina, V. A., 1970, *On the Knowledge of the Early Cambrian of Tien-Shan,* FAN Publishing House, Tashkent, USSR [in Russian].

Chapter 8

Lower Cambrian of Southeastern Newfoundland

Epeirogeny and Lazarus Faunas, Lithofacies–Biofacies Linkages, and the Myth of a Global Chronostratigraphy

ED LANDING

1. Introduction

The Avalon Zone in the coastal northeastern Appalachian Mountains, in Wales, and in central England preserves a particularly significant record of the evolutionary radiations, immigration history, and provincialism of latest Precambrian and Early Cambrian metazoans. A "natural" division of the Lower Cambrian of this region into a subtrilobitic and a higher trilobite-bearing interval was recognized during the late nineteenth century (Landing, 1988c, p. 661), and studies through the 1960s documented aspects of the taxonomy and biostratigraphy of trilobites and nontrilobite elements (Holl, 1865; Billings, 1872, 1874; Walcott, 1884, 1890, 1900; Matthew, 1886a, b, 1889, 1895, 1899a–c; Shaler and Foerste, 1888; Lap-

ED LANDING • New York State Geological Survey, Albany, New York 11230.

Origin and Early Evolution of the Metazoa, edited by Jere H. Lipps and Philip W. Signor. Plenum Press, New York, 1992.

worth, 1888, 1891; Foerste, 1899; Burr, 1900; Grabau, 1900; Groom, 1902; Shimer, 1907; Cobbold, 1910, 1919, 1921, 1931, 1936; Cobbold and Pocock, 1934; Howell and Stubblefield, 1950; Shaw, 1950; Hutchinson, 1952, 1962; Bassett and Walton, 1960; Rushton, 1966).

Four key developments in Lower Cambrian paleontology in the Avalon Zone have taken place over the last several decades. These include the following:

1. Recovery of trilobites from the uppermost part of the Lower Cambrian (Rushton, 1966; Fletcher, 1972).

2. Application of bulk disaggregation techniques in the recovery of acid-resistant body fossils (Walliser, 1958; Landing *et al.*, 1980, 1989; Bengtson and Fletcher, 1983; Landing, 1984, 1988c, 1991; Brasier, 1984, 1986; Hinz, 1987).

3. Documentation of a biostratigraphically-significant trace fossil sequence from the uppermost Precambrian through Lower Cambrian (Crimes and Anderson, 1985; Crimes, 1987; Narbonne and Myrow, 1988; Landing, 1988a; Hofmann and Patel, 1989) and designation of a candidate Precambrian–Cambrian boundary stratotype horizon at Fortune Head, southeastern Newfoundland, at the lowest occurrence of Cambrian-aspect traces (Narbonne *et al.*, 1987).

4. Reconstruction of the depositional environments of the siliciclastic-dominated, fossiliferous lithofacies of the Avalon Zone (Brasier *et al.*, 1978; Brasier and Hewitt, 1979; Benus and Landing, 1984; Landing and Benus, 1984, 1988a, b; Myrow, 1987; Myrow *et al.*, 1988; Landing *et al.*, 1988a–c, 1989; Landing, 1988c, 1991).

Data on the lithologic and faunal sequence of the Lower Cambrian of the Avalon Zone, in particular southeastern Newfoundland, has led to necessary re-evaluations of the relative duration of the subtrilobitic Lower Cambrian, the relative completeness of the Lower Cambrian in Soviet East Asian and Tethyan sections, and the practicality or desirability of globally applicable stage- and series-level divisions of the Lower Cambrian. As discussed below, the subtrilobitic Lower Cambrian of Avalon, or Placentian Series, apparently represents a much longer interval of time than the trilobite-bearing Branchian Series (new). Second, interregional correlations of the Placentian and Branchian Series are very problematic; the much thinner subtrilobitic Lower Cambrian in Soviet East Asia ("Manykaian" and Tommotian Stages) and China (Meishucunian Stage) are cut by unconformities, and the lower ranges of key taxa lie at probable unconformities in the latter regions. The seemingly anomalously "long" stratigraphic ranges of many Avalonian small shelly fossils (to 1.2 km) are due to the anticipated resistance to extinction of species that are geographically widespread and occur on several Early Cambrian continents and mirror their attenuated ranges in the condensed Asian sequences. In addition, small shelly faunas were diverse and abundant only in shallow-water paleoenvironments of Avalon, and the occurrence of similar faunas in geographically separated regions may reflect biofacies similarity rather than temporal equivalence. Similarly, Lower Cambrian trilobite faunas are highly provincial and offer little basis for a finely resolved interregional temporal correlation. For example, a number of trilobite species from the middle part of the *Callavia broeggeri* Zone in Avalon do appear in the lower Botomian Stage of Siberia. However, their appearance closely corresponds to a marked lithofacies change, and this biostratigraphic similarity has greater implications for comparisons of depth-related (?) biofacies than temporal correlation. Trilobite first appearances within the *Callavia broeggeri* and *Protolenus* Zones and their diversity in these zones even within southeastern Newfoundland seem to reflect basin evolution. Highest diversity *Callavia broeggeri* Zone faunas seem to correlate with an interval of very widespread calcareous shale and limestone substrates; appearance of the low-diversity faunas of the *Protolenus howleyi* Zone correlates with a loss of these substrates and the deposition of widespread siliciclastic mudstones. Finally, the utility of Siberian Lower Cambrian stages as globally applicable units (e.g., Cowie and Bassett, 1989) is highly suspect. Cambrian series and stages should be provincial and regional divisions that reflect the unity of faunal provinces and their geological development rather than the hopes of paleontologists for a simplified chronostratigraphic terminology.

2. The Avalon Zone and Southeastern Newfoundland

Presence of a thick upper Precambrian, volcanic-rich, rift-to-drift facies, a subsequent late Precambrian compressive and intrusive episode (Avalonian Orogeny), and "Acado-Baltic" trilobites and graptolites in the siliciclastic-dominated Cambrian and Ordovician have been used to recognize the Avalon Zone in the coastal northeastern United States, southern New Brunswick, northern mainland Nova Scotia, Cape Breton Island, southeastern Newfoundland, Wales, and central England (Rast *et al.*, 1976; see reviews by King, 1979, 1986; O'Brien and King, 1982; Landing *et al.*, 1988c, Conway Morris, 1989). Excellent outcrops along the coast of southeastern Newfoundland have resulted in this region becoming a standard of reference for the geological history of the Avalon Zone. Similarly, the areally extensive outcrops of the uppermost Precambrian and Lower Cambrian provide an unparalleled record of the geological context of earliest Cambrian faunas.

3. Distribution and Age of Uppermost Precambrian and Lower Cambrian

Uppermost Precambrian through Lower Ordovician outliers occur in synclines and down-faulted blocks across southeastern Newfoundland (Figure 1). Fossiliferous Lower Cambrian strata occur in all of these outliers. Precise lithostratigraphic correlations and reconstruction of the dynamic stratigraphy are possible because shore sections are commonly completely exposed with minor structural complexity and because marker beds and other distinctive stratigraphic intervals often persist between the outliers (Landing and Benus, 1988a, b; Landing *et al.*, 1988c, 1989; Myrow *et al.*, 1988). A slatey cleavage is often prominent, although it is weak in eastern Conception Bay, western Trinity Bay, and the southwestern Burin Peninsula.

Although a number of reports have claimed a conformable sequence from the upper Precambrian through Cambrian in parts of southeastern Newfoundland (e.g., Jenness, 1963; McCartney, 1969; O'Brien and King, 1982), an angular unconformity or nonconformity always separates the uppermost Precambrian and Lower Cambrian from the underlying upper Precambrian folded volcanic, volcaniclastic, and siliciclastic rocks and erosionally-unroofed granites of the Avalonian Orogen (Landing and Benus, 1988a, b; Landing *et al.*, 1988c). Similar unconformable relationships between the Avalonian Lower Cambrian and the upper Precambrian of the Avalonian Orogen are documented in England (e.g., Brasier *et al.*, 1978) and eastern Massachusetts and Rhode Island (e.g., Landing, 1988c). Recent reports (Landing, 1991; Landing and Murphy, 1991) indicate that the folded and intruded volcanics of the Forchu Group in Cape Breton Island and the Maple Ridge Slate in the northern Antigonish Highlands, mainland Nova Scotia, are unconformably overlain by uppermost Precambrian subaerial red beds comparable the Rencontre Formation of southeastern Newfoundland (Fig. 2).

The significance of enumerating these stratigraphic relationships lies in the fact that they help emphasize the relatively abrupt nature of the earliest Cambrian evolutionary radiation of trace-producing and skeletalized metazoans. Benus (1988) reported a 565 ± 3 My U–Pb age for zircons from an ash that directly overlies fossils of Vendian metazoans of the Mistaken Point Formation, southeastern Avalon Peninsula (Anderson and Misra, 1968; Anderson and Conway Morris, 1982). The geological history of southeastern Newfoundland after burial of this soft-bodied fauna and prior to deposition of the fossiliferous lowest Cambrian included (1) deposition of a 5-km-thick, shoaling up sequence consisting, successively, of base-of-slope turbidites, slope mudstones and sandstones, shelf sandstones, and fanglomerates (upper Conception through Signal Hill Groups), (2) regional compression and intrusion (Avalonian Orogeny), (3) erosional stripping of up to 10 km of rock from the Avalonian Orogen, and (4) regional extension and deposition of 2.75+ km of uppermost Precambrian subaerial to marginal marine facies (Rencontre and lower Chapel Island Formations) (Landing *et al.*, 1988c). These relationships, along with other dates from the Avalon Zone [see review by Conway Morris (1988)] and Sinai and northern Africa (Odin *et al*, 1983; Odin and Dore, 1985), all suggest that published dates of 570–600 My ago for the base of the Cambrian are highly suspect. The Cambrian evolutionary radiation of metazoans is certainly best designated the "Cambrian explosion" if the period had a length of only 30–40 My.

4. Epeirogenic Controls on Sedimentation, Accumulation, and Faunal Immigration

The Lower Cambrian of southeastern Newfoundland has a maximum composite thickness of approximately 1.6 km. This includes 1.4 km of the lower–middle Lower Cambrian [the apparently nontrilobitic Placentian Series (Landing *et al.*, 1989)] and up to 200 m of the upper Lower Cambrian [Branchian Series]. Deposition did not take place on a tectonically passive shelf, and the terrane seems to have been on characterized by long-term extension or transpression that restricted maximum accumulation to three successive depocenters: the Fortune Bay–Burin Basin, the Placentia–Bonavista axis, and the St. Mary's–east Trinity axis (Fig. 1). These depocenters record a dramatic eastward shift, in terms of modern geography, of maximum accumulation through the Early Cambrian. Each depocenter contains the thickest accumulation of the three unconformity-bounded depositional sequences (*sensu* Vail *et al.*, 1977) of the uppermost

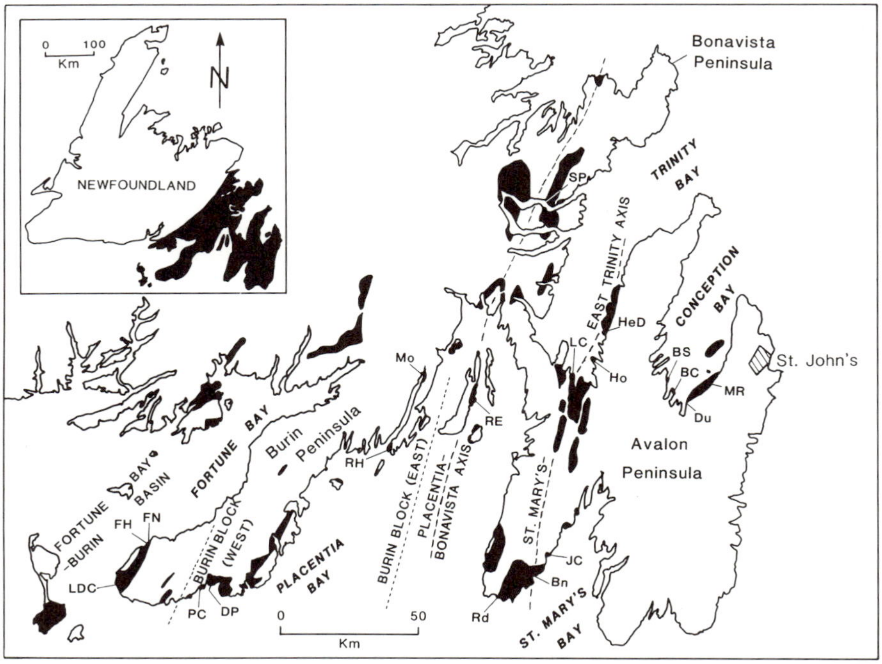

Figure 1. Generalized map showing localities mentioned in the text. Inset map shows the Avalon Zone (in black); uppermost Precambrian (Rencontre and lower part of Chapel Island Formations) through Ordovician outliers are shown in black on the larger-scale map. Abbreviated localities: BC (Bacon Cove), Bn (Branch), BS (Brigus South Point), DP (Duck Point), Du (Duffs), FH (Fortune Head), FN (Fortune North), HeD (Heart's Delight), Ho (Hopeall Head), JC (Jigging Cove), LC (Long Cove), LDC (Little Dantzic Cove), Mo (Monkstown), MR (Manuels River), PC (Pump Cove), Rd (Redlands), RH (Red Head), RE (Rose-au-rue), SP (Smith Point).

Precambrian–Lower Cambrian of southeastern Newfoundland (Landing and Benus 1988a, b). These depositional sequences include the uppermost Precambrian through lowest Cambrian (Rencontre–Chapel Island–Random Formations), the middle Lower Cambrian (Bonavista Group), and upper Lower Cambrian (Brigus Formation). As discussed below, it is probable that key faunal events such as "diversifications" of small skeletalized metazoans through the Placentian Series and the stratigraphically abrupt appearance of the abundantly trilobite-bearing Lower Cambrian reflect changes in regional epeirogenetic activity rather than Early Cambrian eustasy.

4.1. Fortune Bay–Burin Basin

The three conformable formations (Rencontre, Chapel Island, and Random) that compose the uppermost Precambrian through lowest Cambrian are thickest and most completely preserved in the western part of southeastern Newfoundland. These units were deposited in a basin here designated the "Fortune Bay–Burin Basin." The western margin of this basin is not defined, because of limited outcrop in Fortune Bay, and the eastern part of the basin includes the Burin Block, a structural unit that underwent uplift, erosion, and subsidence after deposition of the Random Formation (discussed below).

4.2. Uppermost Precambrian–Lowest Cambrian Faunas and Lithofacies

Initial proximal alluvial fan deposits of the uppermost Precambrian Rencontre Formation (2.8+ km thick) reflect deposition in fault-bounded basins that developed with extension of the upper Precambrian Avalonian Orogen (Smith and Hiscott, 1984; Landing *et al.*, 1988c, pp. 32, 33). The upper one third of the unfossiliferous Rencontre Formation includes marginal marine to peritidal sandstones and siliciclastic mudstones that are transitional into the tidally-influenced, peritidal facies of member 1 of the Chapel Island Formation (to 180 m thick) (Myrow *et al.*, 1988; Landing *et al.*, 1988c).

Uppermost Precambrian trace fossils of the lower *Harlaniella podolica* Zone first appear in lower member 1 of the Chapel Island Formation (Narbonne *et al.*, 1987). The lowest occurrence of body fossils includes flattened fragments of the vendotaenid alga *Tryasotaenia* Gnilovskaya from the upper part of member 1 and transversely annulated, organic-walled tubes of the metazoan (?) *Sabellidites cambriensis* Yanichevsky from the upper 20 m of member 1 (Landing *et al.*, 1989, Fig. 4).

Lowest Cambrian *Phycodes pedum* Zone trace fossils appear 2.4 m above the transition from member 1 into the inner-shelf, storm-influenced, gray-green sandstones and siltstones of member 2A at the candidate Precambrian–Cambrian boundary stratotype at Fortune Head (Fig. 1) (Narbonne *et al.*, 1987). This change from the uppermost Precambrian *Harlaniella podolica* Zone to the lowest Cambrian *Phycodes pedum* Zone does not seem to represent a linked lithofacies–biofacies transition associated with progressive deepening through the lower part of the Chapel Island Formation. The evidence for this interpretation is the occurrence of peritidal red sandstones and shales that show evidence of subaerial exposure through the lower half of member 2A (275 m thick). These intervals include lithofacies representing paleoenvironments comparable to those of member 1 but do not have trace fossils diagnostic of the uppermost Precambrian (Landing *et al.*, 1988c, p. 41, Stop 2D; Narbonne and Myrow, 1988). The second faunal transition in member 2A includes the appearance of *Rusophycus avalonensis* Zone traces 152 m above the base of member 2A at the Fortune Head section (Narbonne *et al.*, 1987).

Body fossils of the lowest Cambrian *Phycodes pedum* through lower *Rusophycus avalonensis* Zones are limited to numerous *Tryasotaenea* fragments and rare specimens of *Sabellidites cambriensis* in darker shales. The *Sabellidites cambriensis* Zone (Landing *et al.*, 1989) straddles the Precambrian–Cambrian boundary, and its top is the highest known occurrence of the eponymous species at the Fortune Head section (170 m). An important recent discovery is the collection of specimens of the soft-bodied macrofossil *Kullingia delicata* (Fedonkin) from the lower *Phycodes pedum* Zone and from the lower part of the barren interval between the *Sabellidites cambriensis* and *"Ladatheca" cylindrica* Zones (Narbonne *et al.*, 1991). These occurrences extend *K. delicata*'s range from the upper Vendian into the Lower Cambrian and support traditional views that Vendian chondrophores are ancestral to Phanerozoic forms.

Thorough search of bedding plane surfaces and acid-disaggregation of all calcareous nodule horizons and sandstone beds through the Chapel Island Formation has not yielded any remains of skeletalized metazoans below the upper part of member 2B (approximately 165 m thick). The problematic metazoan *"Ladatheca" cylindrica* (Grabau; see Landing, 1988c) appears as decalcified, flattened conchs in upper member 2B at Fortune Head (415.9–419.5 m above base of member 2) and as limonite steinkerns in member 3 (180 m thick) at the Little Dantzic Cove and Fortune North sections. The monospecific *"Ladatheca" cylindrica* Zone (Landing *et al.*, 1989) is recognized from this part of the Chapel Island Formation.

"Ladatheca" cylindrica has its lowest local occurrence in gray-green siltstones and thin sandstones (member 2B) and laminated silver-green siltstones (member 3) that represent the deepest depositional environments of the Chapel Island Formation (respectively, subtidal, storm-influenced shelf, and low-energy outer shelf generally below the influence of storm waves) (Myrow *et al.*, 1988; Landing *et al.*, 1988c). This lowest occurrence does not appear to be environmentally controlled and seems to reflect an immigration event that is only coincidental with the deepest facies of the formation. *"Ladatheca" cylindrica* is a eurytopic metazoan that ranges from deep subtidal, level-bottom, siliciclastic mudstones to peritidal limestones in eastern Massachusetts and southeastern Newfoundland (Landing, 1988c; Landing *et al.*, 1989). This habitat range suggests that the species should have occurred in the shallower shelf facies lower in the Chapel Island Formation if it had an earlier time of immigration into the region. In addition, diagenetic environments in lower member 2 seem to have been appropriate for its preservation. Landing *et al.* (1989, pp. 746, 748) determined that *"L." cylindrica* conchs in members 3 and 4 of the Chapel Island Formation were pyrite-infilled and -replaced shortly after burial in the zone of bacterial sulfate reduction. This pyritization took place before the growth of calcareous nodules in the zone of methane formation (e.g., Berner, 1980) during early stages of sediment compaction. Insoluble residues from calcareous nodules through members 2A and 2B of the Chapel Island Formation contain abundant fragments of pyritized burrows, and this suggests that the nodules had a similar diagenetic history. Highly depleted ^{13}C values from nodules from member 2 (S. Grant, personal communication, 1989) further corroborate their growth in the zone of methane formation. Nonrecovery of *"L." cylindrica* in the latter nodules may be evidence for the absence of the species during deposition of the lower part of the Chapel Island Formation.

The top of the *"Ladatheca" cylindrica* Zone is defined by the appearance of low-diversity assemblages of the lower *Watsonella crosbyi* Zone (Landing, 1989) [= lower part of "*Aldanella attleborensis* assemblage" of Bengtson and Fletcher (1983) and "lower *Aldanella attleborensis* assemblage interval" of Landing and Benus (1988a, b)]. These assemblages represent a muddy level-bottom community that includes a rostroconch, *W. crosbyi* Grabau; a snail, *A. attleborensis* (Shaler and Foerste); a tiny monplacophoran, *Helcionella* sp.; and numerous specimens of *"L." cylindrica* (Landing *et al.*, 1989).

The transition from the *"Ladatheca" cylindrica* Zone into the lower *Wat-*

sonella crosbyi Zone is best interpreted as a linked biofacies–lithofacies shift that accompanied the gradual shoaling and habitat change from upper member 3 into the inner-shelf nodular mudstones of member 4 (85 m thick) of the Chapel Island Formation (Landing *et al.*, 1988c; Myrow *et al.*, 1988). The vertical sequence of faunas and lithofacies suggests that the offshore, monospecific "*L.*" *cylindrica* assemblage was replaced laterally by a fauna that included three mollusks on muddy bottoms in intermediate and shallow habitats. The most proximal environments are represented by two condensed limestone beds that form the caps of shoaling cycles in the middle and at the top of member 4 (Myrow *et al.*, 1988). These limestones have features that indicate both the shallowest marine environments of this part of the Chapel Island (i.e., teepee and birdseye structures, planar stromatolites, mud cracks) and the presence of nearby subaerial terrigenous conditions (soil horizons, grains of vein quartz, and clasts of Precambrian argillite and volcanic rocks) (Landing *et al.*, 1989, Figs. 3.3, 3.4). These limestones also have the lowest diversity small shelly faunas in southeastern Newfoundland. The faunas include the eurytopic species of the low-diversity *Watsonella crosbyi* Zone assemblages along with four monoplacophorans (*Anabarella plana* Vostokova, *Archaeospira? avalonensis* Landing, *Bemella? vonbitteri* Landing, *Helcionella* sp.), several conoidal problematica [*Anabarites korobovi* (Missarzhevsky), *Coleoloides typicalis* Walcott, *Plinthokonion arethion* Landing], conodontlike remains (*Fomitchella infundabiliformis* Missarzhevsky, *Maldeotaia bandalica* Singh and Shukla, *Protohertzina anabarica* Missarzhevsky), two tommotiids (*Eccentrotheca kanesia* Landing, Nowlan, and Fletcher, and *Lapworthella ludvigseni* Landing), and a coelosclerotophoran (*Halkieria stonei* Landing).

A number of conclusions and observations are made possible by these data on faunal occurrences in the Chapel Island Formation:

1. The faunal diversity of these peritidal limestones (17 species) is as high as that of various east Asian carbonate platform sequences when the many dubiously and typophyletically defined taxa from the latter regions are evaluated [see Landing (1988b) and systematic evaluations of Landing (1988c, 1991), Qian and Bengtson (1989), and Landing *et al.* (1989)].

2. The stratigraphically abrupt local first appearance of diverse small shelly faunas sometimes has been regarded as a record of evolutionary events (e.g., Cowie and Rozanov, 1974; Luo *et al.*, 1984) when alternative explanations involving unconformities and secular changes in depositional environments are not investigated. The fact that diverse small shelly faunas appear within the siliciclastic mudstone-dominated facies of member 4 only in two thin peritidal limestones demonstrates that local "diversification" can be directly related to the dynamic stratigraphy of a depositional basin.

3. A third observation is that these diverse faunas and their peritidal habits track a lithofacies shift associated with shoaling and offlap. Models that propose that regional deepening and onlap of off-shore facies during eustatic rise could have brought diverse Early Cambrian faunas into a basin (Brasier, 1979, 1980, 1982) are directly contravened by the evidence from the Chapel Island Formation. Briefly stated, peritidal carbonate depositional environments were a locus for immigration and have the most diverse Early Cambrian faunas in Avalon (Landing *et al.*, 1989).

4. Condensed red limestones with hardgrounds from member 4 are lithically comparable to units higher in the Bonavista Group (discussed below) and to the Home Farm Member in Nuneaton, England (e.g., Brasier *et al.*, 1978). These limestones are not deposited in offshore, relatively deep environments (e.g., Brasier, 1989, Fig. 5.3), but are peritidal near- and on-shore lithosomes.

5. Almost all of the genera and species of the diverse *Watsonella crosbyi* faunas have a strongly lithofacies- and environment-specific distribution and do not occur in the shales and carbonate nodules that directly overlie and underlie peritidal limestones (Landing *et al.*, 1989). A number of forms from the peritidal limestone beds of the Chapel Island Formation (species of *Anabarella, Anabarites, Coleoloides, Fomitchella, Maldeotaia, Protohertzina*) are widely distributed on a number of Early Cambrian continents. However, their habitat (including substrate and depth of occurrence) is so restricted in the Chapel Island Formation that they would seem to provide a limited basis for intrabasinal and interregional correlation. Similar habitat restrictions are demonstrable for many of the small skeletalized metazoans from the younger Bonavista Group (discussed above). The likelihood is that small shelly species cannot satisfy the recommendation that a Precambrian–Cambrian boundary stratotype and horizon be established with reference to "diverse small shelly fossils with a good potential for correlation" (Cowie, 1985, p. 93).

6. A final observation is that of the 17 skeletalized metazoans from the *Watsonella crosbyi* Zone, phosphate is used only by the relatively sparse conodontlike forms and tommotiids in sclerite production. Phosphate is limited to a cement in the agglutinated conchs of *Plinthokonion*. Landing *et al.* (1989, pp. 753, 754) reviewed the repeated claims that have been made for the abundance of primary phosphate composition in the construction of skeletons by Early Cambrian metazoans. They concluded that the *Watsonella crosbyi* Zone remains are mineralogically similar to other Early Cambrian and Phanerozoic assemblages. Simply stated, calcium carbonate, probably as aragonite, is the dominant material used in skeletons worldwide in the earliest Cambrian. The earliest known use of calcite in Avalon is by trilobites in the upper Lower Cambrian (Landing *et al.*, 1984).

Deposition of the peritidal limestone cap of member 4 of the Chapel Island Formation was followed by another abrupt biofacies change that accompanied the stratigraphically sharp transition into member 5. The thin-bedded green sandstones, siltstones, and laminated shales of lower member 5 represent a deepening into outer-shelf environments. Member 5 (to 180 m thick) changes upward into storm- and wave-dominated, inner-shelf and shore-face environments and is conformably overlain by macrotidal sand field deposits of the Random Formation (to 250 m thick) [see sedimentological summaries of Hiscott (1982), Myrow *et al.* (1988) and Landing *et al.* (1988c, 1989)].

Body fossils are unknown from member 5 of the Chapel Island Formation and the Random Formation. This reflects a combination of habitat changes that led to regional high-energy conditions and mobile substrates, unfavorable diagenetic conditions that would have resulted in dissolution of calcareous remains in noncalcareous sands and muds, and difficulties involved in bulk processing of siliceous samples (Landing *et al.*, 1989). Trace fossils of the upper *Rusophycus avalonensis* Zone are locally abundant on the numerous bedding soles through this interval, and this sequence has the highest trace fossil diversity of the Avalonian lowest Cambrian (e.g., Narbonne and Myrow, 1988).

4.3. Uplift of the Burin Block

Easterly onlap brought the Chapel Island Formation as far east as the islands and peninsulas in western Placentia Bay, and the feather edge of the Random Formation unconformably overlies the folded, upper Precambrian Musgravetown Group in southeastern Trinity Bay and western St. Mary's Bay (e.g., Landing and Benus, 1988a, b). An important episode of erosion followed deposition of the Random Formation.

Fletcher (1972) noted local erosion of the Random Formation across Cape St. Mary's Peninsula between eastern Placentia and western St. Mary's Bays. This erosion resulted in complete removal of the Random Formation (locally 40 m thick) at Redlands (Fig. 1). This erosion is far more dramatic along a belt extending along the eastern Burin Peninsula region from Pump Cove to Duck Point and along western Placentia Bay at least as far north as Monkstown and Rose-au-rue. In this region, the lowest beds of the Bonavista Group directly overlie intervals as low as member 2A of the Chapel Island Formation (Fig. 2) (Landing *et al.*, 1988c, Fig. 6).

These relationships indicate that an episode of uplift, possibly resulting from block faulting, and erosion removed as much as 750 m from the top of the lowest Cambrian in the Duck Point area. The most dramatic area of uplift involved a region here designated the "Burin Block" in the eastern Burin Peninsula and western Placentia Bay. Movement of a second horst (?) resulted in local removal of all of the Random Formation on Cape St. Mary's.

4.4. Placentia–Bonavista Axis

The tectonic activity of the Burin Block helps emphasize the relative duration of the subtrilobitic Lower Cambrian of southeastern Newfoundland. The lowest Cambrian *Phycodes pedum* Zone through *Rusophycus avalonensis* Zones include as much as 1.1 km of section, some of which represents the product of slow rates of accumulation [members 3 and 4 of the Chapel Island Formation (Myrow *et al.*, 1988, Landing *et al.*, 1988c)]. This lowest Cambrian depositional sequence was then subjected to dramatic block faulting, erosion, and subsequent basin reorganization that led to marine incursion along the Placentia–Bonavista axis, an elongate, northerly-trending depocenter defined by the thickest and most complete development of the Bonavista Group. Fresh-appearing, erosion-unmodified fractures up to 3.0 m deep are filled with initial onlap sediments of the Bonavista Group in Trinity and southern Conception Bays (Fig. 3). These fissures suggest that regional extension accompanied and may be responsible for definition of the Placentia–Bonavista axis and for easterly and westerly onlap of the Bonavista Group across southeastern Newfoundland [see Landing and Benus (1988a, b) and Landing *et al.* (1988c) for a review of tectonic history and a formal definition of formational and member subdivisions of the Bonavista Group].

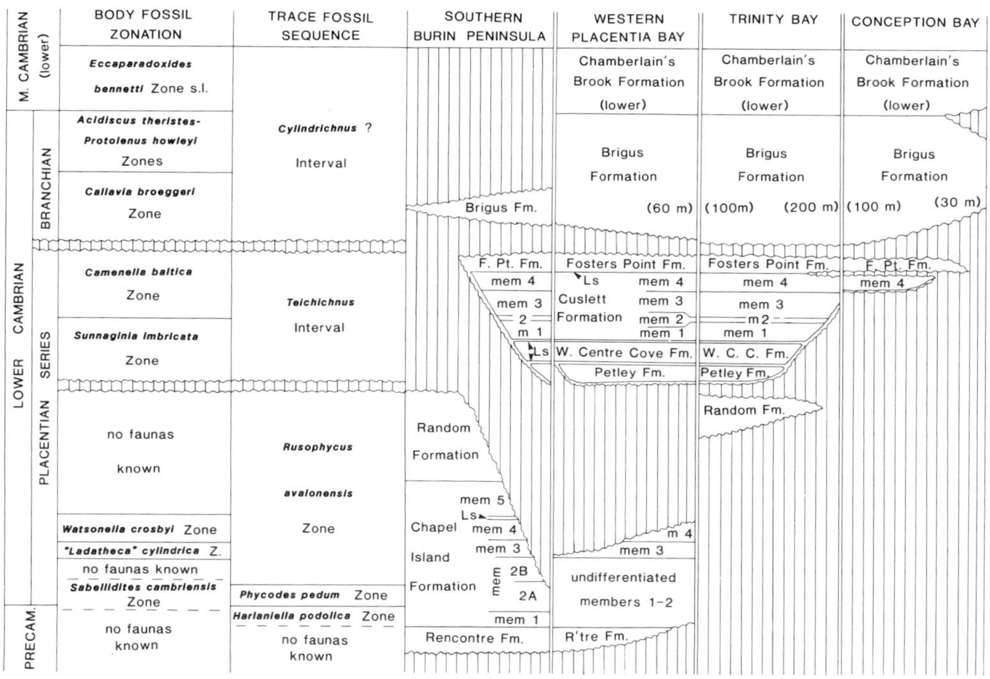

Figure 2. Uppermost Precambrian through Lower Cambrian of southeastern Newfoundland. Members of the Chapel Island Formation as emended by Narbonne *et al.* (1987) and Landing *et al.* (1988c); subdivisions of the Bonavista Group (Petley through Fosters Point Formations and members of Cuslett Formation) from Landing and Benus (1988b). Trace fossil sequence from Narbonne *et al.* (1987) and Landing (1988a); body fossil zonation of Placentian Series after Landing *et al.* (1989) and of Branchian Series modified from Hutchinson (1962) and Fletcher (1972). "Ls" indicates regionally extensive peritidal limestone lithosomes. Vertical rulings indicate hiatus. Stratigraphic columns: Southern Burin Peninsula from Little Dantzic Cove (far left side of column) east to Duck Point (far right), West Placentia Bay column from Monkstown (far left) south to Red Head (far right), Trinity Bay column from Smith Point (far left) east to Heart's Desire (far right), Conception Bay column from Brigus South Point (far left) southeast to Duffs (right) and northeast to Manuels River (far right).

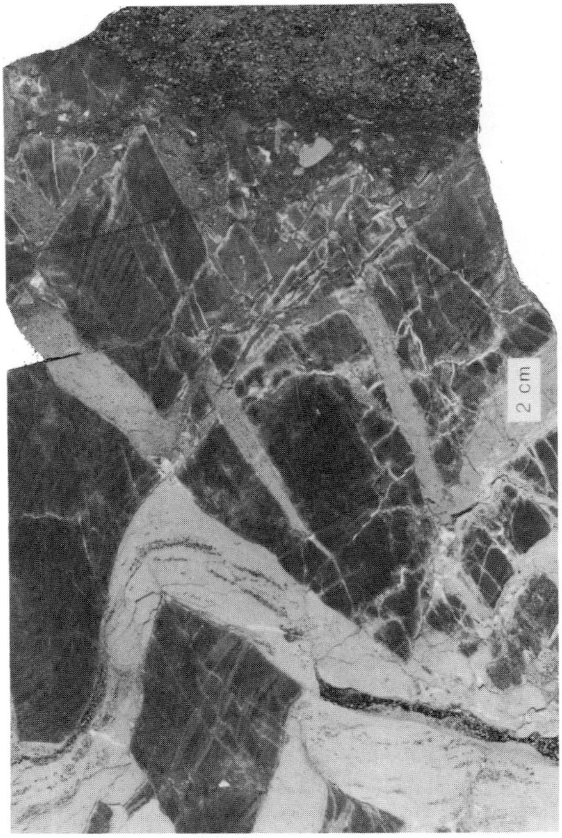

Figure 3. Erosion-unmodified fissures in tightly folded, upper Precambrian, purple argillite (Conception Group) overlain by coarse-grained, reddish-purple, calcareous, feldspathic quartz arenite (at top) and filled by pink calcisiltite with quartz arenite laminae of member 4 of the Cuslett Formation. The contact suggests that regional extension accompanied and was responsible for onlap of the Bonavista Group. Interpreted sequence of events includes (1) development of northerly trending, easterly and westerly dipping conjugate fissure set in folded Conception Group argillire, (2) onlap of member 4 of the Cuslett Formation and initial fill of open fissures with pink carbonate silt and minor quartz sand in low-energy, shoreline setting, (3) continued onlap and concentration of well-sorted and winnowed *remanie* quartz sand with clasts of Precambrian argillite, granite, and vein quartz as thin layer (to 0.75 m) in depressions and stromatolitic, hyolith-rich fossil wackestones on low rises along shoreline (Anderson, 1987), (4) deposition of siliciclastic mudstone-dominated facies of member 4, and (5) compaction of sediments and deformation of sediment in fissure fills with partial closure of fractures. Bacon Cove, southwestern Conception Bay.

4.5. Middle Lower Cambrian Faunas and Dynamic Stratigraphy

The lower part of the Bonavista Group (Petley Formation through member 1d of the Cuslett Formation) is assigned to the *Sunnaginia imbricata* Zone (Landing *et al.*, 1989) [= upper part of "*Aldanella attleborensis* assemblage" of Bengtson and Fletcher (1983) and "upper *Aldanella attleborensis* assemblage interval" of Landing and Benus (1988a, b) and Landing (1988c)]. Faunal assemblages of the

Sunnaginia imbricata Zone are dominated by "Lazarus taxa" (Jablonski, 1980) that reappeared following a protracted absence from the local sequence. With the exception of several rare forms (*Archaeospira? avalonensis*, *Maldeotaia bandalica*, *Protohertzina anabarica*, and, possibly, *Bemella? vonbitteri*), all species of the significantly lower and older *Watsonella crosbyi* Zone appear *de novo* in the lower part of the Bonavista Group. Their reappearance has a relatively simple explanation that reflects two factors. The red, purple, and green siliciclastic-dominated facies of the Bonavista Group represent low-energy, inner-shelf deposits that are comparable to those of member 4 of the Chapel Island Formation. Reestablishment of these depositional environments in southeastern Newfoundland was tracked by key species of the *Watsonella crosbyi* Zone and constituted a major lithofacies–biofacies event in the region. Second, the species that reappear in the Bonavista Group are also those that were noted above as being widely distributed both in the Avalon Zone or on a number of Early Cambrian continents. The well-developed dispersal capacity of these species and a wide habitat tolerance in cool- and warm-water, shallow-water environments meant that they were resistant to extinction and exhibit a virtual evolutionary stasis (Landing, 1988c; see Jackson, 1977).

The *Sunnaginia imbricata* Zone has the lowest occurrences of the tommotiid *S. imbricata* Missarzhevsky, the phosphatic conoidal remains of the problematicum *Torellella laevigata* (Linnarsson), and the monplacophoran *Bemella bella* Chen and Zhang in eastern Massachusetts and southeastern Newfoundland (Landing, 1988c; Landing *et al.*, 1989). As in the *Watsonella crosbyi* Zone, the majority of metazoan species, including mollusks, most conoidal problematica, hyoliths, and coelosclerotophorans, have calcareous and not phosphatic skeletons (Landing, 1988c, Table 1).

Siliciclastic mudstone–sandstone alternations are rare in the Bonavista Group, and bedding soles are infrequently found. As a result, trace fossil diversity is much lower than in the Chapel Island Formation. This low-diversity "*Teichichnus* assemblage" (Landing, 1988a) is named for the most abundant trace in the Bonavista Group.

Associated small shelly faunas of the *Sunnaginia imbricata* Zone show the same lithofacies-specific diversity and occurrence patterns of the *Watsonella crosbyi* Zone. For example, regionally extensive peritidal limestones, such as the 0.5- to 1.0-m thick bed forming the top of the Petley Formation (up to 38 m) and the 2-m-thick limestones at the top of the West Centre Cove Formation (up to 79+ m) (Fig. 2), have faunas that include abundant orthothecids ("*Allatheca*" *degeeri sensu lato*), low helcionaceans, conodontlike remains (*Fomitchella infundabiliformis*), and almost all of the tommotiid specimens (*Eccentrotheca kanesia*, *Lapworthella ludvigseni*, *S. imbricata*) known from this part of the sequence. In contrast, red, purple, and green siliciclastic mudstones with calcareous nodules which can directly underlie and overlie these limestones have faunas identical to the much older low-diversity *Watsonella crosbyi* Zone assemblages (Landing, 1989, p. 570).

The on-shore/off-shore distribution of communities and species of the *Sunnaginia imbricata* Zone is similar to that of the *Watsonella crosbyi* Zone, and the appearance in the Placentia–Bonavista axis of the peritidal limestones with high-diversity faunas that define the tops of the Petley and West Centre Cove Formations is the result of regional shoaling (Landing and Benus, 1988a, b). These limestones are lithically and faunally comparable to limestones deposited as initial onlap facies of the Bonavista Group to the east and west of the Placentia–Bonavista axis (Anderson, 1987). Indeed, these peritidal limestones can be demonstrated to converge and coalesce with the initial onlap limestones of the Bonavista Group along both margins of the Placentia–Bonavista axis (Landing and Benus, 1988b, Figs. 35, 36). This suggests that regional shoaling and consequent rapid progradation of near-shore limestone facies belts explain the presence of these peritidal limestones with their diverse faunas in the middle part of the Placentia–Bonavista axis.

A similar linked lithofacies–biofacies shift that followed long-term siliciclastic mudstone-dominated deposition of member 1 of the Cuslett Formation is associated with the appearance of faunas of the *Camenella baltica* Zone (Landing *et al.*, 1989) [= "*Coleoloides typicalis* assemblage" of Bengtson and Fletcher (1983) and "*Camenella baltica* assemblage interval" of Landing and Benus (1988a, b) and Landing (1988c)]. The base of this zone corresponds to the regionally extensive but relatively short-term change in substrate represented by member 2 of the Cuslett Formation. The lowest occurrence of the tommotiid *C. baltica* (Bengtson) and the rostroconch (?) *Watsonella? terranovica* (Matthew) are in the distinctive interval of alternating thin fossil packstones and thin purple-to-red siliciclastic mudstones of members 2 (3.0–7.5 m thick). *Coleoloides typicalis* undergoes an abrupt and presently inexplicable increase in abundance in the *Camenella baltica* Zone. Its first occurrence is in the much older *Watsonella crosbyi* Zone (Landing *et al.*, 1989), and this biostratigraphic interval should not be named for it.

Member 2 is lithologically comparable to the four successive, regionally traceable, 5- to 8-m-thick, limestone–shale intervals that form the middle and upper parts of member 3 of the Cuslett Formation (up to 50 m thick). Each of these intervals features an upward increase in thickness, sorting, and winnowing of the fossiliferous limestones and an associated change from purple to red coloration of the intervening shales. These limestone–shale intervals are regarded as sixth-order, shoaling-up cycles that are terminated by deepening events which reestablished purple shale-dominated deposition (Landing and Benus, 1988b).

The dynamic stratigraphy of member 3 of the Cuslett Formation is particularly significant in understanding the "diversification" through this interval. The upper shoaling cycles of member 3 feature the lowest occurrence of a number of species: the clam *Fordilla troyensis* Barrande, the hyoliths *Gracilitheca bayonet* (Matthew) and *Hyolithes tenuistriatus* Linnarsson *sensu lato,* a number of conodontlike forms (rare *Rhombocorniculum insolutum* Missarzhevsky and Mambetov, very rare *R. cancellatum* (Cobbold), and common *Hertzina elongata* Muller *sensu lato*), the calcareous problematicum *Anabaritellus hexsulcatus* Missarzhevsky, the high helcionellacean "*Ginella*" *acutacosta* (Walcott) and the oldest-known brachiopods of Avalon [*Obolus? groomi* (Matley *in* Groom) and *Micromitra phillipsi* (Holl)]. With the exception of *Fomitchella infundabiliformis* and *Watsonella crosbyi,* all species from the underlying *Sunnaginia imbricata* Zone persist through this part of the *Camenella baltica* Zone.

These diverse *Camenella baltica* Zone faunas disappear in the upper part of the Cuslett Formation with the onset of siliciclastic mudstone-dominated deposition of member 4 (up to 35 m thick). The *Teichichnus*-churned purple, green, and, locally, red mudstones of member 4 are regionally extensive in southeastern Newfoundland from southeastern Conception Bay through all of the Lower Cambrian outliers in Trinity, Placentia, and northern St. Mary's Bays (Landing and Benus, 1988b).

High-diversity *Camenella baltica* Zone faunas reappear with the onset of regional deposition of the peritidal Fosters Point Formation (Landing and Benus, 1988b) (= lower part of the "Smith Point Limestone" of many earlier reports). This is a second "Lazarus episode" in the Placentian Series.

The Fosters Point Formation records a major shoaling event that terminates the subtrilobitic Lower Cambrian of southeastern Newfoundland. This shoaling event, interestingly, is not associated with offlap, because the Fosters Point unconformably overlies the Random Formation at Jigging Cove, west-central St. Mary's Bay, and the eroded, upper Precambrian Holyrood granite in southern Conception Bay (Hutchinson, 1962; Landing and Benus, 1988b). The seemingly contradictory situation of continued onlap with regional shoaling is best explained as a consequence of reduced rate of relative sealevel rise rather than an absolute fall (Landing and Benus, 1988b; see Pitman, 1978). Shoaling-up cycles with limestone caps in the Chapel Island Formation and lower part of the Bonavista Group (discussed above) may have had a similar history of development. The deposition of the Fosters Point Formation directly on the paleoshoreline in parts of Conception and St. Mary's Bays conclusively demonstrates that Avalonian condensed red limestones are the shallowest facies of "Avalonian shale basins" and appear within more offshore sites of deposition with shoaling and progradation (Landing *et al.*, 1989). They are not an offshore, deeper marine facies and do not appear in shale basins from off-shore locations with onlap and deepening; they cannot be interpreted as as facies that moved shoreward with eustatic rise (compare Brasier, 1980, 1989). If long-term regional extension or transpression controlled onlap of the Bonavista Group, these shoaling events may have been caused by decreases in rate of epeirogenic subsidence rather than by eustatic changes.

A well-developed lithofacies and biofacies spectrum is preserved in the Fosters Point Formation. This includes oncolitic, stromatolitic fossil wackestones to packstones with abundant broken *Coleoloides* and hardgrounds in peritidal environments of Conception, St. Mary's, and southern Trinity Bays. Much different, shallow subtidal algal mud mound limestones with *in situ Coleoloides* are present throughout Placentia and western Trinity Bays. Collections from peritidal faciesare almost completely dominated by tommotiids (*Sunnaginia imbricata* and *Eccentrotheca kanesia*). The diverse *Camenella baltica* Zone assemblages of upper member 3 of the Cuslett Formation reappear in the algal mud mound facies belt (Landing and Benus, 1984; Landing *et al.*, 1984). The latter facies is not well developed in the upper part of member 3, and this may explain the local abundance of *Hyolithes tenuistriatus sensu lato* and common occurrence of *Rhombocorniculum insolutum* in the Fosters Point Formation. The large monoplacophoran *Randomia aurorae* Matthew is a conspicuous element in algal mud mound facies of the Fosters Point Formation. However, shell fragments from the upper part of the Cuslett Formation may indicate an earlier occurrence of the species.

4.6. St. Mary's–East Trinity Axis

A third episode of basin reorganization in southeastern Newfoundland was initiated during the latter stages of deposition of the Placentian Series and controlled lithofacies and faunal distribution and accumulation of the trilobite-bearing Lower Cambrian. This reorganization featured a third easterly shift of Lower Cambrian depocenter to the St. Mary's–east Trinity axis (Landing and Benus, 1988a, b).

Definition of this axis may have taken place as early as the interval corresponding to deposition of the Fosters Point Formation. The Fosters Point is somewhat thicker (8–12 m) along southeastern Trinity Bay than at western localities in western Trinity and Placentia Bays (0.9–7.5 m) and at eastern sites in Conception and St. Mary's Bays (up to 6 m). This local thickening did not result from stratigraphic condensation of bedded limestones in the Cuslett Formation with the peritidal limestones that form the upper part of the Fosters Point Formation (*fide* Fletcher, 1972). The evidence for this is the minor development of small calcareous nodules in member 4 of the Cuslett Formation. In addition, a phosphatized diastem or soil horizon with calcrete occurs at the top of member 4 throughout Conception and eastern Trinity Bays. These two features indicate that member 4 and the Fosters Point are not conformably deposited units in Conception and eastern Trinity Bays (Landing and Benus, 1988a, b).

The maximum thickness of the Brigus Formation is about 200 m in southwestern St. Mary's Bay (e.g., Fletcher, 1972) and at several localities in southeastern Trinity Bay (Long Cove and Heart's Delight). It thins to the east and west and measures approximately 100 m in western Conception (Brigus South Point) and Trinity (Smith Point) Bays and falls to 10–30 m at localities in the southeastern Burin Peninsula and eastern Conception Bay (Hutchinson, 1962; Landing and Benus, 1988b).

The Brigus Formation represents the reestablishment of Avalonian shale basin deposition (Landing *et al.*, 1988c, 1989). It consists of red, purple, and green siliciclastic mudstones with relatively minor bedded and nodular limestones. Detailed field work with A. P. Benus and S. R. Westrop over nine field seasons has served to demonstrate that the Brigus Formation has a very localized distribution of facies. Major facies belts are asymmetrically distributed in the St. Mary's–east Trinity basin. For example, red shale and limestone deposition persisted into the *Protolenus howleyi* Zone only in the northeast (Conception Bay), while correlative strata in western Trinity Bay are green mudstones as low as the lower *Serrodiscus bellimarginatus* Subzone. Stratigraphic correlations in the middle and upper parts of the Brigus Formation feature very abrupt, red-to-green color changes that cross-cut the stratigraphy. These color changes cannot be used for purposes of lithostratographic correlation (compare Fletcher, 1972), and reliance on color as an aid in stratigraphic correlation can result in completely inaccurate correlations. However, thin sandstone event beds and nodular intervals provide a basis for correlation between outliers of the Brigus Formation.

An unconformity is demonstrable locally between the Brigus Formation and the Middle Cambrian Chamberlain's Brook Formation. This results in unconformable contracts of the Chamberlain's Brook Formation on strata as low as the middle *Callavia broeggeri* Zone at Little Dantzic Cone in the southwestern Burin

Peninsula (Hutchinson, 1962). However, we have not found credible evidence that demonstrates the protracted hiatus illustrated by Bengtson and Fletcher (1983, Fig. 5) in western Trinity and Conception Bays. Indeed, depositional continuity in the latter areas is suggested by the occurrence of scattered rhodochrosite crystals and small manganese nodules comparable to those of the Chamberlain's Brook-Formation in unfossiliferous strata as low as 20 m below the top of the Brigus Formation. Stratigraphic condensation of both formations and absence of biostratigraphically significant faunas may better explain the stratigraphic relationships of the two formations to the east and west of the St. Mary's–east Trinity axis.

4.7. Trilobite-Bearing (Upper Lower Cambrian) Faunas

A regionally extensive unconformity lies at the base of the trilobite-bearing Lower Cambrian throughout southeastern Newfoundland (Bengtson and Fletcher, 1983). Landing and Benus (1988b) detailed the physical evidence for this unconformity at all localities in the region. They referred all trilobite-bearing Lower Cambrian strata to the Brigus Formation and referred the subtrilobitic part of the "Smith Point Limestone" to the Fosters Point Formation (Fig. 4).

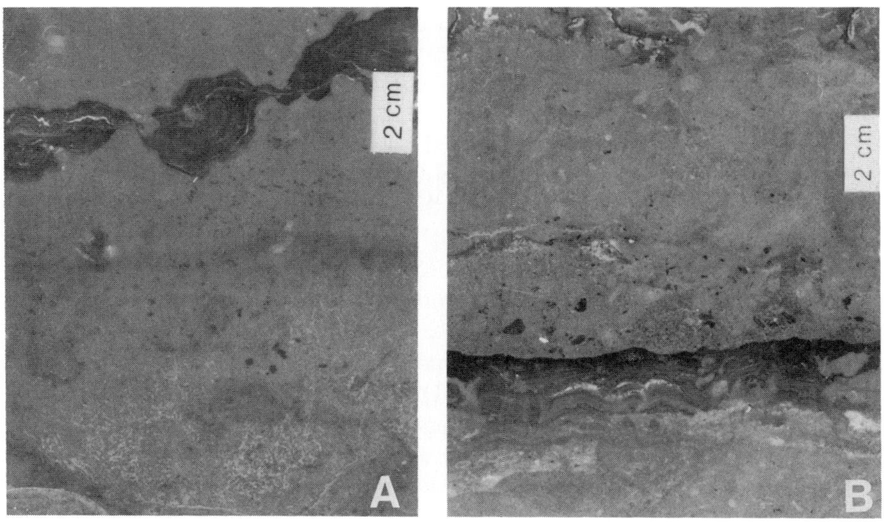

Figure 4. Unconformable contacts between the Fosters Point Formation (uppermost Placentian Series) and overlying Brigus Formation (Branchian Series). (A) Pink trilobite packstone (Brigus Formation, trilobite packstone 1.33 m thick) overlies pink algal mud mound limestone (Fosters Point Formation, 7.0 m thick) at the embayed contact at the top of the figure; the unconformity reaches as low as the limonitic oncolite horizon, darker patches in underlying limestone are neomorphically recrystallized calcite areas that weather as a "blue bed" (Landing and Benus, 1984). Smith Point section, western Trinity Bay (Landing *et al.*, 1988c, Fig. 31). (B) Green-gray, *Linguella*-rich, neomorphic fossil wackestone with oncolites forms the base of the Brigus Formation and overlies the phosphate- and limonite-impregnated algal mat cap of the Fosters Point Formation (1.3 m thick; gray-brown fossil wackestones with fossil packstone veneers on hardgrounds). Note the angular dark clasts of the algal mat in the lowest part of Brigus Formation limestone. Duffs, southeastern Conception Bay (Landing and Benus, 1988a, pp. 30–32).

The relative duration of the hiatus between the Fosters Point and Brigus Formations is undetermined, but may be as significant as that between the Random Formation and Bonavista Group. Landing *et al.* (1989) emphasized that this is a biostratigraphically significant contact that features the local disappearance of a number of species (*Watsonella? terranovica, Aldanella attleborensis, Obolus? groomi, Micromitra phillipsi, Rhombocorniculum insolutum,* and, possibly, *Anabarites korobovi*). Perhaps more significant is the appearance of at least six trilobite genera in the lower part of the *Callavia broeggeri* Zone (Hutchinson, 1962; Fletcher, 1972) in association with several brachipods (*Linguella viridus* Cobbold, *Obolella atlantica* Walcott), *Pelagiella atlantoides* Matthew and a "naticiform" snail, *Scenella reticulata* Billings, *Rhombocorniculum spinosus* (Hinz), *Microdictyon effusum* Bengtson, Missarzhevsky and Matthews, and a large meniscate trace-former [*Cylindrichnus?* sp. of Landing (1988a)]. All of these animals are unknown in the similar (if not identical) lithofacies of the underlying Bonavista Group, and an undetermined interval of time corresponding to the evolution and immigration of these forms seems to be unrepresented at the hiatus (Landing *et al.*, 1989).

The trilobite sequence of the Brigus Formation and coeval units elsewhere in Avalon is relatively poorly known. Faunal zones and subzones have been proposed and used in a number of reports. However, a number of apparent vertical transitions in faunas seem to reflect collecting biases and, more importantly, progressive but subtle changes in environment and habitat area during onlap of the Brigus Formation.

For example, the *Acanthomicmacca* Subzone (Fig. 5) which was proposed by Fletcher (1972) and used by Bengtson and Fletcher (1983) is based on only two small collections from the lower 17 m of the Brigus Formation at Redlands. Two of the three reported species of this "subzone" (*Callavia broeggeri* (Walcott) and a *Comuella*) occur in overlying strata with more diverse faunas of the *Dipharus attleborensis* Subzone of this report [*D. attleborensis* (Shaler and Foerste), *Acanthomicmacca walcotti* (Matthew), a new species of *Serrodisucs,* and a *Kjerulfia?*] (Fletcher, 1972). This subzone, in turn, is succeeded, in Fletcher's (1972) and our collections, by a higher diversity interval with *Dipharus attleborensis* Subzone trilobites in association with *Serrodiscus bellimarginatus* (Shaler and Foerste), *Diapharus planus* Hutchinson, *Ladadiscus llarenai* Richter and Richter, *Strenuella strenua* (Billings), *Calodiscus lobatus* (Hall), and rare *Triangulaspis vigilans* (Matthew). Fletcher's (1972; Bengtson and Fletcher, 1983) "*Serrodiscus bellimarginatus* Subzone" is based, in part, on a perceived "acme" of that species, and the base of a *Serrodiscus bellimarginatus* Subzone should be drawn at its lowest occurrence (Fig. 5). *Bonnia? bombifrons* (Matthew) and *Pseudatops reticulatus* (Walcott) appear in the middle part of the *Serrodiscus bellimarginatus* Subzone of this report, but seem to be so rare that they are not practical aids for finely resolved biostratigraphy in this area.

Key species of the lower and middle *Serrodiscus bellimarginatus* Subzone (*Dipharus attleborensis, D. plana,* and *Strenuella strenua*) occur in initial onlap limestones of the Brigus Formation in western Trinity Bay (e.g., Hutchinson, 1962). This suggests that the high diversity of the lower and middle parts of the subzone may actually reflect an increased habitat area that included areally-extensive deposition of thin-bedded limestones with onlap of the Brigus Formation.

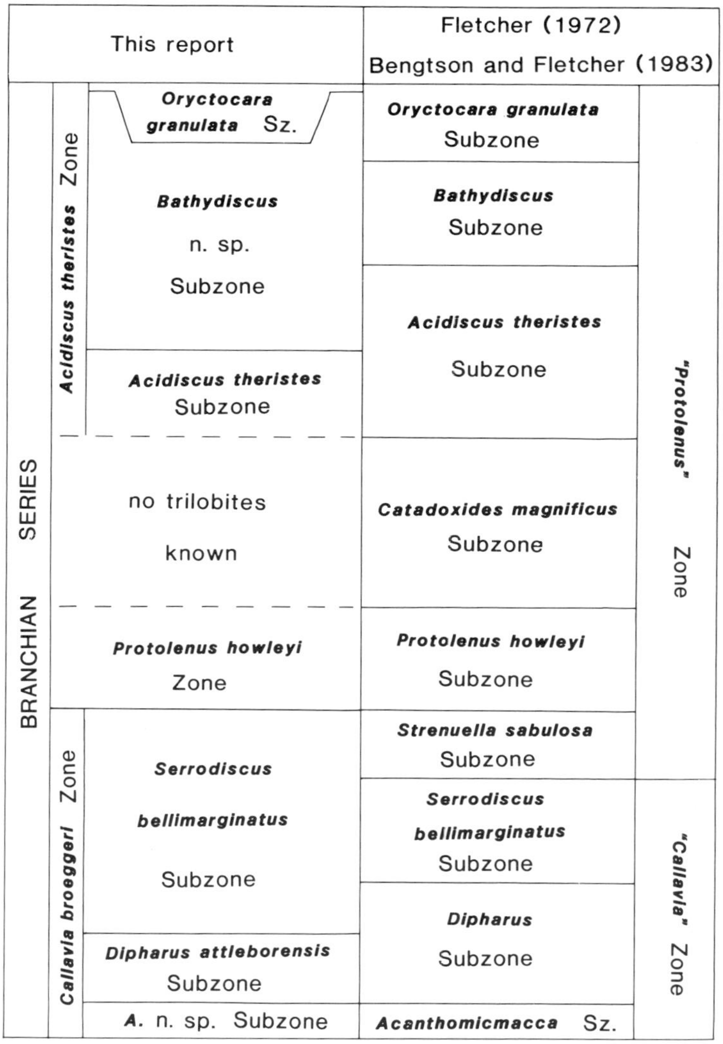

Figure 5. Trilobite sequence of the upper Lower Cambrian Branchian Series in southeastern Newfoundland. Zones and subzones modified from Fletcher (1972) and Bengtson and Fletcher (1983) as discussed in text.

Regional stratigraphic analysis (E. Landing and S. R. Westrop, unpublished data) suggests that Fletcher's (1972; Bengtson and Fletcher, 1983) somewhat lower diversity "*Strenuella sabulosa* Subzone" is coeval with the appearance of areally extensive green mudstone-dominated deposition with monospecific (*S. strenua*) trilobite faunas on the west side of the St. Mary's–east Trinity axis. *Strenuella sabulosa* Rushton appears and *Callavia broeggeri* apparently disappears from the Brigus Formation at this time. However, the overall composition of the lower diversity faunas of the upper *Serrodiscus bellimarginatus* Subzone (*S. bellimarginatus, Ladadiscus llareni, Dipharus planus, Strenuella strenua*) is comparable to that of the lower part of the Subzone, and these strata should be assigned to the *Callavia broeggeri* Zone. These associated faunal changes may really reflect a decreased area of deposition of shallow carbonate and calcareous siliciclastic mudstone substrates through this interval. Low-diversity *Protolenus howleyi* Zone faunas occur as wave-disarticulated assemblages in scattered nodules and weakly calcareous shale intervals in overlying strata through the middle part of the Brigus Formation in Trinity Bay. The associated appearance of *Protolenus howleyi* Zone faunas, disappearance of carbonate substrates, and onset of regional deposition of red, purple, and pyritiferous green siliciclastic mudstones suggest that the transition from the upper *Callavia broeggeri* Zone into the *Protolenus howleyi* Zone is a linked biofacies–lithofacies shift.

Nodules and disseminated carbonate are rare higher in the formation, and trilobite remains apparently underwent complete dissolution at most localities. The "*Catadoxides magnifiucs* Subzone" in this part of the Brigus Formation (Fletcher, 1972; Bengtson and Fletcher, 1983) is completely devoid of trilobites. It was proposed because *C. magnificus* (Matthew) is the only trilobite that occurs in the *Protolenus howleyi* Zone and the upper part of the Brigus Formation. It should be designated a "barren interval."

A window into the uppermost Lower Cambrian trilobites of southeastern Newfoundland was provided by Fletcher's (1972) discovery and thorough collecting of faunas in the upper part of the Brigus Formation at Branch Cove, southwestern St. Mary's Bay. These faunas include species reported by Rushton (1966) from the upper part of the Purley Shale in central England. The base of Fletcher's (1972; Bengtson and Fletcher, 1983, Fig. 5) "*Bathydiscus* Subzone" seems to have been drawn at an arbitrary distance above the highest collection of *Acidiscus therestes* Rushton, and a more "conventional" definition at the base of the local range of *Bathydiscus* is used herein (Fig. 5). Similarly, Fletcher (1972) recovered an assemblage that consists of *Bathydiscus* n. sp. Subzone faunas but with such additional forms as *Condylopyge, Peronopsis,* and *Oryctocara granulata* Tchernyscheva. This assemblage occurs in a collection just below the top of the Brigus Formation. The base of this subzone should reflect its known distribution in the Brigus Formation (Fig. 5).

5. Placentian and Branchian Series

The Placentian Series (Landing *et al.*, 1989) is the interval corresponding to the *Phycodes pedum* Zone through *Camenella baltica* Zone of Avalon. The faunal succession of the series is a distinct biostratigraphic unit that corresponds to a long-term increase in diversity with the disappearance of relatively few early-appearing species in the upper part of the Placentian Series. Five ichnogenera are known in the lowest part of the *Phycodes pedum* Zone, while 28 are recorded in the Random Formation (upper *Rusophycus avalonensis* Zone) (Narbonne and Myrow, 1988). Although *Sabellidites cambriensis* is the only body fossil in the Precambrian–Cambrian boundary interval, 18 species occur in the diverse *Watsonella crosbyi* Zone assemblages (Landing *et al.*, 1989). Seventeen small shelly fossil species have been illustrated from the *Sunnaginia imbricata* Zone of eastern Massachusetts (Landing, 1988c), and this compares to known diversity in the zone in southeastern Newfoundland. Faunal lists from the upper *Camenella baltica*

Zone in England (Cobbold, 1919; Brasier, 1984, 1986), eastern Massachusetts (Landing, 1988c), and Cape Breton Island and southeastern Newfoundland include as many as 37 species.

The interval from the base of the *Callavia broeggeri* Zone through the *Acidiscus theristes* Zone is a second distinct faunal unit. This interval is the upper Lower Cambrian in southeastern Newfoundland and other regions of the Avalon Zone. Although a few small shelly fossil species persist from the underlying Placentian Series, this overlying interval commonly features the abrupt appearance of relatively diverse trilobite faunas and the lowest occurrence of a number of mollusks, several brachiopods, and a number of phosphatic problematica (noted above). The designation "Branchian Series" is proposed for supra-*Camenella baltica* Zone and sub-Middle Cambrian strata in southeastern Newfoundland and the Avalon Zone. The designation "Branchian" refers to the village of Branch, southeastern Newfoundland, and to the section along northeastern Branch Cove where Fletcher (1972) collected the first reported trilobites from the upper part of the Brigus Formation. The reference horizon for the base of the Branchian Series is the base of the Brigus Formation [as revised by Landing and Benus (1988b)] at Redland Point on the west side of Redland Cove, St. Mary's Peninsula. The base of the Brigus Formation at this section is the base of Fletcher's (1972) "*Acanthomicmacca* Subzone." The top of the Branchian Series is defined by strata at least as old as those bearing a Middle Cambrian *Eccaparadoxides bennetti* Zone *sensu lato* fauna in the lowest part of the Chamberlain's Brook Formation at Branch Cove (Fletcher, 1972). The Branchian Series includes trilobite-bearing Lower Cambrian strata of eastern Massachusetts (Landing, 1988c), southern New Brunswick (Matthew, 1895), northern Nova Scotia (Landing *et al.*, 1980) and Cape Breton Island (Hutchinson, 1952), northern Wales (Howell and Stubblefield, 1950; Bassett and Walton, 1960), and central and western England (Cobbold, 1921; Cobbold and Pocock, 1934; Rushton, 1966).

The definition of the Branchian Series means that it may prove to include post-*Camenella baltica* Zone Lower Cambrian strata which do not have trilobites locally. This situation seems to be present in the Lower Comley Sandstone of Shropshire. Landing *et al.* (1989) noted that the *Obolus? groomi* Beds (Ab1) of the Lower Comley Sandstone have brachiopod species that are unknown either below the middle *Camenella baltica* Zone or from the trilobite-bearing Lower Cambrian. Brasier (1989) has reported *Camenella* from the interval. None of these brachiopods are reported in strata in the middle part of the Lower Comley Sandstone (Ab2) which have the lowest occurrence of bradoriid ostracodes in Avalon (*Bradoria benepuncta*, *B.* sp. cf. *obesa*, *B. robusta*, *B. tinea?*) (Cobbold and Pocock, 1934). All of these bradoriid species were first described from the much younger "*Protolenus* Zone" in New Brunswick (Matthew, 1895) and their presence in unit Ab2 may indicate that these strata should be referred to the Branchian Series. Trilobite fragments in the overlying Ab3 Beds which have been variously referred to *Holmia*, *Fallotaspis*, or *Kjerulfia?* (Brasier, 1989, p. 94) and which occur below the lowest specimens of *Callavia broeggeri* and *Dipharus attleborensis* in the *Callavia* Sandstone (Ac1) may represent an interval unrepresented at the unconformity below the Brigus Formation. However, their position above apparent *Camenella baltica* Zone faunas (Ab1) and ostracodes known elsewhere higher in the Lower Cambrian suggests that they occur in strata best referred to the Branchian Series.

The Placentian Series seems to represent a much longer interval of Early Cambrian time than the trilobite-bearing Branchian Series. The Placentian Series is much thicker (1400 m) than the Branchian Series (200 m). More importantly, strata that represent slow rates of accumulation in the Placentian Series (member 4 of the Chapel Island Formation and Bonavista Group) have a composite thickness which is greater than the lithologically comparable Brigus Formation.

6. Discussion

In proposing the Placentian Series, Landing *et al.* (1989) discussed how all earlier named, series-level subdivisions for the Lower Cambrian of Avalon ("Etcheminian," "Handfordian," "Comleyan," and "Caerfaian") have ambiguous upper and/or lower boundaries, type sections, diagnostic faunas, and definitions. In addition, such units as the Tommotian Stage of Siberia and Meischucunian Stage of south China are not appropriate for use in Avalon. These stages are defined on the basis of species that are not present in Avalon (i.e., Tommotian and Atdabanian archaeocyathans), and many small shelly species common to Avalon and these east Asian regions have "anomalously long" stratigraphic ranges in southeastern Newfoundland.

Correlations based on trace fossils and small shelly fossils indicate that the Placentian Series is best regarded as bracketing an interval of time corresponding to the Nemakit–Daldyn Formation through lower (?) Atdabanian Stage of Siberia (Narbonne *et al.*, 1987; Landing *et al.*, 1989). However, interregional correlations based on the relatively diverse small shelly faunas of southeastern Newfoundland are problematic even at the stage level. For example, Landing *et al.* (1989) argued that the faunal composition of the lower *Camenella baltica* Zone indicates a correlation with the lower Tommotian of Siberia. They assigned the underlying *Sunnaginia imbricata* and *Watsonella crosbyi* Zones to an interval seemingly unrepresented at the regional unconformity at the base of the Tommotian and referred the upper part of the *Camenella baltica* Zone to an undifferentiated upper Tommotian–Atdabanian equivalence.

These proposed correlations of the Placentian Series and evaluations of the duration of the sub-Tommotian unconformity are far less significant than the demonstration that the Tommotian and Atdabanian are irrelevant concepts in Avalon. A sub-Tommotian unconformity simply means that faunas from the middle part of the Placentian Series that include species known in the Siberian Tommotian can be wholly older than the Tommotian. Use of "Tommotian" in southeastern Newfoundland, as seems to be urged by Cowie and Bassett's (1989) draft chart, would require attempts to correlate an unconformity into a fossiliferous sequence and to find the base of the Atdabanian in the seemingly undifferentiable upper Tommotian–Atdabanian equivalent in the upper Cuslett Formation. As another example, one of the most frequent cited bases for interregional correlation of the trilobite-bearing Lower Cambrian is the presence of one or more *Serrodiscus bellimarginatus* Subzone species (*Calodiscus lobatus, Dipharus attleborensis,* and *Triangulaspis vigilans*) in the *Elliptocephala asaphoides* fauna of eastern New York, the Botomian Stage of Siberia, and Gislöv Formation in Scania (e.g., Berström and Ahlberg, 1981). However, as Lochman (1956) stressed, there is a lower and upper *E. asaphoides* assemblage, and these are now known to be separated by over 100 m of strata deposited on the continental slope in the Taconic allochthon, eastern New York (Landing, 1988d). *Calodiscus lobatus*' very long range in eastern New York is unlikely to be recorded in Scania and Avalon; the species has a very short stratigraphic range and, apparently, was a short-term immigrant in the latter areas. Its presence there does not provide biostratigraphic resolution for correlation at the zone or stagelevel. The base of the Botomian Stage in northern Siberia features a wholesale introduction of trilobites at a lithologic change into flaggy limestones and shales. The habitat change seems to have been appropriate to allow immigration of species (*T. vigilans*, *D. planus*) that are known in the Avalonian *S. bellimatginatus* Subzone. A conservative evaluation is that the shared taxa are probably more important for illustrating biofacial rather than temporal similarity of Avalonian sequences with the Botomian Stage. The occurrence of these species in the lower Botomian Stage of Siberia is not evidence for concluding that their seemingly much longer stratigraphic range in Avalon is also in lower Botomian-equivalent rock (*fide* Brasier, 1989); Avalon occurrences of these species are as likely to be in upper Atdabanian- to middle Botomian-equivalent strata of the Branchian Series. The pronounced lithofacies–biofacies shift defining the base of the Botomian indicates that this Siberian stage is also inappropriate for the worldwide use suggested by Cowie and Bassett's (1989) chart.

Siberian Lower Cambrian stages reflect a long-term evolutionary development and immigration history of the species and communities that define this warm-water faunal province. Regional and vertical distribution and even preservation of these stages are a product of the unique geological history of depositional basins in Siberia. Siberian Lower Cambrian stages and Avalonian series should be considered to be appropriately defined and recognizable by the faunal sequences specific to each region. One can hope that interregional correlations of the Lower Cambrian will mature beyond the simple comparison of faunal lists and claims of "first appearance datums." The faunal record should come to be understood as a product not only of evolution, but of diachronous immigration events that are controlled by epeirogenic and, at times, eustatic processes that determine the distribution of ancient biofacies.

ACKNOWLEDGMENTS. Field support from the National Geographic Society (grants 2453-82 and 2619-83) and the National Science Foundation (grants EAR83-12939 and EAR-8805774) is gratefully acknowledged. A. P. Benus, P. Myrow, G. M. Narbonne, and S. R. Westrop have been field collaborators for a number of years. W. B. Rogers and R. H. Fakundiny reviewed the preliminary manuscript. Publication Number 643 in the Journal Series of the New York State Geological Survey.

References

Anderson, M. M., 1987, Stratigraphy of Cambrian rocks at Bacon Cove, Duffs, and Manuels River, Conception Bay, Avalon Peninsula, eastern Newfoundland, in: *Northeastern Section of the Geological Society of America Centennial Field Guide* (D. C. Roy, ed.), Geological Society of America, Boulder, Colorado, Vol. 5, pp. 467–472.

Anderson, M. M., and Conway Morris, S., 1982, A review, with descriptions, of four unusual forms, of the soft-bodied fauna of the Conception and St. John's Groups (late Precambrian), Avalon Peninsula, in: *Proceedings. 3rd North American Paleontological Convention* (B. Mamet and M. J. Copeland, eds.), Vol. 1, pp. 1–8.

Anderson, M. M., and Misra, S. B., 1968, Fossils found in pre-Cambrian Conception Group of south-eastern Newfoundland, *Nature* **220:**680–681.

Bassett, D. A., and Walton, E. K., 1960, The Hell's Mouth Grits: Cambrian greywackes in St. Tudwal's Peninsula, North Wales, *J. Geol. Soc. Lond.* **116:**85–119.

Bengtson, S., and Fletcher, T. P., 1983, The oldest sequence of skeletal fossils in the Lower Cambrian of southeastern Newfoundland, *Can. J. Earth Sci.* **20:**525–536.

Benus, A. P., 1988, Sedimentological context of a deep-water Ediacaran fauna (Mistaken Point Formation, Avalon Zone, eastern Newfoundland), in *Trace Fossils, Small Shelly Fossils and the Precambrian–Cambrian Boundary* (E. Landing, G. M. Narbonne, and P. Myrow, eds.), New York State Museum Bulletin **463,** pp. 8–9.

Benus, A. P., and Landing, E. 1984, Depositional environments and biofacies of the Bonavista Formation (Early Cambrian, eastern Newfoundland), *Geol. Soc. Am. Abstr. Progr.* **16:**3.

Bergström, J., and Ahlberg, P., 1981, Uppermost Lower Cambrian biostratigraphy in Scania, Sweden, *Geol. Fören. Stockh. Förh.* **103:**192–214.

Berner, R. A., 1980, *Early Diagenesis: A Theoretical Approach*. Princeton University Press, Princeton, New Jersey.

Billings, E., 1872, On some fossils from the Primordial rocks of Newfoundland, *Can. Nat.* **6:**465–479.

Billings, E., 1874, *Palaeozoic Fossils,* Vol. 2, Geological Survey of Canada, Ottawa, Canada.

Brasier, M. D., 1979, The Cambrian radiation event, in: *The Origin of Major Invertebrate Groups* (M. R. House, ed.), The Systematics Association Special, Vol. 12, Academic Press, New York, pp. 103–159.

Brasier, M. D., 1980, The Lower Cambrian transgression and glauconite-phosphate facies in western Europe, *J. Geol. Soc. Lond.* **137:**695–703.

Brasier, M. D., 1982, Sea-level changes, facies changes, and the late Precambrian-Early Cambrian evolutionary explosion, *Precambrian Research* **17:**105–132.

Brasier, M. D., 1984, Microfossils and small shelly fossils (especially conoidal fossils) from the Lower Cambrian *Hyolithes* Limestone at Nuneaton, English Midlands, *Geol. Mag.* **121:**229–253.

Brasier, M. D., 1986, The succession of small shelly fossils (especially conoidal microfossils) from English Precambrian–Cambrian boundary beds, *Geol. Mag.* **123:**237–256.

Brasier, M. D., 1989, Section in England and their correlation, in: *The Precambrian–Cambrian Boundary* (J. W. Cowie and M. D. Brasier, eds.), Clarendon Press, Oxford, pp. 82–104.

Brasier, M. D., and Hewitt, R. A., 1979, Environmental setting of fossiliferous rocks from the uppermost Proterozoic–Lower Cambrian of central England, *Palaeogeogr. Palaeoclimatol. Palaeoecol.* **27:**35–57.

Brasier, M. D., Hewitt, R. A., and Brasier, C. J., 1978, On the late Precambrian–Early Cambrian Hartshill Formation of Warwickshire, *Geol. Mag.* **115:**21–36.

Burr, H. T., 1900, A new Lower Cambrian fauna from eastern Massachusetts, *Am. Sci.* **25:**42–50.

Cobbold, E. S., 1910, On some small trilobites from the Cambrian rocks of Comley (Shropshire), *J. Geol. Soc. Lond.* **66:**19–37.

Cobbold, E. S., 1919, Cambrian Hyolithidae, etc., from the Hartshill in the Nuneaton District, Warwickshire, *Geol. Mag.* **6:**149–158.

Cobbold, E. S., 1921, The Cambrian horizons of Comley (Shropshire) and their Brachiopoda, Pteropoda, Gasteropoda, etc., *J. Geol. Soc. Lond.* **76:**325–386.

Cobbold, E. S., 1931, Additional fossils from the Cambrian rocks of Comley, Shropshire, *J. Geol. Soc. Lond.* **87:**459–511.

Cobbold, E. S., 1936, The Conchostraca from the Cambrian area of Comley, Shropshire, with a note on a new variety of *Atops reticulatus* (Walcott), *J. Geol. Soc. Lond.* **92:**221–229.

Cobbold, E. S., and Pocock, R. W., 1934, The Cambrian area of Rushton (Shropshire), *Philos. Trans. R. Soc. Lond. B* **223:**305–409.

Conway Morris, S., 1988, Radiometric dating of the Precambrian–Cambrian boundary in the Avalon Zone, in: *Trace Fossils, Small Shelly Fossils and the Precambrian–Cambrian Boundary* (E. Landing, G. M. Narbonne, and P. Myrow, eds.), New York State *Museum Bulletin* 463, pp. 53–58.

Conway Morris, S., 1989, South-eastern Newfoundland and adjacent areas (Avalon Zone), in: *The Precambrian–Cambrian Boundary* (J. W. Cowie and M. D. Brasier, eds.), Clarendon Press, Oxford, pp. 7–39.

Cowie, J. W., 1985, Continuing work on the Precambrian–Cambrian boundary, *Episodes* **8**:93–98.

Cowie, J. W., and Bassett, M. G., 1989, International Union of Geological Sciences 1989 global stratigraphic chart, *Episodes* (Suppl.) **12**(2):unnumbered page.

Cowie, J. W., and Rozanov, A. Yu., 1974, I.U.G.S. Precambrian/Cambrian Boundary Working Group in Siberia, 1973, *Geol. Mag.* **111**:237–242.

Crimes, T. P., 1987, Trace fossils and correlation of Late Precambrian and Early Cambrian strata, *Geol. Mag.* **124**:97–119.

Crimes, T. P., and Anderson, M. M., 1985, Trace fossils from late Precambrian–Early Cambrian strata of southeastern Newfoundland (Canada): Temporal and environmental implications, *J. Paleontol.* **59**:310–343.

Fletcher, T. P., 1972, Geology and Lower to Middle Cambrian trilobite faunas of southeast Avalon, Newfoundland, Ph.D. thesis, University of Cambridge, Cambridge.

Foerste, A. F., 1899, The Cambrian strata of the Attleboro District, in: N. S. Shaler, J. B. Woodworth, and A. F. Foerste, *Geology of the Narragansett Basin* U.S. Geological Survey Monograph 33, Washington, D. C.

Grabau, A. W., 1900, Palaeontology of the Cambrian terranes of the Boston Basin, *Occas. Pap. Bost. Soc. Nat. Hist.* **4**:601–694.

Groom, T., 1902, The sequence of Cambrian and associated beds of the Malvern Hills, *J. Geol. Soc. Lond.* **58**:89–135.

Hinz, I., 1987, The Lower Cambrian microfauna of Comley and Rushton, Shropshire/England, *Palaeontogr. Abt.* **198**:41–110.

Hiscott, R. N., 1982, Tidal deposits of the Lower Cambrian Random Formation, eastern Newfoundland: Facies and paleoenvironments, *Can. J. Earth Sci.* **19**:2028–2046.

Hofmann, H. J., and Patel, I. M., 1989, Trace fossils from the type 'Etcheminian Series' (Lower Cambrian Ratcliffe Brook Formation), Saint John area, New Brunswick, Canada, *Geol. Mag.* **126**:139–157.

Holl, H. B., 1865, On the geological structure of the Malvern Hills and adjacent districts, *J. Geol. Soc. Lond.* **21**:72–108.

Howell, B. F., and Stubblefield, C. J., 1950, A revision of the fauna of the North Welsh *Conocoryphe viola* beds impling a Lower Cambrian age, *Geol. Mag.* **87**:1–24.

Hutchinson, R. D., 1952, The stratigraphy and trilobite faunas of the Cambrian sedimentary rocks of Cape Breton Island, Nova Scotia, *Geol. Surv. Can. Mem.* **263**:1–124.

Hutchinson, R. D., 1962, Cambrian stratigraphy and trilobite faunas of southeastern Newfoundland, *Geol. Surv. Can. Bull.* **88**:1–156.

Jablonski, D., 1980, Apparent versus real biotic effects of transgressions and regressions, *Paleobiology* **6**:398–407.

Jackson, J. B. C., 1977, Some relationships between habitat and biostratigraphic potential of marine benthos, in: *Concepts and Methods of Biostratigraphy* (E. J. Kauffman and J. E. Hazel, eds.), Dowden, Hutchinson, and Ross, Stroudsburg, Pennsylvania, pp. 65–72.

Jenness, S. E., 1963, Terra Nova and Bonavista Map-Areas, Newfoundland, *Geol. Surv. Can. Mem.* **327**:1–184.

King, A. F., 1979, The birth of the Caledonides: Late Precambrian rocks of the Avalon Peninsula, Newfoundland, and their correlatives in the Appalachian-orogen, in: *The Caledonides in the USA* (D. R. Wones, ed.), Virginia Polytechnic Institute and State University Memoir 3, Blacksburg, Virginia, pp. 3–9.

King, A. F., 1986, Geology of the St. John's area, Newfoundland, *Nfld. Dep. Mines Energy Rep.* **86-1**:209–218.

Landing, E., 1984, Skeleton of lapworthellids and the suprageneric classification of tommotiids (Early and Middle Cambrian phosphatic problematica), *J. Paleontol.* **58**:1380–1398.

Landing, E., 1988a, Bioturbation and infaunal diversity, Lower Cambrian, southeastern Newfoundland, *Geol. Soc. Am. Abstr. Progr.* **20**:31.

Landing, E., 1988b, Trace fossils, small shelly fossils and the Precambrian–Cambrian boundary: A premeeting viewpoint, in: *Trace Fossils, Small Shelly Fossils and the Precambrian–Cambrian Boundary* (E. Landing, G. M. Narbonne, and P. Myrow, eds.), New York State Museum Bulletin 463, pp. 1–3.

Landing, E., 1988c, Lower Cambrian of eastern Massachusetts: Stratigraphy and small shelly fossils, *J. Paleontol.* **62**:661–695.

Landing, E., 1988d, Depositional tectonics and biostratigraphy of the western portion of the Taconic allochthon, eastern New York State, in: *The Canadian Paleontology and Biostratigraphy Seminar* (E. Landing, ed.), New York State Museum Bulletin, 462, pp. 96–110.

Landing, E., 1989, Paleoecology and distribution of the Early Cambrian rostroconch *Watsonella crosbyi* Grabau, *J. Paleontol.* **63**:566–573.

Landing, E., 1991, Upper Precambrian through Lower Cambrian of Cape Breton Island: Faunas, paleoenvironments, and stratigraphic revision, *J. Paleontol.* **65**:570–595.

Landing, E., and Benus, A. P., 1984, Lithofacies belts of the Smith Point Limestone (Lower Cambrian, southeastern Newfoundland) and the lowest occurrence of trilobites, *Geol. Soc. Am. Abstr. Progr.* **16**:45.

Landing, E., and Benus, A. P., 1988a, Cambrian depositional history and stratigraphy, Avalon-Bonavista region, southeastern Newfoundland, in: *Field Trip Guidebooks* (P. H. Davenport, R. N. Hiscott, P. P. O'Neill, and L. W. Nolan, eds.), Geological Association of Canada Annual Meeting, Trip A3, Geological Association of Canada, Newfoundland Section, St. John's, Newfoundland.

Landing, E., and Benus, A. P., 1988b, Stratigraphy of the Bonavista Group, southeastern Newfoundland: Growth faults and the distribution of the sub-trilobitic Lower Cambrian, in: *Trace Fossils, Small Shelly Fossils and the Precambrian–Cambrian Boundary* (E. Landing, G. M. Narbonne, and P. Myrow, eds.), New York State Museum Bulletin 463, pp. 59–71.

Landing, E., and Murphy, J. D., 1991, Uppermost Precambrian (?)–Lower Cambrian of mainland Nova Scotia: Faunas, depositional environments, and stratigraphic revision, *J. Paleontol.* **65**:382–396.

Landing, E., Nowlan, G. S., and Fletcher, T. P., 1980, A microfauna associated with Early Cambrian trilobites of the *Callavia* Zone, northern Antigonish Highlands, Nova Scotia, *Can. J. Earth Sci.* **17**:400–418.

Landing, E., von Bitter, P. H., Benus, A. P., and Albanese, J. R., 1984, Oldest "tube worm"–algal mud mound associations (lower Cambrian, eastern Newfoundland), *Geol. Soc. Am. Abstr. Progr.* **16**:568.

Landing, E., Narbonne, G. M., Benus, A. P., and Myrow, P., 1988a, Avalonian sub-trilobitic small shelly fossils: Facies and diagenetic restrictions on the correlation of the lowest Cambrian, in: *Trace Fossils, Small Shelly Fossils and the Precambrian–Cambrian Boundary* (E. Landing, G. M. Narbonne, and P. Myrow, eds.), New York State Museum Bulletin 463, pp. 13–14.

Landing, E., Narbonne, G. M., Myrow, P., Benus, A. P., and Anderson, M. M., 1988b, A sub-trilobitic lowest Cambrian series for Avalon: The Etcheminian, in: *The Canadian Paleontology and Biostratigraphy Seminar* (E. Landing, ed.), New York State Museum Bulletin 462, pp. 9–10.

Landing, E., Narbonne, G. M., Myrow, P., Benus, A. P., and Anderson, M. M., 1988c, Faunas and depositional environments of the Upper Precambrian through Lower Cambrian, southeastern Newfoundland, in: *Trace Fossils, Small Shelly Fossils and the Precambrian–Cambrian Boundary* (E. Landing, G. M. Narbonne, and P. Myrow, eds.), New York State Museum Bulletin 463, pp. 18–52.

Landing, E., Myrow, P., Benus, A. P., and Narbonne, G. M., 1989, The Placentian Series: Appearance of the oldest skeletalized faunas in southeastern Newfoundland, *J. Paleontol.* **63**:739–769.

Lapworth, C., 1888, On the discovery of the *Olenellus* fauna in the Lower Cambrian rocks of Great Britain, *Geol. Mag.* **5**(N. S.):480.

Lapworth, C., 1891, On *Olenellus callavei* and its geological relationships, *Geol. Mag.* **8**(N. S.):529–536.

Lochman, C., 1956, Stratigraphy, paleontology, and paleogeography of the *Elliptocephala asaphoides* strata in Cambridge and Hoosick Quadrangles, New York, *Bull. Geol. Soc. Am.* **67**:1331–1396.

Luo H., Jiang Z., Wu X., Song X., Ouyang L., Xing Y., Liu G., Zang S., and Tao Y., 1984, *Sinian–Cambrian Boundary Stratotype Section at Meishucun, Jinning, Yunnan, China*, People's Publishing House, Kumming, Yunnan, China.

McCartney, W. D., 1969, Geology of the Avalon Peninsula, southeast Newfoundland, in: North Atlantic Geology and Continental Drift (M. Kay, ed.), *Am. Assoc. Pet. Geol. Mem.* **12**:115–129.

Matthew, G. F., 1886a, Note on the occurrence of *Olenellus? kierulfi* in America, *Am. J. Sci.* **31**:472–473.

Matthew, G. F., 1886b, On the Cambrian faunas of Cape Breton and Newfoundland, *Trans. R. Soc. Can.* **4**(4):147–157.

Matthew, G. F., 1889, On Cambrian organisms in Acadia, *Trans. R. Soc. Can.* **7**(4):135–162.

Matthew, G. F., 1895, The *Protolenus* fauna, *Trans. N. Y. Acad. Sci.* **14**:101–153.

Matthew, G. F., 1899a, A Palaeozoic terrane beneath the Cambrian, *Ann. N. Y. Acad. Sci.* **12**:41–56.

Matthew, G. F., 1899b, The Etcheminian fauna of Smith Sound, Newfoundland, *Trans. R. Soc. Can.* **5**(4):97–119.

Matthew, G. F., 1899c, Preliminary notice of the Etcheminian fauna of Newfoundland, *Bull. Nat. Hist. Soc. N. Brunswick* **18:**189–196.

Myrow, P., 1987, Sedimentology and depositional history of the Chapel Island Formation (late Precambrian to Early Cambrian), southeast Newfoundland, Ph.D. thesis, Memorial University, St. John's, Newfoundland, Canada.

Myrow, P., Narbonne, G. M., and Hiscott, R. N., 1988, Storm-shelf and tidal deposits of the Chapel Island and Random Formations, Burin Peninsula: Facies and trace fossils, in: *Field Trip Guidebooks* (P. H. Davenport, R. N. Hiscott, P. P. O'Neill, and L. W. Nolan, eds.), Geological Association of Canada Annual Meeting, Trip B6, Geological Association of Canada, Newfoundland Section, St. John's, Newfoundland, Canada.

Narbonne, G. M., and Myrow, P., 1988, Trace fossil biostratigraphy in the Precambrian–Cambrian boundary interval, in: *Trace Fossils, Small Shelly Fossils and the Precambrian–Cambrian Boundary* (E. Landing, G. M. Narbonne, and P. Myrow, eds.), New York State Museum Bulletin 463, pp. 72–76.

Narbonne, G. M., Myrow, P., Landing, E., and Anderson, M. M., 1987, A candidate stratotype for the Precambrian–Cambrian boundary, Fortune Head, Burin Peninsula, southeastern Newfoundland, *Can. J. Earth Sci.* **24:**1277–1293.

Narbonne, G. M., Myrow, P., Landing, E., and Anderson, M. M., 1991, A chrondrophorine (medusoid hydrozoan) from the basal Cambrian (Placentian) of Newfoundland, *J. Palentol.* **65:**186–191.

O'Brien, S., and King, A. F., 1982, The Avalon Zone in Newfoundland, in: *Guidebook for Avalon and Meguma Zones* (A. F. King, ed.), The Caledonide Orogen, IGCP Project 27, NATO Advanced Study Institute, Atlantic Canada, Publication 50, pp. 1–27.

Odin, G. S., and Dore, F., 1985, Radiometric dating of late Precambrian times, in: The Chronology of the Geological Record (N. J. Snelling, ed.), *Geol. Soc. Lond. Mem* **10:**65–72.

Odin, G. S., Gale, N. S., Auvray, B., Bielski, M., Dore, F., Lancelot, J.-R., and Pasteels, P., 1983, Numerical dating of Precambrian–Cambrian boundary, *Nature* **301:**21–23.

Pitman III, W. C., 1978, Relationship between eustasy and stratigraphic sequences of passive margins, *Bull. Geol. Soc. Am.* **89:**1389–1403.

Qian Y. and Bengtson, S., 1989, Palaeontology and biostratrigraphy of the Early Cambrian Meishucunian Stage in Yunnan Province, South China, *Fossils Strata* **24:**1–56.

Rast, N., O'Brien, B. H., and Wardle, R. J., 1976, Relationships between Precambrian and Lower Paleozoic rocks of the Avalon Platform in New Brunswick, the northeast Appalachians and the British Isles, *Tectonophysics* **30:**315–338.

Rushton, A. W. A., 1966, *The Cambrian trilobites from the Purley Shales of Warwickshire,* Palaeontology Society Monographs.

Shaler, N. S., and Foerste, A. F., 1888, Preliminary description of North Attleboro fossils, *Harvard Mus. Comp. Zool. Bull.* **16:**27–41.

Shaw, A. B., 1950, A revision of several Early Cambrian trilobites from eastern Massachusetts, *J. Paleontol.* **24:**577–590.

Shimer, H. W., 1907, An almost complete specimen of *Strenuella strenua* (Billings), *Am. J. Sci.* **23**(N. S.):199–201.

Smith, S. A., and Hiscott, R. N., 1984, Latest Precambrian to Early Cambrian basin evolution, Fortune Bay, Newfoundland: Fault-bounded basin to platform, *Canadian J. Earth Sci.* **21:**1379–1392.

Vail, P. R., Mitchum, R. M., and Thompson III, S., 1977, Seismic stratigraphy and global changes of sea level, Part 3: Relative changes of sea level from coastal onlap, in: Seismic Stratigraphy—Applications to Hydrocarbon Exploration (E. Payton, ed.), *Am. Ass. Pet. Geol. Mem.* **26:**63–81.

Walcott, C. D., 1884, Review of the fauna of the Saint John Formation contained in the Hartt Collection at Cornell University, *Bull. U. S. Geol. Surv.* **10:**9–54.

Walcott, C. D., 1890, *The Fauna of the Lower Cambrian or Olenellus Zone of North America,* 10th Annual Report of the U.S. Geological Survey.

Walcott, C. D., 1900, Lower Cambrian terrane in the Atlantic Province, *Proc. Wash. Acad. Sci.* **1:**301–339.

Walliser, O. H., 1958, *Rhombocorniculum comleyensis* n. gen. n. sp. (*Incertae sedis,* Unterkambrium, Shropshire), *Paläontol. Z.* **32:**176–180.

Chapter 9

The Lower Cambrian Fossil Record of China

JIANG ZHI-WEN

1. Introduction

The upper Proterozoic and Lower Cambrian strata of China preserve a series of unique fossil faunas that provide unique opportunities for studying and understanding the evolution of early Metazoa. These faunas include a late Proterozoic metazoan fauna, diverse pretrilobite shelly assemblages, and an Early Cambrian

JIANG ZHI-WEN • Yunnan Institute of Geological Sciences, Kunming 650011, Yunnan, China.

Origin and Early Evolution of the Metazoa, edited by Jere H. Lipps and Philip W. Signor. Plenum Press, New York, 1992.

soft-bodied fauna. The purpose of this chapter is to review briefly these faunas, which are not well known outside China.

2. Tectonics and Paleogeography

Prior to the Sinian, three separate, lithologically distinct land masses (North China–Tarim, Yangtze, and South Tibet) that now form China were probably separated by considerable paleogeographic distances. These land masses arelithologically, paleoecologically, and paleontologically distinct. These masses have been tectonically stable throughout the late Proterozic and Early Cambrian (Wang *et al.*, 1985), and facies changes were largely controlled by fluctuations in eustatic sea level. The Early Cambrian faunas originated and flourished mainly in marginal and shelf seas over these land masses. As many as ten Early Cambrian provinces can be recognized (Xiang *et al.*, 1981) and they are strongly controlled by paleogeography (Fig. 1 and Table I). Among these different provinces, the faunal record in southern China (Yangtze region) is particularly well developed.

3. The Fossil Fauna of China

3.1. Shelly Fauna

The evolution of skeletons that defines the beginning of the Cambrian is an important and essential qualitative event in the history of life. A wide variety of different organisms appear simultaneously in the fossil record, and each skeletonized group is characterized by fundamentally different types of skeletal growth and materials. The newly skeletonized animals radiated quickly to fill a multitude of roles in newly forming marine communities and, as a result, the major metazoan phyla and classes were established as distinct clades from the earliest Cambrian.

In China, Cambrian metazoan communities are mainly preserved on stable cratons, such as the Yangtze, Tarim, and North China, whereas they tend to be uncommon within tectonically active areas. The oldest communities occur within the Yangtze area, and slightly later in North China and Tarim to the west. In the Meishucunian, the "nonskeletonized trilobite" zone is the time when this fauna reached its acme, only to be reduced to a secondary role following the appearance of trilobites and other arthropoda in the subsequent Qiongzhusian Stage.

Based upon the available data, seven Early Cambrian assemblage zones can be recognized in China:

VII. *Dorsolinevites–Parakorilithes–Yochelcionella* Zone, Late Tsanglangpuan Stage

VI. *Obliquatheca–Majatheca* Zone, Early Tsanglangpuan Stage

V. *Sulcavitus–Linevitus–Aimitus* Zone, Late Qiongzhusian Stage

IV. *Cambroclovus–Rhombocorniculum–Microconus* Zone, Early or Early–Middle Qiongzhusian Stage

III. *Sinosachites–Tannuolina* Zone, Late Meishucunian Stage

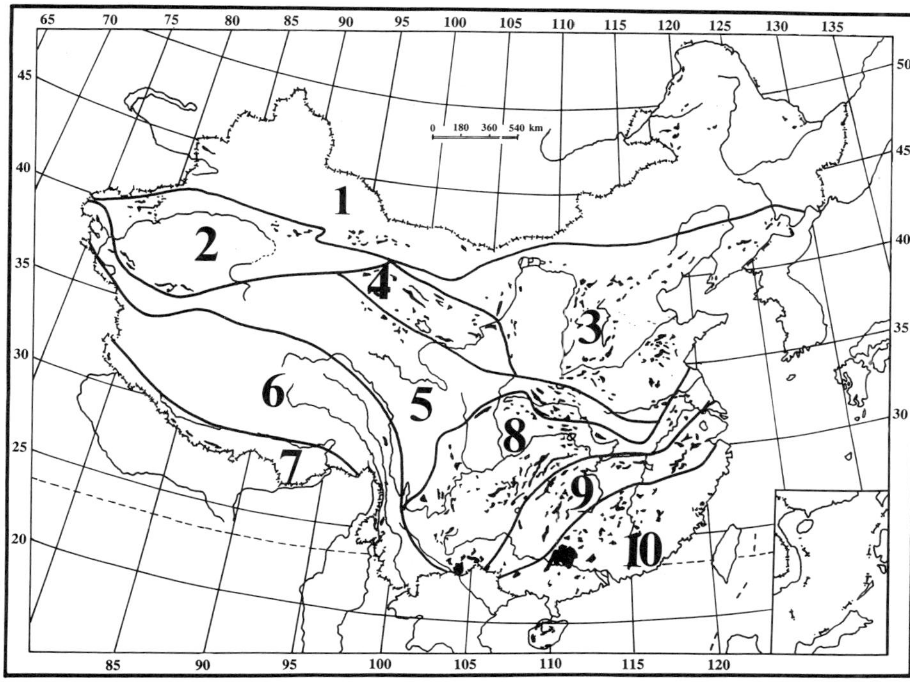

Figure 1. Stratigraphic regions and Cambrian outcrops, China. (1) Tianshan–Hingan, (2) Tarim, (3) N. China, (4) Qilian, (5) Kunlun–Qinling, (6) N. Tibet–W. Yunnan, (7) Himalaya, (8) Yangtze, (9) Jiangnan, and (10) S. China.

II. *Siphogonuchites–Paragloborilus–Heraultipegma* Zone, Middle Meishucunian Stage

I. *Anabarites–"Circotheca"–Protohertzina* Zone, Early Meishucunian Stage.

3.1.1. Yangtze Region

The Yangtze region has the longest history of scientific investigation and is the best understood of the Early Cambrian Chinese provinces. This region, the Upper Yangtze paraplatform, includes eastern Yunnan, western Sichuan, southern Shaanxi, the south-central part of Guizhou, the east gorge part of Hubei, and western Hunan. During the earliest Cambrian (Meishucunian) this area was a vast, shallow, carbonate-dominated continental shelf that was most suitable as an environment for shelly organisms. Later, in the middle to late Early Cambrian (Qiongzhusian and Tsanglangpuan) this self was the site of shallow marine detrital sedimentation, with occasional phosphate beds. Dolomites and oolitic limestones representative of high-energy environments accumulated during the late Early Cambrian (Longwangmiaoan), when the fossil faunas greatly decreased under harsh environmental conditions.

Four assemblages of abundant fossils, designated I–III and V, occur in the Lower Cambrian of China in considerable abundance and in a consistent sequence. These are interpreted as four distinct evolutionary stages in the development of the subtrilobitic shelly faunas.

TABLE I. Correlation for Dividing Zones of Main Fauna, Lower Cambrian, China

Stage	Shelly fossil zones	Trilobite zones	Soft-bodied metazoa	Archaeocyathan assemblages
Longwangmiaoan		8 *Redlichia (Pteroredlichia)*		
		7 *Hoffetella*		
Tsanglangpuan	VII *Dorsolinevitus–Parakorilithes–Yochelcionella*	6 *Megapalaeolenus*		
		5 *Palaeolenus*		Tianheban
	VI *Obliquatheca–Majitheca*	4 *Drepanuroides*		Yingzuiyan
		3 *Yiliangella–Yunnanaspis*		Liangshuijing
Qiongzhusian	V *Sulcavites–Linevitus–Aimitus*	2 *Eoredlichia*	*Alalcomenaeus* *Brachiocaris* *Cambinivalvula* *Fuxianhuia* *Hanchiangella* *Heliomedusa* *Isoxys* *Jianfengia* *Kuamaia* *Kunmingella* *Liangshanella* *Nanchengella* *Naroia* *Nianianshana* *Perspicaris* *Rhombicalvaria* *Rotadiscus* *Stellostomites* *Vetulicola* *Yunnanomedusa*	Qiangzhusian
	IV *Cambroclavus–Rhombocorniculum–Microconus*	1 *Parabadiella–Mianxiandiscus*		
Meishucunian	III *Sinosachites–Tannuolina*			
	II *Siphogonuchites–Paragloborilus–Heraultipegma*			
	I *Anabarites–"Circotheca"–Protohertzina*			

Assemblage I represents the initial stage of the shelly fauna's development, and is characterized by small, simple forms with low diversities and densities. The primary members of this assemblage are orthothecid hyoliths, primitive monoplacophorans, brachiopods, globomorphs, hyolithelminthes, possible conulariids, and others. Genera of shelly fossils in Assemblage I include "*Circotheca*," *Conotheca, Turcutheca, Rogatotheca, Anabarites, Paranabarites, Protohertzina, Spinulitheca, Artimycta, Eoconularia, Hexangularia, Archaeooides, Cassidina, Emeiconus* and the ichnofossils *Sellaulichnus, Chondrites, Phycodes, Cladichnus,* and *Palaeodictyon* (Fig. 2).

Assemblage II is the most diverse and abundant of the Meishucunian faunas. Some stratigraphic horizons are composed entirely of small shelly fossils, including hyoliths, brachiopods, bryozoans, archaeocyathans, chancellorids, camenerids, conodontomorphans, and siphonoguchitids. Not only are the fossils abundant and represent a large number of genera and species, but there is a marked increase in the complexity of the fossils, in their ornament, interior structure, and microtexture. Most of the elements that might prove useful for global

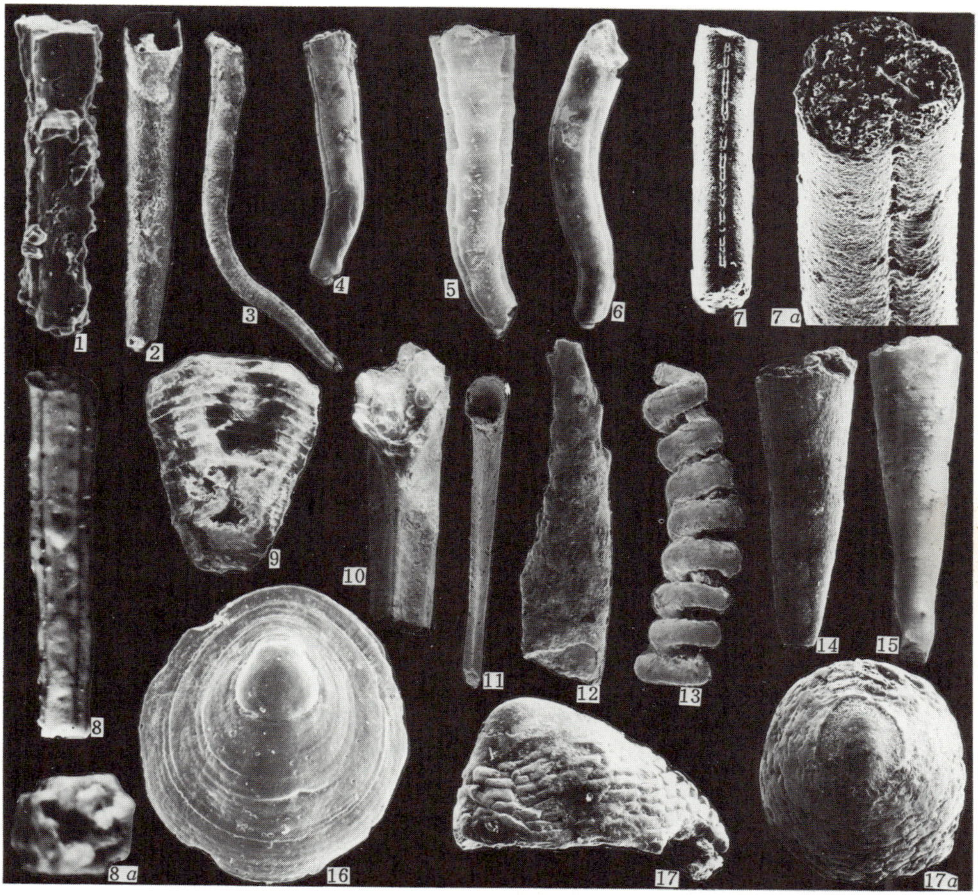

Figure 2. Assemblage I of the shelly fossils in the Meishucunian, Yangtze region. (1–6, 9–10, 12–15, 17) Sampled in the Meishucun section, Yunnan. (1, 2, 8) Located from Xiaowaitoushan member; (3–6, 9, 10, 12–17) located from Zhongyicun member, Yuhucun fm. (1) Protebulichitiida,×67. (2) "*Circotheca*" *longiconica* Qian, 1978, ×27. (3–6) *Anabarites trisulcatus* Missarzhevsky, 1969, (3) ×7, (4) ×19, (5) ×19, (6) ×19. (7) *Anabarites gracilis* Cheng, 1984, (7) ×53, (7a) ×168, apertural view, Yanjiahe, Yichang, Hubei. (8) *Paranabarites tetrosulcatus* Jiang, 1982, (8) ×16; (8a) ×24, apertural view, Annin, Yunnan. (9) *Barbitositheca ansatus* Qian & Jiang, 1980, ×96. (10, 11) *Protohertzina anabarica* Missarzhevsky, 1973, (10) ×104, (11) ×144, Nanjiang, Sichuan, Mofengyan mem., Dengying fm. (12) *Salonacus meishucunensis* Jiang, 1984, ×80. (13) *Spirella columnorus* Jiang, 1984, ×72. (14, 15) *Conotheca ansidata* Qian & Jiang, 1980, (14) ×80, (15) ×31. (16) *Aegides places* Jiang, 1982, ×45, dorsal view, Yulo, Yunnan. (17) *Canopoconus calvatus* Jiang, 1982, (17) ×64, lateral view; (17a) ×52, dorsal view.

correlations are found in this fauna. The genera occurring in this assemblage include *Siphonoguchites, Paleosulcachites, Lopochites, Carinachites, Paracarinachites, Paragloborilus, Leconogloborilus, Yunnanotheca, Ovalitheca, Heterosculpotheca, Bucanotheca, Tiksitheca, Leibotheca, "Lapworthella", Heraultipegma, Conularia, Sunnaginia, Zhijinites, Fomitchella, Scoponodus, Brushenodus, Protoconus, Rostroconus, Igorella, Latouchella, Pelagiella, Maidipingoconus, Yangtzespira, Tannuella, Obtusoconus, Remanta, Parapunctella,* and others (Fig. 3). The ichnofossils *Didymaulichnus, Rusophycus, Cavalichnus, Cruziana, Neonereites, Monomorphichnus, Paleodictyon, Asteriacites, Scoyenia, Cochlichnus,* and *Meishucunichnus* occur with Assemblage II.

Assemblage III corresponds to the latest stage of development of the Meishucunian fauna prior to the advent of trilobites and archaeostracoda in the local fossil record. With the appearance of the arthropods, most of the Meishucunian genera disappear from the record and are replaced by new forms. The major components of Assemblage III include tannuolinidans, camenellidans, chancellorids, and a small number of hyoliths. The genera occurring in this assemblage include *Sinosachites, Sachites, Eonovitatus, Allatheca, Ebianotheca, Scenella, Chancelloria, Onychia, Allonnia, Dimidia, Adversella, Archiasterella, Camenella, Tannuolina, Coleoloides, Thambetolepis, Palaeoacmaea, Lapworthella,* and the ichnogenera *Plagiogmus, Gordia, Taphrhelminthopsis, Skolithos, Diplocraterion, Arenicolites, Cochlichnus,* and others (Fig. 4).

It is worth mentioning that the north-central part of the Upper Yangtze Platform—Zhenxiong of northeast Yunnan, Jinsha of northwestern Guizhou, and Xiannutong of northern Sichuan—had accumulated thick carbonate deposits at this time. The north-central platform has a characteristic fauna characterized by tommotids occurring in association with the archaeocyathans *Aldanocyathus, Ajacicyathus, Rotundocyathus, Coscinocyathus, Conannulofungia, Sibirecyathus,* and others. The shelly fauna includes *Tommotia, Fomitchella, Anabarella, Bemella, Igorella, Latouchella, "Circotheca", Trapezotheca, Chancelloria,* and others. This assemblage, perhaps, corresponds to the upper part of Assemblage III, but the possibility that the assemblage might be somewhat younger cannot be ruled out.

The three Meishucunian Assemblages can tentatively be correlated to other parts of the world, and the area is an ideal standard on which to base a detailed reconstruction of the evolutionary sequence of shelly fossils. Taxa occurring in Assemblage II have been discovered throughout the world, including the Siberian Platform, the Karatau ridge of central Kazakhstan, the Altai of Mongolia, the Lesser Himalayas of India, the Alborz Mountains of Iran, the Mackenzie Mountains of northwest Canada, Sonora, Mexico, and in the Heraulte of southeastern France. This broad geographic distribution of occurrences makes preliminary biostratigraphic correlation feasible.

The appearance of diverse trilobites, archaeostracodes, and inarticulate brachiopods marks the beginning of the Qiongzhusian. The typical Meishucunian shelly fauna apparently became extinct, and was replaced by a younger fauna. One aspect of this younger fauna is another group of hyoliths, characterized by distinctive ventral and dorsal sides, a large apical angle, a central groove, and an operculum. These hyolithids include *Linevitus, Sulcovitus, Ambrolinevitus, Aimitus, Burithes,* and *Dolivitus*. The hyolithids are associated with the trilobites *Eoredlichia, Wutingaspis,* and the famous "Chengjiang Fauna." These are well-

Figure 3. Assemblage II of the shelly fossils in the Meishucunian, Yangtze region. (11–15, 17, 18) Sampled in the Meishucun section; (7–10, 12, 16, 19, 20) Sampled in Huize, Yunnan. (3, 7–9, 12–16, 18, 21) Located from Zhongyicun member; (4, 6, 10, 11, 17, 19, 20) located from Dahai member, Yuhucun fm. (1, 2) *Aldanella yanjiaheensis* Cheng, 1984, (1) ×24, (2) ×22, Yanjiahe, Yichang, Hubei, Yanjiahe fm. (3) *Mobergella? bella* (He and Yang, 1982) ×48, Xianfeng, Yunnan. (4) *Latouchella korobkovi* (Vostokova, 1962), ×24, Annin, Yunnan. (5) *Camenella plana* Missarzhevsky, 1969, ×80, Yuanshan member. (6) *Sachites preinguis* Jiang, 1982, ×56, Yongshan, Yunnan. (7) *Haraultipegma varensalensis* (Cobbold, 1935), ×14. (8) *Lapworthella cancellata* Jiang, 1982, ×32. (9) *Obtusoconus multicostatus* Yu, 1979, ×24. (10) *Paragloborilus mirus* (He, 1977), ×12. (11) *Paragloborilus subglobosus* (He, 1977), ×80. (12) *Spinulitheca secunda* Sysoiev, 1968, ×24. (13) *Yangtzespira exima* Yu, 1979, ×34. (14) *Yunnanospira multiribis* Jiang, 1980, ×24. (15) *Paracarinachites sinensis* Qian & Jiang, 1980, ×48. (16) *Palaeosulcachites biformis* Qian, 1977, ×24. (17) *Rostraconus sinensis* Jiang, 1980, ×38. (18) *Yunnanodus doleres* Wang & Jiang, 1980, ×88. (19) *Zhijinites minutus* Qian, 1978, ×36. (20) *Scoponodus renustus* Jiang, 1982, ×28, Annin, Yunnan. (21) *Brushenodus prionodes* Jiang, 1982, ×19. (22) *Hexaconularia multicostata* He & Yang, 1983, ×34, Nanjiang, Sichuan, Mofengyan member, Dengying fm. (23) *Microlites shotanensis* He & Yang, 1983, ×45, Nanjiang, Sichuan, Mofengyan member, Dengying fm.

Figure 4. Assemblage III of the shelly fossils in the Meishucunian, Yangtze region. All samples for this figure were located at the basal part of the Yu'anshan member, Qiongzhusi formation, Meishucun section (except for sample 9). (1, 2) *Lapworthella rete* Yue, 1986, (1) ×90, (2) ×58. (3) *Lapworthella* cf. *hubeiensis* Qian & Zhang, 1983, ×67. (4) *Burithes* cf. *erum* Missarzhevsky, 1969, ×44. (5) *Tannuolina multifora* Fonin & T. Smirnova, 1967, (5) ×14, dorsal view; (5a) ×12, inner view; (5b) ×16, lateral view. (6) *Lapworthella tortuosa* Missarzhevsky, 1966, ×90. (7) *Lapworthella* cf. *dentata* Missarzhevsky, 1969, ×49. (8) *Thambetolepis delicata* Jell, 1980, ×36. (9) *Sinosachites triangularis* He, 1980, ×38, Huize, Yunnan. (10) *Rhabdochites exaspertus* He & Yang, 1984, ×54. (11) *Calcihexatina isophyllus* Jiang, 1984, ×96. (12) *Archiasterella pentactina* Sdzuy, 1969, ×40. (13) *Igorella* cf. *ungulata* Missarzhevsky, 1969, ×29. (14) *Coleoloides trigeminatus* Missarzhevsky, 1969, ×96. (15) *Coleoloides typicalis* Walcott, 1889, ×120. (16) *Paragloborilus circulatus* Qian & Zhang, 1983, ×52. (17) *Sinosachites fllabeliformis* He, 1968, ×40. (18) *Sachites* cf. *longus* Qian, 1971, ×45.

known components of Assemblage V, which have been employed to establish the *Sulcavitus–Linevitus–Aimitus* Zone.

Within this region, another assemblage has been discovered in the Mingxinsi Formation of northern Guizhou. The taxa of the Mingxinsi Formation include *Obliquatheca, Yankongvitus, Orthotheca, Exilitheca, Parakorilithes, Linevitus, Burithites, Dolivitus, Majatheca, Allatheca, Tommotia,* and *Fomitchella*. Assemblage VI and the *Obliquatheca–Majatheca* Zone have been established on the basis of this fauna.

3.1.2. Tarim Region

Lower Cambrian strata are exposed only along the northwestern and the northeastern margins of this region. The Lower Cambrian Yultus Formation crops out along the northwestern margin of the Tarim in the Koping–Wushi–Aksu area. The following taxa have been recovered from the middle and upper parts of the formation (Fig. 5): *Cambroclavus, Sachites, Isoclavus, Zhijinites, Acrosouama, Lophotheca, Aksuglobolus, "Paragloborilus,"* and *Lopochites*. These are preserved in association with the stromatolite *Linocollenia* and the phosphatocopid bradoriida *Xinjiangella*. The age of the assemblage corresponds to the latest Meishucunian and early Qiongzhusian and has been assigned to Assemblage IV (*Cambroclavus–Rhombocorniculum–Microconus* Zone). Most of these fossils have been coated with a thin film of phosphate that allows them to be recovered from acid digestions of limestone.

3.1.3. Kunlun–Qinling Region

Although this region incorporates a large area, Lower Cambrian strata occur only in the central-northern and eastern portions. The strata and their faunas have not been studied in detail, and the shelly fossils are not well known. In the Shennongjia area of eastern Hubei province, the following taxa have been collected from the Xihaoping Formation (Fig. 6): *Cambroclavus, Isoclavus, Cambrothyra, Rhombocorniculum, "Paragloborilus," Neogloborilus, Lapworthella, Microconus, Adjshevitheca, Pseudorthotheca, Cambrotubulus, Sachites, Paraformitchella, Ovalitheca, Chancelloria,* and *Archiasterella*. These are recognized components of Assemblage IV, and have an age of latest Meishucunian and early Qiongzhusian. At the base of the Shuigoukou Formation in south Shaanxi a few rare *"Circotheca" subcurvata* and *Hyolithellus minus* have been found in siliciclastic rocks. Other Meishucunian taxa might also have been present.

3.1.4. North China Region

Geotectonically, the North China Region corresponds to the area of the North China Craton, including southern northeast China, southern Inner Mongolia, and parts of Hebei, Henan, and Anhui provinces. This region has not been tectonically active since the middle Early Cambrian. Sediments bearing abundant faunas accumulated there in a stable platform setting. Lower Cambrian fossils from this area include (see Fig. 7) *Dorsolinevitus, Linevitus, Microconus, Parakorilithes, Anhuiconus, Anabarella, Tannuella, Spatuloconus, Bemella, Mellopegma, Yochel-*

Figure 5. Typical shelly fossils of the Lower Cambrian, Tarim region, Xinjiang. All samples were located from the Yurtus fm. of the Lower Cambrian, Aksu-Wushi county, Xinjiang. (1, 2) *Sachites mira* J. X. Qian & Xiao, 1984, (1) ×98, (2) ×67. (3) *Tianshandiscus verrucosus* J. X. Qian & Xiao, 1984, ×80. (4) *Xiadongtubulus punctatus* (Qian, 1978), ×72. (5, 6) *Polycladium yurtusensis* J. X. Qian and Xiao, 1984, (5) ×134, (6) ×80. (7, 12) *Cambroclavus soleiformis* (J. X. Qian and Xiao, 1984), (7) ×98, (12) ×107. (8, 13) *Cambroclavus bicornis* (J. X. Qian and Xiao, 1984), (8, 8a) ×90; (13) ×134. (9, 10) *Wushichites minutus* J. X. Qian & Xiao, 1980, ×72. (11) *Aurisella tarimensis* J. X. Qian & Xiao, 1984, ×107.

cionella, Scenella, Pojetaia, Oryzoconcha, Allonnia, Henaniodus, Bioistrodina, and others. Also, the prolenid trilobites *Bergeroniella* and *Hsuaspis* and the archaeostracods *Ophiosema* and *Houlongdonella* occur in this fauna. This fauna is recognized as Assemblage VII (the *Dorsolinevitus–Parakorilithes–Yochelcionella* Zone). Assemblage VII corresponds to the late Tsanglangpuan stage, established in the Xinji Formation of Fangchen County, Henan, in the Houjiashan Formation of Houqui County of Anhui, in the Wendurmina Group (in the narrow sense) of Inner Mongolia, and the Ligou Formation of south Liengling. This distribution roughly corresponds to the north and south margins of the North China Craton. Eastwardly, it may extend into the north part of the Korean peninsula and may be represented in the *Redlichia* or *Protolenus* shales of the Zhonghe Formation (Saito, 1936).

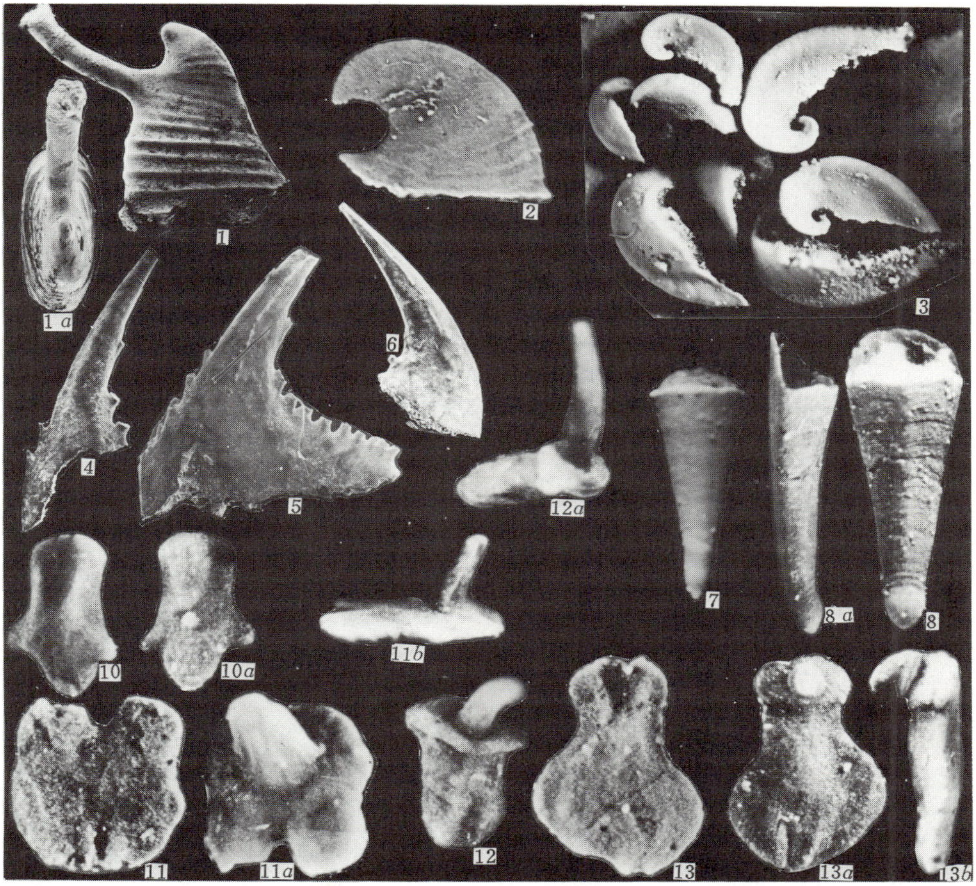

Figure 6. The chief shelly fossils of Lower Cambrian, North China and Kunlun-Qinling Region. (1–7) located from Henan, North China region: (1) Yexian county; (2–7) Fangchen county. Xinji fm., Є$_1^3$. (8–13) Located from Fangxian county of Kunlung–Qinling region, Xihaoping fm., Є$_1$. (1) *Yochelcionella chinensis* Pei, 1985, (1) ×22; (1a) ×24. (2) *Anabarella drepanoida* He & Pei, 1984, ×20. (3) *Auriculatespira andunca* He & Pei, 1984, ×13. (4–6) *Henaniodus magicus* He & Pei, 1984, ×20. (7) *Parakorilithes mammillatus* He & Pei, 1984, ×20. (8) *Microcornus eximius* Duan, 1983, (8, 8a) ×28. (10) *Phyllochites shawanensis* Duan, 1983, (10, 10a) ×40. (11) *Tanbaoites porosus* Duan, 1983, (11, 11a, 11b) ×40. (12) *Cambroclavus dolicanthus* Duan, 1983, (12, 12a) ×48. (13) *Cambroclavus olecranus* Duan, 1983, (13, 13a, 13b) ×40.

3.1.5. Some Basic Problems

3.1.5a. Time–Space Distribution. In a broad sense, evolutionary studies should include documentation of both the local diversification and the geographic expansion of newly evolved groups. This requires analysis of both temporal and biogeographic aspects of diversification and the systematic relationships of the biota. As one evolutionary and biogeographic component of the Cambrian fauna, the Early Cambrian shelly fauna of China has temporal and biogeographic contexts. It is these relationships that are needed to construct global biostratigraphic correlations and to understand the early radiation of Metazoa.

It is well known that the Yangtze region, particularly the area including eastern Yunnan and western Sichuan, is the site of the oldest shelly fauna. The first appearances of shelly fossils are increasingly younger to the north and east. It can be shown that in the southern part of east Yunnan, on the western margin of the Yangtze region, Assemblage I is well developed and characterized by small forms and primitive *Anabarites primitivus* (possibly the primitive *A. trisulcatus*) in its lowest part. Northwardly, toward the Sichuan–Yunnan, the lowest form is the younger and more advanced form, *Anabarites trisulcatus*. Further north toward the Emei area of southwest Sichuan, *Anabarites* has not been found and the fauna includes younger forms such as *Rugatotheca* and *Spinulitheca*. Still further north, near Ningqiang in Shaanxi Province, *A. trisulcatus* and the siphonoguchitids that normally occur at the top of Assemblage I and bottom of Assemblage II are the lowest occurring fossils. This pattern suggests a tendency toward somewhat later first occurrences of fossils to the north. A similar pattern can be seen toward the central and eastern parts of the Yangtze region.

From the Yangtze region northward to North China and western to the Kunlun–Qinling and Tarim region, the first occurrence of somewhat younger shelly fossils than found in eastern Yunnan is a regular pattern. In North China the first shelly fossils correlate with the late Qiongzhusian–Tsanglangpuan stages, and Assemblages I–V are not found. To the west, in the Tarim region, Assemblages I and II have not been found.

Analyzing the constitutents of the faunas, it can been seen that Assemblage I consists mainly of circothecid hyoliths (Orthothecida). These are characterized by long shells with small apical angles. Less common but also present are primitive tryblidiod monoplacophora, inarticulate brachiopoda, protoconondonts, primitive conuariids (Hexangulaconulariidae), globomorphs and thin-shelled worm-tubes.

The constituents of Assemblage II are much more complex than earlier fossils. The Circothecidae (Hyolitha) of Assemblage I lack the bilateral symmetry often developed in the Orthothecidae and Allathecidae of Assemblage II. Also, some hyoliths develop a labrum for the first time. Planispiral and helicospiral gastropods and rostroconchs (*Heraultipegma, Rostroconus,* etc.), including some forms that might be transitional between the rostroconchs and monoplacophora, appear for the first time. The first bivalve mollusks occur in this assemblage. There is a much greater diversity of inarticulate brachiopods than is found in Assemblage I, including ocurranid, scambiciid, tianzhushanellid, and plicatolingulid forms. Symmetrical and asymmetrical chancellorids diversify in this assemblage. South Asian endemics, including camenids with simple ornament, siphogonuchitiids, and conodontomorphs, are common in this assemblage. The first occurrences of Archaeocyatha are in this assemblage.

Assemblage III is the last of the Meishucunian Stage faunas. The number of taxa is reduced, but the abundances remain large. Among the hyoliths, only a few representatives of the Allathecidae and Globoriliida remained and the Hyolitha with broad divergent angles began to appear. The Chancellorida reaches a peak in its morphological diversification, and all combinations of branches can be found in this assemblage. These can be differentiated on the basis of the number of central (H) and lateral (V) branches in each cluster. Using this notation, the following taxa can be recognized (Fig. 7):

7–8H	+ 1V	*Chancelloria* Walcott, 1920
6H	+ 1V	*Eiffellia* Walcott, 1929
2H	+ 1V	*Adversella* Jiang, 1980
5H	+ 0V	*Archiasterella* Sdzuy, 1969
4H	+ 0V	*Onychia* Jiang, 1982
3H	+ 0V	*Allonnia* Doré and Reid, 1965
2H	+ 0V	*Dimidia* Jiang, 1982

The taxonomic relationship of *Sachites* is the subject of dispute. It first appears in Assemblage II, and reaches its peak diversity in Assemblage III. Typical Chinese taxa such as *Lapworthella, Camenella,* and *Coleoloides* only occur in this assemblage. *Tannuolina* occurs at higher levels in other regions, but is a member of Assemblage III in Meishucun. Also, a few gastropods and brachiopods occur in this assemblage.

Assemblage IV is characterized by the joint occurrence of *Cambroclavus* and *Rhombocorniculum;* the age of this assemblage is early Qiongzhusian. This assemblage is distributed from Tarim eastward through Shennongjia of Hubei Province and southward to Hainan Island (and possibly into Australia).

3.1.5b. Comparability and Sequences. Several shelly fossils might serve as excellent index fossils for regional and global correlations. These include *Anabarites, Protohertzina, Aldanella, Igorella, Bemella, Latouchella, Obtusoconus, Tommotia, Sachites, Siphogonuchites,* and *Paragloborilus,* among others. There are also many taxa that apparently have appearances that are diachronous in different regions. These include the following:

(i) *Tannuolina* Fonin and Smirnova, 1967. This genus has only been found within a narrow zone extending north–south at the longitudes 94–140° E. Seven sites have produced *Tannuolina,* including the Sivelik-Hem and Ulg-Sangan sections of Tuvonsk (USSR) to the north (the type locality for *Tannuolina*). In the central part of this zone are the Meishucun and Dahai sections of eastern Yunnan, a borehole in Zhengong of neighboring Kunming, and the Maidiping section of Omei, Sichuan province. In the southern part of the zone they are found in the Flinders Ranges and in the Ajax Limestone of the Beltane District of Australia. During the Cambrian this zone was situated at a similar latitude, reaching north–south in orientation (Jiang, 1984b).

In all Chinese sections the genus occurs in Assemblage III of the Meishucunian shelly fauna and the oldest trilobite, *Parabadiella,* first occurs above them. In southern Siberia, it is associated with archaeocyathans, trilobita, brachiopods, and hyoliths. Daily (1956) believed that the Ajax Limestone represents the second assemblage of the Lower Cambrian; perhaps it should correspond to the latest Meishuchunian–early Qiongzhusian. Many specimens of *Tannuolina* from this district have been illustrated by Bischoff (1976). Thus, there is a tendency for *Tannuolina* to occur later to the north and south of the Yangtze Platform.

(ii) *Cambroclavus* Mambetov, 1979. It was mentioned above that *Cambroclavus* shows a geographic range extending east–west, in contrast to the north–south distribution of *Tannuolina* (Fig. 8). It first appeared in the Minor Karatau of Talasky Ala-Tau, Kazakhstan, Central Asia, and the northern Tianshan region, and expanded into China through Aksu, Wushi, and Loping over the north margin of Tarim, passing through Shennongjia of western Hubei to the Yaxian of Hainan Island. This distribution forms a "T" pattern of distribution across the central-

Figure 7. The main kinds of Chancelloriida in China. (1–8) Located in assemblage III of Yu'anshan member, Qiongzhusian fm: (1, 2, 5, 6) Sampled from Meishucun section, Yunnan; (3, 4, 7, 8) sampled from Huize, Yunnan. (9, 10) Sampled from Yurtus fm., Aksu-Wushi region, Xinjiang. (11, 12) Sampled from Damao group of Middle Cambrian, Hainan Island. (1) *Dimidia simpleca* Jiang, 1982, ×58. (2) *Allonnia erromenosa* Jiang, 1982, (2) ×54; (2a) ×45. (3) *Onychia tetrathellis* Jiang, 1982, (3) ×32, (3a) ×34. (4) *Archiasterella pentactina* Sdzuy, 1969, (4, 4a) ×32. (5, 6) *Adversella montanoides* Jiang, 1982, (5) ×50; (6) ×40. (7, 8) *Eiffella? bispanica* Sdzuy, 1969, (7, 7a) ×40, (8) ×28. (9, 10) *Chancelloria altaica* Romanenko, 1968, (9) ×40, (10) ×27. (11, 12) *Miriella hainanensis* Jiang & Huang, 1986, (11) ×98, (12) ×54.

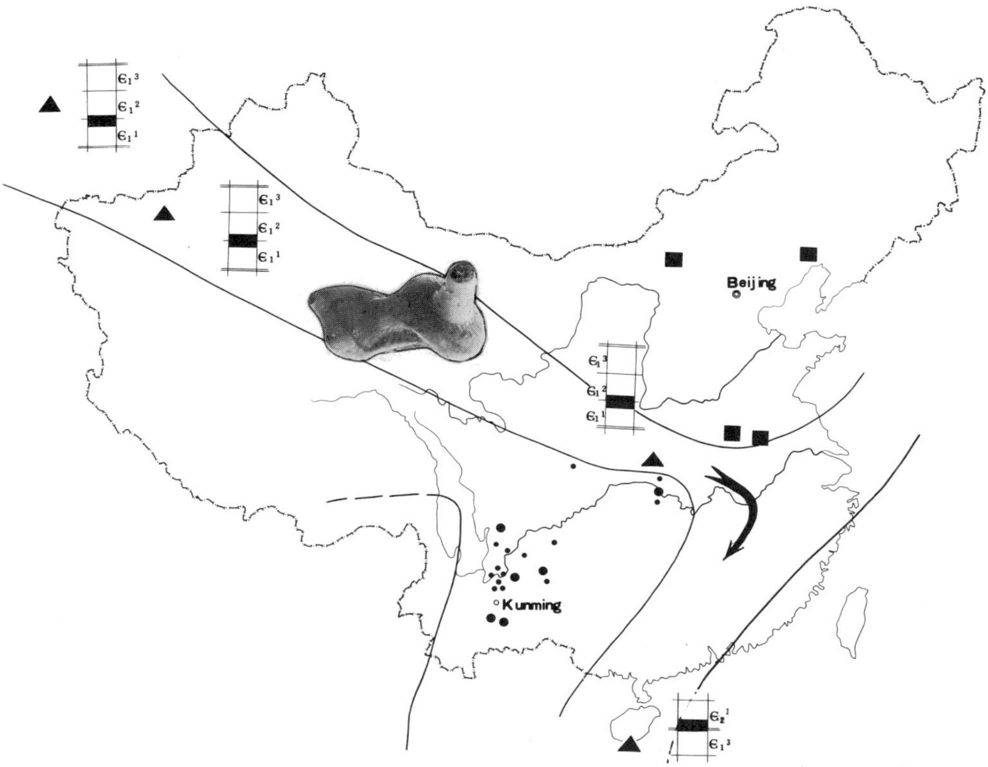

Figure 8. Biogeographic province of *Cambroclavus*. Circles, shelly fossil locations of Meishucun type. Triangles, locations of *Cambroclavus*. Squares, shelly fossil locations in North China.

southeast parts of China. Thus far, it is the only region where *Cambroclavus* occurs, and might indicate a distribution between the northern and southern cratons of China. The occurrences in the Minor Kalatau are in strata of Tommotian age. The Tarim occurrences are in trilobite-bearing (*Shizhudiscus*) strata. To the south, the occurrences on Hainan Island are in strata of Ordian or Templetonian age (early Middle Cambrian).

A better understanding of the temporal distribution of taxa will definitely lead to the recognition of additional index fossils and previously unrecognized diachronous occurrences.

3.2. Trilobites

The appearance of trilobites defines the beginning of the Qiongzhusian stage. Five orders of trilobites have been discovered in the Lower Cambrian of China: Agnostida, Redlichiida, Corynexochiida, Ptychopariida, and Nektaspida. They replace the small shelly fauna as the primary index fossils. The most important of these are the Redlichiida, which are the typical constituents of the "oriental trilobite" faunal province.

3.2.1. Redlichiida

The Redlichiidae are generally found in the Lower Cambrian and possess two phases of development. The earlier phase is represented by the genera *Eoredlichia, Pararedlichia, Wutingaspis, Chanaspis, Kunyangia, Parabadiella,* and others. The later phase includes the genera *Redlichia, Metaredlichia, Neoredlichia,* and *Spinoredlichia*. Most of them possessed genal spines, opisthoparian sutures, and clearly divided glabellas.

The Drepanopygiidae occur only in the southwest region of China. Except for *Paokania,* which occurs in the upper middle Lower Cambrian, most genera (*Drepanopyge, Drepanuroides, Yunnanaspis, Yunnanaspidella, Yinites,* etc.) occur in the lower portion of the middle Lower Cambrian.

Protoleniids occur exclusively in the upper Lower Cambrian throughout the world. In China, they usually occur in the Tsanglongpuan, especially in the Yangtze region. The commonest genera are *Hsuaspis, Bergeroniellus, Ichangia, Pseudoichangia,* and *Hupeia*. Most of these genera also occur along the west and south margins of the Sino-Korean Platform, as far as the vicinity of Pyongyang.

During the Early Cambrian, representatives of the Ellipsocephalidae are uncommon. *Palaeolenus* occurs in the upper part of the Tsanglangpuan throughout North China and in the southwestern part of the Yangtze region, while *Hoffetella* occurs only in eastern Yunnan. These two genera are useful for long-range correlations to other regions.

3.2.2. Agnostida

There are many representatives of the Eodiscina in the Early Cambrian of China. In the early middle stage, these are mostly members of the Metapagetiidae and include *Tsunyidiscus, Shizhudiscus, Mianxiandiscus, Guizhoudiscus, Sinodiscus, Neocobboldia,* and *Pagetia*. Later faunas in the regions of Tianshan of Xingjiang and Beishan of Gangsu are dominated by *Calodiscus, Serrodiscus, Tannudiscus,* and other representatives of the Eodiscidae.

3.2.3. Naktaspida

These are distinctive trilobites represented by the single genus *Naraoia* Walcott. It was first discovered in the Burgess Shale fauna of Canada in 1912, and has been found more recently to the south in the states of Idaho and Utah. In China, it first appears in the *Eoredlichia–Wutingaspis* Zone of the Chengjiang Fauna of Yunnan. This is the earliest known occurrence of *Naraoia*.

3.3. Chengjiang Fauna

An extraordinary fauna has recently been discovered in the Chengjiang Prefecture, in the vicinity of Kunming, Yunnan, China (Fig. 9). The fauna occurs

Figure 9. Important representatives of the Chengjiang fauna. (1–6) Located from Maotianshan Mt. of Chengjiang county. (7) Located from Meishucun section. All were sampled in Qiongzhusi fm., *Eoredlichia–Wutingaspis* Zone. (1) *Naraoia spinosa* Zhang & Hou, 1985, ×.8. (2) *Yunnanomedusa eleganta* Sun & Hou, 1987, ×.8. (3) *Fuxianhuia protensa* Hou, 1987, ×3.2. (4) *Jianfengia multisegmentalis* Hou, 1987, ×6. (5) *Maotianshania cylindrica* Sun & Hou, 1987, ×4. (6) *Alaleomenaeus? illecebrosus* Hou, 1987, ×8. (7) *Odaraia? eurypetala* Hou & Sun, 1988, ×6.

within strata of the Qiongzhusian Stage. This is the oldest Cambrian occurrence of abundant, well-preserved soft-bodied metazoan fossils. The fauna includes the trilobites *Eoredlichia, Yunnanocephalus, Kuanyangia,* and the unusual trilobite *Naraoia,* which are useful for correlating the fauna with other areas. Other members of the fauna are the large bivalved arthropods *Branchiocaris, Cambinivalvula, Isoxys, Perspicaris,* and *Vetulicola,* the bradoriids *Kuanyangia, Kunmingella, Neokunmingella, Tsunuiella, Wutingella,* and *Yaoyingella,* and the arthropods *Jiangfengia, Alalacomenaeus,* and *Fuxianhuia. Maotianshania, Cricocosmia,* and *Palaeoscolex* are worms that have been described from the fauna. Several medusae occur in the Chengjiang fauna, including *Rotadiscus, Stellostomites,* and *Yunnanomedusa.* In addition, the fauna included poorly preserved nontrilobite arthropods, brachiopods, and algae. The fossil priapulid *Maotienshania* is morphologically distinct from the priapulids of the more famous Burgess Shale (*Ottoia, Selkirkia,* and *Louisella*). *Maotienshania* also is smaller than those later genera.

The preservation of the Chengjiang fauna is exquisite. In fossil medusae, the structure of the tentacles, radial canals, annular muscle, central cavity, and mouth can be seen clearly. The proboscis and annulations of worms are perfectly preserved. The arthropods are preserved with eyes, digestive organs, appendages, and fine surface texture intact. This is the first convincing occurrence of fossil medusae in the Lower Cambrian.

The age of the Chengjiang fauna is younger than Australia's "Ediacaran Fauna" and older than the Middle Cambrian "Burgess Shale" fauna, which both include significant numbers of soft-bodied organisms. Each fauna occupied a different continental block separated by oceans. The relationship of the three faunas, with the intermediate temporal position of the Chengjiang fauna, is important for investigations of the early evolutionary history of invertebrates.

3.4. Archaeocyatha

The geographic and stratigraphic distribution of archaeocyathans is restricted by these organisms' paleoecological requirements. Most Chinese Archaeocyatha occur in the Tsanglangpuan Stage, in south Shaanxi, north Sichuan, southwest Sichuan, north Guizhou, northeast Yunnan, and southeast Yunnan. Four assemblages can be recognized at successively higher stratigraphic levels.

3.4.1. Qiongzhusian Archaeocyathan Assemblage

This assemblage occurs in the *Eoredlichia* Zone. Only a few primitive Septoidea, characterized by small size and simple structure, are found. The primary genera are *Hemithamalocyathus, ?Nalivkinicyathus,* and *Dictyocyathus.*

3.4.2. Liangshuijing (or Xiannitong) Archaeocyathan Assemblage

The Liangshuijing assemblage occurs in the *Malungia* (Trilobita) Zone in west Sichuan and south Shaanxi. The primary members are septoids and a few genera that possess features transitional with the Taenioidea. The primary genera are *Ajacicyathus, Archaeofungia, Cambrocyathus, Coscinocyathus, Chengkoucyathus, Cononnulofungia, Dictyocyathus, Protopharetra, Rotundocyathus, Sibirecyathus,* and *Taylorcyathus.*

3.4.3. Yingzuiyan Archaeocyathan Assemblage

This assemblage occurs in the *Drepanuroides* Zone of the Tsanglangpuan Stage. While the septoids are still dominant, the taenioids occur in increasingly large numbers and can occasionally be codominant. Coscinocyathids are common, as are irregular archaeocyathans. The genera *Agastrocyathus, Aldanocyathus, Archaeofungia, Coscinocyathus, Protopharetra,* and *Robustocyathus* are common members of the Yingzuiyan assemblage. It occurs in north Sichuan and central Xingjiang.

3.4.4. Tianheban Archaeocyathan Assemblage

The Tianheban assemblage is sometimes referred to as the Jinolingshan archaeocyathan assemblage. It occurs in the *Palaeolenus* Zone of the Tsanglangpuan Stage. Septoids are absent from this assembage and coscinocyathids are uncommon. The assemblage consists almost entirely of taenioid archaeocyathans, including *Archaeocyathus, Archaeofungia, Coscinocyathus, Dictyocyathus, Protopharetra, Rotundocyathus, Retecyathus,* and *Sanxiacyathus.* It has been reported in north Sichuan, north Guizhou, northeast Yunnan, and the Yangtze Gorge area of Hupei.

In addition to these four assemblages, there might be a Jingsha archaeocyathan assemblage that includes the genera *Aldanocyathus, Ajacicyathus, Coscinocyathus, Conannulafungia,* and *Rotundocyathus.* This assemblage, if its separate recognition proves to be warranted, is overlain by the Liangshuijing assemblage and occurs in the *Yunnanaspis–Yiliangella* Zone of the Tsanglangpuan.

4. Organismal Evolution in the Precambrian–Cambrian Transition in China

The Precambrian–Cambrian transition has an enormous significance in the evolutionary history of life. The emergence of the Ediacaran fauna marks the appearance of sexually reproducing organisms with differentiated organ tissues. The advance from colonies consisting of relatively few cell types to integrated organisms allowed the organisms to adapt to new modes of life in many environments. At the beginning of the Cambrian, the diversification of shelly organisms constitutes a second explosive radiation adding to diversity, complexity, and size, and thus allowing new body forms and modes of life. In the middle Early Cambrian, the continuing development and eventual dominance of the arthropods is a third stage of the metazoan radiation.

In the course of early animal evolution, three distinctive lineages can be recognized in the Precambrian–Cambrian transition of China (Fig. 10).

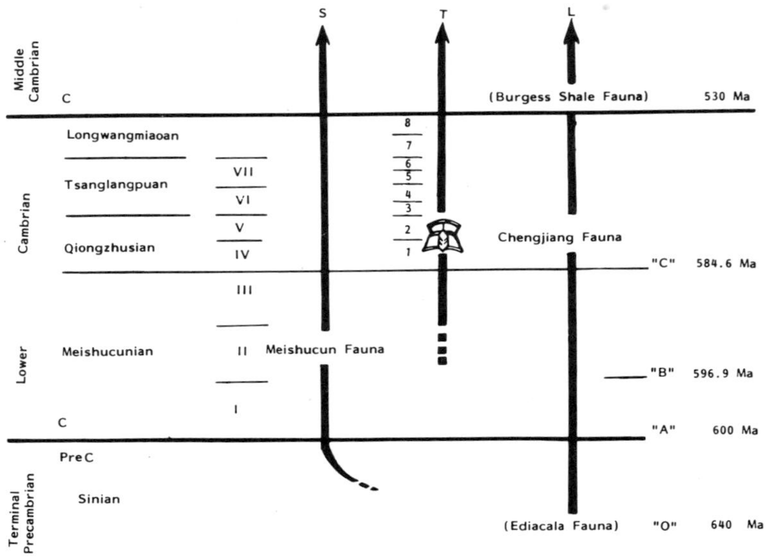

Figure 10. Evolutionary lineages of the main fauna at the Precambrian–Cambrian transition in China. S, Shelly fauna. T, trilobites. L, Large, soft-bodied metazoans. I–VII, shelly fossil assemblages. 1–8, trilobite assemblages. "A"–"C," reference points of the Precambrian–Cambrian boundary in China.

4.1. The Evolutionary Pedigree of Shelly Organisms

The shelly organisms that appear in the Lower Cambrian of China are known collectively as the Meishucunian fauna (Jiang, 1980b). Seven successive assemblages are known from the Meishucunian, Qiongzhusian, and Tsanglangpuan Stages, each with distinctive faunal characteristics but sufficiently similar with other faunas to form clear evolutionary successions. There is some geographic variation among the early faunas of China. The typical Meishucunian faunas occur in east Yunnan–west Sichuan, and to the east the early faunas become more similar to the Early Cambrian Tommotian faunas of the Siberian platform. These eastern faunas are known as the Xiadong (eastern Yangtse Gorge) faunas.

4.2. The Evolutionary Pedigree of Arthropods

By inference, trace fossils such as *Rusophycus, Cruziana, Didymaulichnus,* and *Monomorphichnus* probably reflect the activity of bilaterally symmetrical, limb-bearing Metazoans, probably unmineralized trilobites. Previously, none of these traces were known to occur prior to the appearance of trilobites in the fossil record. In China, the earliest trilobites were *Parabadiella,* found at the base of the Qiongzhusian Stage. Trilobites immediately began to diversify and dominate the benthic fauna. Eight trilobite range zones have been established in the interval from the Qiongzhusian to the Longwangmiaoan Stages and are useful for long-range correlations. The Early Cambrian trilobite faunas of the Yangtze region are particularly complete; the Qiongzhusian Stage is completely missing in North China and the upper Qiongzhusian is missing from north of the 35th parallel. The best-preserved and complete sequences occur in east Yunnan.

4.3. The Evolutionary Pedigree of the Chengjiang Fauna

Known from only one occurrence in China, the Chengjiang fauna is the first Early Cambrian assemblage of its kind in the world. It documents the presence of a diverse unique fauna well before the later, better-known Burgess Shale of the Middle Cambrian. Many forms, especially the medusae, closely resemble some members of the Ediacaran fauna. The diversification of this fauna must have occurred prior to the time when the fossils were actually deposited, concurrent with the radiation of shelly fossils.

5. Summary

With many years of accumulating research, the evolutionary record of fossil Metazoa in China clearly demonstrates the composition and patterns of distribution of early faunas, and the evolutionary patterns that led to the temporal and geographical distribution of those faunas. There is no question that the Early Cambrian Metazoa of China will play an increasingly important role in the study and interpretation of the early evolutionary history of animals.

The sequence of Chinese faunas is unique in the world. The Proterozoic occurrences of an Ediacaran fauna, including *Charnia*, *Sinotubulites*, and *Vendotaenia*, and the algal fossils *Chuaria* and *Tawuia*, are succeeded by the Early Cambrian Meishucunian fauna. The Meishucunian fauna of body fossils is supplemented by a diverse trace fossil fauna. The trilobite-dominated Qiongzhusian fauna follows the Meishucunian, and includes the soft-bodied Chengjiang fauna. These three types of fauna, successive in time and closely interrelated, is a unique representation of the evolutionary transition from the Proterozoic to the Cambrian.

ACKNOWLEDGMENTS. I am greatly indebted to Yu Tzuendean and Amelia M. McNamara for their generous help on the English version of this chapter.

References

Bischoff, G. C. O., 1976, *Dailyatia*, a new genus of the Tommotiidae from Cambrian strata of S.E. Australia (crustacea, cirripedia), *Senckenb. Lethaea* **57**(1):1–33.

Chen Menge, 1979, Some skeletal fossils from the phosphatic sequence, early lower Cambrian, S. China, *Scientia Geol. Sinica* **1979**(2):188–189.

Cobbold, E. S., 1935, Lower Cambrian fossils from Herault, *Quart. J. Geol. Soc. Lond.* **91**:25–49.

Crimes, T. P., and Jiang Zhi-wen, 1986, Trace fossils from the Precambrian–Cambrian boundary candidate at Meishucun, Jinning, Yunnan, China, *Geol. Mag.* **123**(6):641–649.

Daily, B., 1956, The Cambrian of S. Australia, in: *El sistema Cambrico, su paleogeografia y el problema de su base*, 20th International Geological Congress, Mexico, Part 2, pp. 91–147.

Duan Cheng-hua, 1983, Small shelly fossils from the lower Cambrian Xihaoping Formation in the Shennongjia district, Hubei province—Hyoliths and fossil skeletons of unknown affinities, *Bull. Tianjin Inst. Geol. Min. Res.* **7**:143–188.

Endo, R., and Resser, C. E., 1937, *The Sinian and Cambrian Formation and Fossils of S. Manchuria*, Manchurian Science Museum Bulletin 1.

Gao Zhen-jia *et al.*, 1984, *Precambrian Geology in Xinjiang, China*, Xinjiang Peoples Publishing House.

He Ting-qui and Yang Xianhe, 1982, Lower Cambrian Meishucun Stage of the western Yangtze stratigraphic region and its small shelly fossils. *Bulletin of the Chengdu Institute of Geology* **1982**:69–95, Chinese Academy of Geological Sciences [in Chinese].

He Ting-qui, Pei Fang, and Fu Guang-hong, 1984, Some small shelly fossils from the lower Cambrian Xinji Formation in Fangcheng county, Henan province, *Acta Palaeontol. Sinica* **25**(3):349–357.

Jiang Zhi-wen, 1980a, Monoplacophorans and gastropods fauna of the Meishucun stage from the Meishucun section, Yunnan, *Acta Geol. Sinica* **1980**(2):112–123.

Jiang Zhi-wen, 1980b, The Meishucun stage and fauna of Jinning County, Yunnan, *Bull. Chin. Acad. Geol. Sci. Ser. I* **1980**:75–92.

Jiang Zhi-wen, 1981, Shelly fossil assemblage and the trace fossil in the Meishucun stage, China, in: *Short Papers Section, International Symposium on the Cambrian System*, pp. 98–99.

Jiang Zhi-wen, 1982, Small shelly fossil assemblage and sequences of trace fossils in the Meishucun stage of China, *Yunnan Geol.* **1**(3):256–264.

Jiang Zhi-wen, 1984a, Global distribution of the early shelly metazoans, *Geol. Mag.* **121**(3):185–188.

Jiang zhi-wen, 1984b, Evolution of early shelly metazoans and basic characteristics of Meishucun fauna, *Prof. Pap. Strat. Palaeontol.* **13**:1–22.

Jiang Zhi-wen, 1985a, On the genus *Tannuolina*, *Acta Micropalaeontol. Sinica* **2**(3):232–238.

Jiang Zhi-wen, 1985b, The shelly metazon fauna of early and middle Cambrian in China, *Yunnan Geol.* **4**(4):388–397.

Jiang Zhi-wen, 1987, The nonisochroneity of phosphate deposition in earliest Cambrian of S. Asia, *Yunnan Geol.* **6**(3):201–210.

Jiang Zhi-wen and Huang Zhao-bi, 1986, Middle Cambrian small shelly fauna in Yaxian county, Hainan Island, *Geol. Rev.* **32**(4):323–329.

Jiang Zhi-wen, Luo Hui-lin, and Zhang Shi-shan, 1982, Trace fossils of the Meishucun stage (Lowermost Cambrian) from the Meishucun section, China, *Geol. Rev.* **1982**(1):7–13.

Liang Zhong-fa, 1982, Early and middle Cambrian monoplacophorans from southern northeast China, *Bull. Shenyang Inst. Geol. Min. Res.* **3**:1–6.

Liu Di-yong, 1979, Earliest Cambrian brachiopods from S. W. China, *Acta Palaeontol. Sinica* **18**(5):505–511.

Luo Hui-lin, Jiang Zhi-wen, *et al.*, 1982, *The Sinian–Cambrian Boundary in Eastern Yunnan, China*, Peoples Publishing House, Kunming, China.

Luo Hui-lin, Jiang Zhi-wen, Wu Xi-che, Song, Xue-liang, and Qu-yang Lin, 1984, *Sinian–Cambrian Boundary Stratotype Section at Meishucun, Jinning, Yunnan, China*, Peoples Publishing House, Kunming, Yunnan, China.

Lu Yenhao *et al.*, 1974, Bio-environmental control hypothesis and its application to Cambrian biostratigraphy and palaeozoogeography, *Mem. Nanking Inst. Geol. Palaeontol.* **5**:27–116.

Pei Fang, 1985, First discovery of *Yochelcionella* from the lower Cambrian in China and its significance, *Acta Micropalaeontol. Sinica* **2**(4):395–400.

Qian Jianxin and Xiao Bing, 1984, An Early Cambrian small shelly fauna from Aksu-Wushi Region, Xinjiang, *Prof. Pap. Strat. Paleontol.* **13**:65–90, Geological Publishing House, Beijing [in Chinese].

Qian Yi, 1977, Hyolitha and some problematica from the lower Cambrian Meishucun stage in central and S. W. China, *Acta Palaeontol. Sinica* **16**(2):256–275.

Qian Yi, 1978, The early Cambrian hyolithids in central and S. W. China and their stratigraphical significance, *Mem. Nanjing Inst. Geol. Palaeontol.* **11**:1–38.

Saito, K., 1936, Older Cambrian brachiopoda, gastropoda, etc. from N. W. Korea, *J. Fac. Sci. Imp. Univ. Tokyo* **IV**(3).

Vostokuva, V. A., 1962, Cambrian gastropods from the Siberian Platform and Tajmyr, *Sbornik Stetei po Paleontologii i Stratigrafii* **28**:51–74 [in Russian].

Walcott, C. D., 1913, Cambrian faunas of China, in: *Reserch in China*, Carnegie Institution, Washington, D. C., Vol. 3.

Wang Hong-zhen *et al.*, 1985, *Atlas of the Palaeogeography of China*, Cartographic Publishing House, Beijing.

Wang Yan-geng, Yin Gongzheng, Zheng Shufeng, and Qian Yi, 1984, *The Upper Precambrian–Cambrian Boundary in Guizhou*, Peoples Publishing House, Guizhou, China.

Xiang Li-wen *et al.*, 1981, *The Cambrian System of China. Stratigraphy of China*, Vol. 4, Geological Publishing House, Beijing.

Yin Ji-cheng, *et al.*, 1980, *The Palaeontology and Sedimentary Environment of the Sinian System in Emei-Ganluo Area, Sichuan*, Printer of the Peoples Republic of China, Cheng Duo, China.

Yuan Ke-xing and Zhang Sen-qui, 1980, Lower Cambrian archaeocyatha of central and S. W. China, *Acta Palaeontol. Sinica* **19**(5):380–392.

Yuan Ke-xing and Zhang Sen-qui, 1983, Discovery of the *Tommotia* fauna in S. W. China, *Acta Palaeontol. Sinica* **22**(1):31–40.

Yu Wen, 1979, Earliest Cambrian monoplacophorans and gastropods from W. Hubei with their biostratigraphical significance, *Acta Palaeontol. Sinica* **18**(3):233–266.

Zhang Sen-Qui, 1982, Oldest trilobite in the world and correlation of some archaeocyathid assemblage, *Acta Palaeontol. Sinica* **21**(5):516–525.

Zhang Wen-tang, 1987, Early Cambrian Chengjiang fauna and its trilobite, *Acta Palaeontol. Sinica* **26**(3):223–235.

Zhang Wen-tang and Hou Xian-guang, 1985, Preliminary notes on the occurrence of the unusual trilobite *Naraoia* in Asia, *Acta Palaeontol. Sinica* **24**(6):591–595.

Chapter 10

The Early Cambrian Radiation of Arthropods

DEREK E. G. BRIGGS and RICHARD A. FORTEY

DEREK E. G. BRIGGS • Department of Geology, University of Bristol, Bristol BS8 1RJ, England. RICHARD A. FORTEY • Department of Palaeontology, Natural History Museum, London SW7 5BD, England.

Origin and Early Evolution of the Metazoa, edited by Jere H. Lipps and Philip W. Signor. Plenum Press, New York, 1992.

1. Introduction

The Phylum Arthropoda includes the overwhelming majority of living animal species, and arthropods are among the earliest body fossils of which we have record. Their importance as members of all the major biotic communities throughout the Phanerozoic would be hard to overestimate. Arthropods have variously become deposit feeders, filter feeders, scavengers, predators, or parasites. Their generally small size means that they can derive a livelihood from any environment, from the interstitial habitat between sedimentary particles to the nutrient-poor desert. They can occupy positions low in the food chain as direct plankton feeders, such as copepods, or they can be important predators, limited only by the fact that arthropod construction places an upper limit on size. Arthropods are united by having an exoskeleton, which is molted to allow size increase. The exoskeleton is usually made of chitin, but with calcium carbonate, or even calcium phosphate, as an important component in some groups. The characteristically jointed limbs give the phylum its name, but there are other uniting characters, such as the possession of a hemocoel, and the common occurrence of compound eyes. There is considerable debate about whether the arthropods constitute a monophyletic group. They share bilateral symmetry and serial segmentation with a larger group including Annelida.

The fossil record of arthropods is, in general, somewhat sporadic. Most arthropods are small, and they can be overlooked even when preserved. Many are fragile, and their hard parts do not survive the attentions of scavengers or the processes of sedimentation *post mortem*. Even when sedimentary conditions are favorable, their exoskeletons tend to disarticulate. The groups with the best fossil record are those marine or freshwater arthropods with mineralized cuticles: trilobites, cirripedes, decapods, ostracodes. Even among the trilobites, of which thousands of species have been described from dorsal exoskeletons, there are less than 20 in which the unmineralized appendages are preserved in the fossil state. If this proportion is an approximate measure for the marine arthropods as a whole, it suggests that we may know as little as one in a thousand of the unmineralized species as fossils. Conway-Morris (1986) estimated that about 14% of the total Burgess Shale fauna (genera) had hard parts capable of being preserved in a "normal" Cambrian fossil fauna. Most of our knowledge of unmineralized forms comes from a few exceptional preservations, Konservat-Lagerstätten. There is no guarantee that such special deposits provide a representative sample of the arthropods of the past: for example, they may be biased toward organisms which could live in dysaerobic environments. In spite of these restrictions, it is clear that by the Cambrian, arthropods had already diversified into a number of distinct types, including representatives of the major living groups. Other Cambrian arthropods are more puzzling and are not readily compared with living taxa, and the problem of interpreting their relationships is one of the themes of this chapter. All the Cambrian arthropods were, of course, marine, but there is evidence that the arthropods were the first animals to accompany the plants in the colonization of the land. This adaptive breakthrough stimulated the greatest radiation in the phylum, taking the number of species in the living fauna as a measure of diversification. However, it is also the case that an earlier diversification must have taken place to account for the variety of arthropods present in the Cambrian. This diversification was the more important in establishing the major arthropod clades.

2. Taphonomy

It is first necessary to discuss the factors that influence the arthropod fossil record, because we see the Cambrian history of the group only through a taphonomic filter.

The average Cambrian shale or limestone yields only the hard parts of trilobites. There is no reason to suppose that other arthropods were not living in the same paleoenvironments—it is just that they are not preserved. The fossilization potential of unmineralized cuticle is low. Even for the trilobites, there are important taphonomic influences on what is recovered from the rocks. Localities which yield whole, articulated specimens are rather rare: many Cambrian species are described from cranidia alone. The fragmentary nature of many trilobite occurrences is not only the result of the disarticulation and sorting of dead animals; many of the fossils are exuviae. It is not known how many times trilobites molted between the larval stage (protaspis) and the largest adult. Hunt (1967) recognized seven or eight molts in an agnostid with an adult size of less than 1 cm, and it is certain that large trilobites, such as paradoxidids, which often attain a length of 20 cm or so, must have had many more molts than this. So the trilobites generated several times more potential fossils than the standing population during the course of their ontogeny. Such exuviae were readily reworked by currents to form coquinas. In shales which accumulated under quieter conditions, however, it is not unusual to find the various parts of the trilobite exoskeleton associated together in a molt configuration. Several different arrangements of such exuviae have been described in Cambrian trilobites (Henningsmoen, 1975; McNamara and Rudkin, 1984). The commonest technique for molting resulted in the inversion of the free cheeks; the "soft-shell trilobite" then crawled away through the anterior of the old carapace leaving the molted thorax and pygidium right way up. So far, as we know, all Cambrian arthropods grew by molting at more or less regular intervals, which means that absolute size is a character which must be used with caution in systematics. However, the majority of species of unmineralized arthropods in Cambrian Konservat-Lagerstätten appear to be recovered as dead animals rather than molts, and this can probably best be explained by the peculiar conditions necessary to preserve soft-bodied faunas.

Many of the Cambrian deposits which yield trilobite remains accumulated on shallow-water shelves. The Cambrian included times of widespread immersion of cratonic areas, which produced patterns of facies belts related to the paleocontinental configurations of the time (Palmer, 1972; Lu, 1974). Each of these belts tended to have genera endemic to it. The occurrences of Konservat-Lagerstätten are not randomly distributed through these belts, but tend to occur in deeper water and more exterior sites where transport and rapid burial in poorly oxygenated conditions pertained for the preservation of unmineralized cuticle. Thus the soft-bodied faunas may only preserve a part of the whole environmental profile of the Cambrian marine realm, although some examples from the Lower Cambrian represent more shallow water (Conway Morris, 1989). Aglaspids were apparently part of the inshore fauna of the Upper Cambrian, otherwise typified by dikelocephalacean trilobites. Cambrian Konservat-Lagerstätten occur at low paleolatitudes (Conway Morris, 1985, Fig. 1), and we simply do not know whether the faunas at high paleolatitudes were comparable or different.

The preservation of unmineralized arthropods requires rapid burial under anaerobic conditions. Such conditions are found in the oceanic oxygen minimum layer where it abuts the continental shelf, or within basins having a high input of organic material in the bottom sediments. The decay of the soft parts may have stimulated authigenic mineralization replacing or surrounding soft tissues. The major Cambrian Konservat-Lagerstätten fall into two categories which are distinct in terms of both taphonomy and the nature of the arthropod fauna: the Burgess Shale type and the "Orsten."

Burgess Shale-type faunas are now known from many locations and the similarities between them extend beyond the arthropod component (Conway Morris, 1989). Conway Morris *et al.* (1987) referred to a "broadly comparable Cambrian biota with a distinct identity" for faunas of this type having taxa in common (Table I) even if they differ in relative abundance from one locality to another. The major examples are as follows (Conway Morris, 1985, 1989):

1. Chengjiang Fauna, Chiungchussu Formation, Eastern Yunnan, southwestern China. This fauna is the oldest Konservat-Lagerstätten known with a substantial number of lightly skeletized arthropods; some 17 such taxa (excluding bradoriids) have been described thus far (Hou and Bergström, 1991). It occurs in the basal part of the *Eoredlichia–Wutingaspis* Zone of the Early Cambrian Chiungchussu Stage (Zhang and Hou, 1985) and is known from two localities, Chengjiang and Meishucun (Hou and Sun, 1988) about 40 km apart. It is currently under intensive study and a number of papers have treated the arthropods (Zhang and Hou, 1985; Hou, 1987a–c; Hou and Sun, 1988; Hou *et al.*, 1989; Hou and Bergström, 1991).

2. Buen Formation, Peary Land, northern Greenland. The fauna is mid-to-late Early Cambrian in age, and is thought most likely to occur within the lower part of the *Bonnia–Olenellus* Zone, although a position within the underlying *Nevadella* Zone is conceivable (Conway Morris *et al.*, 1987). Five lightly skeletized arthropod taxa have been described so far from a preliminary sample of material (Conway Morris *et al.*, 1987), but more await description.

3. Kinzers Formation, Pennsylvania, United States. The fauna occurs in the early Cambrian *Olenellus* Zone. The arthropod component is dominated by trilobites, but is otherwise not particularly diverse. Lightly skeletized forms are confined to *Tuzoia* and two other rare bivalved arthropods, and tergite fragments of an unidentified form.

4. Emu Bay Shale, Kangaroo Island, South Australia. The fauna is late Early Cambrian in age, but has not been assigned to a zone. A variety of trilobites occurs (e.g., Conway Morris and Jenkins, 1985). Glaessner (1979) also recorded *Isoxys communis, Tuzoia australis,* and an unidentified "crustacean."

5. Stephen Formation, British Columbia, Canada. This includes the most diverse Cambrian fauna of lightly skeletized and soft-bodied invertebrates, that of the Phyllopod Bed, Burgess Shale, from which about 40 lightly skeletized arthropods are known. It occurs in the *Bathyuriscus–Elrathina* Zone (Middle Cambrian). The Burgess Shale fauna has been the subject of recent investigation [major reviews relevant to the arthropod component are provided by Briggs and Whittington (1985), Conway Morris (1986), and Whittington (1985)]. Several other localities in the Stephen Formation represent different assemblages (Collins *et al.*, 1983), most notably that on Mount Stephen (Briggs and Collins, 1988), which occurs in the preceding *Glossopleura* Zone.

TABLE I. Occurrence of Arthropods in Major "Burgess Shale-Type" *Konservat-Lagerstätten*

	British Columbia	Utah	China	Greenland	Pennsylvania (Kinzers Fm.)	South Australia (Emu Bay)
Acanthomeridion			×			
Actaeus	×	?				
Alalcomenaeus	×	×	?			
Aysheaia	×	×				
Branchiocaris	×	×	?			
Burgessia	×					
Canadaspis	×	×				
Carnarvonia	×					
Ecnomocaris		×				
Emeraldella	×	?				
Fuxianhuia			×			
Habelia	×					
Helmetia	×					
Houghtonites	×					
Hurdia	×					
Isoxys	×		×	×		×
Jianfengia			×			
Kuamaia			×			
Leanchoilia	×	?		cf.		
Molaria	×					
Mollisonia	×	×				
Naraoia	×	×	×			
Odaraia	×		?			
Perspicaris	×	?	?			
Plenocaris	×					
Priscansermarinus	×	?				
Proboscicaris	×	×				
Pseudoarctolepis		×	cf. *Vetulicola*			
Retifacies			×			
Rhombicalvaria			×			
Sanctacaris	×					
Sarotrocercus	×					
Sidneyia	×	?			cf.	
Skania	×					
Sunella				×		
Tegopelte	×			cf.		
Thelxiope	×					
Tuzoia	×	×			×	×
Urokodia			×			
Utahcaris		×				
Waptia	×			cf.		
Yohoia	×					

6. Wheeler, Marjum, Spence Shale Formations, Utah, United States. Middle Cambrian localities in Utah have so far yielded about 20 lightly skeletized arthropods (Briggs and Robison, 1984; Robison, 1984; Robison and Richards, 1981; Conway Morris and Robison, 1988; Robison, 1991). They occur in the early Middle Cambrian Spence Shale (*Peronopsis bonnerensis* Assemblage Zone = *Glos-*

sopleura Assemblage Zone) of northern Utah, and the late Middle Cambrian Wheeler and Marjum Formations (*Ptychagnostus atavus* and *P. punctuosus* Interval Zones, respectively) of west-central Utah.

The Burgess Shale fossils are preserved in aluminosilicate films, but at least in some cases kerogenized organic carbon is also present (Butterfield, 1990). Limbs stand out from the surrounding rock because of a difference in color and reflectivity. How this preservation came about, and why it is not more common, have not yet been explained satisfactorily. It is clear, however, that decay was inhibited, possibly by the intimate association of clay particles (Butterfield, 1990). Pyrite does occur in the Burgess Shale (Conway Morris, 1986), but it appears to have been confined mainly to hard parts. The exoskeleton of the trilobite *Olenoides* was originally calcium carbonate, and there is no evidence that it suffered dissolution. The cuticle was evidently sufficiently robust to fracture under later overburden pressure (figures in Whittington, 1985). The exoskeleton has been largely replaced by chlorite and illite (Whittington, 1980). The Burgess animals were engulfed in periodic sediment slides into the anoxic basin, which left them in random orientation with respect to the subsequent compaction of the sediment. Now this is reflected in the varied foreshortening of the fossils depending on their original orientation with respect to bedding. Some are flattened dorsoventrally, some laterally, others are even vertically flattened, head-to-tail, as it were. This was first elegantly described by Whittington (1971; see also Whittington, 1985) in *Marrella*.

In the Upper Cambrian anthraconitic "Örsten" of southern Sweden the fossils are from nodular phosphatic limestones, and mineral replacement consists of calcium phosphate. The fauna of the Örsten [reviewed by Müller and Walossek (1985b)] includes ostracodes and trilobites (some with preserved appendages) and some ten genera of tiny crustaceans and crustacean-like arthropods (Walossek and Müller, 1990). Examples have been collected from at least five trilobite zones in the Upper Cambrian. The small size and phosphatic mineralization of the fossils indicates a taphonomic history very different from that of the Burgess Shale-type faunas. The replication of tissue was so rapid that it may almost have proceeded *in vivo* (Müller and Walossek, 1985a,b). Such preservation is three-dimensional, and allows even the finest setae and sensory organs to be seen. In some cases of phosphatic preservation the mineral precipitates onto the cuticle rather than replacing the tissue itself. In the Upper Cambrian Örsten the size of the fossils recovered is invariably small, 1–2 mm at most, and it is possible that in this case phosphatization works well only on animals of small size, which would provide an additional taphonomic filter to what we see of the fauna. It is not known whether the predominance of crustaceans and very low proportion of "problematic" arthropods represent a real contrast between the arthropods of the late Cambrian and those which preceded them (Briggs, 1985). Similarly phosphatized Cambrian arthropods have been discovered in a large area of southern Sweden, in erratic boulders in northern Germany, in a borehole in northern Poland, and even in southern Britain.

To summarize, the Cambrian record of arthropods is dominated by the hard parts of trilobites; we may know only 1% or even less of the total arthropod fauna from fossils. Nor do the Konservat-Lagerstätten which give us our knowledge of unmineralized arthropods necessarily sample the whole range of habitats likely to have been colonized by Cambrian arthropods. Rapid burial under anoxic conditions seems to be necessary for the preservation of unmineralized cuticle. The correct circumstances for preserving unmineralized arthropods are not only sporadic in the Cambrian paleogeography, but are also scattered through the Cambrian geological column (Conway Morris, 1989). What we know about the early history of arthropods must be only a small part of what there is to know. This caution should be borne in mind in any speculation about the early diversification of the group.

3. Cuticle

The acquisition of the sclerotized exoskeleton was a key event in the process of "arthropodization"—whether or not this process happened more than once in phylogeny. The cuticles of unmineralized living arthropods share many similarities. The major part is composed of a nitrogenous polysaccharide ("chitin"), the long-chain-like molecular structure of which accounts for much of the cuticle's strength and flexibility. The cuticle is usually divided into several layers: a thick lamellar endocuticle, usually thinner exocuticle, and a very thin, exterior, nonchitinous, waxy epicuticle (terrestrial arthropods). The cuticle is penetrated by ducts for sensory organs, and by minute pore canals which are implicated in the hardening of the cuticle after molting. Wherever they are known in living arthropods mineralized cuticles are secondarily derived from chitinous ones, and there is no reason to suppose that anything else pertained in the Cambrian organisms that concern us here. If the Crustacea, Chelicerata, and Trilobita had a common arthropod ancestor, it is likely to have had a chitinous exoskeleton.

The cuticle of the unmineralized arthropods from the Burgess Shale was presumably of this type. The preservation does not retain the fine structure, but the fine phosphatized preservation of the Upper Cambrian Örsten means that many more details can be appreciated. The cuticles of the small arthropods from that deposit are very thin (of the order of 5–10 μm thick); some specimens show wrinkled carapaces that have been interpreted (e.g., Müller, 1982) as the "soft-shelled" condition before cuticle hardening has proceeded far after molting. Details of the fine pores in the dorsal surface of the cuticle are preserved in several Örsten species, showing that these Cambrian arthropods had similar kinds of sensory organs to those of living marine arthropods. Because of the low fossilization potential of chitinous cuticle, our knowledge of these early arthropods is largely limited to Konservat-Lagerstätten. However, the cephalic carapaces of phyllocarid crustaceans such as *Hymenocaris* are not uncommon fossils (although little studied) especially in deeper-water shales of Cambrian age. Their relative abundance may indicate that they were comparatively robust compared with other chitinous arthropod cuticles.

Some Cambrian arthropods had cuticles which were unique in the history of the phylum: they were composed of calcium phosphate. This is of particular interest in view of the common occurrence of phosphatic shells in other phyla in the Cambrian. There are no Recent arthropods with a cuticle composed entirely of calcium phosphate, although a phosphatic component is present in some cuticles (e.g., the crab *Cancer*), especially in their outer part. Phosphatic cuticle has been demonstrated in two groups in the Cambrian: the primitive ostracod group Phosphatocopina (Müller, 1979) has phosphatic valves, and the aglaspid dorsal exoskeleton (Briggs and Fortey, 1982) is similarly apatitic. These two groups of arthropods are most unlikely to be closely related. It is probable, therefore, that the phosphatic cuticle was derived from separate chitinous ancestors in these two groups. The cuticle of the phosphatocopines was less than half the thickness of that of the aglaspids, but the latter are large arthropods. Sections through the aglaspid cuticle clearly show the laminated endocuticular layer. It is tempting to assume that phosphatic cuticles were at some kind of selective disadvantage compared with calcareous ones, because subsequent groups of arthropods that became mineralized invariably used calcium carbonate for the purpose. However, Runnegar and Bengtson (1990) comment that of the early Cambrian clades (all phyla) "42% of the carbonate ones survive until the present, as compared to 25% of the phosphatic ones. . . . Both categories include clades which are today very successful and diverse . . . . there is no reason to assume that the shell mineral was the particular factor that decided the survival of each of the early lineages."

The trilobite cuticle is composed of calcite with a subsidiary organic matrix. Most of the histological studies of the trilobite exoskeleton have been carried out on taxa younger than Cambrian in age, but there is sufficient similarity between them all to suggest a common pattern. Two layers appear to be present in many species, and where the outer, thin exocuticle is not seen on some specimens this is usually attributed to deficiencies in preservation—but perhaps it was lacking in certain species. The thick endocuticle is composed of fibrous calcite, but a laminated structure of typical arthropod type is not always present. The outer cuticular layer often comprises prismatic calcite; Teigler and Towe (1975) recorded a phosphatic outer layer in two species, and a "pigmented" outer layer in the Cambrian trilobite *Paradoxides*. The optical axes of the calcite crystals in the cuticle are more or less normal to the surface, and this is invariably the case in the lenses of the eyes, in which such *c*-axis orientation is an intrinsic part of the optical mechanism. Eyes of Cambrian trilobites are rarely seen because the visual surface was surrounded by a circumocular suture. If *Olenellus* is taken to represent the primitive trilobite as far as cuticular structure is concerned, then it, too, had an inner fibrous and outer prismatic layer in the cuticle (Teigler and Towe, 1975). Many small trilobite growth stages show a pattern of "cell polygons" on the surface, a fine reticulum which has been taken to represent the pattern of epidermal cells secreting the cuticle. These are often shown on protaspides, where they are preserved in the original calcite (Fortey and Chatterton, 1988, pl. 18, Fig. 7), but may not survive phosphatization or silicification. A much coarser reticulum is present on certain Cambrian olenellids such as *Wanneria*, but its scale is such that it is not likely to represent the traces of secretory cells.

The trilobite cuticle can get very thick—more than 1 mm in some large, inner-shelf forms from the Ordovician. Early Cambrian forms seem to have had much thinner cuticles, but there is no detailed information on this. Agnostid and olenid trilobites from later in the Cambrian certainly retain very thin cuticles. It is not uncommon to find redlichiids with the cephalic shield in relief and the posterior part of the thorax and pygidium flattened, which suggests that the cuticle was progressively thinner along the length of the animal. Trilobite cuticle can be perforated by a variety of canals suggesting the presence of sensory organs: it is very likely that there were tactile or chemosensory setae on the dorsal surface, for example. Even in the Early Cambrian, redlichiids have ventral terrace ridges on the cephalic doublure (Hupé, 1953, pl. 9, Fig. 9). These fine, cuesta-shaped ridges are typical of many later trilobites, and the scarp slopes have been claimed to be associated with regular, canal-like perforations through the cuticle.

4. Trilobites

4.1. Early Cambrian Stratigraphic Record

Trilobites are the only arthropod group for which the record of early diversification can possibly be read directly from the sequence of fossils. Indeed, Cambrian trilobite taxonomists have tended to use stratigraphic occurrence as an implicit part of their taxonomic treatment. Trilobites occur in nearly all the sections that span the Lower Cambrian, and provided one of the first methods for biotratigraphic subdivision of these rocks. The type Cambrian succession is that of Wales and the Welsh borderlands (Rushton, 1974), and the fullest Early Cambrian succession there is that of the Comley area in the county of Shropshire. However, classic though it is, the Comley section is not well exposed, is certainly condensed, and is probably incomplete; better exposed and more complete sections are known from other parts of the world. In these successions the trilobites do *not* appear among the first fossils; they are underlain by assemblages of small shelly fossils usually assigned to the Tommotian Stage (Cowie and Brasier, 1988). Because the first trilobites that appear are often different in the various sections—olenelloids in some, redlichioids in others—there is no doubt that their sudden appearance does not reflect the origin of the group. The trilobites as a whole must have had an earlier history, and the appearance of the fossils in any given section is therefore a matter of extrinsic influences—such as a change in sedimentary facies, or the sufficient reduction of current activity to ensure the survival of the usually fragile trilobite exoskeleton.

Important sections recording the early history of the trilobite exoskeleton are found in Morocco, Sweden, the Siberian platform, North America (especially California), and several provinces of China. Most of these are "platform" successions. Figure 1 summarizes the record of trilobites in some of these sections. In most a single genus appears initially, to be followed by a rather rapid increase to four or five. However, this general similarity of pattern is not matched by a comparable succession of taxa in the sections. Five major groups of trilobites are concerned: Olenellida, Redlichiida, Eodiscina, Corynexochida, and Ptychopariina. All of these appear in the earlier part of the Cambrian in one or more of the sections, which implies that the major taxonomic differentiation of the trilobites happened before or within the Lower Cambrian, depending on the view taken of the completeness of the record. The earliest trilobites to appear in the most complete successions in North America, Siberia, and most sections in the Anti-Atlas, Morocco, are olenellids belonging to the genera *Profallotaspis* or *Fallotaspis*. In China (and Australia) such olenellids are not known, and the earliest trilobites are redlichiids (*Parabadiella*). Hence, even in the earliest Cambrian there is paleogeographic differentiation of trilobite faunas: these major divisions

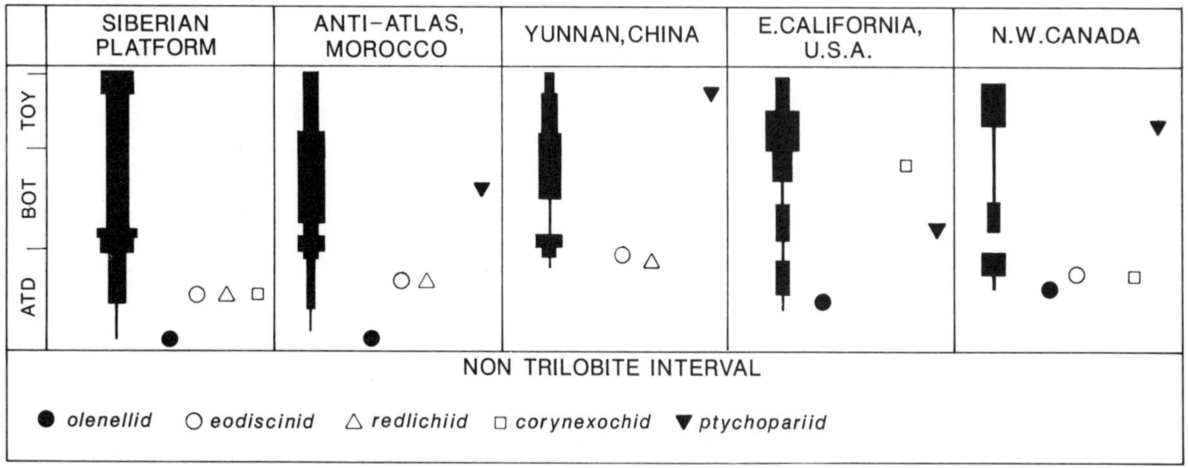

Figure 1. Patterns of diversification of trilobites in basal Cambrian successions in important biogeographic regions, showing horizon of first appearance of major taxa. Where they occur, olenellids appear first (which is consistent with the interpretation of this group as including the most primitive trilobites), but there is no generally applicable order of appearance of other orders of trilobites. Correlation between sections is still controversial; here after Brasier (1989).

were dubbed respectively the olenellid and redlichiid faunal "realms" by Cowie (1971). The redlichiids so common in China and Australia are not present in the North American sections. Some workers believe that the first appearance of trilobites in the North American olenellid realm may antedate their first appearance in the Chinese redlichiid realm, but the opposite view has also been maintained. There are many other indications of endemicity; for example, the Chinese sections have yielded a variety of endemic eodiscids as well as redlichiids. This provinciality has two important implications for the early evolutionary history of trilobites:

1. Worldwide correlation becomes difficult, causing debate about the biostratigraphy/chronostratigraphy, which means that it is difficult to compare time of appearance of taxa in different provinces.

2. If the trilobites are a monophyletic group, as they appear to be (Fortey and Whittington, 1989), there must have been prior evolutionary history of which we know nothing, to account for dispersal and vicariance events.

In Morocco and Siberia there are mixed olenellid and redlichiid faunas. Although the olenellids apparently usually predate the redlichiids, Sdzuy (1978) has described a Moroccan section in which the earliest fallotaspid olenellids (*Eofallotaspis*, claimed to be ancestral to *Fallotaspis*) are accompanied by trilobites with functional opisthoparian facial sutures which are presumably redlichiid (*Hupetina*). Hence the stratigraphic evidence provides only equivocal support for the contention that olenellids are more primitive than redlichiids. In the most complete sections the appearance of eodiscids postdates the earliest appearance of olenellids or redlichiids, but in some Chinese sections, and in Sardinia (Rasetti, 1972), redlichiids and eodiscids appear together at the local base of the Cambrian. The appearance of eodiscids always predates the appearance of the first agnostids, of which they are the primitive sister group. The first ptychopariids may appear before (California) or after (Siberia) the first Corynexochida, but both have first occurrences within a zone or two of the appearance of the trilobites as a whole. The major Cambrian groups of trilobites all appear low in the Cambrian, and olenellids and/or redlichiids predate the other groups. The order of appearance of corynexochids, eodiscids, and ptychopariids is not always consistent from section to section. Stratigraphic sequence does not present an unequivocal narrative of early trilobite evolution, and there must have been events in that evolution of which we have no record in the rocks.

4.2. Initial Evolutionary Radiation

If the stratigraphic record does not provide an account of the *relationships* of the early Trilobita, it is nonetheless obviously true that within a zone or two of their first appearance there is already a remarkable diversity of trilobites belonging to groups with a long subsequent history (Fig. 2). The olenellids lack dorsal facial sutures. So do many later trilobites, but in these cases the lack of functional sutures is known to be the result of secondary loss. However, it has been contended that in olenellids this absence is primitive (Bergström, 1973). This is supported by the fact that olenellids lack such sutures throughout their ontogeny. We would also expect the sister group of the trilobites as a whole to lack dorsal sutures and it

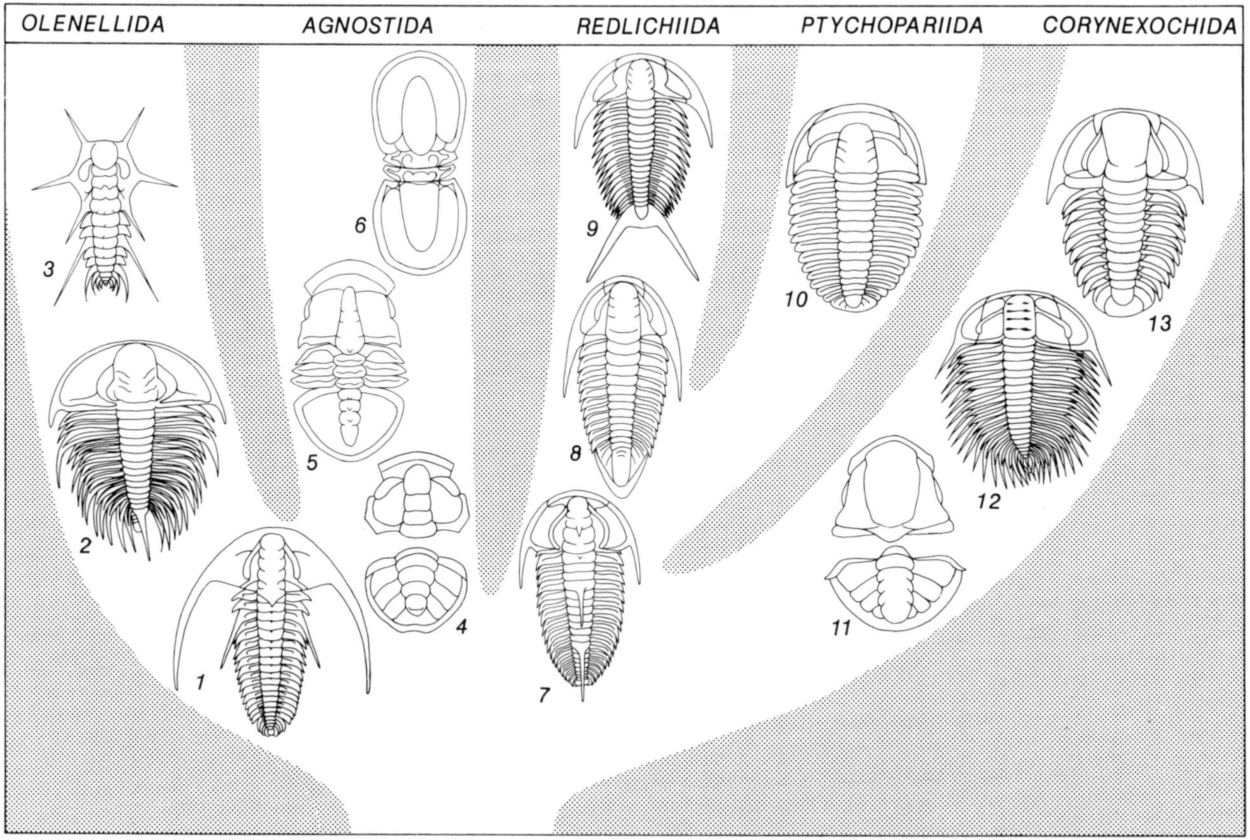

Figure 2. Early radiation of the Trilobita in terms of morphological diversity of major taxa in the Early Cambrian. Agnostida, Ptychopariida, and Corynexochida are probably natural groups, but Redlichiida and Olenellida as conventionally employed may include species ancestral to other clades. (1) *Fallotaspis*, (2) *Olenellus*, (3) *Olenelloides*, (4) *Sinodiscus*, (5) *Hupeidiscus*, (6) *Eoagnostus*, (7) *Redlichia*, (8) *Parapaokannia*, (9) *Yunnanaspidella*, (10) ptychopariid, (11) *Bonnia*, (12) *Lancastria*, (13) *Poliella*.

is not unreasonable to suppose that the most primitive trilobites—as defined by the other autapomorphies of the group, which olenellids share—might retain the primitive condition of the ecdysial sutures. This would identify the acquisition of dorsal sutures as a synapomorphy of all "higher" trilobites, and would place the redlichiids as the primitive sister group of all other nonolenellid trilobites. It is not in question that the possession of dorsal sutures is a character unique to trilobites among the arthropods considered here.

The interpretation of olenellids as the most primitive trilobites may be doubted by those who believe that the lack of dorsal sutures in the group is secondary (Hupé, 1953)—the result of ankylosis in the dorsal shield in the manner of some later trilobites such as harpids or brachymetopids. Nonetheless, there are other features which are consistent with a view of olenellid primitiveness (Fortey and Whittington, 1989). On the early growth stages the fixed cheeks inside the eyes show distinct segmentation (Palmer, 1957), which is a primitive feature, and loss of such segmentation is characteristic of the early growth stages of all "higher" trilobites. Also, the thoracic articulation is characterized by a pleural flange which extends beneath the segment in front, a feature which we have noted in several of the primitive arthropod sister groups in our cladistic analysis. So we think it likely that the olenellids are genuinely primitive trilobites (including the condition of the sutures) which have not yet acquired some of the advanced characters of redlichiids and all later trilobites. This raises the question of whether the olenellids are trilobites at all! They have been regarded (Lauterbach, 1980) as stem group Arachnata, more closely related to *Limulus* and its allies, especially on the basis of the division of the thorax into a prothorax and opisthothorax, which were homologized with the main body divisions in arachnates. We do not dispute that Arachnata and trilobites ultimately constitute a monophyletic group (below). Lauterbach did not fully list the synapomorphies linking the olenellids with the rest of the trilobites—which render his rejection of a closer linkage of the olenellids with the rest of the trilobites than with chelicerates the less parsimonious estimate of relationships. These characters include: (1) compound eyes composed of calcium carbonate lenses (unique in the animal kingdom); (2) eye ridges (a structure not known in other arthropods having dorsal eyes on the carapace, but ubiquitous in all primitive trilobites, although secondarily lost in advanced forms); (3) the pygidium, a dorsoventrally flattened shield including more than one segment and underlain by doublure (olenellid pygidia have been misinterpreted as comprising a true telson, but, although they are small, they actually include two or more axial rings); (4) a rostral plate on the venter; (5) a distinctive calcareous cuticle structure, otherwise matched only by advanced crustaceans. Lauterbach did not realize that olenellids also have typical antennae, a primitive character, but one which is lacking in Arachnata *sensu* Lauterbach. We conclude that the trilobites are a natural group, and olenellids their most primitive known representatives.

The olenellids underwent a modest diversification, with at least 30 genera in the Lower Cambrian. Although one might assume that the presence of large numbers of thoracic segments is a primitive character, there is wide variation among olenellids: most have more than 17, but *Olenellus* (*Olenelloides*) only has nine. This form is regarded as having originated progenetically (McNamara, 1983). Although some olenellids have the thorax divided into pro- and opisthothorax, others do not; many have the third thoracic segment macropleural and prolonged into a long spine, but others have a normally tapering thorax. *Olenellus* and allied genera have a unique kind of reticulate surface sculpture. The hypostome position varies in the group, as does the point of origination of the genal spine. The significance of these characters has not been assessed phylogenetically, but it is likely that some of the group will prove to be closer to redlichiids than others.

Redlichiids (and "higher" trilobites) acquired functional ecdysial sutures on the dorsal surface. The circumocular suture which allowed the corneal surface of the eye to be molted separately was certainly present on redlichiids, but may have been present on olenellids also. Redlichiids acquired normal trilobite thoracic articulation, and terrace ridges on the cephalic doublure. More than 100 redlichiid genera have been described, many of them from China, in recent years. Primitive redlichiids, such as *Redlichia* itself, retain a strong likeness to olenellids. Even allowing for "splitting," there is clearly a Cambrian diversification of the group of some magnitude (Fig. 2). Redlichiids include the first trilobites to adopt certain morphological specializations which are subsequently repeated many times in trilobite history. Genal spines may become reduced; eyes may migrate across the genae away from the glabella; glabellar furrows may become deep and cross the glabella, or become effaced; caudalization may proceed further to produce macropygous forms; various types of surface sculpture may be prominent; anterior branches of facial sutures may become highly divergent. It is likely that the ancestors of agnostids, corynexochids, and ptychopariids would be classified currently in Redlichiida, which means that it may prove to be a paraphyletic group.

The Order Agnostida comprises one of the most distinctive groups of Cambrian trilobites. The description of the unmineralized appendages of *Agnostus* (Müller and Walossek, 1987) means that we know a great deal about agnostid morphology, and it is perhaps not surprising that there are many autapomorphies peculiar to the group. Agnostids appear first high in the Lower Cambrian. They are usually regarded as related to eodiscids, which, as mentioned above, first appear low in the Lower Cambrian. Eodiscids are less specialized than agnostids. Many retain both eyes and dorsal sutures (agnostids are secondarily blind), and normal trilobite thoracic structure, but like agnostids they are small isopygous trilobites with few thoracic segments. Agnostids have two thoracic segments with a distinctive pleural structure, and the articulating half ring on the first segment is lost. Because of their peculiarities, agnostids + eodiscids are often regarded as "separate" from the rest of the Trilobita, but if the presence of dorsal sutures is a synapomorphy as we suggest, they belong in the postolenellid trilobite clade. It has been suggested that agnostids had a planktic mode of life (Robison, 1972), and this might account for their specializations. The agnostid–eodiscid clade probably arose progenetically from a redlichiid ancestor. Because thoracic segments are normally released from the front of the pygidium during trilobite ontogeny, the small size, the few thoracic segments of Agnostida, and their large pygidium with numerous segments seem to be explicable by "arresting" development at an early stage, with accompanying precocious sexual maturity.

Whatever the explanation of their origin, the Agnostida quickly diversified in the lower part of the Cambrian. For example, there are some 13 eodiscoid genera in the Taconic fauna described by Rasetti (1967), and a comparable number from the Early Cambrian of China—and few genera in common between the two. From an early time the Agnostida show a propensity to be most diverse in marginal or off-shelf situations, which may support a planktic habit. Even accepting this, it is not easy to account for their diversity. Robison (1975) suggested that agnostids with different adult dimensions may have specialized by consuming different particle sizes, a limitation on competition found in living animals having comparable use of resources. Agnostids increased in diversity at the expense of eodiscoids in the mid-Cambrian, and persisted until the end of the Ordovician.

Corynexochida are a major group of medium-sized, mostly macropygous trilobites with a characteristic pestle- or club-shaped glabella, and the rostral plate typically fused to the hypostome. Although the maximum diversity of the group was within the Middle Cambrian, the main morphological radiation was over by the end of the Lower Cambrian. The early corynexochid *Bonnia* is a widespread and characteristic genus in the later Lower Cambrian of North America. Suvorova (1964) recorded no fewer than 19 genera of corynexochids from the Lower Cambrian of the Siberian platform. The appendages of the corynexochid *Olenoides* are known from the Burgess Shale (Whittington, 1975). The presence of caudal cerci is unique among trilobites with known appendages. The spinose inner edges of the coxae on the legs suggested to Whittington that *Olenoides* was "a benthonic predator on small and soft bodied animals as well as a scavenger." This observation is of interest because other authors (Bergström, 1973) have suggested that olenellids, for which appendages are not fully known, were also at least partly predatory in their habits. If these observations are correct, it implies that the primitive life habit of the trilobites as a whole included predation.

The Ptychopariida are the largest group of trilobites. The ancestors of some of the major later trilobite orders are currently classified with ptychoparioids, which is almost certainly a paraphyletic group. Proetida, for example, which include the youngest of all trilobites and all post-Devonian ones, probably originated in the Upper Cambrian from some kind of generalized ptychoparioid. Nonetheless, the definition of the group has been imprecise. We regard the Ptychopariida as including those trilobites in which the hypostome has become detached *in vivo* from the cephalic doublure [natant hypostomal condition of Fortey and Chatterton (1988)]. They have a preglabellar field underlain neither by doublure nor hypostome. Most ptychoparioids also have a tapering glabella, anteriorly somewhat truncate, medially placed eyes well removed from the glabella, and many thoracic segments (which may vary in number even within a genus), and a comparatively small pygidium which is more transverse than that of redlichiids. They remain, as a whole, a remarkably conservative group, and it is possible to find broadly similar morphologies throughout the Cambrian. They diversified mainly in the Middle to Late Cambrian, although early representatives extend to the Lower Cambrian. Certain Protolenidae, currently classified with redlichiids, would be regarded as ptychopariids on the basis of a detached hypostome, and thus extend the record of the group into the earliest zone or two of the trilobite record.

The early trilobite history may be summarized by emphasizing that the major groups all extend far back to the Lower Cambrian, by the end of which they were already highly diverse at the generic level. While the order of their appearance does not help greatly in deducing their phylogenetic relationships, it is still probable that the olenellids, which appear first in many sections, were the most primitive trilobites, and that redlichiids were the primitive sister group of most "higher" trilobites. From very early on trilobites were able to live in a variety of habitats from inshore to open sea. Most of the larger forms were benthic; the Agnostida may have included planktic species. A case can be made for predatory/scavenging habits in the primitive forms.

5. Approaches to the Taxonomy of the Early Arthropods

The evidence of Konservat-Lagerstätten shows that there was a considerable diversity of arthropods in the Early Cambrian in addition to trilobites, testifying to a substantial radiation. How this radiation is described depends on the taxonomic interpretation of these arthropods. The taxonomic treatment is not merely a question of pigeon-holing the taxa; it colors our perception of the nature of this early radiation and what can be deduced on the basis of early fossils belonging to a major group. The taxonomy of these early forms has been approached in three main ways in the last few decades:

1. The "lumping" approach. This is exemplified by Størmer's (1944) treatment of Burgess Shale and other arthropods [included in modified form in Størmer (1959)]. Størmer erected a Subphylum Trilobitomorpha based largely on "appendages of trilobitan type." This subphylum, given equal status with the Chelicerata, included an array of Paleozoic forms, most from the Burgess Shale. The assumption that the Trilobitomorpha is a natural group implied that the early Cambrian witnessed an explosive diversification of arthropods which have little to do with the subsequent history of the group. It emphasized the peculiarities of these early forms rather than their morphological connections with other arthropods.

2. The "multiple polyphyly" approach. This approach was influenced by the work of Manton (1977) and Anderson (1973), who argued separate origins for the major living groups of arthropods. The various Cambrian arthropods were likewise surmised to have had separate "soft-bodied" ancestors (Whittington, 1979, Fig. 2). No phylogenetic links were drawn between the several evolutionary lineages recognized (they were illustrated as parallel lines crossing the Precambrian–Cambrian boundary, a phylogenetic "lawn"). No specific soft-bodied ancestors (e.g., particular annelid "worms") were proposed either; hence there was no indication of whether any one arthropod is more closely related to any other.

3. The cladistic approach. A preliminary cladistic analysis was made by Briggs and Whittington (1981), and Briggs (1983) presented two conflicting cladograms and considered the implications of these results. This approach rejects polyphyly, i.e., it regards those features shared by early arthropods (jointed limbs, biramous appendages, etc.) as sufficient to indicate a monophyletic origin. It diminishes the significance of the peculiarities of individual Cambrian taxa (e.g., the "great appendage" of *Leanchoilia*) in favor of those characters which are shared between taxa. The most important distinction is between characters which are primitive (plesiomorphic) and those which are derived (apomorphic). Such a distinction can either be made from *a priori* assumptions [as did Briggs (1983)], or

by identifying a particular arthropod (e.g., *Marrella*) as exhibiting the primitive character states. Any of the Cambrian arthropods is, potentially at least, a sister group to a living group.

A cladistic approach provides the most constructive way of analyzing the Cambrian radiation of the arthropods, and is adopted here. Approach 1 (above) does not discriminate between primitive and derived character states. The trilobitomorphs are united by symplesiomorphies (essentially a very broadly defined trilobite limb), and they are at best a paraphyletic assemblage. The taxon Trilobitomorpha simply recognizes the presence of many primitive arthropods in the Cambrian. Approach 2 merely pushes the problem of relationships back into the Precambrian (and denies the possibility of solving it by the scientific method). It seems improbable that those characters uniting the known Cambrian arthropods evolved on more than one occasion. Although the acquisition of an arthropod character may have happened more than once, this should not be assumed to have been the case without good evidence.

6. The PAUP Analysis

In this chapter we present the result of a cladistic analysis based on the PAUP (Phylogenetic Analysis Using Parsimony) program (Version 2.4.1). Since this chapter was submitted for publication, the results of this analysis have been published in outline elsewhere (Briggs and Fortey, 1989; Briggs, 1990) and the data base revised and expanded (Briggs, Fortey, and Wills, in preparation). The analysis considered 29 taxa using 46 characters (Tables II and III). Only those taxa preserved in sufficient detail to allow a high proportion of the character states to be coded were selected for inclusion. The phosphatocopine ostracodes were omitted, as their peculiar combination of primitive and derived characters results in a number of different possible placements, and hence increases the number of equally parsimonious trees without greatly clarifying relationships (see discussion below). Representatives of the major clades of living aquatic arthropods were added to the list of taxa in order to establish which Cambrian arthropods appear most closely related to them. Those selected are taxa which have been regarded as the most plesiomorphic: the crustaceans *Hutchinsoniella* (Sanders, 1957) and *Speleonectes* (Schram *et al.*, 1986), and the aquatic chelicerate *Limulus*. *Baltoeurypterus* was also included to represent the extinct Paleozoic eurypterids. In later runs those clades which demonstrated a consistent relationship (the trilobites; *Canadaspis*, *Perspicaris*, and *Odaraia*; *Baltoeurypterus*, *Limulus*, *Sarotrocercus*, and *Yohoia*) were represented by just one of their member taxa, thus reducing the number under consideration and enhancing the resolving power of the program.

6.1. The Data Matrix

The data matrix (Table II) was constructed using, where possible, the morphology of *Marrella* (Fig. 4) as the primitive state or "root" in establishing polarity. The selection of *Marrella* to play this role is based on the position it

TABLE II. Coding of Characters (Table III) Used in the PAUP Analysis[a]

	1	2	3	4	5	6	7	8	9	10	11	12	13	14	15	16	17	18	19	20	21	22	23	24	25	26	27	28	29	30	31	32	33	34	35	36	37	38	39	40	41	42	43	44	45	46
Aglaspis	1	2	1	2	1	1	0	1	1	1	1	0	1	3	0	0	1	1	1	0	0	0	0	0	0	0	0	1	2	1		0	0	0	1	0	1	1	1	0	0	4	0	0	1	0
Agnostus	1	3	1	2	1	2	1	1	1	1	1	1		3	1	0	1	1	1	0	1	0	0	0	1	1	0	1	2	1	0	1	0	1	0	0	1	0		0	0	5	0	0	1	1
Alalcomenaeus	1	1	1	2	0	0	0	1	1	1	0	1		3	0	0	0	0	1	0	1	0	0	0	0	0	0	1	2	1	1	0	0	1	1	0	1	0	2	0	0	4	0	0	1	0
Aysheaia	0	0	0	0	0	0	0	0	0	0	0	0	0	0	0	0	0	0	0	0	0	0	0	0	0	0	0	0	0	0				0	0	0	0	0	0	0	0	0	0		1	0
Baltoeurypterus	1	1	1	2	0	0	0	1	1	1	1	1	0	5	1	0	0	1	1	0	1		0	0	0	1	0	0	0	2				1	2	0	1	0	1	1	0	3	1	0	1	1
Branchiocaris	2	1	1	1	0	0	0	2	1	1	0	0		1	0	0	0	0	0	0	0	0	0	0	0	0	0	1	0	2	1	1	0	0	0	1	0	0	0	0	0	0	0	1	0	0
Burgessia	1	1	1	1	0	0	0	2	1	0	0	0		3	0	0	0	0	0	0	0	0	0	0	1	1	0	1	0	1	1	1	0	1		0	1	0	1	0	0	5	0	0	1	0
Canadaspis	1	1	1	1	0	0	0	2	1	1	0	0		3	0	0	0	0	1	0	0	1	1	1	0	1	1	1	2	1	1	1	0	1	0	0	0	0	0	0	0	2	1	1	1	0
Emeraldella	1	1	1	2	0	0	0	1	1	1	0	0		5	0	0	0	0	1	0	0	0	0	0	0	1	0	1	2	1	1	0	1	0	1	0	1	1	1	1	0	3	0	0	1	0
Habelia	1	1	1	2	1	1	0	2	1		0	0		2	0	0	0	1		0	0	0	0		0	0	0	1		1	1	0	0	1	2	0	1	0	1	0	0	3	0	0	1	0
Hutchinsoniella	1	1	1	2	0	1	0	1	1	1	0	0		4	0	0	0	0	1	0	0	0	1	1	0	1	0	1	2	1	0	1	0	0	0	0	0	0	0	1	1	1	1	0	1	0
Leanchoilia	1	1	1	2	1	0	0	2	1	0	0	0		2	0	0	0	0	0	0	1	0	0	0	0	0	0	1	0	1	1	0	0	1	1	0	1	0	1	0	0	4	0	0	1	0
Limulus	1	1	1	2	1	0	0	1	1	1	1	1	1	5	1	0	1	1	1	0	1		0	0	0	0	0	0	0	2	1	1	0	1	0	1	1	0	1	0	0	5	0	0	1	1
Marrella	1	1	1	1	0	0	0	1	1	1	0	0		1	0	0	1	0	0	0	0	0	0	0	0	0	0	1	0	1	0	0	0	1	0	0	0	0	0	0	0	0	0	0	0	0
Martinssonia	1	1	1	1	0	0	0	2	1	1	0	0		4	0	0	0	0	1	0	0	0	0	0	0	1	0	1	1	1	1	0	0	1	0	0	1	0	0	1	0	5	1	0	1	0
Molaria	1	1	1	2	1	1	0	1	1	1	0	0		3	0	0	0	1	1	0	0	0	0	0	1	0	0	1	2	1	1	0	0	0	1	0	1	0	1	1	0	5	0	0	1	0
Naraoia	1	1	1	2	1	0	1	1	1	1	0	0		3	0	0	0	1	1	0	0	0	0	0	1	0	0	1	2	1	1	0	1	1	1	0	1	0	0	0	0	1	0	0	0	0
Odaraia	2	1	1	1	0	0	0	2	1	1	0	1		4	0	0	0	0	0	0		1	1	1	0	1	0	1	2	1	1	1	0	0	0	0	0	0	0	0	0	0	0	1	0	0
Olenoides	1	3	1	2	1	2	1	1	1	1	1	0	1	3	2	1	1	1	1	0	0	0	0	0	1	1	0	1	2	1	1	0	1	1	2	0	1	0	0	0	1	3	0	0	1	1
Perspicaris	1	1	1	1	0	0	0	2	1	1	0	0			0	0	0	0		0	0	1	1	1	0	1	1	1		2	1	1	0	1	0	0	0	0	0	1	1	2	1	1	1	0
Plenocaris	1	1	1	1	0	0	0	1	1	1	0	0		0	0	0	0	0	0	0	0	0		0	0	1	0	0		1				0	0	0	0	0	0	1	1	3	1	1	1	0
Sanctacaris	1	1	1	2	1	1	0	1	1	1	0	0		5	0	0	0	1	1	0	1	0	0	0	0	1	0	1	2	1	1	0	0	0	1	0	1	0	2	1	0	4	0	0	1	0
Sarotrocercus	1	1	1	2	1	1	0	1	1	1	0	0		1	0	0	0	0	0	0	0	0	0	0	0	0	0	0	0	2	1	0	0	0	1	0	1	0	1	1	0	5	0	0	1	0
Sidneyia	1	1	1	2	0	0	0	1	1	0	0	0		0	0	0	0	0	0	0	0	0		0	0	2	0	1	2	1	1	0	0	0	1	0	1	0	2	0	0	3	1	0	1	0
Skara	1	1	1	2	0	0	0	1	1	1	0	0		4	0	0	0	0	1	0	0	0	0	0	0	1	0	1	2	1	0	0	0	1	0	0	0	0	0	1	1	3	1	0	1	0
Speleonectes	1	1	1	2	0	0	0	1	1	1	0	0		4	0	0	0	0	1	1	0	0	0	1	0	1	0	1	0	1	0	0	0	0	1	0	0	0	0	0	1	0	0	0	0	0
Triarthrus	1	3	1	2	1	2	1	1	1	1	1	0	1	3	2	1	1	1	1	0	0	0	0	0	1	1	0	1	2	1	1	0	0	1	2	0	1	0	0	0	0	0	0	0	0	1
Waptia	1	1	1	1	0	0	0	2	1	1	0	0		4	0	0	0	0		0	0	1	1	0	0	2	0	0		1	1	0	0	0	0	0	0	0	0	1	1	2	1	1	1	0
Yohoia	1	1	1	2	1	2	0	1	1	1	0	0		3	0	0	1	0	0	0	1	0	0	0	0	1	0	0	0	2	1	1	0	0	1	1	1	0	2	1	0	3	1	0	1	0

[a]A blank space indicates that the character could not be coded (i.e., it was inapplicable, not preserved, etc.). Characters were coded with varying degrees of certainty, but space does not permit discussion of details here.

TABLE III. Character States Used in Constructing Table II and in the PAUP Analysis[a]

1. Trunk annulated (0) (the condition in *Aysheaia*, whose position consistently resolves as a primitive sister group of all the other arthropods under consideration), clearly segmented (1), showing no clear correspondence between segments and limbs (2)
2. Exoskeleton soft (0) (the condition in *Aysheaia*), sclerotized (1), calcium phosphate (2); calcium carbonate (3) (both 2 and 3 derive independently from 1)
3. Carapace absent (0) (the condition in *Aysheaia*), present (1)
4. Cephalic shield absent (0) (the condition in *Aysheaia*), present (1), present with tergites (2)
5. Trilobation absent (0), present (1)
6. Pleura absent (0), unfurrowed (1), grooved (2)
7. Pygidium absent (0), present (1)
8. Dorsoventral division of the body absent—hence no possibility of a doublure (0), present with doublure (1), lacking doublure (2)
9. Mouth terminal (0) (the condition in *Aysheaia*), ventral (1)
10. Labrum absent (0), present (1)
11. Eyes absent or "free"/pedunculate (0), on carapace (1) (the polarity between 0 and 1 is uncertain)
12. Median eye absent (0), present (1)
13. Eyes superposition (0), apposition (1) (the nature of the eyes of the Cambrian animals is uncertain—it is treated as constant)
14. Number of cephalic limbs 1 (0), 2 (1), 3 (2), 4 (3), 5 (4), 6 (5)
15. Ecdysial sutures absent (0), marginal (1), dorsal (2)
16. Eye ridges absent (0), present (1)
17. Genal spines absent (0), present (1)
18. Marginal rim absent (0), present (1)
19. Cephalic gnathobases absent (0), present (1)
20. First cephalic appendage uniramous (0), biramous (1)
21. First cephalic appendage antenniform (0), otherwise (1)
22. Antenniform first cephalic appendage "standard" (0), reduced (1)
23. Second cephalic appendage not antenniform (0), antenniform (1) (here "antenniform" is regarded as implying an appendage made up of many similar short podomeres; the second cephalic appendage in *Marrella* is excluded; this concept has the drawback that it also excludes the second antenna in a number of crustacean groups; it is preferable, however, to the alternative, which would involve some sort of subjective definition of antenna)
24. Mandible absent (0), present (1)
25. Gut diverticula absent (0), present (1)
26. Number of trunk tagmata (excluding telson) 1 (0), 2 (2), 3 (2)
27. Malacostracan tagmosis absent (0), present (1)
28. Trunk appendages uniramous (0), biramous (1) (this polarity is based on *Aysheaia* as the root; using *Marrella* would reverse it)
29. Trunk gnathobases absent (0), on some (1), on all (2)
30. Inner ramus lobate (0) (the condition in *Aysheaia*), segmented (1), reduced or absent (2)
31. Outer ramus segmented (0), otherwise (1)
32. Outer ramus bearing narrow filaments (0), otherwise (1)
33. Distal lobe on outer ramus absent (0), present (1)
34. Body flexure lateral and dorsoventral (0), dorsoventral only (1)
35. Postcephalic articulation without overlap (0), with overlapping pleurae (1), articulating half-ring (2)
36. Grasping claw absent (0), present (1)
37. Anus posterior (0), ventral (1)
38. Postventral plate absent (0), present (1)
39. Posterior extremity rounded (0), styliform articulated at base (1), paddle-shaped articulated at base (2) (similar structures in crustaceans are not articulated at the base, or are incorporated into tail-fans, and hence are rather different from those in chelicerates)
40. Pretelson appendages present (0), absent (1)

(*continued*)

TABLE III. *(Continued)*

41. Movable telson appendages absent (0), present (1)
42. Number of trunk somites 25+ (0), 16–24 (1), 15 (2), 12–14 (3), 11 (4), less than 11 (5)
43. Appendages on posteriormost tagma (excluding telson) present (0), absent (1)
44. Cephalic shield entire (0), bivalved (1)
45. Trunk limbs diminishing in size (0), not diminishing (1)
46. Ecdysial sutures absent (0), present (1)

[a]As far as possible, the increasing scores indicate progressively more derived states. *Aysheaia* was not included in the analysis which resulted in the cladogram in Fig. 3. Hence characters 3, 9, and 13 were constant.

consistently occupied at the base of PAUP cladograms when the Onychophora were the designated outgroup. Previous studies resulted in a similar placement for *Marrella* (Briggs and Whittington, 1981; Briggs, 1983). As Briggs (1983, p. 19) pointed out, "*Marrella* has an unusual cephalic shield and filamentous outer rami but in most other respects the morphology is primitive enough to give rise to any of the three major groups" (i.e., trilobites, crustaceans, and chelicerates). The program allows polarity reversals as synapomorphies. The program does not rank characters in any hierarchy or sequence; all are regarded as of equal significance, and the position or level in the cladogram at which they may define nodes is not constrained.

The analysis was run several times with slight variations in the taxa included, the number and coding of characters, the polarity of character states, etc. While the cladogram presented here (Fig. 3) is the most parsimonious solution for the taxa considered (there are no alternative equally parsimonious cladograms using this set of data), it must be emphasized that it is not presented here as a definitive solution to the question of relationships among the Cambrian arthopods. There is obviously uncertainty regarding some characters (their preservation and the identification of homologies), and the consistency index (0.384) of

Figure 3. PAUP-generated cladogram of Cambrian and selected Recent arthropods. Autapomorphies for each node on the cladogram, as identified by PAUP, as follows (character states are listed in Table III):

N1: 34 1>0 (i.e., node 1, character 34, changes from state 1 to state 0).
N2: 4 *1>2; 14 *1>4; 19 *0>1; 41 *0>1.
N3: 26 *0>1; 29 *0>2; 40 *0>1; 42 *0>1; 43 *0>1; 45 *0>1.
N4: 42 *1>3.
N5: 31 *0>1.
N6: 4 *2>1; 44 *0>1.
N7: 8 *1>2; 22 *0>1; 23 *0>1; 42 3>2.
N8: 24 *0>1; 32*0>1.
N9: 40 1>0; 41 1>0.
N10: 37 *0>1; 41 *1>0.
N11: 35 *0>1; 39 *0>1.
N12: 43 1>0.
N13: 5 *0>1; 6 *0>1; 18 *0>1.
N14: 14 4>3; 26 1>0.
N15: 40 1>0.
N16: 34 *0>1.
N17: 6 1>0; 21 *0>1.

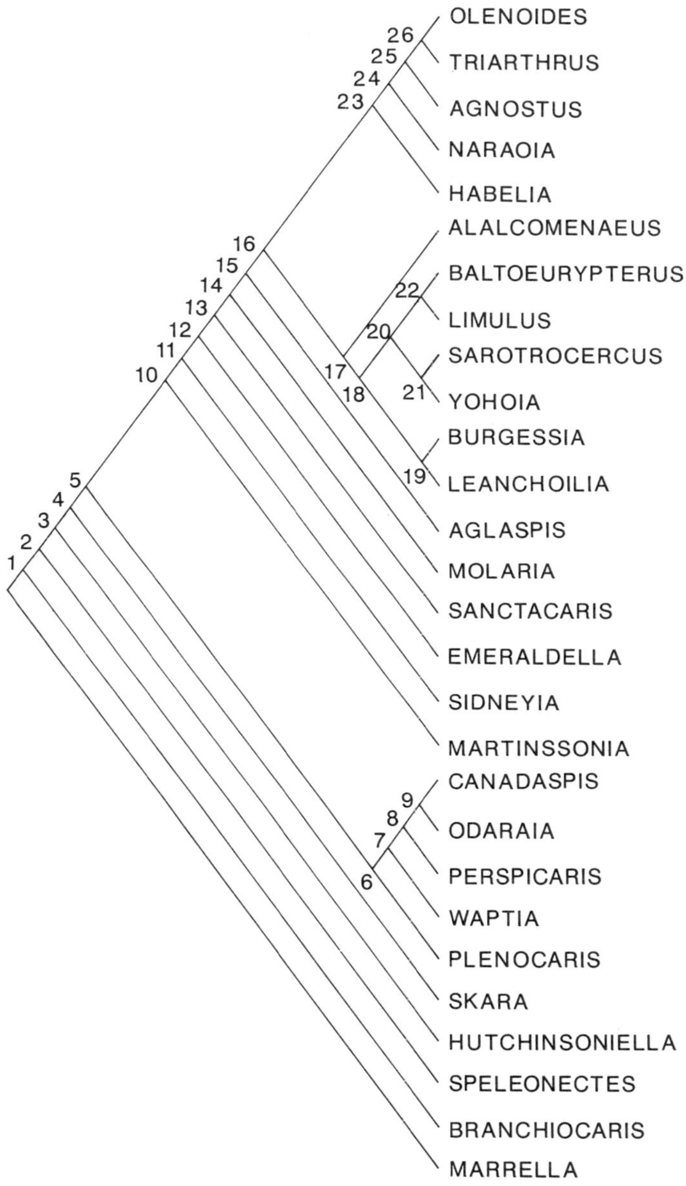

N18: 29 2>0.
N19: 8 1>2; 10 1>0; 18 1>0; 19 1>0.
N20: 28 *1>0; 30 *1>2.
N21: 6 0>1; 19 1>0; 34 1>0; 40 0>1.
N22: 11 *0>1; 12 *0>1; 14 *3>5; 15 *0>1
N23: 42 4>3.
N24: 7 *0>1; 25 *0>1; 33 *0>1; 39 1>0.
N25: 2 *1>3; 11 *0>1; 15 *0>1; 17 *0>1; 26 *0>1.
N26: 15 *1>2; 16 *0>1; 35 *1>2.

Character changes identified by an asterisk are considered robust, as compared to those not so identified, which include reversals, losses, etc.

character transformations is relatively low. The cladogram does, however, illustrate that a reasonable series of steps can account for relationships between groups, without recourse to solutions involving polyphyly. It also provides a framework for discussion of the relationships of the Cambrian arthropods.

A detailed description of the application of the PAUP program to an analysis of the phylogenetic history, and hence radiation, of the early arthropods is beyond the scope of the present treatment. Some of the problems inherent in applying the PAUP technique in this investigation, however, require comment here.

6.1.1. Primitive Character Complexes

Living arthropod groups are defined with reference to character complexes in an advanced state, yet these must have been derived by several stages. It is therefore likely that the Cambrian representatives of these groups will display an early stage in the differentiation of these character complexes—the synapomorphies will be incompletely expressed. Cephalization is the obvious example. Living arthropod groups are characterized by particular configurations of cephalic segments and corresponding appendages. Crustaceans, for example, have five pairs: two pairs of antennae, a mandible, and two pairs of maxillae.

It is certainly reasonable to assume that arthropods with fewer cephalic appendages preceded those with more; such a character polarity is endorsed if we regard Onychophora as a sister group of the arthropods. Some of the Cambrian arthropods have few cephalic appendages (e.g., *Branchiocaris, Marrella;* Fig. 4) and are obviously primitive with regard to this character. As the cladistic method does not attach importance to shared primitive features, however, the classification of these arthropods relies on the recognition of other, advanced characters. To identify homologies without evidence—to assume, for example, that the first three postcephalic limbs of *Marrella* were "destined" to become part of the crustacean head—would be to prejudge the classificatory outcome, by assuming a transformational hypothesis. Here we have avoided such assumptions in an attempt to be as objective as possible. This may, however, impair the result in some cases by eliminating from the analysis characters which may be robust synapomorphies. The problem of coding the crustacean mandible is a case in point: to identify a relatively undifferentiated appendage as a mandible would be to align the arthropod in question with the crustaceans.

6.1.2. Peripatetic Taxa

With the help of P. L. Forey, we produced a variety of cladograms, making different assumptions about outgroups, for example. A discussion of all these versions is beyond the scope of this chapter. Certain consistent groupings of arthropods were revealed, but other taxa occupy different positions according to the assumptions made. These are essentially plesiomorphic forms and none of their placings is particularly well supported by the distribution of characters.

In all our analyses using an onychophoran outgroup, *Marrella* came out consistently as the most primitive of the arthropods under consideration. Hence we felt justified in using it as the outgroup in subsequent cladograms, including that presented here.

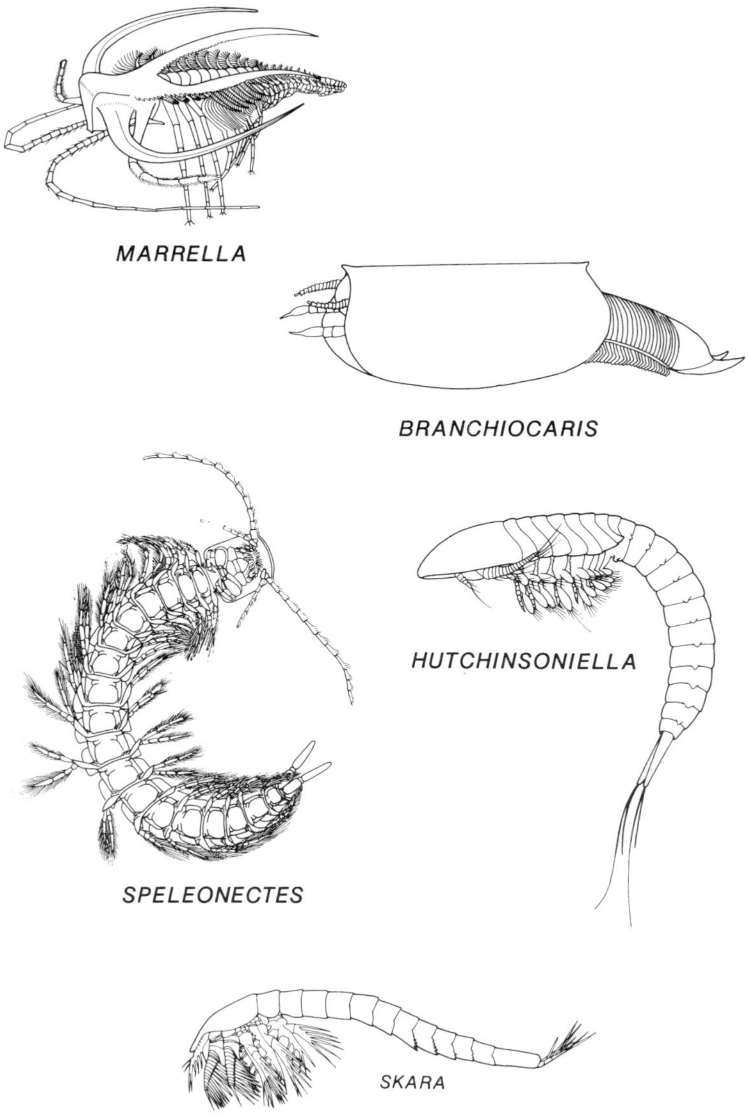

Figure 4. Cambrian (*Marrella, Branchiocaris, Skara*) and living (*Speleonectes, Hutchinsoniella*) arthropods occupying the most primitive positions on the cladogram (Fig. 3).

6.2. The Resultant Pattern

As a result of a preliminary cladistic analysis, Briggs (1983) presented two conflicting hypotheses. One (Briggs, 1983, Fig. 4), based on numbers of cephalic appendages, a character which readily separates the major groups of living arthropods, resulted in obvious convergences of taxa with low numbers of head appendages. The other [Fig. 3 of Briggs (1983), a modified version of that published by

Briggs and Whittington (1981)] separated arthropods with a cephalic shield and tergites from those with a large carapace. This divided crustaceans and crustaceanlike arthropods in one clade from the remainder, which included the trilobites. Briggs (1983) concluded that while the crustaceans were part of a separate group, the difficulty in arriving at a definitive solution to the problem of relationships among the Cambrian arthropods was a function of the wide range of morphologies resulting from the first major metazoan radiation.

In the PAUP analyses carried out in the course of this study, different combinations of taxa and variations in the list of characters used resulted in cladograms which differed in detail. The basic pattern remained the same, however: the trilobites are a derived group and most closely related to the chelicerates and "cheliceratelike" arthropods, the crustaceans and "crustaceanlike" arthropods form a PRIMITIVE sister group to this clade. The demonstration of the derived status of the trilobites and the primitive status of the crustaceans among the arthropods is one of the most significant results of this study.

Previous considerations of the phylogeny of the major arthropod groups and their relationship to early Paleozoic forms identified other relationships. Cisne (1974, p. 16) regarded the trilobites as suggestive of "the very primitive and generalized plan from which the crustacean and chelicerate conditions can be derived" (see also Cisne, 1982). Hessler and Newman (1975) reached a similar conclusion. Schram (1978) argued for four arthropodous phyla: Uniramia, Crustacea, Cheliceriformes, and Trilobitomorpha, but did not consider relationships between them. Briggs (1983; see also Briggs and Whittington 1981) showed that the crustaceans and crustaceanlike arthropods formed a sister group to the rest of the Cambrian arthropods, but did not identify their more primitive status. Schram (1986, pp. 24, 28) acknowledged the possibility that trilobitomorphs and cheliceriforms might be sister groups, leaving the crustaceans in an "isolated phyletic position" (presumably more derived, although this was not explicitly stated). This is a return to Størmer's (1944) concept of the Arachnomorpha, which groups Trilobitomorpha and Chelicerata, separating them from the Crustacea. Størmer (1944, p. 143) considered that these two different branches of the Arthropoda possibly "developed independently from different annelid ancestors." Bergström (in press) also regards trilobites and chelicerates as more closely related than either is to the crustaceans. He acknowledges that trilobitomorphs are unlikely to have given rise to the crustaceans.

With the exception of the trilobites, the "higher" chelicerates, and the Burgess Shale bivalved arthropods (*Branchiocaris* excepted), the majority of the arthropods in our analysis do not group into clades (Fig. 3). Rather, the general pattern for the fossil taxa is one of a series of plesions which ascend from crustaceans (and crustaceanlike arthropods), through cheliceratelike arthropods, to the "higher" chelicerates plus trilobites which together comprise a clade. This pattern is presumably a reflection of the predominance of primitive character complexes among the Cambrian arthropods. There is an absence of synapomorphic character combinations to define additional clades, although the addition of primitive fossil taxa [such as *Mimetaster* from the Devonian Hunsrück Slate (Stürmer and Bergström, 1976)] might augment some of these "steps" to small clades known only from fossils.

6.3. The Position and Arrangement of Taxa on the Cladogram

6.3.1. *Branchiocaris*

Briggs (1976, 1983) argued that *Branchiocaris* (Fig. 4) is not a crustacean. It is eliminated by the presence of only two limbs in the head, and a single antenna followed by a large, apparently chelate appendage. There is no evidence that any of the anteriormost trunk limbs are in any respect differentiated—i.e., that they show any tendency to cephalization. The position of *Branchiocaris* and *Odaraia* (Fig. 5) in the cladogram suggests that the similarities between them, including the bivalved carapace, the large number of trunk annulations which do not correspond 1:1 to the limbs, and the "fixed" telson processes, are homoplastic.

6.3.2. *Speleonectes* and *Hutchinsoniella*

These two genera (Fig. 4) were selected as representatives of two orders of living crustaceans, the Remipedia and Cephalocarida, which have provided the basis for major hypotheses of crustacean origins and early evolution. Their position on the cladogram reflects their apparently primitive nature. For some 30 years following the discovery of *Hutchinsoniella* (Sanders, 1955, 1957, 1963), the Cephalocarida were considered to reflect the primitive morphology of crustaceans. The Remipedia have usurped this position in some recent treatments of crustacean taxonomy and phylogeny. Schram (1983, 1986) argued that the mode of feeding in the remipedes, grappling with the head appendages (cephalophagy), is more likely to be primitive than the method of passing food along a thoracic "groove" (thoracophagy) employed by cephalocarids. The primitive limb is then biramous, as in remipedes, rather than polyramous, as in cephalocarids. This analysis lends some support to Schram's hypothesis; the remipede *Speleonectes* (and *Branchiocaris*, which is also cephalophagous) lies below *Hutchinsoniella* on the cladogram (Fig. 3). This polarity is not consistent, however—*Skara* (Fig. 4), which follows *Hutchinsoniella*, is again cephalophagous in that it lacks thoracic appendages with the exception of the one maxilliped (Müller and Walossek, 1985a).

Speleonectes and *Hutchinsoniella* both have a mandible, as does the clade comprised of *Canadaspis*, *Perspicaris*, and *Odaraia* (Fig. 5) (node 8). This acquisition therefore appears as a homoplasy. We do not code the appropriate appendage as a mandible in *Skara* [although it is identified as such by Müller and Walossek (1985a)], because it is not significantly differentiated from the other cephalic appendages. This may simply be a function of the larval nature of the specimens (Schram, 1986, p. 523). The preservation is inadequate to show whether a mandible is present in *Plenocaris* and *Waptia* (Fig. 5). If these taxa were coded for a mandible, the taxa from *Speleonectes* to *Canadaspis* would constitute a separate clade equivalent to the Crustacea, rather than a paraphyletic group.

6.3.3. The Burgess Shale Bivalved Arthropods

The bivalved arthropods from the Burgess Shale (with the exception of *Branchiocaris*) are united in a separate monophyletic clade on the basis of the

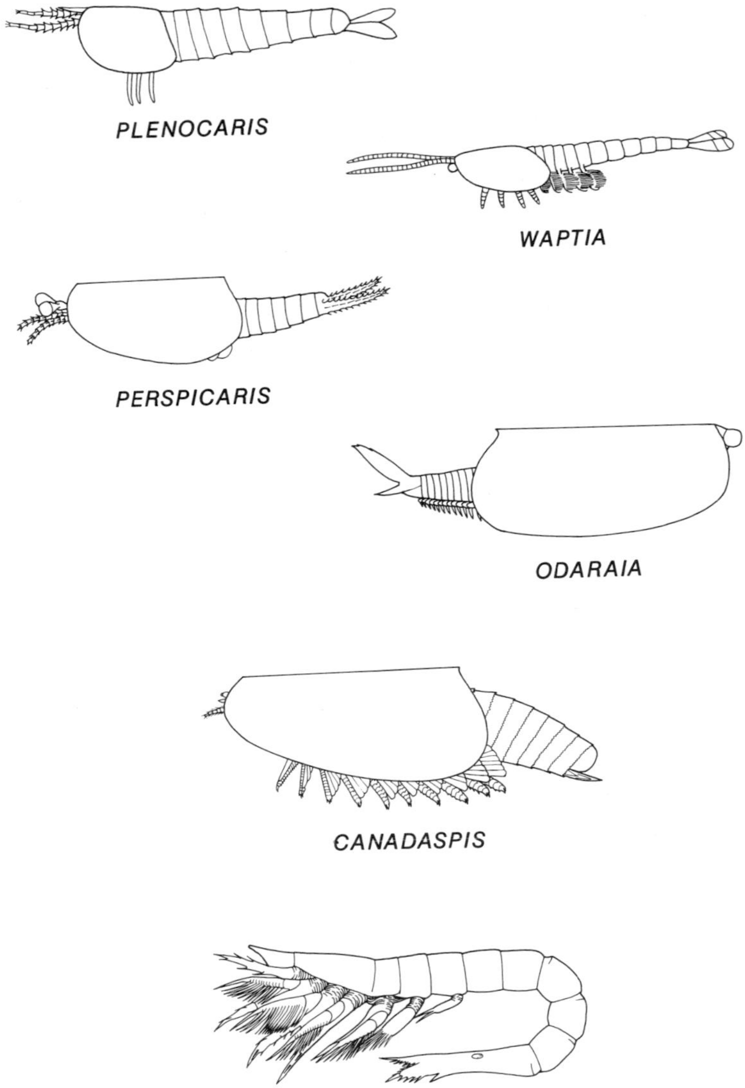

Figure 5. Bivalved Burgess Shale arthropods, and *Martinssonia* from the Upper Cambrian of Sweden.

possession of a single bivalved cephalic shield and the absence of dorsal tergites (node 6). *Plenocaris* (Fig. 5) is the primitive sister group of the rest of these bivalved arthropods, a position which may be partly a reflection of the lack of evidence for the nature of most of its head appendages (Whittington, 1974).

The 15 trunk somites of *Waptia* (node 7 on the cladogram) may not be homologous with those of *Canadaspis* and *Perspicaris*—the arrangement of tagmata is not the same.

Canadaspis, Perspicaris, and *Odaraia* are separated from *Waptia* (Fig. 5) by the presence of a mandible, characteristic of the crustaceans, and by the loss of filaments on the trunk exites (node 8). Both *Odaraia* and *Canadaspis* preserve relief traces of what Briggs (1978, 1981) interpreted as denticles on the mandible, implying that the cuticle on this surface was more robust than that of the rest of the limbs. A mandible is presumed to have been present in *Perspicaris*, and there are other grounds for arguing a close relationship between it and *Canadaspis* (Briggs, 1977, 1978). All three arthropods have large flattened lobate outer rami. In other respects, however, the trunk appendages of *Canadaspis* (the poorly preserved limbs of *Perspicaris* were probably similar) differ from those of *Odaraia*. The PAUP program usually identifies a closer relationship between *Canadaspis* and *Odaraia* than between either of these arthropods and *Perspicaris*. This grouping of *Canadaspis* and *Odaraia* is based on the presence of pre-telson appendages, and the lack of appendages on the telson itself (node 9). It reduces the similar (malacostracan) number of trunk somites and division of tagmata in *Canadaspis* and *Perspicaris* to a homoplasy. We prefer to emphasize the likely significance of this character and to identify *Odaraia* as the sister taxon of the phyllocarids (*sensu* Briggs, 1978) *Canadaspis* and *Perspicaris*.

Dahl (1983, 1984) argued against extending the concept of Phyllocarida (and hence Malacostraca) to include *Canadaspis* and *Perspicaris*. His approach emphasizes the differences rather than the similarities between these Cambrian arthropods and later phyllocarids. Some major differences—the segmentation/boundary between the cephalon and thorax, the structure of the thoracic limbs—are arguably primitive characters and thus of limited significance in determining relationships [Briggs (1978) with reference to *Canadaspis;* Briggs and Collins (1988) on *Sanctacaris*]. Others, such as the apodous abdomen, may imply functional contrasts.

The grouping of these Cambrian arthropods in a monophyletic clade including *Canadaspis*, which displays diagnostic crustacean features (Briggs, 1978, 1983), suggests that all should be considered "Crustacea." Where the diagnostic arrangement of cephalic appendages cannot be confirmed in some of these taxa, this may be taphonomic.

6.3.4. *Martinssonia*

The morphology of *Martinssonia* (Fig. 5) is suggestive of a primitive crustacean (Müller and Walossek, 1986a), but our analysis positions it as more advanced than the rest of the crustaceans under consideration. This is due to the ventral anus, and the lack of an articulation at the base of the telson appendages (node 10). The largest known specimens of this arthropod, however, are only 1.5 mm and may still be immature. As Müller and Walossek (1986a, p. 91) pointed out, the morphology of the posterior extremity of the trunk may be larval. They regarded *Martinssonia* as "crustaceanlike" rather than a crustacean—an early offshoot before the true crustacean "level of evolution" was reached. Its position on the cladogram may be spurious.

6.3.5. Status of the Crustaceans

The Crustacea do not emerge as a natural group; they are paraphyletic rather than a true clade. The existence of a natural group Crustacea is a basic tenet among neontologists; its failure to emerge in our analysis may reflect the problem of primitive character complexes outlined above. The crustaceans do, however, form a sequence on the cladogram (Fig. 3) and separate, even if only paraphyletically, from an "arachnomorph" clade.

The possibility that the Crustacea, as presently understood, is not a true clade, should not be ruled out. Schram (1986, p. 3) pointed out that "aside from the features of the head, it is impossible to characterize crustaceans except by noting tendencies toward certain conditions or states." Certain characters regarded as "crustacean" may be primitive for the arthropods under consideration here. The possession of a nauplius larva, for example, might be a symplesiomorphy, not a synapomorphy. A discussion of the reality of the Crustacea is beyond the scope of this chapter: here the taxa from *Martinssonia* to *Speleonectes* on the cladogram can be referred to as Crustacea. *Marrella* and *Branchiocaris* (Fig. 4) are primitive sister groups of the rest of the arthropods.

6.3.6. An Arachnomorph Clade

Node 11 separates the crustaceans, which lie at the primitive end of the cladogram, from the remainder of the arthropods, which form an "arachnomorph" clade including the chelicerates and trilobites. This node is defined by a styliform telson and overlap in the pleural area of the tergites.

6.3.7. *Sidneyia*

Sidneyia (Fig. 6) is the primitive sister group of the rest of the arachnomorph clade. PAUP identifies a number of autapomorphies which separate *Sidneyia* from the ancestral node. These include a reduction of the cephalic appendages to one pair of antennae, an increase in the number of trunk tagmata to three, and modification of the telson from the primitive styliform morphology to a paddle-shaped structure incorporated into a tail fan. A dramatic reduction of the number of cephalic appendages (from five to one) seems improbable. However, if the first four appendages of the trunk in *Sidneyia* (which differ from those following in lacking outer rami), together with the pair of antennae, were considered equivalent to the head, this would increase the number of head limbs to five and reduce the number of trunk tagmata to two!

6.3.8. *Emeraldella* to *Alalcomenaeus*

The morphology of *Emeraldella* (Fig. 6) is largely primitive. The analysis interprets the postventral plate, which is otherwise known only in the aglaspids, as homoplastic.

Node 13 is marked by trilobation (and the presence of pleurae), and by the presence of a marginal rim. Nodes 14 and 15, on the other hand, are defined by small numbers of reversals, and hence are not well constrained: the relative positions of *Sanctacaris*, *Molaria*, and *Aglaspis* (Fig. 6) are somewhat equivocal.

Node 17 defines a clade on the basis of a first cephalic appendage which is specialized, not antenniform. This naturally includes the chelicerates *Limulus* and *Baltoeurypterus* (Fig. 7), in which this appendage is the chelicera. The inclu-

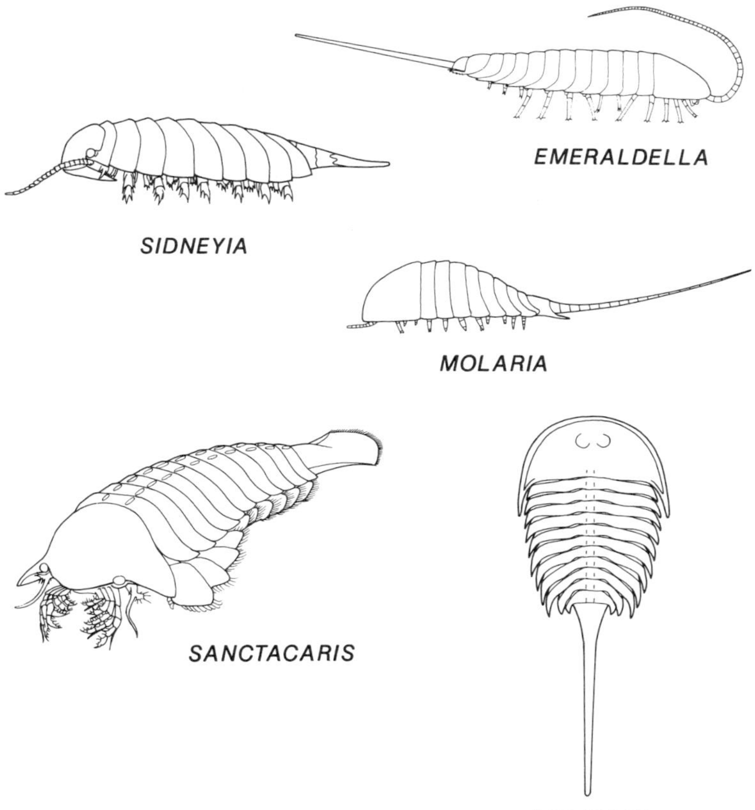

Figure 6. Cambrian "arachnomorph" arthropods, including the chelicerate *Sanctacaris.*

sion of *Burgessia* (Fig. 7) within this clade is anomalous, and involves a transformation of the first cephalic appendage back into an antenna. Indeed, the position of *Burgessia* as a sister group of *Leanchoilia* is based exclusively on the loss of characters (node 19).

In the several PAUP analyses performed, *Baltoeurypterus* and *Limulus* occur consistently in a clade with *Sarotrocercus* and *Yohoia* (Fig. 7), genera which have previously been considered neither closely related to each other nor to the chelicerates (Whittington, 1974, 1981; Briggs and Whittington, 1981; Briggs, 1983). This clade is united by a reduction of the trunk limbs to an outer ramus (node 20). Within it *Yohoia* and *Sarotrocercus* plot as sister taxa (node 21).

6.3.9. Status of the Chelicerates

The absence of chelicerates in the Burgess Shale appeared anomalous, not least because the two other major aquatic arthropod groups (the trilobites and crustaceans) are represented, and the chelicerates are known from Ordovician rocks. This gap has recently been filled by the description of *Sanctacaris* (Fig. 6) (Briggs and Collins, 1988) from the Stephen Formation on Mount Stephen. A

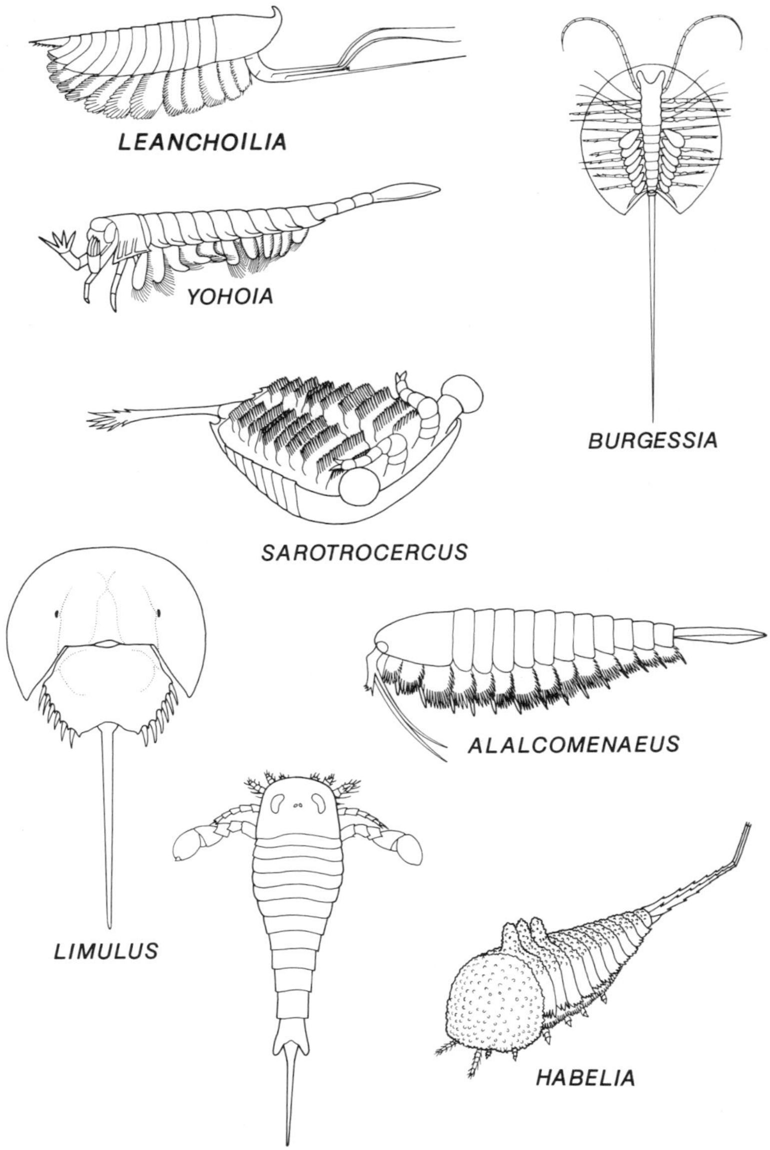

Figure 7. Members of the clade including the living chelicerate *Limulus* and the fossil eurypterid *Baltoeurypterus* (Fig. 3). *Habelia* emerges as a primitive sister taxon of the trilobites (Fig. 3).

larval chelicerate is also known from the Upper Cambrian Örsten deposits (Müller and Walossek, 1986b).

There are few characters that can be observed readily which are diagnostic of the chelicerates: six or seven head appendages, first head appendage not anten- niform (chelicerae are normally small and lie under the head, and consequently are rarely evident), and the presence of a telson [primitively styliform, otherwise paddle-shaped (Weygoldt and Paulus, 1979)] articulating at the base and lacking appendages. This last character, which occurs in the aquatic chelicerates, is potentially one of the most useful, as it is readily preserved, and absent in crustaceans and trilobites.

The acquisition of the six cephalic appendages characteristic of the living chelicerates occurs twice in the cladogram, in the *Baltoeurypterus* plus *Limulus* clade, and in *Sanctacaris,* and is by implication homoplastic. Other chelicerate-like arthropods under consideration here have fewer. All those taxa occupying positions in the cladogram between *Sanctacaris* [identified as a chelicerate by Briggs and Collins (1988)] and the clade including *Limulus* and *Baltoeurypterus* should have chelicerate affinities (i.e., all taxa from *Sanctacaris* to *Alalcomenaeus*).

6.3.10. *Habelia*

The position of *Habelia* (Fig. 7) as the most primitive member of a clade including the trilobites is equivocal. Node 23 is based on a single character reversal—an increase in trunk somites from 11 to 12–14. The styliform telson articulating at the base suggests that *Habelia* should probably be grouped with the cheliceratelike forms.

6.3.11. Trilobites

Olenoides, Triarthrus, Agnostus, and *Naraoia* (Fig. 8) consistently form a monophyletic clade. While this was an expected outcome, it supports the inclusion of the agnostids (Müller and Walossek, 1987), and forms like *Naraoia* with an unmineralized cuticle (Whittington, 1977), in the trilobites. The synapomorphies defining the group in the cladogram are the pygidium (to which may be linked the absence of any telson paddle or spine), the gut diverticula, and the distal lobe on

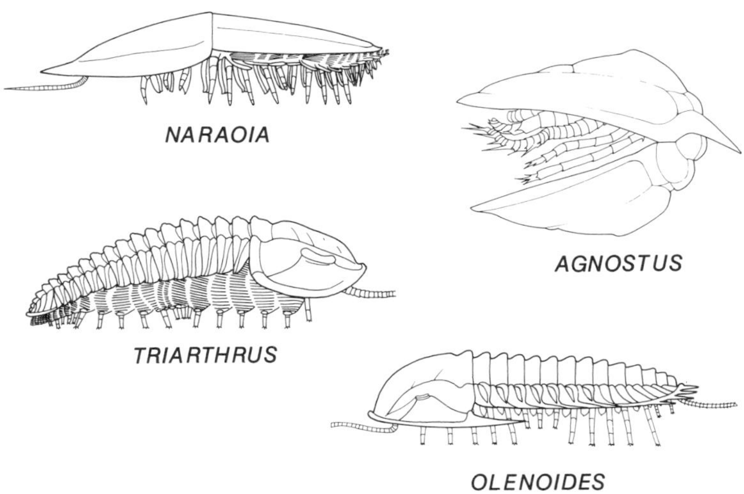

Figure 8. Cambrian and Ordovician (*Triarthrus*) trilobites.

the outer ramus of the limbs. Recognition of the presence of a pygidium in *Naraoia* relies on the interpretation of the posterior shield as equivalent to the thorax and pygidium of other trilobites. A pygidium, dorsal ecdysial sutures, eye ridges, and an articulating half-ring do not occur in any of the arthropods other than the higher trilobites.

Elsewhere in this chapter we argue on both morphologic and stratigraphic grounds that the olenellids are the most primitive of those trilobites with mineralized exoskeletons. The earliest faunas containing trilobites which, like *Naraoia*, have unmineralized cuticles are Lower Cambrian—the Buen Formation (lower part of the *Bonnia–Olenellus* or even the underlying *Nevadella* Zone) of North Greenland (Conway Morris *et al.*, 1987) and the Chiungchussu Formation (lower part of the *Eoredlichia–Wutingaspis* Zone) of Maotianshan (Zhang and Hou, 1985). In both cases, nonmineralized trilobites with few or no articulating tergites and a large posterior shield are associated with olenellids or redlichiids. The analysis identifies *Naraoia* as the primitive sister group of the rest of the trilobites. It lacks a number of synapomorphies characterizing the group, including a mineralized exoskeleton, eyes and eye ridges, and ecdysial sutures. Whether or not the taxon Trilobita is extended, as advocated by Whittington (1977), to accommodate forms like *Naraoia* is a matter of convention.

7. The Diversity of the Cambrian Arthropods

7.1. Morphology

The multiple polyphyly approach to classifying the Burgess Shale arthropods (see Section 5) focused on the differences between them, rather than the similarities, and tended to exaggerate their morphologic disparity. Thus, Whittington (1980, p. 146) wrote that the Burgess Shale shows "a spectrum of not closely related animals, even among those we group as structurally similar, such as . . . arthropods," and later (Whittington, 1982, p. 12) he again emphasized "the wide morphological gaps between kinds" of arthropods. This view was prompted at least in part by some of the more obvious bizarre attributes of these arthropods, such as the spinose cephalic shield of *Marrella*. It led to a perception of the Cambrian arthropods as displaying significantly greater morphologic disparity as well as taxonomic diversity than the living arthropod fauna (see Gould, 1989).

A cladistic analysis is concerned with identifying shared characters in order to determine relationship. A preliminary study of this kind led Briggs (1983, p. 20) to conclude that "it seems unlikely than many of the groups . . . are of separate origin." The possession of unique attributes—autapomorphies—is not relevant to a cladistic analysis, and hence their significance is not given undue emphasis. Characters unique to individual Cambrian arthropods, however, provide an indication of the morphologic separation between them and the major groups (including the trilobites), and therefore some measure of the morphologic disparity.

It should be noted that there are two distinct categories of autapomorphy in the context of the PAUP analysis: (1) characters unique to Cambrian taxa and (2) characters which change in terminal taxa on the cladogram.

1. Characters which are essentially unique to Cambrian taxa are largely omit-

ted from the PAUP analysis because they do not help in determining relationship. They are significant, however, in a consideration of morphologic disparity. This category includes the following:

(i) *Aglaspis:* Postventral plate. A similar structure, homoplastic according to the PAUP analysis, occurs in *Emeraldella* (Bruton and Whittington, 1983).

(ii) *Alalcomenaeus:* Large first cephalic appendage, with three distal (as opposed to originating on a basal podomere), segmented projections.

(iii) *Branchiocaris:* Large clawlike second appendage. Claws, with an opposable fixed and movable finger, are known in other groups (chelicerates, decapods), but they differ in structure and position.

(iv) *Burgessia:* Single shield covering both head and trunk. Spinelike posteriormost trunk limbs. Single shields are known in other groups (e.g., some branchiopods, *Vachonisia*), but they are different in character.

(v) *Habelia:* Telson spine with single articulation.

(vi) *Leanchoilia:* First cephalic appendage with three distal annulated projections.

(vii) *Marrella:* Deeply divided spinose cephalic shield.

(viii) *Molaria:* Multisegmented telson.

(ix) *Odaraia:* Tubular carapace, three-fluked tail.

(x) *Sidneyia:* Tail fan, consisting of a telson and flanking uropods borne by the last abdominal somite. Similar structures occur in the malacostracan crustaceans, but are unknown in the Lower Paleozoic.

(xi) *Yohoia:* Large ?second appendage of two segments with four terminal spines.

Few of these characters are strictly unique. They represent varying degrees of morphologic separation. The potential for functional diversification of limbs, for example, even among closely related arthropods, is well known [see, for example, McLaughlin (1982) on crustacean appendages], and differences in limb morphology alone presumably indicate separation at a relatively low level in the taxonomic hierarchy. The postventral plate of aglaspids (and possibly the three-fluked tail of *Odaraia*) is perhaps the only attribute significant at a relatively high level in the hierarchy and apparently confined to the Cambrian.

2. The PAUP solution inevitably involves changes in character state in terminal taxa (characters in which a taxon differs from the nested set in which it occurs). These autapomorphies may be reversals and frequently occur as homoplasies elsewhere on the cladogram.

The number of such autapomorphies is listed and computed by PAUP for each taxon as a distance from the adjacent node. In the current analysis *Olenoides* and *Limulus*, representatives of clades which are major arthropod groups, are further from the origin of the cladogram (in terms of changes in coded character states) than any other taxa. These somewhat crude measures reinforce the conclusion that those taxa which do not fall into any of the major living groups, or the trilobites, do not show any unusual morphological separation.

This survey identifies only a very small number of characters which are unique to the Cambrian arthropods, but clearly only a minute fraction of those arthropods is sufficiently well preserved (indeed preserved at all) to be considered here. The sample may be reasonably representative at higher taxonomic levels, however, and it is probable that the discovery/addition of further arthropods would decrease the distances between taxa, by introducing intermediates, rather than add to the morphologic disparity. Thus, what has previously been perceived as a much greater morphological disparity among Cambrian arthropods than in those living today (see Gould, 1989) results from combinations of characters in the Cambrian forms which are unfamiliar with reference to the living fauna and which have prevented us assigning the fossil arthropods to living taxa. Their morphologic disparity appears no greater than that of living crustaceans, for example, which range from barnacles to brachyurans, and Schram (1986, p. 3) argued that "crustaceans exhibit a greater degree of diversity in form than that seen in any other animal phylum."

7.2. Affinities and Relationships

The morphologic disparity of the Burgess Shale arthropods is more apparent than real, and is merely a reflection of the fact that few of them conform to our concepts of the major living groups. Applying the criteria used to classify living arthropods, many of the Burgess Shale arthropods would rank as separate orders, even classes. In the recent series of redescriptions of Burgess Shale arthropods, Whittington and his co-workers balked at this solution. "To place each in an extinct group of high rank, an order, or even a class, one known only from the Burgess Shale, is a rather desperate solution to classification" (Whittington, 1985, p. 68). Only taxa which could be assigned to living groups were assigned to higher taxonomic categories. This approach was appropriate as long as there were neither sufficient data, nor suitable methods, to investigate relationships.

Briggs (1983), for example, argued that the arrangement of cephalic appendages was the only reliable character which could be used in the Cambrian fossils to identify representatives of the major groups of living arthropods. This was based on a perceived requirement to identify in the fossils characters known to be diagnostic of living crustaceans. There are drawbacks in this approach (Briggs, 1983, p. 17), in addition to the fact that the head appendages are frequently not sufficiently well preserved to reveal the necessary detail. Homologies are difficult to identify. The nature and arrangement of head appendages may not have fully evolved in Cambrian arthropods. This is the case in the crustacean *Canadaspis* (Briggs, 1978) and the chelicerate *Sanctacaris* (Briggs and Collins, 1988), although in both cases there is evidence to identify the diagnostic compliment of cephalic limbs, and other characteristic traits are well developed. Cephalic morphology does not provide the same clear indication of affinity in the Cambrian arthropods as it does in the Recent fauna. This is particularly the case with taxa that have fewer cephalic appendages than any of the major groups. *Branchiocaris*, *Sarotrocercus*, and *Marrella* all have two cephalic appendages, but this clearly cannot be interpreted as evidence of relationship. The unsuitability of the arrangement of head appendages as a basis for grouping these Cambrian arthropods is emphasized by the low consistency index for transformations of this character in the PAUP analysis; only 7 of the 46 characters used gave a lower consistency index.

In taxa which are largely complexes of primitive characters, highly evolved diagnostic attributes are not available to indicate relationships. The PAUP analysis provides an alternative approach. How the groups of taxa revealed by the cladogram are ranked and named is a matter of convention (e.g., Ax, 1985), but no generally accepted solution has been proposed. Living taxa may or may not be expanded to include primitive fossil forms. Matters are simplified if single taxa in an ascending series are designated plesions, implying monophyletic groups of unspecified rank (Patterson and Rosen, 1977). A further difficulty is the fact that the position of some of the taxa is equivocal.

7.3. Conclusions

No formal classification is proposed here, but the following observations can be offered (many of which summarize extended discussions above):

1. *Marrella* and *Branchiocaris* can be regarded as primitive sister groups of all the taxa that succeed them.

2. The genera *Speleonectes* to *Martinssonia* correspond to the Crustacea, which this analysis reveals to be a paraphyletic group. A wider identification of the mandible might have resulted in a holophyletic taxon. The Crustacea are primitive with respect to the rest of the arthropods. The position of *Martinssonia* may be spurious.

3. *Sidneyia* and those taxa which succeed it form a holophyletic taxon which corresponds closely to the concept of Arachnomorpha (Størmer, 1944).

4. *Limulus* and *Baltoeurypterus* occur within a monophyletic clade from *Leanchoilia* to *Alalcomenaeus* (the position of *Burgessia* within this clade is equivocal). This provides a basis for enlarging the concept of the Chelicerata to include primitive relatives. If all the taxa between *Sanctacaris* [assigned to the Chelicerata by Briggs and Collins (1988)] and *Alalcomenaeus* are included, the taxon becomes paraphyletic.

5. The Trilobita form a holophyletic clade. The position of *Habelia* is probably spurious.

The cladogram provides no justification for identifying additional taxa at a taxonomic rank equivalent to the Crustaceans, Chelicerates, and Trilobites. Nor do the arthropods of the Cambrian show greater morphological disparity than those living today.

ACKNOWLEDGMENTS. We are grateful to P. L. Forey for running the PAUP program and generously providing comments and interpretations of the method and results. M. D. Brasier kindly made available unpublished Early Cambrian correlations. Pam Baldaro drafted the figures.

References

Anderson, D. T., 1973, *Embryology and Phylogeny of Annelids and Arthropods*, Pergamon Press, Oxford.

Ax, P., 1985, *Das Phylogenetische System*, Gustav Fischer Verlag, Stuttgart.

Bergström, J., 1973, Organisation, life and systematics of trilobites, *Fossils Strata* **2**:1–69.

Bergström, J., In press, The oldest arthropods and the origin of the Crustacea. *Acta Zool., Stockh.*

Brasier, M. D., 1989, China and the Palaeotethyan belt, in: *The Precambrian-Cambrian Boundary* (J. W. Cowie and M. D. Brasier, eds.), Clarendon Press, Oxford, pp. 40–74.

Briggs, D. E. G., 1976, The arthropod *Branchiocaris* n. gen., Middle Cambrian, Burgess Shale, British Columbia, *Bull. Geol. Surv. Can.* **264:**1–29.

Briggs, D. E. G., 1977, Bivalved arthropods from the Cambrian Burgess Shale of British Columbia, *Palaeontology* **20:**595–621.

Briggs, D. E. G., 1978, The morphology, mode of life, and affinities of *Canadaspis perfecta* (Crustacea: Phyllocarida), Middle Cambrian, Burgess Shale, British Columbia, *Philos. Trans. R. Soc. Lond. B.* **281:**439–487.

Briggs, D. E. G., 1981, The arthropod *Odaraia alata* Walcott, Middle Cambrian, Burgess Shale, British Columbia, *Philos. Trans. R. Soc. Lond. B* **291:**541–584.

Briggs, D. E. G., 1983, Affinities and early evolution of the Crustacea: The evidence of the Cambrian fossils, in: *Crustacean Phylogeny* (F. R. Schram, ed.), Balkema, Rotterdam, pp. 1–22.

Briggs, D. E. G., 1985, Les premiers arthropodes, *La Recherche* **16:**340–349.

Briggs, D. E. G., 1990, Early arthropods: Dampening the Cambrian explosion, *Short Courses in Paleontology* **3:**24–43, Paleontological Society.

Briggs, D. E. G., and Collins, D., 1988, A Middle Cambrian chelicerate from Mount Stephen, British Columbia, *Palaeontology* **31:**779–798.

Briggs, D. E. G., and Fortey, R. A., 1982, The cuticle of aglaspid arthropods, a red herring in the early history of the vertebrates, *Lethaia* **15:**25–29.

Briggs, D. E. G., and Robison, R. A., 1984, Exceptionally preserved nontrilobite arthropods and *Anomalocaris* from the Middle Cambrian of Utah, *Univ. Kans. Paleontol. Contrib.* **111:**1–23.

Briggs, D. E. G., and Fortey, R. A., 1989, The early radiation and relationships of the major arthropod groups, *Science* **246:**241–243.

Briggs, D. E. G., and Whittington, H. B., 1981, Relationships of arthropods from the Burgess Shale and other Cambrian sequences, in: *Short Papers for the Second International Symposium on the Cambrian System* (M. E. Taylor, ed.), U. S. Department of the Interior, Geological Survey Open-File Report 81-743, pp. 38–41.

Briggs, D. E. G., and Whittington, H. B., 1985, Modes of life of arthropods from the Burgess Shale, British Columbia, *Trans. R. Soc. Edinb. Earth Sci.* **76:**149–160.

Bruton, D. L., and Whittington, H. B., 1983, *Emeraldella* and *Leanchoilia,* two arthropods from the Burgess Shale, Middle Cambrian, British Columbia, *Philos. Trans. R. Soc. Lond. B* **300:**553–585.

Butterfield, N. J., 1990, Organic preservation of non-mineralizing organisms and the taphonomy of the Burgess Shale, *Paleobiology* **16:**272–286.

Cisne, J. L., 1974, Trilobites and the origin of arthropods, *Science* **186:**13–18.

Cisne, J. L., 1982, Origin of the Crustacea, in: *The Biology of Crustacea* (D. E. Bliss, ed.), Academic Press, New York, Vol. 1, pp. 65–92.

Collins, D., Briggs, D., and Conway Morris, S., 1983, New Burgess Shale fossil sites reveal middle Cambrian faunal complex, *Science* **222:**163–167.

Conway Morris, S., 1985, Cambrian Lagerstätten: Their distribution and significance, *Phil. Trans. R. Soc. Lond.* **B311:**49–65.

Conway Morris, S., 1986, The community structure of the Middle Cambrian Phyllopod Bed (Burgess Shale), *Palaeontology* **29:**423–468.

Conway Morris, S., 1989, The persistence of Burgess Shale-type faunas: Implications for the evolution of deeper-water faunas, *Trans. R. Soc. Edinb. Earth Sci.* **80:**271–283.

Conway Morris, S., and Jenkins, R. J. F., 1985, Healed injuries in Early Cambrian trilobites from South Australia, *Alcheringa* **9:**167–177.

Conway Morris, S., and Robison, R. A., 1988, More soft-bodied animals and algae from the Middle Cambrian of Utah and British Columbia, *Univ. Kans. Paleontol. Contrib.* **122.**

Conway Morris, S., and Whittington, H. B., 1979, The animals of the Burgess Shale, *Sci. Am.* **241:**122–133.

Conway Morris, S., Peel, J. S., Higgins, A. K., Soper, N. J., and Davis, N. C., 1987, A Burgess shale-like fauna from the Lower Cambrian of North Greenland, *Nature* **326:**181–183.

Cowie, J. W., 1971, Lower Cambrian faunal provinces, in: Faunal Provinces in Space and Time (F. A. Middlemiss, P. F. Rawson, and G. Newall, eds.), *Geol. J. Spec. Issue* **4:**31–46.

Cowie, J. W., and Brasier, M. D. (eds.), 1988, *The Precambrian–Cambrian Boundary,* Oxford University Press, Oxford.

Dahl, E., 1983, Phylogenetic systematics and the Crustacea Malacostraca—A problem of prerequisites, *Verh. Naturwiss. Ver. Hamburg* **26:**355–371.

Dahl, E., 1984, The subclass Phyllocarida (Crustacea) and the status of some early fossils: A neontologist's view, *Vidensk. Medd. Dan. Naturhist. Foren.* **145:**61–76.

Fortey, R. A., and Chatterton, B. D. E., 1988, Classification of the trilobite suborder Asaphina, *Palaeontology* **31:**165–222.

Fortey, R. A., and Whittington, H. B., 1989, The trilobites as a natural group, *Hist. Biol.* **2:**125–138.

Glaessner, M. F., 1979, Lower Cambrian Crustacea and annelid worms from Kangaroo Island, South Australia, *Alcheringa* **3:**21–31.

Gould, S. J., 1989, *Wonderful life, the Burgess Shale and the nature of history,* Norton, New York.

Henningsmoen, G., 1975, Moulting in trilobites, *Fossils Strata* **4:**179–200.

Hessler, R. R., and Newman, W. A., 1975, A trilobitomorph origin for the Crustacea, *Fossils Strata* **4:**437–459.

Hou Xian-guang, 1987a, Two new arthropods from Lower Cambrian, Chengjiang, Eastern Yunnan, *Acta Palaeontol. Sinica* **26:**236–256.

Hou Xian-guang, 1987b, Three new large arthropods from Lower Cambrian, Chengjiang, Eastern Yunnan, *Acta Palaeontol. Sinica* **26:**272–285.

Hou Xian-guang, 1987c, Early Cambrian large bivalved arthropods from Chengjiang, Eastern Yunnan, *Acta Palaeontol. Sinica* **26:**286–298.

Hou Xian-guang, and Bergström, J., 1991, The arthropods of the Lower Cambrian Chenjiang fauna, with relationships and evolutionary significance, in: *The Early Evolution of Metazoa and the Significance of Problematic Taxa* (A. M. Simonetta and S. Conway Morris, eds.), Cambridge University Press, pp. 179–187.

Hou Xian-guang and Sun Wei-guo, 1988, Discovery of Chengjiang fauna at Meishuchun, Jinning, Yunnan, *Acta Palaeontol. Sinica* **27:**1–12.

Hou Xian-guang, Chen Junyan, and Lu Haozhi, 1989, Early Cambrian new arthropods from Chengjiang, Yunnan, *Acta Palaeontol. Sinica,* **28:**42–57.

Hunt, A. S., 1967, Growth, variation and instar development of an agnostid trilobite, *J. Paleontol.* **41:**203–208.

Hupé, P., 1953, Contribution à l'étude du Cambrien inferieur et du Precambrien III de l'Anti Atlas marocain, *Notes Mém. Service Carte Géol. Maroc* **103:**1–402.

Lauterbach, K.-E., 1980, Schlusselereignisse in der Evolution des Grundplans der Arachnata (Arthropoda), *Abh. Naturwiss. Ver. Hamburg* **23:**163–327.

Lu Yan-hao *et al.,* 1974, The bio-environmental control hypothesis and its application to the Cambrian biostratigraphy and palaeozoogeography, *Mem. Nanking Inst. Geol. Palaeontol.* **5:**27–110 [in Chinese].

Manton, S. M., 1977, *The Arthropoda, Habits, Functional Morphology and Evolution,* Clarendon Press, Oxford.

McLaughlin, P. A., 1982, Comparative morphology of crustacean appendages, in: *The Biology of Crustacea* (D. E. Bliss, ed.), Academic Press, New York, Vol. 2, pp. 197–256.

McNamara, K. J., 1983, Progenesis in trilobites, *Spec. Pap. Palaeontol.* **30:**59–68.

McNamara, K. J., and Rudkin, D. M., 1984, Techniques of trilobite exuviation, *Lethaia* **17:**153–173.

Müller, K. J., 1979, Phosphatocopine ostracods with preserved appendages from the Upper Cambrian of Sweden, *Lethaia* **12:**1–27.

Müller, K. J., 1982, *Hesslandona unisulcata* sp. nov. (Ostracoda) with phosphatised appendages from the Upper Cambrian "Orsten" of Sweden, in: *A Research Manual of Fossil and Recent Ostracods* (R. H. Bate, ed.), Ellis Horwood, Chichester, England, pp. 276–307.

Müller, K. J., and Walossek, D., 1985a, Skaracarida, a new order of Crustacea from the Upper Cambrian of Vastergotland, Sweden, *Fossils Strata* **17:**1–65.

Müller, K. J., and Walossek, D., 1985b, A remarkable arthropod fauna from the Upper Cambrian "orsten" of Sweden, *Trans. R. Soc. Edinb. Earth Sci.* **76:**161–172.

Müller, K. J. and Walossek, D., 1986a, *Martinssonia elongata* gen. et sp. n., a crustacean-like euarthropod from the Upper Cambrian "Orsten" of Sweden, *Zool. Scripta* **15:**73–92.

Müller, K. J., and Walossek, D., 1986b, Arthropod larvae from the Upper Cambrian of Sweden, *Trans. R. Soc. Edinb. Earth Sci.* **77:**157–179.

Müller, K. J., and Walossek, D., 1987, Morphology ontogeny and life habit of *Agnostus pisiformis* from the Upper Cambrian of Sweden, *Fossils Strata* **19:**1–123.

Palmer, A. R., 1957, Ontogenetic development of two olenellid trilobites, *J. Paleontol.* **31:**105–128.
Palmer, A. R., 1972, Problems of Cambrian biogeography, in: *Proceedings. 24th International Geological Congress, Montreal,* Vol. 7, pp. 310–315.
Patterson, C., and Rosen, D. E., 1977, Review of ichthyodectiform and other Mesozoic teleost fishes and the theory and practice of classifying fossils, *Bull. Am. Mus. Nat. Hist.* **158:**81–172.
Rasetti, F., 1967, Lower and Middle Cambrian trilobite faunas from the Taconic sequence of New York, *Smithson. Misc. Collect.* **152:**1–111.
Rasetti, F., 1972, Cambrian trilobite faunas of Sardinia, *Atti Acad. Naz. Lincei Mem. Ser. 8* **9:**1–100.
Robison, R. A., 1972, Mode of life of agnostid trilobites, in: *Proceedings. 24th International Geological Congress, Montreal,* Vol. 7, pp. 33–40.
Robison, R. A., 1975, Species diversity among agnostoid trilobites, *Fossils and Strata* **4:**219–226.
Robison, R. A., 1984, New occurrences of the unusual trilobite *Naraoia* from the Cambrian of Idaho and Utah, *Univ. Kans. Paleontol. Contrib.* **112:**1–8.
Robison, R. A., 1991, Middle Cambrian biotic diversity: Examples from four Utah Lagerstätten, in: *The Early Evolution of Metazoa and the Significance of Problematic Taxa* (A. M. Simonetta and S. Conway Morris, eds.), Cambridge University Press, pp. 77–93.
Robison, R. A., and Richards, B. C., 1981, Larger bivalve arthropods from the Middle Cambrian of Utah, *Univ. Kans. Paleontol. Contrib.* **106:**1–19.
Runnegar, B., and Bengtson, S., 1990, Origin of the hard parts—Early skeletal fossils, in: *Palaeobiology: A Synthesis* (D. E. G. Briggs and P. R. Crowther, eds.), Blackwell Scientific Publications, London, pp. 24–29.
Rushton, A. W. A., 1974, The Cambrian of Wales and England, in: *Lower Palaeozoic Rocks of the World* (C. H. Holland, ed.), Wiley, Chichester, England, Vol. 2, pp. 43–121.
Sanders, H. L., 1955, The Cephalocarida, a new subclass of Crustacea from Long Island Sound, *Proc. Natl. Acad. Sci. USA* **41:**61–66.
Sanders, H. L., 1957, The Cephalocarida and crustacean phylogeny, *Syst. Zool.* **6:**112–128.
Sanders, H. L., 1963, The Cephalocarida, functional morphology, larval development, comparative external anatomy, *Mem. Conn. Acad. Arts Sci.* **25:**1–80.
Schram, F. R., 1978, Arthropods: A convergent phenomenon, *Fieldiana Geol.* **39:**61–108.
Schram, F. R., 1983, Remipedia and crustacean phylogeny, in: *Crustacean Phylogeny* (F. R. Schram, ed.), Balkema, Rotterdam, pp. 23–28.
Schram, F. R., 1986, *Crustacea,* Oxford University Press, Oxford.
Schram, F. R., Yager, J., and Emerson, M. J., 1986, Remipedia. Part 1. Systematics, *San Diego Soc. Nat. Hist.* **15:**1–60.
Sdzuy, K., 1978, The Precambrian–Cambrian boundary beds in Morocco (preliminary report), *Geol. Mag.* **115:**83–94.
Størmer, L., 1944, On the relationships and phylogeny of fossil and Recent Arachnomorpha, *Skr. Nor. Vidensk. Akad. Mat. Naturvidensk. Kl.* **5:**1–158.
Størmer, L., 1959, Trilobitoidea, in: *Treatise on Invertebrate Paleontology,* Part O, *Arthropoda 1* (R. C. Moore, ed.), Geological Society of America and University of Kansas Press, Lawrence, Kansas, pp. 23–37.
Størmer, W., and Bergström, J., 1976, The arthropods *Mimetaster* and *Vachonisia* from the Devonian Hunsrück Shale, *Palaontol. Z.* **50:**78–111.
Suvarova, N. P., 1964, Corynexochoid trilobites and their historical development, *Trudy Palaeontol. Inst. SSSR* **103:**1–319 [in Russian].
Teigler, D. J., and Towe, K. M., 1975, Microstructure and composition of the trilobite cuticle, *Fossils Strata* **4:**137–150.
Walossek, D., and Müller, K. J., 1990, Upper Cambrian stem-lineage crustaceans and their bearing upon the monophyletic origin of Crustacea and the position of *Agnostus, Lethaia* **23:**409–427.
Weygoldt, P., and Paulus, H. F., 1979, Untersuchungen zur Morphologie, Taxonomie und Phylogenie der Chelicerata, *Z. Zool. Syst. Evol.* **17:**85–200.
Whittington, H. B., 1971, Redescription of *Marrella splendens* (Trilobitoidea) from the Burgess Shale, Middle Cambrian, British Columbia, *Bull. Geol. Surv. Can.* **209:**1–24.
Whittington, H. B., 1974, *Yohoia* Walcott and *Plenocaris* n. gen., arthropods from the Burgess Shale, Middle Cambrian, British Columbia, *Bull. Geol. Surv. Can.* **231:**1–27.
Whittington, H. B., 1975, Trilobites with appendages from the Middle Cambrian Burgess Shale, British Columbia, *Fossils Strata* **4:**97–136.

Whittington, H. B., 1977, The Middle Cambrian trilobite *Naraoia*, Burgess Shale, British Columbia, *Philos. Trans. R. Soc. Lond. B* **280:**409–443.
Whittington, H. B., 1979, Early arthropods, their appendages and relationships, in: *The Origin of Major Invertebrate Groups* (M. R. House, ed.), Academic Press, New York, pp. 253–268.
Whittington, H. B., 1980, The significance of the fauna of the Burgess Shale, Middle Cambrian, British Columbia, *Proc. Geol. Assoc.* **91:**127–148.
Whittington, H. B., 1981, Rare arthropods from the Burgess Shale, Middle Cambrian, British Columbia, *Philos. Trans. R. Soc. Lond. B* **292:**329–357.
Whittington, H. B., 1982, The Burgess Shale fauna and the early evolution of metazoan animals, in: *Palaeontology, Essential of Historical Geology* (E. M. Gallitelli, ed.), STEM Mucchi, Modena, pp. 11–24.
Whittington, H. B., 1985, *The Burgess Shale*, Yale University Press, New Haven, Connecticut.
Zhang Wen-tang and Hou Xian-guang, 1985, Preliminary notes on the occurrence of the unusual trilobite *Naraoia* in Asia, *Acta Palaeontol. Sinica* **24:**591–595.

Chapter 11

Radiation of Echinodermata

JAMES SPRINKLE

1. Introduction

1.1. The Early to Middle Cambrian Echinoderm Record

Echinoderms underwent an initial radiation into marine shelf environments in the Early to Middle Cambrian. Within about 45 million years, nine new classes of echinoderms (as recognized in an evolutionary classification) appeared suddenly in the fossil record, many of these without obvious ancestors. This radiation then tapered off during the Late Cambrian, and no additional classes of echinoderms appeared in the record until the Early Ordovician, about 20 million years later, when a secondary radiation of echinoderms began.

JAMES SPRINKLE • Department of Geological Sciences, University of Texas, Austin, Texas 78713-7909.

Origin and Early Evolution of the Metazoa, edited by Jere H. Lipps and Philip W. Signor. Plenum Press, New York, 1992.

This initial radiation of Cambrian echinoderms produced a large number of classes with quite different body forms and ways of life, but relatively low diversity. Only 34 genera and about 60 species of Early and Middle Cambrian echinoderms have been named and described (see Fig. 1), or about 3.8 genera and 6.7 species per class on average. Even the largest echinoderm class during this time (eocrinoids) had a total of only 12 named genera. This pattern agrees with the model developed by Valentine (1969) for low-diversity generalized ecosystems at the beginning of the Paleozoic.

Complete Cambrian echinoderms appear to be much less widespread and usually less common than trilobites, brachiopods, and hyolithids; only at a few favorable localities are echinoderms very abundant. Relatively few collecting localities have produced complete Early and Middle Cambrian echinoderm specimens. For example, Sprinkle (1976, p. 65) listed only 21 localities from the Rocky Mountains that had produced complete echinoderms, based on five summers of field collecting. Probably fewer than 75 localities worldwide have yielded complete Early and Middle Cambrian echinoderms. Most Cambrian echinoderm genera are still known from only a single locality and stratigraphic unit, although some genera are more widespread on a single continent and a few are known from two or more continents.

Echinoderm debris (separate plates and columnals) is more widely distributed and makes up a large percentage of some Cambrian calcarenites. However, it is often difficult to identify which echinoderm genera or even classes are present in these occurrences if no complete specimens have been found (e.g., Donovan and Paul, 1982).

Echinoderms show a somewhat different diversity pattern than other members of the Cambrian fauna, such as arthropods. Trilobites dominate the Cambrian record of arthopods with about 815 genera or 94%; the five other arthropod classes in the Cambrian are relatively to very small, fewer than 50 genera (Robison and Kaesler, 1987). No single class of echinoderms dominates the Early and Middle Cambrian record, and the two largest classes in an evolutionary classification, eocrinoids (about 15 genera total) and edrioasteroids (at least eight genera), make up only about 37% and 20%, respectively, of the total diversity (see Fig. 1). Only much later, in the Middle Ordovician, did a single echinoderm class (crinoids) rise to dominance over the 16 other echinoderm classes present at that time (Sprinkle, 1980).

1.2. Morphologic Features of Cambrian Echinoderms

Most echinoderms occurring in the Cambrian have generalized and primitive morphology. The thecae of Cambrian echinoderms have numerous adjacent or imbricate plates, sometimes in multiple layers (Derstler, 1985); incompletely-developed symmetry, such as pentameral symmetry affecting only the ambulacral system, incomplete bilateral symmetry, or even asymmetry in such groups as helicoplacoids and some stylophorans; sutural pores (epispires) between the thecal plates for respiration; short, unbranched arms or brachioles for erect feeding appendages; and use of a multiplated (stelar) holdfast (Sprinkle, 1973; Brett, 1981) to elevate the theca above the sea floor.

Although Middle and Late Cambrian echinoderm genera retain many of these primitive features, some echinoderm genera show considerable improvements in morphology and the development of new and advanced features. The first columnal-bearing stem was developed in Middle Cambrian eocrinoids (Prokop, 1962; Sprinkle, 1973; Prokop and Fatka, 1985; Berg-Madsen, 1986). Some eocrinoids and edrioasteroids drastically reduced the number of thecal plates and began to organize them into alternating circlets (Ubaghs, 1953; Prokop and Fatka, 1985), an arrangement that was standardized in most later crinozoan and blastozoan genera. Pentameral symmetry spread in those genera with fewer thecal plates so that the thecal plates and other structures were organized into a fivefold arrangement. Middle Cambrian ctenocystoids already had relatively well-developed bilateral symmetry, and later members of several classes repeatedly developed this feature independently. Some Middle Cambrian eocrinoids and edrioasteroids lost the sutural epispires that characterized most earlier members of these groups (Sprinkle, 1973).

1.3. Ways of Life Present in Cambrian Echinoderms

Six major ways of life have commonly been used by fossil and living echinoderms (Sprinkle, 1980), and three of these were apparently developed during the Early to Middle Cambrian. Several echinoderm groups in the Cambrian became mobile, sessile, or attached, low-level epifaunal suspension feeders; these included edrioasteroids, homosteleans, small helicoplacoids, and perhaps some eocrinoids and homoiosteleans. These groups used tube feet on recumbent ambulacra, short brachioles, or short arms to capture small food particles drifting by in the lowest 5 cm above the sea floor.

Other groups in the Cambrian developed ways of elevating the feeding structures 5–12 cm above the sea floor to capture small food particles in higher bottom currents, thus becoming mobile or attached, medium-level epifaunal suspension feeders. Most Cambrian eocrinoids and crinoids having a multiplated stelar holdfast or columnal-bearing stem (Sprinkle, 1980), large upright helicoplacoids, and perhaps stylophorans using an erect aulacophore (Parsley and Caster, 1975) used this way of life. Tiering of Cambrian epifaunal suspension feeders was not very well developed yet (Ausich and Bottjer, 1985), and only the low to intermediate levels were occupied.

Several Cambrian echinoderm groups became mobile epifaunal, or shallow-burrowing infaunal, detritus feeders, including homoiosteleans, ctenocystoids, and perhaps homosteleans and holothurians (Sprinkle, 1980). These forms used short arms, disclike teeth, tube feet, or other appendages to sift through or ingest wholesale the top layer of organic-rich sediment. Most Cambrian forms using this way of life probably stayed within the top 2 cm of sediment.

No echinoderm herbivores, carnivores, or deep infaunal deposit feeders are known from the Cambrian. Herbivores such as regular echinoids and perhaps ophiocistioids and carnivores such as asteroids, ophiuroids, and some regular echinoids first appeared in the Early or Middle Ordovician, and deep-burrowing, deposit-feeding holothurians and irregular echinoids did not evolve until much later in the Late Paleozoic or Early Mesozoic.

1.4. Grouping into Evolutionary Faunas

The nine classes of echinoderms present in the Early to Middle Cambrian can be assigned to Sepkoski's three Phanerozoic evolutionary faunas (Sepkoski, 1981, 1984). Four classes appear to belong to the Cambrian evolutionary fauna because they have a peak in diversity within the Cambrian (Fig. 1). These are eocrinoids, the largest group of Cambrian echinoderms, helicoplacoids, only found in the Early Cambrian, and homosteleans and ctenocystoids, two small groups known only from the Middle Cambrian. These groups seem to be among the earliest echinoderms to appear in the fossil record and either are very successful in the Cambrian, thus reaching their peak of diversity here, or are present for only a short time before becoming extinct, so that their entire record is within the Early or Middle Cambrian (Sprinkle, 1981).

Four additional echinoderm classes that originated in the Cambrian appear to belong to Sepkoski's Paleozoic evolutionary fauna, having a peak in diversity during the Ordovician or later in the Paleozoic. These include edrioasteroids, the second most common echinoderm class in the Cambrian, stylophorans, and homoiosteleans, all of which reached their maximum diversity in the Early or Middle Ordovician (Ubaghs, 1968c; Caster, 1968; Sprinkle, 1980; Derstler, 1985). The fourth class was crinoids, a group that was rare in the Cambrian (one species from the Burgess Shale), became the dominant Paleozoic echinoderm in the Middle Ordovician, and peaked in generic diversity in the Mississippian (Sprinkle, 1980). Except for edrioasteroids, which already had several genera in the Early Cambrian, these other three classes may have appeared somewhat later in the record, or were more persistent and successful than members of the Cambrian evolutionary fauna.

Only one possible member of Sepkoski's Modern evolutionary fauna, the questionable holothurian *Eldonia*, has been found in the Middle Cambrian (Durham, 1974). Holothurians do not appear again until the Middle Ordovician (based on microscopic sclerites) and are uncommon during the rest of the Paleozoic (Sprinkle, 1980), probably because their sclerite-bearing skeletons disarticulate more easily than those of other echinoderms. However, Paul and Smith (1984, p. 469) express doubt that *Eldonia* was a holothurian or even an echinoderm; if this view is correct, then no member of Sepkoski's Modern evolutionary fauna was present in Cambrian echinoderms.

1.5. Geographic and Stratigraphic Distribution of Echinoderm Faunas

Most Early and Middle Cambrian echinoderm faunas based on complete specimens are known from eastern and western North America, northwestern Europe (Baltica), central and southwestern Europe, northwestern Africa, and eastern Australia and Tasmania. Although scattered occurrences of disarticulated plates have been reported from other areas such as Iran (Wolfart, 1974) and North Korea (Durham, 1978), Cambrian echinoderms appear to be uncommon and perhaps even absent from faunas in most other parts of the world. This apparent distribution pattern may be strongly biased by: (1) the distribution, exposure, and

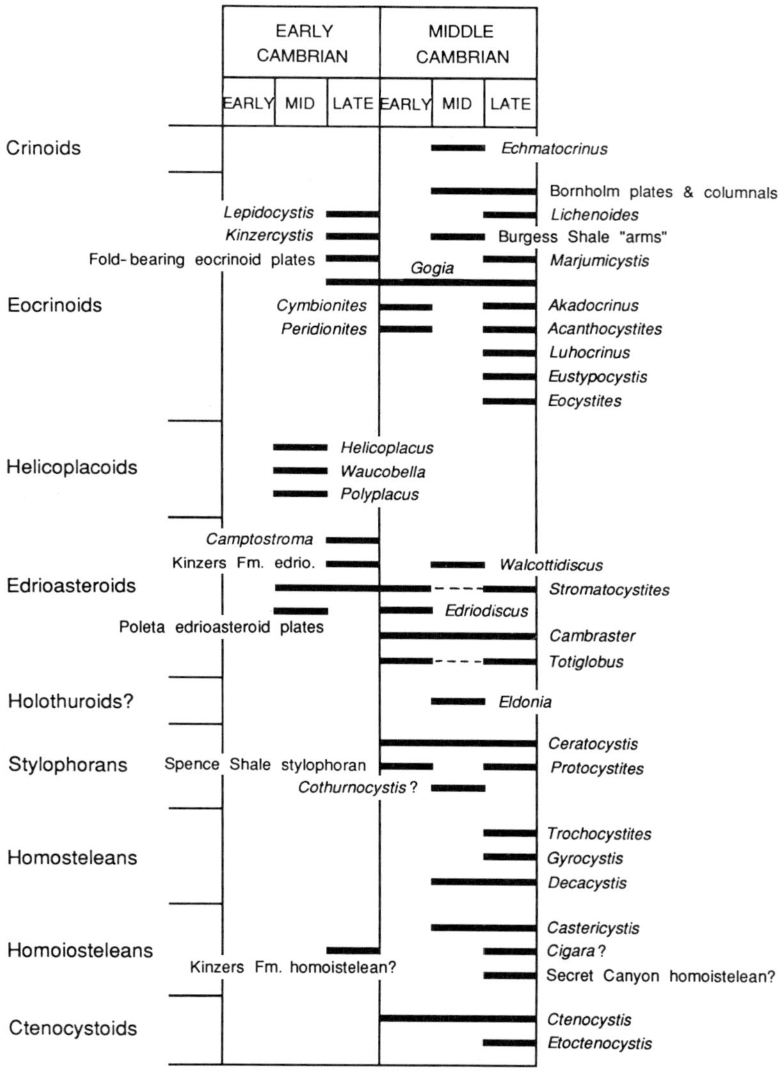

Figure 1. Stratigraphic ranges of all known Early and Middle Cambrian echinoderm genera grouped into four subphyla and nine classes using an evolutionary classification (Sprinkle, 1976, 1980) and plotted on a three-part zonation of both the Early and Middle Cambrian from information in Briggs and Robison (1984, Fig. 2), Berg-Madsen (1986, Fig. 2), and R. A. Robison (personal communication). Early Early Cambrian = *Fallotapis* zone or Tommotian; middle Early Cambrian = *Nevadella* Zone or Atdabanian and lower Botomian; late Early Cambrian = *Bonnia–Olenellus* Zone or upper Botomian and Toyonian; early Middle Cambrian = *Plagiura* through *Glossopleura* Zones or nearly to the top of the *Peronopsis bonnerensis* Zone; middle Middle Cambrian = top half of the *Oryctocephalus* to lower third of the *Bolaspidella* Zone or *Ptychagnostus praecurrens* to *P. atavus* Zones; late Middle Cambrian = upper two thirds of the *Bolaspidella* Zone into the lower part of the *Cedaria* Zone or *Ptychagnostus punctuosus* to *Lejopyge laevigata* Zones.

accessibility of Cambrian sedimentary rocks in which these echinoderms occur, (2) the presence in these areas of outer-shelf shales, the facies in which the most complete and best-preserved echinoderms have been found, (3) the location of tropical to low-temperate climates in the Cambrian where echinoderms apparently had their greatest abundance and diversity, and (4) the distribution of amateur and professional collectors, colleges with geology departments, and natural history museums to provide workers to collect and describe these echinoderms.

The following sections briefly review the occurrence and morphology of all echinoderm genera known from the Early and Middle Cambrian. These genera are considered by subphylum and class, using an evolutionary classification (Ubaghs, 1968a; Sprinkle, 1980) to group together genera that appear to be closely related. The stratigraphic occurrence of these echinoderm genera is plotted in Fig. 1; both the Early and Middle Cambrian have been arbitrarily divided into three parts (early, middle, and late), using correlations based on European and North American trilobite zones shown by Briggs and Robison (1984) and Berg-Madsen (1986). Echinoderm ranges were plotted as they were originally reported in the literature or by correlating to better-known areas, but some of these ranges or correlations may be incorrect (R. A. Robison, personal communication).

2. Crinozoa. Class Crinoidea (One Genus)

Only a single apparent crinoid genus has been described from the Early or Middle Cambrian (Fig. 1). This genus is *Echmatocrinus* (Sprinkle, 1973), described from seven specimens occurring in the middle Middle Cambrian Burgess Shale Member of the Stephen Formation in British Columbia, western Canada. *Echmatocrinus* has a conical theca with numerous irregularly-arranged thecal plates, 6–10 relatively short uniserial arms with small soft appendages (tube feet?) attached to them, and a long conical stelar holdfast used for attachment to objects on the sea floor (Fig. 2F). It appears to differ considerably from all later crinoids in lacking the well-organized theca with alternating rows of five plates each (radials, basals, etc.), and in lacking even a partly organized four- or five-part stem such as found in the next-youngest crinoids that appeared in the Early Ordovician.

3. Blastozoa. Class Eocrinoidea (Fifteen Genera)

Twelve named genera of eocrinoids have been described from Early and Middle Cambrian rock units in North America, Europe, and Australia. Several other questionably named or informal genera are based on fragmentary remains (plates,

Figure 2. Early and Middle Cambrian echinoderms from western North America. (A–E) Latex casts of natural or artificial molds; (F) a slab surface immersed in water; white bar scale in each figure equals 5 mm. (A) New stylophoran similar to *Cothurnocystis*, Spence Shale (Middle Cambrian), Calls Fort, northern Utah. (B) *Helicoplacus* sp., lower theca showing spiraling pleats and two ambulacra (top right), Poleta Formation (Lower Cambrian), Westgard Pass, eastern California. (C) *Totiglobus nimius* Bell and Sprinkle, oral surface of holotype MCZ 983, Chisholm Shale (Middle Cambrian), just northwest of Pioche, eastern Nevada. (D) *Ctenocystis utahensis* Robison and Sprinkle, aboral (top) and side views,

Spence Shale (Middle Cambrian), Antimony Canyon, northern Utah. (E) *Gogia* n. sp. showing theca with few epispires, brachioles, and stelar holdfast, Spence Shale (Middle Cambrian), Antimony Canyon, northern Utah. (F) *Echmatocrinus brachiatus* Sprinkle, complete holotype GSC 25962B showing soft tube feet on arms (top), irregular theca, and stelar holdfast attached to worm tube, Burgess Shale Member of the Stephen Formation (Middle Cambrian), north of Field, British Columbia, Canada. Slightly modified from Sprinkle (1976).

arms, or thecal bases). Four eocrinoid genera occur in the Early Cambrian, while 12 genera occur in the Middle Cambrian.

Most Cambrian eocrinoids are characterized by a globular to conical theca with numerous imbricate or adjacent plates (or some combination of these plate types) and epispires (sutural pores) between the adjacent plates either over much of the theca or near the summit (Broadhead, 1982). A recumbent ambulacral system is present on the flattened or domed upper side of the theca, usually consisting of five ambulacral grooves showing a "2–1–2" branching pattern (Sprinkle, 1973) at the central mouth and erect biserial brachioles attached alongside the ambulacra. The theca is attached to the substrate in most genera using either a cylindrical or inflated, multiplated, stelar holdfast with imbricate or adjacent plates (Sprinkle, 1973; Brett, 1981) or a longer columnal or fusular ring-bearing stem sometimes with a terminal cemented disc.

Two genera of lepidocystid eocrinoids occur in the Kinzers Formation of late Early Cambrian age (lower *Olenellus* Zone) in southeastern Pennsylvania in the eastern United States (Durham, 1968; Sprinkle, 1973; Paul and Smith, 1984). These two genera seem closely related, differing in thecal and holdfast proportions, living height above the sea floor, and development of the ambulacral system, especially brachiole size. *Lepidocystis* is represented by nearly 200 specimens, many of them unstudied. It has a short conical theca, imbricate plates, and a domed oral surface with numerous small epispire-bearing adjacent plates and an ambulacral system with erect brachioles that greatly increase in size away from the mouth. *Kinzercystis* is represented by 8–9 specimens in the Kinzers Formation. It has a larger conical theca with small imbricate plates, a short, cylindrical stelar holdfast made of small imbricate plates, and a wide, domed, oral surface with large, polygonal, epispire-bearing adjacent plates and bearing long ambulacra with erect brachioles that do not increase much in size away from the central mouth. These two genera are the earliest complete eocrinoids known from the fossil record, and are the only eocrinoids having most of the theca and holdfast covered by imbricate plates.

Donovan and Paul (1982) described but did not name several types of possible eocrinoid thecal plates from the Protolenus Limestone of late Early Cambrian age from the Comley district of Shropshire, England. Some of these plates have folds on the exterior surface and possibly epispires on the sutured margins. They do not match thecal plates from other eocrinoids, and may belong to an undescribed genus.

Gogia is known from rocks in western North America, southern France, and perhaps Australia, ranging in age from latest Early Cambrian (Durham, 1978) to latest Middle Cambrian (Robison, 1965; Sprinkle, 1973). Twelve species have been described, based on several thousand complete specimens, and at least three undescribed species are known. *Gogia* has a globular, conical, or barrel-shaped theca with numerous adjacent plates, sutural epispires over all or at least the upper part of the theca, numerous straight or spiraling brachioles attached to the edge of the summit at the end of 3–5 ambulacral grooves, and a cylindrical to inflated stelar holdfast for attachment (Fig. 2E). *Gogia* is the most widespread, diverse, and morphologically variable Cambrian eocrinoid known.

Marjumicystis is a small eocrinoid based on seven specimens recently described by Ubaghs and Robison (1985) from the Marjum Formation of late Middle

Cambrian age in western Utah in the western United States. It has a theca with numerous polygonal plates lacking epispires, a basal attachment disc on the lower theca instead of a holdfast, and short ambulacra on the summit leading to 2–7 short brachioles. *Marjumicystis* seems most closely related to *Gogia*, which occurs in the same region.

Akadocrinus (Prokop, 1962) and *Acanthocystites* (Ubaghs, 1968b; Fatka and Kordule, 1984) are two similar eocrinoids from the late Middle Cambrian of Jince, Czechoslovakia. Both genera have numerous irregular thecal plates with epispires, brachioles arranged around the summit at the end of short ambulacra, and a long columnal-bearing stem with incomplete wedge-shaped columnals proximally and an attachment disc at the distal tip. These two genera differ in thecal size and in the type of epispires present between the thecal plates, *Akadocrinus* having simple round epispires and *Acanthocystites* having elongate, partly closed epispires.

Eocystites is the name given to isolated thecal plates from the Middle Cambrian of New Brunswick, eastern Canada (Ubaghs, 1968b), and a disarticulated theca and plates of late Middle Cambrian age from southern France (Ubaghs, 1987). This genus has numerous thecal plates with radiating ridges but no epispires, high crests on some plates, and possibly a large periproct based on new material studied by Ubaghs (1987). No other distinctive features are known, and *Eocystites* seems most closely related to the Late Cambrian genus *Cambrocrinus* from Poland or the Early Ordovician widespread genus *Macrocystella*.

Luhocrinus was recently described by Prokop and Fatka (1985) from a single specimen in the late Middle Cambrian of Czechoslovakia. It is a small genus with 4–5 mostly alternating circlets of thecal plates lacking epispires, 6–10 long brachioles attached around the summit, and a large, columnal-bearing stem of unknown length having a large central lumen.

Lichenoides was redescribed by Ubaghs (1953, 1968b) based on about 10–12 specimens from the late Middle Cambrian of Czechoslovakia. *Lichenoides* is an unusual stemless eocrinoid having three major circlets of five alternating, epispire-bearing plates, 10–21 long brachioles mounted on small facets on the top two plate circlets, and the lowest plate circlet strongly inflated or bulged, probably to add weight to the lower theca. *Lichenoides* had no stem, but several small plates fill in the center of the aboral end where the specimens apparently sat on the sea floor. It is one of the earliest eocrinoids to develop a reduced number of thecal plates in alternating circlets with five plates each.

Cymbionites and *Peridionites* are two unusual genera originally described by Whitehouse (1941) from numerous disarticulated plates in the early Middle Cambrian of Queensland, Australia. They were more recently redescribed by Ubaghs (1968e), briefly discussed by Sprinkle (1973, pp. 135–136), and extensively reviewed by Smith (1982). Smith has clearly shown that *Cymbionites* is a heavily calcified, radially symmetrical, basal cup made of five plates; these cups are closely associated with epispire-bearing, polygonal plates that were located higher in the theca, and small U-shaped plates that probably supported brachioles on the summit. *Peridionites* is a similar, heavily calcified, but bilaterally symmetrical, basal cup made of five plates; these cups are also associated with polygonal, epispire-bearing plates. Smith (1982) has argued that these basal cups and associated plates are most likely stemless eocrinoids that sat upright (*Cym-*

bionites) or lay flat (*Peridionites*) on the sea floor. *Cymbionites* may have been closely related to *Lichenoides*, which also had a heavily calcified lower theca, no stem, and epispires between most of the plates. *Peridionites* may have been similar to flattened rhipidocystid eocrinoids (?) from the Middle Ordovician, but these forms differ somewhat in morphology and are widely separated in time.

Eustypocystis is a small eocrinoid from the Secret Canyon Formation of late Middle Cambrian age in central Nevada in the western United States described by Sprinkle (1973) based on about 20 specimens. It has a barrel-shaped theca with numerous small plates lacking epispires, usually three short brachioles branching off the summit, and a short stem with disclike columnals. It is the earliest eocrinoid having a columnal-bearing stem known from North America.

Berg-Madsen (1986) recently reported the presence of ridged thecal plates with epispires and columnals with nodes and crenulae of apparent eocrinoid origin from the Exsulans Limestone/Kalby Marl of middle Middle Cambrian age and the Andrarum Limestone of late Middle Cambrian age at Bornholm, Denmark. No taxa were named, but these columnals are the oldest ones known from Europe.

Finally, several "arm" fragments from an otherwise unknown eocrinoid have been described by Sprinkle (1973) from the Burgess Shale Member of the Stephen Formation of middle Middle Cambrian age in British Columbia, western Canada. These segments have apparent biserial plating with two rows of smaller biserial brachioles branching off along the length. Unfortunately, no other thecal or stem features of this taxon are known. These "arms" appear to be erect ambulacra with brachioles similar to the armlike appendages of the Late Cambrian eocrinoid *Trachelocrinus* from nearby Wyoming and Montana in the northwestern United States (Sprinkle, 1973, pp. 123–125).

4. Echinozoa

4.1. Class Helicoplacoidea (Three Genera)

Helicoplacoids are now the most diverse echinoderms in the Early Cambrian, with three named genera and six named species (Durham and Caster, 1963; Durham, 1967); several other species await description. Helicoplacoids only occur in the middle Early Cambrian in eastern California and adjacent western Nevada in the western United States and from the same age at one locality in northwestern Alberta, western Canada; between 500 and 600 complete specimens have been collected from these areas. Two closely related genera (*Helicoplacus* and *Waucobella*) have a fusiform theca with diagonally spiraling pleats each made up of three rows of plates forming an interambulacral column. Smaller-plated ambulacra with pores for the tube feet on each side and tiny cover plates protecting a central food groove spiral around the theca every 10–13 pleats (Fig. 2B). The third genus (*Polyplacus*) has all of the spiraling pleats composed of tiny plates that resemble the ambulacra of the other two helicoplacoid genera; the ambulacra in *Polyplacus* have not been identified for certain.

The original authors inferred that the mouth was at the upper rounded pole of the theca and that the long single ambulacrum branched once about halfway down the theca (Durham and Caster, 1963; Durham, 1967, 1990). However, Derstler (1981, 1985) [followed by Paul and Smith (1984)] has convincingly argued that the mouth was located where the ambulacra split and that three ambulacral branches lead away from it, designated A (spiraling downward), B/C, and D/E (both spiraling upward) in the standard system. The anal opening or pyramid has not been found. Helicoplacoids were apparently attached, low- to medium-level suspension feeders that buried the lower theca in the sediment (Durham, 1990) or attached to objects lying on the substrate (Derstler, 1985).

4.2. Class Edrioasteroidea (Eight Genera)

Eight genera of edrioasteroids are known from the Early and Middle Cambrian, four genera (two named, two unnamed) occurring in the Early Cambrian, and five named genera in the Middle Cambrian (Smith, 1985).

Most Cambrian edrioasteroids are characterized by a biscuit-shaped to domed theca with a pentagonal to circular outline, imbricate or adjacent plates on both the upper (oral) and lower (aboral) surfaces, five long, straight or curved ambulacra radiating out in a "2–1–2" arrangement (Sprinkle, 1973) from the central mouth on the oral surface, biserial flooring plates with pores for tube feet making up the grooved bottom of each ambulacrum with 1–4 series of smaller cover plates protecting this ambulacral groove, a lateral anal pyramid in the CD interray, and an apparent attachment area with larger flat plates on the aboral side or forming a peripheral rim around the margin of this side. Most Cambrian edrioasteroids appear to have been sessile or attached, low-level epifaunal suspension feeders.

Camptostroma is an early edrioasteroid (?) known from the Kinzers Formation of late Early Cambrian age from southeastern Pennsylvania in the eastern United States. Nearly 200 specimens have been collected and studied. *Camptostroma* is a large, domed, multilayered edrioasteroid with an imbricately pleated lower theca and five, recumbent, curved ambulacra leading to a central mouth on the upper surface. The thecal morphology is complex and difficult to interpret, as shown by the different interpretations that various authors have made. Ruedemann (1933) named and described *Camptostroma* using the original natural molds, but interpreted this genus as a hydrozoan. Durham (1966) recognized *Camptostroma* as an echinoderm because of its complex calcite plates, but could not find a class to assign it to and instead set up the new class Camptostromatoidea. The most recent authors to study this genus (Derstler, 1981, p. 72; 1985; Paul and Smith, 1984, p. 451) agree that *Camptostroma* is an early (stem) echinoderm or edrioasteroid, but have reconstructed the thecal shape, especially the pleated and imbricately-plated lower theca, much differently.

A second, small, unnamed edrioasteroid represented by about ten specimens occurs with *Camptostroma* in the Kinzers Formation in Pennsylvania; it was briefly described by Derstler (1981, p. 72), but has not been illustrated or described in any detail.

Stromatocystites is a large edrioasteroid (?) from the late Early Cambrian of Newfoundland, eastern Canada, the early Middle Cambrian of western Queensland, Australia (Jell *et al.*, 1985; Smith and Jell, 1990), the late Middle Cambrian of Czechoslovakia, and possibly the middle Early Cambrian of Nevada (Durham, 1967). *Stromatocystites* has a completely plated theca with a rounded pentagonal outline and a domed oral surface, five narrow, straight ambulacra that reach the thecal margin and have biserial flooring plates and several series of cover plates, epispires between the adjacent oral surface plates located between the ambulacra, and somewhat larger thecal plates at the margin of the oral surface and in the center of the aboral surface.

Disarticulated thecal plates that may belong to an edrioasteroid have been described from the middle Early Cambrian of California as "eocrinoid (?) plates from the Poleta Formation" (Sprinkle, 1973). Some of these plates with a half or full ambulacrum are very similar to plates found in later edrioasteroids such as *Cambraster* or *Totiglobus* from the Middle Cambrian. However, these plates cannot at present be assigned with certainty to any edrioasteroid genus.

Cambraster has been described from the late Middle Cambrian of southern France (Cabibel *et al.*, 1958; Termier and Termier, 1969; Ubaghs, 1971; Smith, 1985) and from the early and middle Middle Cambrian of eastern Australia and Tasmania (Jell *et al.*, 1985). *Cambraster* has a fully plated domed theca with adjacent polygonal plates, 15–18 stout marginal ossicles, epispires between the adoral surface plates, relatively wide, pore-bearing, straight ambulacra that extend out just over the marginal ossicles, and larger elongate plates around the center of the aboral surface. Some species of *Cambraster* may have been attached aborally.

Edriodiscus, known from three specimens from the early Middle Cambrian of western Queensland, eastern Australia, was originally described as a cyclocystoid by Henderson and Shergold (1971), but was redescribed as an edrioasteroid and transferred to a new genus by Jell *et al.* (1985) and further redescribed by Smith and Jell (1990). It has fully plated aboral and oral surfaces with adjacent plates and no epispires, five straight to slightly curved ambulacra that just cross a thin marginal rim that appears on both the oral and aboral surfaces, a well-developed peripheral skirt of small plates outside the rim, and the anal pyramid just inside the rim between the C and D ambulacra. *Edriodiscus* seems closely related to *Cambraster*, which also occurs in Australia.

Walcottidiscus is a poorly known edrioasteroid based on two specimens from the middle Middle Cambrian Burgess Shale Member of the Stephen Formation in eastern British Columbia, western Canada; it has recently been redescribed by Smith (1985). *Walcottidiscus* has a domed, fully plated, oral surface with imbricate plates and no epispires, a mostly unplated aboral surface with a tiny-plated peripheral rim, five strongly curved (clockwise) ambulacra with flooring plates and several series of smaller cover plates, and an anal pyramid low on the theca. *Walcottidiscus* seems most closely related to *Edrioaster* and *Edriophus* from the Middle Ordovician, but better-preserved specimens will be needed to confirm this.

Totiglobus is a globular edrioasteroid known from the Chisholm Shale of early Middle Cambrian age near Pioche, eastern Nevada, based on over 900 specimens (Bell and Sprinkle, 1978), and questionably from the Marjum Formation of late Middle Cambrian age in western Utah (one specimen described by Sprinkle, 1985), both in the western United States. It has a fully plated globular to ellipsoidal theca with imbricate to small adjacent plates lacking epispires, long straight ambulacra with pore-bearing flooring plates and 2–4 series of cover plates, no peripheral rim but a central basal disc on the aboral surface with ridged plates internally and a tiny-plated membrane externally that was probably used for attachment, and an anal pyramid nearly on the aboral surface. *Totiglobus* appears to be closely related to *Walcottidiscus* and later edrioasterid genera, but differs from all of these in its thecal shape, straight ambulacra, and unique basal disc (Fig. 2C).

4.3. Class Holothuroidea(?) (One Genus)

Only a single possible holothurian, the genus *Eldonia,* is known from the Cambrian, from about 100 specimens occurring in the middle Middle Cambrian Burgess Shale Member of the Stephen Formation in eastern British Columbia, western Canada. *Eldonia* is a soft-bodied organism with an umbrella-shaped body, short oral tentacles, and a coiled gut suspended by apparent mesenteries within a coelom. Durham (1974) restudied the type material and concluded that *Eldonia* was most likely a holothurian echinoderm and not a siphonophore as suggested by some other authors; however, Paul and Smith (1984, p. 469) strongly disagreed with this assignment and argued that *Eldonia* is not even an echinoderm.

5. Homalozoa

5.1. Class Stylophora (Four Genera)

Stylophorans are puzzling asymmetrical homalozoans or "carpoids" that have been considered as primitive chordates by some authors (Jefferies, 1967, 1969, 1979, 1986; Jefferies *et al.*, 1987; Paul, 1977) and true echinoderms by others (Ubaghs, 1968c, 1971, 1975; Regnéll, 1975; Sprinkle, 1980, 1983; Derstler, 1985; Campbell and Marshall, 1987; Parsley, 1988, 1991). Stylophorans are considered as echinoderms here, convergent on a mobile, "flat-fish," body design, even though they apparently had a low- to medium-level suspension-feeding way of life (Parsley and Caster, 1975; Parsley, 1991). Four genera (three named) belonging to the more primitive Order Cornuta are known from the Middle Cambrian of Europe and North America.

Ceratocystis is a primitive stylophoran from the middle Middle Cambrian of Czechoslovakia (Ubaghs, 1967; Jefferies, 1969), the late Middle Cambrian of southern France (Ubaghs, 1987), and the early Middle Cambrian of Sweden (see comment in Berg-Madsen, 1986, p. 68). *Ceratocystis* has an asymmetrical, boot-shaped, flattened theca with relatively few large plates on both surfaces. A small pyramid (anal pyramid?) occurs at one end of the theca, and a row of small sutural pores resembling epispires occurs on one side of the "toe region" of the "boot." A long moveable appendage called an aulacophore is attached to the edge of the theca opposite the pyramid; the aulacophore has a large-diameter, hollow, flexible region with imbricate plates near the theca, a conical solid stylocone medially, and a long, slowly-tapering, distal region of uniserial, hemispherical ossicles protected on the flattened side by a double series of cover plates. This appendage appears to have been a moveable feeding structure bearing a protected food groove along its length with tube feet that could be exposed by opening the cover plates.

Protocystites from the late Middle Cambrian of southern Wales has recently been redescribed by Jefferies *et al.* (1987), based on six poorly preserved but complete thecae and many separate plates. This genus is similar to *Ceratocystis* in most features and apparently closely related to it, but has a slightly different thecal shape, smaller and more numerous plates on one of the flattened thecal sides, fewer pores on this side, and a shorter but better organized proximal aulacophore with distinct rings of four plates each.

A species of *Cothurnocystis* (?) has recently been described from the Wheeler Shale of middle Middle Cambrian age from western Utah in the western United States based on about 200 specimens (Ubaghs and Robison, 1988); several other species of this genus occur in the Ordovician in Europe (Ubaghs, 1968c). *Cothurnocystis* has a flattened boot-shaped theca, a marginal frame of elongate U-shaped plates supporting upper and lower membranes bearing numerous polygonal plates, long spines or knobs on some of the marginals, and a long, well-organized aulacophore similar to that found in *Protocystites*.

Another similar stylophoran that may be related to *Cothurnocystis* or belong to a new genus was reported by Robison and Sprinkle (1969) and figured by Sprinkle (1976) from the Spence Shale of early Middle Cambrian age in northern Utah (see Fig. 2A). This form, represented by only four specimens, has a flattened theca with a rounded boot- to heart-shaped frame, similar small-plated upper and lower membranes, and a long, well-organized aulacophore.

5.2. Class Homostelea (Three Genera)

Homosteleans are homalozoans having a flattened, heavily-framed theca, smaller-plated central areas that contain epispires in one genus, marginal food grooves on the thecal rim leading to a central mouth, a hinged operculum that may have covered the anal opening, and a long tail-like stele attached to one margin of the frame (Ubaghs, 1968d; Schroeder, 1973). Three genera are known from the Middle Cambrian of Europe and North Africa. *Trochocystites* is known from the late Middle Cambrian of Czechoslovakia, Germany, and southern France; it has long ambulacral grooves on the thecal margin and is the only genus known to have epispires. *Gyrocystis* is known from the late Middle Cambrian of France, Spain, and Morocco; it has shorter ambulacral grooves and larger centrals. *Decacystis* occurs in the middle and late Middle Cambrian of Spain and southern France; it is a small form with thick marginals and only one ambulacral groove. Several other genera described by Cabibel *et al.* (1958) and Termier and Termier (1973) appear to be synonyms. Homosteleans were apparently slow-moving, benthic detritus or suspension feeders.

5.3. Class Homoiostelea (Four Genera)

Two named genera and two undescribed forms that may be homoiosteleans occur in the Early or Middle Cambrian. Homoiosteleans have a somewhat asymmetrical ovoid to triangular theca with numerous small plates, a long stele with an enlarged part near the theca, a smaller biserial arm attached to one side of the theca, apparently with the mouth at its base, and bearing an ambulacral groove protected by biserial cover plates, and an anal pyramid or hinged plate at one corner of the theca. Homoiosteleans were apparently mobile, epifaunal suspension or detritus feeders.

Derstler (1981, 1985, pp. 400–404) briefly described but did not name a possible homoiostelean based on a single specimen from the Kinzers Formation of late Early Cambrian age in Pennsylvania in the eastern United States. This specimen has a rounded triangular theca with numerous poorly-preserved plates that appear to be imbricate, a relatively long arm attached to one side, and a relatively short undifferentiated stele attached to the lower corner of the theca. It is the only homoiostelean without an enlarged proximal stele where this structure is attached to the theca.

Castericystis is now a well-known homoiostelean based on about 1200 specimens from the Wheeler and Marjum Formations of middle and late Middle Cambrian age in western Utah in the western United States (Ubaghs and Robison, 1985, 1988). It has an elongate ovoid theca with numerous adjacent plates, an anal pyramid on the lower theca, a long stele divided into an enlarged imbricately-plated proximal part and a tapering spiny distal part, and a smaller arm with cover plates that is attached to the opposite end of the theca. One unusual feature is that small juvenile specimens in this genus are found attached to all parts of adults using the distal tip of the stele at one occurrence, indicating that this structure may have temporarily served for attachment.

Cigara is a puzzling echinoderm that may be a homoiostelean from the Jince Formation of late Middle Cambrian age (Ubaghs, 1968b; Sprinkle, 1973) in Czechoslovakia and from the Oelandicus Shale of early Middle Cambrian age in Sweden (Berg-Madsen, 1986, p. 67). About ten specimens have been collected from these two occurrences, most of them poorly preserved or incomplete. *Cigara* has a bent theca (?) with one end covered by numerous polygonal epispire-bearing plates, the bent middle part made up of X-braced plates arranged in a latticelike pattern, and the other tapering end by large, imbricate, rugose plates lacking epispires. *Cigara* could belong to almost any echinozoan or homalozoan group, and is only provisionally included here with the homoiosteleans.

The final possible homoiostelean occurs with the small, stemmed eocrinoid *Eustypocystis* in the Secret Canyon Formation of late Middle Cambrian age in central Nevada in the western United States; about 20 specimens are known, and a partial stele (?) was figured by Sprinkle (1973, pl. 28, Fig. 4). The remains consist of curved, elongate, biserially-plated structures that resemble the wide distal steles of *Castericystis*, which is about the same age. No other parts of this echinoderm are known, and it has not been described in detail or named.

5.4. Class Ctenocystoidea (Two Genera)

Two genera and at least five species of ctenocystoids have been named and described since the first occurrences were found and recognized as new in 1968 (Robison and Sprinkle, 1969). Species of *Ctenocystis* occur in the early and middle Middle Cambrian in northern Utah and southern Idaho (Sprinkle and Robison, 1978) and in western Utah (Ubaghs and Robison, 1988) in the western United States, in the middle Middle Cambrian of Poland (Orlowski, 1959; Derstler, 1985), in the late Middle Cambrian of southern France (Ubaghs, 1987) and southern Wales (Jefferies *et al.*, 1987), and in the middle Middle Cambrian of Tasmania (Jell *et al.*, 1985). The related genus *Etoctenocystis* occurs in the late Middle Cambrian of Czechoslovakia (Fatka and Kordule, 1985). Both of these genera have a small flattened theca with near bilateral symmetry, a double-layered marginal frame having elongate plates, a grill-like set of ctenoid plates on one margin around the mouth and short food grooves, small plates in a membrane covering the top surface and larger pointed to polygonal plates on the bottom, and a conelike anal pyramid between the frame layers opposite the mouth (Fig. 2D). No appendages (aulacophore, stele, or arm) are present, unlike other homalozoans. Ctenocystoids were apparently mobile, epifaunal or barely infaunal detritus feeders that could probably swim short distances because of their streamlined shape and good bilateral symmetry.

6. Classification of Early and Middle Cambrian Echinoderms

6.1. Evolutionary Classification

Most classifications of fossil echinoderms have used a classical evolutionary philosophy as their basis, combining information about the development of new features and the branching sequence of the fossil groups along with information on overall similarity, level of development, and persistence and success of the fossil groups (Bather, 1900; Jaekel, 1918; Regnéll, 1945; Ubaghs, 1968a; Nichols, 1969; Sprinkle, 1973, 1976, 1980; Paul, 1977, 1979). As pointed out by Smith (1984), evolutionary systematists working on fossil echinoderms have tended to emphasize the differences between taxa, and have produced somewhat ambiguous classifications in which one cannot reconstruct the exact sequence of branching directly from the classification. Evolutionary systematists have typically used "spindle diagrams" to show species or genera grouped into higher taxa, along with information on stratigraphic range, taxon diversity, and some degree of relationship (how the solid or dashed lines between groups are connected), all together in a single diagram. Evolutionary systematists cannot go directly from a diagram of this type to a Linnaean classification, or vice versa.

Some of these criticisms by Smith (1984) may be valid, but others are probably overdone or inappropriate. For example, Linnaean classifications were never intended to be perfectly dichotomous or to have maximum "information content" about relationships. In many ways this ambiguity in classification may mirror the complexity found in nature.

Figure 3 is a "spindle diagram" showing the Early and Middle Cambrian echinoderm groups from Fig. 1 based on an evolutionary classification with some of their inferred relationships. Groups have been established both on a detailed analysis of their morphology and on their overall similarity and way of life. This means that genera such as *Camptostroma* and *Stromatocystites* are considered as early and primitive members of the edrioasteroids and not split off as stem echinoderms or stem eleutherozoans. Some homologous characters used to link together groups in Paul and Smith (1984) have also been included here to indicate possible relationships. Some classes such as eocrinoids were (strictly) mono-

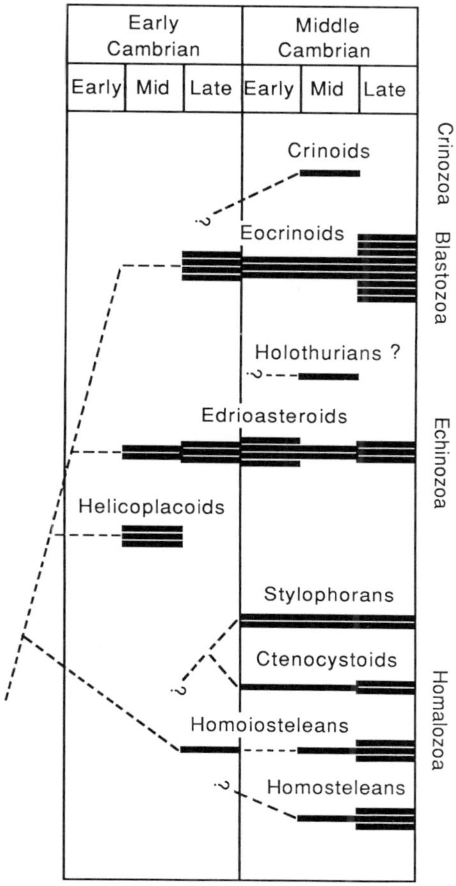

Figure 3. "Spindle diagram" showing the diversity, stratigraphic ranges, and inferred relationships of the Early and Middle Cambrian echinoderm genera plotted in Fig. 1, using an evolutionary classification. Note that two classes (helicoplacoids and edrioasteroids) appeared almost simultaneously in the middle part of the Early Cambrian, and all others had appeared by the middle part of the Middle Cambrian. Stratigraphic units are the same as in Fig. 1.

phyletic at the time of this diagram (end of the Middle Cambrian), even though they later became paraphyletic (see later discussion) when more advanced groups such as rhombiferans subsequently branched from them. Several classes cannot easily be linked to others, indicating a long previous history probably extending some distance back into the Early Cambrian or latest Precambrian.

6.2. Cladistic Classification

Several recent classifications of fossil echinoderms or some of their included groups have been proposed using a cladistic (also phylogenetic or genealogical) approach based only on the development of new homologous characters to indicate branching sequence. Paleontologists using these cladistic methods have tried either to fit fossil groups into a phylogeny derived from characters in living echinoderms (Smith, 1984, 1988a) or to work out a cladistic methodology (stem and crown groups) for analyzing the relationship and history within extinct fossil echinoderm groups (Jefferies, 1979, 1986; Haugh and Bell, 1980; Paul and Smith,

1984; Derstler, 1985; Smith, 1985, 1988b; Donovan, 1988; Paul, 1988). Although I disagree with some of the basic cladistic philosophy and some of the details (see below), I consider these recent papers important and well-written contributions to working out the relationships, phylogeny, and classification of these echinoderms. The cladograms produced for particular groups by these analyses (for example, Smith, 1984, Text-Fig. 9; Derstler, 1985, Fig. 2-2) appear to be quite reasonable and informative. However, the cladistic classifications produced directly from these cladograms are awkward to use and much less satisfactory (Smith, 1984, Table 3; Jefferies *et al.*, 1987, p. 436); transferring a cladistic analysis into a reasonable cladistic-based Linnaean classification seems to be a major problem.

In contrast, a few cladistic workers have taken the extreme position that the fossil record is useless in trying to work out the relationships and overall phylogeny of a group of organisms, instead relying only on features shown by monophyletic living representatives (Patterson, 1981). I believe that this extreme cladistic viewpoint is clearly incorrect, and that authors who ignore the fossil record have much less chance of correctly reconstructing the relationships and phylogeny of any group with a long past history. Fossils are historical data, a record of what has happened to organisms through several hundred million years of geologic time. Even with an incomplete and somewhat biased fossil record, the data available from this source are probably as good for reconstructing relationships and phylogeny as the data from any other source, such as embryology of living forms or DNA sequences. The present day with its living fauna is a "slice of time," just like any other single time plane in the fossil record, such as the Middle Cambrian or the Late Permian. Because it is a summation over many time planes, the fossil record has geologic time on its side, yielding historical information on what has happened in particular groups that cannot always be reconstructed by analyzing much-removed living groups and their morphologic features. Both the present-day fauna with its more complete and easily-obtained data and the fossil record with its historical information on morphology and timing of events need to be used together to work out the relationship and phylogeny of complex and long-lived groups of organisms such as echinoderms or vertebrates [see similar statements in Paul (1985, p. 8) and Gauthier *et al.* (1988, p. 191)].

One area of disagreement between cladists and evolutionary systematists involves the definition and use of monophyletic and paraphyletic groups. Cladists use a strict definition of a monophyletic (holophyletic) group to include the latest common ancestor of the group plus *all* of its fossil and living descendants. The living descendants and their latest common ancestor represent the (monophyletic) crown group, while the now-extinct fossil representatives down to their latest common ancestor represent a paraphyletic stem group (Jefferies, 1979). Groups that develop similar morphology by convergence represent artificial polyphyletic groups when classified together. Cladists reject the use of both polyphyletic and paraphyletic groups in classification; evolutionary systematists reject polyphyletic groups but accept paraphyletic groups, at least provisionally. In fossil echinoderms, groups such as eocrinoids and edrioasteroids would be paraphyletic groups because they were apparently ancestral to later, more-advanced groups (Sprinkle, 1973; Paul and Smith, 1984; Smith, 1988b; Paul,

1988). However, there is nothing inherently "wrong" with using paraphyletic groups in classification; they are monophyletic (wide definition) in that they come from one ancestor, are often distinctive morphologically, but have given rise to other groups that developed new morphological features. They sometimes represent "ragbag" groups that retain primitive features and usually become extinct before their more advanced descendants, but this may be typical for groups undergoing a series of adaptive radiations. Of special interest, if paraphyletic fossil groups are rejected in the classification of organisms, then very few strictly monophyletic groups become extinct in the fossil record, and some of the biggest mass extinctions almost disappear when these groups are plotted against geologic time (Patterson and Smith, 1987). Paraphyletic groups apparently carry much of the information about the magnitude and timing of extinctions in the past (Sepkoski, 1987).

I have reservations about several of the proposed homologies used by cladistic workers to link together different groups of Cambrian echinoderms; two of these are discussed below. Derstler (1981, p. 73; 1985, Fig. 3-7) and Paul and Smith (1984, Fig. 13) have argued that brachioles in Cambrian eocrinoids were derived from the aligned cover plate sheets found in Early Cambrian forms (edrioasteroids?) such as *Camptostroma* and *Stromatocystites*. This derivation is a brilliant idea and may even be true, but more evidence is needed to support it and several problems with this proposed origin need to be resolved. First, cover plates are not well known in most Cambrian edrioasteroids, and it is not known how extensive aligned cover-plate sheets really are. If brachioles originated from aligned cover plates, where did the brachiolar cover plates that protect them originate? The "lateral branches" (= tube feet?) in the ambulacral grooves of *Camptostroma* lie in between the aligned cover plates that became brachioles in this model; this location would make it difficult for these branches to continue up the newly-formed brachioles of eocrinoids to produce multiple tube feet for feeding. Paul and Smith (1984, p. 464) and Derstler (1981, p. 72; 1985) disagree on whether these newly-formed brachioles did in fact contain tube feet (see also Sprinkle, 1983, pp. 10–12). Finally, *Camptostroma* and *Lepidocystis* have large pores for ampullae or tube feet through their ambulacral plates, in contrast to later eocrinoids and other blastozoans that do not; what happened to these water vascular structures and tube feet in the ambulacra of these later blastozoans?

Paul and Smith (1984, Fig. 13) argued that the ambulacra in *Lepidocystis* (Early Cambrian) extended off the thecal surface to produce an erect biserial arm that was the source for the erect uniserial arms found in *Echmatocrinus* (Middle Cambrian) and later crinoids (see Fig. 4). Previously, this structure in *Lepidocystis* had been interpreted as an enlarged terminal brachiole (Sprinkle, 1973, p. 65) and not a projecting ambulacrum. I doubt that the erect arm hypothesis is correct, for the following reasons. (1) The morphology of the terminal "arm" in *Lepidocystis* is nearly identical to the other brachioles in this genus, except it is larger, longer, and has two sets of cover plates instead of just one. (2) All of the brachioles gradually increase in size distally on the ambulacrum; the terminal one is just the largest (about 1½ times the length of the next proximal one). (3) In some specimens the terminal "arm" appears to attach alongside the ambulacral groove and not directly at the end of it (see Sprinkle, 1973, Text-Fig. 21B). (4) This terminal "arm" is attached to small, curved, ambulacral mounting plates as are the other

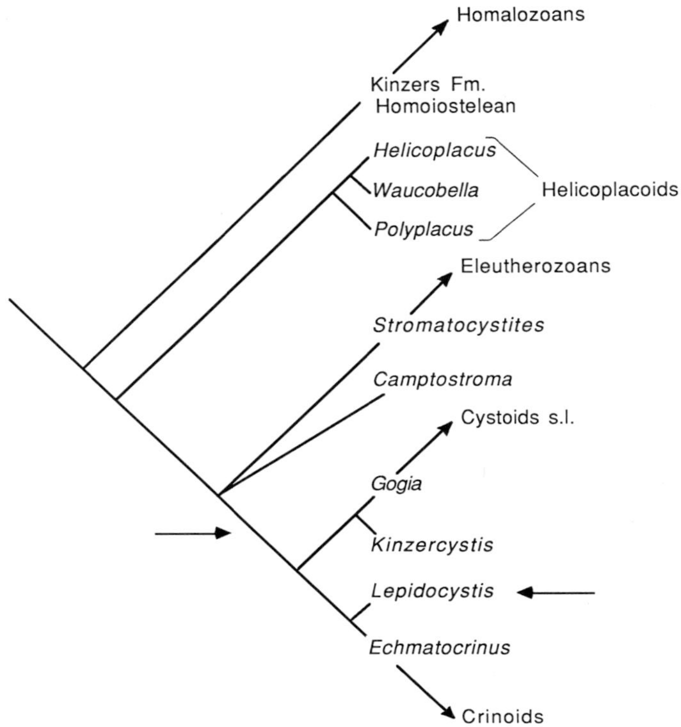

Figure 4. Modified cladogram derived from Paul and Smith (1984, Fig. 12) showing their inferred relationships of Early (and a few Middle) Cambrian echinoderms and the lineages leading to later groups. Their proposed development of brachioles from cover plate sheets (left arrow) and development of arms from erect ambulacra (right arrow) are discussed further in the text.

brachioles. (5) If this structure were an erect ambulacrum projecting off the edge of the theca, it might be expected to have smaller brachioles attached to it as are present alongside the recumbent ambulacrum, but it has no smaller appendages. (6) *Lepidocystis* and *Kinzercystis* appear to be closely related genera in the same family having very similar morphology and occurring together in the same formation; an "erect arm" would represent a major difference between these genera, whereas an enlarged terminal brachiole would not. For these reasons, I think it is unlikely that *Lepidocystis* has "erect arms" or that it is closely related or ancestral to *Echmatocrinus* and later crinoids, as inferred by Paul and Smith (1984).

7. Conclusions

1. Echinoderms underwent a major radiation at the beginning of the Cambrian, producing 8–9 classes by the end of the Middle Cambrian. Diversity was relatively low (about 34 genera), and no one class dominated the echinoderm record until later, in the Ordovician.

2. Most Early Cambrian echinoderms had primitive morphologic features, but later forms gradually improved the morphology and produced better designs for feeding, attachment, and other activities.

3. Mobile, sessile, or attached, low- and intermediate-level suspension feeders plus mobile, epifaunal detritus feeders dominated the Cambrian record of echinoderms. Other ways of life were not developed until the Late Cambrian or Early Ordovician.

4. Cambrian echinoderms are about evenly divided between groups belonging to the Cambrian evolutionary fauna and others belonging to the Paleozoic evolutionary fauna. Only a single member of the Modern evolutionary fauna may be present in the Cambrian.

5. An evolutionary classification of Early and Middle echinoderms assigns them to four subphyla (several originating in the Late Precambrian) and to 8–9 classes (several minor or short-lived), many of which do not have obvious connections or known ancestors at present. This classification is preferred by the present author.

6. Several authors have proposed cladistic classifications of Early and Middle Cambrian echinoderms that connect many more groups together into two strictly monophyletic subphyla and about four large but widely varying classes, all of which originated within the Early or Middle Cambrian. Although this approach has merit, some of the inferred connections may not be correct.

ACKNOWLEDGMENTS. I thank Richard A. Robison, University of Kansas, Lawrence, for information on Cambrian correlations, Tim Rowe, University of Texas at Austin, for help interpreting the cladistic classifications, C. R. C. Paul, University of Liverpool, for reviewing the completed manuscript, and Rosemary Brant and Linda Carlson, University of Texas at Austin, for assistance in typing various drafts of this manuscript. The Geology Foundation, University of Texas at Austin, provided financial assistance at various stages of this project.

References

Ausich, W. I., and Bottjer, D. J., 1985, Echinoderm role in the history of Phanerozoic tiering in suspension-feeding communities, in: *Echinodermata (Proceedings of the Fifth International Echinoderm Conference, Galway)* (B. F. Keegan and B. D. S. O'Connor, eds.), Balkema, Rotterdam, pp. 3–11.

Bather, F. A., 1900, The Echinoderma, in: *A Treatise on Zoology*, Part 3 (E. F. Lankester, ed.), Adam and Charles Black, London, pp. 1–216.

Bell, B. M., and Sprinkle, J., 1978, *Totiglobus*, an unusual new edrioasteroid from the Middle Cambrian of Nevada, *J. Paleontol.* **52**:243–266.

Berg-Madsen, V., 1986, Middle Cambrian cystoid (*sensu lato*) stem columnals from Bornholm, Denmark, *Lethaia* **19**:67–80.

Brett, C. E., 1981, Terminology and functional morphology of attachment structures in pelmatozoan echinoderms, *Lethaia* **14**:343–370.

Briggs, D. E. G., and Robison, R. A., 1984, Exceptionally preserved nontrilobite arthropods and *Anomalocaris* from the Middle Cambrian of Utah, *Univ. Kans. Paleontol. Contrib. Pap.* **111**:1–23.

Broadhead, T. W., 1982, Reappraisal of class Eocrinoidea (Echinodermata), in: *Echinoderms: Proceedings of the International Conference, Tampa Bay* (J. M. Lawrence, ed.), Balkema, Rotterdam, pp. 125–131.

Cabibel, J., Termier, H., and Termier, G., 1958, Les echinodermes Mesocambriens de la Montagne Noire (Sud de la France), *Ann. Paleontol.* **44**:281–294.

Campbell, K. S. W., and Marshall, C. R., 1987, Rates of evolution among Palaeozoic echinoderms, in: *Rates of Evolution* (K. S. W. Campbell and M. F. Day, eds.), Allen and Unwin, London, pp. 61–100.

Caster, K. E., 1968, Homoiostelea, in: *Treatise on Invertebrate Paleontology*, Part S, *Echinodermata 1* (R. C. Moore, ed.), Geological Society of America, New York, and University of Kansas Press, Lawrence, Kansas, pp. S581–S623.

Derstler, K., 1981, Morphological diversity of Early Cambrian echinoderms, in: *Short Papers for the Second International Symposium on the Cambrian System* (M. E. Taylor, ed.), U. S. Geological Survey Open-File Report 81-743, pp. 71–75.

Derstler, K., 1985, Studies on the morphological evolution of echinoderms, Ph.D. thesis, University of California, Davis.

Donovan, S. K., 1988, The early evolution of the Crinoidea, in: *Echinoderm Phylogeny and Evolutionary Biology* (C. R. C. Paul and A. B. Smith, eds.), Clarendon Press, Oxford, pp. 235–244.

Donovan, S. K., and Paul, C. R. C., 1982, Lower Cambrian echinoderm plates from Comley, Shropshire, England, *Geol. Mag.* **119:**611–614.

Durham, J. W., 1966, *Camptostroma,* an Early Cambrian supposed scyphozoan, referable to Echinodermata, *J. Paleontol.* **40:**1216–1220.

Durham, J. W., 1967, Notes on the Helicoplacoidea and early echinoderms, *J. Paleontol.* **41:**97–102.

Durham, J. W., 1968, Lepidocystoids, in: *Treatise on Invertebrate Paleontology,* Part S, *Echinodermata 1* (R. C. Moore, ed.), Geological Society of America, New York, and University of Kansas Press, Lawrence, Kansas, pp. S631–S634.

Durham, J. W., 1974, Systematic position of *Eldonia ludwigi* Walcott, *J. Paleontol.* **48:**750–755.

Durham, J. W., 1978, A Lower Cambrian eocrinoid, *J. Paleontol.* **52:**195–199.

Durham, J. W., 1990, More on the helicoplacoids (Echinodermata) (Abst.). *Paleobios* **13**(Suppl.)**:**3.

Durham, J. W., and Caster, K. E., 1963, Helicoplacoidea: A new class of echinoderms, *Science* **140:**820–822.

Fatka, O., and Kordule, V., 1984, *Acanthocystites* Barrande, 1887 (Eocrinoidea) from the Jince Formation (Middle Cambrian) of the Barrandian area, *Vestnik Ustredniho Ustavu Geol.* **59:**299–302.

Fatka, O., and Kordule, V., 1985, *Etoctenocystis bohemica* gen. et sp. nov., new ctenocystoid from Czechoslovakia (Echinodermata, Middle Cambrian), *Vestnik Ustredniho Ustavu Geol.* **60:**225–230.

Gauthier, J., Kluge, A. G., and Rowe, T., 1988, Amniote phylogeny and the importance of fossils, *Cladistics* **4:**105–209.

Haugh, B. N., and Bell, B. M., 1980, Classification schemes, in: *Echinoderms, Notes for a Short Course* (T. W. Broadhead and J. A. Waters, eds.), University of Tennessee Department of Geological Sciences, Studies in Geology, No. 3, pp. 94–105.

Henderson, R. A., and Shergold, J. H., 1971, *Cyclocystoides* from Early Middle Cambrian rocks of northwestern Queensland, Australia, *Palaeontology* **14:**704–710.

Jaekel, O., 1918, Phylogenie und system der Pelmatozoen, *Palaeontol. Z.* **3**(1)**:**1–128.

Jefferies, R. P. S., 1967, Some fossil chordates with echinoderm affinities, *Symp. Zool. Soc. Lond.* **20:**163–208.

Jefferies, R. P. S., 1969, *Ceratocystis perneri* Jaekel—A Middle Cambrian chordate with echinoderm affinities, *Palaeontology* **12:**494–535.

Jefferies, R. P. S., 1979, The origin of chordates—A methodological essay, in: *The Origin of Major Invertebrate Groups,* (M. R. House, ed.), Academic Press, New York, pp. 443–477.

Jefferies, R. P. S., 1986, *The Ancestry of Vertebrates,* British Museum (Natural History), London.

Jefferies, R. P. S., Lewis, M., and Donovan, S. K., 1987, *Protocystites menevensis*—A stem-group chordate (Cornuta) from the Middle Cambrian of South Wales, *Palaeontology* **30:**429–484.

Jell, P. E., Burrett, C. F., and Banks, M. R., 1985, Cambrian and Ordovician echinoderms from eastern Australia, *Alcheringa* **9:**183–208.

Nichols, D., 1969, *Echinoderms,* 4th ed., Hutchinson University Library, London.

Orlowski, S., 1959, Archaeocyatha from the Middle Cambrian of the Holy Cross Mountains, *Bull. Acad. Polon. Sci.* **7**(5)**:**363–368.

Parsley, R. L., 1988, Feeding and respiratory strategies in Stylophora, in: *Echinoderm Phylogeny and Evolutionary Biology* (C. R. C. Paul and A. B. Smith, eds.), Clarendon Press, Oxford, pp. 347–361.

Parsley, R. L., 1991, Review of selected North American mitrate stylophorans (Homalozoa: Echinodermata), *Bull. Am. Paleontol.* **100:**1–57.

Parsley, R. D., and Caster, K. E., 1975, Zoological affinities and functional morphology of the Mitrata (Echinodermata) (Abst.), *Geol. Soc. Am. Abstr. Progr.* **7**(7)**:**1225–1226.

Patterson, C., 1981, Significance of fossils in determining evolutionary relationships, *Annu. Rev. Ecol. Syst.* **12:**195–223.

Patterson, C., and Smith, A. B., 1987, Is the periodicity of extinctions a taxonomic artifact?, *Nature* **330:**248–251.

Paul, C. R. C., 1977, Evolution of primitive echinoderms, in: *Patterns of Evolution as Illustrated by the Fossil Record* (Hallam, ed.), Elsevier, Amsterdam, pp. 123–158.

Paul, C. R. C., 1979, Early echinoderm radiation, in: *The Origin of Major Invertebrate Groups* (M. R. House, ed.), Academic Press, New York, pp. 415–434.

Paul, C. R. C., 1985, The adequacy of the fossil record reconsidered, in: Evolutionary Case Histories from the Fossil Record (J. C. W. Cope and P. W. Skelton, eds.), *Spec. Pap. Palaeontol.* **33**:7–15.

Paul, C. R. C., 1988, The phylogeny of the cystoids, in: *Echinoderm Phylogeny and Evolutionary Biology* (C. R. C. Paul and A. B. Smith, eds.), Clarendon Press, Oxford, pp. 199–213.

Paul, C. R. C., and Smith, A. B., 1984, The early radiation and phylogeny of echinoderms, *Biol. Rev.* **59**:443–481.

Prokop, R. J., 1962, *Akadocrinus* nov. gen., nova lilijice z jineckeho kambria (Eocrinoidea), *Ustredniho Ustavu Geol. Sbornik (Paleontol.)*, **27**:31–39.

Prokop, R. J., and Fatka, O., 1985, *Luhocrinus monicae* gen. et sp. n. (Eocrinoidea) from the Middle Cambrian of Bohemia, *Vestnik Ustredniho ustavu Geol.* **60**:231–234.

Regnéll, G., 1945, Non-crinoid Pelmatozoa from the Paleozoic of Sweden, *Lunds Geol. Mineral. Inst. Medd.* No. 108.

Regnéll, G., 1975, Review of recent research on "pelmatozoans", *Palaeontolol. Z.* **49**:530–564.

Robison, R. A., 1965, Middle Cambrian eocrinoids from western North America, *J. Paleontol.* **39**:355–364.

Robison, R. A., and Kaesler, R. L., 1987, Phylum Arthropoda, in: *Fossil Invertebrates* (R. S. Boardman, A. H. Cheetham, and A. J. Rowell, eds.), Blackwell, Palo Alto, California, pp. 205–269.

Robison, R. A., and Sprinkle, J., 1969, Ctenocystoidea: New class of primitive echinoderms, *Science* **166**:1512–1514.

Ruedemann, R., 1933, *Camptostroma*, a Lower Cambrian floating hydrozoan, *U. S. Natl. Mus. Proc.* **82**(13):1–8.

Schroeder, R., 1973, Carpoideen aus dem Mittelkambrian Nordspaniens, *Paleontogr. Abt. A* **141**(5–6):119–142.

Sepkoski, J. J., 1981, A factor analytic description of the Phanerozoic marine fossil record, *Paleobiology* **7**:36–53.

Sepkoski, J. J., 1984, A kinetic model of Phanerozoic taxonomic diversity. III. Post-Paleozoic families and mass extinctions, *Paleobiology* **10**:246–267.

Sepkoski, J. J., Jr., 1987, Reply to Patterson and Smith, *Nature* **330**:251–252.

Smith, A. B., 1982, The affinities of the Middle Cambrian Haplozoa (Echinodermata), *Alcheringa* **6**:93–99.

Smith, A. B., 1984, Classification of the Echinodermata, *Palaeontology* **27**:431–459.

Smith, A. B., 1985, Cambrian eleutherozoan echinoderms and the early diversification of edrioasteroids, *Palaeontology* **28**:715–756.

Smith, A. B., 1988a, Fossil evidence for the relationship of extant echinoderm classes and their times of divergence, in: *Echinoderm Phylogeny and Evolutionary Biology* (C. R. C. Paul and A. B. Smith, eds.), Clarendon Press, Oxford, pp. 85–97.

Smith, A. B., 1988b, Patterns of diversification and extinction in Early Palaeozoic echinoderms, *Palaeontology* **31**:799–828.

Smith, A. B., and Jell, P. A., 1990, Cambrian edrioasteroids from Australia and the origin of starfishes, *Mem. Queensl. Mus.* **28**:715–778.

Sprinkle, J., 1973, *Morphology and Evolution of Blastozoan Echinoderms*, Harvard University, Museum of Comparative Zoology Special Publication.

Sprinkle, J., 1976, Biostratigraphy and paleoecology of Cambrian echinoderms from the Rocky Mountains, *Brigham Young Univ. Geol. Stud.* **23**(2):61–73.

Sprinkle, J., 1980, An overview of the fossil record, in: *Echinoderms, Notes for a Short Course* (T. W. Broadhead and J. A. Waters, eds.), University of Tennessee Department of Geological Sciences, Studies in Geology No. 3, pp. 15–26.

Sprinkle, J., 1981, Diversity and evolutionary patterns of Cambrian echinoderms, in: *Short Papers for the Second International Symposium on the Cambrian System* (M. E. Taylor, ed.), U. S. Geological Survey Open-File Report 81-743, pp. 219–221.

Sprinkle, J., 1983, Patterns and problems in echinoderm evolution, in: *Echinoderm Studies* (M. Jangoux and J. M. Lawrence, eds.), Balkema, Rotterdam, Vol. 1, pp. 1–18.

Sprinkle, J., 1985, New edrioasteroid from the Middle Cambrian of western Utah, *Univ. Kans. Paleontol. Contrib. Pap.* **116**:1–4.

Sprinkle, J., and Robison, R. A., 1978, Ctenocystoids, in: *Treatise on Invertebrate Paleontology,* Part T, *Echinodermata 2* (R. C. Moore and Curt Teichert, eds.), Geological Society of America, Boulder, Colorado, and University of Kansas Press, Lawrence, Kansas, pp. T998–T1002.

Termier, H., and Termier, G., 1969, Les Stromatocystitoides et leur descendance, Essai sur l'evolution des premiers echinoderms, *Geobios* **2**:131–156.

Termier, H., and Termier, G., 1973, Les echinodermes Cincta du Cambrien de la Montagne Noire (France), *Geobios* **6**:243–266.

Ubaghs, G., 1953, Notes sur *Lichenoides priscus* Barrande, eocrinoide du Cambrien Moyen de la Tchecoslovaquie, *Inst. R. Sci. Nat. Belg. Bull.* **29**(34):1–24.

Ubaghs, G., 1967, Le Genre *Ceratocystis* Jaekel (Echinodermata, Stylophora), *Univ. Kans. Paleontol. Contrib. Pap.* **22**:1–16.

Ubaghs, G., 1968a, General characters of Echinodermata, in: *Treatise on Invertebrate Paleontology,* Part S, *Echinodermata 1* (R. C. Moore, ed.), Geological Society of America, New York, and University of Kansas Press, Lawrence, Kansas, pp. S3–S85.

Ubaghs, G., 1968b, Eocrinoidea, in: *Treatise on Invertebrate Paleontology,* Part S, *Echinodermata 1* (R. C. Moore, ed.), Geological Society of America, New York, and University of Kansas Press, Lawrence, Kansas, pp. S455–S495.

Ubaghs, G., 1968c, Stylophora, in: *Treatise on Invertebrate Paleontology,* Part S, *Echinodermata 1* (R. C. Moore, ed.), Geological Society of America, New York, and University of Kansas Press, Lawrence, Kansas, pp. S495–S565.

Ubaghs, G., 1968d, Homostelea, in: *Treatise on Invertebrate Paleontology,* Part S, *Echinodermata 1* (R. C. Moore, ed.), Geological Society of America, New York, and University of Kansas Press, Lawrence, Kansas, pp. S565–S581.

Ubaghs, G., 1968e, *Cymbionites* and *Peridionites*—Unclassified Middle Cambrian echinoderms, in: *Treatise on Invertebrate Paleontology,* Part S, *Echinodermata 1* (R. C. Moore, ed.), Geological Society of America, New York, and University of Kansas Press, Lawrence, Kansas, pp. S634–S637.

Ubaghs, G., 1971, Diversite et specialisation des plus anciens echinodermes que l'on connaisse, *Biol. Rev.* **46**:157–200.

Ubaghs, G., 1975, Early Paleozoic echinoderms, *Annu. Rev. Earth Planet. Sci.* **3**:79–98.

Ubaghs, G., 1987, Echinodermes nouveaux du Cambrien Moyen de la Montagne Noire (France), *Ann. Paleontol.* **73**:(1):1–27.

Ubaghs, G., and Robison, R. A., 1985, A new homostelean and a new eocrinoid from the Middle Cambrian of Utah, *Univ. Kans. Paleontol. Contrib. Pap.* **115**:1–24.

Ubaghs, G., and Robison, R. A., 1988, Homalozoan echinoderms of the Wheeler Formation (Middle Cambrian) of western Utah, *Univ. Kans. Paleontol. Contrib. Pap.* **120**:1–17.

Valentine, J. W., 1969, Patterns of taxonomic and ecological structure of the shelf benthos during Phanerozoic time, *Palaeontology* **12**:684–709.

Whitehouse, F. W., 1941, The Cambrian faunas of north-eastern Australia. Part 4: Early Cambrian echinoderms similar to the larval stages of Recent forms, *Mem. Queensl. Mus.* **12**(1):1–28.

Wolfart, R., 1974, Die fauna (Brachiopoda, Mollusca, Trilobita) aus dem Unter-Kambrium von Kerman, sudost-Iran, *Geol. Jahrb. B* **8**:5–70.

Chapter 12

The Cambrian Radiation of Brachiopods

LEONID YE. POPOV

1. Introduction

Traditionally, Brachiopoda (*sensu lato*) is regarded as a distinct phylum of lophophorate animals and is divided into two major taxa, Class Inarticulata (or Ecardines) and Class Articulata (or Testicardines). Both taxa are polythetic, but their diagnoses are based mainly on two features: the presence or absence of the articulation of the valves, and an open or blind digestive tract. The known differences between classes are possibly persistent from the beginning of the Ordovician. However, the earliest Cambrian brachiopods display various combinations of unusual morphological features, which were not inherited in further evolution. During the past two decades considerable progress has been made toward understanding the morphology and phylogenetic relationships between major taxa of Cambrian brachiopods, but our knowledge of the initial Cambrian radiation is still incomplete.

LEONID YE. POPOV • Scientific Research Geological Institute, St. Petersburg, Russia.
Origin and Early Evolution of the Metazoa, edited by Jere H. Lipps and Philip W. Signor. Plenum Press, New York, 1992.

The most distinctive feature of brachiopods as a grade, which separates them from the other lophophore-bearing animals, is a filtration system in an isolated cavity. Affinities of the earliest brachiopods and lingulates, expressed in the mineralized bivalved shell, in the relative position of the mantle and visceral cavities, in the shelly musculature, as well as in lophophore morphology, seem to have resulted from coadaptations of the shell-opening mechanisms and the unique pattern of the feeding system. There are essential differences in the chemical composition and morphology of the shell, as well as in the anatomy of soft parts of the body between Early Cambrian lingulate and brachiopod stocks. Filtration in the isolated cavity may have originated two or three times during the history of primeval lophophorates.

The oldest known phosphate-shelled "brachiopods," or lingulates, are the paterinids of the family Cryptotretidae. They are recorded from the pretrilobite Tommotian Stage in Siberia. Cryptotretidae are characterized by biconvex shells with a straight posterior margin and well-defined ventral pseudointerarea bearing the homeodeltidium. The relatively large visceral cavity is disposed close to the posterior shell margin, but the data on the shell musculature and mantle canal system are inadequate. The well-defined muscle scars and mantle canals are known only for the Family Paterinidae, which first appeared in the Atdabanian (Laurie, 1987). The homology of the main muscle scars of paterinids and other brachiopods is still obscure. The shell ultrastructure of closely packed vertical prisms, which is characteristic of paterinids, is extremely different from those of the other phosphate-shelled stocks (Popov and Ushatinskaya, 1987).

The Orders Lingulida and Acrotretida are also represented in the Early Cambrian. Both are recorded beginning with the Atdabanian. The first ancient acrotretids and lingulids are characterized by a peripherally placed shell musculature and a baculate type of mantle canal system with well-developed dorsal vascula media. Their shells are composed of alternating mineralized and organic layers. Their valves are opened hydraulically. The dermal muscles are probably the most important in the hydraulic opening system of lingulates, given the absence of any kind of oblique muscles inserted by one end into the body wall. The pedicle foramen, which was regarded for a long time as the distinctive feature of acrotretids, has been discovered recently in the aberrant lingulid family Lingulellotretidae.

The calcareous-shelled brachiopods also were represented among the earliest skeletonized animals. The most ancient orders, Obolellida and Kutorginida, are recorded first in the late Tommotian (Grigorjeva *et al.*, 1983; Ushatinskaya, 1987). The earliest orthids of the Superfamily Protorthacea are known since Botomian time. True pentamerids have been recorded not earlier than the Middle Cambrian.

The majority of the Cambrian calcareous-shelled brachiopods are characterized by variable primitive articulation. The deltidiodont type of cardinalia originated in the Middle Cambrian. The earliest obolellids lack articulation. The first "articulate" members of the order with a primitive strophic articulation and direct shell-opening mechanism regulated by the diductor muscles originated in the late Atdabanian.

The Kutorginids and Nisusiids, belonging to the same order, possess an extremely primitive articulation lacking teeth. The distinctive features of those stocks are the large convex pseudodeltidium and the minute apical foramen.

Nisusiids probably possessed an open digestive tract with a posteriorly placed anus (Rowell and Caruso, 1985).

The only known Cambrian inarticulate stock is the Family Matutellidae, which is first recorded in the middle Early Cambrian. This family is characterized by a straight posterior margin lacking articulation, and hemiperipheral growth of the brachial valve. The species also possess a pseudodeltidium and an additional aperture near the beak of the pedicle valve, as well as a large, posteriorly placed visceral cavity.

The shell structure of Cambrian brachiopods is poorly known, but evidently there are two main patterns of shell ultrastructure in the Early Cambrian—one with fibrous and the other with laminar secondary shell layers.

At least two major stocks can be distinguished among brachiopods (*sensu lato*): Phylum Brachiopoda (*sensu stricto*) and Class Lingulata. They display differences which sometimes affect their body plans. Their content does not correspond to the class divisions as traditionally used, but generally reflects the differences in the chemical composition of the shell.

Phylum Brachiopoda embraces the articulate brachiopods, and living and extinct inarticulate taxa with calcareous shells. Class Lingulata includes only inarticulate taxa with a calcium phosphate shell. The ancestors of brachiopods may have been derived from the protolophophorates (Gorjansky and Popov, 1986) before the radiation of the other lophophorate stocks. The taxonomic position of the Class Lingulata and its relations to other lophophorates is unclear. The origin of lingulates, however, is associated with the subsequent radiation of lophophorates.

2. Major Taxa and Their Diagnostic Characteristics

2.1. Phylum Brachiopoda (*sensu stricto*)

Brachiopoda are animals with bilaterally symmetrical, bivalved calcareous shells secreted by the dorsal and ventral mantle lobes and composed of periostracum, a primary layer secreted by the edges of the mantle, and a secondary layer secreted by the entire surface of the outer epithelium. A lophophore with or without skeletal supports occupies the mantle cavity. It has one row of filaments at the trocholophe stage and two rows later in ontogeny. The nervous system is subepithelial. An anus lies in the midpart of the posterior body wall or ends blindly. The circulatory system is open with one or, less commonly, two pairs of metanephridia that act as gonoducts. Gonads are located in the coelomic canals of the mantle. An enterocoeleous coelom with a large mesocoel is connected with the lophophore and mantle. A small metacoel is present in the pedicle segment. In Recent articulate brachiopods, this division of the coelomic space is present only in larvae and disappears during metamorphosis. The larva of brachiopods (cephalula) is lecithotrophic. It settles on the peduncular lobe, which is transformed in articulate stocks into a cartilaginous stalk. After settlement the larva undergoes a metamorphosis, which is expressed in the inversion of the mantle folds, reduction of the aboral nervous center, and the formation of shell and lophophore. Most of these features are characteristic of all calcareous-shelled brachiopods, except for craniids, in which the initial stages of ontogeny are unknown.

Brachiopods feed on strongly transformed organic matter and bacterial aggregates. Both *Crania* and articulate brachiopods react quite negatively to increased suspended phytoplankton, as confirmed by numerous experiments and the decreased diversity of Recent brachiopod faunas in highly productive oceanic zones (Zezina, 1976).

The subdivision of the Phylum Brachiopoda into two classes, Inarticulata and Articulata, remains reasonable. The ordinal classification of Inarticulata used here differs significantly from that used by Williams and Rowell (1965). Only three orders are assigned to this class: the Craniida, ranging from the Ordovician to the Recent without any appreciable change, and the extinct Craniopsida and Trimerellida. Previously these stocks were included in the Order Lingulida.

The traditional opinion that the Class Articulata is represented in the Cambrian by only two orders must be revised. In addition to the Orthida and Pentamerida, two short-lived stocks, Obolellida and Kutorginida, may also be assigned to the Class Articulata.

The Obolellida is commonly included in the Class Inarticulata. However, renewed study of *Alisina* and *Trematobolus* has shown that the systematic position of this order needs to be reconsidered. Both genera have a primitive strophic articulation wherein their valves were opened by diductors (Gorjansky and Popov, 1985). The Order Kutorginida also has an articulation that is the most primitive among the known brachiopods.

The Cambrian Orthida, in fact, represents a grade. Moreover, the Family Nisusiidae, which usually has been regarded as an ancestral articulate stock, lacks the characteristic hinge mechanism of the deltidiodont type (Rowell and Caruso, 1985), and at the same time is much more closely related to kutorginids than was previously thought. Obolellids and kutorginids are apparently direct descendants of the ancestral stock of the Class Articulata.

2.2. Class Lingulata

The Class Lingulata is characterized by a bilaterally symmetrical, bivalved calcium phosphate shell, secreted by the ventral and dorsal mantle lobes. The lophophore, with large and small brachial canals, has no skeletal supports, and bears two rows of filaments at the trocholophe stage and medial tentacle (epistome). The nervous system is located in the basal part of the epidermis, forming only one subenteric ganglion. The digestive tract is open with an anterior anus near the right nephropore. Gonads occupy the free edges of the peritoneal bands. The circulatory system is open. There is one pair of metanephridia. The coelom is schizocoeleous with a mesocoel (brachiocoel) and a large metacoel entering the pedicle and mantle. The planktotrophic larva settles on the pedicle, which is an outgrowth of the ventral part of the body. The mantle lobes are persistent in their definitive position.

Carbonogydratolytic enzymes, such as cellulase, have been found in the digestive tract of lingulids (Chuang, 1959; McCammon, 1981) and an abundance of phytoplankton does not adversely affect their distribution (Zezina, 1976).

Six orders can be assigned to the Class Lingulata with a greater or lesser certainty. Four of them, the Acrotretida, Lingulida, Paterinida, and Heosomocelyphida, are known from the Early Cambrian, the Siphonotretida appear close to the Middle Cambrian, whereas Discinida make their appearance in the Ordovician.

The siphonotretids and discinids are ranked differently by different workers. As a rule, both stocks are included in the Order Acrotretida at the rank of superfamilies. However, strong arguments confirm the origin of both directly from obolids (Holmer, 1987; Popov and Nõlvak, 1987). They will be discussed in more detail below, when evolution of the Class Lingulata is analyzed.

3. Some Aspects of Morphology

3.1. The Inhalant and Exhalant Current System in the Mantle Cavity and Its Connection with a Shell-Opening Mechanism of the Prototypic Brachiopod

Filtration in a separate cavity likely appeared in ancestors of brachiopods (*sensu stricto*) before they had acquired a mineralized exoskeleton, but after they had expanded mantle lobes. The function of the hydroskeleton, which imparted a relative rigidity to the soft covers surrounding the filtration cavity, or brachiocoel, was primary for coelomic canals in the mantle. At this level of organization none of the shell-opening mechanisms known in brachiopods would work because they require the presence of rigid valves. Perhaps the shell-opening mechanism of protobrachiopods was inherited by some Early Cambrian brachiopods, such as the Family Matutellidae and related inarticulate brachiopods found recently in the Lower Cambrian of South Kirgizia.

The most unusual feature of Matutellidae is the presence of two apertures on one of the valves. The first is a triangular delthyrium on a flat pseudointerarea that most likely served for the emergence of a pedicle. A small convex pseudodeltidium in the apical part of the delthyrium covered the proximal end of a pedicle. Another large aperture is located in front of the umbo (Fig. 1I) and is partly covered anteriorly by a plate. Along with a traditional interpretation of this aperture as a pedicle opening, it could also have served as an additional aperture, connected with the canals of the mantle cavity located along the dorsal body wall and similar to those observed in Recent *Crania*. The possible function of this additional aperture was shown by experiments on living *Crania* with a drilled dorsal valve (Chuang, 1974). In the closed shells, after the lophophore starts working, water is admitted into the mantle cavity by suction through the drilled hole, but when the brachia and filaments relaxed, it continued to function as an additional exhalant aperture. Thus, the foramen of Matutellidae may have served as part of a "hydrodynamic" shell-opening mechanism. In a closed shell, water was sucked into the mantle cavity through the foramen and thus created an initial pressure difference enabling the valves to open halfway. Such a mechanism could also function in brachiopod ancestors lacking a mineralized shell.

The relation of the valve homology in Matutellidae to those of other brachiopods is an important problem. According to the above reconstruction, the

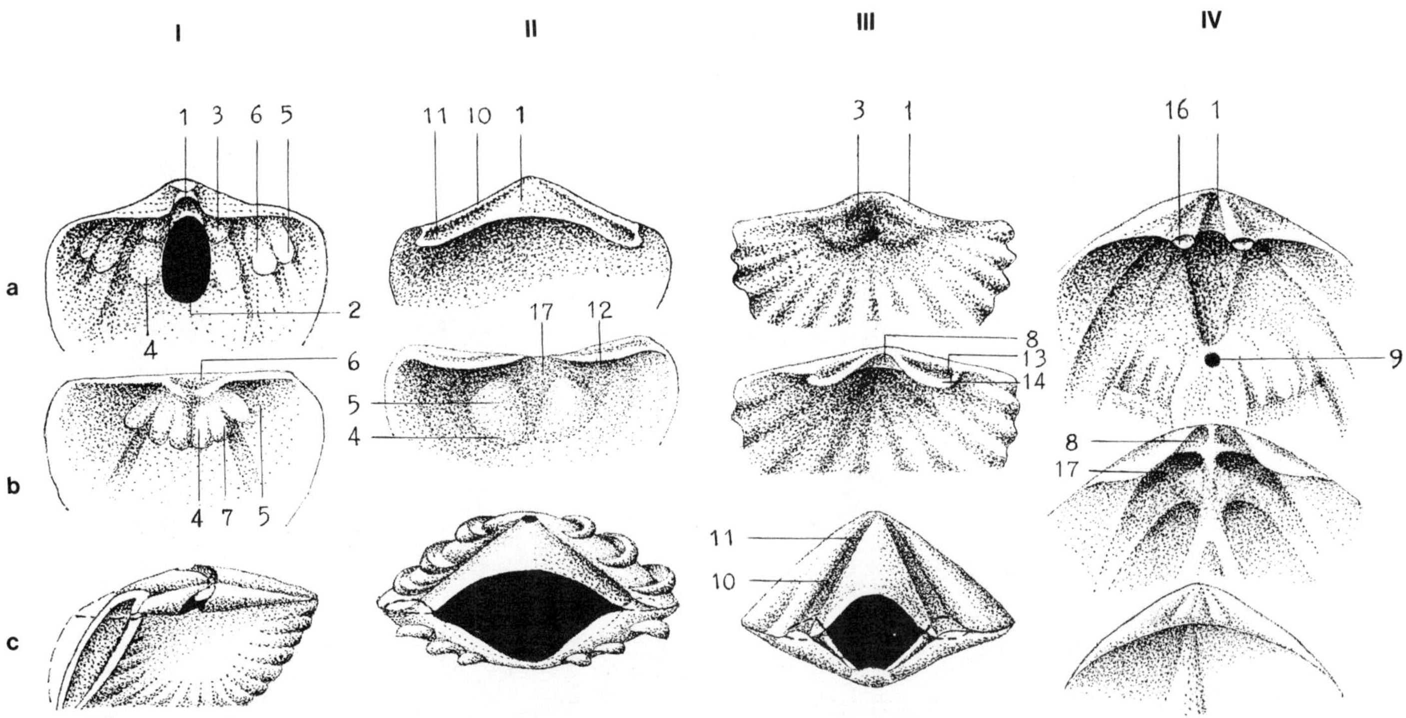

Figure 1. Internal morphology of (a) pedicle and (b) brachial valves, and (c) posterior view of the shell of selected Cambrian calcareous-shelled brachiopods. (I) Gen. et sp. indet. Matutellidae, Lower Cambrian, South Kirgizia; (II) *Kutorgina catenata* Koneva, Lower Cambrian, South Kirgizia; (III) *Nisusia* sp., Lower Cambrian, South Kirgizia; (IV) *Trematobolus pristinus* bicostatus Gorjansky, Lower Cambrian, Siberia; (V) *Bajarinovia* (?) sp., Lower Cambrian, South Kirgizia; (VI) *Glyptoria* sp., Middle Cambrian, South Kirgizia; (VII) *Arctohedra pyramidalis* Aksarina, Middle Cambrian, South

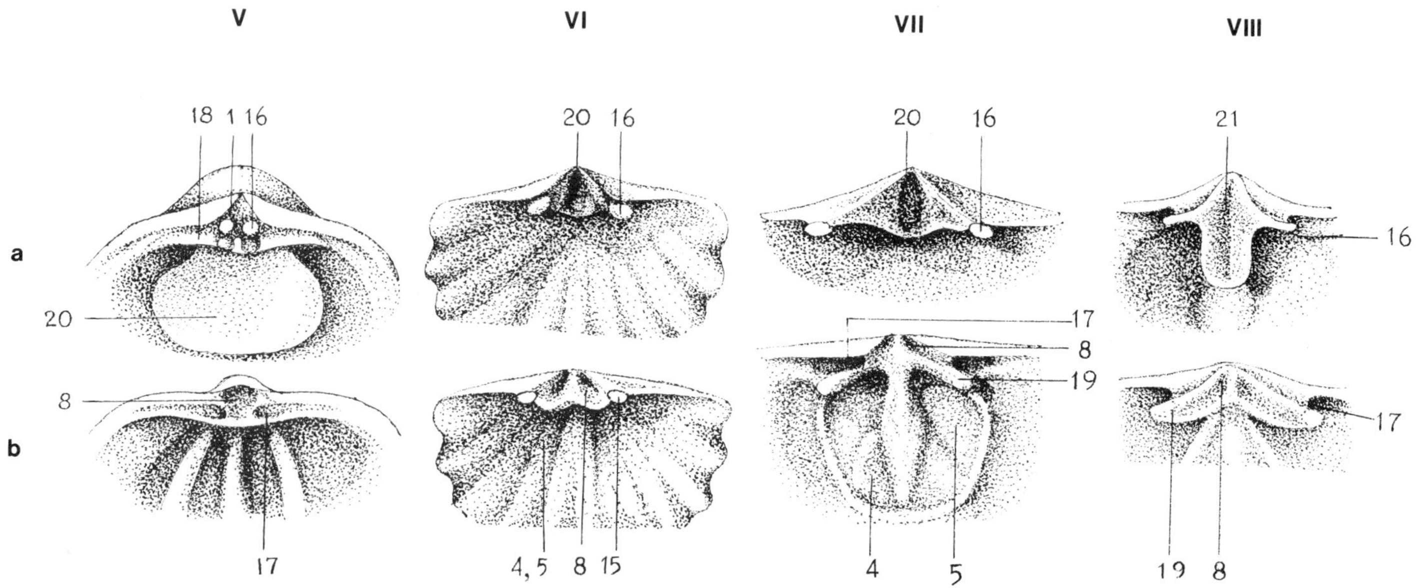

Kirgizia; (VIII) *Palaeostrophia fibrillosa* Nikitin & Popov, Upper Cambrian, South Kazakhstan. (1) Pseudodeltidium, (2) additional aperture, (3–8) principal muscle scars (3, adjustors; 4, anterior adductors; 5, posterior adductors; 6, oblique internal; 7, oblique lateral; 8, diductors), (9) pedicle foramen, (10) cardinal ridges, (11) cardinal furrows, (12) triangular propareas, (13) cardinal sockets, (14) socket plates, (15) "brachiophore" nubs, (16) teeth, (17) hinge sockets, (18) arcuate plate (anteris), (19) brachiophores, (20) free apical plate, (21) spondylium.

perforated valve is a homologue of the dorsal valve of *Crania,* whereas the opposite valve, lacking an interarea, is a homologue of the ventral valve, which in *Crania* attaches to the substrate. The position of the delthyrial opening, through which the pedicle protrudes, makes it possible to homologize the perforated valve with the pedicle (or ventral) valve of articulate brachiopods. At any rate, the problem of valve homology in *Crania* and articulate brachiopods deserves special study and will not be considered herein.

Matutellids lack effective articulation structures, though they are characterized by a straight posterior margin. The second characteristic feature is shell resorption around the foramen, which is a quite rare phenomenon in Early Paleozoic brachiopods. It is recorded only in the obolellid *Trematobolus* and was possibly also characteristic of *Trematosia.* Apart from these exceptions, Cambrian brachiopods do not seem to have acquired the ability to use resorption for the construction of their shells. The ability to resorb shells in articulate brachiopods appears again only in the Ordovician, simultaneously with appearance of a cyrtomatodont dentition in Rhynchonellida and also probably in *Porambonites* (Jaanusson, 1971). At the same time shell resorption is not at all typical of orthaceans and billingsellaceans.

3.2. The Pedicle and Possible Course of Appearance of a Blind Intestine

Rowell and Caruso (1985) showed convincingly that the nisusiids, one of the most ancient articulate brachiopods, had an open intestine with the anus in the middle part of the posterior body wall, indicating a close affinity between craniids and articulate brachiopods. Nevertheless, their information also indicates that the apical foramen in nisusiids and, thus, kutorginids and related stocks of articulate brachiopods, may not have been an aperture through which the pedicle protruded. In fact, the small size of the pedicle required by the diameter of this foramen probably did not provide a reliable anchor under conditions of the high hydrodynamic activity issuing from Rowell and Caruso's reconstructions of the burial conditions of their *Nisusia.* Also, the specimens, illustrated in Fig. 5 of Rowell and Caruso (1985) were most likely attached to the substrate by a short, massive pedicle passing between the valves. During the life of the animal, this pedicle was covered with pellets, which promoted its fossilization. Rowell and Caruso (1985) rejected this idea because of the difficulties of reconstructing the function of the apical foramen. Nevertheless, a different interpretation of the origin of an apical foramen is also possible, assuming that it is a rudiment of a hydrodynamic shell-opening mechanism. Its preservation indicates that the ancestors of nisusiids and kutorginids evolved a levator shell-opening mechanism skipping the stage of hydraulic opening of the shell.

A unique find by Rowell and Caruso (1985) confirms the earlier proposed homology of the valves of *Crania* and articulate brachiopods, because in Recent *Crania,* the anus has a dorsoterminal position near the proximal end of the anal papilla, a homologue of the pedicle in articulate brachiopods (Starobogatov, 1979, 1983). According to the reconstruction by Rowell and Caruso (1985, Fig. 7), the anus in *Nisusia* was nearer the brachial valve; however, judging by the photograph in their Fig. 6, the distribution of pellets suggests that it was quite near pseudodeltidium. This problem deserves further study. Presumably, a pedicle in nisusiids developed from the posterior larval segment, and preserved coelomic cavities in an adult stage, whereas the cartilaginous tissue was as yet absent. However, the rhynchonellids, the oldest stock of Recent brachiopods, which appeared at the beginning of the Middle Ordovician, already had a blind digestive tract and cartilaginous tissue in the pedicle. Coelomic cavities in pedicle segment disappear in the adult stage, most likely due to the appearance of new supporting elements. Thus, a combination of features characterizing Recent representatives of the Class Articulata was present before the Middle Ordovician. The origin of a blind digestive tract in articulate brachiopods possibly has a phyletic correlation with the appearance of a deltidiodont dentition and cartilaginous tissue in the pedicle. It can be regarded as a pedomorphic alternation (Gorjansky and Popov, 1986). A small gape between valves, the closely packed diductor muscles, and the pedicle itself replaced by a cartilage served as obstacles for the normal functioning of the intestine opened in the middle part of the posterior body wall.

In obolellids the appearance of articulation and the levator shell-opening mechanism was accompanied by the migration of the pedicle (Fig. 1IV), which was possibly due to preservation of an open digestive tract. In *Alisina* and *Trematobolus* there is no gape along the posterior shell margin and the pedicle protrudes through the foramen. The small dimensions of the foramen would not be an obstacle to the normal functioning of the open intestine if one presumes that in obolellids the pedicle resembled the extended anal papilla of craniids, and the main supporting rôle was played by coelomic cavities, and not by cartilaginous tissue.

3.3. Comparative Morphology of the Earliest Lingulates

Homology of the muscle systems of Recent lingulids and discinids with those of the extinct Early Paleozoic lingulate stocks is important but far from being solved. An asymmetric shell musculature with a nonpaired posterior adductor is unknown among the Cambrian lingulids, acrotretids, and siphonotretids, whereas the shell musculature of the Cambrian lingulids repeated with minor variations the type described for the genus *Obolus* (Fig. 2). When establishing the homology of muscular impressions, account should be taken of their arrangement relative to the base of the main mantle canals. The point is that the baculate mantle canal system is remarkably constant in most of the Early Paleozoic lingulates, except paterinids. On the ventral valves of lingulids, acrotretids, and siphonotretids, three sectors are distinguished within the visceral area: a median one between the bases of vascula lateralia and two posterolateral sectors. In all three orders, muscle fields are confined to the anterior part of the median sector. In the Family Obolidae they served as attachment areas of three paired muscles: central, middle lateral, and outside lateral. Similar muscle scars are also recorded on the ventral valve of the siphonotretid genus *Dysoristus* (Fig. 2IV). In acrotretaceans, the median sector of the ventral visceral area corresponds to the apical process, which formed an elevated muscular platform (Fig. 2VI). Sometimes in its anterior portion impressions of six paired muscles are discernible, which can be regarded as

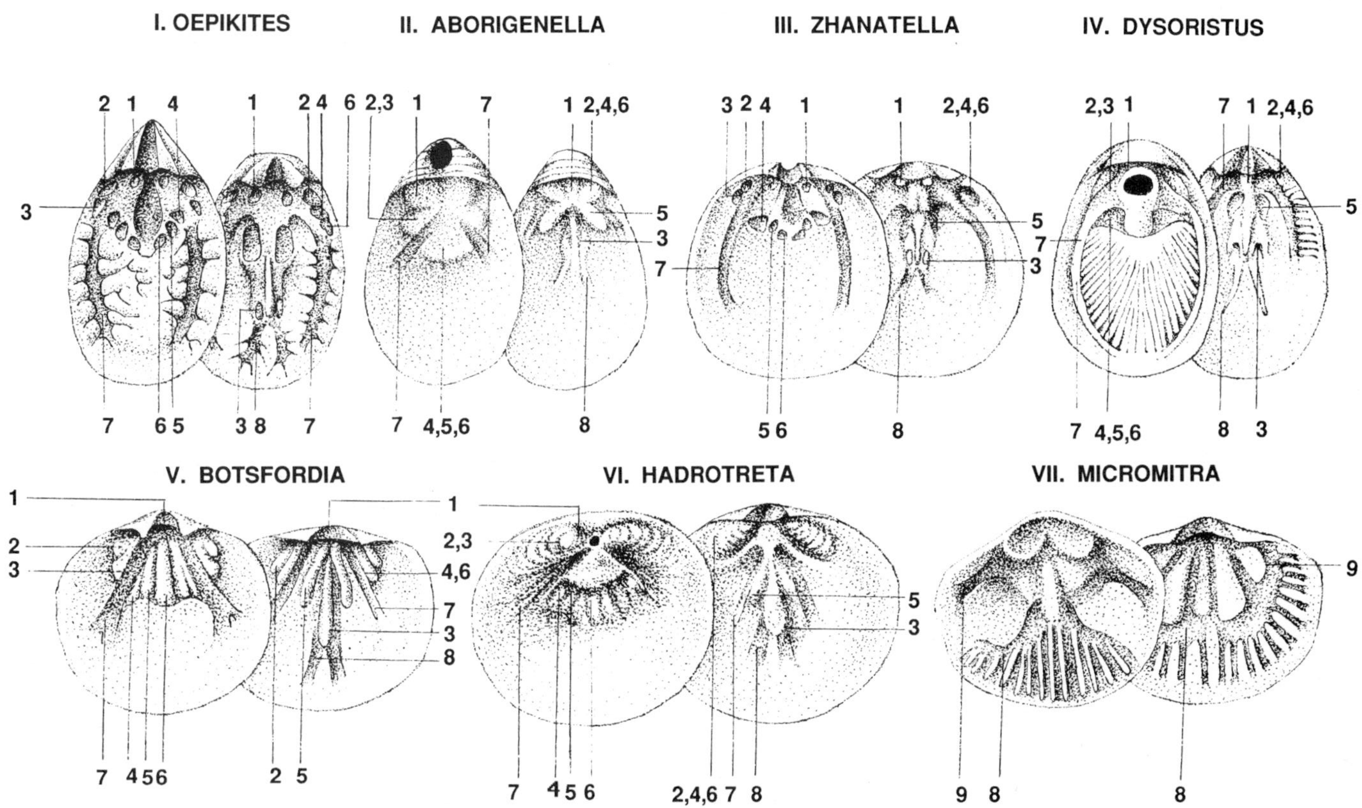

Figure 2. Internal morphology of ventral (left) and dorsal (right) valves of selected Cambrian lingulates. (1–6) principal muscle marks (1, posterior adductors, or umbonal muscles of lingulids; 2, transmedian muscles; 3, anterior lateral muscles; 4, outside lateral muscles; 5, anterior adductors, or central muscles; 6, middle lateral muscles); (7–9) mantle canal impressions (7, vascula lateralia; 8, vascula media; 9 (?) vascula genetalia).

homologues of similarly arranged muscles of *Obolus*. Between the proximal ends of the vascula lateralia and the margins of propareas in obolids and *Dysoristus*, there are posterolateral ventral muscle fields, which served as areas of attachment of anterior laterals and transmedian muscles. They are homologous to cardinal muscle impressions in acrotretaceans, which are located on the internal side of the ventral propareas. Thus, the "ventral propareas" in acrotretids are not homologous to the ventral propareas of lingulids and siphonotretids, but are homologous to the posterolateral areas of their ventral valves (Popov and Nõlvak, 1987). This conclusion is confirmed by the results of observations on the ontogeny of *Hadrotreta* (Rowell, 1980) and *Scaphelasma* (Nazarov and Popov, 1980). At the early postlarval stages these genera have a broad delthyrium, which occupies most of the narrow posterior margin. The slightly concave lateral margins of the delthyrium most likely represent strongly reduced propareas. Further growth of the shell results in intergrowth of the distal margins of the delthyrium and the appearance of a pedicle foramen. Thus, the ventral valve of acrotretaceans can be represented as an obolid valve, which is rolled up and joined along the reduced propareas (Fig. 3). The area secreted by the posterior part of the ventral mantle lobe was reduced and corresponded only to a small part of the palintrope around and behind the pedicle foramen in the Family Curticiidae, the Subfamilies Linnarssoniinae and Scaphelasmatinae, as well as some mainly Cambrian Acrotretinae. In most of the other Acrotretidae it is not discernible because the foramen is developed at earlier stages of ontogeny; in larval shells, prior to calcium phosphate deposition.

Changes in the geometry of ventral valve growth in acrotretaceans resulted in the displacement of the ventral impressions of posterior adductors, which acquired a subcentral position. The paired apical pits in acrotretacean ventral valves likely corresponded to them (Fig. 2VI).

The dorsal visceral area of Early Paleozoic lingulates has a characteristic median projection. The posterior boundaries of this projection correspond to places of attachment of the gastroparietal bands. Large paired impressions in front of them are recorded in lingulids, siphonotretids, and certain Cambrian acrotretids. They are homologous in all three orders and correspond to the anterior adductors (or central muscles). Other paired muscle scars are disposed behind the bases of the vascula media. They correspond to the anterior lateral muscles of *Obolus*. Posterolateral dorsal muscle fields, lying between the bases of vascula lateralia and propareas in lingulids and *Dysoristus*, served as the attachment areas of three pairs of muscles: transmedian, outside lateral, and middle lateral. Their homologues in acrotretaceans are the large cardinal muscle impressions in a similar position. The attachment areas of separate muscles on them are commonly indiscernible. However, the arrangement of muscle impressions on the ventral valve suggests that the dorsal cardinal muscle fields in the oldest acrotretaceans also served as attachment areas of three pairs of muscles, homologous to the transmedian, outside lateral, and middle lateral (Fig. 2).

Thus, the Cambrian representatives of the orders Acrotretida, Lingulida, and Siphonotretida have a similar shell musculature, though in the Superfamily Acrotretacea the relative arrangement of ventral muscle impressions underwent appreciable changes. Further evolution of acrotretaceans resulted in a gradual simplification of shell musculature. This process proceeded simultaneously in

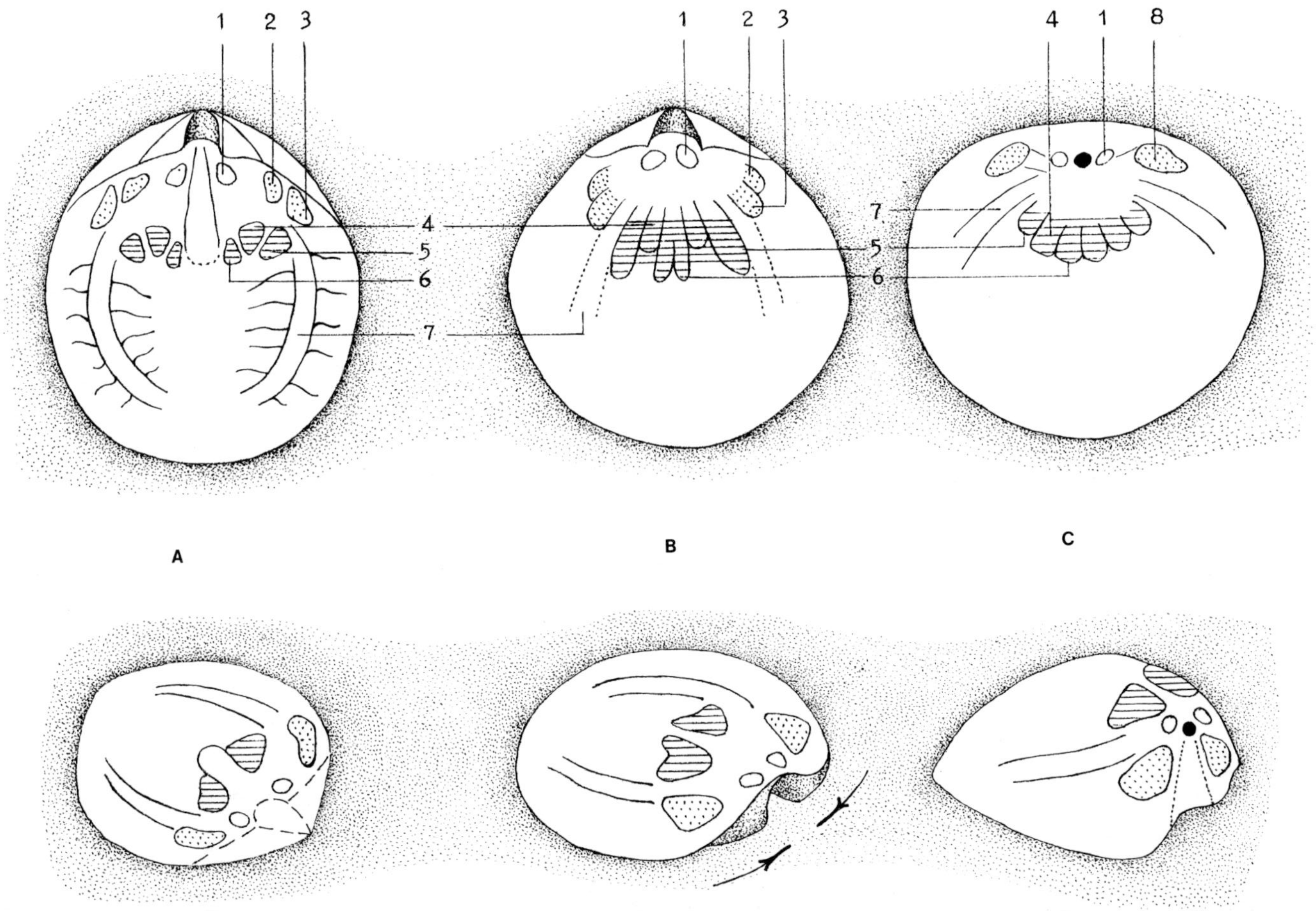

Figure 3. Main steps toward the origin of the pedicle foramen in the Order Acrotretida. (A) Obolid stage, (B) botsfordiid stage, (C) acrotretid stage. Principal muscle marks: (1) umbonal muscles, (2) transmedian muscles, (3, 8) anterior lateral muscles, (4) central muscles, (5) outside lateral muscles, (6) middle lateral muscles. (7) Mantle canal impressions.

several phylogenetic lineages. The earliest was the reduction of anterior lateral muscles, which are recorded only in Linnarssoniinae and the most ancient Cambrian Acrotretinae. In the majority of other Cambrian and Ordovician acrotretaceans, the anterior boundary of the dorsal visceral field is located immediately in front of the adductor scars. The disappearance or appreciable reduction of the apical process on the ventral valve in the subfamilies Ephippelasmatinae and Torynelasmatinae correlates distinctly with the disappearance of dorsal impressions of anterior adductors.

4. Debate about Brachiopod Origin

4.1. Monophyly or Polyphyly?

No model concerning the origin of brachiopods, their initial radiation, and phylogenetic relations between the major taxa is generally accepted. The existing models can be reduced to three major hypotheses: monophyletic, polyphyletic, and diphyletic.

4.1.1. Monophyly

The traditional notions of a monophyletic origin of brachiopods are well stated by Williams and Rowell (1965), Williams and Hurst (1977), Rowell (1982), and Emig (1984). I summarize only the most important aspects.

Brachiopods (*sensu lato*) are commonly regarded as having risen from a trimerous, tubicolous, coelomate, phoronidlike ancestor (Clark, 1979). According to the hypothetical reconstruction proposed by Williams and Hurst (1977), the prototypic brachiopod had a recurved gut with the anus outside the lophophore consisting of a single row of filaments, as in the tricholophe of the articulate brachiopods, and a pedicle, which is an outgrowth of the ventral body wall. The animal had a pair of unarticulated completely organic valves, secreted by ventral and dorsal mantle lobes. Calcium-phosphatic mineralization of the shell seems to be most ancient, whereas calcareous-shelled brachiopods radiated repeatedly from different ancestors with apatite shells.

Articulate brachiopods are presumably a monophyletic taxon. Williams and Rowell (1965) assumed that orthaceans were the ancestral stock for the lineage in which the pedicle was formed from the posterior larval segment, whereas the mantle reversal during larval stage of growth was first recorded in Order Rhynchonellida. The appearance of a blind intestine is associated with the origin of deltidiodont teeth and sockets. This feature could have been developed simultaneously in several phyletic lineages of the articulate brachiopods (Williams and Hurst, 1977).

Recently, Rowell (1982) considered the arguments in favor of a monophyletic origin of brachiopods. The most important of these is the great similarity in morphology of the lophophore in all Recent brachiopods and its significant difference from that of other lophophorates. Among synapomorphic features is also the presence of a bivalved shell secreted by dorsal and ventral mantle lobes with coelomic canals and setae along the margin of the mantle lobes (Fig. 4A). How-

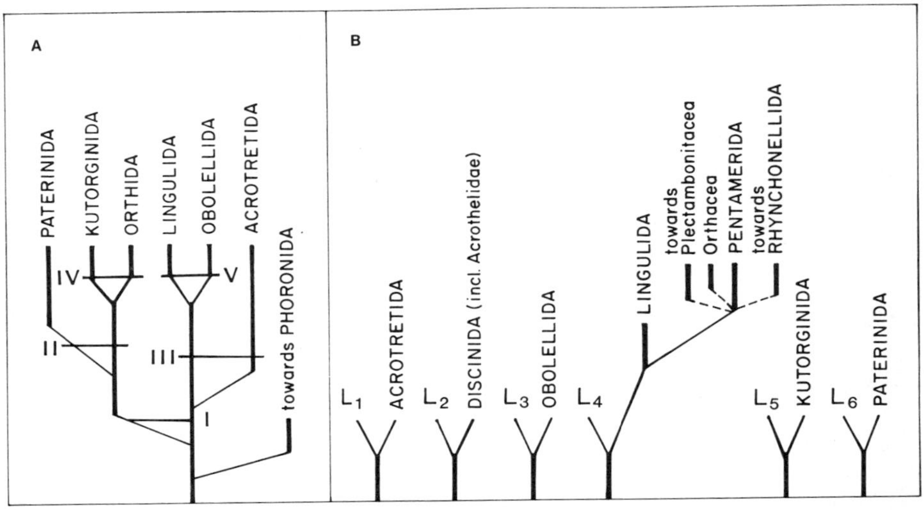

Figure 4. Monophyletic and polyphyletic hypotheses of the origin of brachiopods. (A) Monophyletic origin of brachiopods from a phoronidlike ancestor [after Rowell (1982), modified]. Numerals and bars connecting taxa show synapomorphies, uniquely derived evolutionary novelties: (I) development of ventral and dorsal mantles and setae at mantle margin. (II) Medially located muscle scars, straight posterior margin of shell, and open delthyrium may be partially closed apically. (III) Peripheral location of shell muscles in body cavity. (IV) Pseudodeltidium and apical foramen. (V) Large anterior adductors and marginal beak in both valves. (B) Polyphyletic origin of brachiopods from several phoronidlike ancestors (L_1 to L_6). The Class Articulata is a holophyletic taxon [after Wright (1979), modified].

ever, the homology of these characteristics in the articulate brachiopods and craniids on the one hand and the "inarticulate brachiopods" with a calcium phosphate mineralization of the shell on the other hand has been questioned on the basis of ontogeny and biochemistry (Gorjansky and Popov, 1985, 1986). The differences in the structure of the lophophore at the trocholophe stage also needs to be explained. In Craniida and articulate brachiopods it has only one row of filaments, but in lingulids and discinids it has two rows. Possibly this difference has a rather ancient origin.

4.1.2. Polyphyly

The polyphyletic hypothesis of the origin of brachiopods was supported with serious arguments by Valentine (1975, 1977) and Wright (1979). They consider brachiopods as a grade, and that the level of organization characterizing this taxon appeared independently in several lineages of the phoronidlike infaunal lophophorates (Fig. 4B). The proponents of the polyphyletic origin of the brachiopods indicate that inarticulate and articulate stocks have several essential anatomical differences and very distinctive ontogeny. Valentine did not rule out the possibility that the articulate brachiopods are a polyphyletic group; however, until recently, there were not enough data to check this assumption. Wright believes that the articulate brachiopods could have a monophyletic origin directly from a *Lingula*-like ancestor. The advantages and flaws of this hypothesis were critically analyzed by Rowell (1982).

4.1.3. Diphyly

According to the diphyletic hypothesis, which I accept, brachiopods in the traditional view include two major lineages of the lophophorate animals of quite different origin (Fig. 5). One of them diverged from the main lophophorate stock prior to the radiation of other lophophorates. In the Recent fauna it is represented by articulate brachiopods and calcareous-shelled inarticulates (Craniids). This stock retained certain features inherited from the protolophophorates, which were lost by other lophophorate animals. They lack a metasomal outgrowth of the ventral body wall, and the intestine has retained its initial position, which is confirmed by a dorsoterminal anus in craniids and the oldest articulate brachiopods of the Family Nisusiidae. According to data on the ontogeny of articulate brachiopods (Percival, 1944), most of the brachiopod body, including the lophophore and mantle, is a derivative of the mesosome. With a transition to a benthic mode of life, the larva settles down on the posterior end of a pedicle segment.

The second major lineage is lingulates (or "inarticulate brachiopods" with an apatitic shell), which originated due to a later radiation of lophophorates (Starobogatov, 1979; Gorjansky and Popov, 1986) at the stage when they already had a metasomal pouch of the ventral body wall, into which the median parts of the coelom and intestine submerged, whereas a derivative of the metasome accounted for most of the body. In lingulates the larvae settle on the ventral side of body as in phoronids.

Both stocks independently reached the brachiopod level of organization, and a majority of similar features appeared convergently. Thus, the reconstruction of the prototype is hardly possible at a prebrachiopod level of organization.

In contrast to what is held in monophyletic and polyphyletic models, brachiopods (*sensu lato*) and lingulates apparently evolved independently from epibentic protolophophorate ancestors, and the phoronids are represented as a sister stock for lingulates. A similar model of lophophorate phylogeny was proposed earlier by Jägersten (1972). The origin of various calcareous-shelled inarticulate and articulate stocks from ancestors with a calcium phosphate mineralization of the shell is completely rejected. All calcareous-shelled brachiopods are a single monophyletic taxon. Phylogenetic relations between the main lineages of lingulates and articulate brachiopods also has been revised.

The differences in monophyletic, polyphyletic, and diphyletic conceptions of brachiopod evolution are based on the alternative interpretations of the origin and successive evolution of such important features as the subdivision of the coelomic space, shell-opening mechanisms, and articulation of the valves. Clark's hypothesis on the important role of peristaltic burrowing in the origin of the coelom has to be modified in light of the recent data on the Precambrian and early Paleozoic paleoichnology. All these problems will be discussed below in detail.

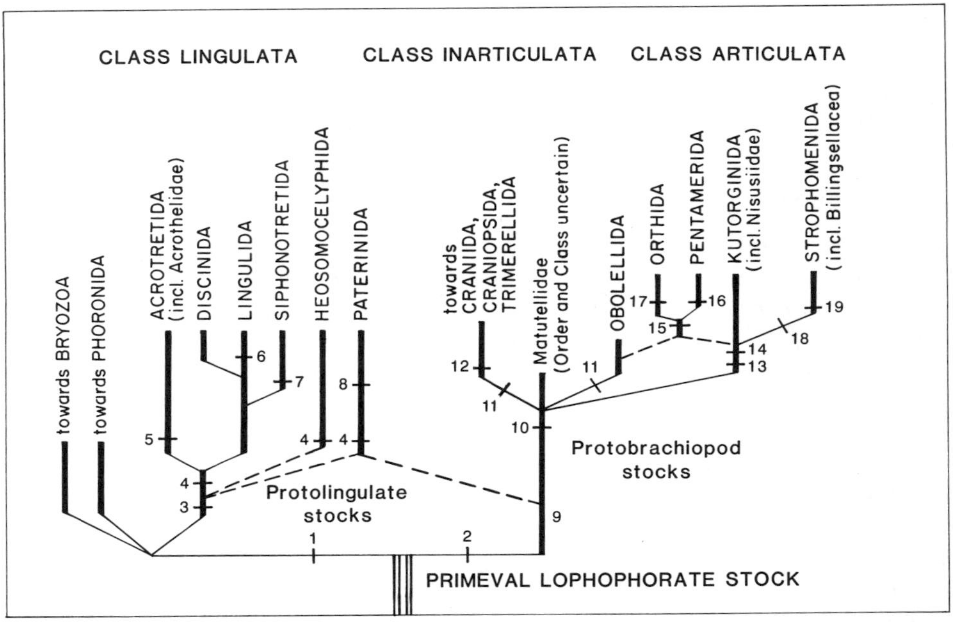

Figure 5. The probably phylogenetic relations among the oldest brachiopod and lingulate stocks. Numerals show points at which the most important evolutionary events took place: (1) Metasomal pouch and recurved digestive tract originate, larva settles on the ventral body wall. (2) Initial orientation of a digestive tract with the posteriorly placed anus, and settlement of larva on the posterior part of the body persists. (3) Mantle lobes and filtrating chamber appears. (4) Calcium phosphate mineralization of a shell and hydraulic shell-opening mechanism originate. (5) Pedicle foramen and organic larval shell with a vesicular microstructure originate. (6) Infaunal mode of life originates. (7) Hollow spines and pedicle foramen originate. (8) Straight posterior margin, probably with fused mantle lobes originates. (9) Soft mantle lobes, filtering chamber, and hydrodynamic shell-opening mechanism originate. Coelomic space divides into a large mesocoel with mantle and lophophore, and a metacoel in the pedicle segment. (10) Calcareous shell constructed of periostracum, with primary and secondary layers, originates. Mantle lobes fused along the posterior margin of a shell. (11) Hydraulic shell-opening mechanism and two generative zones along the posterior margin of a shell originate. (12) Pedicle is reduced. (13) Two generative zones along fused posterior margin of the mantle lobes originate. (14) Primitive articulation and levator shell-opening mechanism originate. (15) Deltidiodont teeth, dental sockets, blind intestine, and cartilaginous tissue in the pedicle originate. (16) Dental plates and brachiophorium appear. (17) Dental plates and brachiophores originate. (18) Deltidiodont teeth, dental sockets, and laminar secondary layer originate. (19) Dental plates and pseudopunctation originate.

4.2. Interpretation of Relationships of Cambrian Brachiopods and Lingulates

The direct ancestors of brachiopods (*sensu stricto*) were not burrowing lophophorates similar in their organization to phoronids, as accepted in the majority of hypotheses proposed in recent years, but sedentary epifaunal filter-feeding animals as Jägersten (1972) assumed. The organization of ancient as well as of Recent representatives of this phylum does not show any features pointing to a burrowing mode of life for their ancestors, if the presence of a coelom itself is not regarded as such a feature. The proposed interpretation does not conform to Clark's well-known hypothesis in which the coelom appeared as a hydroskeleton primarily adapted to peristaltic burrowing in a substrate.

The hydroskeleton for the body-wall musculature is undoubtedly the most important primary function of a coelom. However, its origin cannot be unambiguously associated with a peristaltic burrowing into the sediment. Paleoichnological data indicate that Precambrian Metazoa inhabited only the near-surface part of the bottom (Fedonkin, 1985), and this contradicts a burrowing mode of life. Infaunal organisms are recorded only at the beginning of the Cambrian, simultaneously with the appearance of the first coelomates. The contradictions can be eliminated if Clark's theory is supplemented by several assumptions put forward by Starobogatov (1983) in his hydrocoel (brachiocoel) theory. In it the coelom originated as a hydroskeleton of the tentacular apparatus, and its use in peristaltic burrowing into the sediment was an important early, but nevertheless secondary, function. Hypothetical ancestors of the Phylum Brachiopoda (*sensu stricto*) may have been epibenthic, sedentary, trimerous lophophorates devoid of a shell, but having mantle lobes bounding the filtration cavity. A pedicle, homologous to the posterior segment of a larva of the articulate brachiopods and anal papilla of craniids, had musculature and coelomic cavities. The intestine was open with a dorsoterminal anus located near the proximal end of a pedicle.

The most ancient brachiopods retained these characters, but acquired a calcareous shell. They inherited the hydrodynamic shell-opening mechanism. However, for the normal functioning of the additional inhalant aperture, resorption of the surrounding shell during ontogenetic development was necessary. The most primitive of the known brachiopods, such as the Family Matutellidae, had mantle lobes fused along the posterior margin. In this case, only the pedicle valve was accreted along the posterior margin by a single generative zone. The brachial valve with a mixoperipheral growth was accreted only along the anterior and lateral commissures.

The presence of well-developed mantle lobes along the posterior margin, characteristic of *Crania* and extinct calcareous-shelled inarticulate brachiopods, is not likely to represent the initial stage. They could rather appear later in the lineage which give rise to class Inarticulata simultaneously with the formation of a hydraulic shell-opening mechanism. The latter imposes several limitations on the shape of the shell and the relations of the mantle and visceral cavities (Gutmann *et al.*, 1978) which led to the conservation of many archaic features, also retained in Recent craniids. Apparently at the pre-Ordovician evolutionary stage of inarticulate brachiopods (*sensu stricto*), a reduction of the pedicle took place. The main events in the evolution of the Class Inarticulata, whose reliable representatives are first known in the Ordovician, were considered by Gorjansky and Popov (1985, 1986).

Another trend in brachiopod evolution, which gave rise to the class Articulata, is associated with the appearance of an effective articulation of the valves and a levator mechanism for opening them. The oldest articulate brachiopods belong to the Order Kutorginida (Fig. 6). The earliest representative of this order is *Khasagtina* from the late Tommotian of Mongolia (Ushatinskaya, 1987).

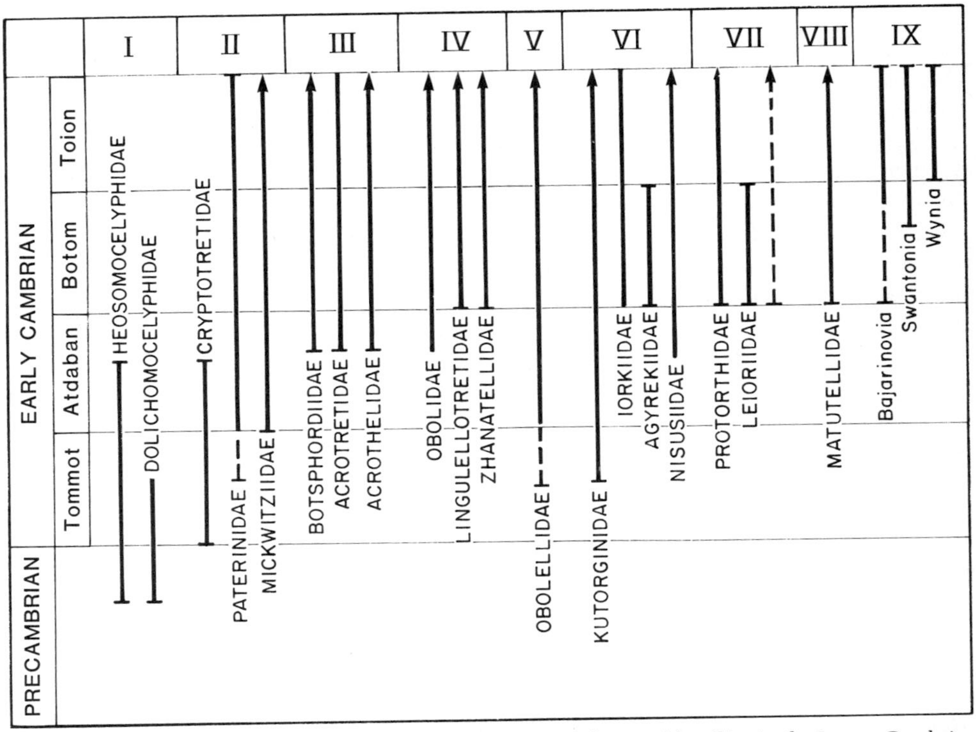

Figure 6. Stratigraphic ranges of brachiopod and lingulate orders and families in the Lower Cambrian. Numerals show orders: (I) Heosomocelyphoda, (II) Paterinida, (III) Acrotretida, (IV) Lingulida, (V) Obolellida, (VI) Kutorginida, (VII) Orthida, (VIII) *incertae ordinis*, (IX) genera of uncertain affinities.

Kutorginaceans already had a very primitive hinge mechanism (Fig. 1II). The axis of rotation along the hinge line is fixed by short triangular interareas on the brachial valve that fit into cardinal furrows along the lateral sides of the pseudodeltidium on the pedicle valve. No doubt the Family Nisusiidae is the direct descendant of kutorginaceans. Both stocks have a large pseudodeltidium and a minute apical foramen. Nisusiids in a rudimentary state have structures that fixed the axis of rotation in kutorinaceans (Fig. 1III). Nisusiids clearly lack teeth and true hinge sockets, dental plates, and brachiophores (Rowell and Caruso, 1985). The axis of rotation of the valves, fixed by margins of the interarea of the pedicle valve immediately lateral of the delthyrium, sits on the lateral plates bounding the notothyrium.

Possibly, the Superfamily Billingsellacea also descended from nisusiids, but this needs proof because a considerable morphological hiatus exists between the two taxa. Billingsellaceans have typical deltidiodont teeth and hinge sockets, whereas their secondary shell layer has a laminar structure (Williams and Hurst, 1977).

Thus, a major evolutionary trend can be envisaged, having its roots at the beginning of the Cambrian and giving rise to numerous taxa of articulate brachiopods, whose origin is associated with that of Superfamily Billingsellacea.

The second evolutionary trend is represented by the Order Obolellida. The oldest representative of this order is *Nocharoiella*, which appears in the second half of the Tommotian (Grigorjeva *et al.*, 1983). However, its taxonomic affinities are not quite certain, for the pseudointerareas are not yet developed and details of the interiors of both valves are unknown. Reliable obolellids are widespread starting with Atdabanian time (Ushatinskaya, 1986). They have pseudointerareas on both valves, which are indicative of the existence of two generative zones along the posterior margin of the shell. The oldest obolellids lack a hinge mechanism, and their valves possibly opened by means of a hydraulic mechanism. As early as the Atdabanian the first obolellids appear with a primitive articulation of a strophic type (Fig. 1IV). They have closely paired teeth along the margins of a delthyrium and hinge sockets on the notothyrial chamber floor (Gorjansky and Popov, 1986). Simultaneously a levator shell-opening mechanism developed in articulate obolellids.

The origin of the peculiar genus *Bajarinovia* (Fig. 1V) could possibly be connected with obolellids. It is characterized by massive teeth composed of secondary shell material on the arcuate platform (*anteris*), and a hanging visceral platform (or free plate) on the pedicle valve.

The origin of the Orders Orthida and Pentamerida remains problematic. In the second half of the Early Cambrian, the Orthida was represented by the superfamilies Orthacea and Protorthacea (Fig. 6). The oldest and most primitive protorthacean genera, such as *Glyptoria, Israeleria,* and *Psiloria,* have well-developed interareas on both valves and an open delthyrium, with paired teeth along the margin (Fig. 1VI). However, they have no dental sockets or well-developed brachiophores (Cooper, 1976). How the valves were fixed is unclear because no complete shells are known. The Middle Cambrian Protortacean genus *Arctohedra* (Fig. 1VII) already had hinge sockets and brachiophores. A unique feature of this superfamily is a free plate, which represents the hanging visceral platform.

The Early Cambrian monotypic family Leioridae also had quite well-developed brachiophores. Cooper (1976) believed that it was among the most ancient orthaceans. However, the free plate in the pedicle valve strongly indicates that it should be placed among protorthaceans. The taxonomic affinity of the Early Cambrian Orthacea cannot be considered reliable, since either the detailed structure of their cardinalia remains unknown (Rowell, 1977), or the teeth, brachiophores, and hinge sockets are absent (for instance, in "*Diraphora*" *venzlaffi* Wolfart). True orthaceans with deltidiodont dentition and rudimentary brachiophores, but lacking dental lamellae, are known only beginning with the Middle Cambrian (Havlíček, 1971).

Which of the two oldest superfamilies was the first orthid is not clear. There are no transitional groups between orthaceans and protorthaceans which would allow tracing the development of a free plate on the pedicle valve. A close affinity between protorthaceans and nisusiids is not likely, since they have an essentially different hinge structure; furthermore, protorthaceans lack a pseudodeltidium, and nisusiids have no free plate.

If the family Bohemiellidae is the most primitive stock within the order, presumably it separated from nisusiids by a series of pedomorphic alternations, resulting in a loss of the pseudodeltidium and the cardinalia that characterizes nisusiids, and, possibly, also the appearance of blind intestines. Initially, the valves were fixed to each other only by mantle lobes fused along the posterior margin. "*Diraphora*" *venzlaffi* (Wolfart, 1974) from the Lower Cambrian of Iran may represent this stage, the first for orthids.

The oldest Middle Cambrian pentamerids have deltidiodont teeth and hinge sockets, well-developed dental plates, and brachiophores. They are also characterized by a tendency to form a sessile spondylium. Thus, the dental plates of the earliest pentamerids might have formed by intergrowth of the free plate of protorthaceans and the valve's floor. At any rate, among the Early Cambrian articulate brachiopods, only the protorthaceans had a complex cardinalia, from which the cardinalia of the oldest pentamerids can be deduced.

Ancestors of the class Lingulata diverged during the latest radiation of protolophophorates, at the stage when they already had a well-developed metasomal outgrowth of the ventral body wall. The recurved digestive tract and anteriorly placed anus in Recent lingulids and discinids are very ancient features, inherited by both orders from the common Cambrian ancestor. On the dorsal valve of the Cambrian lingulids, siphonotretids, and acrotretids there are traces of attachment of the gastroparietal bands, which enables an assumption about a similar position of the digestive tract in the oldest representatives of the class. Apparently, as far back as the Precambrian, the lingulates acquired an organic shell and hydraulic shell-opening mechanism. The dermal muscles are important in this mechanism, but their equivalents are not recorded in calcareous-shelled brachiopods. No convincing evidence exists that the ancestors of lingulates were burrowing animals. The oldest calcium phosphate-shelled "brachiopods" are the Order Paterinida and the enigmatic Heosomocelyphida. They can be assigned to the Class Lingulata with some restraint, since they have not been adequately studied, and the known features of the shell morphology distinguish them from other orders within the Class Lingulata.

The Order Heosomocelyphida was erected by Liu (1987) on the basis of Late Sinian material from China. It contains small (about 1 mm) shells with bilaterally symmetrical, calcium–phosphatic valves. A broad gape along the posterior margin served for the emergence of a pedicle, according to Liu. Sometimes, on one of the valves there are structures resembling the homeodeltidium and a ventral pseudointerarea of paterinids. The data on the shell structure given by Liu indicate that the shell was composed of alternating organic and mineralized layers. No data are available on the relative position of the mantle and visceral cavities and on the structure of shell musculature, without which the affinities between the Order Heosomocelyphida and other lingulates cannot be determined.

The Order Paterinida displays, along with peculiar features of the shell morphology, also some characteristics which are diagnostic on one hand for lingulates and on the other hand for typical brachiopods. It is similar to lingulates in chemical composition of its shell; however, the shell structure is quite distinctive (Popov and Ushatinskaya, 1987). The pattern of the shell structure in paterinids was apparently the most primitive one among all the described lingulates and brachiopods. Their shells are composed of several lamellae disposed subparallel to the valve surface. Each lamella is composed of closely packed calcium phosphate prisms. The length of the prisms coincided with the thickness of the lamellae, and their width is about 7–8 μm. Each prism was, probably, initially enclosed in an organic sheet and was accreted by a single cell or a compact group of cells. Thus, paterinids had a cellular mode of shell secretion similar to the articulate brachiopods and craniids. The disposition of the main muscle scars and the mantle canal system is known only for the latest Family Paterinidae (Laurie, 1987), which is characterized by a straight posterior margin, along which the valves were accreted by fused mantle lobes. At any rate, no other mode of fixation of the rotation axis has been recorded in paterinids. Muscular impressions, located on the inner side of the homeodeltidium, were positioned beyond the axis of rotation of the valves. The muscles could function as diductors. At the same time, the homologization of the shell musculature of paterinids with that of lingulates and brachiopods presents considerable difficulties. The arrangement of the main mantle canals on both valves of the Family Paterinidae (Fig. 2VII) differs markedly from those characteristic of lingulates, but is somewhat similar to the pattern observed in some ancient articulate brachiopods. Paterinids may display impressions resembling *vascula genitalia* of articulate brachiopods (Laurie, 1987). If confirmed in the future, this interpretation will provide strong evidence for a common origin of paterinids and articulate brachiopods. Possibly the paterinids independently acquired a mineralized shell, and their alternative origin from the ancestors common with brachiopods (*sensu stricto*) or lingulates should be regarded as equally possible.

The common origin of the four other lingulate orders, Lingulida, Acrotretida, Siphonotretida, and Discinida, is presently undoubted. The oldest lingulids and acrotretids appear almost simultaneously in the second half of the Atdabanian (Rowell, 1977; Ushatinskaya, 1986). Beginning with the Botomian, the marine fauna is represented by three lingulid families (Obolidae, Lingulellotretadae, and Zhanatellidae) of the four families known in the Cambrian, as well as the major lineages of acrotretids (the families Acrotretidae, Botsfordiidae, and Acrothelidae). All this is indirect evidence that both orders had an eventful history in the first half of the Early Cambrian.

Presumably the Family Obolidae retained most of the features of the lingulate prototype. The family is ancestral to the Lingulellotretidae and Zhanatellidae, which appeared in the middle part of the Early Cambrian (Fig. 6). This radiation of lingulids at the beginning of the Botomian was associated with transformations in morphology of the posterior shell's margin. The Family Lingulellotretidae is characterized by the pedicle foramen and internal pedicle tube, which arise as a result of the knitting of the ventral propareas above the pedicle groove. Representatives of the Family Zhanatellidae acquired an ability to resorb the apical part of the ventral valve (Koneva, 1986).

The earliest known lingulids were epibenthic and most likely inhabited offshore environments. They started colonizing unstable sandy bottoms in the onshore zone beginning in the Middle Cambrian (Khazanovitch *et al.*, 1984). By this time lingulids must already have had the ability to anchor in soft sandy bottoms, an effective rejection mechanism, and a ciliary-cleansing mechanism of the mantle and body wall. Morphological characteristics of obolids such as a subcircular outline and an unequivalved shell with well-developed pseudointerareas, as well as a posterior visceral cavity, restricted the mobility of the valves, and were unlikely to be adaptations to burrowing, as described by Thayer and Steele-Petrovich (1975) for Recent *Glottidia*.

The anterior body wall in obolids continues far forward along the dorsal valve and separates the right and left parts of brachiocoel, indicating that the inhalant aperture was located along the anterior commissure line and the exhalant ones along the lateral margins of the shell. The burrowing mechanism of lingulids is unique because the valves are used as shovels to excavate a burrow. This method is primarily used by epibenthic obolids for escaping from the sandy sediment. It may have appeared as a response to periodic burial of the animal in the sandy sediment in unstable onshore environments.

The divergence of the orders Lingulida and Acrotretida is possibly associated with changes in the life cycle of the ancestors of the latter (Rowell 1982). Acrotretids acquired a characteristic organic larval shell, possibly with a vesicular structure. Its negative cast shows pitted apical areas on both valves (Popov *et al.*, 1982). The growth pattern of the ventral vale has also changed in acrotretids in comparison to lingulids. It is correlated with changes in the relative arrangement of some internal organs, among them shell musculature.

The shell morphology of the Family Botsfordiidae displays a mosaic combination of features characterizing both acrotretids and obolids. It may represent a slightly modified descendant of the ancestral acrotretid stock. However, this suggestion needs confirmation through study of the shell structure and microornament of the larval shell.

The origin of siphonotretids and discinids is commonly associated with the Order Acrotretida (Williams and Rowell, 1965). However, recent studies suggest that this order, as defined by Williams and Rowell, does not represent a monophyletic clade. The Family Lingulellotretidae shows that the pedicle foramen cannot be regarded as a unique feature of acrotretids. Also, siphonotretids and discinids have a smooth larval shell that is much closer to the lingulacean type than to the pitted larval shell of acrotretaceans. The origin of a foramen in the most primitive siphonotretids, such as *Dysoristus* and *Acanthambonia*, is not accompanied by the displacement of the ventral posterolateral muscle fields onto the posterior slope of the valve, and most likely proceeded by means of knitting of the outer sides of a pedicle groove, as in Lingulelotretidae (Popov and Nõlvak 1987). The resorption of the apical part of the ventral vale in the lingulid family Zhanatellidae (Koneva 1986) can be considered as an intermediate stage in the formation of the migrating pedicle foramen of *Dysoristus*.

The shell structure of the Recent and extinct discinids has a close similarity to that of the lingulid genus *Glottidia*, but differs significantly from the shell structure of the Early Paleozoic acrotretaceans (Ushatinskaya *et al.*, 1988). Most probably the origin of siphonotretids is connected with the Family Zhanatellidae. The origin of discinids is more problematic, and requires further study.

5. Problems Requiring Further Study

There are different, sometimes diametrically opposite, points of view on the origin and initial radiation of brachiopods, as well as on phylogenetic relations between the main taxa. Any hypothesis concerning the origin of brachiopods and their Cambrian radiation, if reliable, cannot ignore certain essential characteristics of the Cambrian lingulate and brachiopod fauna. The most significant among them are the following.

1. In the Tommotian and first half of the Atdabanian, "brachiopods" with calcium phosphate shells are represented only by paterinids and problematic heosomocelyphids.

2. The Orders Lingulida and Acrotretida make a relatively late appearance, and reliable occurrences are known only from the second half of the Atdabanian. A rapid radiation of both orders apparently began at the beginning of the Botomian.

3. The earliest calcareous-shelled brachiopods appeared much earlier than lingulids and acrotretids, as far back as the end of the Tommotian, which is already characterized by reliable kutoginids and problematic obolellids.

4. An early appearance and rapid radiation of the articulate brachiopods in the Early Cambrian and an independent origin of the deltidiodont type of cardinalia occurred in at least two phylogenetic lineages.

5. At least partly, the Cambrian articulate brachiopods possessed an open digestive tract, arranged similarly to that of Recent craniids.

6. The burrowing mode of life of the oldest ancestors of the Cambrian brachiopods and lingulates cannot be established.

7. At the beginning of the Cambrian, brachiopods, initially without an articulation and a levator shell-opening mechanism, were characterized by a hemiperipheral growth of one of the valves, and possibly by the perforation of the dorsal body wall by the canals of the mantle cavity, but at the same time similar to the oldest articulate brachiopods in shell shape (Family Matutellidae).

The difference in the ontogeny and chemical composition of brachiopod and lingulate shells, as well as the known differences in the structure and position of the digestive tract in lingulates, craniids, and articulate brachiopods, must also be accounted for.

The solution of many vague aspects of the origin and evolution of brachiopods can be promoted by further study of the ontogeny, anatomy, and morphology of Recent brachiopods such as craniids, thecideaceans, and the peculiar genus *Taumatosia*. Studies on the ontogeny of craniids are of utmost significance for checking the hypothesis of their common origin with articulate brachiopods, and confirming the interpretation of the subdivision of a coelomic space in brachiopods, accepted in the present chapter. The studies on the ontogeny of thecideaceans seem important, since they are the only possible descendants of the strophomenid clade.

A particular emphasis should be placed on the extremely rare and poorly known genus *Taumatosia*. It is included among terebratulids; however, the shell shape and type of articulation of the valves resemble some orthids (Cooper, 1973). Such features as the absence of calcified branchidial supports and a schizolophe lophophore are also archaic. If the assumption of the affinities between *Taumatosia* and orthids is confirmed, its careful study could shed light on the time of appearance of such features, common to all living brachiopods of the Class Articulata, as blind intestines, inversion of mantle lobes in ontogeny, reduction of the subdivision of the coelomic space during metamorphosis, and the appearance of cartilaginous tissue in a pedicle.

Parallel and convergent evolution are common both among the Cambrian brachiopods and lingulates. The above examples illustrate the independent formation of a strophic articulation with teeth and dental sockets in obolellids, orthaceans, and billingsellaceans, as well as a simultaneous appearance of a pedicle foramen in the four orders of lingulates. Thus, important features in the existing brachiopod classifications had an independent origin. This convergence and parallelism create considerable difficulties in using cladistic methods to analyze the processes of appearance of the main brachiopod and lingulate taxa.

ACKNOWLEDGMENTS. I am grateful to T. N. Koren (All-Union Research Geological Institute, St. Petersburg) and G. T. Ushatinskaya (Paleontological Institute, Moscow) for critical reading of the manuscript. I thank A. Rowell (University of Kansas), who corrected the language and offered comments on the manuscript. The excellent drawings of lingulates and brachiopods are the work of G. A. Bulaievskaya (St. Petersburg). I also thank T. I. Vassilieva (All-Union Research Geological Institute, St. Petersburg) for help in translation, and Jere H. Lipps (University of California, Berkeley) for editorial assistance.

References

Chuang, S. H., 1959, The structure and function of the alimentary canal in *Lingula unguis* (L.) (Brachiopoda), *Proc. Zool. Soc. Lond.* **132**:283–311.

Chuang, S. H., 1974, Observations on the ciliary feeding mechanism of the brachiopod *Crania anomalia, J. Zool. Lond.* **173**:441–449.

Clark, R. B., 1979, Radiation of the Metazoa, in: *The Origin of Major Invertebrate Groups* (M. R. House, ed.), Academic Press, New York, pp. 55–101.

Cooper, G. A., 1973, New Brachiopoda from the Indian Ocean, *Smithson. Contrib. Paleobiol.* **16**:1–43.

Cooper, G. A., 1976, Lower Cambrian brachiopods from the Rift Valley (Israel and Jordan), *J. Paleontol.* **50**:269–289.

Emig, C., 1984, On the origin of the Lophophorata, *Sonderdruck Z. Zool. Syst. Evolutionsforsch.* **22**(2):91–94.

Fedonkin, M. A., 1985, Palaeoichnology of the Vendian Metazoa, in: *The Vendian System*, Vol. 1, *Paleontology*, Nauka, Moscow, pp. 112–117 [in Russian].

Gorjansky, V. Ju., and Popov, L. Ye., 1985, The morphology, systematic position and origin of inarticulate brachiopods with carbonate shells, *Paleontol. Zh.* **1985**(3):3–14 [in Russian].

Gorjansky, V. Ju., and Popov, L. Ye., 1986, On the origin and systematic position of the calcareous shelled inarticulate brachiopods, *Lethaia* **19**:233–240.

Grigorjeva, N. V., Melnikova, L. M., and Pelman, Ju. L., 1983, Brachiopods, ostracods (Bradoriida) and problematics from the type sections of the Lower Cambrian stages, *Paleontol. Zh.* **1983**(3):54–58 [in Russian].

Gutman, W. F., Vogel, K., and Zorn, H., 1978, Brachiopods: Biomechanical interdependences governing their origin and phylogeny, *Science* **199**:890–893.

Havlíček, V., 1971, Brachiopods of the order Orthida in Czechoslovakia, *Rozpravy Ustredniho Ustavu Geol.* **44**:1–327.

Holmer, L. E., 1987, Discinacean brachiopods from the Ordovician Kullsberg and Boda limestones of Dalarna, Sweden, *Geol. För. Stockh. Förh.* **109**(4):317–326.

Jaanusson, V., 1971, Evolution of the brachiopod hinge, in: Paleozoic Perspectives, a Paleontological Tribute to G. A. Cooper (J. T. Dutro, ed.), *Smithson. Contrib. Paleobiol.* **3**:33–46.

Jägersten, G., 1972, *Evolution of the Metazoan Life Cycle,* Academic Press, New York.

Khazanovitch, K. K., Popov, L. Ye., and Melnikova, L. M., 1984, First occurrences of faunal remains in the Sablinka Formation (Middle Cambrian) in the northwestern part of Russian Platform, *Paleontol. Zh.* **1984**(4):33–47.

Koneva, S. P., 1986, A new family of Cambrian inarticulate brachiopods, *Paleontol. Zh.* **1986**(1):49–55 [in Russian].

Laurie, J., 1987, The musculature and vascular systems of two species of Cambrian Paterinidae (Brachiopods), *Bur. Miner. Resour. J. Aust. Geol. Geophys.* **10**:261–265.

Liu Di-yong, 1987, Brachiopods and tommotiids near Precambrian–Cambrian boundary in SW China, in: *Stratigraphy and Palaeontology of Systemic Boundaries in China, Precambrian–Cambrian Boundary 1*, Nanjing University, Nanjing, China, pp. 346–400.

McCammon, H. M., 1981, Physiology of the Brachiopod digestive system, in: *Lophophorates, Notes for a Short Course* (T. W. Broadhead, ed.), University of Tennessee, Department of Geological Sciences, Studies in Geology, No. 5, pp. 170–189.

Nazarov, B. B., and Popov, L. Ye, 1980, Stratigraphy and fauna of Ordovician siliceous-carbonate deposits of Kazakhstan, *Trudy Geol. Inst. SSSR Akad. Nauk* **331**:3–190 [in Russian].

Percival, E., 1944, A contribution to the life-history of the brachiopod *Terebratella inconspicula*, *Trans. R. Soc. N. Zealand* **74**:1–23.

Popov, L. Ye., and Nõlvak, J., 1987, Revision of the morphology and systematic position of the genus *Acanthambonia* (Brachiopoda, Inarticulata), *Eesti NSV Tead. Akad. Toimetsed. Geol.* **36**(1):14–19.

Popov, L. Ye., and Ushatinskaya, G. T., 1987, New data on the shell of the inarticulate brachiopods of the order Paterinida, *Doklady Akad. Nauk SSSR* **293**(5):1228–1230 [in Russian].

Popov, L. Ye., Zezina O. N., and Nõlvak, J., 1982, Microstructure of the apical region of the shell of inarticulate brachiopods and its ecological significance, *Byull. Mosk. Ovo. Ispyt. Prir. Otd. Biol.* **87**(6):94–104.

Rowell, A. J., 1977, Early Cambrian brachiopods from the south-western Great Basin of California and Nevada, *J. Paleontol.***51**:68–85.

Rowell, A. J., 1980, Inarticulate Brachiopods of the Lower and Middle Cambrian Pioche Shale of the Pioche District Nevada, *Univ. Kans. Paleontol. Contrib.* **98**:1–26.

Rowell, A. J., 1982, The monophyletic origin of the Brachiopoda, *Lethaia* **15**:299–307.

Rowell, A. J., and Caruso, N. E., 1985, The evolutionary significance of *Nisusia sulcata*, an early articulate brachiopod, *J. Paleontology* **59**:1227–1242.

Starobogatov, Ya. I., 1979, Evolution of the pelagic larvae of Protostomia and problem of the main body components, *Zool. Zh.* **78**(2):149–160 [in Russian].

Starobogatov, Ya. I., 1983, Brachiocoelic (hydrocoelic) hypothesis of the origin of coelom, *Trudy Zool. Inst. Akad. Nauk SSSR* **109**:111–122 [in Russian].

Thayer, C. W., and Steele-Petrovich, H. M., 1975, Burrowing of the lingulid brachiopod *Glottidia pyramidata*: Its ecologic and paleoecologic significance, *Lethaia* **8**:209–221.

Ushatinskaya, G. T., 1986, The Cambrian brachiopods (Review of localities and some principles of geographical distribution), in: *Trans. Joint Soviet–Mongolian Paleontol. Expedition* **29**:8–34 [in Russian].

Ushatinskaya, G. T., 1987, Unusual inarticulate brachiopods from the Lower Cambrian of Mongolia, *Paleontol. Zh.* **1987**(2):62–68 [in Russian].

Ushatinskaya, G. T., Zetina, O. N., Popov, L. E., and Putivleva, N. B., 1988, On the microstructure and mineralogy of phosphato-calcium shells of brachiopods, *Paleontol. Zh.* **1988**(3):45–55 [in Russian].

Valentine, J. W., 1975, Adaptive strategy and the origin of grades and ground plans, *Am. Zool.* **15**:391–404.

Valentine, J. W., 1977, General patterns of metazoan evolution, in: *Patterns of Evolution as Illustrated by the Fossil Record* (A. Hallam, ed.), Elsevier, Amsterdam, pp. 27–57.

Williams, A., and Hurst, J. D., 1977, Brachiopod evolution, in: *Patterns of Evolution as Illustrated by the Fossil Record* (A. Hallam, ed.), Elsevier, Amsterdam, pp. 79–121.

Williams, A., and Rowell, A. J., 1965, Evolution and phylogeny, in: *Treatise on Invertebrate Paleontology*, Part H, *Brachiopoda* (R. C. Moore, ed.), Geological Society of America, New York, and University of Kansas Press, Lawrence, Kansas, pp. H164–H199.

Wolfart, R., 1974, Die Fauna aus dem Unter-Kambrium von Kerman, Sudost Iran, *Geol. Jahrb. B* **8**:5–70.

Wright, A. D., 1979, Brachiopod radiation, in: *The Origin of Major Invertebrate Groups* (M. R. House, ed.), Academic Press, New York, pp. 232–252.

Zezina, O. N., 1976, *Ecology and Distribution of Recent Brachiopods*, Nauka, Moskow [in Russian].

Chapter 13

Diversification of Archaeocyatha

FRANÇOISE DEBRENNE

1. Introduction

1.1. Definition of Archaeocyatha

Archaeocyaths are an extinct group of sessile marine organisms that lived on carbonate shelves in reef environments during the Cambrian. The basic skeleton is composed of a calcareous cup made up of two coaxial inverted porous cones, the "walls" of which are connected in the "intervallum" by various more or less horizontal elements. Solitary or rarely modular or pseudocolonial, their gross morphology can range from cylindrical to discoid; the intervallum can present horizontal bulging or longitudinal grooves and fluting. Exo- and endostructures, together with secondary development of the skeleton, can develop inside and outside of the cups.

FRANÇOISE DEBRENNE • Institute of Paleontology, 75005 Paris, France.

Origin and Early Evolution of the Metazoa, edited by Jere H. Lipps and Philip W. Signor. Plenum Press, New York, 1992.

1.2. Archaeocyath Significance for Cambrian Biotas

Archaeocyaths are typically Early Cambrian in age, but some taxa persist through the Middle Cambrian into the Upper Cambrian [one locality in Antarctica (Debrenne *et al.*, 1984)].

Archaeocyaths in conjunction with calcimicrobes are significant producers of carbonate sediments. During middle and late Early Cambrian time they are associated with bioherm constructions; they represent a biotic and structural prototype for later reef development in the Phanerozoic (James and Debrenne, 1980; Debrenne and James, 1981).

2. Morphological Diversification

2.1. Characteristics of Archaeocyatha

When the number of described archaeocyaths reached 400, it became obvious to specialists that a major natural subdivision existed among archaeocyaths, and the taxon was divided into two groups: the Regulares and the Irregulares (Vologdin, 1937), considered since that time as classes. The basis for this subdivision was, at first, the general features of the intervallum: radial rods or radial porous or aporous septa, in the Regulares, versus obliquely curved, more or less regularly oriented porous taeniae and vesicular tissue in the Irregulares. However, in the most derived forms of Irregulares, regular septa may be found. Horizontal structures, the tabulae, are present in both classes as porous flat plates (*Erismacoscinus* in Regulares, *Anthomorpha* in Irregulares) or as curved porous plates linked with one or both walls (*Coscinocyathus* versus *Clarucoscinus*). Hexagonal porous tubes building a honeycomb intervallar structure is known only in Irregulares (*Syringocnema*). All these structures are represented in Figure 1.

2.2. Ontogeny

Ontogenetic study of individual archaeocyath cups is based on longitudinal or serial transverse sections. Archaeocyaths have a worldwide distribution of

Figure 1. Intervallum structures of Archaeocyatha. (a) Irregulares with platelets (folii) in the intervallum (*Protopharetra*), Sardinia, Matoppa Member, M84040 MNHN, ×4. (b) One-wall Regulares (Monocyathina), South Australia, Sellick Hill Formation, M82027 MNHN, ×10. (c) Regulares with radial rods (Dokidocyathina), South Australia, Sellick Hill Formation, M82027 MNHN, ×10. (d) Regulares with septa and tabulae (Erismacoscinina), Maroc, Tasousekhtian, M80043 MNHN, ×4. (e) Irregulares with radial nonporous pseudosepta (*Anthomorpha*), Sardinia, Matoppa Member, M84248 MNHN, ×7.5. (f) Irregulares with twisted taeniae (*Archaeocyathus*), Western Canada, Rosella Formation, M83129 MNHN, ×3. (g) Irregulares with arched tabulae (*Archaeosycon*), Newfoundland, M83048 MNHN, ×3. (h) Regulares with normal septa (Ajacicyathina), South Australia, faunal Assemblage II, M82004 MNHN, ×4. (i) Catenulate colony (*Erbocyathus*), Kuznetsk Alatau (Toyonian), coll. Rozanov PIN, ×3. (j) Intervallum with hexagonal porous tubes (Syringocnemidida), South Australia, Ajax Limestone, M82027 MNHN, ×3. (k) Globular Archaeocyatha (Capsulocyathina), Kuznetsk Alatau (Atdabanian), coll. Rozanov PIN, ×10.

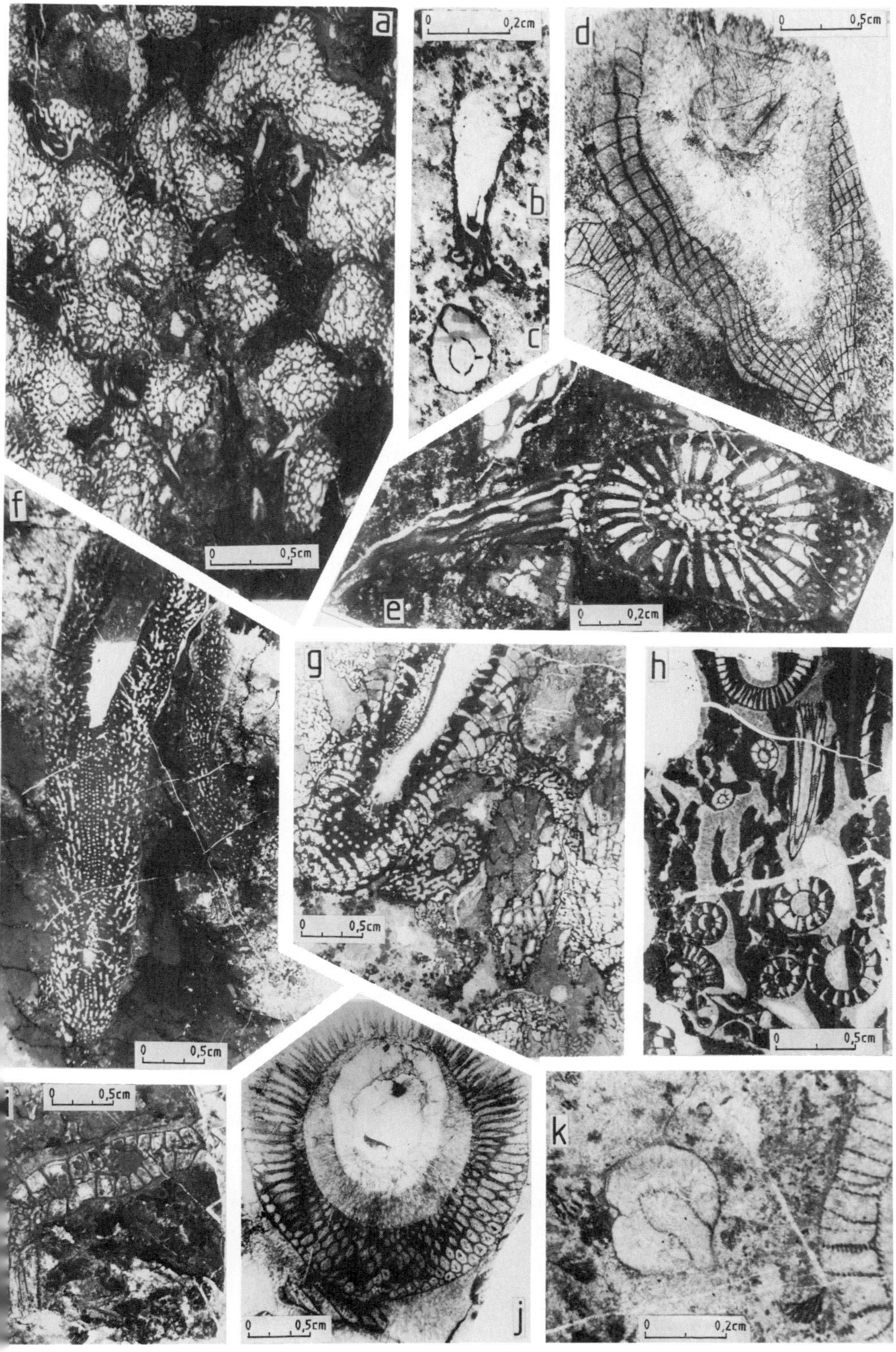
a
0 0,5cm
b
c
0 0,2cm
d
0 0,5cm
e
0 0,2cm
f
0 0,5cm
g
0 0,5cm
h
0 0,5cm
0 0,5cm
j
0 0,5cm
k
0 0,2cm

genera and species, so that a planktonic nonskeletonized larval stage is inferred. The postlarval development begins with the formation of an aporose curved cup which corresponds to the tip of all archaeocyaths, followed by a stage with two walls connected with radial elements. After these common stages, development differs in Regulares and in Irregulares.

In two-walled Regulares the porosity of the outer wall appears at a diameter of 0.15–0.20 mm, as do radial rods (*Dokidocyathus* stage) which later may be transformed into septa. The inner wall differentiates from the inner edge of the rods, so that the direction of initial development is toward the center (centripetal type). In forms with arched tabulae, the tabulae appear before the inner wall and septa. The inner wall differentiates by the invagination of curved tabulae (invaginal type).

In Irregulares the duration of all the ontogenetic stages is greatly lengthened. First to appear is an empty subcylindrical or narrow conical one-walled cup. The second step is the development of a two-walled cup with taeniae pierced by one row of pores per intervallum, very similar to the Regulares *Dokidocyathus* stage; the inner wall is developed from the inner part of the taeniae, as in Regulares, from the septa. In the third step the taeniae become typically multiporous; these taenial structures distinguish Irregulares from Regulares. Taeniae could persist in mature stages or evolve into pseudosepta (porous or nonporous), pseudotaeniae (with synapticulae), a dictyonal framework, and vertical tube-like elements (calicles) or subhorizontal porous polyhedral tubes (syrinx). Tabulae may occur, formed from the upturning of the outer wall. The outer wall is primarily nonporous at early stages; then the different types occur immediately above, without intermediate stages. The inner wall is always porous. Specific characters are stabilized very late in Irregulares for a diameter of 7–10 mm instead of 1–2 mm in Regulares.

Besides this type of development, another group of Irregulares has a thalamid growth pattern.

2.3. Histology of the Skeleton

The archaeocyath skeleton is composed of calcite grains. This structure was first described by Taylor (1910). Only modern techniques—use of ultrathin sections of 0.3 μm thick and the scanning electron microscope, allowing observations at high magnification—have given a more precise picture of the shape and ornamentation of the grains. The results are not diagenetic artifacts, because tests have been made within the same genus from the Tommotian to the Botomian and in different localities (*Nochoroicyathus*) and within different genera (*Acanthopyrgus, Cordilleracyathus, Archaeocyathus*) coming from the same calcareous block. The microstructure of algae (*Epiphyton, Renalcis*) and a coelenterate (*Tabulaconus*), all from the same sample, show significant differences (Debrenne, Gangloff, and Lafuste, 1987).

The archaeocyath skeleton is made of interlocking granules of calcite (Lafuste and Debrenne, 1970). These elements come together to build a compact structure of well-geared elements which cannot be mistaken for independent modified spicules or sclerites. The style of preservation along with the small size of the crystals suggest that the original skeletal mineralogy was calcite. Microstructure is consistent throughout the archaeocyaths; there are no differences between Regulares and Irregulares. Nevertheless, smaller (1.7–1.9 μm) and less-embossed grains are observed within the one-walled Regulares as well as in the vesicular tissue and stereoplasma of Irregulares (Lafuste and Debrenne, 1977, 1982).

Because of strong similarities with some sphinctozoans, Debrenne and Vacelet (1984) assumed that the skeleton was secreted within an organic template, as in the Recent sponge *Vaceletia*, and not the result of biologically induced mineralization (Lowenstam, 1981). The assumption that the crystals are of organic and not mineralogical origin is based on the presence of hollows and bumps on the surfaces, which would be smooth if inorganic.

2.4. Skeletal Diversity within the Regulares

Ontogenetic variations are used to subdivide the Regulares taxonomically into orders and suborders (Table I).

Since the differentiation of the various outer wall structures occurs earlier than that of the inner wall, outer wall structures are taken as the basis for the definition of superfamilies and inner wall structures for families. The structures are represented Figures 2A–2J for the outer wall, and Figures 2K–2P for the inner wall.

Minor structures such as synapticulae, pectinate tabulae, dissepiments, and the upper covering of the cup (pelta) have no high taxonomic significance.

2.5. Skeletal Diversity within the Irregulares

Until recently, the Irregulares had not been as thoroughly investigated as the Regulares. Debrenne and Zhuravlev are now preparing a revision of the Irregulares; their first results are used here.

Ontogenic variations are now known for most of the genera and allow the distinction between two orders, one with taenial development, the other with a thalamid type of growth (Table II).

Intervallar structures (used for the definition of suborders) are much more diverse and complicated than their analogues in the Regulares; while septa are developed in the same radial plane, taeniae diverge in different directions from the main radial plane, giving in section an image of disoriented rods or "folia." The development of synapticulae leads to a dictyonal, a calicular, or syringoid type of intervallum. Taeniae could evolve in radial porous or nonporous pseudosepta. Tabulae are generally not independent and are formed by the outer wall covering the intervallum, except in *Anthomorpha* or related genera, in which they develop independently.

Different types of wall have been recognized. In the outer wall the subdivisions proposed are derived from Gravestock (1984): simple (rudimentary), centripetal, compound, with canals simple, subdivided, or geniculate, and tabellar. The inner wall is simple (rudimentary) with bracts and canal pores, compound, not diverse as in the Regulares. One of the most typical features of the Irregulares is the abundance of dissepiments and the development of a secondary skeleton. Unlike most of the Regulares, the living processes occur within the uppermost part of the cup, whereas the lower part is no longer functional.

TABLE I. Regulares Suborders

One wall	Two walls					
Monocyathida	Radial elements Ajacicyathida		+ Plate tabulae Erismacoscinida	Subspherical chambers Capsulocyathida		Horizontal partitions Tabulacyathida
	Radial rods Dokidocyathina	Radial rods Ajacicyathina		Separate chambers Capsulocyathina	Successive chambers Coscinocyathina	

Figure 2. Types of walls of Archaeocyatha. Outer wall: (A) Simple pores (A1: round; A2: slitlike; A3: diaphragmed). (B) Pores with independent microporous sheath (*Pretiosocyathus* type). (C) Pores with attached microporous sheath (*Erbocyathus* type). (D) Simple tumuli. (E) Multiperforate tumuli. (F) Pores with bracts. (G) Pore canals. (H) Canals covered by microporous sheath. (I) Clathrate pores. (J) Annuli. Inner wall: (K) Simple pores. (L) Pores with bracts and scales. (M) Annuli. (N) Nonporous canals. (O) Porous canals (O1: simply porous; O2: highly porous—spongiose wall). (P) Pores with screen.

TABLE II. Irregulares Suborders[a]

Taenial type of development					Thalamid type of development
Archaeocyathida					Kazachstanicyathida
Pseudosepta		Taeniae	Calicle	Syrinx	
No tabulae	Tabulae				
Loculicyathina	Anthomorphina	Archaeocyathina	Dictyofavina	Syringocnemidina	

[a] All with two walls.

2.6. Significance of Morphological Diversity

The above account of the different morphological skeletal features was done not to give a key for archaeocyath classification, but to show how diverse the archaeocyaths are, and to introduce the diversification pattern and the time of feature appearance of all archaeocyaths, from a systematic point of view.

Interpretation of the systematic value of the skeletal structures is verified by the study of ontogenetic stages and, by the law of homologous variability: the existence of homologous series at all levels, including superfamilies and suborders, was shown by Rozanov (1973) in Regulares and by Debrenne and Zhuravlev (in press) in Irregulares. It allows us to estimate the significance of numerous parallelisms in the development of the group. This law has been recently applied to Irregulares (Debrenne and Zhuravlev, in press).

These principles, used by archaeocyath specialists, may lead one to overestimate the rank of the systematic subdivisions. Sepkoski (1979) did not include the archaeocyaths in his kinetic model because the number of families of the group was so high that an unexplicable peak disturbed the graph (Sepkoski, 1979, Fig. 4). He noted that (1) the taxonomy of archaeocyaths may be oversplit relative to other early Phanerozoic groups, and (2) their morphology may be sufficiently distinct to warrant separation from the Metazoa.

More recent work (Debrenne and Vacelet, 1984; Zhuravlev, 1985; Debrenne, Rozanov, and Zhuravlev, 1990) has concluded undoubtedly that Archaeocyatha are part of the sponges, so the assertion that archaeocyaths are not Metazoa is not tenable. On the contrary, the first proposal has to be considered: if Sepkoski has used superfamilies instead of families as defined by archaeocyath specialists, the result should have fit with his kinetic model of early Phanerozoic taxonomic diversity.

3. Diversification of Archaeocyatha in Time

3.1. Appearance and Disappearance of Archaeocyath Genera

Rozanov (1973) established four subdivisions within the Lower Cambrian on the basis of the steps of Regulares archaeocyath development, using mainly the changes in the porosity of the skeleton: the number of pores in the outer wall (and septa) tends to become reduced (oligomerization). Subsequently, a new process takes place, compensation, with formation on the outer wall of an additional porous sheath, thereby returning the porosity to the preoligomerization state.

The four-stage biostratigraphic zonation as defined above will be used in this chapter: the Tommotian (T1–T4), the stage of simple Regulares with simple porosity; the Atdabanian (A1–A4), the appearance of a great number of new features; the Botomian (B1–B3), a new combination between former characteristic features; and the Toyonian (Ty1–Ty3), few archaeocyaths survive.

The distribution of genera through time is presented here on three graphs. Figure 3 shows the range of each genus, Regulares and Irregulares together. Few genera extend from the beginning of the Tommotian to the end of the Botomian, and none from the Tommotian to the Toyonian. The vast majority are limited to

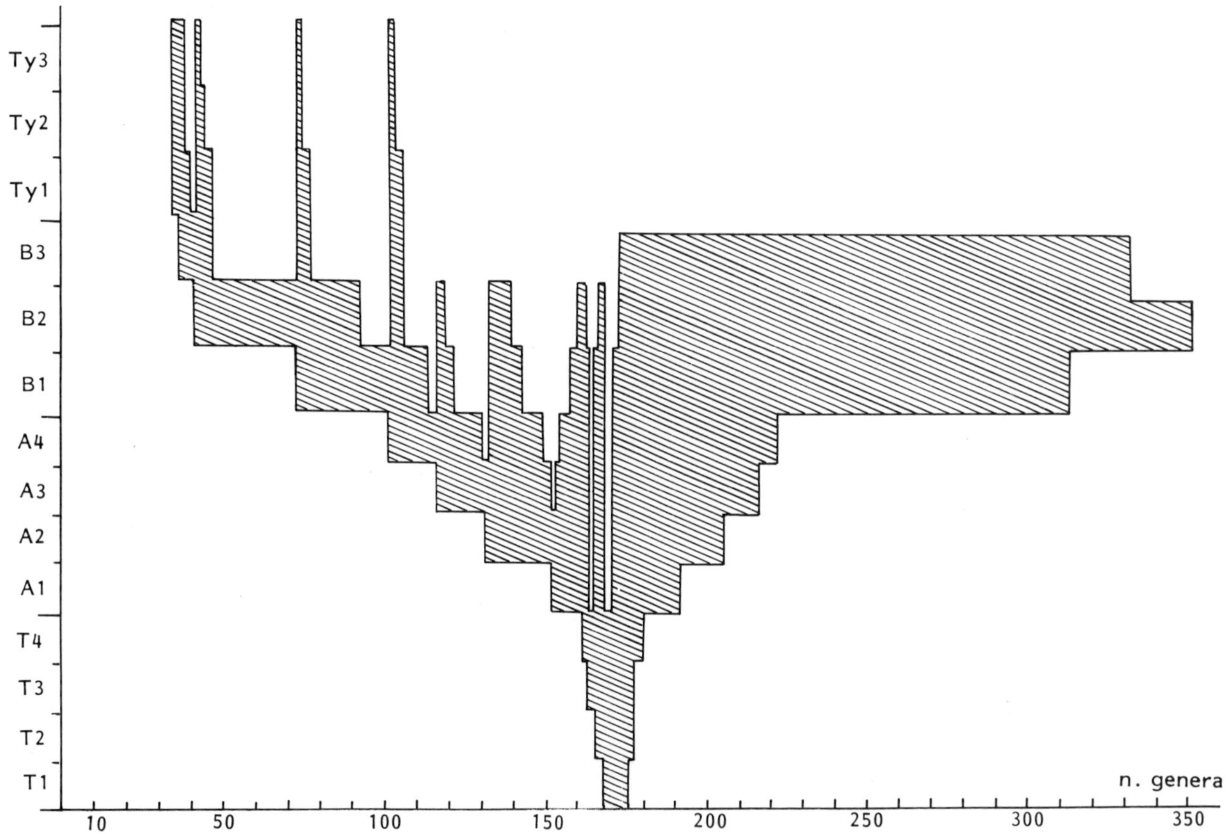

Figure 3. Appearance and disappearance of archaeocyath genera. Horizontal axis, number of genera; vertical axis, stratigraphic column. Age: Tommotian T1–T4, Atdabanian A1–A4, Botomian B1–B3, Toyonian Ty1–Ty3.

the Botomian, but a significant number persist from the Atdabanian to the Botomian. No Atdabanian genera are represented in the Toyonian, while some Botomian genera are present in the Toyonian. If the Regulares were separated from the Irregulares on the graph, the Regulares would show shorter stratigraphic ranges than the Irregulares.

Figures 4 and 5 show the number of appearances and disappearances of Regulares and Irregulares genera represented separately; a cumulative curve summarizes the results. It should be noted that there are no major differences between Regulares and Irregulares diversification. However, the diversification of Regulares begins earlier, in the *Dokidocyathus lenaicus* Zone (T3), than the diversification of Irregulares, which takes place in the *Fansycyathus lermontovae* Zone (A4). The rate of diversification of Regulares speeds up from A4 to B1 and is still relatively high in B1, then slows down. For Irregulares a high rate of diversification is observed as in Regulares in A4–B1, but begins to slow down earlier, as early as the beginning of the Botomian. During the Toyonian both groups remain almost constant with no or few new genera.

Disappearance of genera begins at the end of the Tommotian (Fig. 5). The rate of disappearance is relatively and constantly low for the Irregulares from the beginning of the Atdabanian (A1) up to the upper Botomian (beginning of B3), when it increases significantly; then the decrease slows down to reach a new peak at the beginning of the upper Toyonian (T3). The disappearance of Regulares increases from the upper Atdabanian (A4) with a considerable crisis during the end of the Botomian (upper part of B2 and B3); this phenomenon appears earlier

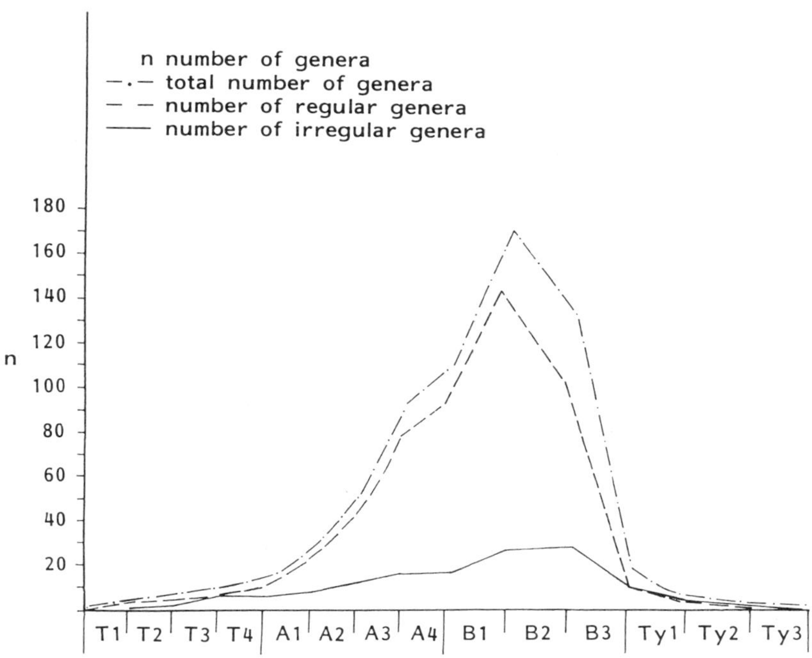

Figure 4. Number *n* of new genera through time. (– –) Regulares, (—) Irregulares, (– • –) total.

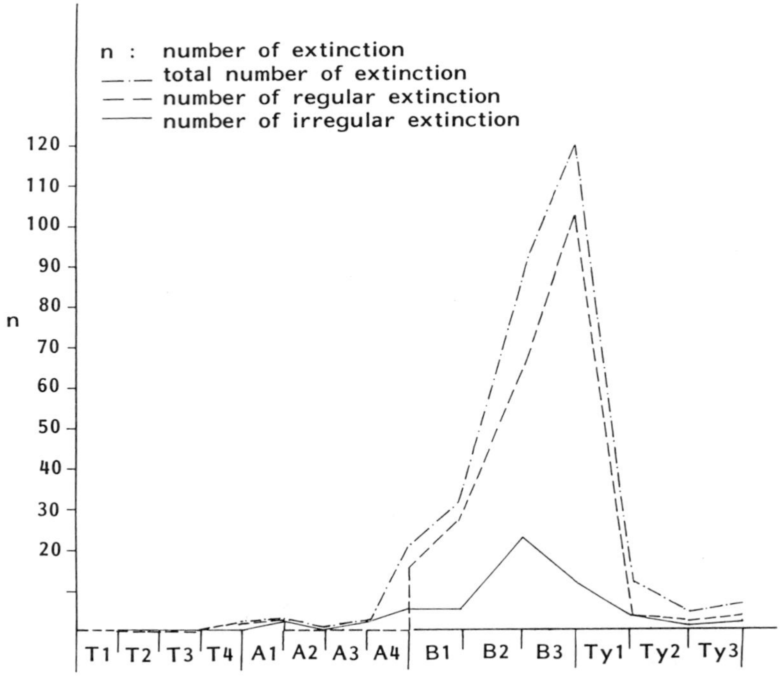

Figure 5. Number *n* of lost genera through time. (– –) Regulares, (—) Irregulares (– • –) total.

in Regulares than in Irregulares. Then the rate of disappearance for the Regulares slows down until the end of the Toyonian, parallel to the Irregulares.

In conclusion, Regulares show a higher rate of appearance and disappearance than Irregulares. They begin their diversification earlier, at the late Tommotian; they are better zone indexes, as their stratigraphic range is more restricted in time than that of the Irregulares. The Irregulares are more globally distributed and are useful for intercontinental biostratigraphic correlation.

3.2. Appearance of Morphological Features

3.3.1. The Timing of First Appearance of Regulares Skeletal Features

Shape of cup	
Conical to cylindroconical	T1
Discoid	T2
Dendroid colonies	T2
With transverse bulging of both walls	T4
Subspherical	T4
Asteroid outer wall	A2
Asteroid inner wall	A2
Catenulate colonies	B1

One-walled cup	T1
Two-walled cup	T1
Primary intervallar elements	
Radial rods (Dokidocyathina)	T1
Radial septa: (Ajacicyathina) normally porous	T1
Reticular	T2
Nonporous	A1
Plate tabulae: (Erismacoscinina) normally porous	T2
With slitlike pores	T2
Heterogeneous	A3
Subspherical chambers (Capsulocyathina)	A1
Arched tabulae linked with one or two walls (Coscinocyathina)	A2
Only tabulae without septa (Tabulocyathina)	A3
Secondary intervallar elements	
Pectinate tabulae	T1
Redimiculi (vertical platelets) in outer wall	T1
Spines	T1
Synapticulae	T4
Stirrup-pores	A3
Redimiculi in inner wall	B1

Table III summarizes the order of appearance of the outer and inner wall porosity types in the various suborders, which, in turn, reflect the structure of the intervallum. When the different structures first appear in the outer wall, they always are associated with a simple inner wall except in the most complicated outer wall cases (marked by an asterisk): an erbocyathoid outer wall with canals

TABLE III. Order of Appearance of the Outer and Inner Wall Porosity Types in Regulares[a]

	M	D	A	E	Ca	Co	T
Outer wall							
A Simple	T1	T1	T1	T1	A1	A2	B1
B Pretiosocyathoid		A2	A2	A2		B1	
C Erbocyathoid	B1		B1	B1	B1		
D Simple tumuli	T2	A1	T4	A2	A1		A3
E Multiperforate tumuli	B1	A3	A2*	B1	A1	A2	
F Bracts	A2	A4	A1	B1	Ty2		
G Canals	B1	B1	A2	A1			
H Canal + sheath			A4*	B1*	Ty1		
I Clathrate			B1	B1*		A4	
J Annular			B1*				
Inner wall							
K Simple		T1	T1	T2	A1	A2	A3
L Bracts		A2	A1	B2			
M Annuli		A3	A1	A1			
N Nonporous canals			A1	A2*		B1	B1
O Porous canals			A1	A4		B1	
P Sheath			A1	A3*		B1*	

[a]M, Monocyathina; D, Dokidocyathina; A, Ajacicyathina; E, Erismacoscinina; Ca, Capsulocyathina; Co, Coscinocyathina; T, Tabulocyathina. See text for explanation of the asterisk.

and sheath first appears with nonporous canals in the inner wall; a clathrate outer wall is associated with an inner wall with bracts; an annular outer wall first appears in genera with annuli or nonporous canals in the inner wall.

The first appearance of inner wall structures follows the same pattern. When they first appear, they are associated with a simple porosity of the outer wall. Some exceptions also occur: Erismacoscinina with nonporous canals on the inner wall present nonporous canals also in the outer wall, not simple pores; forms with an inner wall sheath have an outer wall with sheath, too. It is commonly observed that when the associated wall is not simple, it has the same basic structure as the wall considered.

In both walls the more complex morphologies appear sequentially later than the simple morphologies, following the general trends of evolution defined within the Regulares. One exception is observed in Tabulocyathina: the genus with a simple outer wall appears later (B1) than the genus with outer wall tumuli. There are no clear rules for the appearance of the different types of walls within the suborders: they appear earlier, simultaneously, or later without any order.

3.2.2. The Timing of First Appearance of Irregulares Skeletal Features

Feature	First appearance
Shape of cup	
Domal	B2
Plate-like	T3
Bowl-shaped	T3
Conical to cylindroconical	T1
Discoid	T1
Dendroid pseudocolonies	T2
Massive modular forms	A2
Intervallar elements	
Porous pseudosepta	T1
Porous taeniae	T1
Dictyonal network	T2
Nonporous pseudosepta	T3
Chambers and pillars	B2
Pseudotaenial network	T3
Hexagonal calicles	A2
Tetragonal syrinx	A4
Tetragonal calicles	B1
Hexagonal syrinx	B2
Outer wall	
Simple	T1
Compound	T3
Pustular	T3
Canals	A3
Centripetal	A1
Tabellar	B1
Subdivided canals	A3
Geniculate	A4

Inner Wall
- Simple T1
- Bracts and canals A2
- Compound A4

Most features appeared during the Tommotian and Early Atdabanian, the most complicated forms at the beginning of Botomian. No new features appeared after the Middle Botomian (Figs. 4 and 5). But, contrary to Regulares, a significant number of new characters appear even at the ordinal level in the early Botomian (also see Table IV). Genera in Irregulares have a longer range than in Regulares: the genus *Protopharetra* is known as early as in the Tommotian (T3), while a genus close to *Archaeocyathus* is the only representative surviving in the Upper Cambrian: the suborder Archaeocyathina assures the survival of the group through the Cambrian.

Irregulares have fewer diagnostic features and genera than Regulares, but as far as the number of individuals or the biovolume within a given stratum or bioherm is considered, Irregulares are far more abundant than Regulares. Irregulares are the main contributors in the frameworks of buildups, whereas Regulares are scattered within the construction or live in the immediate vicinity of the reefs.

4. Diversification of Archaeocyatha in Space

At the beginning of the Cambrian, the interaction of various factors affected the development of the biota.

Paleogeographic reconstructions based on recent paleomagnetic data together with paleontological and paleoclimatic information place the Siberian Platform,

TABLE IV. Order of Appearance of Outer and Inner Wall Porosity Types in Irregulares[a]

L	An	Ar	D	S	K	
Outer wall						
A Simple	A1	B1	T1	A2		B1
B Centripetal			A2		B1	
C Compound			A2	B3		
D Pustular	B3				B3	
G Canal			A3			
H1 Subdivided canals			A3			
H2 Geniculate canals			A4			
I Tabellar			B1			
Inner wall						
I Simple			T1			B1
II Bracts–canals	A2		B1			
III Compound			B2			

[a]L, Loculicyathina; An, Anthomorphina; Ar, Archaeocyathina; D, Dictyofavina; S, Syringocnemidina; K, Kazachstanicyathida.

Mongolia, and Australia in the equatorial zone, and North Africa and North America in middle southern latitudes from the Tommotian to the Botomian (Rozanov, 1984).

Archaeocyath distribution is linked with the deposition of calcium carbonate and with the development of buildups. The distribution of archaeocyaths within the different Cambrian basins is plotted on Fig. 6, together with the times of appearance and disappearance in each basin.

The oldest archaeocyaths are found on the Siberian Platform (Aldan and Igarka regions) within a narrow belt of shallow marine bioherms. This is the first center of origin and diversification of Archaeocyatha, during the Tommotian. Archaeocyaths are not found outside of Siberia at that time. The restriction of archaeocyaths to this region in the Tommotian may be due to the fact that only dolomites and terrigenous sediments are deposited elsewhere in the world, conditions unfavorable to the development of Archaeocyatha.

The largest Lower Cambrian transgression started in the Middle Atdabanian and reached its maximum in the Botomian, resulting in broad carbonate shelves. The diversification and biogeographic extension of archaeocyaths occur concomitantly. By the end of the Atdabanian in all basins of North Africa, southern Europe, the Middle East, Mongolia–Okhotsk, China, Australia, and the Siberian Platform similar assemblages of archaeocyaths were established. If the hypothesis of a Palaeopangea is correct (Kirschvink *et al.*, 1984), then the basins occupied by archaeocyaths probably were not isolated.

Two main dispersal routes may be proposed: from Siberia (Siberian Platform and South Fold Belt) (1) through the Far East to Australia and (2) through Iran, the Caucasus, and the Balkans to western Europe and North Africa.

Between the late Atdabanian and the early Botomian there is again an important increase in the number of archaeocyaths (see Figs. 3–5). The beginning of a significant participation of archaeocyaths as frame builders in reefs also took place at this period of time (James and Debrenne, 1980). A new province is created in a narrow belt from Mexico to Alaska and Koryakiya, with a high percentage of endemic regular forms [61.5% according to Zhuravlev (1986)]. The other Botomian province is the Afro-Siberian–Antarctic province, in which it is possible to define subprovinces: (1) the Afro-European subprovince extending to China and Kazakhstan, characterized by Anthomorphidae, (2) the Australo-Antarctic subprovince, with Flindersicyathidae, Metacyathidae, and Syringocnemididae, and (3) the Siberian subprovince, where all these families are represented.

The provincial differentiation as the main basins are still in low latitudes is probably due to greater geographic isolation of each area.

At this time four centers of dispersion may be recognized: Mongolia–Okhotsk, North Africa, Australia, and North America. The main routes, based upon the presence of common genera, started through Siberia and the Mongolia–Okhotsk Fold Belt (1) to China, Australia, and Antarctica (2) to Europe and North Africa, and (3) through the Far East to Koryakiya and North America.

The Toyonian marks the beginning of a regression and the general reduction of the Early Cambrian basins. Archaeocyaths drastically decrease in number, but nevertheless reach new regions, Greenland and Virginia, coming from the Altai-Sayan Fold Belt and the Siberian Platform area, all of which are characterized by

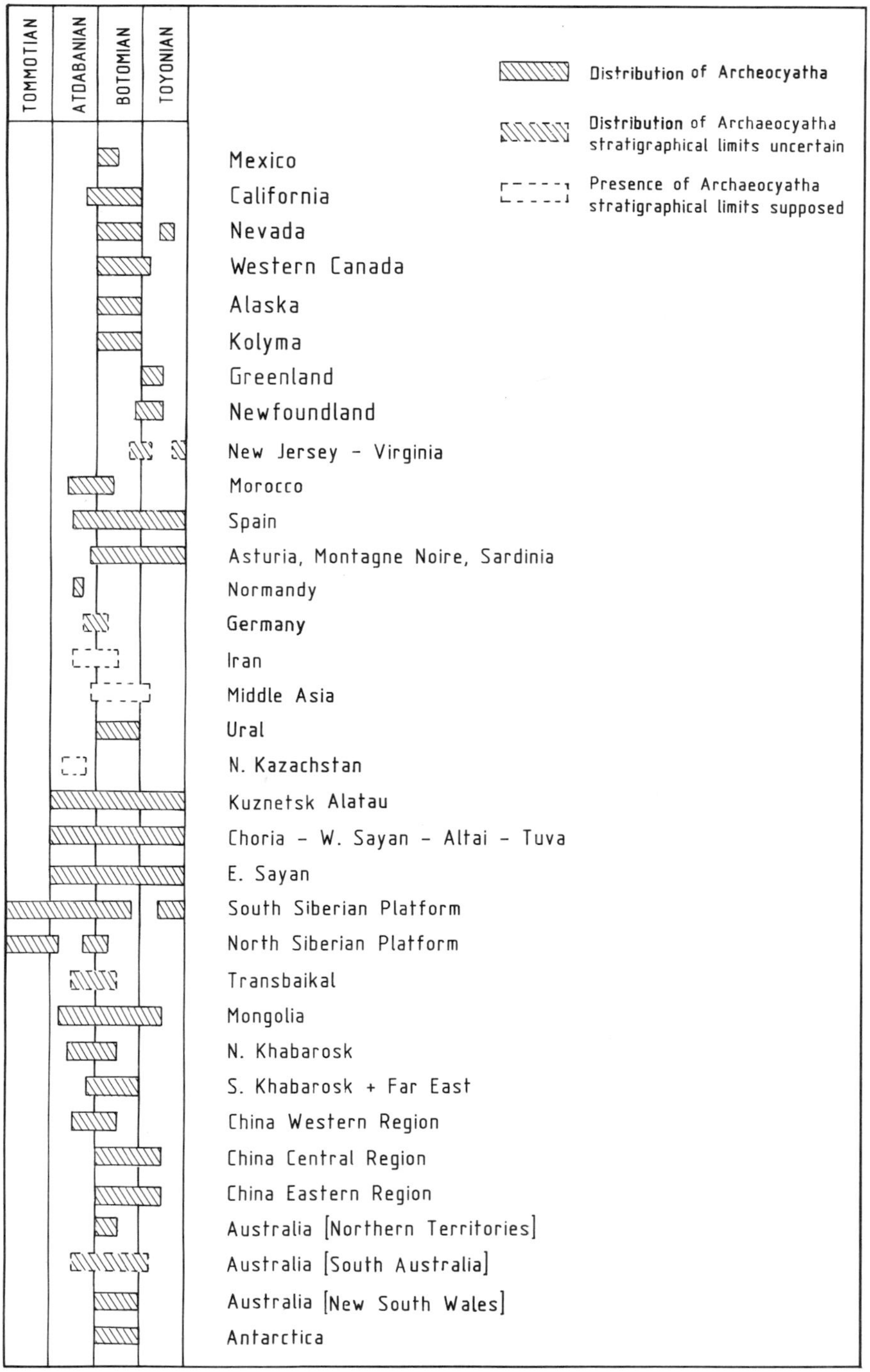

Figure 6. Geographic distribution of Archaeocyaths.

Tegerocyathidae and Erbocyathidae. The other basins—South Australia, China, Spain, Sinai, Newfoundland and Labrador, and western North America—contain Archaeosyconidae and Archaeocyathidae. The links between these basins were apparently broken and the faunas evolved in place.

In conclusion, Regulares, mostly endemic, serve to define the provinces, while Irregulares, with a wider geographic distribution, delineate the migration routes.

References

Debrenne, F., and James, N. P., 1981, Reef-associated archaeocyathans from the Lower Cambrian of Labrador and Newfoundland, *Paleontology* (Lond.) **24**(2):343–378.

Debrenne, F., and Vacelet, J., 1984, Archaeocyatha: Is the sponge model consistent with their structural organisation? *Paleontogr. Amer.* **54**:358–369.

Debrenne, F., and Zhuravlev, A. Yu., in press, Archéocyathes Irréguliers. Cah. Paléont., Ed. du CNRS, Paris.

Debrenne, F., Gangloff, R. A., and Lafuste, J., 1987, *Tabulaconus* Handfield: Microstructure and its implication in the taxonomy of primitive corals. *J. Paleont.* **61**(1):1–9.

Debrenne, F., Rozanov A. Yu., and Webers, G. F., 1984, Upper Cambrian Archaeocyatha from Antarctica, *Geol. Mag.* (Lond.) **121**(4):291–299.

Debrenne, F., Rizanov, A. Yu., and Zhuravlev, A. Yu., 1990, Archéocyathes Réguliers, *Cah. Paléont.* 5Ed. du CNRS, Paris, pp. 1–218.

Gravestock, D. I., 1984, Archaeocyatha from lower parts of the Lower Cambrian carbonate sequence in South Australia. *Mem. Ass. Austral. Paleont.* **2**:1–139.

Hill, D., 1972, Archaeocyatha, in: *Treatise on Invertebrate Paleontology*, Part E, *Archeocyatha*, 2nd ed., Geological Society of America and University of Kansas Press, Lawrence, Kansas, pp. E1–E158.

James, N. P., and Debrenne, F., 1980, Lower Cambrian bioherms: Pioneer reefs of Phanerozoic, *Acta Palaeont. Polonica* **25**(3/4): 655–668.

Kirschvink, J. L., Barr, T. D., and Rozanov, A. Yu., 1984, The destruction of Palaeospangaea (?) in the Early Cambrian, in: *Abstracts., 27th International Geological Congress, Moscow, 1984*, Nauka, Moscow, Vol. 47, p. 113.

Lafuste, J., and Debrenne, F., 1970, Observations en lames ultra-mince de la microstructure d'archéocyathes, *C.R. Som. Société Géologique de France,* **56**:224–225.

Lafuste, J., and Debrenne, F., 1977, Présence de deux types de microstructures chez *Archaeocyathus atlanticus* Billings (Cambrien inférieur, Labrador, Canada). *Géobios* **510**(1):103–107.

Lafuste, J., and Debrenne, F., 1982, Différences microstructurales entre formes à une muraille et formes à deux murailles chez les archéocyathes réguliers: Faut-il "couper" les Monocyathida des Archaeocyatha? 9° RAST, Paris, Société Géologique de France, p. 350.

Löwenstam, H. A., 1981, Minerals formed by organisms, *Science* **211,** 4487, pp. 1126–1131.

Rozanov, A. Yu., 1973, Zakonomernosti Morfologicheskoy Evolutsii Arkheotsiat i Voprosy Yarusnogo Raschleneniya Nizhnego Kembriya. (Regularities of Archaeocyathan Morphological Evolution and the Problems of the Lower Cambrian Stage Subdivision), *Trudy Geol. Inst. Akad. Nauk SSSR* **241**:1–164. [In Russian].

Rozanov, A. Yu., 1984, Some aspects of studies on bio- and paleogeography of the Cambrian, in: *Proceedings of the 27th Geological Congress, Moscow, 1984,* Vol. 2, *Paleontology,* VNU Science Press, Ubrecht, pp. 143–157.

Sepkoski, J. J., Jr., 1979, A kinetic model of Phanerozoic taxonomic diversity. II—Early Phanerozoic families and multiple equilibria, *Paleobiology* **5**(3):222–251.

Taylor, T. G., 1910, The Archaeocyathinae from the Cambrian of South Australia, *Mem. R. Soc. S. Austral.* **2**(2):55–188.

Vologdin, A. G., 1937, Arkheotsiaty i rezultaty ikh izucheniya v SSSR (kratkaya svodka). [Archaeocyaths and the results of their study in USSR (a brief review)], *Problemy Paleont.* **2/3**:453–481 [in Russian].

Zhuravlev, A. Yu., 1985, Sovremennye arckheotsiaty? (Recent archaeocyaths ?), in: Problematiki

Pozdnogo Dokembriya i Paleozoya. (Problematics of the Late Precambrian and Palaeozoic), B. S. Sokolov and I. T. Zhuravleva, eds, *Trudy Inst. Geol. Geofiz. Sibirsk. Otdel. Akad. Nauk SSSR*, **632**:24–33. [In Russian].

Zhuravlev, A. Yu., 1986, Evolution of archaeocyaths and palaeobiogeography of the Early Cambrian, *Geol. Mag.* (Lond.) **123**(4):377–385.

V

Overview of the Proterozoic/ Cambrian Radiations

Chapter 14

Early Radiation of Biomineralizing Phyla

STEFAN BENGTSON and SIMON CONWAY MORRIS

STEFAN BENGTSON • Institute of Palaeontology, S-751 22 Uppsala, Sweden. SIMON CONWAY MORRIS • Department of Earth Sciences, Cambridge University, Cambridge CB2 3EQ, England.

Origin and Early Evolution of the Metazoa, edited by Jere H. Lipps and Philip W. Signor. Plenum Press, New York, 1992.

1. Introduction

Phanerozoic rocks, from the Cambrian onward, teem with fossils of biomineralizing organisms. The long Proterozoic sequence, up to its very contact with the Phanerozoic, reflects an almost total absence of biologically controlled mineralization. Indications of calcareous incrustations of Proterozoic photosynthesizing organisms (Klein *et al.*, 1987) are at most questionable, and in any case do not imply more than that mineralization may have been indirectly induced by biological activity. Records of biogenic magnetite in the Proterozoic (Chang and Kirschvink, 1989) suggest more direct biological control and are thus of considerable interest with regard to the early evolution of biomineralization. In terms of structural and mineralogical diversity, however, the transition to the Phanerozoic nevertheless represents a quantum jump.

The dramatic difference in visible fossil content between the Proterozoic and the Phanerozoic therefore has often been seen as mainly due to the advent of biomineralization, increasing the fossilization potential in already existing lineages. It is now evident, however, that biomineralization was just one facet of the biotic revolution during the Proterozoic–Phanerozoic transition. This explosive diversification involved all kinds of organisms, some with, but many without hard skeletons. At the moment, however, the skeletalized taxa are the only ones with a complete enough record to yield reliable information about morphology and distribution. Because they include all levels of organization, they may serve as a useful subset of the total biota for investigation of the anatomy of the Precambrian–Cambrian radiation.

Biostratigraphic correlation and resolution, however, are so poor that a sequence of appearances among skeletal groups can be suggested only in outline. Moreover, significant divergence in the interpretation of radiometric data means that the rate at which hard parts appeared is difficult to assess; estimates of as little as 1 or 2 million years (or even less) stand in contrast to proposals that the acquisition of skeletons stretched over tens of millions of years.

It remains also to be determined to what extent hard-part development represents actual evolutionary originations as against changing taphonomic potential. This question arises because of evidence that primitive skeletons were predominantly organic, with the mineral salts present either in low abundance or as scattered granules. In either event, the preservation potential of such skeletons would have been exceptionally low, so that even if individual granules survived to accumulate in the sediment, their recognition would be almost impossible (Lowenstam and Margulis, 1980a). In principle, therefore, the onset of skeletogenesis may have preceded the widespread appearance of fossil hard parts by millions of years, with the latter event marking the breaching of a taphonomic threshold resulting from the evolution of more robust skeletons.

In the absence of soft-bodied Lagerstätten of the appropriate age it is difficult to assess whether skeletogenesis was abrupt or gradual. Certainly the oldest known animal hard parts, those of the enigmatic tubicolous taxon *Cloudina* from Ediacaran limestones of Namibia (Germs, 1972), appear to consist of organic matter that is weakly impregnated with calcite (Grant, 1990). Similarly, in Early Cambrian faunas there are a number of skeletal remains whose flexibility, as evident from such features as folding and crumpling, argue for only weak biomineralization. The conspicuously thin exoskeletons of early trilobites have been remarked upon by various workers. Indeed, if comparisons with modern faunas are any guide, it would be unreasonable to expect there to be a rigid definition of hard parts; rather a graduation between lightly mineralized to robust skeletons would be variously represented according to shifting taphonomic circumstances. It is important also to stress that not only may local conditions favor (or militate against) preservation of the more delicate skeletons, but that the positive aspects of taphonomy whereby skeletons are replaced during diagenesis by phosphate also play a crucial role in our perception of the fossil record of early skeletal organisms. Phosphogenic episodes close to the Precambrian–Cambrian boundary seem to indicate major paleoceanographic changes, some of which are considered to have had a direct influence on the development of skeletons and the wider diversification of metazoans (Cook and Shergold, 1984). Much less emphasis, however, has been given to the widespread phosphatization of calcareous skeletons after burial. The taphonomic mechanisms of this replacement are poorly understood, but in favored circumstances tens of thousands of skeletal remains can be recovered per kilogram of sample by acid digestion of the carbonate matrix. Inevitably this leads to a taphonomic bias in our collections, but there has been no systematic attempt to see whether particular assemblages are favored or discriminated against according to their position on the shelf transsect and relative to centers of phosphogenesis.

It will be clear, therefore, that the onset of skeletogenesis is neither well constrained in terms of time nor facies. Similarly, the biogeographic distribution of this event has received only simplistic attention. Nevertheless, recent advances have been considerable and it seems assured that skeletons developed independently in many unrelated groups, although further information on the underlying genetic coding for skeletogenesis is required urgently.

2. Types of Skeletal Biomineralization

2.1. Minerals

The production of minerals by organisms is a considerably broader phenomenon than the formation of skeletons. Of the approximately 60 minerals known to be formed by living organisms (Lowenstam and Weiner, 1989), only calcite, magnesian calcite, aragonite, apatite, and opal have any extensive use for skeleton-building. The fossil record suggests that skeletons made of these different biominerals appeared during a narrow interval of time at the Proterozoic–Phanerozoic transition (Fig. 1). Hard skeletons appear to have evolved separately in at least around 40 clades, as discussed below [for details, see Schopf and Klein (in press)]. These clades represent a number of different mechanisms for the formation of skeletons.

Due to the influence of taphonomic and diagenetic factors, the original structure and mineralogical composition of fossil skeletons is often problematic. Interpretations may be based on nothing better than observed diagenetic composition in a few lithologies and on comparisons with known and presumably related taxa. Few detailed studies have been carried out on the detailed petrography and geochemistry of fossils and their surrounding rocks in this part of the stratigraphical sequence (e.g., James and Klappa, 1983; Runnegar and Bentley, 1983; Grant, 1990), and the mineralogy of early mollusk shells has in some cases been investigated from crystal morphologies replicated on the surface of internal molds (Runnegar, 1985a; Bengtson *et al.*, 1990). More such studies are needed before we can reach beyond the fairly general observations reported herein.

2.1.1. Opal

Biogenic opal (a hydrated silica gel) is diagenetically unstable and only rarely preserved in the fossil state. Measurements in modern equatorial oceans suggest that only about 0.1% of the opal produced becomes incorporated in the sediment (Hurd, 1973), and over a longer time period the opal almost always recrystallizes, eventually to quartz. Opal is widespread among protists and sponges; it occurs in a number of multicellular plants (Sangster and Parry, 1981), but is not formed by the so-called higher metazoans. This pattern of distribution among protists and animals is already evident in the earliest record of skeletal fossils: there seem to be no extinct phyla among the early siliceous fossils. On the other hand, utilization of silica for skeleton formation appears to have increased dramatically at several later stages during the Phanerozoic, for example, with the radiation of diatoms in the Late Mesozoic. The present undersaturation of sea water with respect to silica is believed to be a late effect due to the extensive biomineralization by diatoms and radiolarians in particular (Broecker, 1971; Lowenstam and Weiner, 1989, pp. 244–247). Thus, during the early history of skeletal opal, sea water may have been saturated with silica.

Why does skeleton-forming opal have this particular distribution within the protist, animal, and plant kingdoms? Silicon is the second most abundant element in the earth's crust, and organisms have physiological means of handling it. In vertebrates it is known to be an essential trace element for growth and for the apatite mineralization of bone (Carlisle, 1981). But opal may have its limitations in the construction of larger skeletons. Although opal skeletons are often characterized by an elaborate morphogenesis (e.g., radiolarian or diatom tests), morphogenetic control is essentially limited to the intracellular environment (Simpson and Volcani, 1981, Table 1-1), where the opal is formed within a cytoplasmic membrane (silicalemma).

From its wide distribution among protists of different phylogenetic lineages (Round, 1981; Bovee, 1981), intracellular opal biomineralization would appear to

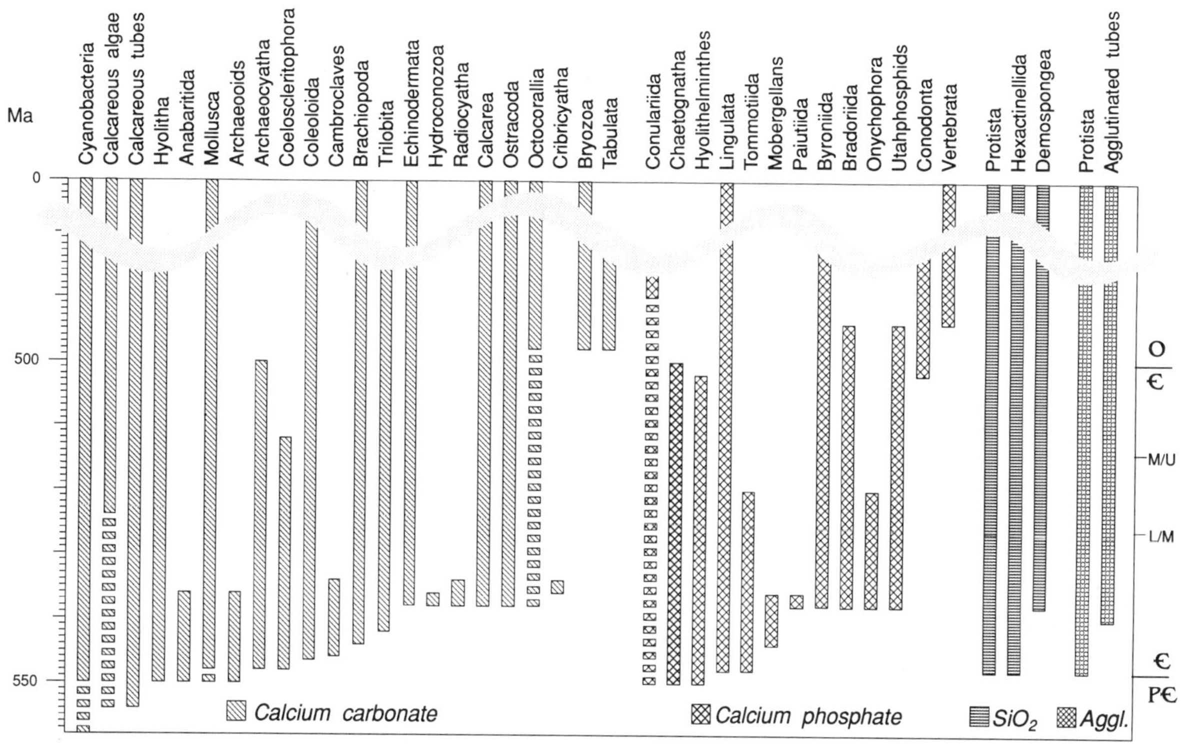

Figure 1. Distribution of major skeleton-forming (biomineralizing and agglutinating) taxa during the latest Precambrian to earliest Ordovician, as well as the Recent (wavy line indicates omitted time in diagram). "Wastebasket" taxa (calcareous tubes, agglutinating tubes) are taxonomically poorly defined, but contain at least one separate skeleton-forming clade. Broken bars indicate uncertain identity of fossils.

be a primitive character that evolved early in the history of eukaryotes. Yet no clear evidence for siliceous skeletons older than the Cambrian exists in the fossil record, the record of Cambrian silica-producing protists is very scarce, and some major groups of silica-producers, such as diatoms and silicoflagellates, did not make their appearance until considerably later. Thus, it is possible that silica biomineralization evolved separately in several lineages of protists and sponges. Nevertheless, the recent report of diverse siliceous microfossils in the Lower Cambrian (?) of Yukon Territory (Allison and Hilgert, 1986) (cf. below, p. 458) shows that the fossil record still has surprises in store and that silica biomineralization may be less sporadic than appears in the literature. If it is indeed a plesiomorphic trait among protists and multicellular plants and animals, the lack of biogenic opal in nonporiferan metazoans would suggest that the ability to form opal, once lost, is not easily regained; there is good evidence from physiology, development, histology, and biochemistry for the close phylogenetic proximity of poriferans with other metazoans (Bergquist, 1985; Runnegar, 1985b) [but see Bergström (1991) for arguments regarding a separate derivation of the sponges], and there would seem to be a number of uses for silica also in higher metazoans.

2.1.2. Calcium Salts

The other major skeleton-forming minerals are calcium salts, that is, carbonates or phosphates. This fact has inspired a number of hypotheses in which calcium plays a central role in the early evolution of biomineralization and of metazoans in general (Lowenstam and Margulis, 1980a, b; Degens *et al.*, 1985; Kazmierczak *et al.*, 1985; Kazmierczak and Degens, 1986). Calcium is essential in eukaryote cell metabolism, interacting with a variety of proteins, including structural proteins such as collagen. The two main polymorphs of calcium carbonate, calcite and aragonite, are partly interchangeable as biominerals. Mollusks, for example, may form skeletons consisting of one or the other, or a mixture of both; in several other groups the mineral form may vary from one taxon to the other, and may even vary within one species due to environmental factors, mainly temperature (Lowenstam and Weiner, 1989).

The early history of calcite versus aragonite mineralization is still largely obscure, because in many cases the original mineralogy of calcareous fossils is not known. Aragonite is diagenetically unstable and only exceptionally preserved in pre-Mesozoic rocks. Calcite and magnesian calcite are more stable, although the latter commonly is diagenetically transformed into low-magnesian calcite. The original aragonitic composition usually has to be inferred from cooccurrences of recrystallized skeletons and originally calcitic ones. A high strontium and low magnesium content may also indicate original aragonite, since the calcite lattice allows replacement of Ca by Mg ions, whereas in aragonite Sr is the preferred substitution.

Calcium carbonates are the dominant skeletal mineral group in extant animals, and appear to have been so from the onset of skeletal biomineralization. The first mineral skeletons known from the fossil record are of calcium carbonate, and among the diverse skeletal faunas in the earliest Cambrian, the majority of higher taxa used this material (Fig. 1).

Calcium phosphates as biominerals occur mainly as crystalline carbonate apatite or amorphous calcium phosphate (ACP). Among recent organisms, only vertebrates and some inarticulate brachiopods are known to produce apatitic skeletons. Apatite occurs also in the periostracum of some mytilid bivalves (Waller, 1983; Carter *et al.*, 1990), teeth of certain chitons (Lowenstam and Weiner, 1985), and possibly the sclerites of a few stolidobranch ascidians (Lambert *et al.*, 1990). Calcium phosphate has thus been considered a less common biomineral in spite of its quantitatively large occurrence in the diverse and widespread vertebrates. Recent investigations have shown, however, that ACP occurs in no less than 20 living phyla (Lowenstam and Weiner, 1989; Watabe, 1990).

Early Phanerozoic rocks contain a number of problematic fossils with apatitic skeletons, the diversity of which has been taken to suggest a wider distribution of this shell mineral among the earliest skeletal fossils (e.g., Rhodes and Bloxam, 1971; Lowenstam and Margulis, 1980a, b; Cook and Shergold, 1984, 1986; Lowenstam and Weiner, 1989). Phosphorus, being an essential element for vital functions, is a limiting factor for biomass in modern oceans. During the Proterozoic–Phanerozoic transition, however, major phosphate deposits were formed in various parts of the world (Cook and Shergold, 1986), suggesting that phosphorus may have been more readily available than now. Rhodes and Bloxam (1971) proposed that phosphate biomineralization evolved as a detoxification process in a phosphorus-rich environment. A milder version of this hypothesis is that if phosphorus was abundant, it would have been physiologically inexpensive to use it rather than carbonate for calcification (Lowenstam and Margulis, 1980a, b; Cook and Shergold, 1984, 1986; Lowenstam and Weiner, 1989), particularly as phosphate is also known to act as a crystal poison of carbonate mineralization (Simkiss, 1964).

Calcium phosphate is diagenetically the most stable of the major shell-building minerals, and when the original crystallography has been preserved, it is usually not difficult to recognize primary phosphate in the fossil record. Nevertheless, a number of Cambrian fossils are known only from phosphate deposits where secondary impregnation or overgrowths of skeletal matter have taken place. In those cases it is often not easy to see whether the shells were originally calcareous or phosphatic, or even conceivably consisted of some other mineral.

Although there is no question that apatite in the Cambrian occurred in a larger number of major animal groups than today, this appears to be only a reflection of the higher diversity of phylum-level taxa at that time. Of the skeletalized clades recognized in Fig. 1, 11 out of 36 recognized in the Early Cambrian appear to have been phosphatic, as against 2 out of 17 in today's biota. In view of the higher probability of recovering the chemically stable phosphatic shells from the fossil record, this difference may well be an artifact.

2.2. Skeletal Types

Before addressing the taxonomic profile of skeletogenesis, it is necessary to define their principal forms and mineralogy, the latter of which are coded as C for calcareous, O for resistant organic biopolymers, P for phosphatic, and S for siliceous.

2.2.1. Spicules

These are a widespread, but relatively neglected component of the early skeletal faunas. They are defined as internally secreted elements that are usually embedded in soft tissue. The isolated nature of spicules means that on death they disperse, so that taxonomically many may have to be treated as sciotaxa (cf. Bengtson, 1985), especially as the morphological range in a species may be considerable. In many cases spicules of modern organisms are composed of diagenetically unstable minerals and cannot be expected to be readily fossilized (Rieger and Sterrer, 1975; Lowenstam and Weiner, 1989). Even if the spicular component of many forms is still poorly understood, their assignment to known groups is possible in some cases. In the lower part of the Lower Cambrian of South Australia (Bengtson *et al.*, 1990), spicules with well-preserved mineralogical and morphological characters can be referred to several extant major taxa: hexactinellid (S) and calcareous (C) sponges, echinoderms (C), and probably also cnidarian octocorals (C). Claims for ascidian (P; although in living forms more commonly C and A) affinity of utahphosphids (Bendix-Almgren and Peel, 1988; cf. below) remain more speculative. However, at present many spicules defy ready assignment to known groups, and may represent extinct classes or even phyla.

Based largely on comparisons with living relatives, the early spicules are envisaged as being embedded in the body wall, primarily serving for support and protection. As so little is known about the functional morphology of modern spiculate taxa, as well as the significance (if any) of particular spicule morphs, so there has been little detailed speculation about the paleoecology of their earliest relatives. In addition, the taphonomy of such taxa awaits proper investigation. Certainly, the high surface area-to-volume ratio of spicules can encourage dissolution, but there appear to have been no systematic investigations of the taphonomic behavior of different groups.

2.2.2. Tubes

Tubicolous taxa are very widespread and also account for some of the earliest records in the form of *Cloudina* (C) and the sabelliditids (O). In the early Cambrian faunas the tubes tend to be morphologically simple, and it is difficult to ascertain the true diversity that may be masquerading behind the facade of apparent tubular uniformity. Cambrian tubes may be phosphatic, calcareous, siliceous, agglutinating, or purely organic. There are at least seven major groups of tubicolous taxa, the more distinct of which are discussed below.

2.2.3. Conchs

Included under the category of conchs are shells that are secreted by marginal accretion and so typically bear growth lines. Univalve forms are dominated by the molluscan monoplacophorans, rostroconchs, and gastropods (all C), although possible ambiguity resides in various cap-shaped shells of unknown origin and tubicolous worms that adopt a spiral configuration. Bivalved conchs are represented by bivalve mollusks (C), stenothecoids (C), brachiopods (C, P), and an informal group of so-called pseudobrachiopods (C). In addition, the hyoliths (C) are included by virtue of their conical conch closed by an operculum. As with tubes, the primary function of conchs was presumably protective; with adpression of the shell, closure of valves or operculum serves to seal off the soft parts. However, although many conch-bearing taxa are inferred to have been deposit feeders, in at least the brachiopods (and presumably pseudobrachiopods) the bivalved arrangement serves to define a feeding chamber in which the lophophore operates.

2.2.4. External Sclerites

Taxa bearing this type of skeleton seem to have been of unusual importance in early faunas. Like the spicules, they suffer the disadvantage of disaggregation of the scleritome on the animal's death, so that reconstruction of original arrangements may be difficult. Because sclerites from different regions of the body may appear rather different, there has been unwarranted enthusiasm in some quarters to erect a large number of form taxa. Diversity estimates based on an uncritical tally of taxa are likely to be wildly awry. Difficulties in scleritome reconstruction may be offset by discovery of fused sclerites and in Lagerstätten even articulated specimens, as well as recognition of articulatory facets and other evidence for mutual accommodation. Such data, in combination with those lines of evidence that have been adopted routinely by conodont workers (mutual cooccurrence, similar ultrastructure, etc.) and additional criteria [articulation facets, ontogenetic fusion of sclerites, etc. (Qian and Bengtson, 1989, p. 13)], are now leading to more plausible scleritome arrangements.

Several unrelated groups bore sclerites. These include the coeloscleritophorans (chancelloriids, halkieriids, sachitids, and siphogonuchitids; all C), tommotiids (including camenellids, tannuolinids, eccentrothecids, lapworthellids, and sunnaginiids; all P) and the cambroclaves (C). In the case of halkieriids and chancelloriids, discoveries of articulated specimens have provided indications of sclerite arrangements (Walcott, 1920; Bengtson and Conway Morris, 1984; Conway Morris and Peel, 1990). The sclerites of *Microdictyon* have recently been shown to be serially repeated, paired elements situated along the bodies of onychophoranlike animals (Chen *et al.*, 1989a; Ramsköld and Hou, 1991). In remaining groups, however, knowledge of sclerite disposition is at a more preliminary stage. It seems clear that the appearance of the scleritome varied widely. In some taxa sclerites were closely articulated to form a continuous covering, whereas in others probably they were studded in a more haphazard fashion. In any event, the scleritome would have conferred protection, especially when the sclerites bore spines. However, the functional explanations of variations in sclerite shape are largely unaddressed.

2.2.5. Toothlike Structures

Teeth are often a good source of paleontological information, because they tend to be impregnated with substances that make them resistant to wear and, consequently, to destruction before fossilization. There are numerous toothlike fossils in Cambrian rocks. Elongate, thornlike objects grade into slender and gracile teeth, while yet others have more robust shapes. Most of them, however, are recurved to varying degrees. Many seem to be derived from grasping apparatuses, perhaps of chaetognaths [protoconodonts, all P (Szaniawski, 1982; Bengtson, 1983)]. Others, however, may have been more similar to grasping mandibles [*Cyrtochites*, ?P (Qian and Bengtson, 1989)]. Although tooth-shaped objects may also have other functions, this category of skeletal fossils is important because it provides one of the few direct lines of evidence for predators in these early metazoan communities.

2.2.6. Carapaces

The group encompasses the arthropods with a cuticular exoskeleton that usually is conspicuously more mineralized than the underlying appendages. In Lower Cambrian assemblages the principal groups are the trilobites (all C) and bradoriids (all P), a group that is usually included with the ostracodes. In addition there is a wide variety of arthropods, many of uncertain systematic position, with exoskeletons more or less minimally mineralized.

2.2.7. Calcareous Reinforcements

Calcareous reinforcements of bodies are often massive and have good fossilization potential. Archaeocyathans (all C) and echinoderms (all C), characterized by internal such skeletons, have an excellent fossil record which may closely correspond to their actual distribution in time and space (this volume, Chapters 11 and 14). There are also a number of Cambrian fossils that appear to represent massive basal skeletons of spongelike or cnidarianlike animals (Jell, 1984; Debrenne *et al.*, 1990).

2.2.8. Others

Although the seven categories encompass the majority of early skeletal fossils, there remains a potpourri of taxa which are less easy to define in the context of skeletal type because of continuing uncertainties of their biology and ecology. In most cases they are plate- or cap-shaped structures that have not yet been adequately explained in terms of functional morphology.

3. Lineages of Early Skeletal Fossils

3.1. Attempts to Identify Independent Origins of Mineralization

Biomineralization in various lineages may ultimately be seen to have one common origin in the evolution of physiological control mechanisms that would serve for ion concentration and nucleation for a number of biominerals. Nonetheless, in the fossil record a number of skeleton-bearing taxa appeared suddenly without any evidence of one type of skeleton having evolved from the other. Although independent origins can never be definitely proven unless the origin can be identified in each specific case, the profound differences in structure and composition between various early skeletons make direct homologies unlikely. In the following we shall therefore treat separately all clades of biomineralizing organisms that do not show evidence of having a common biomineralizing ancestor with any other clade.

3.2. Early Biomineralizing Lineages

3.2.1. Calcareous Cyanobacteria

Recent calcified cyanobacteria mostly occur in nonmarine environments and have their mucilagenous sheaths impregnated with carbonate crystals, perhaps as a by-product of the photosynthetic removal of CO_2 from the water in which they lived (Riding, 1977). Although arguments have been made for biologically induced mineralization by cyanobacteria in deposits as far back as 2.5–2.3 Gy old (Klein *et al.*, 1987), this is clearly anomalous with respect to the total Proterozoic record, in which cyanobacteria are almost exclusively nonmineralized. Close to the Precambrian–Cambrian transition, however, cyanobacteria provide one of the best examples of the appearance of mineralization in previously nonmineralized taxa. The genus *Obruchevella* Reitlinger, 1948, is represented by helically coiled threads, about 5–25 μm in diameter, in rocks of late Precambrian to Devonian age. They are particularly abundant around the Precambrian–Cambrian boundary (e.g., Voronova and Radionova, 1976; Riding and Voronova, 1984; Song, 1984), and in the lowermost Cambrian similar forms appear with a calcified outer sheath (Peel, 1988; Qian and Bengtson, 1989; Bengtson *et al.*, 1990). Impressions of the cellular patterns of the trichomes leave little doubt about the cyanobacterial nature of the fossils (Peel, 1988). Other filamentous cyanobacteria also became calcified at this time (Danielli, 1981; Riding and Voronova, 1982; Bengtson *et al.*, 1990).

3.2.2. Calcareous Eukaryotic Algae

Multicellular eukaryotic algae show a similar pattern of calcification to cyanobacteria, but in the early Paleozoic there may be considerable difficulties in determining the affinity of calcareous "algae." Riding and Voronova (1984) compiled stratigraphic ranges of various calcareous algae (including cyanobacteria) across the Precambrian–Cambrian boundary in Siberia and Mongolia. Whereas some members of their earliest assemblages may indeed be true calcareous algae, and even a late Precambrian candidate has been presented (Grant *et al.*, 1991), more convincing such fossils do not appear until the Middle Cambrian (Schopf and Klein, in press).

3.2.3. Protists

The contribution of this kingdom to the early skeletal record is remarkably scanty, which is all the more surprising given that skeletogenesis during the "Cambrian explosion" was not confined to metazoans but encompassed cyanobacteria and a variety of eukaryotic algae.

The Lower Cambrian *Platysolenites* is widely accepted as an agglutinating foraminiferan (Glaessner, 1963; Føyn and Glaessner, 1979), although the possibility that the tubes housed some sort of metazoan seems to be worth reconsidering. A Lower Cambrian *Ammodiscus* has recently been figured (Culver, cited in Huber, 1991), and by Middle Cambrian times a moderate diversity of foraminiferans had been reached (Cherchi and Schroeder, 1985). Most claims for early foraminiferans, including supposed examples from the Precambrian, either have not withstood scrutiny or require reappraisal to determine their dubious affinities. In any case, all early examples were agglutinating; calcareous foraminiferans did not appear until the Devonian (Boersma, 1978).

Scarce radiolarianlike fossils occur throughout the Cambrian (Nazarov, 1973, 1975; White, 1986; Bengtson, 1986; Conway Morris and Chen, 1990a), but not until the Ordovician did the group radiate to encompass more than just a few genera (Fortey and Holdsworth, 1971; Nazarov and Ormiston, 1985).

Allison and Hilgert (1986) reported a diverse association of siliceous microfossils from chert-containing portions of the Tindir Group in northwestern Canada, of probable earliest Cambrian age. Although their exact affinity is unknown, they show considerable resemblance to skeletal parts of living groups of protists (in particular, chrysophytes), and it is likely that they represent early protists. This discovery suggests that it is possible to obtain much more information on early biomineralizing protists if suitable techniques are applied.

3.2.4. Hexactinellid Sponges

Hexactinellids differ from other living sponges to such an extent that they have been proposed to be a separate phylum (Bergquist, 1985). Spicules of hexact type occur in sediments as early as the lower Meishucunian of China (Ding and Qian, 1988) and the approximately equivalent lower Tommotian of the Siberian Platform (Sokolov and Zhuravleva, 1983). Higher in the Cambrian, hexactinellids are abundant (Finks, 1983; Rigby, 1986; Bengtson, 1986; Bengtson *et al.*, 1990).

Recent hexactinellid spicules are composed of opal, deposited intracellularly around an organic axis (e.g., W. C. Jones, 1969). The Meishucunian spicules are reported to have a siliceous composition (Ding and Qian, 1988). In the Atdabanian of South Australia, the spicules now consist of quartz and show the characteristic central canal (Bengtson *et al.*, 1990). Reports of younger calcareous spicules of hexactinellid type [see Sdzuy (1969), and subsequent papers referring to "Calcihexactina") probably represent diagenetic replacement of originally opaline spicules.

3.2.5. Demosponges

Demosponges are a large and diverse group of sponges that have the ability to form several types of mineral skeletons (in addition to organic ones), the main polymorphs being siliceous spicules and, in the coralline sponges and the "sphinctozoans," a calcareous nonspicular skeleton. The first definitive demosponges occur in the Atdabanian (Chen *et al.*, 1989b, 1990; Bengtson *et al.*, 1990), but it is not unlikely that earlier siliceous spicules also belong to this group. The later fossil record of demosponges is very rich (Rigby, 1983, 1986).

Although there are constructional as well as morphological differences between demosponge and hexactinellid spicules (Reiswig, 1971), they have the same opaline composition and intracellular mode of formation. In view of the fact that the two groups appear to be very clearly differentiated even from their first appearance, it is likely that they acquired their biomineralizing habit independently, although the underlying physiological mechanism may have been inherited from a common ancestor.

3.2.6. Calcareous Sponges

Calcareous sponges have spicules of magnesian calcite, each one consisting of a single crystal (W. C. Jones, 1970, 1979). The spicules are typically mono- or triradiate and formed extracellularly by interaction of sclerocytes, generally two per ray. These sponges are also known to produce calcareous (calcitic) coralline skeletons similar to those of some demosponges (Vacelet, 1985).

Fossil spicules of calcareous sponges are not always identifiable as such, because diagenetic effects may have obscured the original composition. The oldest spicules demonstrated to consist of original calcite are Atdabanian in age, and these Australian examples have a strict triradial symmetry (Bengtson *et al.*, 1990).

3.2.7. Archaeocyathans

The first representatives of this group are Tommotian in age. Long thought to be confined to the Early Cambrian, the archaeocyathans have recently been shown to have survived at least into the Late Cambrian (Debrenne *et al.*, 1984). The apparent rate of diversification, outstripping all other groups, is probably an artifact of taxonomy due to an abundance of describable characteristics that has encouraged oversplitting. Nevertheless, there is no doubt that this is a morphologically diverse and successful group that in many parts of the Lower Cambrian world must have provided the most conspicuous skeletal organisms. In addition, the contribution of archaeocyathans to reef systems, together with stromatolites and calcareous algae, is a landmark in marine paleoecology, although their ecological complexity is limited in comparison to later examples. Two principal groups are identified, the regular and irregular archaeocyathans. Although the textbook example of a double-walled form with connecting septa and prominent pores is regarded as typical, a wide diversity of other forms, including encrusters, also occur. The principal investigations of evolution have been by Soviet and French workers (e.g., Zhuravleva, 1960; Rozanov, 1973; Debrenne *et al.*, 1989).

Archaeocyathan affinities have been the subject of some debate, opinions generally focusing on either a poriferan relationship or a status as an extinct phylum. Recent research on living and fossil coralline sponges (e.g., Vacelet, 1985) suggests that the archaeocyathan type of skeleton may well have been formed by a sponge (Debrenne and Vacelet, 1984; Debrenne and Wood, 1990). Cogent additions to the debate may depend on further investigations into functional morphology, especially with respect to proposed modes of water filtering, and skeletal ultrastructure. However, a place within the sponges seems reasonable for at least some archaeocyathans.

3.2.8. Radiocyathans

Radiocyathans are spongelike fossils with walls reinforced by dumbbell-shaped calcareous sclerites having a central shaft and a rosette at the outer and inner end (Debrenne *et al.*, 1970, 1971; Nitecki and Debrenne, 1979; Zhuravlev, 1986). They have been referred to sponges or archaeocyathans, but have also been interpreted as a separate group related to these phyla or to receptaculitid algae. They are confined to the Lower Cambrian and are mainly known from Australia (Bedford and Bedford, 1934, 1936) and Siberia (Zhuravleva and Myagkova, 1981).

3.2.9. Cnidarians

Although numerous lines of evidence hint that cnidarians existed during the Early Cambrian, identifiable skeletal remains are exceedingly scarce. Calcareous sclerites reminiscent of octocoral spicules have been found in the Lower Cambrian of South Australia (Bengtson *et al.*, 1990). A small number of colonial and solitary calcareous skeletons [reviewed by Scrutton (1979) and Jell (1984)], including the Hydroconozoa Kordeh, 1963, have been referred to the Cnidaria, but their exact affinity is unknown (Debrenne *et al.*, 1990; Lafuste *et al.*, 1990). The phosphatic byroniids, including some Early Cambrian representatives, have been interpreted as representing polyp stages of scyphozoan medusae (Bischoff, 1989) (see also remarks on conulariids, Section 3.2.22).

Cnidarians are important biomineralizers throughout most of the Phanerozoic, producing a considerable portion of the skeletal carbonates in modern sediments. The apparent rarity of biomineralizing cnidarians in the Cambrian poses a puzzle, in view of the ancient origin and alleged primitiveness of the group.

3.2.10. Tubicolous Organisms

These forms probably encompass a variety of major groups, although assignments to known phyla are seldom secure. At least seven independently biomineralizing clades can be discerned; in addition, there are numerous insufficiently known tubular fossils (e.g., Fig. 2B) that may or may not be related to any of these clades.

The sabelliditids occur in Vendian–Cambrian boundary deposits and had a flexible tube of organic composition. Although usually broken, the available lengths and lack of tapering point to very elongate tubes, and comparisons have been drawn frequently with the pogonophorans (Sokolov, 1965, 1967). Ultrastructural comparisons between the two types suggest a simpler laminar structure in the sabelliditids (Urbanek and Mierzejewska, 1977), although taphonomic studies on pogonophoran tubes could also be instructive to determine if this structure is the result of preferential decay of parts of the tube.

The phosphatic hyolithelminths have also been compared with pogonophorans (Poulsen, 1963) and with serpulids (Grigor'eva and Zhegallo, 1979; Grigor'eva, 1980), although none of these suggestions appears solidly founded. Preliminary observations of a crossed-fibrous wall structure (Bengtson *et al.*, 1990) need to be followed up.

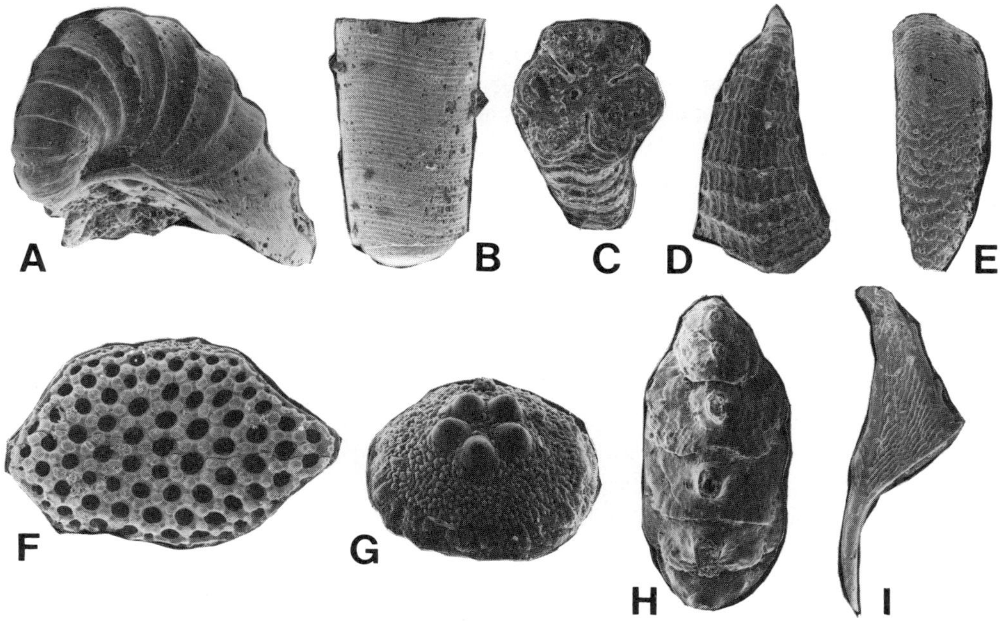

Figure 2. Examples of early skeletal fossils, all, except G, from the Lower Cambrian. (A) *Archaeospira ornata* Yu, 1979, a mollusk, ×27; (B) *Actinotheca* n. sp., a calcareous tubular fossil, ×27; (C) *Anabarites* n. sp., an anabaritid, ×50; (D) *Lapworthella* n. sp., a tommotiid, ×50; (E) Sachitidae n. gen., n. sp., a coeloscleritophoran, ×27; (F) *Microdictyon rhomboidale* Bengtson, Matthews, and Missarzhevskij, 1986, ×50; (G) *Hadimopanella* n. sp., a utahphosphid, ×200; (H) *Paracarinachites spinus* (Yu, 1984), a paracarinachid, ×27; (I) *Scoponodus renustus* Jiang, 1982, ×17. (A, H, I) From the Middle Meishucunian Stage of Yunnan, China (Qian and Bengtson, 1989); (B–E) from the Parara Limestone (Atdabanian Stage equivalents) of South Australia (Bengtson *et al.*, 1990); (F) from the Atdabanian Stage of Tamdytau, Kazakhstan (Bengtson *et al.*, 1986); (G) from the Amgian Stage of the Siberian Platform (unpublished picture; collection of Yu L. Pel'man).

The coleolids (Fisher, 1962) had a resistant calcareous tube with longitudinal striations spiraling along the length. In some instances they lived embedded vertically in the sediment, forming local thickets that trapped sediment into a moundlike structure (Braiser and Hewitt, 1979; Landing *et al.*, 1988).

The paiutiids are an exclusively Lower Cambrian group having phosphatic tubes with up to seven longitudinal septa arranged in bilateral symmetry (Tynan, 1983). They have been suggested to be cnidarians related to anthozoan corals (Tynan, 1983). Similarities with the polyp stages of coronate scyphozoans, including the fossil byroniids, are more compelling, but certain basic differences in tube formation appear to exist (Bischoff, 1989).

The anabaritids (Fig. 2C) had calcareous tubes, distinguished by a prominent triradial symmetry (Voronova and Missarzhevskij, 1969; Rozanov *et al.*, 1969; Missarzhevskij, 1974; Val'kov and Sysoiev, 1970; Abaimova, 1978; Val'kov, 1982; Conway Morris and Chen, 1989; Bengtson *et al.*, 1990). The internal keels, which may be notched or whole, are of uncertain functional significance. In some taxa there were also elongate flanges extending from the three external corners (e.g.,

Abaimova, 1978), and these may have helped to stabilize the tube in shifting sediment. Anabaritids are characteristic components of the first assemblages of skeletal fossils to appear close to the Precambrian–Cambrian boundary in various parts of the world. Anabaritids have been proposed to belong to the Annelida (Voronova and Missarzhevskij, 1969; Glaessner, 1976), Cnidaria (Missarzhevskij, 1974), and specifically the Scyphozoa (Val'kov, 1982; Fedonkin, 1985), but the unusual triradial symmetry of the whole body makes such assignments problematic. The presence of discoidal fossils with triradial symmetry in Ediacaran rocks may be a significant clue to their ancestry (Fedonkin, 1985).

The status of the late Precambrian *Cloudina* is also unresolved, and comparisons with serpulid annelids (Glaessner, 1976) are not supported. *Cloudina* tubes consisted of loosely stacked cones of flexible organic lamellae, probably impregnated with high-magnesium calcite (Grant, 1990). Comparisons have been made with the Lower Cambrian cribricyathans (Germs, 1972; Glaessner, 1976), but no convincing similarities have been demonstrated.

The cribricyathans have calcareous tubes of complex microstructure, with an outer wall composed of short, stacked, concentric lamellae, apparently representing incremental growth, and sometimes a thin inner porous wall. They are almost exclusively associated with archaeocyathans, and have mostly been reported only from the Altaj-Sayan Region, Tuva, USSR. Cribricyathans have been regarded as planktic larval stages (Vologdin, 1932) or other life stages (Zhuravleva and Okuneva, 1981) of archaeocyathans, but most commonly they are interpreted as a separate problematic group (Hill, 1972), sometimes at the level of an independent phylum (Jankauskas, 1972).

The agmatans are a small group of Early Cambrian fossils having tubes with or without external calcareous wall and with the interior filled with agglutinated material deposited in a funnel-shaped pattern around a central tube. They have been compared with a number of animal groups, including annelids and protozoans [see Lipps and Sylvester (1968) and Yochelson (1977) for reviews], but none of these suggestions have been well supported. Yochelson (1977) proposed that they be regarded as a separate phylum, the Agmata.

3.2.11. Mollusks

Understanding of molluscan relationships in the Cambrian has advanced considerably during the last decades. Despite the general accord that most of the principal groups of mollusks (Fig. 2A) can be derived either directly or indirectly from the monoplacophorans, there is reason to believe that yet more primitive forms have spicular coats and were derived from turbellarianlike creatures.

At some time near the beginning of the Cambrian these spicules are envisaged as having fused into a coherent shell, raising considerably the preservation potential of the early mollusks. Given that spicules and sclerites are abundant in early skeletal fossils, it might seem reasonable to identify the precursors of mollusks with a coherent conch. Thus far, however, they have not been positively identified, although the paucity of information on the soft parts of many sclerite-bearing groups limits speculation. Of possible significance is the recognition in the Middle Cambrian wiwaxiids, which are believed to be closely related to halkieriids (Bengtson and Conway Morris, 1984) [this connection, however, is queried by

Butterfield (1990)], of a possible radula and soft ventral area comparable to a mollusk foot (Conway Morris, 1985). It seems possible that the molluscan ancestors, equipped with a muscular foot and feeding radula, included forms that secreted skeletal elements in a variety of ways, and only those leading to the spicular coat of aplacophorans and to the cap-shaped shells of monoplacophorans were to enjoy long-term success.

Although the diversity of mollusks in the Cambrian was considerably lower than during later Phanerozoic periods, the number of Cambrian fossil taxa referable to mollusks is relatively high: about 160 genera occur in the Lower Cambrian alone (Schopf and Klein, in press, Chapter 23.5). In addition, there are a number of fossil shells and sclerites that may conceivably be mollusks. Many of the uncertainties in the interpretation are due to lack of information about structure and composition of the skeleton. Recent mollusks construct their shells or spicules of aragonite or calcite. Even by the Middle Cambrian, a substantial proportion of the known varieties of shell ultrastructure had evolved. These include spherulitic prismatic aragonite, aragonitic nacre, tangentially arranged fibrous aragonite, cross-lamellar aragonite, and foliated calcite (Runnegar, 1985a). Furthermore, there are a few examples of apatite being used for constructional purposes by living mollusks (Lowenstam and Weiner, 1985; Watabe, 1990). Although these cases are exceedingly rare, they suggest that Cambrian phosphatic-shelled organisms should not be automatically excluded from mollusk affinities.

Of particular relevance to the question of early molluscan shell evolution is the unresolved question of the nature and affinities of those monoplacophoranlike shells composed of closely packed cellular units. These occur among the earliest assemblages of shelly fossils, and include the distinctive genus *Maikhanella* (Kerber, 1988; Qian and Bengtson, 1989). These shells are reminiscent of fused sclerites and may help to explain the origin of the molluscan shell (Missarzhevskij, 1989; Bengtson, 1990).

3.2.12. Coeloscleritophorans

This is an exclusively Cambrian group of metazoans characterized by calcareous, probably aragonitic, hollow sclerites which may have been molted during ontogenetic growth (Bengtson and Missarzhevsky, 1981). They include vagrant, bilaterally symmetrical forms with scale- and spine-shaped sclerites (halkieriids and wiwaxiids), probably sedentary forms with bag-shaped bodies and covered with spiny, usually composite, sclerites (chancelloriids), and several apparently intermediate forms with spine-shaped sclerites and unknown body shape (sachitids, Fig. 2E; siphogonuchitids). Detailed investigations of the group have been presented by Jell (1981), Bengtson and Conway Morris (1984), Conway Morris (1985), Qian and Bengtson (1989), and Bengtson *et al.* (1990). Coeloscleritophorans appear to constitute a major Cambrian clade, possibly representing a stock of metazoans of platyhelminth grade close to the mollusks.

3.2.13. Hyoliths

Hyoliths are bilaterally symmetrical metazoans with a cone-shaped aragonitic conch carrying an operculum and, in some forms, a pair of helens, curved appendages that protrude through slits between conch and operculum. They usually have one flatter and one more convex side, presumed to be ventral and dorsal. They were probably benthic deposit feeders, because a sediment-filled U-shaped gut is sometimes preserved, consisting of a convoluted ventral and a straight dorsal limb. The shell structure resembles molluscan crossed-lamellar fabric (Runnegar *et al.*, 1975). The two main groups, orthothecids and hyolithids, differ in that the latter have a ventral lip (ligula) that is expressed in both conch and operculum. Hyoliths are exclusively Paleozoic and may have been present in the earliest skeletal assemblages.

The systematic affinity of hyoliths has been hotly debated. They are often included in the mollusks (e.g., Marek and Yochelson, 1976), but have also been suggested to constitute an extinct phylum related to the Sipunculida (Runnegar *et al.*, 1975).

3.2.14. Brachiopods

Studies of this group have lagged somewhat, although it is clear that the early radiation produced a variety of short-lived and morphologically distinct groups. The two principal questions in the context of the present chapter are: (1) Are brachiopods monophyletic in that all known taxa share a common shelled ancestor? (2) What is the significance of the various short-lived groups and how are they best classified?

The first question has been dealt with in some detail by Wright (1979), Rowell (1982), and Goryansky and Popov (1986); Popov (this volume, Chapter 12) addresses both of them in an innovative fashion. There is agreement that a mineralized shell evolved several times in the history of the group, the differences focusing on the phylogenetic relationships between the various groups and whether or not the last common ancestor had a brachiopodlike mantle. Popov (this volume, Chapter 12) regards the calcareous-shelled brachiopods ("Phylum Brachiopoda") and the phosphatic-shelled ones ("Class Lingulata") to be sister groups that evolved independently from a lophophorate ancestor. Both these groups appeared in the earliest Cambrian [note that the two families appearing in the late Precambrian in Popov's diagram refer to Middle Meishucunian occurrences that are probably Tommotian equivalents, and consequently of Early Cambrian age as currently understood (Qian and Bengtson, 1989)].

3.2.15. Arthropods

The early history of arthropods is dealt with in detail elsewhere (Briggs and Fortey, this volume, Chapter 10). Trace fossils attributable to arthropods occur in probable Tommotian-equivalent strata, but body fossils of arthropods are not definitely known until the Atdabanian. Mineralized carapaces in the Lower Cambrian are only known from two or possibly three groups. Trilobites have a calcareous skeleton, probably originally consisting of low-magnesian calcite (Wilmot and Fallick, 1989). Lower Cambrian ostracodes mostly belong to the Order Bradoriida, which have a carapace more or less impregnated with calcium phosphate. The origin of calcareous ostracodes is not known. P. J. Jones and McKenzie (1980) suggest that the Atdabanian genus *Ushkarella* is a palaeocope ostracode, but this possibility needs to be more closely investigated. Although the contrary has often been claimed, there is no evidence of a phylogenetic transition from a phosphatic to a carbonatic ostracode carapace. Another example of an early arthropod with mineralized carapace is provided by Atdabanian representatives of the problematic bivalved arthropod *Isoxys*, in which the long anterior and posterior spines (but possibly not the remainder of the carapace) appear to have been originally impregnated with calcium phosphate (Bengtson *et al.*, 1990).

3.2.16. Echinoderms

Echinoderms radiated into a number of distinct classes after their appearance in the Atdabanian (Paul and Smith, 1984). Of the characteristic features of living echinoderms—pentaradial symmetry, a water vascular system, and a mesodermal skeleton consisting of calcite in a stereom meshwork—only the two latter may be found in all early echinoderms. One of the first groups to appear, the helicoplacoids, lacked pentameral symmetry, but they had a three-rayed ambulacral system that appears to correspond to the three main branches of the five-rayed system in other echinoderms (Derstler, 1981). The carpoids, probably early offshoots of the echinoderm branch, similarly had no plane of symmetry.

The echinoderm skeletal stereom consists of magnesian calcite with small amounts of organic matter (Lowenstam and Weiner, 1989, pp. 132–134). It is intermeshed with the organic stroma, with which it is in constant physiological interaction, providing a tight control on skeletal growth and morphogenesis. This type of skeleton was present in the earliest echinoderms.

3.2.17. Protoconodonts

The genus *Protohertzina* is part of the earliest assemblages of skeletal fossils (e.g., Missarzhevskij, 1973; Nowlan *et al.*, 1985; Brasier and Singh, 1987; Qian and Bengtson, 1989) and ranges at least into the upper Atdabanian (Bengtson *et al.*, 1990). It is a representative of the so-called protoconodonts, slender, spine-shaped, phosphatic elements with a lamellar structure indicating accretionary growth on the inner side and basal margin (Bengtson, 1976, 1983). They are regarded as grasping hooks of predatory animals similar to and perhaps related to recent chaetognaths (Szaniawski, 1982). The relationship of protoconodonts to the para- and euconodonts (Bengtson, 1983) is obscure, although a direct phyletic connection may exist (Bengtson, 1976; Szaniawski, 1987).

3.2.18. Utahphosphids

This group is represented by button-shaped, phosphatic sclerites, about 50–250 μm in diameter and consisting of a fibrous core and an enamel-like capping (Fig. 2G). Occasional specimens preserved several sclerites set in a matrix of finely granular apatite. Utahphosphids are known from the Atdabanian to the Lower Ordovician. In the Upper Cambrian *Utahphospha sequina*, the sclerites are fused into polygonal platelets with no intervening matrix, the whole forming a hollow cone with an open tip (Müller and Miller, 1976). Histological similarities with vertebrate dermal skeletons (Bengtson, 1977) have favored hypotheses of vertebrate affinity of utahphosphids (Wrona, 1982, 1987; Dzik, 1986), and a proposal of affinity with tunicates has also been made (Bendix-Almgren and Peel, 1988). Kraft and Mergl (1989), Boogard (1989), and Hinz *et al.* (1990) identified utahphosphidlike structures in the cuticle of palaeoscolecid worms, suggesting that utahphosphids belong to the Palaeoscolecida Conway Morris and Robison, 1986, of uncertain systematic affinity.

3.2.19. Tommotiids

This is a diverse group represented by sclerites built up of phosphatic growth laminae externally expressed as distinct growth lines (Fig. 2D). The sclerites presumably mantled part or all of the surface of a sluglike animal (Bengtson, 1970), but the scleritome structure may vary from simple packing of more or less irregular sclerites (*Eccentrotheca*) to a distinctly bimembrate (*Camenella, Tannuolina, Micrina*) or polymembrate (*Dailyatia*) organization. The classification is problematic and demonstrates problems of taxonomic analysis of dispersed skeletal elements where no complete skeletons or even associated sclerites have been found [see recent discussions by Landing (1984), Bengtson (1986), and Laurie (1986)]. The only direct evidence for multisclerite composition is provided by the occasional ontogenetic merging of sclerites during growth (Landing, 1984; Qian and Bengtson, 1989).

Tommotiids are typical of the lower part of the Lower Cambrian, but range into the Middle Cambrian. Although ideas were initially expressed that the phosphate in the sclerites was diagenetic (Fonin and Smirnova, 1967), the primary phosphatic composition has been amply demonstrated by a consistent phosphatic composition in a variety of lithologies and through microstructural investigations (Landing *et al.*, 1980; Bengtson, 1983; Landing, 1984; Conway Morris and Chen, 1990b). A proposal (Bengtson, 1970) that the Ordovician–Carboniferous machaeridians, worm-shaped animals with platelike calcareous sclerites, are descendants of the bimembrate tommotiids (mitrosagophorans) has won some support (Jell, 1979; Dzik, 1986), but still needs to be critically tested.

3.2.20. Cambroclaves

Cambroclaves are a little known and exclusively Lower Cambrian group. They are mostly represented by disarticulated sclerites, although a few more complete sets of articulated sclerites have been found (Mambetov and Repina, 1979; Bengtson *et al.*, 1990; Yue, 1991; Conway Morris and Chen, 1991). The sclerites have a circular to elongated basal shield and an elongate spine, and were probably of calcareous composition. They formed a coat of interlocking sclerites, sometimes in the shape of a tube, but the shape of the complete skeleton is not known.

A relationship with the endoparasitic acanthocephalans has been proposed (Qian and Yin, 1984), but this is based only on the morphological similarities of individual sclerites and hooks, disregarding the nature of the articulated cambroclave scleritome. The cambroclaves should be regarded as a clade of uncertain affinities, but possibly related to the paracarinachitids.

3.2.21. Paracarinachitids

Paracarinachitids had elongated sclerites with a row of spines along the convex side, usually one spine per growth increment (Fig. 2H). The mineralogical composition is not firmly established, but may be calcareous. Paracarinachitids have been interpreted as polyplacophoran mollusks (Qian, 1983, 1984; Yu, 1984, 1987), but this interpretation is not tenable (Kerber, 1988; Qian and Bengtson, 1989). They may be related to the cambroclaves, but the further affinities are unknown. Hypothetical reconstructions of the sclerite arrangement in the scleritome (Qian and Bengtson, 1989; Conway Morris and Chen, 1991) are tentative. The group is seemingly restricted to the lower part of the Lower Cambrian.

3.2.22. Conulariids

Conulariids are known mainly from the Ordovician–Triassic, as tubular or pyramidal fossils with double bilateral symmetry and quadratic to rectangular transverse cross section. The conchs are built up of fine organic–phosphatic laminae periodically thickened into stiff transverse rods that typically form a chevron pattern on the flat faces. The flat faces are usually delimited by longitudinal corner furrows. Recent finds from the Lower Cambrian of China (e.g., Chen, 1982; He, 1987; Qian and Bengtson, 1989) and India (Brasier and Singh, 1987) appear to be referable to this group.

Conulariids have been interpreted as sessile scyphozoans on account of their alleged tetraradiate symmetry and the preservation of tentacles in the Ordovician presumed relative *Conchopeltis*. However, the latter relationship does not withstand scrutiny (Oliver, 1984), and the conulariids may alternatively be regarded as a separate clade of uncertain affinities (Babcock and Feldmann, 1986).

3.2.23. Onychophorans

Microdictyon, known from the Lower Cambrian to the basal part of the Middle Cambrian, had phosphatic netlike plates with a crudely hexagonal meshwork (Fig. 2F) constructed of several layers of apatite (Bengtson *et al.*, 1986). Recent discoveries in the Chengjiang fauna of Yunnan (Chen *et al.*, 1989a; Ramsköld and Hou, 1991) show that *Microdictyon* belongs to a group of onychophoranlike animals that also contains a number of other Cambrian fossils with or without mineralized skeletons, including the spiny *Hallucigenia* from the Middle Cambrian Burgess Shale (Conway Morris, 1977).

3.2.24. Other Groups

In addition to the representatives of known phyla and major problematic groups occurring in the earliest skeletal biotas, there are a number of groups that may indeed represent phylum-level taxa on their own, but that are either too poorly known or too poorly understood biologically for their nature to be properly assessed. Only a few of these will be mentioned below.

Apart from the protoconodonts (see above, Section 3.2.17), Cambrian rocks abound with toothlike fossils that may or may not be related to conodonts (Bengtson, 1983). The taxonomy of these forms is mostly unmanageable, as most named taxa are based on fragmented spines with undetermined structure and composition. Some forms are better known, however, such as the distinctly ornamented *Rhombocorniculum* (Walliser, 1958; Landing *et al.*, 1980). Some have a morphology that strongly suggests a function as jaws in a predatory animal, for example, the mandiblelike hooks of *Cyrtochites* (Qian and Bengtson, 1989). Others are of more obscure function, such as the funnel-shaped *Fomitchella* with phosphatic lamellae apparently deposited externally (Bengtson, 1983). Several major taxa are probably represented among these fossils.

Mobergellans (Fisher, 1962; Bengtson, 1968) are a small and exclusively Lower Cambrian group represented by rounded phosphatic shells with distinct paired muscle impressions on the inside. The shells are usually convex, but in some forms they are flat discs whose convex part is restricted to the center. Mobergellans have been interpreted as monoplacophorans, on account of their paired muscle prints, but the phosphatic composition is unknown among monoplacophorans. In addition, the generally flat shape of the shells suggests that they were opercula of a larger animal rather than conchs.

The laminated phosphatic sclerites of *Tumulduria*, for a while misinterpreted as early Tommotian trilobites (Fedorov *et al.*, 1979), so far defy biological understanding, but appear to represent an incompletely mineralized shell (Bengtson *et al.*, 1987).

Other examples of early skeletal fossils of potential significance as representing separate biomineralizing clades may be culled almost at will from the literature. It suffices here to list alphabetically some genera that appear to be particularly worthy of further study: *Aldanolina* Pel'man, 1976, *Archaeooides* Qian, 1977, *Bioistodina* He and Pei in He, Pei and Fu, 1984, *Cambrocassis* Missarzhevskij, 1977, *Cambrothyra* Qian and Zhang, 1983, *Cowiella* Hinz, 1987, *Gonamella* Val'kov and Karlova, 1984, *Koksodus* Missarzhevskij in Missarzhevskij and Mambetov, 1981, *Kuanchuanella* Yue in Xing *et al.*, 1984, *Mongolitubulus* Missarzhevskij, 1977, *Parapunctella* Jiang, 1982 in Luo *et al.*, 1982, *Petalites* [*nomen nudum* in Zhong (1977)], *Pyrgites* Yue in Xing *et al.*, 1984, *Rushtonites* Hinz, 1987, *Salanacus* Grigor'eva in Voronin *et al.*, 1982, *Scoponodus* Jiang in Luo *et al.*, 1982, (Fig. 2I), *Stefania* Grigor'eva in Voronin *et al.*, 1982, and *Stictoconus* Qian and Bengtson, 1989.

4. Ecology

The complex taphonomic history of many assemblages of early skeletal fossils have not only led us into a taxonomic quagmire, but have retarded informed speculation on the paleoecology of these diverse faunas. Our understanding of this area is still in its infancy, and even the traditional lines of enquiry of feeding types and position relative to the sediment–water interface are still obscure in many cases.

In terms of trophic specialization it seems that many of the tubicolous taxa were suspension feeders, and recurrent comparisons to groups such as the sedentary annelids would have us believe that the tube-dweller extended a feeding crown from the aperture of the tube. The spongelike nature of the chancelloriids also suggests that they may have mimicked sponges in their feeding behavior, even though there is little evidence for a close relationship. The undoubted sponges (hexactinellids, demosponges, and calcareous sponges) certainly were suspension feeders then as now, and whether or not the enigmatic archaeocyathans are included in the sponge phylum, they also most probably represented this mode of feeding. Among the conch-bearing taxa, both the brachiopods and the pseudobrachiopods may be identified with some confidence as suspension feeders, and in at least one species of the latter informal group, there is compelling evidence for separation of feeding currents by sulci (Bengtson *et al.*, 1990).

Many of the sluglike metazoans armored with sclerites, including the halkieriids and perhaps the tommotiids, may be assigned to the deposit feeders, although the tentative nature of the proposal needs to be stressed. More definite is the deposit-feeding habit of the hyoliths and many of the mollusks.

Convincing evidence for predatory habits of particular taxa appears to be available from the protoconodonts, as well as the more enigmatic *Cyrtochites*. Direct evidence of predation on shelly fauna includes examples of small boreholes in mobergellans, tubicolous species, inarticulate brachiopods, and other shelly fossils. The makers of these boreholes are not known. Increasing evidence for tightly articulated scleritomes, especially in the halkieriids and cambroclaves, suggests that they would have been effective in deterring predators. In a similar fashion, the abundance of bivalved forms and operculate tubes presumably served to confer protection.

The position of these taxa with regard to the sediment–water interface is often open to more than one interpretation. The majority of tubicolous species presumably were semi-infaunal, and *Coleoloides* formed local thickets, effective in baffling sediment and building small bioherms. In some anabaritids, extensive flanges may have acted as stabilizers in shifting sediment.

The vagrant epifauna is assumed to have included the mollusks, hyoliths, and many of the scleritome-bearing taxa, such as halkieriids, tommotiids, and cambroclaves. The tentative nature of these assignments, especially in the last two cases, needs emphasis, and alternative proposals that some belong either to the roster of vagrant infauna or sessile epifauna merit consideration.

The pelagic component is imagined to include the protoconodonts, largely because of their functional and structural similarity to the Recent chaetognaths, a phylum whose representatives, with few exceptions, are pelagic.

The sketchiness of the above proposals will make it clear how much remains to be established, and how insecure some proposals remain. In general it appears that the incoming of the earliest Cambrian faunas involved an abrupt rise in the utilization of organic matter near the sediment–water interface, mostly by sessile epifauna and infauna. An increase in diversity of the mobile epifauna, mainly arthropods, followed in the Atdabanian. For further refinement and corrections of this general picture, improved understanding of the autecology of the various enigmatic organisms is needed.

5. Radiation of Skeletal Fossils as a Radiation of Phyla

5.1. Introduction

Although the record of skeletal fossils is not an unbiased subset of the total biota, it can be used for certain extrapolations and deductions regarding also the role of nonskeletal organisms. It is particularly fortunate for such analyses that the Cambrian sequence indeed contains a number of Lagerstätten with preserved soft-bodied fossils, in particular the Lower Cambrian Chengjiang fauna (e.g., Hou, 1987a–d; Sun and Hou, 1987a–b; Hou and Chen, 1989; Chen *et al.*, 1989a, b), the Middle Cambrian Burgess Shale biota (e.g., Whittington, 1985), and the Upper Cambrian *Orsten* fauna of phosphatized arthropods (e.g., Müller, 1983; Müller and Walossek, 1985, 1987, 1988).

5.2. How Large a Percentage of the Phyla Were Fossilized?

Present-day phyla are generally clearly circumscribed, but in practice the phyla have tended to be defined by the limits of our phylogenetic understanding (Bengtson, 1986). Whichever phylum definition is chosen, it inevitably breaks down when dealing with the early phases of metazoan radiation. Evolutionary distances to the common ancestor were smaller, many short-lived lineages existed, and the limited nature of data available from fossils further obscures phylum divisions. Nevertheless, phyla have a crude reality as measures of body-plan diversity, and it may be worthwhile to consider how much of this diversity we may expect to discover through the record of early skeletal fossils.

Johnson (1964) estimated the proportion of species with fossilizable skeletons in modern marine metazoan communities to be about 30%. This estimate is likely to be high, as new discoveries continue to be made of small metazoans without fossilization potential, particularly in the interstitial habitats. According to Nicol (1977), the total proportion of normally fossilizable species in all Recent habitats is less than 10%. In the Middle Cambrian Burgess Shale only about 20% of the metazoan genera had normally preservable hard parts (Conway Morris, 1986). As most Burgess Shale genera are monospecific, this number is approximately valid for species as well, but it may again be too high, because nonskeletal species of millimeter size are not likely to have been preserved even in the Burgess Shale.

On the other hand, the species that do have a good fossilization potential tend to be well-represented in the fossil record. Valentine (1989) found that 77% of the gastropod and bivalve mollusk species now living in southern California are represented in the Pleistocene record. The recovery is likely to become poorer with increasing age because of less complete preservation of the sedimentary sequence, but for the present argument we may use this number to estimate the maximum probability of finding a readily fossilizable species in the fossil record of a shallow-water environment. This probability may then be set at 0.77. Accepting the Burgess Shale value of 20% as a maximum value for the proportion of skeletal taxa in the Cambrian, we see that the number of skeletal species known from the Cambrian would correspond to no more than 15% of the total number of metazoan species. As the fossil record in the Cambrian is certainly poorer than that of the Californian Pleistocene, and the Burgess Shale number is very probably too high, we may conclude that the true value indeed cannot be higher than 15%.

The matter is different with regard to recovery of phyla, however. The probability of a higher taxon being recovered must be higher than for a species, because it is easier to hit a large than a small taxon in sampling [see Raup (1975) for rarefaction plots at different taxonomic levels]. For easily fossilized taxa in a homogeneous environment the probability of recovery approaches certainty as soon as the number of species is larger than just a few. Using Valentine's figure of 0.77 for the probability of species recovery, we have that the probability of finding the larger taxon is $1 - (1 - 0.77)^n$, where n is the number of species in the taxon. This probability exceeds 0.95 already at an n value between 2 and 3, but the calculation obviously assumes that there is no correlation between the abundances of different species within a taxon. This is not necessarily true, but is presently not possible to quantify. In Valentine's study, 89% of the families were recovered from the fossil record. The families had a species content of 1–55, the average being 6. Of the 12 families that were not found as fossils, one contained two species and the remainder one (Valentine, 1989, Appendix).

The risk of missing all of the 698 species of Phylum Mollusca in Valentine's study is obviously nil. Due to the generally lower diversities in the Cambrian, however, phyla of that time are probably more similar in diversity terms to Recent mollusk families than to the entire Phylum Mollusca, or even Californian bivalves and gastropods. Thus, given the nearly ideal circumstances of the Californian Pleistocene and assuming that the Cambrian phyla had the same diversity distribution as the families of Valentine's study, we might expect to be able to retrieve around 90% of the fully skeletalized phyla from the Cambrian rocks.

But all phyla are not skeletalized, and even the ones that are may have a substantial number of nonskeletalized members. The pattern of distribution of skeletal taxa is also important. If fossilizable skeletons occur in 20% of the species, it makes a substantial difference whether skeletalization is restricted to a few phyla or whether it is spread out among a larger number. In the latter case more phyla should be recovered from the fossil record than in the former case.

Figure 3 shows a calculation of probabilities of recovery of a hypothetical set of 1000 "phyla" with various numbers of species, from 1 to 30, distributed along an exponential decline curve. [The exact shape of this curve is not crucial for the argument, although it has been set to resemble actual frequency distributions of lower taxa among higher ones in the fossil record (cf. Sepkoski, 1978, Fig. 11).] The probability of recovery of skeletal species has been set to 0.77, following Valentine's (1989) results. It was assumed that one third of the phyla did not contain any skeletal species at all (approximately the proportion among living phyla) and that the remainder each had 30% skeletal species. As could be expected, most nonrecovery is among the low-diversity phyla. The probability of recovering skeletal phyla reaches the 0.95 level at a species number of 12.

These calculations suggest that if the probability of recovery of skeletal spe-

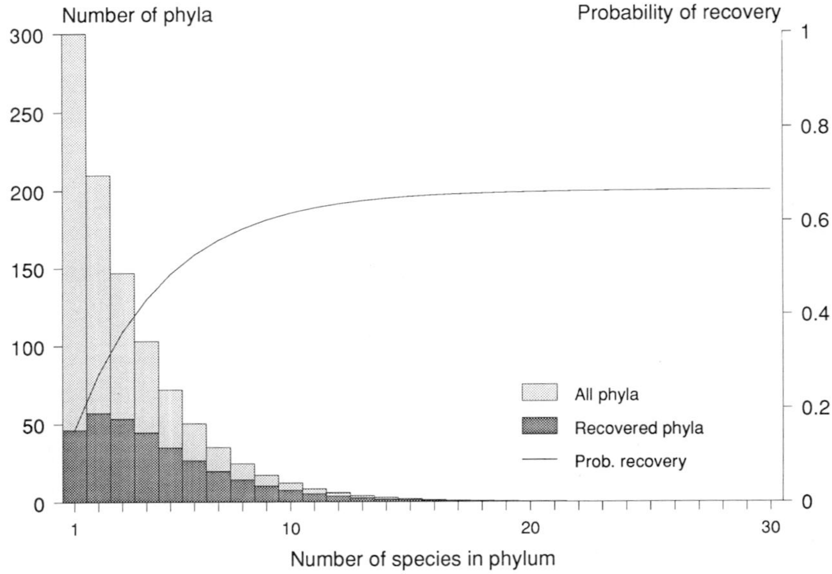

Figure 3. Plot of ideal recovery of phyla of low diversities (1–30 species), assuming that 77% of the skeletal species are recovered, 67% of the phyla contain 30% skeletal species, and the remainder of the phyla were nonskeletal (see text). The curve shows the probability of phylum recovery, calculated as $\{1 - [1 - (0.77 \times 0.30)^n]\} \times 0.67$, where n is the number of species within the phylum. Bars show the frequency distribution of artificial sets of "phyla," following the equation $y = 300 \times 0.7^{x-1}$. The darker shaded bottom parts of bars represent the number of phyla recovered according to the probability curve.

cies is comparable to that in the Californian Pleistocene, the higher taxa eventually overlooked in the census of skeletal fossils would be those that either lack skeletal species or have low diversities, less than about ten species. For small taxa the chance of recovery decreases rapidly. The question of how many of the original Cambrian phyla we may be able to recognize boils down—not unexpectedly—to how low a diversity we accept in a phylum. If one-species phyla are accepted, the question is meaningless, because there is no way to define or recognize such phyla consistently. If we speak about phyla larger than a dozen species, however, we would expect to find nearly all that contained at least their share of skeleton-producing species.

When have we found all skeletal phyla in the Cambrian? Looking at the number of taxa described in the literature over the last century (Fig. 4), it appears that we are in the steep phase of an exponential growth curve. The curve is likely to flatten out as we approach the level of species recovery that is possible in practice. This level should ideally mean that 70–80% of the species have been retrieved. There will always be species that are so rare in terms of individuals that they will probably never be found, but provided that this rarity is not the trait of a whole phylum, this will not matter. We may at that point be reasonably secure that we have found practically all of the skeletal phyla containing at least a dozen species in the environments preserved in the fossil record. But it is obvious from Fig. 4 that we are far from that goal.

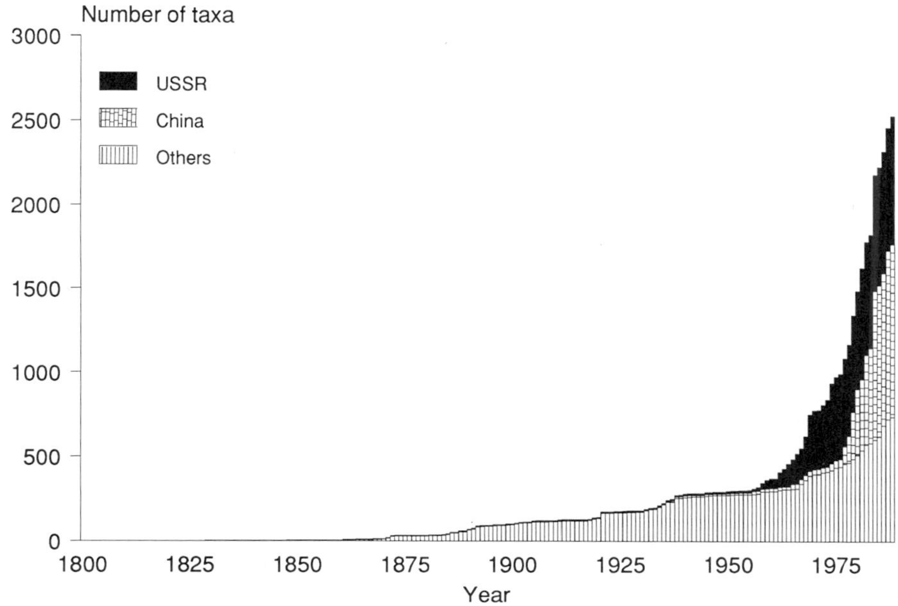

Figure 4. Cumulative plot of taxa of Early Cambrian skeletal fossils (except trilobites and archaeocyathans) described in the literature from 1800 to 1988. Data are from the ESF (Early Skeletal Fossils) database compiled by Bengtson (unpublished data).

References

Abaimova, G. P., 1978, Anabaritids—Ancient fossils with carbonate skeleton, in: New Material on the Stratigraphy and Paleontology of Siberia (S. V. Sukhov, ed.), *Trudy Sibirsk. Nauchno-Issled. Inst. Geol. Gefiz. Mineral.* **260:**77–83.

Allison, C. W., and Hilgert, J. W., 1986, Scale microfossils from the Early Cambrian of Northwest Canada, *J. Paleontol.* **60:**973–1015.

Babcock, L. E., and Feldmann, R. M., 1986, The phylum Conulariida, in: *Problematic Fossil Taxa* (A. Hoffman and M. H. Nitecki, eds.), Clarendon Press, Oxford, pp. 135–147.

Bedford, R., and Bedford, W. R., 1934, New species of Archaeocyathinae, *Mem. Kyancutta Mus.* **1:**1–7.

Bedford, R., and Bedford, J., 1936, Further notes on Cyathospongia (Archaeocyathi) and other organisms from the Lower Cambrian of Beltana, South Australia, *Mem. Kyancutta Mus.***3:**21–26.

Bendix-Almgren, S. E., and Peel, J. S., 1988, *Hadimopanella* from the Lower Cambrian of North Greenland: Structure and affinities, *Bull. Geol. Soc. Den.* **37:**83–103.

Bengtson, S., 1968, The problematic genus *Mobergella* from the Lower Cambrian of the Baltic area, *Lethaia* **1:**325–351.

Bengtson, S., 1970, The Lower Cambrian fossil *Tommotia, Lethaia* **3:**363–392.

Bengtson, S., 1976, The structure of some Middle Cambrian conodonts, and the early evolution of conodont structure and function, *Lethaia* **9:**185–206.

Bengtson, S., 1977, Early Cambrian button-shaped phosphatic microfossils from the Siberian Platform, *Palaeontology* **20:**751–762.

Bengtson, S., 1983, The early history of the Conodonta, *Fossils Strata* **15:**5–19.

Bengtson, S., 1985, Taxonomy of disarticulated fossils, *J. Paleontol.* **59:**1350–1358.

Bengtson, S., 1986, The problem of the Problematica, in: *Problematic Fossil Taxa* (A. Hoffman and M. H. Nitecki, eds.), Clarendon Press, Oxford, pp. 3–11.

Bengtson, S., 1990, *Maikhanella* and *Siphogonuchites:* Problems of skeleton formation in Cambrian metazoans, in: *Palaeontological Association, Durham 1990, Abstracts,* p. 3.

Bengtson, S., and Conway Morris, S., 1984, A comparative study of Lower Cambrian *Halkieria* and Middle Cambrian *Wiwaxia, Lethaia* **17**:307–329.

Bengtson, S., and Missarzhevsky, V. V., 1981, Coeloscleritophora—A major group of enigmatic Cambrian metazoans, in: *Short Papers for the Second International Symposium on the Cambrian System 1981* (M. E. Taylor, ed.), U. S. Geological Survey Open-File Report 81-743, pp. 19–21.

Bengtson, S., Matthews, S. C., and Missarzhevsky, V. V., 1986, The Cambrian netlike fossil *Microdictyon*, in: *Problematic Fossil Taxa* (A. Hoffman and M. H. Nitecki, eds.), Clarendon Press, Oxford, pp. 97–115.

Bengtson, S., Fedorov, A. B., Missarzhevsky, V. V., Rozanov, A. Yu., Zhegallo, E. A., and Zhuravlev, A. Yu., 1987, *Tumulduria incomperta* and the case for Tommotian trilobites, *Lethaia* **20**:361–370.

Bengtson, S., Conway Morris, S., Cooper, B. J., Jell, P. A., and Runnegar, B. N., 1990, Early Cambrian fossils from South Australia, *Mem. Australas. Assoc. Palaeontol.* **9**:1–364.

Bergquist, P. R., 1985, Poriferan relationships, in: *The Origins and Relationships of Lower Invertebrates* (S. Conway Morris, J. D. George, R. Gibson, and H. M. Platt, eds.), Clarendon Press, Oxford, pp. 14–27.

Bergström, J., 1991, Metazoan evolution around the Precambrian–Cambrian transition, in: *The Early Evolution of Metazoa and the Significance of Problematic Taxa* (A. M. Simonetta and S. Conway Morris, eds.), Cambridge University Press, Cambridge, pp. 25–34.

Bischoff, G. C. O., 1989, Byroniida new order from early Palaeozoic strata of eastern Australia (Cnidaria, thecate scyphopolyps), *Senckenb. Lethaea* **69**:467–521.

Boersma, A., 1978, Foraminifera, in: *Introduction to Marine Micropaleontology* (B. U. Haq and A. Boersma, eds.), Elsevier, New York, pp. 19–77.

Boogard, M., van den, 1989, Isolated tubercles of some Palaeoscolecida, *Scripta Geol.* **90**:1–12.

Bovee, E. C., 1981, Distribution and forms of siliceous structures among Protozoa, in: *Silicon and Siliceous Structures in Biological Systems* (T. L. Simpson and B. E. Volcani, eds.), Springer, New York, pp. 233–279.

Brasier, M. D., and Hewitt, R. A., 1979, Environmental setting of fossiliferous rocks from the uppermost Proterozoic–lower Cambrian of central England, *Palaeogeogr. Palaeoclimatol. Palaeoecol.* **27**:35–57.

Brasier, M. D., and Singh, P., 1987, Microfossils and Precambrian–Cambrian boundary stratigraphy at Maldeota, Lesser Himalayas, *Geol. Mag.* **124**:323–345.

Broecker, W. S., 1971, A kinetic model for the chemical composition of sea water, *Quaternary Research* **1**:188–207.

Butterfield, N. J., 1990, A reassessment of the enigmatic Burgess Shale fossil *Wiwaxia corrugata* (Matthew) and its relationship to the polychaete *Canadia spinosa* Walcott, *Paleobiology* **16**:287–303.

Carlisle, E. M., 1981, Silicon in bone formation, in: *Silicon and Siliceous Structures in Biological Systems* (T. L. Simpson and B. E. Volcani, eds.), Springer, New York, pp. 69–94.

Carter, J. G., Lutz, R. A., and Tevesz, M. J. S., 1990, Shell microstructural data for the Bivalvia. Part VI. Orders Modiomorphoida and Mytiloida, in: *Skeletal Biomineralization: Patterns, Processes and Evolutionary Trends* (J. G. Carter, ed.), pp. 391–411, Van Nostrand Reinhold, New York.

Chang, S.-B. R., and Kirschvink, J. L., 1989, Magnetofossils, the magnetization of sediments, and the evolution of magnetite biomineralization, *Ann. Rev. Earth Planet. Sci.* **1989**(17):169–195.

Chen Junyuan, Hou Xianguang, and Lu Haozhi, 1989a, Early Cambrian netted scale-bearing worm-like sea animal, *Acta Palaeontol. Sinica* **28**(1):1–16 [in Chinese].

Chen Junyuan, Hou Xianguang, and Lu Haozhi, 1989b, Lower Cambrian leptomitids (Demospongae), Chengjiang, Yunnan, *Acta Palaeontol. Sinica* **28**(1):17–31 [in Chinese].

Chen Junyuan, Hou Xianguang, and Li Guoxiang, 1990, New Lower Cambrian demosponges—*Quadrolaminiella* gen. nov. from Chengjiang, Yunnan, *Acta Palaeontol. Sinica* **29**(4):402–414.

Chen Menge, 1982, The new knowledge of the fossil assemblages from Maidiping section, Emei County, Sichuan with reference to the Sinian–Cambrian boundary, *Scientia Geol. Sinica* **1982**:253–262 [in Chinese].

Cherchi, A., and Schroeder, R., 1985, Middle Cambrian Foraminifera and other microfossils from SW Sardinia, *Boll. Soc. Paleontol. Ital.* **1985**(23):149–160.

Conway Morris, S., 1977, A new metazoan from the Cambrian Burgess Shale of British Columbia, *Palaeontology* **20**:623–640.

Conway Morris, S., 1985, The Middle Cambrian Metazoan *Wiwaxia corrugata* (Matthew) from the Burgess Shale and *Ogygopsis* Shale, British Columbia, Canada, *Philos. Trans. R. Soc. Lond. B* **307**:507–586.

Conway Morris, S., 1986, The community structure of the Middle Cambrian Phyllopod Bed (Burgess Shale), *Palaeontology* **29**:423–467.

Conway Morris, S., and Chen Menge, 1989, lower Cambrian anabaritids from South China, *Geol. Mag.* **126**:615–632.

Conway Morris, S., and Chen Menge, 1990a, *Blastulospongia polytreta* n. sp., an enigmatic organism from the Lower Cambrian of Hubei, China, *J. Paleontol.* **64**:26–30.

Conway Morris, S., and Chen Menge, 1990b, Tommotiids from the Lower Cambrian of South China, *J. Paleontol.* **64**:169–184.

Conway Morris, S., and Chen Menge, 1991, Cambroclaves and paracarinachitids, early skeletal problematica from the Lower Cambrian of South China, *J. Paleontol.* **34**:357–397.

Conway Morris, S., and Peel, J. S., 1990, Articulated halkieriids from the Lower Cambrian of north Greenland, *Nature* **345**:802–805.

Cook, P. J., and Shergold, J. H., 1984, Phosphorus, phosphorites and skeletal evolution at the Precambrian–Cambrian boundary, *Nature* **308**:231–236.

Cook, P. J., and Shergold, J. H. (eds.) 1986, *Phosphatic Deposits of the World*, Vol. 1, *Proterozoic and Cambrian Phosphorites*, Cambridge University Press, Cambridge.

Danielli, H. M. C., 1981, The fossil alga *Girvanella* Nicholson & Etheridge, *Bull. Br. Mus. Nat. Hist. (Geol.)* **35**:79–107.

Debrenne, F., and Vacelet, J., 1984, Archaeocyatha: Is the sponge model consistent with their structural organization?, in: Recent Advances in the Paleobiology and Geology of the Cnidaria (W. A. Oliver, Jr., W. J. Sando, S. D. Cairns, A. G. Coates, *et al.*, eds.), *Palaeontogr. Am.* **54**:358–369.

Debrenne, F., and Wood, R., 1990, A new Cambrian sphinctozoan sponge from North America, its relationship to archaeocyaths and the nature of early sphinctozoans, *Geol. Mag.* **127**:435–443.

Debrenne, F., Termier, H., and Termier, G., 1970, *Radiocyatha*. Une nouvelle classe d'organismes primitifs du Cambrien inférieur, *Bull. Soc. Géol. Fr. 7 Ser.* **12**:120–125.

Debrenne, F., Termier, H., and Termier, G., 1971, Sur de nouveaux représentans de la classe des *Radiocyatha*. Essai sur l'évolution des Métazoaires primitifs, *Bull. Soc. Géol. Fr. 7 Ser.* **13**:439–444.

Debrenne, F., Rozanov, A. Yu., and Webers, G. F., 1984, Upper Cambrian Archaeocyatha from Antarctica, *Geol. Mag.* **121**:291–299.

Debrenne, F., Zhuravlev, A. Yu., and Rozanov, A. Yu., 1989, Regular archaeocyathans, *Trudy Paleontol. Inst. Akad. Nauk SSSR* **233**:1–196.

Debrenne, F., Lafuste, J., and Zhuravlev, A., 1990, Coralomorphes et spongiomorphes à l'aube du Cambrien, *Bull. Mus. Nat. Hist. Nat. Paris 4e Sér.* **12**:17–39.

Degens, E. T., Kazmierczak, J., and Ittekott, V., 1985, Cellular response to Ca^{2+} stress and its geological implications, *Acta Palaeontol. Polon.* **30**(3–4):115–135.

Derstler, C., 1981, Morphological diversity of Early Cambrian echinoderms, in: *Short Papers for the Second International Symposium on the Cambrian System 1981* (M. E. Taylor, ed.), U. S. Geological Survey Open-File Report 81-743, pp. 71–75.

Ding Weiming and Qian Yi, 1988, Late Sinian to Early Cambrian small shelly fossils from Yangjiaping, Shimen, Hunan, *Acta Micropalaeontol. Sinica* **5**(1):39–55 [in Chinese].

Dzik, J., 1986, Chordate affinities of the conodonts, in: *Problematic Fossil Taxa* (A. Hoffman and M. H. Nitecki, eds.), Clarendon Press, Oxford, pp. 240–254.

Fedonkin, M. A., 1985, The non-skeletal fauna of the Vendian: A promorphological analysis, in: *The Vendian System*, Vol. 1, *Paleontology* (B. S. Sokolov and A. B. Ivanovskij, eds.), Nauka, Moscow, pp. 10–69 [in Russian].

Fedorov, A. B., Egorova, L. I., and Savitskij, V. E., 1979, First find of oldest trilobites in the lower parts of the Lower Cambrian Tommotian Stage (R. Aldan), *Doklady Akad. Nauk SSSR* **249**:1187–1190 [in Russian].

Finks, R. M., 1983, Fossil Hexactinellida, in: *Sponges and Spongiomorphs—Notes for a Short Course* (T. Broadhead, ed.), University of Tennessee, Department of Geological Sciences, Studies in Geology, No. 7, pp. 101–115.

Fisher, D. W., 1962, Small conoidal shells of uncertain affinities, in: *Treatise on Invertebrate Paleon-*

tology, Part W, *Miscellanea* (R. C. Moore, ed.), American Geological Society and University of Kansas Press, Lawrence, Kansas, pp. W98–W143.

Fonin, V. D., and Smirnova, T. N., 1967, A new group of problematic Early Cambrian organisms and some methods of preparing them, *Paleontol. Zh.* **1967**(2):15–27 [in Russian].

Fortey, R. A., and Holdsworth, B. K., 1971, The oldest well-preserved Radiolaria, *Boll. Soc. Paleontol. Ital.* **10**(1):35–41.

Føyn, S., and Glaessner, M. F., 1979, *Platysolenites*, other animal fossils, and the Precambrian–Cambrian transition in Norway, *Norsk Geol. Tidsskrift* **59**:25–46.

Germs, G. J. B., 1972, The stratigraphy and paleontology of the Lower Nama Group, South West Africa, *Univ. Cape Town Dept. Geol. Precambrian Res. Unit Bull.* **12**:1–250.

Glaessner, M. F., 1963, Major trends in the evolution of the Foraminifera, in: *Evolutionary Trends in Foraminifera* (G. H. R. von Koenigswald, ed.), Elsevier, Amsterdam, pp. 9–24.

Glaessner, M. F., 1976, Early Phanerozoic annelid worms and their geological and biological significance. *J. Geol. Soc. Lond.* **132**:259–275.

Gorjansky, W. Ju., and Popov, L. Ye., 1986, On the origin and systematic position of the calcareous-shelled inarticulate brachiopods, *Lethaia* **19**:233–240.

Grant, S. W. F., 1990, Shell structure and distribution of *Cloudina*, a potential index fossil for the terminal Proterozoic, *Am. J. Sci.* **290**(A):261–294.

Grant, S. W. F., Knoll, A. H., and Germs, G. J. B., 1991, Probable calcified metaphytes in the latest Proterozoic Nama Group, Namibia: Origin, diagenesis, and implications, *J. Paleontol.* **65**:1–18.

Grigor'eva, N. V., 1980, On the question of microstructural investigations of hyolithelminths, in: *26th International Geological Congress. Paleontological and Stratigraphic Reports of Soviet Geologists*, Nauka, Moscow, pp. 49–55 [in Russian].

Grigor'eva, N. V., and Zhegallo, E. A., 1979, To the study of the microstructure of some Tommotian fossils, *Paleontol. Zh.* **1979**(2):142–144 [in Russian].

He Tinggui, 1987, Early Cambrian conulariids from the Yangtze Platform and their early evolution, *J. Chengdu College Geol.* **14**(2):7–18 [in Chinese].

He Tinggui, Pei Fang, and Fu Guanghong, 1984, Some small shelly fossils from the Lower Cambrian Xinji Formation in Fangcheng County, Henan Province, *Acta Palaeontol. Sinica* **23**:350–357 [in Chinese].

Hill, D., 1972, Archaeocyatha, in: *Treatise on Invertebrate Paleontology*, Part E1, 2nd ed. (C. Teichert, ed.), Geological Society of America and University of Kansas Press, Lawrence, Kansas, pp. E1–E158.

Hinz, I., 1987, The Lower Cambrian microfauna of Comley and Rushton, Shropshire/England, *Palaeontogr. A* **198**(1–3):41–100.

Hinz, I., Kraft, P., Mergl, M., and Müller, K. J., 1990, The problematic *Hadimopanella*, *Kaimenella*, *Milaculum* and *Utahphospha* identified as sclerites of Palaeoscolecida, *Lethaia* **23**:217–221.

Hou Xianguang, 1987a, Two new arthropods from Lower Cambrian, Chengjiang, eastern Yunnan, *Acta Palaeontol. Sinica* **26**:236–256 [in Chinese].

Hou Xianguang, 1987b, Three new large arthropods from Lower Cambrian, Chengjiang, eastern Yunnan, *Acta Palaeontol. Sinica* **26**:272–285 [in Chinese].

Hou Xianguang, 1987c, Early Cambrian large bivalved arthropods from Chengjiang, eastern Yunnan, *Acta Palaeontol. Sinica* **26**:286–298 [in Chinese].

Hou Xianguang, 1987d, Oldest Cambrian bradoriids from eastern Yunnan, in: *Stratigraphy and Palaeontology of Systemic Boundaries in China—Precambrian–Cambrian Boundary* 1, Nanjing University Publishing House, Nanjing, China, pp. 537–545.

Hou Xianguang and Chen Junyuan, 1989, Early Cambrian tentacled worm-like animals (*Facivermis* gen. nov.) from Chengjiang, Yunnan, *Acta Palaeontol. Sinica* **28**:32–41 [in Chinese].

Huber, B., 1991, Micropaleontology, *Geotimes* **1991**(2):34–35.

Hurd, D. C., 1973, Interactions of biogenic opal, sediment and seawater in the Central Equatorial Pacific, *Geochim. Cosmochim. Acta* **37**:2257–2282.

James, N. P., and Klappa, C. F., 1983, Petrogenesis of Early Cambrian reef limestones, Labrador, Canada, *J. Sedment. Petrol.* **53**:1051–1096.

Jankauskas, T. V., 1972, Lower Cambrian cribricyathans from Siberia, in: *Problems in the Biostratigraphy and Paleontology of the Lower Cambrian of Siberia* (I. T. Zhuravleva, ed.), Nauka, Moscow, pp. 161–183 [in Russian].

Jell, J. S., 1984, Cambrian cnidarians with mineralized skeletons, in: Recent Advances in the Paleobiology and geology of the Cnidaria (W. A. Oliver, Jr., W. J. Sando, S. D. Cairns, A. G. Coates, *et al.*, eds.), *Palaeontogr. Am.* **54:**105–109.

Jell, J. S., 1979, *Plumulites* and the machaeridian problem, *Alcheringa* **3:**253–259.

Jell, P. A., 1981, *Thambetolepis delicata* gen. et sp. nov., an enigmatic fossil from the Early Cambrian of South Australia, *Alcheringa* **5:**85–93.

Johnson, R. G., 1964, The community approach to paleoecology, in: *Approaches to Paleoecology* (J. Imbrie and N. Newell, eds.), Wiley, New York, pp. 107–134.

Jones, P. J., and McKenzie, K. G., 1980, Queensland Middle Cambrian Bradoriida (Crustacea): New taxa, palaeobiogeography and biological affinities, *Alcheringa* **4:**203–225.

Jones, W. C., 1970, The composition, development, form and orientation of calcareous sponge spicules, *Symp. Zool. Soc. Lond.* **25:**91–123.

Jones, W. C., 1979, The microstructure and genesis of sponge biominerals, in: Biologie des spongiaires, *Colloq. Int. CNRS* **291:**425–447.

Kazmierczak, J., and Degens, E. T., 1986, Calcium and the early eukaryotes, *Mitt. Geol. Palaeontol. Inst. Univ. Hamburg* **61:**1–20.

Kazmierczak, J., Ittekott, V., and Degens, E. T., 1985, Biocalcification through time: Environmental challenge and cellular response, *Palaeontol. Z.* **59:**15–33.

Kerber, M., 1988, Mikrofossilien aus unterkambrischen Gesteinen der Montagne Noire, Frankreich, *Palaeontogr. Abt. A* **202:**127–203.

Klein, C., Beukes, N. J., and Schopf, J. W., 1987, Filamentous microfossils in the Early Proterozoic Transvaal Supergroup: Their morphology, significance, and paleoenvironmental setting, *Precambrian Res.* **36:**81–94.

Kraft, P., and Mergl, M., 1989, Worm-like fossils (Palaeoscolecida; ?Chaetognatha) from the Lower Ordovician of Bohemia, *Sb. Geol. Ved. Paleontol.* **30:**9–36.

Lafuste, J., Debrenne, F., and Zhuravlev, A., 1990, Les fuscinules, type nouveau de biocristaux dans le squelette d'*Hydroconus* Korde 1963, Coralomorphe du Cambrien inférieur, *C. R. Acad. Sci. Paris II* **310:**1553–1559.

Lambert, G., Lambert, C. C., and Lowenstam, H. A., 1990, Protochordate biomineralization, in: *Skeletal Biomineralization: Patterns, Processes and Evolutionary Trends* (J. G. Carter, ed.), pp. 461–469, Van Nostrand Reinhold, New York.

Landing, E., 1984, Skeleton of lapworthellids and the suprageneric classification of tommotiids (Early and Middle Cambrian phosphatic problematica), *J. Paleontol.* **58:**1380–1398.

Landing, E., Nowlan, G. S., and Fletcher, T. P., 1980, A microfauna associated with Early Cambrian trilobites of the *Callavia* Zone, northern Antigonish Highlands, Nova Scotia, *Can. J. Earth Sci.* **17:**400–418.

Landing, E., Narbonne, G. M., Myro, P., Benus, A. P., and Anderson, M. M., 1988, Faunas and depositional environments of the upper Precambrian through Lower Cambrian, southeastern Newfoundland, in: *Trace Fossils, Small Shelly Fossils and the Precambrian–Cambrian Boundary* (E. Landing, G. M. Narbonne, and P. Myrow, eds.), New York State Museum Bulletin 463, pp. 18–52.

Laurie, J. R., 1986, Phosphatic fauna of the Early Cambrian Todd River Dolomite, Amadeus Basin, central Australia, *Alcheringa* **10:**431–454.

Lipps, J. H., and Sylvester, A. G., 1968, The enigmatic Cambrian fossil *Volborthella* and its occurrence in California, *J. Paleontol.* **42:**329–336.

Lowenstam, H. A., and Margulis, L., 1980a, Calcium regulation and the appearance of calcareous skeletons in the fossil record, in: *The Mechanisms of Biomineralization in Animals and Plants* (M. Omori and N. Watabe, eds.), Tokai University Press, Tokyo, pp. 289–300.

Lowenstam, H. A., and Margulis, L., 1980b, Evolutionary prerequisites for early Phanerozoic calcareous skeletons, *Biosystems* **12:**27–41.

Lowenstam, H. A., and Weiner, S., 1985, Transformation of amorphous calcium phosphate to crystalline dahllite in the radular teeth of chitons, *Science* **227:**51–53.

Lowenstam, H. A., and Weiner, S., 1989, *On Biomineralization*, Oxford University Press, Oxford.

Luo Huilin, Jiang Zhiwen, Wu Xiche, Song Xueliang, Ouyang Lin, *et al.*, 1982, *The Sinian–Cambrian Boundary in Eastern Yunnan, China*, Kunming, China [in Chinese].

Mambetov, A. M., and Repina, L. N., 1979, The Lower Cambrian of Talasskij Ala-Too and its correlation with the sections of Malyj Karatau and the Siberian Platform, *Trudy Inst. Geol. Geofiz. Sibirsk. Otd. Akad. Nauk SSSR* **406:**9–158 [in Russian].

Marek, L., and Yochelson, E. L., 1976, Aspects of the biology of Hyolitha (Mollusca), *Lethaia* **9:**65–82.

Missarzhevskij, V. V., 1973, Conodont-shaped organisms from the Precambrian–Cambrian boundary beds of the Siberian Platform and Kazakhstan, in: Problems in the Paleontology and Stratigraphy of Siberia and the Far East (I. T. Zhuravleva, ed.), *Trudy Inst. Geol. Geof. SO Akad. Nauk SSSR* **49:**53–57 [in Russian].

Missarzhevskij, V. V., 1974, New data on the oldest Lower Cambrian fossils of the Siberian Platform, in: *Lower Cambrian Biostratigraphy and Paleontology of Europe and Northern Asia* (I. T. Zhuravleva and A. Yu. Rozanov, eds.), Nauka, Moscow, pp. 179–189 [in Russian].

Missarzhevskij, V. V., 1977, Conodonts (?) and phosphatic problematica from the Cambrian of Mongolia and Siberia, in: Paleozoic Invertebrates of Mongolia (L. P. Tatarinov *et al.*, eds.), *Trans. Joint Soviet–Mongolian Paleontol. Expedition* **1977:**10–19 [in Russian].

Missarzhevskij, V. V., 1989, The oldest skeletal fossils and stratigraphy of the Precambrian–Cambrian boundary beds, *Trudy Geol. Inst. Akad. Nauk SSSR* **443:**1–237 [in Russian].

Missarzhevskij, V. V., and Mambetov, A. M., 1981, Stratigraphy and fauna of the Precambrian–Cambrian boundary beds in Malyj Karatau, *Trudy Geol. Inst. Akad. Nauk SSSR* **326:**1–92 [in Russian].

Müller, K. J., 1983, Crustacea with preserved soft parts from the Upper Cambrian of Sweden, *Lethaia* **16:**93–109.

Müller, K. J., and Miller, J. F., 1976, The problematic microfossil *Utahphospha* from the Upper Cambrian of the western United States, *Lethaia* **9:**391–395.

Müller, K. J., and Walossek, D., 1985, Skaracarida, a new order of Crustacea from the Upper Cambrian of Västergötland, Sweden, *Fossils Strata* **17:**1–65.

Müller, K. J., and Walossek, D., 1987, Morphology, ontogeny, and life habit of *Agnostus pisiformis* from the Upper Cambrian of Sweden, *Fossils Strata* **19:**1–124.

Müller, K. J., and Walossek, D., 1988, External morphology and larval development of the Upper Cambrian maxillopod *Bredocaris admirabilis, Fossils Strata* **23:**1–70.

Nazarov, B. B., 1973, Radiolarians from the Lower Cambrian beds of Bateny Hills, in: *Problems in the Paleontology and Biostratigraphy of Siberia and the Far East* (I. T. Zhuravleva, ed.), *Trudy Inst. Geol. Geofiz. Sibirsk. Otd. Akad. Nauk SSSR* **49:**1–13 [in Russian].

Nazarov, B. B., 1975, Lower–Middle Palaeozoic radiolarians of Kazakhstan, *Trudy Geol. Inst. Akad. Nauk SSSR* **275:**1–203 [in Russian].

Nazarov, B. B., and Ormiston, A. R., 1985, Evolution of Radiolaria in the Paleozoic and its correlation with the development of other marine fossil groups, *Senckenb. Lethaea* **66:**203–215.

Nicol, D., 1977, The number of living animal species likely to be fossilized, *Florida Sci.* **40**(2)**:**135–139.

Nitecki, M. H., and Debrenne, F., 1979, The nature of radiocyathids and their relationship to receptaculitids and archaeocyathids, *Geobios* **12**(1)**:**5–27.

Nowlan, G. S., Narbonne, G. M., and Fritz, W. H., 1985, Small shelly fossils and trace fossils near the Precambrian–Cambrian boundary in the Yukon Territory, Canada, *Lethaia* **18:**233–256.

Oliver, W. A., 1984, *Conchopeltis:* Its affinities and significance, in: Recent Advances in the Paleobiology and Geology of the Cnidaria (W. A. Oliver, Jr., W. J. Sando, S. D. Cairns, A. G. Coates, *et al.*, eds.), *Palaeontogr. Am.* **54:**141–147.

Paul, C. R. C., and Smith, A. B., 1984, The early radiation and phylogeny of echinoderms, *Biol. Rev. Camb. Philos. Soc.* **59:**443–481.

Peel, J. S., 1988, *Spirellus* and related helically coiled microfossils (cyanobacteria) from the Lower Cambrian of North Greenland, *Grønlands Geol. Undersøgelse Rapp.* **137:**5–32.

Pel'man, Yu. L., 1976, Early–Middle Cambrian stenothecoids and new skeletal remains of uncertain systematic position from the stratotypic region of the Aldan and Lena Rivers, *Trudy Inst. Geol. Geofiz. Sibirsk. Otd. Akad. Nauk SSSR* **296:**176–179 [in Russian].

Poulsen, V., 1963, Notes on *Hyolithellus* Billings, 1871, class Pogonophora Johannson, 1937, *Biol. Medd. K. Dan. Vidensk. Selsk.* **23**(12)**:**1–15.

Qian Yi, 1977, Hyolitha and some problematica from the Lower Cambrian Meishucunian Stage in central and southwestern China, *Acta Palaeontol. Sinica* **16:**255–275 [in Chinese].

Qian Yi, 1983, Sinian–Cambrian boundary in China, in: *Studies on Stratigraphic Boundaries in China,* p. 1–11 [in Chinese].

Qian Yi, 1984, Several groups of bizarre sclerte fossils from the earliest Cambrian in eastern Yunnan, *Bull. Nanjing Inst. Geol. Paleontol. Acad. Sinica* **1983**(6)**:**85–99 [in Chinese].

Qian Yi and Bengtson, S., 1989, Palaeontology and biostratigraphy of the Early Cambrian Meishucunian Stage in Yunnan Province, South China, *Fossils Strata* **24:**1–156.

Qian Yi and Yin Gongzheng, 1984, Zhijinitidae and its stratigraphical significance, *Acta Palaeontol. Sinica* **23**:215–223 [in Chinese].

Qian Yi and Zhang Shi-ben, 1983, Small shelly fossils from the Xihaoping member of the Tongying Formation in Fangxian County of Hubei Province and their stratigraphical significance, *Acta Palaeontol. Sinica* **22**:82–94 [in Chinese].

Ramsköld, L., and Hou Xianguang, 1991, New early Cambrian animal and onychophoran affinities of enigmatic metazoans, *Nature* **351**:225–228.

Raup, D. M., 1975, Taxonomic diversity estimation using rarefaction, *Paleobiology* **1**:333–342.

Reiswig, H. M., 1971, The axial symmetry of sponge spicules and its phylogenetic significance, *Cahiers Biol. Mar.* **12**:505–514.

Reitlinger, E. A., 1948, Cambrian foraminifers of Yakutia, *Byull. Mosk. Obshch. Ispyt. Prir. Otd. Geol.* **23**(2):77–81.

Rhodes, F. H. T., and Bloxam, T. W., 1971, Phosphatic organisms in the Paleozoic and their evolutionary significance, in: Phosphate in fossils (E. L. Yochelson, ed.), *Proceedings of the North American Paleontological Convention, Chicago, 1969,* K:1485–1513.

Riding, R., 1977, Calcified *Plectonema* (blue-green algae), a Recent example of *Girvanella* from Aldabra Atoll, *Palaeontology* **20**:33–46.

Riding, R., and Voronova, L. G., 1982, Calcified cyanophytes and the Precambrian–Cambrian transition, *Naturwissenschaften* **69**:498–499.

Riding, R., and Voronova, L. G., 1984, Assemblages of calcareous algae near the Precambrian/Cambrian boundary in Siberia and Mongolia, *Geol. Mag.* **121**:205–210.

Rieger, R. M., and Sterrer, W., 1975, New spicular skeletons in Turbellaria, and the occurrence of spicules in marine meiofauna, *Z. Zool. Syst. Evolutionsforsch.* **13**:207–278.

Rigby, J. K., 1983, Fossil Demospongia, in: *Sponges and Spongiomorphs—Notes for a Short Course* (T. Broadhead, ed.), *University of Tennessee, Department of Geological Sciences, Studies in Geology,* No. 7, pp. 12–39.

Rigby, J. K., 1986, Sponges of the Burgess Shale (Middle Cambrian), British Columbia, *Palaeontogr. Can.* **2**:1–105.

Round, F. E., 1981, Morphology and phyletic relationships of the silicified algae and the archetypal diatom—Monophyly or polyphyly, in: *Silicon and Siliceous Structures in Biological Systems* (T. L. Simpson and B. E. Volcani, eds.), Springer, New York, pp. 97–128.

Rowell, A. J., 1982, The monophyletic origin of the Brachiopoda, *Lethaia* **15**:299–307.

Rozanov, A. Yu., 1973, Regularities in the morphological evolution of archaeocyathans and questions regarding the stage division of the Lower Cambrian, *Trudy Geol. Inst. Akad. Nauk SSSR* **241**:1–164 [in Russian].

Rozanov, A. Yu., Missarzhevskij, V. V., Volkova, N. A., Voronova, L. G., Krylov, I. N., Keller, B. M., Korolyuk, I. K., Lendzion, K., Michniak, R., Pykhova, N. G., and Sidorov, A. D., 1969, The Tommotian Stage and the problem of the lower boundary of the Cambrian, *Trudy Geol. Instituta Akad. Nauk SSSR* **206**:1–380 [in Russian].

Runnegar, B., 1985a, Shell microstructure of Cambrian molluscs replicated by calcite, *Alcheringa* **9**:245–257.

Runnegar, B., 1985b, Collagen gene construction and evolution, *J. Molec. Evol.* **22**:141–149.

Runnegar, B., and Bentley, C., 1983, Anatomy, ecology and affinities of the Australian Early Cambrian bivalve *Pojetaia runnegari* Jell, *J. Paleontol.* **57**:73–92.

Runnegar, B., Pojeta, J. Jr., Morris, N. J., Taylor, J. D., Taylor, M. E., and McClung, G., 1975, Biology of the Hyolitha, *Lethaia* **8**:181–191.

Sangster, A. G., and Parry, D. W., 1981, Ultrastructure of silica deposits in higher plants, in: *Silicon and Siliceous Structures in Biological Systems* (T. L. Simpson and B. E. Volcani, eds.), Springer, New York, pp. 383–407.

Schopf, W. J., and Klein, C. (eds.), In press, *The Proterozoic Biosphere—A Multidisciplinary Study,* Cambridge University Press, Cambridge.

Scrutton, C. T., 1979, Early fossil cnidarians, in: *The Origin of Major Invertebrate Groups* (M. R. House, ed.), Academic Press, New York, pp. 161–207.

Sdzuy, K., 1969, Unter- und mittelkambrische Porifera (Chancelloriida und Hexactinellida), *Palaeontol. Z.* **43**(3/4):115–147.

Sepkoski, J. J., Jr., 1978, A kinetic model of Phanerozoic taxonomic diversity. I. Analysis of marine orders, *Paleobiology* **4**:223–251.

Simkiss, K., 1964, Phosphates as crystal poisons of calcification, *Biol. Rev.* **39**:487–505.

Simpson, T. L., and Volcani, B. E., 1981, Introduction, in: *Silicon and Siliceous Structures in Biological Systems* (T. L. Simpson and B. E. Volcani, eds.), Springer, New York, pp. 3–12.

Sokolov, B. S., 1965, The oldest Early Cambrian deposits and sabelliditids, in: *Reports. All-Union Symposium on Precambrian and Cambrian Paleontology,* Institute of Geology and Geophysic, Siberian Branch, USSR Academy of Sciences, Novosibirsk, USSR, pp. 78–92 [in Russian].

Sokolov, B. S., 1967, The oldest pogonophorans, *Doklady Akad. Nauk SSSR* **177**:201–204 [in Russian].

Sokolov, B. S., and Zhuravleva, I. T. (eds.), 1983, Stage division of the Lower Cambrian of Siberia. Atlas of fossils, *Trudy Inst. Geol. Geofiz. Sibirsk. Otd. Akad. Nauk SSSR* **558**:1–216.

Song Xueliang, 1984, *Obruchevella* from the early Cambrian Meishucun Stage of the Meishucan section, Jinning, Yunnan, China, *Geol. Mag.* **121**:179–183.

Sun Weiguo and Hou Xianguang, 1987a, Early Cambrian medusae from Chengjiang, Yunnan, China, *Acta Palaeontol. Sinica* **26**:257–271 [in Chinese].

Sun Weiguo and Hou Xianguang, 1987b, Early Cambrian worms from Chengjiang, Yunnan, China: *Maotianshania* gen. nov., *Acta Palaeontol. Sinica* **26**:299–305 [in Chinese].

Szaniawski, H., 1982, Chaetognath grasping spines recognized among Cambrian protocondonts, *J. Paleontol.* **56**:806–810.

Szaniawski, H., 1987, Preliminary structural comparisons of protoconodont, paraconodont, and euconodont elements, in: *Palaeobiology of Conodonts* (R. J. Aldridge, ed.), Ellis Horwood, Chichester, England, pp. 35–47.

Tynan, M. C., 1983, Coral-like microfossils from the Lower Cambrian of California, *J. Paleontol.* **57**:1188–1211.

Urbanek, A., and Mierzejewska, G., 1977, The fine structure of zooidal tubes in Sabelliditida and Pogonophora with reference to their affinity, *Acta Palaeontol. Polon.* **22**:223–240.

Vacelet, J., 1985, Coralline sponges and the evolution of the Porifera, in: *The Origins and Relationships of Lower Invertebrates* (S. Conway Morris, J. D. George, R. Gibson, and H. M. Platt, eds.), Clarendon Press, Oxford, pp. 1–13.

Valentine, J. W., 1989, How good was the fossil record? Clues from the Californian Pleistocene, *Paleobiology* **15**:83–94.

Val'kov, A. K., 1982, *Lower Cambrian Biostratigraphy of the East Siberian Platform,* Nauka, Moscow [in Russian].

Val'kov, A. K., and Karlova, G. A., 1984, The fauna of the transitional Vendian–Cambrian beds in the lower reaches of the River Gonam, in: *Late Precambrian and Early Paleozoic Stratigraphy, Middle Siberia* (V. V. Khomentovsky, ed.), pp. 12–41 [in Russian].

Val'kov, A. K., and Sysoiev, V. A., 1970, Cambrian angustiochreids from Siberia, in: *Stratigraphy and Paleontology of the Proterozoic and Cambrian of the East Siberian Platform* (A. K. Bobrov, ed.), Yakutsk Publishing House, Yakutsk, USSR, pp. 94–100 [in Russian].

Vologdin, A. G., 1932, *Siberian Archaeocyathans, Part 2, The Fauna of the Cambrian Limestones of Altaj,* State Scientific–Technical Geologic Exploratory Publishing House, Moscow [in Russian].

Voronin, Yu. I., Voronova, L. G., Grigor'eva, N. V., Drozdova, N. A., Zhegallo, E. A., Zhuravlev, A. Yu., Ragozina, A. L., Rozanov, A. Yu., Sayutina, T. A., Sysoev, V. A., and Fonin, V. D., 1982, The Precambrian–Cambrian boundary in the geosynclinal regions (reference section Salany-Gol, MPR), in: *Joint Soviet–Mongolian Paleontol. Expedition* **18**:1–150 [in Russian].

Voronova, L. G., and Missarzhevskij, V. V., 1969, Finds of algae and worm tubes in the Precambrian–Cambrian boundary beds in the northern part of the Siberian Platform, *Doklady Akad. Nauk SSSR* **184**(1):207–210.

Voronova, L. G., and Radionova, Eh. P., 1976, Palaeozoic algae and microphytolites, *Trudy Geol. Inst. Akad. Nauk SSSR* **294**:1–220 [in Russian].

Walcott, C. D., 1920, Cambrian geology and paleontology IV:6—Middle Cambrian Spongiae, *Smithson. Misc. Collect.* **67**(6):261–364.

Waller, T. R., 1983, Dahllite in the periostracum of *Lithophaga nigra* (Mollusca: Bivalvia) and its taxonomic and functional implications, *Am. Malacol. Bull.* **1**:101.

Walliser, O. H., 1958, *Rhombocorniculum comleyense* n. gen., n. sp. (Incertae sedis, Unterkambrium, Shropshire), *Palaeontol. Z.* **32**:176–180.

Watabe, N., 1990, Calcium phosphate structures in invertebrates and protozoans, in: *Skeletal Biomineralization: Patterns, Processes and Evolutionary Trends* (J. G. Carter, ed.), pp. 35–44, Van Nostrand Reinhold, New York.

White, R. D., 1986, Cambrian Radiolaria from Utah, *Journal of Paleontology* **60**:778–780.
Whittington, H. B., 1985, *The Burgess Shale*, Yale University Press, New Haven, Connecticut.
Wilmot, N. V., and Fallick, A. E., 1989, Original mineralogy of trilobite exoskeletons, *Palaeontology* **32**:297–304.
Wright, A. D., 1979, Brachiopod radiation, in: *The Origin of Major Invertebrate Groups* (M. R. House, ed.), Academic Press, New York, pp. 235–252.
Wrona, R., 1982, Early Cambrian phosphatic microfossils from southern Spitsbergen (Hornsund Region), *Palaeontol. Polon.* **43**:9–16.
Wrona, R., 1987, Cambrian microfossil *Hadimopanella* Gedik from glacial erratics in West Antarctica, *Palaeontol. Polon.* **49**:37–48.
Xing Yusheng, Chen Yiyuan, Zhang Shusen, Liu Guizhi, Xiong Xingwu, Chen Ping, Luo Huilin, Jiang Zhiwen, Wu Xiche, *et al.*, 1984, The Sinian–Cambrian boundary of China, *Bull. Inst. Geol. Chin. Acad. Geol. Sci. Spec. Issue* **10**:1–260 [in Chinese].
Yochelson, E. L., 1977, Agmata, a proposed extinct phylum of Early Cambrian age, *J. Paleontol.* **51**:437–454.
Yu Wen, 1984, Early Cambrian molluscan faunas of Meishucun Stage with special reference to Precambrian–Cambrian boundary, in: *Developments in Geoscience. Contributions to the 27th International Geological Congress, 1984 Moscow*, Academia Sinica, Science Press, Beijing, pp. 21–35.
Yu Wen, 1987, Yangtze Micromolluscan Fauna in Yangtze Region of China with notes on Precambrian–Cambrian boundary, in: *Stratigraphy and Palaeontology of Systemic Boundaries in China—Precambrian–Cambrian Boundary 1*, Nanjing University Publishing House, Nanjing, China, pp. 19–344.
Yue Zhao, 1991, Discovery of fused sclerites of Early Cambrian *Phyllochites* and its relation with zhijinitids, *Kexue Tongbao* **1991**(1):47–50 [in Chinese].
Zhong Hua (Chen Menge), 1977, Preliminary study of the ancient fauna of south China and its stratigraphic significance, *Scientia Geol. Sinica* **1977**:118–128 [in Chinese].
Zhuravlev, A. Yu., 1986, Radiocyathids, in: *Problematic Fossil Taxa* (A. Hoffman and M. H. Nitecki, eds.), Clarendon Press, Oxford, pp. 35–44.
Zhuravleva, I. T., 1960, *Archaeocyathans of the Siberian Platform*, USSR Academy of Sciences, Moscow [in Russian].
Zhuravleva, I. T., and Myagkova, E. I., 1981, Material for the study of Archaeata, in: Problematics of the Phanerozoic (B. S. Sokolov, ed.), *Trudy Inst. Geol. Geofiz. Sibirsk. Otd. Akad. Nauk SSSR* **481**:41–74 [in Russian].
Zhuravleva, I. T., and Okuneva, O. G., 1981, On the nature of cribricyathans, in: Problematics of the Phanerozoic (B. S. Sokolov, ed.), *Trudy Inst. Geol. Sibirsk. Otd. Akad. Nauk SSSR* **481**:23–30 [in Russian].

Chapter 15

Paleoceanography and Changes in the Biological Cycling of Phosphorus across the Precambrian–Cambrian Boundary

M. D. BRASIER

1. Introduction

This chapter has two main aims. First, to draw attention to the evidence for changing oceanography and atmospheres coincident with the "Cambrian explo-

M. D. BRASIER • Department of Earth Sciences, Oxford University, Oxford OX1 3PR, England.

Origin and Early Evolution of the Metazoa, edited by Jere H. Lipps and Philip W. Signor. Plenum Press, New York, 1992.

sion." And second, to discuss the profound changes in nutrient cycling likely to have been brought about by this remarkable biological revolution.

The first part of the chapter therefore reviews several approaches to paleoceanography, relying heavily upon analogues from Mesozoic to Quaternary paleoceanography. An attempt has been made to draw together the fragmentary evidence, and the reader will note how much remains to be done. Evidence is inescapable, however, for the importance of nutrient-enriched and oxygen-depleted water masses over this interval.

The second part of the chapter initially attempted to answer the central question: what was the impact of changing ocean–atmosphere chemistry on the biota, e.g., with respect to biomineralization? This proved a highly elusive and unsatisfactory exercise, however, and the question has therefore been turned around to one that may be more rewarding: what was the likely impact of metazoan evolution upon the biological phosphorus cycle? The conclusions are, of course, entirely tentative.

Both paleoceanography and the biological phosphorus cycle have enormous implications for our understanding of carbon isotope stratigraphy. This new technique lies at the core of an endeavor to improve the stratigraphic resolution of this important evolutionary event (IGCP Project 303 on Late Precambrian and Cambrian Event Stratigraphy). If successful, it could revolutionize the time frame and render much of the vague and loosely structured discussion of past decades entirely obsolete. This chapter is dedicated to that effort.

1.1. Stratigraphic Controls

In the following discussion, some of the better known strata are referred to chronostratigraphic and informal biostratigraphic intervals. The writer has elsewhere discussed at length the problems of biostratigraphy and correlation in the Precambrian–Cambrian boundary interval (Brasier, 1989a; Cowie and Brasier, 1989), while recent work has also emphasized the utility of trace fossils (Narbonne, *et al.*, 1987; Narbonne and Myrow, 1988; Crimes, 1987, 1989). It is by no means certain that successive small shelly fossil assemblages are everywhere contemporaneous, nor has a satisfactory correlation been demonstrated between these and trace fossil stratigraphy. To break this deadlock, an attempt is being made to test the use of nonbiostratigraphic techniques for correlation, through the work of IGCP Project 303 on Precambrian-Cambrian Event Stratigraphy. This work will involve the integrated analysis of biological events, chemostratigraphy (isotopes of carbon, oxygen, strontium, and trace element anomalies), paleomagnetism, and sequence stratigraphy. Provisional attempts at correlation have already been made from strata in India to Siberia and Morocco (Aharon *et al.*, 1987), from southern China to Iran (Brasier *et al.*, 1990) and from Siberia to Morocco (Kirschvink *et al.*, 1991). Current work on the Siberian and Newfoundland sections indicates the existence of many carbon isotope excursions, however, and work proceeds with caution.

Stratigraphy is therefore in a state of transition. The correlation scheme followed here is that previously employed by the writer (Fig. 1) (Brasier, 1989a; Brasier *et al.*, 1990), but uses the familiar nomenclature of the Siberian Platform wherever possible. The basic terms (as used in this chapter) are defined below.

Era	Period	Epoch	Age (U.S.S.R)	Glaciations	Evaporites	Phosphorites	Black shales	Age (China)
Paleozoic	Cambrian	Lenian	Toyonian		◇ China, Australia, Siberia	● Thorntonia		Maozhuangian + Longwangmiaoan
			Botomian			● Xinji	■ Sinska	Canglanpuian
		Aldanian	Atdabanian		◇ Siberia		■ Shuijngtuo, Yuhucun	Qiongzhusian
			Tommotian		◇ Yudoma	● Tal, Badaowan	■ Badaowan	Meishucunian
		–	Nemakit–Daldynian		◇ Salt Range ◇ Hormuz	● Zhongyicun		
Sinian	Vendian	Ediacarian	Kotlinian	?▲ Late Sinian	◇ Yudoma		■ Jiucheng, Hormuz, Chopoghlu	Dengyingxian
			Redkinian		◇ ◇ Hormuz	● ● Doushantuo		–
		Varangian	–	▲ Varangian (Nantuo)				
	Sturtian	–		▲ Sturtian				

Figure 1. Stratigraphic table showing major lithological and oceanographic events against a preliminary time scale. The position of the Precambrian–Cambrian boundary is extrapolated from the boundary point in the Burin peninsula, southeastern Newfoundland; epoch names are provisional; stage names from Siberia and China are shown.

Varangian (or Varangerian): The interval that includes the late Precambrian glaciation, below the appearance of most (but not all) of the Ediacara fauna, included at the base of the Vendian System of the USSR, and within the Sinian System of China.

Redkinian: A Neoproterozoic interval above the Varangerian but below the Kotlinian. It contains the Ediacara soft-bodied fauna (e.g., *Charnia*) and simple trace fossils. Examples include the Valdaian Stage of the Vendian, the Conception Group of Newfoundland, and the Pound Quartzite of South Australia.

Kotlinian: This Neoproterozoic interval is characterized by the *Harlaniella podolica* trace fossil assemblage, above the Ediacara fauna with *Charnia* and below the appearance of the *Phycodes pedum* ichnofossil assemblage. A few early skeletal fossils may occur in this interval. Examples include the Kotlinian Stage of the Vendian on the East European Platform and the lower Chapel Island Formation of Newfoundland.

Nemakit-Daldynian: This lies above the last certain appearance of Ediacaran fauna and *Harlaniella podolica*. It spans the lower range of the *Anabarites trisulcatus–Protohertzina anabarica* skeletal assemblage (Brasier, 1989a). Organic tubes of *Sabellidites cambriensis* and the *Phycodes pedum* ichnofossil assemblage are thought to appear at a similar level, and the latter is taken to mark the base of the Cambrian in Newfoundland and elsewhere (Narbonne *et al.*, 1987; Landing *et al.*, 1989). Cap-shaped shells such as *Canopoconus*, *Purella*, and *Latouchella* may mark a later stage of this assemblage in Asia and Newfoundland. The lower part of the *Aldanella attleborensis* assemblage may also belong here. Examples include the Rovno Formation of the East European Platform, the Nemakit-Daldyn Formation of the Yudoma Suite in Siberia, Zones I and II of the Meishucunian Stage (i.e., the upper Dengying Formation) in south China, and Members 2 (to 5?) of the Chapel Island Member in Newfoundland.

Tommotian: A time of major flooding of the cratons, which brought in the first archaeocyathans, inarticulate brachiopods, triangulate hyoliths, and a variety of phosphatic problematica such as tommotiids. The upper part of the *Aldanella attleborensis* assemblage (Brasier, 1989a, p. 155) belongs here. Examples include the type Tommotian of Siberia, the Lontova and Mazowsze Formations of the East European Platform, and the Petley to Cuslett Formations of Newfoundland. Less certain is the inclusion of parts of Zone II to Zone III of south China in this interval.

Early Atdabanian: This contains the greater part of the range of the pseudoconodont *Rhombocorniculum insolutum* and its associated assemblages (Brasier, 1989a), including the earliest olenellid trilobites (*Profallotaspis*, *Fallotaspis*, *Schmidtiellus*). Examples are the *lenaicus* to *pinus* or *anabarus* Zones of Siberia, the upper *Camenella baltica* Zone (Fosters Point Formation) in Newfoundland, and the Home Farm Member in England. Zone III of the Meishucunian in South China may also belong here.

Late Atdabanian: This contains the lower part of the range of the *Rhombocorniculum cancellatum* assemblage, together with trilobites such as *Callavia*, *Judomia*, *Holmia*, and *Parabadiella* (Brasier, 1989a). Examples include the upper *Pagetiellus anabarus* to *Judomia* Zones of Siberia, the lower *Callavia* Zone of England and Newfoundland, the *Nevadella* Zone of North America, and probably Zone IV (the *Parabadiella* Zone) of south China. The *Lapworthella cornu* assemblage appeared in the upper part of this interval in many regions (Brasier, 1989a).

Botomian: This is characterized by the continuation of the *Lapworthella cornu* assemblage, plus a succession of distinctive trilobites such as *Serrodiscus bellimarginatus, Calodiscus lobatus,* and *Pseudatops* sp. and the appearance of Protolenidae or Redlichiidae sp. Examples include the latest Atdabanian–Botomian of Siberia, the upper *Callavia* to Strenuellid–Protolenid Zones of England and equivalent zones of southeastern Newfoundland, and the *Eoredlichia* to *Palaeolenus* Zones of the Qiongzhusian Stage in south China.

1.2. The Neoproterozoic and Cambrian Radiations

Speculative models abound at this level in the stratigraphic column. This study therefore begins by outlining the paleobiological characteristics of the Neoproterozic and Cambrian as discovered from the fossil record itself. Many of these were reviewed by the writer over a decade ago (Brasier, 1979a) and have been supported by further research, as outlined below.

The Neoproterozoic and Cambrian radiations of invertebrate phyla together resulted in a great increase in diversity that reached equilibrium near the Middle Cambrian (e.g., Sepkoski, 1979; Brasier, 1979b, 1982).

Ancient stromatolitic communities suffered a decline in abundance and diversity during the Varangian glaciation(s) and the Redkinian time interval, perhaps in response to grazing pressures (Walter and Heys, 1985). Stromatolites are said to have recovered slightly during the ensuing Kotlinian Stage, but declined again with the appearance of skeletons (Sokolov and Fedonkin, 1986). Structureless, calcified thrombolites of *Renalcis* appeared in the Nemakit-Daldynian Stage, contemporaneous with the earliest tubular invertebrate skeletons (Riding and Voronova, 1984) and with some of the first complex infaunal traces (Fedonkin, 1988).

Organic-walled microplankton (acritarchs) suffered a marked decline in the Varangian glaciation(s), locally remained of low diversity through the Redkinian, and became more diverse through the Cambrian with diversification of spinose and sculptured vesicles (Vidal and Knoll, 1983; Knoll and Swett, 1987). Early stages of this adaptive radiation are well seen in type Tommotian strata (Moczydlowska and Vidal, 1988) but began in the Nemakit-Daldynian Stage (G. Vidal, personal communication 1990).

The soft-bodied Ediacara fauna mostly occurs above deposits of the Varangerian glaciation(s). They have been regarded mainly as impressions of cnidarian polyps and medusoids, annelid worms, and a few arthropods (Glaessner, 1984). Others argue that this fauna represents a heterogeneous group of trace fossils and other unidentified organisms, perhaps of phyla unique to the Ediacaran (e.g., Seilacher, 1984) or of forms that hosted symbiotic algae (McMenamin, 1987). This assemblage appears to have suffered a diminution in size and diversity during the following Kotlinian Stage (containing the *Harlaniella podolica* trace fossil assemblage) and worldwide disappearance near the end of the Precambrian. This disappearance coincided with major physical and taphonomic changes in the environment (Brasier, 1989b).

Small, shallow-burrowing trace fossils of Kotlinian type, such as *Harlaniella* and *Palaeopascichnus*, were rather abruptly replaced by larger, more penetrative and complex traces, such as deposit-feeding *Phycodes pedum*, plus arthropod traces, U-tube dwellings of *Diplocraterion*, and many other "Phanerozoic" types (Crimes and Anderson, 1985; Crimes, 1987, 1989; Narbonne *et al.*, 1987) in strata of presumed Nemakit-Daldynian age. The depth and intensity of bioturbation did not increase dramatically, however, until trilobitic strata of presumed Atdabanian age, being greatest in shallow-water facies (Droser and Bottjer, 1988).

The "mass radiation" of invertebrate phyla probably began with the development of soft-bodied invertebrates. This is not only indicated by the trace fossils, but is also hinted at by late Atdabanian–Botomian Burgess Shale-type Lagerstätten fossils of Yunnan, plus younger occurrences in Greenland and British Columbia (e.g., Conway Morris *et al.*, 1982, 1987). Groups present include ?chordates, arthropods, polychaetes, priapulids, sponges, cnidarians [including fronds of ?Ediacara affinity (S. Conway Morris, personal communication 1991)], and problematica. The appearance of this kind of fauna in the fossil record may owe much to changing paleoceanographic and taphonomic processes.

This radiation was accompanied by several major episodes of biomineralization. The following groups made their first major skeletal appearance at this time: eukaryotic algae, protists, parazoans (including archaeocyathans), brachiopods, mollusks, hyoliths, arthropods, echinoderms, ?chaetognath worms, ?pogonophoran worms, numerous tubular problematica (including some possible annelids), tommotiids, and other multisclerite problematica (Brasier, 1979a; Sepkoski, 1979). As mentioned above, mineralization also developed among cyanobacteria at this time (e.g., Riding and Voronova, 1984; Pentecost and Riding, 1986).

These new forms of skeleton employed a wide variety of materials: organic and chitinoid, arenaceous, siliceous, phosphatic, aragonitic, and calcitic (Brasier, 1979a). To these may be added the domichnial "skeletons" of infaunal trace-making organisms, the fluid "skeletons" of many coelomate groups, and even internal sand-skeletons (A. Seilacher, personal communication 1991). Biomineralization was therefore part of a wider phenomenon, which was "biologically controlled" in most invertebrates and "biologically induced" in cyanobacteria (*sensu* Lowenstam, 1986). Various kinds of multielement skeletons of problematica appeared at this time (e.g., Landing, 1984; Bengtson and Conway Morris, 1984; Qian and Bengtson, 1989; Bengtson *et al.*, 1990), as did a range of phosphatic skeletons, the latter being locally important and diverse in phosphogenic facies (Brasier, 1979a, 1990a; Brasier and Hewitt, 1979). Indeed, phosphogenesis appears to have opened a taphonomic window for the preservation of small shelly fossils across the boundary interval (Brasier, 1990b).

From here, we must move into the more uncertain territory of ecological changes and evolutionary patterns. The biological revolution that distinguishes Cambrian from Ediacaran faunas may have been accompanied by changes in feeding, respiration, and metabolism. For example, Ediacaran faunas lack evidence for gills (Seilacher, 1984; Runnegar, 1982) and few have been proven to possess either mouth or gut, so that some have speculated upon the presence of endosymbionts (McMenamin, 1987). In contrast, the Cambrian radiations involved many gill-bearers, filter-feeders and scavengers, protected by burrows or true skeletons (Brasier, 1979a).

A reported decrease in the size of body fossils, from larger to smaller Ediacaran faunas to relatively microscopic small shelly fossils, was accompanied by an apparent increase in organic preservation of macroalgal remains (Sokolov and Fedonkin, 1986). There followed a general increase in the maximum body diameter of trace fossils (see Fig. 8).

A dramatic increase in cropping pressures (Stanley, 1976) is supported by bioturbation and grazing traces as well as by evidence for predation (e.g., Brasier, 1979a; Conway Morris and Jenkins, 1985; Droser and Bottjer, 1988). Rapid expansion into vacant niches (Brasier, 1979a, 1982) was accompanied by rather plastic variations of skeletal morphotype, with many intermediates. These initially show little evidence for directional natural selection (e.g., Conway Morris and Fritz, 1984) and suggest the existence of an "expansive phase" in the adaptive radiation at this time.

Three successive phases of diversification (trace fossils—small shelly fossils—trilobites) were recognized by Brasier (1979a, 1980, 1982) and resemble kinetically driven radiations in successively further offshore communities, in phase with increases in habitat through the Cambrian transgressions. The evolutionary pattern was much more complex than this, of course, with paleoceanography, paleogeography, and climate probably affecting the time and place of biological events. As will be discussed below, the development of deeper nutrient-enriched waters appears to have favored the early appearance, spread, and diversification of certain "phosphatophilic" taxa, while the formation of surface nutrient-depleted waters may parallel the later dramatic rise of archaeocyathan communities (e.g., Brasier, 1990a, in press a).

2. Precambrian–Cambrian Paleoceanography

What is the evidence for paleoceanographic conditions during the Precambrian–Cambrian boundary interval? This is a timely question, since the history of the biosphere in relation to the nutrient and carbon cycles, oxygen levels, and ocean–atmosphere interactions is of much current interest.

Broadly progressive patterns have been argued for oxygenation of the atmosphere (Berkner and Marshall, 1965; Cloud, 1976) and ventilation of the oceans (Berry and Wilde, 1978) from Precambrian times onward, though these views have not gone uncontested. The prevailing view for the Phanerozoic, perhaps, is that the world's oceans and atmosphere have not undergone radical, long-term changes in composition. Instead, there appear to have been rhythmic to abrupt changes in chemistry and climate of the oceans and atmosphere (e.g., Fischer, 1984; Holser, 1984; Holser *et al.*, 1986; Wilde and Berry, 1986). There is evidence, perhaps, for two Phanerozoic supercycles, each c. 300 My long and related to mantle convection. These long cycles may have been perturbed by shorter cycles of c. 30 My duration, in which pCO_2 is suggested to have had a major influence on climate and the carbon cycle (e.g., Fischer and Arthur, 1977; Fischer, 1984; Berner, 1990). Evidence for yet shorter cyclic variations in pCO_2 include ?orbitally forced periodicities of 2.0, 1.25, 0.800, 0.400, 0.115, and 0.93 My for the Miocene (Tiwari, 1987) and orbitally forced 100,000- and 21,000-year rhythms in pCO_2, sea level, and oceanic circulation through the later Quaternary (Pisias and Shackleton, 1984; Shackleton and Pisias, 1985; Genthon *et al.*, 1987). Even pO_2 may have fluctuated since the Cretaceous (Berner and Landis, cited in Anderson, 1987), according to the evidence from gaseous inclusions in amber.

Paleoceanography of the late Precambrian and Cambrian is in its infancy and there has been little attempted synthesis of the evidence for oceanic and atmospheric change until now. Some observers considered that this interval was accompanied by an increase in atmospheric oxygen (e.g., Cloud, 1976). Most of the arguments for this have been based on evidence from comparative physiology of living or fossil taxa (e.g., Berkner and Marshall, 1965; Towe, 1970; Runnegar, 1982) or generalized modeling from the ecology of Recent anoxic basins (e.g., Rhoads and Morse, 1971; Degens and Stoffers, 1976). Models based on sedimentological and geochemical criteria, however, have implied that the oceans became more stagnant in the latest Precambrian and Cambrian (e.g., Holser, 1984; Fischer 1984) with episodes of overturn from stagnant to mixed states near the boundary interval (Cook and Shergold, 1984; Aharon *et al.*, 1987) as well as at higher stratigraphic levels (Notholt and Brasier, 1986; Shergold and Brasier, 1986).

The scenario to be considered here is summarized in Section 4. Models and examples for the development of anoxic, nutrient-enriched water masses are examined below. This is followed by a discussion of criteria used to piece together the chemistry of late Precambrian–Cambrian seawater.

2.1. Models for Anoxia

Oxygenation of seawater is related primarily to the degree of bottom water circulation, to temperature and climate (Fig. 2), and to organic productivity. In the present oceans, increased input of carbon to the sea floor is mainly related, in a complex way, to increased productivity of primary producers at the surface (Berger *et al.*, 1989). The latter respond positively to temperature and light, but are limited by the availability of nutrients, especially phosphate, nitrate, iron, and trace elements such as molybdenum (e.g., Berger *et al.*, 1989; Codispoti 1989; Martin and Fitzwater, 1988). After death, this organic matter settles through the water column, consuming oxygen as it decays, producing a lower layer of oxygen-deficient to anoxic water. If consumption remains fixed, increased organic input will result in a rise in the boundary between oxygenated and oxygen-deficient waters. Because of the decreased solubility of oxygen with temperature, this boundary will tend to lie higher in warmer or low-latitude water masses. Salinity has a similar effect, with decreasing solubility of oxygen in more saline (usually warmer) water masses.

Such oxygen-deficient water masses tend to be enriched in P, Si, Mn, Fe, trace metals, H_2S, ammonia, and dissolved organic matter, while the pH tends to be lower than that of normal seawater. Furthermore, the activities of heterotrophs (especially sulfate-reducing and methanogenic bacteria) causes an excess of respiration over photosynthesis, resulting in locally raised concentrations of CO_2, particularly in cooler waters where this gas is more soluble.

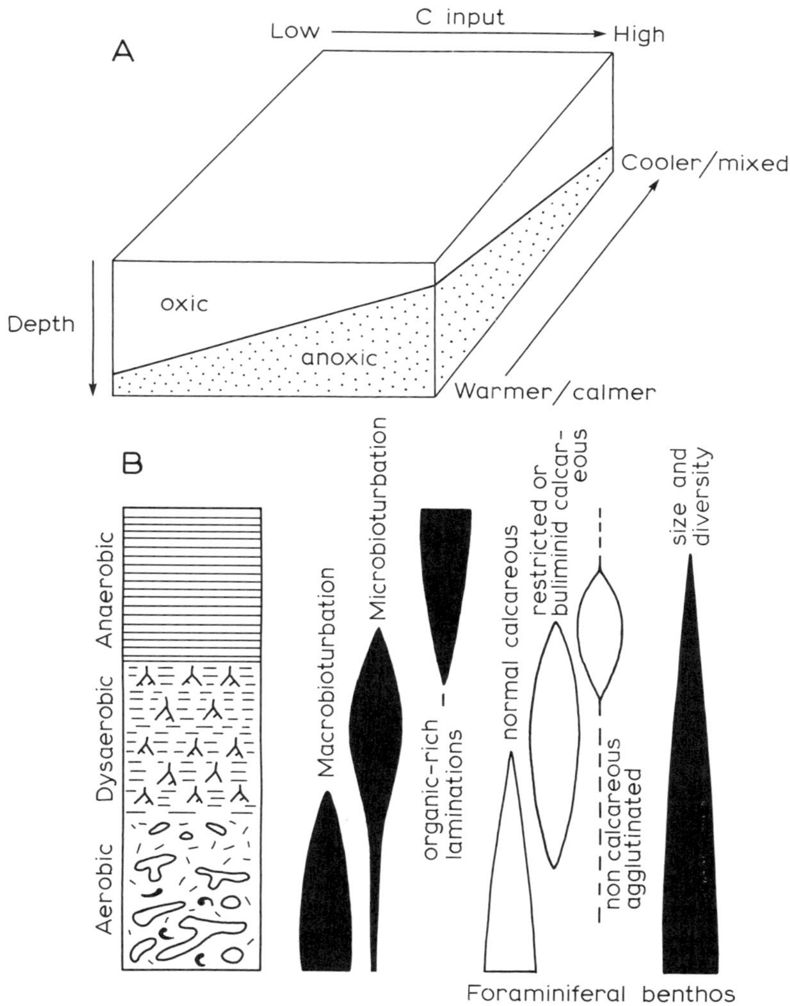

Figure 2. Models of oxygenation in the oceans. (A) Block diagram illustrating some of the factors (other than basin configuration) that influence the development of anoxic water masses in a stratified ocean. (B) Generalized lithological column showing how bioturbation and benthos can be used to track changes from aerobic to anaerobic condition (based on the writers' studies of CTBBS and TBS events). Comparable approaches need to be undertaken for Precambrian–Cambrian anoxic and phosphogenic events.

At present, such conditions tend to arise in two distinct situations: with upwelling or in barred basins. In the former case, strong upwelling causes vertical advection of nutrients such as phosphates, leading to the stimulation of primary production in surface layers and the consumption of oxygen in midwater (Baturin, 1982). This oxygen minimum zone can also be driven upward by vertical advection of H_2S, since oxidation of the latter to sulfate consumes oxygen (Southam *et al.*, 1982). Wind-driven coastal upwelling is often, but not invariably, associated with the deposition of "bioproductites" such as phosphorites, siliceous diatomites, radiolarites, and organic-rich facies (Kruijs and Barron, 1990) and appears to have been strongest during glacial episodes, when atmospheric circulation was stronger and pCO_2 was low (Barnola *et al.*, 1987), consumed by nutrient stimulation of the primary producers

In barred basins, anoxia is related more to poor bottom-water circulation, resulting in greater density stratification of the waters. This results in marked boundaries (thermocline, halocline, or the density-related pycnocline) that also separate anoxic from oxygenated waters (Degens and Stoffers, 1980). In euxinic basins such as the present Black Sea, the pycnocline comprises a narrow zone of wide-ranging redox and Eh, where chemicals tend to become concentrated, and either dissolved or precipitated according to vertical movements of the boundary layer (Degens and Stoffers, 1976). Quaternary cores through such basins therefore show cyclic alternations between strata laid down under stratified and non-stratified states (Degens and Stoffers, 1980). Seasonally varved sapropels also develop during warm interglacials, following spillover of saline brines into the barred and brackish basin of the Black Sea. The rising pycnocline tends to concentrate dissolved ions of Fe, Mn, and CO_3, while its fall during cooler periods results in the precipitation of hydroxides of Fe and Mn, plus carbonates of Fe, Mn, and Ca around the pycnocline. This can result in centimeter- to decimeter-scale rhythms of laminated pyritic sapropel, passing up into bioturbated levels and burrowed carbonates (Degens and Stoffers, 1976, 1980).

Anoxic conditions are largely restricted to continental margins and land-locked basins today, but were widespread across shelves and even ocean floors during the Cambrian to Devonian and Jurassic to Cretaceous. The latter have been referred to as oceanic anoxic events (OAEs) (Jenkyns, 1980). Below are considered those criteria that point to the development of anoxic and nutrient-enriched waters on the margins of late Precambrian–Cambrian cratons.

2.2. Occurrence of Organic-Rich Shales

Much of our understanding of "black shales" comes from studies of Mesozoic OAEs, such as those that laid down the Toarcian black shales (TBS) and the Cenomanian–Turonian boundary black shales (CTBBS). The total organic carbon (TOC) content of such black shales tends to be relatively high [average 1.2–13.7% (Summerhayes, 1987)], but the color may relate more to that of the organic matter than to organic content (Spears, 1980) and is not a safe criterion.

Depositional models for these black shales are still controversial, but a strong link has been noted between anoxia, rising sea levels, the formation of oceanic rift basins, and salinity stratification (e.g., Arthur *et al.*, 1987). Current evidence certainly seems to favor bottom-water stagnation rather than raised productivity for the Toarcian and Kimmeridgian black shales. In the TBS, this developed as a result of spillover of evaporative or oceanic ridge brines into restricted, opening oceanic basins, locally with brackish surface waters (Reigel *et al.*, 1986; Farrimond *et al.*, 1989; Baudin *et al.*, 1990). This event seems to correspond with an opening phase of the central Atlantic Ocean (Jenkyns, 1980) and coincided with a major rise in sea level (Hallam, 1981). Oxygen-deficient conditions reappeared during the Callovian–Oxfordian transgression (Duff, 1975; Morris, 1980) coincident with the opening of the North Atlantic, while the formation of warm saline bottom waters led to oil-bearing sapropels in the Kimmeridgian at a time of rifting and sea level rise in the North Sea Basin (Tyson *et al.*, 1979; Miller, 1990).

The CTBBS is a particularly widespread organic-rich facies, coincident with a major rise in sea level and a postulated rise in the oxygen minimum zone to within 100–200 m of the surface (Jenkyns, 1980; de Graciansky *et al.*, 1985; Schlanger *et al.*, 1987). There is some evidence that freshwater and evaporitic runoff in restricted basins promoted periodic salinity stratification and stagnation (e.g., Arthur *et al.*, 1987), but the upwelling-productivity model appears to be supported by evidence for high rates of organic carbon accumulation, organic geochemistry, biogenic silica deposits, opportunistic plankton assemblages, and climatic modeling (Kuhnt *et al.*, 1990; Kruijs and Barron, 1990; Farrimond *et al.*, 1990).

Although black shales and associated rocks (alum shales, stinkstones, stone coals) are regionally common in the Cambrian, they are not widely encountered in earlier Varangerian or Redkinian strata (Fig. 1). Shallow-water sediments of relatively light or even reddish color seem to be more common at this time of low sea levels, when tidal currents scoured the shelves of Laurentia and Baltica. Carbonaceous preservation of "algal" *Vendotaenia* and *Chuaria* is, at least, known from Redkinian rocks of the East European Platform (Urbanek and Rozanov, 1983), the Wernecke Mountains of northwest Canada (Narbonne and Hoffman, 1987), the Yudoma Suite of the Siberian Platform (Sokolov and Fedonkin, 1986), and the Doushantuo Formation of the Yangtze Gorges in south China (Xing *et al.*, 1985). More importantly, carbonaceous shales are typical of the late Redkinian to Kotlinian of the East European Platform (Felitsyn *et al.*, 1989), while bituminous limestones and dolomites are known from the Yukanda to Sakhara and Khatyspyt Formations of the Yudoma Suite in Siberia (Khomentovsky, 1984; Sokolov and Fedonkin, 1986) of Redkinian age. Black shales, cherts, and carbonates also accumulated in the coeval Doushantuo Formation (Xing *et al.*, 1985).

Carbonaceous deposits that include vendotaeniids and chuariamorphids are more widely known from successive Kotlinian and Nemakit-Daldynian strata. Examples include the Kotlin and Rovno stages of the East European Platform (Urbanek and Rozanov, 1983; Felitsyn *et al.*, 1989), Chapel Island Members 1 and 2, southeastern Newfoundland (Narbonne *et al.*, 1987), the Pusa Shales of central Spain (Brasier *et al.*, 1979), the Dengying Formation of south China (Luo *et al.*, 1984; Xing *et al.*, 1985), and the Lower (Chopoghlu) Shale of the Soltanieh Formation, Elburz Mountains of Iran (Hamdi *et al.*, 1989). Vendotaeniid ribbons could even represent the envelopes of anoxic bacteria (Vidal, 1989).

Commercially important hydrocarbons were also generated during this Kotlinian to Nemakit-Daldynian interval, especially in the Oman–Persian Gulf region (Edgell, 1989; Husseini and Husseini, 1990) and on the Siberian Platform (Kontorovitch *et al.*, 1990). In both cases, very low $\delta^{13}C$ in hydrocarbons may indicate bacterial blooms associated with hypersaline facies (see pp. 503–506).

Black shales occur especially along the Arabic–Asian "Palaeotethyan" carbonate margin, associated with transgressive facies of Kotlinian to Botomian age. Black shales or dolomites and phosphorites are found particularly in Meishucunian Zones I (c. Nemakit-Daldynian). III (c. Tommotian), and IV–V (late Atdabanian–Botomian). The Chengjiang Lagerstätten biota of Yunnan was preserved during the latter interval of anoxia. Such black shales can be traced from the center of the Yangtze Platform toward the southeastern slopes of Jiangnan, where anoxia prevailed throughout most of this interval (e.g., Brasier and Gao, in press). On the Siberian Platform, black shale facies of the Sinsk Formation appeared during the peak transgression of Botomian times (Zhuravleva *et al.*, 1990).

Black shales did not become widespread in Avalonia and Baltica until Middle and Upper Cambrian times (Leggett, 1980; Brasier, 1980; Notholt and Brasier, 1986; Thickpenny and Leggett, 1987), of which the most notable was a widespread episode of anoxic deposition initiated during the *Ptychagnostus gibbus* Zone (Shergold and Brasier, 1986). The Burgess Shale fossil Lagerstätten was deposited around Laurentia during early Middle Cambrian times (e.g., Conway Morris, 1986); at the type locality, this involved anoxic conditions at the foot of a steep reef scarp, about 100 m below the surface of the carbonate bank (McIlreath, 1977).

In a survey of carbonaceous shales through time, Raiswell and Berner (1986) commented on the scarcity of carbonaceous shales from normal marine settings in the Cambrian. That is to say, the majority proved to be euxinic. This led them to suggest that normal marine shales were rare because oxygenated waters were restricted to shallow regions with turbulent circulation. This picture is also supported by sedimentological studies of the Middle–Upper Cambrian Alum Shales of Scandinavia (Thickpenny, 1985; Thickpenny and Leggett, 1987). High sea-level stands are inferred at this time, with low detrital influx, stratification of the water column, relatively low organic carbon production, and euxenic conditions below wave base, leading to high preservation rates for organic carbon.

The evidence therefore suggests an increase in the stratigraphic incidence of anoxic facies through the Early to Middle Cambrian, appearing first on the low-latitude margins of Gondwana (i.e., China to Arabia, Siberia) and later at higher latitudes in Avalonia and Baltica. Although these conditions spread over the drowned cratons, like the Mesozoic OAEs discussed above, these anoxic events appear to have lasted much longer.

Much more needs to be learned about the nature and lateral extent of these anoxic facies and of oceanic sedimentation. Did black shales extend onto the continental slope, as with an oxygen minimum zone, or even onto the deep ocean floor?

2.3. Degree of Pyritization and C/S Ratios

The degree of pyritization of iron (DOP) can be used to differentiate normal from restricted or inhospitable bottom conditions (Raiswell and Berner, 1985). This is because anaerobic respiration by sulfate-reducing bacteria results in the formation of pyrite. This ratio may be preferable to that of organic carbon to sulfur (C/S), though both may be limited by the availability of iron rather than organic carbon (Raiswell and Berner, 1985; R. Raiswell, personal communication 1988).

Evidence is not available for the C/S and DOP of late Precambrian sediments, but normal marine Cambrian to Silurian shales analyzed by Raiswell and Berner (1986) appeared to show a higher proportion of pyrite sulfur to carbon than in Devonian to Recent examples. These authors suggested a number of possible explanations: greater organic productivity of the Devonian and younger terrestrial ecosystems; lower metabolic reactivity of terrestrial-derived carbon from Devonian times onward; increasing bioturbation through the Paleozoic, irrigating the sediment and depleting the carbon; and rising oxygen levels through the Paleozoic, converting reactive organic matter to relatively unreactive humic matter. Donnelly *et al.* (1988b) have, however, questioned these anomalous C/S ratios, which were not found in their studies of Middle Cambrian shales from Australia. They suggested that the influence of geothermal fluids, and the burial rates of both carbon and sulfur, need to be better understood before the signal of C/S ratios can be understood. A more comprehensive study by Raiswell and Al-Biatty (1989) has since confirmed that the DOP of Cambrian shales does compare with Devonian to Cretaceous values.

2.4. Sulfur Isotopes and Evaporites

Although the isotopes of sulfur in evaporitic sulfates are more or less in equilibrium with seawater, pyritic sulfur is more enriched in the lighter isotope of ^{32}S. This is because of preferential fractionation by dissimilarity sulfate-reducing bacteria (Schidlowski *et al.*, 1983). Global excursions of $\delta^{34}S$ in evaporites, in particular, may therefore be interpreted as evidence for major episodes of sulfide deposition and anoxia (Holser, 1977; Holser *et al.*, 1986). Such changes in the balance between sulfates and sulfides is likely to have had short-term repercussions on the operation of the oxygen cycle (Schidlowski *et al.*, 1983): times of excess evaporite formation may have been times of excess oxygen storage in sedimentary sulfates, while times of excess sulfide formation may have led to oxygen release.

Evaporites of latest Precambrian and Cambrian age occur extensively in Siberia and across the northern margin of Gondwana (Figs. 1 and 3). The $\delta^{34}S$ values of middle Proterozoic evaporites are mostly less than 20‰ but a gradual increase took place during the late Proterozoic, culminating in a maximum [the "Yudomski event" of Holser (1977)]. This was first reported from the Yudomski and Motskaya Suite, around the southern edge of the Irkutsk Amphitheater in Siberia, where gypsiferous and saliferous rocks occur in Nemakit-Daldynian to Tommotian dolomites (Khomentovsky, 1986; Kontorovitch *et al.*, 1990). Another evaporitic deposit of about Nemakit-Daldynian age is the Hormuz Salt Formation of Iran and the Persian Gulf (e.g., Wolfart, 1981); this shows similarly high $\delta^{34}S$ values (Holser, 1977). This remarkable unit accumulated up to 3000 m of interbedded evaporites, marine sediments, hydrocarbons, and igneous rocks, and extended from southern Oman to central Iran. It appears to pass laterally into the Soltanieh Formation of northern Iran (Husseini and Husseini, 1990) of which the middle dolomite can be correlated with the Nemakit-Daldynian of Siberia and China (Hamdi *et al.*, 1989). Up to 2000 m of evaporites accumulated at about the same time in the Salt Range Formation of Pakistan (Yeats and Lawrence, 1984).

A scenario for the Yudomski event was suggested by Holser (1977) as follows. Bacterial sulfides accumulated in reducing conditions, perhaps stratified beneath evaporative brines in rift basins such as those forming the Hormuz Salt. This led to massive extraction of light ^{32}S and enrichment of brines in ^{34}S. The concept of a

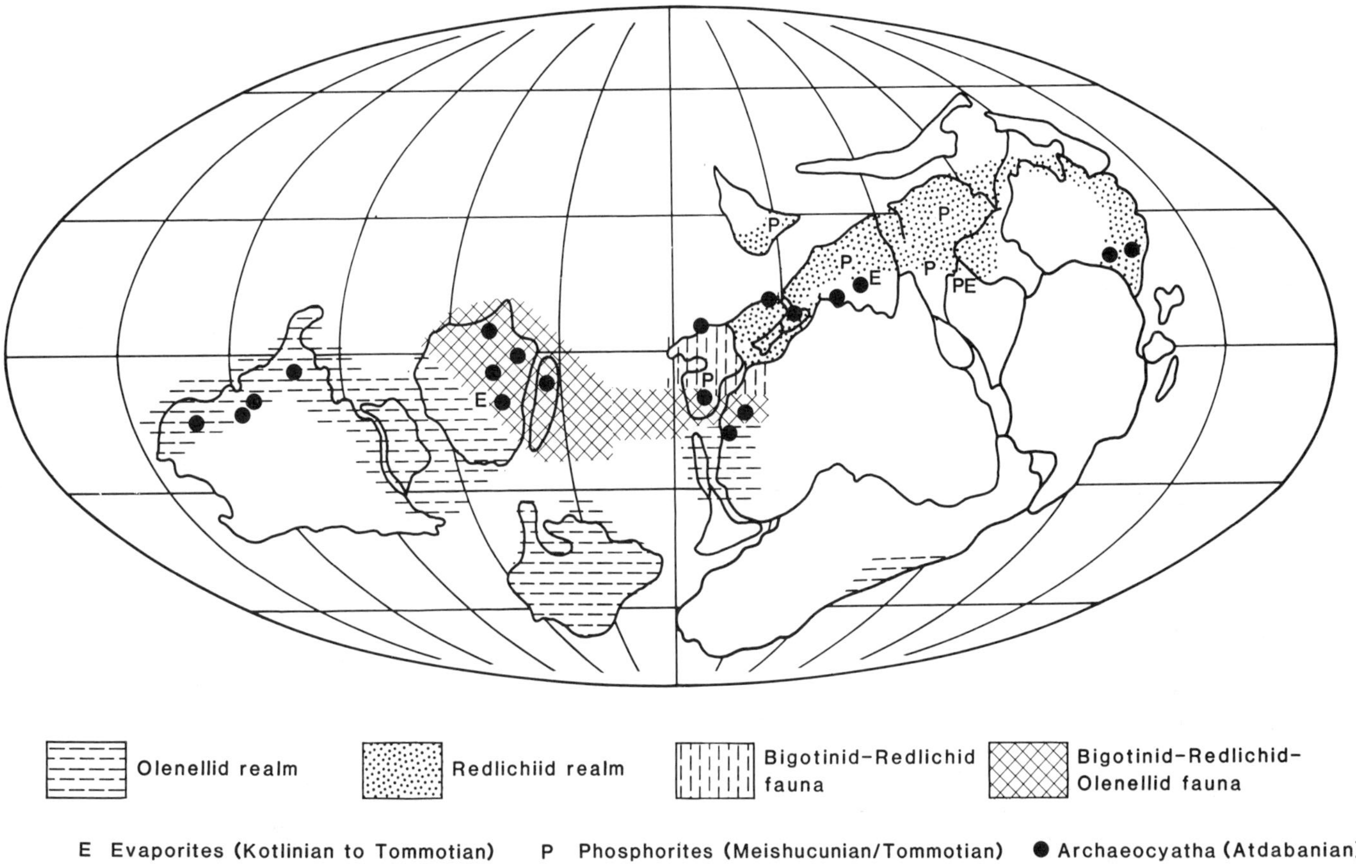

Figure 3. Paleogeographic reconstruction for Early Cambrian continents, based on data in Pillola (1990) and McKerrow *et al.* (in press). Trilobite faunal realms indicate the possible existence of two major supercontinental assemblies, with transitional faunas in Siberia, southern Europe, and Morocco. Archaeocyathans give an indication of tropical to subtropical latitudes, while phosphorites and evaporites occur particularly in the arid subtropics of Gondwana.

sharp excursion in $\delta^{34}S$ (Holser, 1977) now appears incorrect, however, as shown by high $\delta^{34}S$ in organic-rich shales of Kotlinian age on the East European Platform (Vidal, 1989) and in evaporites throughout the Cambrian (Holser, 1984). High values have also been obtained from lattice sulfate in francolites of Nemakit-Daldynian to Tommotian age (Cook and Shergold, 1984) and in baritic black cherts of Atdabanian–Botomian age in Qinling (Fan *et al.*, 1989), but these may have been affected by fractionation within sulfate-reducing conditions (e.g., Benmore *et al.*, 1983).

The widespread anoxia implied by these signals should, in theory, have been balanced by an increase in the flow of oxygen into shallow waters and the atmosphere, to be consumed in the oxidation of carbonaceous matter, resulting in a negative shift in $\delta^{13}C$ (e.g., Veizer *et al.*, 1980; Donnelly *et al.*, 1990). This could explain negative shifts in $\delta^{13}C$ associated with transgressive pulses through the Cambrian (see below, pp. 503–506), but comparative data on $\delta^{13}C$ and $\delta^{34}S$ profiles are not available to confirm this.

2.5. Evidence from Metals, Trace Elements, and Rare Earth Elements

Metallic ions, especially iron and manganese, tend to concentrate close to the pycnocline in euxinic basins (Degens and Stoffers, 1976, 1980; Frakes and Bolton, 1984). This may explain Mn enrichment in shelf carbonates coincident with the positive $\delta^{13}C$ excursion of both the TBS and the CTBBS (Jenkyns and Clayton, 1986; Schlanger *et al.*, 1987) and the association between berthierinitic (chamositic)–sideritic ironstones and times of falling sea level (e.g., Hallam and Bradshaw, 1979).

Trace metals are adsorbed onto organic matter (e.g., fecal pellets) during descent through the water column, with concentrations that tend to reflect the residence time of organic matter in the water column or on the sea floor (e.g., Cherry *et al.*, 1978). This carries the implication that the trace element composition of seawater may largely be modulated by adsorption of ions onto fecal matter. Trace element concentrations in rocks may vary in relation to the abundance of clay minerals. It follows that values should be normalized to common clay elements such as Al, Fe, Ti, or Rb, or to a standard clay sample (e.g., Middleton and Grant, 1990), a calculation often missing from published accounts. The relative enrichment or depletion of trace elements may then be used to indicate conditions of formation in carbonates or mudrocks. For example, raised concentrations of Cd, Ag, Mo, Sb, Zn, V, and Cu may indicate low rates of sedimentary accumulation in stagnant conditions, as during the CTBBS (Brumsack, 1986).

Particulate Ba in the water column is an excellent modern indicator of high productivity and upwelling, apparently being taken up secondarily by decaying organic matter in the water column, in constant proportion to phosphorus. Where phosphorus is depleted, its concentrations cease to change over the surface of the warmer ocean (Broecker and Peng, 1982; Bishop, 1989; Bruland *et al.*, 1989). Another useful indicator is Cd, which is depleted in normal surface waters but may become concentrated in the oxygen minimum zone, in a pattern which precisely matches that of PO_4 (Broecker and Peng, 1982). Thus, lattice-bound Cd/Ca ratios in the aragonite of Recent scleractinian corals may indicate episodes of upwelling (Shen *et al.*, 1987), as may an increase in skeletal Cd/Ca of foraminifera, especially when combined with a decrease in $\delta^{13}C$ (e.g., Delaney and Boyle, 1987). Both Ni and Zn may also serve as nutrient indicators, the latter having a distribution close to that of dissolved silica (Broecker and Peng, 1982).

Normal marine sediments and associated biogenic apatite are both markedly deficient in the rare earth element Ce when compared with its neighbors La and Nd. In anoxic regions such as the Black Sea, however, the Ce anomaly is reduced and the Ce(anom) in apatite provides a useful measure of the oxidation/reduction system prevailing in surrounding waters (Holser *et al.*, 1986). Rare earth elements can also provide important evidence for the influence of volcanic or extraterrestrial influence in sediments (Orth 1989). Light and heavy rare earth elements tend to be concentrated in the upper and lower parts of the Earth, respectively. Thus, normal shales have La and Ce abundances about 80–150 times that of chondrites, while Yb and Lu are depleted by a factor of about 15–25 (Xu *et al.*, 1989).

A variety of studies have looked at trace elements and rare earth elements (REEs) through the Precambrian–Cambrian boundary interval. In many cases, high concentrations of metals have been found in association with phosphorites, black shales, and sapropelic "stone coals," as, for example, near the base of the Qiongzhusi Formation in south China (i.e., Meishucunian Zones II to III, of inferred early Tommotian age). Here occur high concentrations of V, Co, Mo, Ni, U, Ba, Cu, Ir, Ta, Hf, Th, Rb, Ir, Os, and Au associated with a negative $\delta^{13}C$ signature in carbonates (Hsü *et al.*, 1985; Zhang *et al.*, 1987; Xu *et al.*, 1989). The ratio of Th/U rises markedly in the "boundary clay" at the base of this unit, which overlies a major disconformity surface and could represent a paleosol. Absolute abundances and La/Yb ratios resemble those of chondrites rather than of normal marine shales (Xu *et al.*, 1989). Chinese authors have therefore inferred the impact of an extraterrestrial bolide, pointing to the drastic reduction in faunal diversity (although the latter can be explained in purely ecological terms). These trace element and REE anomalies, however, have chalcophile affinities according to studies by X.-Y. Mao, which does not suggest an extraterrestrial origin (Orth, 1989). Association with the diachronous, anoxic phosphatic base of the Qiongzhusi Formation, and reappearance within the "Ni–Mo polyelement" facies of the Nuititang Formation (Xu *et al.*, 1989; Brasier and Gao, in press), all suggest very slow deposition beneath the pycnocline, where metallic ions from the water column (perhaps bearing the imprint of submarine vents) became concentrated in organic matter, diagenetic sulfide, and phosphate on the sea floor.

Trace element anomalies (especially Ir) have also been traced on the East European Platform (Felitsyn *et al.*, 1989). Here there was a gradual rise to 7.0 ppb of iridium, broadly coincident with the spread of argillaceous, carbonaceous, and volcaniclastic facies. However, the supposed Ir anomalies from the basal Tommotian glauconitic dolomites of Ulukhan-Sulugur in Siberia (Nazarov *et al.*, 1983) have not been substantiated by Orth (1989).

Middle to Upper Cambrian alum shales of Scandinavia are rich in alum and U (Martinsson, 1974), while associated stinkstone carbonates have anomalously high levels of radiogenic U (J. Russell, personal communication 1991). Both Mo-depleted and Mo-enriched carbonates are reported from the Middle Cambrian phosphogenic province of Australia (Donnelly *et al.*, 1988a). In fact the metal concentrations and the ratios of both Mo/C_{org} and U/C_{org} of Early Palaeozoic shales may be up to ten times higher than in the present Black Sea, confirming the higher metal content of many seas at that time (Holser, 1984).

The Ce(anom) has not yet been studied in apatite from late Precambrian to Early Cambrian apatite skeletons. Evidence for widespread local anoxia is indicated, however, from REEs in Late Cambrian apatite (Wright *et al.*, 1986), a trend which seems to have declined gradually through the Paleozoic (Holser, 1984; Holser *et al.*, 1986).

Nor are data available on trace element concentrations in early Cambrian skeletons. Such work would provide a useful guide to nutrient pathways and upwelling events. The potential of Ba is indicated, perhaps, by its very high concentrations in phosphogenic carbonates of Zones I and II (c. Nemakit-Daldynian) in south China (Xu *et al.*, 1989). When traced northward into the basinal deposits of south Qinling, barite is enormously concentrated in sponge spicule cherts and black slates, and conceivably had its origin in hot volcanic brines (Fan *et al.*, 1989).

2.6. Phosphatic Occurrences

The biolimiting nutrients of phosphate, nitrate, and silica tend to behave in a similar manner, being depleted in the photic zone and more concentrated with depth, especially in oxygen-deficient conditions, as during episodic expansion of the oxygen minimum zone (Baturin, 1982; Piper and Codispoti, 1975; Berger *et al.*, 1989). Calcium phosphate then appears to be liberated from decaying organic matter, concentrated mainly through the mechanisms of west coast or equatorial upwelling (Cook and McElhinny, 1979; Sheldon and Burnett, 1981). Major phosphorites therefore tend to be found in facies associations that indicate increasing anoxia with depth (carbonates–cherts–phosphorites–black shales) and are correlated with times of maximum flooding and condensation (Cook *et al.*, 1990).

It has often been inferred that phosphogenesis occurs at the onset of vertical mixing, after episodes of stagnation such as oceanic anoxic events (e.g., Cook and Shergold, 1984, 1986a). There are several reasons to be cautious about the implication of upwelling, however. First, these facies may reappear in reverse order during regression (e.g., Eganov, 1979). Where this occurs, pycnocline movement of a stratified water mass is also a possible factor in phosphogenesis. Second, most Precambrian–Cambrian phosphorites are peritidal and occur in restricted lagoonal embayments, hundreds of kilometers inboard of the ocean margin. Here, the necessary nutrient supply may also have been assisted by high evaporation rates over peritidal flats.

The geographic and stratigraphic distribution of late Precambrian and Cambrian phosphorites (Figs. 1 and 3) has been reviewed more fully by Shergold and Brasier (1986). Phosphorites of Redkinian age are very restricted. They occur in the Doushantuo Formation of south China and, questionably, in the Volta and Kedougou regions of west Africa. Phosphorites of the Cambrian appear to be distributed in six phosphogenic cycles, each relating to major transgressions and the spread of pandemic, cooler-water faunas (Notholt and Brasier, 1986; Shergold and Brasier, 1986). Two major peaks of phosphogenesis span the Nemakit-Daldynian to basal Tommotian, and can be traced across Mongolia, the Altay-Sayan, and the north and eastern margins of the Siberian Platform, into southern Kazakhstan, and thence from Iran, through Pakistan, India, and the Tarim Platform of China, to the Yangtze Platform of south China (Cook and Shergold 1984, 1986b; Shergold and Brasier 1986; Brasier, 1987, 1989a, 1990b; Brasier *et al.*, 1990). Although these all appear to have formed along the Palaeotethyan seaway, their spread does not correspond well with any predicted patterns of upwelling (McKerrow *et al.*, in press), but may be related to the formation of rift basins, evaporites, and anoxic water masses.

A second major phosphogenic interval can be traced across the Lower–Middle Cambrian boundary in south China, Tarim, and Queensland, Australia (Shergold and Brasier, 1986; Donnelly *et al.*, 1990) roughly coincident with the "Hawke Bay" regression. Minor phosphatic occurrences are widespread through the Lower to Middle Cambrian and show a clear relationship with bituminous and glauconitic facies and oxygen-deficient water masses (e.g., Brasier, 1980; Notholt and Brasier, 1986).

In each of these cases, phosphorites appear to have formed on the fringes of oxygen-depleted, trace-element-enriched water masses, especially when these impinged on calcium-enriched waters of carbonate shelves. Thus, models that involve glacially-driven upwelling (Cook and Shergold, 1984) or equatorial upwelling (Cook and McElhinny, 1979; Sheldon and Burnett, 1981) no longer seem to comply with all the evidence.

Whatever the mechanism, it appears to have declined after the Middle Cambrian, when black shales remain common locally but commercial phosphorites are lacking. This phenomenon has been interpreted as indicating the final elimination of old, deep, anoxic waters, present since late Precambrian times (Donnelly *et al.*, 1990). This seems questionable, however, given the phosphatic character of Late Cambrian Alum Shales of Baltica and Avalonia. Climatic change is an alternative explanation, in which prolonged greenhouse conditions led to reduced convective circulation of the oceans (e.g., Fischer and Arthur, 1977; Fischer, 1984a; but see Conway Morris and Rushton, 1988) and to nutrient depletion at the ocean surface.

2.7. Nutrient Biominerals

Skeletal fossils that employ P or Si in biominerals may act as markers for nutrient-enriched, oxygen-depleted water masses. Biological opaline silica deposits (mainly formed by diatoms and radiolarians), for example, are associated with high surface productivity, and although these are scarce in some areas of CTBBS (Bralower and Thiersten, 1984; Thiersten, 1989), they are dominant in others (Kuhnt *et al.*, 1990). At earlier times, the abundance and diversity of phosphatic conodonts has been used. In the Silurian, for example, assemblages from stratified (S-state) waters show reduced diversity and abundance in comparison with mixed episodes (P-states) (Jeppsen, 1990).

The potential of nutrient biominerals as paleoceanographic indicators in the late Precambrian and Cambrian has been explored recently by Brasier (1990a, b, 1991 in press a) and will only be mentioned briefly here. The early skeletal fossil record was enhanced by phosphatization of thin calcareous shells, forming "taphonomic windows" as nutrient-enriched water masses (NEW) spread diachronously across shallow shelves at low latitudes. Associated *Coleoloides* and phosphatic shells suggest that these waters may have been relatively cool.

Silica and phosphate nutrients also made an early appearance as biominerals (Kotlinian to Nemakit-Daldynian), often associated with the cherts, phosphorites, and metaliferous black shales referred to above. Sponge spicularites, phosphatized-organic-matter-dominant (POD) skeletal assemblages, and selected primary phosphatic taxa all provide potential markers for the spread of NEW.

The diversity of "phosphatophilic" skeletons may also be used to track the history of these waters onto the Siberian Platform, which peaked during Nemakit-Daldynian to mid Tommotian times, to be replaced later by more nutrient-depleted waters (NDW) and diversification of archaeocyathan communities. Comparative studies along the Gondwana margins of southern Asia indicate successive peaks of phosphatic taxa and repeated incursions of anoxic water masses. In Avalonia, the acme of phosphatic forms was delayed until Atdabanian–Botomian times, associated with the development of glauconite–phosphate facies (Brasier, in press b).

2.8. The Position of the CCD

Some of the factors that influence inorganic precipitation of carbonate minerals (aragonite, high-Mg calcite, low-Mg calcite) are modeled in Figure 4. Supersaturation and precipitation are enhanced by increasing temperature, such as occur in shallow waters and low latitudes. Carbonates become dissolved at greater depths in relation to increasing dissolved CO_2, lowered temperatures, and raised

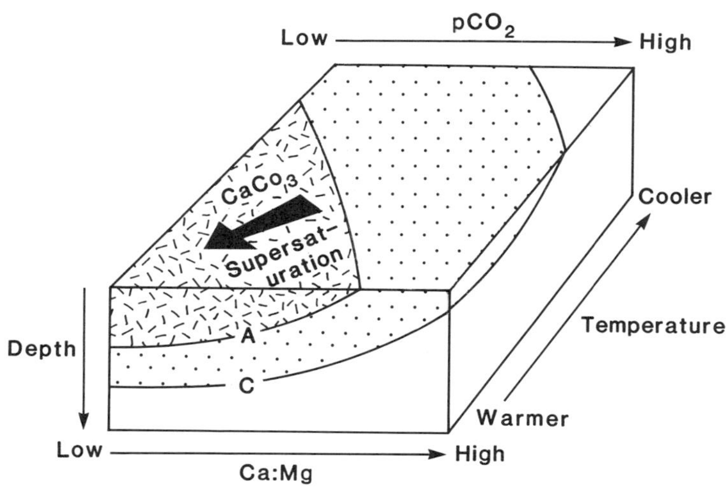

Figure 4. Block diagram illustrating some factors thought to influence the inorganic precipitation and dissolution of calcium carbonate minerals aragonite (A) and calcite (C) in a stratified ocean. Compiled from sources in the text.

hydrostatic pressures, especially below the datum level known as the calcium carbonate compensation depth (CCD).

There is now good evidence that the depth of the CCD has oscillated through time. During orbitally-driven warmer interglacials, and ?tectonically-driven greenhouse states, most carbonate deposition takes place on shelves and the CCD rises in compensation (e.g., Ramsay, 1977; Fischer, 1984a). Although the mechanisms are still far from clear, increased CO_2 in the atmosphere and water column also seems to be indicated at these times (e.g., Pisias and Shackleton, 1984; Kerr, 1984; Fischer, 1984a). During glacials and "icehouse" conditions, both sea levels and CCD appear to be low; the latter may be pumped down by the greater relative input of nannoplanktonic and planktonic foraminiferal carbonate.

As outlined above, the latest Precambrian–Early Cambrian was a time of rising sea level, increasing anoxia, and ameliorating climate. Comparison with similar patterns at later times (e.g., Late Cretaceous, Middle Eocene) might therefore lead one to speculate that the CCD was high at this time, particularly since there was no rain of carbonate detritus from the plankton. If the bulk of CO_2 was derived from bacterial degradation of organic matter, as seems likely, we may agree with Degens *et al.* (1984) and infer from the almost constant $\delta^{13}C$ of Precambrian times that there was a large reservoir of dissolved carbonate.

A high CCD in the oceans has implications for nutrient supply since $CaCO_3$ is the major sink for phosphorus. The latter could therefore accumulate to higher concentrations in the deeper ocean, providing a reservoir for phosphorite formation and phosphatic biomineralization over the boundary interval.

Geochemical changes signaled by changing $\delta^{13}C$ are likely to have involved changes in dissolved CO_2 and a response in the CCD. Deeper-water carbonates of the late Precambrian and Cambrian are likely to have originated as storm or turbidite events and evidence should therefore be sought for discontinuity surfaces and dissolution horizons.

2.9. Carbonate Mineralogy

Field and laboratory studies have shown that the precipitation of aragonite (A), high-Mg calcite (HMC), or low-Mg calcite (LMC) depends largely on the temperature, pCO_2, and Ca:Mg ratio of seawater (Fig. 4), where other factors are equal. Evidence is growing for an oscillating trend in the mineralogy of carbonate ooids through time. Sandberg (1983) has found that A ooids predominated during the late Precambrian, Late Carboniferous–Triassic, and Cenozoic. LMC ooids dominated during Ordovician–Carboniferous and Jurassic–Cretaceous times. These broad trends were inferred by him to compare with broad oscillations in sea level, climate, and granite emplacement. The most likely control is thought to have been plate-tectonically-influenced oscillations in pCO_2 (e.g., Mackenzie and Piggott, 1981; Sandberg, 1983: Tucker 1989), or possibly changes in oceanic Ca:Mg ratio (e.g., Riding, 1985; Tucker, 1989).

The Precambrian–Cambrian transition occupies an interesting place in this story. Tucker (1987, 1989) infers an aragonitic origin for ooids preserved in Vendian carbonates, where calcitic ooids are rare. In the Tommotian of Morocco

(Series Lie de Vin), two-phase calcite–aragonite ooids are most common. Calcitic ooids become more common in Moroccan and Australian limestones of probable Atdabanian age, and they dominate in some high Lower Cambrian carbonates.

This impression of aragonite-dominated carbonate mineralogy has a parallel in the general pattern of dolomite precipitation and distribution (e.g., Tucker, 1982). Dolomites were widespread across Asia in Nemakit-Daldynian times, while calcitic limestones became more common on the Siberian Platform from Tommotian times onward (Rozanov and Sokolov, 1984; Khomentovsky, 1986; Cowie and Brasier, 1989).

The details of these changes have yet to be worked out, but are unlikely to provide a simple pattern. Careful documentation of carbonate mineralogy could, however, provide useful constraints on the speculative models for changes in pCO_2, Ca:Mg, and carbonate biomineralization through the late Precambrian and Early Cambrian.

2.10. The Evidence from Carbon Isotopes

Much interest has focused on the evidence for dramatic fluctuations in $\delta^{13}C$ across the Precambrian–Cambrian boundary interval and their potential for high-resolution stratigraphic correlation (Fig. 5) (e.g., Hsü *et al.*, 1985; Tucker, 1986; Magaritz *et al.*, 1986; Knoll *et al.*, 1986; Aharon *et al.*, 1987, Lambert *et al.*, 1987; Brasier *et al.*, 1990; Kirschvink *et al.*, 1991). This discussion looks briefly at their setting and possible interpretation, beginning with a brief review of work at younger levels.

The carbonaceous matter in Toarcian and Cenomanian–Turonian boundary black shales is, of course, relatively enriched in ^{12}C (the light isotopic fraction preferred by plant photosynthesis). But the $\delta^{13}C$ of contemporaneous organic matter and marine carbonates shows a sharp temporal excursion toward the heavy isotopic fraction, broadly coincident with the maximum extent of anoxia (Jenkyns, 1985; Jenkyns and Clayton, 1986; Schlanger *et al.*, 1987; Arthur *et al.*, 1987). Such "positive" or "heavy" excursions are not easy to interpret, but two major influences are thought to have prevailed: high surface-water productivity (e.g., Broecker and Peng, 1982; Kuhnt *et al.*, 1990) and/or bottom-water anoxia (e.g., Thiersten, 1989; Kuhnt *et al.*, 1990).

Part of the problem of interpretation lies in the recognition of local to regional components of the signal. Fermentative methane generation can bring about extreme fractionation, leading to very light bottom (or pore) waters and heavy residual waters, as can the effects of sulfate-reducing bacteria (e.g., Hudson, 1977; Schidlowski *et al.*, 1983; Tucker and Wright, 1990). Widespread and persistent episodes of evaporite deposition can deplete the ^{12}C reservoir of a basin (e.g., Knoll *et al.*, 1986; Klinkhammer and Lambert, 1989), causing both ^{12}C burial and $^{12}CO_2$ evaporation. Regional stratification in a basin can lead to ^{12}C enrichment of lower layer deposits and ^{13}C enrichment of shallower carbonates (Holser *et al.*, 1986). To these effects must be added fractionation that results from local changes in biota and within microbial mats (e.g., Schidlowski *et al.*, 1983; Degens *et al.*, 1985; Burne and Moore, 1987). These local effects can result, cumulatively, in global isotopic shifts such as those mentioned above during the Mesozoic oceanic anoxic events, or in tandem with the Quaternary climatic oscillations (Shackleton and Pisias, 1985).

The interpretation of "light" or "negative" excursions in $\delta^{13}C$ needs particularly careful consideration. So-called "Strangelove Ocean" perturbations can result from drastically lowered productivity, such as argued for the Cretaceous–Tertiary boundary extinctions (Hsü *et al.*, 1985), but not all negative excursions represent such extraordinary events. For example, increases in the lighter isotopic fraction can be brought about by fractionation within the sulfate-reducing or methanogenic zone (e.g., Hudson, 1977; Benmore *et al.*, 1983), upwelling of light bicarbonate ions and organic matter (Jenkyns and Clayton, 1986), reworking or weathering of organic deposits (Delaney and Boyle, 1987), climate- or ecologically-related changes in fractionation among primary producers (Jenkyns and Clayton, 1986; Benner *et al.*, 1987; Stuiver and Brazuinas, 1987), increased level of carbonate precipitation and/or biomineralization (Fischer and Arthur, 1977; Degens *et al.*, 1985), or changes in volcanic outgassing of CO_2 (Fischer, 1984a).

Carbon isotope stratigraphy must therefore proceed with caution. Wherever possible, the following conditions should be observed: (1) stratigraphic precision (especially biostratigraphy, but also including magnetostratigraphy or other proven method); (2) petrographic, geochemical, and cathodoluminescence (CL) analysis of the sample (effects of bottom waters and diagenesis); (3) basin analysis (effects of facies changes, depositional sequence changes, and basin history), (4) component analysis—comparative analysis of biomineral, cement, and matrix components, or of different kerogen types (effects of local and biological versus global changes; reconstruction of $\Delta\delta^{13}C$ through the water column), and (5) comparative analysis of organic and carbonate components (similar to condition 4). The following data set is sadly deficient according to conditions 3–5, and one of the major aims of IGCP Project 303 must be to address these questions.

Figure 5 presents selected data on $\delta^{13}C$ in carbonates from Precambrian–Cambrian boundary strata. For whatever reason, it is clear that major changes in $\delta^{13}C$ took place over this interval. The signature for the Varangerian ice age tended to be light (Knoll *et al.*, 1986), indicating either conditions of low carbon burial or low productivity. That for the Ediacaran to Nemakit-Daldynian was variable, but included high values, tending toward a maximum in Siberia, Iran, India, the Yangtze Platform, and Morocco.

On the Siberian Platform, this carbon isotope maximum has been well demonstrated at Dvortsy and Ulukhan-Sulugur (Fig. 5) (Magaritz *et al.*, 1986; Magaritz, 1989). The coincidence with shallow-water dolomites and evaporites (e.g., Khomentovsky, 1986) is notable, as is the concordance of adjacent negative excursions with phosphatic preservation of small shelly fossils. The lower Tommotian even yields very light hydrocarbons which may have migrated from underlying evaporite–dolomite facies (e.g., Kontorovitch *et al.*, 1990).

A similar situation is found in the Elburz Mountains of Iran (Brasier *et al.*, 1990). Here, negative values occur associated with hydrocarbon-bearing black shales and with phosphatic carbonates. Maximum values occur near the top of the carbonates, which are broadly contemporaneous with massive evaporites of the Hormuz Salt Formation (e.g., Husseini and Husseini, 1990). The thick succession at Maldeota in the Lesser Himalaya (Fig. 5) also reveals maximum values near the

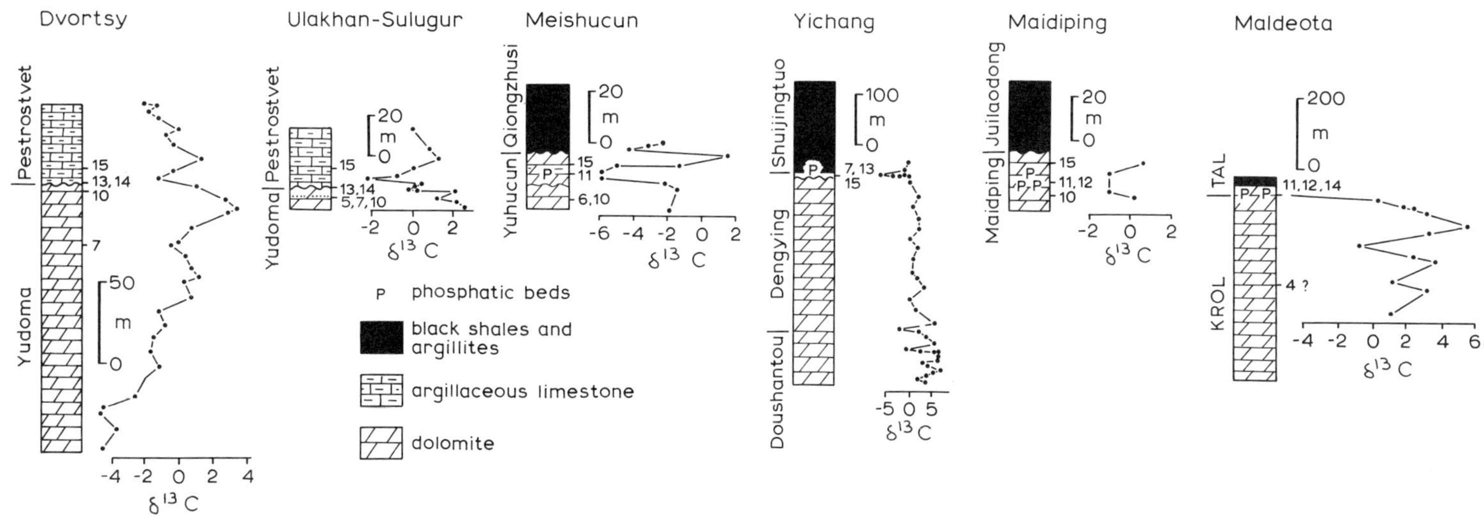

Figure 5. Carbon isotope stratigraphy of Precambrian–Cambrian boundary strata through six major Asiatic sections. note the widespread occurrence of dolomites in the late Nemakit-Daldynian (upper Yudoma, Yuhucun, Dengying, Maidiping, and Krol Formations) and the coincidence of negative signatures with phosphatic or black shale facies. From sources cited in the text. Further and alternative comparisons are given in Brasier *et al.* (1990) and Kirschvink *et* al. (1991).

top of the carbonates, with negative values occurring with phosphorites both before and after (Aharon *et al.*, 1987). Contemporaneous evaporitic facies occur in the Salt Range Formation of Pakistan (Yeats and Lawrence, 1984).

A remarkably similar succession of events can be traced onto the Yangtze Platform in Maidiping, Meishucun, and the Yangtze Gorges (Xu *et al.*, 1989; Brasier *et al.*, 1990). Here the climax of the Dengying dolomites may also include evaporitic pseudomorphs and is seen to contain the maximal values of $\delta^{13}C$. These were punctuated by lower values during transgressions that brought in phosphorites of the Zhongyicun and Badaowan Members, the latter characterized by very low $\delta^{13}C$, trace element anomalies, and organic-rich facies (Hsü *et al.*, 1985; Xu *et al.*, 1989) that suggest the arrival of aged, oxygen-depleted, and nutrient-enriched water masses (Brasier *et al.*, 1990).

The positive excursion within the Dolomie Inferieur of Morocco (Tucker, 1986) appears to lie below Tommotian strata (Latham and Riding, 1990) and may well correlate with the positive excursion at the top of the Nemakit-Daldynian (Latham and Riding, 1990; Brasier *et al.*, 1990; Kirschvink *et al.*, 1991). Less certain is the correlation of carbon isotopes from the Flinders Ranges of South Australia (Tucker, 1989).

Recent work has shown that such excursions continued through the course of the Early Cambrian. Five positive excursions can be traced up to the end of this interval in southeastern Newfoundland (Brasier, Anderson, and Corfield, in press) while a similar number occur in Siberia (M. D. Brasier, and A. Yu. Rozanov, unpublished). If these can be confirmed as of global significance, it follows that similar paleoceanographic factors operated throughout this interval, and no rare event is required to explain them. The challenge, however, will be to confirm their global correlation in the context of facies-controlled biostratigraphic indices.

3. Reconstructing the Biological Phosphorus Cycle

The evidence given above indicates the influence of aging, stratified water masses on shallow platforms through the course of the Cambrian transgressions. Such physical conditions are likely to have caused massive fluctuations in the supply of primary nutrients to the surface layers of the water column (Brasier 1990a, 1991, in press a). How did the biosphere respond to such drastic alternations? What was the impact of biological innovations themselves on these nutrient cycles: did they accelerate or dampen the rate of flow?

The importance of such questions is evident from studies on productivity in the modern ocean, where the concepts of new, regenerated, and export production have developed (e.g., Berger *et al.*, 1989). New production relates to photosynthesis in response to the supply of nutrients (mainly nitrates) brought from below the euphotic zone, as with upwelling. Regenerated production relies on recycling of nutrients within that zone. Export production refers to the amount of production "exported" (by gravitational sinking) below the euphotic zone, and is the factor most applicable to geological questions of carbon and nutrient burial. It tends to be highest in regions of episodic, high new production (e.g., eutrophic conditions) and least in areas of stable, regenerated production (e.g., oligotrophic conditions).

Similar concepts can be applied to the P cycle. In particular, it is necessary to consider the input of "new P" (in this case meaning inorganic sources); the pathways of biologically "regenerated P" through the ecosystem; and "exported P," removed from the ecosystem into sediments. These factors are discussed below for two contrasting conditions of the biosphere: the pre-Redkinian, when Metazoa were scarce or absent (Fig. 6), and the Cambrian, when Metazoa were abundant (Fig. 7). It should be noted from this that three biological factors leading to increased P regeneration or export became common through the Precambrian–Cambrian boundary interval: metazoan biominerals, metazoan activity, and metazoan zooplankton/nekton fecal pellets.

3.1. Terrestrial Runoff

At present, the oceans mainly derive new P from terrestrial weathering and runoff (Figs. 6 and 7) (Froelich *et al.*, 1982). This runoff is likely to be greater during times of low sea level (e.g., Bramlette, 1965; Broecker, 1982). Much of this P is bound to organic matter from terrestrial vegetation (e.g., Vaithiyanathan *et al.*, 1989).

High terrestrial runoff appears to be indicated for the Precambrian–Cambrian boundary interval by heavy strontium isotopes and facies evidence for low sea levels (Donnelly *et al.*, 1990; Brasier, 1989b; Brasier *et al.*, 1990). The relatively

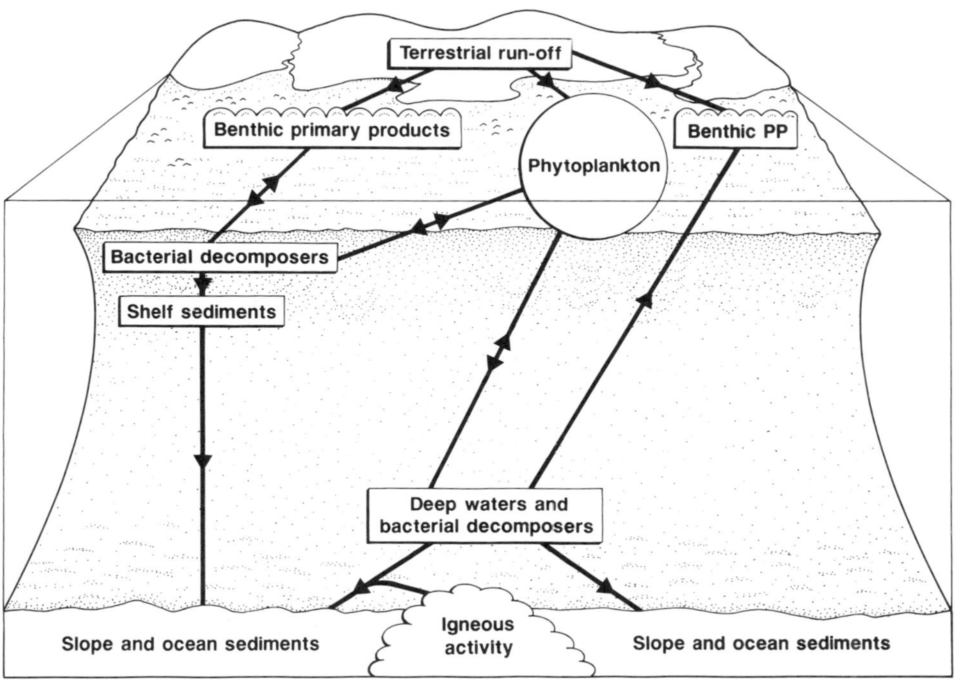

Figure 6. Biogenic pathways of phosphorus, inferred for the biosphere of earlier Proterozoic times, prior to the development of large and abundant invertebrates. See text for discussion.

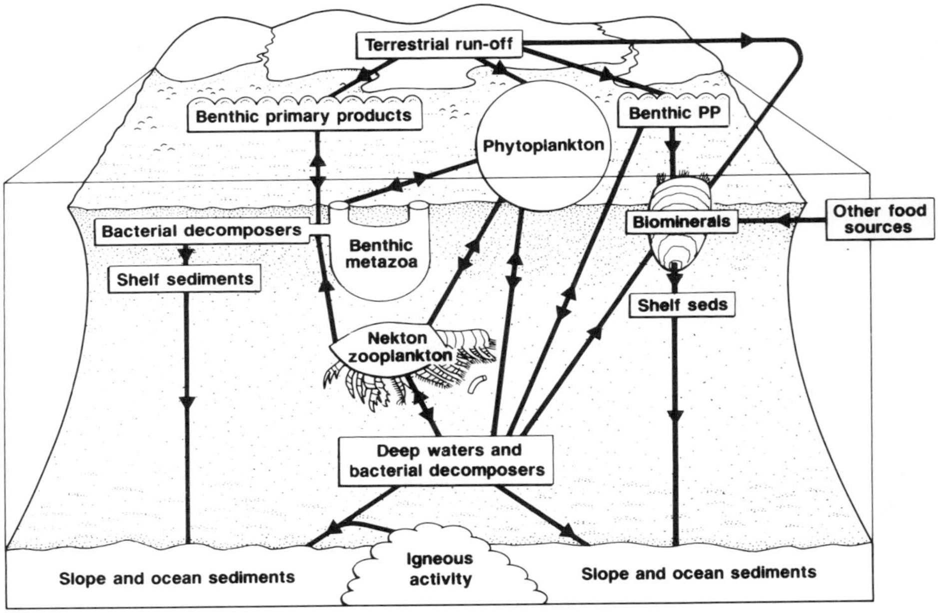

Figure 7. Biogenic pathways of phosphorus inferred for the biosphere after the "Cambrian explosion" of Metazoa. See text for discussion.

poor organic-bound nutrient content of this fluvial detritus is suggested in part by the red coloration of many shallow-water marine sandstones and mudstones (e.g., Ziegler and McKerrow, 1975). With little or no land vegetation in the Precambrian and Cambrian, it seems possible that P mobility from terrestrial sources was considerably less than at present, mainly restricted to inorganic P bound to clays or to iron compounds. If so, P was relatively inaccessible to primary producers in shallow, oxygenated waters unless mobilized within the lower pH and Eh of stagnant sediment or introduced by upwelling from deeper anoxic waters (e.g., Froelich *et al.*, 1982).

3.2. Igneous Removal

Although new Si and Fe may arrive in the oceans via submarine volcanic activity, iron hydroxides formed at the ridges can scavenge P from surrounding seawater, thereby removing it as insoluble compounds (Figs. 6 and 7) (Freely *et al.*, 1990). This process currently results in P depletion of bottom waters, with up to 12% of the terrestrial input to the oceans being removed in this way. Ridge activity and ridge length are therefore factors that may affect removal of dissolved P with time.

It is difficult to demonstrate such a process over the boundary interval. Several supercontinents appear to have been breaking into smaller cratons and forming new ocean basins (e.g., McKerrow *et al.*, in press), notably along the margins of the

Yangtze, Indo-Pak, and Arabian cratons. In the latter case, basalts, evaporites, and black shales occur together in the thick Hormuz Salt Formation of the Persian Gulf (e.g., Wolfart, 1981) and are traceable onshore into phosphatic carbonates (Hamdi *et al.*, 1989). Here it seems that submarine volcanicity erupted into stagnant saline bottom waters whose reducing environment, arguably, would not have depleted the level of dissolved P. Changes in the redox potential of bottom waters would, of course, vary this.

3.3. Benthic Algae and Cyanobacteria

Much P from runoff is taken up by shallow-water primary producers, particularly benthic algae, protists, and cyanobacteria (Figs. 6 and 7). Recent studies of cyanobacteria in hypersaline lagoons show that low P supply encourages mucilaginous benthic mats, while high P supply favors planktonic forms which tend to shade and depress the benthic mats (Javour, 1989).

The existence of comparable late Precambrian and Cambrian microbes is clearly indicated by fossil chert biotas, calcareous algae, stromatolites, thrombolites, and cryptalgal fabrics. By analogy with the present, such communities are likely to have been P-limited because of the availability of nitrates through N fixation by cyanobacteria (e.g., Codispoti, 1989).

A variety of factors can reduce the areal extent of primary producers and N-fixers: increased grazing by metazoans, increased bioturbation, increased turbidity of the water column, reduced extent of the photic zone, and reduced levels of P and other biolimiting nutrients. Of these, the decline of stromatolites in the late Precambrian and Cambrian has been attributed to metazoan activity (Awramik, 1971). It is therefore important to note that both a reduction in cyanobacteria and N fixation, or an increase in bacterial nitrate depletion in surface waters, could lead to the buildup of P in the water column.

3.4. Biominerals

At the present time, much P is extracted from the surface layers of the ocean by the calcitic skeletons of nannoplankton and foraminifera. No such plankton existed in the late Precambrian–Cambrian, but apatitic skeletons were common, especially among benthic Metazoa (Brasier, 1990a). To this form of P extraction must be added that due to phosphatization of calcium carbonate and organic skeletons preserved in and adjacent to phosphogenic environments (Brasier, 1990a). As noted above, this P extraction was associated with the removal of silica nutrients, mainly in the form of sponge spicule cherts.

The role of P in cell biology and biomineralization has been reviewed recently by Simkiss and Wilbur (1989). Phosphate ions are essential for energy metabolism (via ATP), for nucleic acid metabolism, and for buffering ionic chemistry of the cell (via ATP). Cellular storage of P is therefore essential, but a mechanism is needed for Ca^{2+} extraction, to prevent the precipitation of insoluble hydroxyapatite within. Such precipitation is encouraged by the removal of H^+ ions (e.g., from photosynthesis, nitrate reduction, or rising pH) or introduction of excess HPO_4^{2-} ions. Extracellular secretion of apatite could also help with this problem, but requires the secretion of an organic matrix for crystal nucleation, plus organic and chemical modifiers and inhibitors of crystal growth. These inhibitors include a wide range of phosphatic compounds (pyrophosphate, nucleotide polyphosphates, ATP, dentine phosphoproteins, phospholipids, phosphocitrates). Inhibition of hydroxyapatite precipitation is also an essential step in the secretion of calcium carbonate biminerals, especially of calcite (e.g., Simkiss, 1964).

The phosphorus in apatitic skeletons was almost certainly derived from metabolic P, accumulated from the ingestion of coastal plant biomass, phytoplankton, zooplankton or invertebrate benthos (Fig. 7). But the primary source of this P in Cambrian skeletons is uncertain. An association between phosphatic skeletons and ocean-facing habitats (Brasier, 1980), transgressions, and metaliferous alum shale facies (Wright *et al.*, 1986; Brasier, 1990a) is, however, consistent with a source derived from nutrient-enriched bottom waters.

Phosphatic skeletons are relatively insoluble and many are inferred to have become buried in sediments (Fig. 7) with modest change by diagenetic processes. It follows that biomineralization (phosphatic, calcareous, siliceous) provided new pathways for removing nutrients from the water column in the Cambrian.

3.5. Phytoplankton

Modern phytoplankton plays a major role in the uptake of P from terrestrial runoff, from seasonal mixing in the photic zone, or from upwelling of deeper waters (Figs. 6 and 7). These organisms currently include a variety of protists (e.g., diatoms, coccolithophores, dinoflagellates) together with cyanobacterial picoplankton. The latter are likely to have been present in the water column during the late Precambrian and Cambrian, alongside organic-walled acritarchs, interpreted as resting cysts of motile protists (e.g., Downie, 1973). This group evolved rapidly from simple spheroidal vesicles in the late Precambrian to more diverse spinose assemblages in the Cambrian (Moczydlowska and Vidal, 1988).

Modern phytoplankton also plays a part in the export of carbon and nutrients from the photic zone toward bottom waters and deep-sea sediments (Figs. 6 and 7). This export increases with cell size, especially at diameters above 5 μm (e.g., Degens *et al.*, 1985; Legendre and Le Fevre, 1989). In this respect we should note that the latest Precambrian and Cambrian acritarchs are much smaller than their Riphean precursors, with a decline taking place toward the boundary interval (W. J. Schopf and G. Vidal, personal communication 1990). This alone might indicate a lower settling velocity, other factors being equal. In addition, the spiny processes of Early Cambrian acritarchs further increased the drag coefficient. Hence it would appear that phytoplanktonic export of nutrients to the sea floor could have been slight in comparison with earlier Neoproterozoic times.

3.6. Zooplankton and Nekton

Zooplankton and nekton both play a variety of nutrient-related roles in the modern ocean. Their ingestion of phytoplankton controls surface water blooms and recycles P and nitrates in the water column (Fig. 7). Vertical migration in the water column (e.g., through diurnal and breeding cycles) may also recycle nutrients from deeper waters (e.g., Angel, 1989) and release them to phytoplankton.

In the longer term, the production of fecal pellets by zooplankton and nekton assumes considerable importance because it results in the massive removal of carbon and nutrients from surface layers into bottom waters or into deeper sediments (Fig. 7). Here, the size and solubility of fecal particles affects their settling velocity. Thus, small copepod pellets may disaggregate and dissolve before sinking far, while larger fish pellets may sink rapidly without decay (e.g., Peinert *et al.*, 1989). Fecal pellets with a longer residence time in the water column also adsorb greater concentrations of metallic ions from seawater (Cherry *et al.*, 1978).

While some zooplankton and nekton must have existed in the Vendian, evidence for this is scant. The expansion of these niches across the boundary interval is, however, indicated by a variety of evidence: (1) explosive evolution of many extant phyla known to have planktonic larval stages (e.g., Porifera, Mollusca, Annelida, Arthropoda, Brachiopoda, Echinodermata). (2) Pandemic distribution of many extinct taxa (e.g., protoconodonts, pseudoconodonts, mitrosagophorans). (3) Success of agnostoid arthropods of probable planktonic habit (e.g., Robison, 1972). (4) Evidence of phosphatized zooplankton themselves (e.g., Müller and Wallosek, 1986). (5) Evidence for nektonic metazoans in Burgess Shale-type faunas (e.g., Conway Morris *et al.*, 1982). Of these, the agnostoids, at least, would seem to have thrived as adult zooplankton/nekton in outer-shelf to open-ocean habitats.

Less clear is the role of marine invertebrates in the export of nutrients via a rain of fecal pellets to the sea floor. Such pellets are rare in the boundary interval, despite the abundance of suitable phosphatic facies. They abound, however, in phosphatic agnostoid limestones of the Upper Cambrian Alum Shale of Sweden, where arthropods and conodontophorids of planktonic or nektonic life habits were common (e.g., Müller and Walosek, 1986). The zooplanktonic/nektonic origin of these fecal pellets remains untested, but the likely removal of P via agnostoid and other fecal pellets deserves further study.

3.7. Benthic Metazoan Activity

Trace fossils provide the most striking evidence for changes in the behavior of metazoans through the late Precambrian and Cambrian. These activities include burrowing (leading to bioturbation), grazing, scavenging, deposit feeding, suspension feeding, and predation. They are very likely to have affected the pattern and rate of both carbon and nutrient cycling (e.g., Bromley, 1990).

Benthic metazoans derive nutrients from a variety of sources, such as benthic algae, phytoplankton, zooplankton, nekton, and other invertebrate benthos (Fig. 7). Their ingestion of these food items regenerates carbon and nutrients, especially where assisted by bacterial decomposition in the gut and fecal pellets. Burrowing moves these materials between geochemical reaction zones, notably displacing anaerobic zones downward and increasing the amount of faster, aerobic recycling within the upper layers of the sediment (e.g., Aller, 1982). Organic material (including mucus burrow linings and fecal pellets) is also pushed further down into the anaerobic zone by deeper burrows, where regeneration is less efficient, aiding the burial of organic matter (e.g., Reimers, 1989). Thus, the effects of bioturbation on carbon burial will vary according to rates of sedimentation, steepness of the redox gradient, and the depth and intensity of bioturbation (Berger *et al.*, 1989; Jumars *et al.*, 1989).

Bioturbation is also known to alter the flux of different elements within the sediments. Thus, silicon flux is increased by a factor of ten (Schink and Guinasso, 1977; Aller, 1980) owing mainly to an enormous increase in the area of the sediment–water interface represented by burrow walls. The same is likely to be true for P mobility.

The trace fossil record shows that metazoan activity changed dramatically over the Precambrian–Cambrian boundary interval. First, the maximum size of traces tended to increase (Fig. 8). Second, the diversity of traces increased in the Early Cambrian, particularly in shallow-water ichnofacies, to become reduced again during the Middle and Late Cambrian (Crimes 1989; Crimes and Crossley, 1991). Third, the degree of bioturbation shows a stepwise increase in the extent and depth of bioturbation from "pretrilobite" through to Mid Ordovician times, interpreted as the successive occupation of deeper ecological tiers within the sediment (Droser and Bottjer, 1988).

Interesting patterns also emerge from onshore–offshore contrasts in ichnofacies. Deep-water ichnogenera such as *Palaeodictyon* and *Hormosiroidea* are typically found in shallow-water facies of the Early Cambrian (Crimes and Anderson, 1985). This paradox has been interpreted as evidence for a progressive colonization of the deep sea from shallow waters (Crimes, 1974; Crimes and Crossley, 1991), but this no longer seems consistent with evidence for extensive *Nereites* ichnofacies in deeper waters of late Precambrian age (Narbonne and Aitken, 1990).

Did unusual conditions discourage colonization in the deeper offshore muds of Cambrian times? Bioturbation certainly shows a marked reduction offshore in the Early Cambrian (Droser and Bottjer, 1988), while deeper marine sediments

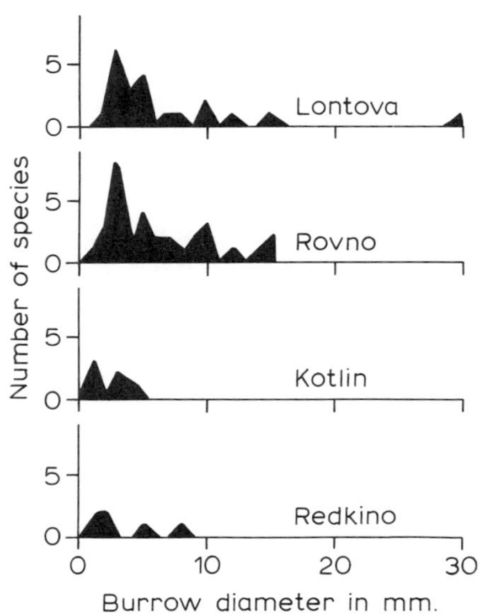

Figure 8. Maximum burrow diameter of invertebrates, obtained from descriptions of taxa at four successive intervals through the Precambrian–Cambrian boundary interval. Based on data from the East European Platform and Avalon Platform in Urbanek and Rozanov (1983), Crimes and Anderson (1985), and Narbonne *et al.* (1987). Kotlinian assemblages appear to record a slight diminution of diameter. The basal Cambrian is taken at the Rovno horizon, where a great increase in diameter occurs.

bear traces that are not only scarcer but smaller. Oxygen depletion, high concentrations of toxic metals, and low food supply are just three possibilities that deserve to be explored further.

3.8. Discussion

The answer to the question, "Did biological innovations accelerate or dampen the flow of nutrients," is, of course, that they did both. A pattern may emerge, however, if we consider the generalized responses of present marine communities to nutrient flux (Table I). Where levels of nutrient supply are low and stable, the minimum amount of P export is encouraged by factors which favor P regeneration. Coral reefs are good examples of such oligotrophic ecosystems, whose health and diversity depend on low nutrient levels (e.g., Hallock and Schlager, 1986). The Early Cambrian equivalent of this ecosystem would appear to be the archaeocyathan–algal reefs of carbonate platforms (e.g., Rowland and Gangloff, 1988).

Where levels of nutrient supply are episodically high, the maximum amount of P leakage is brought about by factors which favor P export. Good examples of such episodically eutrophic ecosystems include diatomaceous and radiolarian blooms (e.g., Kuhnt *et al.*, 1990), leading to biogenic silica deposits. The Cambrian equivalent of this would appear to be assemblages with diverse and abundant phosphatic skeletons, plus siliceous sponges (Brasier, 1990a).

It is therefore tempting to ask whether biological innovations at the Precambrian–Cambrian boundary involved adaptations to improve the biological command of phosphorus. Such a command was surely an essential step toward the increasingly complex tiering of heterotrophic ecosystems (Fig. 7) and these, arguably, marked a huge advance upon the patterns of nutrient flow in the Precambrian (Fig. 6).

TABLE I. Some Generalized Characteristics of Oligotrophic and Eutrophic Ecosystems[a]

	Oligotrophic	Eutrophic
New phosphorus	Low, steady flow	High, unstable flow
Regenerated phosphorus	High	Low
Exported phosphorus	Low	High
Nitrate depletion	Low	High
Cyanobacteria	Benthic mats	Planktonic
Phytoplankton	Low abundance	High abundance
	Cells smaller	Cells larger
Zooplankton	Low abundance	High abundance
	Smaller bodies	Larger bodies
	Smaller fecal pellets	Larger fecal pellets
Bioturbation	Larger, deeper	Smaller, shallower
Reefs	Extensive	Restricted
Oligotrophs	Diverse	Less diverse
Phosphatophiles/silicophiles	Less diverse	More diverse

[a]Mainly from sources in Berger *et al.* (1989), Javour (1989), and Jeppson (1990).

4. Conclusion

A connected sequence of environmental and evolutionary events is therefore beginning to emerge from work that builds upon several decades of international research into the Precambrian–Cambrian boundary (IGCP Project 29) and Proterozoic and Cambrian Phosphorites (IGCP 156). A proper understanding of environmental changes, and their use in high-resolution stratigraphy, is now the responsibility of the project on Precambrian–Cambrian Event Stratigraphy (IGCP Project 303).

Precambrian–Cambrian paleoceanography is in its infancy and the foregoing discussion has therefore concentrated on approaches rather than on speculative models. A huge amount of data awaits to be gathered. Even so, it may be helpful to sketch below a "working hypothesis" that appears to be consistent with these researches.

1. The Precambrian–Cambrian boundary interval straddles a period of rifting between cratons that were formerly clustered in the late Precambrian supercontinents of Gondwana–S. Kazakhastan–China and Laurentia–Baltica–Siberia (e.g., McKerrow *et al.*, in press). Submarine vulcanicity in rifting basins may be one source for the enriched trace elements of Cambrian black shales.

2. Massive deposits of evaporites were forming at low latitudes, the leaching of which provided a possible source for saline bottom waters during this the interval.

3. Stratified water masses, with poorly-oxygenated lower layers, moved episodically over the shelves during the great "Cambrian transgression" (from Kotlinian to Late Cambrian times).

4. Deeper layers of these water masses were greatly enriched in the biolimiting nutrients of phosphorus and silica as well as in trace elements such as Ba, Ni, V, and U.

5. Surface layers of the shelf seas and oceans tended to be relatively depleted in these nutrients, potentially leading toward the buildup of atmospheric carbon dioxide and greenhouse conditions (as suggested by changes in the mineralogy of carbonate ooids). Exceptions to this may relate to regions of coastal upwelling, and to intervals of prolonged stillstand and regressive–transgressive pulses. Huge amounts of nutrients and trace elements were then delivered into shallow-water habitats over large areas.

6. Disappearance of the Ediacara fauna and the ensuing radiation of burrowing traces and early skeletal fossils broadly spanned a major interval of increased stratification and nutrient flux.

7. Siliceous and phosphatic biomineralization was characteristic of these nutrient-enriched waters, as were various taphofacies that favored soft-body preservation. Calcareous archaeocyathan biofacies flourished in more nutrient-depleted waters during the Early Cambrian.

8. It is possible that biomineralization accelerated in response to the effects of transgression and stratification upon nutrient pathways. The primary factor in biomineralization, however, is likely to have been the need for a skeleton for support and protection, in response to increasing predation, grazing, and ecological escalation.

9. The biosphere is inferred to have gained greater command of the phos-

phorus cycle during this interval of highly variable nutrient flux, notably through the innovations of biomineralization, bioturbation, and fecal production.

A major and unresolved question concerns possible changes in atmospheric oxygen over the boundary interval. The evidence reviewed above points to an apparent paradox: that increasing bioturbation and skeletalization took place at times of increasing anoxia. This may seem contrary to the predictive model of Rhoads and Morse (1971), who suggested that the Cambrian radiation compared with changes from anoxic to oxic assemblages in present waters. But it remains a possibility that atmospheric oxygen levels were made to rise by these episodes of greater stagnation, resulting in some very steep gradients through the water column. As Hoffman *et al.* (1990) have pointed out, this possibility may yet be tested by a comparative analysis of the stable isotopes of sulfur and carbon. It is hoped that these and other new data will become available over the next few years through the efforts of IGCP Project 303, allowing this working hypothesis to be tested and refined.

ACKNOWLEDGMENTS. The writer is grateful for the critical comments of those who read through the early drafts of this manuscript, notably John Shergold, Terry Donnelly, Mordecai Magaritz, and Simon Conway Morris. This paper is a contribution to IGCP Project 303 on Precambrian–Cambrian Event Stratigraphy.

References

Aharon, P., Schidlowski, M., and Singh, I. B., 1987, Chronostratigraphic markers in the end-Precambrian carbon isotope record of the Lesser Himalaya, *Nature* **327**:699–702.

Aller, R. C., 1980, Relationships of tube-dwelling benthos with sediment and overlying water chemistry, in: *Marine Benthic Dynamics* (K. R. Tenore and B. C. Coull, eds.), University of South Carolina Press, Chapel Hill, pp. 285–308.

Aller, R. C., 1982, The effects of macrobenthos on chemical properties of marine sediment and overlying water, in: *Animal–Sediment Relationships* (P. L. McCall and M. J. S. Tevesz, eds.), Plenum Press, New York, pp. 53–102.

Anderson, I., 1987, Dinosaurs breathed air rich in oxygen, *New Sci.* **1987**(5 November):25.

Angel, M. V., 1989, Does mesopelagic biology affect the vertical flux?, in: *Productivity of the Ocean: Past and Present* (W. H. Berger, V. S. Smetacek, and G. Wefer, eds.), Wiley, Chichester, England, pp. 155–174.

Arthur, M. A., Schlanger, S. O., and Jenkyns, H. C., 1987, The Cenomanian–Turonian Oceanic Anoxic Event, II. Palaeoceanographic controls on organic matter production and preservation, in: Marine Petroleum Source Rocks (J. Brooks and A. J. Fleet, eds.), *Spec. Publ. Geol. Soc. Lond.* **26**:401–420.

Awramik, S. M., 1971, Precambrian columnar stromatolite diversity: Reflection of metazoan appearance, *Science* **174**:825–827.

Barnola, J. M., Raynaud, D., Korotkevich, Y. S., and Lorius, C., 1987, Vostok ice core provides 160,000-year record of atmospheric CO_2, *Nature* **329**:408–414.

Baturin, G. N., 1982, *Phosphorites on the Seafloor. Origin, Composition and Distribution*, Elsevier, Amsterdam.

Baudin, F., Herbin, J. P., Bassolet, J.-P., Dercourt, J., Lachkar, G., Manivit, H., and Renard, M., 1990, Distribution of organic matter during the Toarcian in the Mediterranean Tethys and Middle East, in: Deposits of Organic Facies (A. Y. Huc, ed.), *Am. Assoc. Pet. Geol. Stud. Geol.* **30**:73–79.

Bengtson, S., and Conway Morris, S., 1984, A comparative study of Lower Cambrian *Halkieria* and Middle Cambrian *Wiwaxia*, *Lethaia* **17**:307–329.

Bengtson, S., Conway Morris, S., Cooper, B. J., Jell, P. A., and Runnegar, B. N., 1990, Early Cambrian fossils from South Australia, *Assoc. Australas. Palaeontol. Mem.* **9**.

Benmore, R. A., Coleman, M. L., and McArthur, J. M., 1983, Origin of sedimentary francolite from its sulphur and carbon isotope composition, *Nature* **302:**516–518.

Benner, R., Fogel, M. L., Sprague, E. K., and Hodson, R. E., 1987, Depletion of ^{13}C in lignin and its implications for stable isotope studies, *Nature* **329:**708–710.

Berger, W. H., Smetacek, V. S., and Wefer, G. (eds.), 1989, *Productivity of the Ocean: Past and Present,* Wiley, Chichester, England.

Berkner, L. V., and Marshall, L. C., 1965, History of major atmospheric components, *Proc. Natl. Acad. Sci. USA* **53:**1215–1225.

Berner, R. A., 1990, Atmospheric carbon dioxide levels over Phanerozoic time, *Science* **249:**1382–1386.

Berry, W. B. N., and Wilde, P., 1978, Progressive ventilation of the oceans—An explanation for the distribution of the Lower Palaeozoic black shales, *Am. J. Sci.* **278:**257–275.

Bishop, J. K. B., 1989, Regional extremes in particulate matter composition and flux: Effects on the chemistry of the ocean interior, in *Productivity of the Ocean: Past and Present* (W. H. Berger, V. S. Smetacek, and G. Wefer, eds.), Wiley, Chicester, England, pp. 117–138.

Bralower, T. J., and Thiersten, H. R., 1984, Low productivity and slow deep-water circulation in mid Cretaceous oceans, *Geology* **12:**614–618.

Bramlette, M. N., 1965, Massive extinction in biota at the end of Mesozoic time, *Science* **148:**1696–2699.

Brasier, M. D., 1979a, The Cambrian radiation event, in: *The Origin of Major Invertebrate Groups* (M. R. House, ed.), Academic Press, New York, pp. 103–159.

Brasier, M. D., 1979b, The early fossil record, *Chem. Br.* **15:**588–592.

Brasier, M. D., 1980, The Lower Cambrian transgression and glauconite-phosphate facies in western Europe, *J. Geol. Soc. Lond.* **137:**695–703.

Brasier, M. D., 1982, Sea level changes, facies changes and the late Precambrian–early Cambrian evolutionary explosion, *Precambrian Res.* **17:**105–123.

Brasier, M. D., 1987, 'Inner Tethyan' Precambrian–Cambrian boundary sequences from China, India, Pakistan and Iran, in: *Abstracts of the International Symposium on the Terminal Precambrian and Cambrian Geology, Yichang,* pp. 1–2.

Brasier, M. D., 1989a, Towards a biostratigraphy of the earliest skeletal biotas, in: *The Precambrian–Cambrian Boundary* (J. W. Cowie and M. D. Brasier, eds.), Clarendon Press, Oxford, pp. 117–165.

Brasier, M. D., 1989b, On mass extinction and faunal turnover near the end of the Precambrian, in: *Mass Extinctions, Processes and Evidence* (S. K. Donovan, ed.), Columbia University Press, New York, pp. 73–88.

Brasier, M. D., 1990a, Nutrients in the early Cambrian, *Nature* **347:**521–522.

Brasier, M. D., 1990b, Phosphogenic events and skeletal preservation across the Precambrian–Cambrian boundary interval, in: Phosphorite research and development (A. J. G. Notholt and I. Jarvis, eds.), *Spec. Pub. Geol. Soc. Lond.* **52:**289–303.

Brasier, M. D., 1991, Nutrient flux and the evolutionary explosion across the Precambrian–Cambrian boundary interval, *Hist. Biol.* **5:** 85–93.

Brasier, M. D., In press a, Nutrient-enriched waters (NEW) and the early skeletal fossil record, *J. Geol. Soc. Lond.* **149.**

Brasier, M. D., and Gao Shuping, In press, *The Stratigraphy of China,* Vol. 4, *The Cambrian System,* Scottish Academic Press, Edinburgh.

Brasier, M. D., and Hewitt, R. A., 1979, Environmental setting of fossiliferous rocks from the uppermost Proterozoic–Lower Cambrian of central England, *Palaeogeogr. Palaeoclimatol. Palaeoecol.* **27:**35–57.

Brasier, M. D., Anderson, M. M., and Corfield, R. M., In press, Oxygen- and carbon-isotype stratigraphy of early Cambrian carbonates in Southeastern Newfoundland and England, *Geol. Mag.* **129.**

Brasier, M. D., Perejon, A., and de San Jose, M. A., 1979, Discovery of an important fossiliferous Precambrian–Cambrian sequence in Spain, *Estudios Geol.* **35:**379–383.

Brasier, M. D., Magaritz, M., Corfield, R., Luo Huilin, Wu Xiche, Ouyang Lin, Jiang Zhiwen, B. Hamdi, He Tinggui, and A. G. Fraser, 1990, The carbon- and oxygen-isotope record of the Precambrian–Cambrian boundary interval in China and Iran and their correlation, *Geol. Mag.* **127:**319–332.

Broecker, W. S., 1982, Ocean chemistry during glacial time, *Geochim. Cosmochim. Acta* **46:**1689–1705.

Broecker, W. S., and Peng, T.-H., 1982, *Tracers in the Sea*, Eldigio Press, Columbia University, Palisades, New York.

Bromley, R. G., 1990, *Trace Fossils, Biology and Taphonomy*, Unwin Hyman, London.

Bruland, K. W., Bienfang, P. K., Bishop, J. K. B., Eglinton, G., Ittekot, V. A. W., Lampitt, R., Sarntheim. M., Thiede, J., Walsh, J. J., and Wefer, G., 1989, Group Report. Flux to the seafloor, in: *Productivity of the Ocean: Past and Present* (W. H. Berger, V. S. Smetacek, and G. Wefer, eds.), Wiley, Chichester, England, pp. 193–216.

Brumsack, H.-J., 1986, Trace metal accumulation in black shales from the Cenomanian/Turonian boundary event, in: *Global Bioevents* (O. H. Walliser, ed.), Springer-Verlag, Berlin, pp. 337–344.

Burne, R. V., and Moore, L. S., 1987, Microbialites: Organosedimentary deposits of benthic microbial communities. *Palaios* **2**:241–254.

Cherry, R. D., Higgo, J. J. W., and Fowler, S. W., 1978, Zooplankton faecal pellets and element residence times in the ocean, *Nature* **274**:246–248.

Cloud, P., 1976, Beginnings of biospheric evolution and their biochemical consequences, *Paleobiology* **3**:351–382.

Codispoti, L. A., 1989, Phosphorus vs. nitrogen limitation of new and export production, in *Productivity of the Ocean: Past and Present* (W. H. Berger, V. S. Smetacek, and G. Wefer, eds.), Wiley, Chichester, England, pp. 377–394.

Conway Morris, S., 1986, The community structure of the Middle Cambrian Phyllopod Bed (Burgess Shale), *Palaeontology* **29**:423–467.

Conway Morris, S., and Fritz, W. H., 1984, *Lapworthella filigrana* n. sp. (incertae sedis) from the Lower Cambrian of the Cassiar Mountains, northern British Columbia, Canada, with comments on possible levels of competition in the early Cambrian, *Palaeontol. Z.* **58**:197–209.

Conway Morris, S., and Jenkins, R. J. F., 1985, Healed injuries in Early Cambrian trilobites from South Australia, *Alcheringa* **9**:167–177.

Conway Morris, S., and Rushton, A. W. A., 1988, Precambrian to Tremadoc biotas in the Caledonides, in: The Caledonian–Appalachian Orogen (A. L. Harris and D. G. Fettes, eds.), *Spec. Pub. Geol. Soc. Lond.* **38**:93–109.

Conway Morris, S., Whittington, H. B., Briggs, D. E. G., Hughes, C. P., and Bruton, D. L., 1982, *Atlas of the Burgess Shale,* Palaeontological Association, London.

Conway Morris, S., Peel, J. S., Higgins, A. K., Soper, N. J., and Davis, N. C., 1987, A Burgess shale-like fauna from the Lower Cambrian of North Greenland, *Nature* **326**:181–183.

Cook, P. J., and McElhinny, M. W., 1979, A re-evaluation of the spatial and temporal distribution of sedimentary phosphate deposits in the light of plate tectonics, *Econ. Geol.* **74**:315–330.

Cook, P. J., and Shergold, J. H., 1984, Phosphorus, phosphorites and skeletal evolution at the Precambrian–Cambrian boundary, *Nature* **308**:231–236.

Cook, P. J., and Shergold, J. H., 1986a, Proterozoic and Cambrian phosphorites—Nature and Origin, in: *Phosphate Deposits of the World,* Vol. 1, *Proterozoic and Cambrian Phosphorites* (P. J. Cook and J. H. Shergold, eds.), Cambridge University Press, Cambridge, pp. 369–386.

Cook, P. J., and Shergold, J. H. (eds.), 1986b, *Phosphate Deposits of the World,* Vol. 1, *Proterozoic and Cambrian Phosphorites* Cambridge University Press, Cambridge.

Cook, P. J., Shergold, J. H., Burnett, W. C., and Riggs, S. R., 1990, Phosphorite research: A historical overview, in: Phosphorite Research and Development (A. J. G. Notholt and I. Jarvis, eds.), *Geol. Soc. Spec. Publ.* **52**:1–22.

Cowie, J. W., and Brasier, M. D., eds., 1989, *The Precambrian–Cambrian Boundary,* Clarendon Press, Oxford.

Crimes, T. P., 1974, Colonization of the early ocean floor, *Nature* **248**:328–330.

Crimes, T. P., 1987, Trace fossils and the Precambrian–Cambrian boundary, *Geol. Mag.* **124**:97–119.

Crimes, T. P., 1989, Trace fossils, in: *The Precambrian–Cambrian boundary* (J. W. Cowie and M. D. Brasier, eds.), Clarendon Press, Oxford.

Crimes, T. P., and Anderson, M. M., 1985, Trace fossils from late Precambrian–early Cambrian strata of southeastern Newfoundland (Canada): Temporal and environmental implications, *J. Paleontol.* **59**:310–343.

Crimes, T. P., and Crossley, J. D., 1991, A diverse ichnofauna from Silurian Flysch of the Aberystwyth Grits Formation, Wales, *Geol. J.* **26**:27–64.

Degens, E. T., and Stoffers, P., 1976, Stratified waters as a key to the past, *Nature* **263**:22–27.

Degens, E. T., and Stoffers, P., 1980, Environmental events recorded in Quaternary sediments of the Black Sea, *J. Geol. Soc. Lond.* **137**:131–138.

Degens, E. T., Kempe, S., and Spitzy, A., 1984, CO_2: A biogeochemical portrait, in: *Handbook of Environmental Chemistry* (O. Hutzinger, ed.), Springer-Verlag, Berlin, pp. 127–215.

Degens, E. T., Kazmierczak, J., and Ittekot, V., 1985, Cellular response to Ca^{++} stress and its geological implications, *Acta Palaeontol. Polonica* **30**:115–135.

De Graciansky, P. C., Deroo, G., Herbin, J. P., Montadert, L., Muller, C., Schaaf, A., and Sigaly, J., 1985, Ocean-wide stagnation episode in the late Cretaceous, *Nature* **308**:346–349.

Delaney, M. L., and Boyle, E. A., 1987, Cd/Ca in late Miocene benethic foraminifera and changes in the global organic carbon budget, *Nature* **300**:156–158.

Donnelly, T. H., Shergold, J. H., and Southgate, P. N., 1988a, Anomalous geochemical signals from phosphatic middle Cambrian rocks in the southern Georgina Basin, Australia, *Sedimentology* **35**:549–570.

Donnelly, T. H., Shergold, J. H., and Southgate, P. N., 1988b, Pyrite and organic matter in normal marine sediments of Middle Cambrian age, southern Georgina Basin, Australia, *Geochim. Cosmochim. Acta* **52**:259–263.

Donnelly, T. H., Shergold, J. H., Southgate, P. N., and Barnes, C. J., 1990, Events leading to global phosphogenesis around the Proterozoic/Cambrian transition, in: Phosphorite Research and Development (A. J. G. Notholt and I. Jarvis, eds.), *Spec. Pub. Geol. Soc. Lond.* **52**:273–287.

Downie, C., 1973, Observations on the nature of the acritarchs, *Palaeontology* **16**:239–259.

Droser, M. L., and Bottjer, D. J., 1988, Trends in depth and extent of bioturbation in Cambrian carbonate marine environments, western United States, *Geology* **16**:233–236.

Duff, K. L., 1975, Palaeoecology of a bituminous shale—the Lower Oxford Clay of Central England, *Palaeontology* **18**:443–482.

Edgell, J. S., 1989, Infracambrian salt basins of Persian Gulf area and their role in prolific hydrocarbon generation, in: *Abstracts. 28th International Geological Congress, Washington, D. C.*, Vol. 1, pp. 433–434.

Eganov, E. A., 1979, The role of cyclic sedimentation in the formation of phosphorite deposits, in: *Proterozoic–Cambrian Phosphorites* (P. J. Cook and J. H. Shergold, eds.), Canberra Publishing and Printing, Canberra, Australia. pp. 22–25.

Fan Delian, Wang Zhonbgchen, Ye Jie, and Zhan Tianwei, 1989, Origin of early Palaeozoic black chert and associated barite-witherite ore deposits in south Qinling, in: *Abstracts. 28th International Geological Congress, Washington, D. C.*, Vol. 1, p. 471.

Farrimond, P., Eglinton, G., Brassell, S. C., and Jenkyns, H. C., 1989, Toarcian anoxic event in Europe: An organic geochemical study, *Mar. Pet. Geol.* **6**:136–147.

Farrimond, P., Eglinton, G., Brassell, S. C., and Jenkyns, H. C., 1990, The Cenomanian/Turonian anoxic event in Europe: An organic geochemical study, *Mar. Pet. Geol.* **7**:75–89.

Fedonkin, M. A., 1988, Paleoichnology of the Precambrian–Cambrian transition in the Russian Platform and Siberia, in: *Trace Fossils, Small Shelly Fossils and the Precambrian–Cambrian boundary* (E. Landing and G. M. Narbonne, eds.), New York State Museum Bulletin 463, p. 12.

Felitsyn, S. B., Sochava, A. V., and Vaganov, P. A., 1989, Ir anomaly at boundary of Ediacara fauna extinction, in: *Abstracts. 28th International Geological Congress, Washington, D. C.*, Vol. 1, p. 478.

Fischer, A. G., 1984, The two Phanerozoic supercycles, in: *Catastrophes and Earth History* (W. A. Berggren and J. A. Van Couvering, eds.), Princeton University Press, Princeton, New Jersey, pp. 129–150.

Fischer, A. G., and Arthur, M., 1977, Secular variations in the pelagic realm, in: Deep Water Carbonate Environments (H. E. Cook and P. Enos, eds.), *Soc. Econ. Paleontol. Min. Spec. Pub.* **25**:19–50.

Frakes, L. A., and Bolton, B. R., 1984, Origin of manganese giants; sea level changes and anoxic–oxic history, *Geology* **12**:83–86.

Freely, R. A., Massoth, G. J., Baker, E. T., Cowen, J. P., Lamb, M. F., and Krogslund, K. A., 1990, The effects of hydrothermal processes on midwater phosphorus distributions in the northeast Pacific, *Earth Planet. Sci. Lett.* **96**:305–318.

Froelich, P. N., Bender, M. L., Luedtke, N. A., Heath, G. R., and Devries, T., 1982, The marine phosphorus cycle, *Am. J. Sci.* **282**:474–511.

Genthon, C., Barnola, N. M., Raynaud, D., Lorius, C., Jouzel, J., Barkov, N. I., Korotkevich, Y. S., and Kotlyakov, V. M., 1987, Vostok ice core: Climatic response to CO_2 and orbital forcing changes over the last climatic cycle, *Nature* **329**:414–418.

Glaessner, M. F., 1984, *The Dawn of Animal Life*, Cambridge University Press, Cambridge.

Hallam, A., 1981, A revised sea-level curve for the early Jurassic, *J. Geol. Soc. Lond.* **138:**735–743.

Hallam, A., and Bradshaw, M. J., 1979, Bituminous shales and oolitic ironstones as indicators of transgressions and regressions, *J. Geol. Soc. Lond.* **136:**157–164.

Hallock, P., and Schlager, W., 1986, Nutrient excess and the demise of coral reefs and carbonate platforms, *Palaios* **1:**389–398.

Hamdi, B., Brasier, M. D., and Jiang Zhiwen, 1989, Earliest skeletal fossils from Precambrian–Cambrian boundary strata, Elburz Mountains, Iran, *Geol. Mag.* **126:**283–289.

Hoffman, A., Gruszczynski, M., and Malkowski, K., 1990, Oceanic delta-13-carbon values as indicators of atmospheric oxygen depletion, *Mod. Geol.* **14:**211–221.

Holser, W. T., 1977, Catastrophic chemical events in the history of the ocean, *Nature* **267:**403–407.

Holser, W. T., 1984, Gradual and abrupt shifts in ocean chemistry during Phanerozoic time, in: *Patterns of Change in Earth Evolution* (H. D. Holland and A. F. Trendall, eds.), Springer-Verlag, Berlin, pp. 123–143.

Holser, W. T., Magaritz, M., and Wright, J., 1986, Chemical and isotopic variations in the world ocean during Phanerozoic time, in: *Global Bioevents* (O. H. Walliser, ed.), Springer-Verlag, Berlin, pp. 63–74.

Hsü, K. J., Oberhansli, H., Gao, J. Y., Sun Shu, Chen Haihong, and Krahenbuhl, U., 1985, 'Strangelove ocean' before the Cambrian explosion, *Nature* **316:**809–811.

Hudson, J. D., 1977, Stable isotopes and limestone lithification, *J. Geol. Soc. Lond.* **133:**637–660.

Husseini, M. I., and Husseini, S. I., 1990, Origin of the Infracambrian salt basins of the Middle East, in: Classic Petroleum Provinces (J. Brooks, ed.), *Spec. Pub. Geol. Soc. Lond.* **50:**279–292.

Javour, B., 1989, *Hypersaline Environments*, Brock/Springer Series in Contemporary Bioscience, Berlin.

Jenkyns, H. C., 1980, Cretaceous anoxic events: From continents to ocean, *J. Geol. Soc. Lond.* **137:**171–188.

Jenkyns, H. C., 1985, The Early Toarcian and Cenomanian–Turonian anoxic events in Europe: Comparisons and contrasts, *Geol. Rundsch.* **74/3:**505–518.

Jenkyns, H. C., and Clayton, C. J., 1986, Black Shales and carbon isotopes in pelagic sediments from the Tethyan Lower Jurassic, *Sedimentology* **33:**87–106.

Jeppson, L., 1990, An oceanic model for lithological and faunal changes tested on the Silurian record. *J. Geol. Soc. Lond.* **147:**663–674.

Jumars, P. A., Altenbach, A. V., De Lange, G. J., Emerson, S. R., Hargrave, B. T., Muller, P. J., Prahl, F. G., Reimers, C. E., Steiger, T., and Suess, E., 1989, Group Report. Transformation of seafloor-arriving fluxes into the sedimentary record, in: *Productivity of the Ocean: Past and Present* (W. H. Berger, V. S. Smetacek, and G. Wefer, eds.), Wiley, Chichester, England, pp. 291–312.

Kerr, R. A., 1984, Carbon dioxide and the control of Ice Ages, *Science* **223:**1053–1054.

Khomentovsky, V. V., 1986, The Vendian System of Siberia and a standard stratigraphic scale, *Geol. Mag.* **123:**334–348.

Kirschvink, J. L., Magaritz, M., Ripperdan, R. L., Zhuravlev, A. Yu., and Rozanov, A. Yu., 1991, The Precambrian/Cambrian boundary: Magnetostratigraphy and carbon isotopes resolve correlation problems between Siberia, Morocco, and South China, *GSA Today* **1:**69–70, 87, 91.

Klinkhammer, G. P., and Lambert, C. E., 1989, Preservation of organic matter during salinity excursions, *Nature* **339:**271–274.

Knoll, A. H., and Swett, K., 1987, Micropalaeontology across the Precambrian–Cambrian boundary in Spitsbergen, *J. Paleontol.* **61:**898–926.

Knoll, A. H., Hayes, J. M., Kaufman, A. J., Swett, K., and Lambert, I., 1986, Secular variation in carbon isotope ratios from Upper Proterozoic successions of Svalbard and East Greenland, *Nature* **321:**832–838.

Kontorovitch, A. E., Mandel'baum, M. M., Surkov, V. S., Trofimuk, A. A., and Zolotov, A. N., 1990, Lena-Tunguska Upper Proterozoic–Palaeozoic petroleum subprovince, in: *Classic Petroleum Provinces* (J. Brooks, ed.), *Spec. Pub. Geol. Soc. Lond.* **50:**473–489.

Kruijs, E., and Barron, E., 1990, Climate model prediction of paleoproductivity and potential source-rock distribution, in: Deposits of Organic Facies (A. Y. Huc, ed.), *Am. Assoc. Pet. Geol. Stud. Geol.* **30:**195–216.

Kuhnt, W., Herbin, J. P., Thurow, J., and Wiedmann, J., 1990, Distribution of Cenomanian–Turonian organic facies in the western Mediterranean and along the adjacent Atlantic margin, in: Deposits of Organic Facies (A. Y. Huc, ed.), *Am. Assoc. Pet. Geol. Stud. Geol.* **30:**133–160.

Lambert, I. B., Walter, M. R., Zang Wenlong, Lu Songnian, and Ma Guogan, 1987, Palaeoenvironment and carbon isotope stratigraphy of the Yangtze Platform, *Nature* **325:**140–142.

Landing, E., 1984, Skeleton of lapworthellids and the suprageneric classification of tommotiids (early and middle Cambrian phosphatic problematica), *J. Paleontol.* **58:**1380–1398.

Landing, E., Myrow, P., Benus, A. P., and Narbonne, G. M., 1989, The Placentian Series: Appearance of the oldest skeletalized faunas in southeastern Newfoundland, *J. Paleontol.* **63:**739–768.

Latham, A., and Riding, R., 1990, Fossil evidence for the location of the Precambrian/Cambrian boundary in Morocco, *Nature* **344:**752–754.

Legendre, L., and Le Fevre, J., 1989, Hydrodynamical singularities as controls of recycled versus export production in oceans, in: *Productivity of the Ocean: Past and Present* (W. H. Berger, V. S. Smetacek, and G. Wefer, eds.), Wiley, Chichester, England, pp. 49–64.

Leggett, J. K., 1980, British Lower Palaeozoic black shales and their palaeooceanographic significance, *J. Geol. Soc. Lond.* **137:**139–156.

Lowenstam, H. A., 1986, Mineralization processes in monerans and protoctists, in: Biomineralization in Lower Plants and Animals (B. Leadbeater and R. Riding, eds.), *Syst. Assoc. Spec. Pub.* **30:**1–18.

Luo Huilin, Jiang Zhiwen, Wu Xiche, Song Xueliang, and Ouyang Lin, 1984, *Sinian–Cambrian Boundary Stratotype Section at Meishucun, Jinning, Yunnan, China,* People's Publishing House, Kunming, Yunnan, China.

Mackenzie, F. T., and Piggott, J. D., 1981, Tectonic controls of Phanerozoic sedimentary rock cycling, *J. Geol. Soc. Lond.* **138:**183–196.

Magaritz, M., 1989, ^{13}C minima follow extinction events: A clue to faunal radiation, *Geology* **17:**337–340.

Magaritz, M., Holser, W. T., and Kirschvink, J. L., 1986, Carbon-isotope events across the Precambrian/Cambrian boundary on the Siberian Platform, *Nature* **320:**258–259.

Martin, J. H., and Fitzwater, S. E., 1988, Iron deficiency limits phytoplankton growth in the north-east Pacific subarctic, *Nature* **331:**341–343.

Martinsson, A., 1974, The Cambrian of Norden, in: *Cambrian of the British Isles, Norden and Spitsbergen* (C. H. Holland, ed.), Wiley, London, pp. 185–283.

McIlreath, I. A., 1977, Accumulation of a Middle Cambrian, deep-water limestone debris apron adjacent to a vertical submarine carbonate escarpment, South Rocky Mountains, Canada, *Spec. Pub. Soc. Econ. Paleontol. Mineral.* (Tulsa) **25:**113–124.

McKerrow, W. S., Scotese, C. R., and Brasier, M. D., In press, Early Cambrian continental reconstructions, *J. Geol. Soc. Lond.* **149.**

McMenamin, M., 1987, The emergence of animals, *Sci. Am.* **255:**94–102.

Middleton, R., and Grant, A., 1990, Heavy metals in the Humber estuary: *Scrobicularia* clay as a pre-industrial datum, *Proc. Yorkshire Geol. Soc.* **48:**75–80.

Miller, R. G., 1990, A palaeoceanographic approach to the Kimmeridge Clay Formation, in: *Deposits of organic facies* (A. Y. Huc, ed.), *Am. Assoc. Pet. Geol. Stud. Geol.* **30:**13–26.

Moczydlowska, M., and Vidal, G., 1988, How old is the Tommotian?, *Geology* **16:**166–168.

Morris, K. A., 1980, Comparison of major sequence of organic-rich mud deposition in the British Jurassic, *J. Geol. Soc. Lond.* **137:**157–170.

Müller, K. J., and Walossek, D., 1986, Arthropod larvae from the Upper Cambrian of Sweden, *Trans. R. Soc. Edinb. Earth Sci.* **77:**157–179.

Narbonne, G. M., and Aitken, J. D., 1990, Ediacaran fossils from the Sekwi Brook area, Mackenzie Mountains, northwestern Canada, *Palaeontology* **33:**945–980.

Narbonne, G. M., and Hoffman, H. J., 1987, Ediacaran biota of the Wernecke Mountains, Yukon, Canada, *Palaeontology* **30:**647–676.

Narbonne, G. M., and Myrow, P., 1988, Trace fossil biostratigraphy and the Precambrian–Cambrian boundary interval, in: *Trace Fossils, Small Shelly Fossils and the Precambrian–Cambrian boundary* (E. Landing, G. M. Narbonne, and P. Myrow, eds.), New York State Museum Bulletin 463, pp. 72–76.

Narbonne, G. M., Myrow, P. M., and Landing, E., 1987, A candidate stratotype for the Precambrian–Cambrian boundary, Fortune Head, Burin Peninsula southeastern Newfoundland, *Can. J. Earth Sci.* **24:**1277–1293.

Nazarov, M. A., Barsukova, L. D., Kolesov, G. M., and Alekseev, A. S., 1983, Iridium abundances in the

Precambrian–Cambrian boundary deposits and sedimentary rocks of Russian Platform, in: *Abstracts of the Lunar and Planetary Science Conference*, Vol. 14, pp. 546–547.

Notholt, A. J. G., and Brasier, M. D., 1986, Proterozoic and Cambrian phosphorites—Regional review: Europe, in: *Phosphate Deposits of the World*, Vol. 1, *Proterozoic and Cambrian Phosphorites* (P. J. Cook and J. H. Shergold, eds.), Cambridge University Press, Cambridge, pp. 91–100.

Orth, C. J., 1989, Geochemistry of bio-event horizons, in: *Mass Extinctions, Processes and Evidence* (S. K. Donovan, ed.), Columbia University Press, New York, pp. 37–72.

Peinert, R., von Bodungen, B., and Smetacek, V. S., 1989, Food web structure and loss rate, in: *Productivity of the Ocean: Past and Present* (W. H. Berger, V. S. Smetacek, and G. Wefer, eds.), Wiley, Chichester, England, pp. 35–48.

Pentecost, A., and Riding, R., 1986, Calcification in cyanobacteria, in: *Biomineralization in lower plants and animals* (B. Leadbeater and R. Riding, eds.), Clarendon Press, London, pp. 73–90.

Pillola, G. L., 1990, Lithologie et trilobites du Cambrien inferieur du SW de la Sardaigne (Italie): Implications paleobiogeographique, *C. R. Acad. Sci. Paris II* **310:**321–328.

Piper, D. Z., and Codispoti, L. A., 1975, Marine phosphorite deposits and the nitrogen cycle, *Science* **188:**15–18.

Pisias, N. G., and Shackleton, N. J., 1984, Modelling the global climate response to orbital forcing and atmospheric carbon dioxide changes, *Nature* **310:**757–759.

Qian Yi and Bengtson, S., 1989, Palaeontology and biostratigraphy of the early Cambrian Meishucunian Stage in Yunnan Province, south China, *Fossil Strata* **24**. [Entire issue]

Raiswell, R., and Al-Biatty, H. J., 1989, Depositional and diagenetic C–S–Fe signatures in early Paleozoic normal marine shales, *Geochim. Cosmochim. Acta* **53:**1147–1152.

Raiswell, R., and Berner, R. A., 1985, Pyrite formation in euxinic and semi-euxinic sediments, *Am. J. Sci.* **285:**710–724.

Raiswell, R., and Berner, R. A., 1986, Pyrite and organic matter in Phanerozoic normal marine shales, *Geochim. Cosmochim. Acta* **50:**1967–1976.

Ramsay, A. T. S., 1977, Sedimentological clues to paleo-oceanography, in: *Oceanic Micropalaeontology* (A. T. S. Ramsay, ed.), Academic Press, London, Vol. 2, pp. 1371–1453.

Reigel, W., Loh, H., Maul, B., and Prauss, M., 1986, Effects and causes in a black shale event—The Toarcian Posidonia Shale of NW Germany, in: *Global Bio-Events* (O. H. Walliser, ed.), Springer-Verlag, Berlin, pp. 267–276.

Reimers, C. E., 1989, Control of benthic fluxes by particulate supply, in: *Productivity of the Ocean: Past and Present* (W. H. Berger, V. S. Smetacek, and G. Wefer, eds.), Wiley, Chichester, England, pp 217–314.

Rhoads, D. C., and Morse, J. W., 1971, Evolutionary and ecologic significance of oxygen-deficient marine basins, *Lethaia* **4:**413–428.

Riding, R., 1985, Calcareous algae near the Precambrian/Cambrian boundary, *NERC News J.* **3:**11–12.

Riding, R., and Voronova, L., 1984, Assemblages of calcareous algae near the Precambrian/Cambrian boundary in Siberia and Mongolia, *Geol. Mag.* **121:**205–210.

Robison, R. A., 1972, Mode of life of agnostid trilobites, *Proceedings. 24th International Geological Congress*, Section 7, pp. 33–40.

Rowland, S., and Gangloff, R. A., 1988, Structure and palaeoecology of Lower Cambrian reefs, *Palaios* **3:**111–135.

Rozanov, A. Yu., and Sokolov, B. S., eds., 1984, *Lower Cambrian Stage Subdivision. Stratigraphy*, USSR Academy of Sciences, Moscow [in Russian].

Runnegar, B., 1982, Oxygen requirements, biology and phylogenetic significance of the late Precambrian worm *Dickinsonia* and the evolution of the burrowing habit, *Alcheringa* **6:**223–239.

Sandberg, P. A., 1983, An oscillating trend in Phanerozoic non-skeletal carbonate mineralogy, *Nature* **305:**19–22.

Schidlowski, M., Hayes, J. M., and Kaplan, I. R., 1983, Isotopic inferences of ancient biochemistries: Carbon, sulfur, hydrogen and nitrogen, in: *Earth's Earliest Biosphere, Its Origin and Evolution* (W. J. Schopf, ed.), Princeton University Press, Princeton, New Jersey, pp. 149–186.

Schink, D. R., and Guinasso, N. L., 1977, Modelling the influence of bioturbation and other processes on calcium carbonate dissolution at the sea floor, in: *The Fate of Fossil Fuel CO_2 in the Oceans* (N. R. Andersen and A. Malahoff, eds.), Plenum Press, New York, pp. 375–399.

Schlanger, S. O., Authur, M. A., Jenkyns, H. C., and Scholle, P. A., 1987, The Cenomanian–Turonian Oceanic Anoxic Event, I. Stratigraphy and distribution of organic carbon-rich beds and the marine

$\delta^{13}C$ excursion, in: Marine Petroleum Source Rocks (J. Brooks and A. Fleet, eds.), *Geol. Soc. Lond. Spec. Pub.* **26:**371–399.

Seilacher, A., 1984, Late Precambrian and early Cambrian Metazoa: Preservational or real extinctions?, in: *Patterns of Change in Earth Evolution* (H. D. Holland and A. F. Trendall, eds.), Springer-Verlag, Berlin, pp. 159–170.

Sepkoski, Jr., J. J., 1979, A kinetic model of Phanerozoic taxonomic diversity. II. Early Phanerozoic families and multiple equilibria, *Paleobiology* **5:**221–251.

Shackleton, N. J., and Pisias, N. G., 1985, Atmospheric carbon dioxide, orbital forcing and climate, *Am. Geophys. Union Geophys. Monogr.* **32:**303–317.

Sheldon, R. P., and Burnett, W. C., 1981, Paleogeography of Phanerozoic phosphorites and extreme upwelling systems, *Am. Assoc. Pet. Geol. Bull.* **65:**991.

Shen, G. T., Boyle, E. A., and Lea, D. W., 1987, Cadmium in corals as a tracer of historical upwelling and industrial fallout, *Nature* **328:**794–796.

Shergold, J. H., and Brasier, M. D., 1986, Biochronology of Proterozoic and Cambrian phosphorites, in: *Proterozoic and Cambrian phosphorites* (P. J. Cook and J. H. Shergold, eds.), Cambridge University Press, Cambridge, pp. 295–326.

Simkiss, K., 1964, Phosphates as crystal poisons of calcification, *Biol. Rev.* **39:**487–505.

Simkiss, K., and Wilbur, K., 1989, *Biomineralization. Cell Biology and Mineral Deposition*, Academic Press, San Diego.

Sokolov, B. S., and Fedonkin, M. A., 1986, *Global Bio-Events* (O. H. Walliser, ed.), Springer-Verlag, Berlin, pp. 105–108.

Southam, J. R., Peterson, W. H., and Brass, G. W., 1982, Dynamics of anoxia, *Palaeogeogr. Palaeoclimatol. Palaeoecol.* **40:**183–198.

Spears, D. A., 1980, Towards a classification of shales, *J. Geol. Soc. Lond.* **137:**125–130.

Stanley, S. M., 1976, Ideas on the timing of metazoan diversification, *Paleobiology* **2:**209–219.

Stuiver, M., and Brazuinas, T. F., 1987, Tree cellulose $^{13}C/^{12}C$ isotope ratios and climate change, *Nature* **328:**58–60.

Summerhayes, C. P., 1987, Organic-rich Cretaceous sediments from the North Atlantic, in: Marche Petroleum Source Rocks (J. Brooks and A. Fleet, eds.), *Geol. Soc. Lond. Spec. Publ.* **26:**301–316.

Thickpenny, A., 1985, "Black Shales" and early Palaeozoic palaeooceanography, *Terra Cognita* **5:**109.

Thickpenny, A., and Leggett, J. K., 1987, Stratigraphic distribution and palaeooceanographic significance of European early Palaeozoic organic-rich sediments, in: Marine Petroleum Source Rocks (J. Brook and A. Fleet, eds.), *Geol. Soc. Lond. Spec. Pub.* **26:**301–316.

Thiersten, H. R., 1989, Inventory of paleoproductivity records. The mid-Cretaceous enigma, in: *Productivity of the Ocean: Past and Present* (W. H. Berger, V. S. Smetacek, and G. Wefer, eds.), Wiley, Chichester, England, pp. 355–376.

Tiwari, R. K., 1987, Higher order eccentricity cycles of the middle and late Miocene climatic variations, *Nature* **327:**219–221.

Towe, K. M., 1970, Oxygen-collagen priority and the early metazoan fossil record, *Proc. Natl. Acad. Sci. USA* **65:**781–788.

Tucker, M. E., 1982, Precambrian dolomites: Petrographic and isotopic evidence that they differ from Phanerozoic dolomites, *Geology* **10:**7–12.

Tucker, M. E., 1986, Carbon isotope excursions in Precambrian–Cambrian boundary beds, Morocco, *Nature* **319:**48–50.

Tucker, M. E., 1987, Changes in carbonate mineralogy across the Precambrian–Cambrian boundary, in: *Abstracts of the International Symposium on the Terminal Precambrian and Cambrian Geology, Yichang*, pp. 78–79.

Tucker, M. E., 1989, Carbon isotopes and Precambrian–Cambrian boundary geology, South Australia: Ocean basin formation, seawater chemistry and organic evolution, *Terra Nova* **1:**573–582.

Tucker, M. E., and Wright, V. P., 1990, *Carbonate Sedimentology*, Blackwell, Oxford.

Tyson, R., Wilson, R. C. L., and Downie, C., 1979, A stratified water column environmental model for the type Kimmeridge Clay, *Nature* **277:**377–380.

Urbanek, A., and Rozanov, A. Yu., 1983, *Upper Precambrian and Cambrian Palaeontology of the East-European Platform*, Wydawnictwa Geologiczne, Warsaw.

Vaithiyanathan, P., Subramanian, V., and Ramanthan, A. L., 1989, Transport and distribution of phosphorus by the rivers of India, in: Phosphorite of India (D. M. Banerjee, ed.), *Geol. Soc. India Mem.* **13:**127–137.

Veizer, J., Holser, W. T., and Wilgus, C. K., 1980, Correlation of $^{13}C/^{12}C$ and $^{34}S/^{32}S$ secular variations, *Geochim. Cosmochim. Acta* **44**:579–587.

Vidal, G., 1989, Are late Proterozoic carbonaceous megafossils metaphytic algae or bacteria?, *Lethaia* **22**:375–379.

Vidal, G., and Knoll, A. H., 1983, Proterozoic plankton, *Mem. Geol. Soc. Am.* **161**:265–277.

Walter, M. R., and Heys, G. R., 1985, Links between the rise of the Metazoa and the decline of stromatolites, *Precambrian Res.* **29**:149–174.

Wilde, P., and Berry, W. B. N., 1986, The role of oceanographic factors in the generation of global bioevents, in: *Global Bio-Events* (O. H. Walliser, ed.), Springer-Verlag, Berlin, pp. 76–91.

Wolfart, R., 1981, Lower Palaeozoic rocks of the Middle East, in: *Lower Palaeozoic Rocks of the Middle East, Eastern and Southern Africa and Antarctica* (C. H. Holland, ed.), Wiley, Chichester, England, pp. 5–130.

Wright, J., Miller, J. F., and Holser, W. T., 1986, Conodont chemostratigraphy across the Cambro-Ordovician boundary: Western United States and southeast China, in: *Conodonts: Investigative Techniques and Applications* (R. L. Austin, ed.), Ellis Horwood, Chichester, England, pp. 256–283.

Xing Yusheng, Duan Chenghua, Liang Yuzuo, Cao Renguan, *et al.*, 1985, *Late Precambrian Palaeontology of China*, Geological Memoirs, Series 2, No. 2, Geological Publishing House, Beijing [in Chinese].

Xu Dao-Yi, Yan Zheng, Sun Yi-Yin, He Jin-Wen, Zhang Qin-Wen, and Chai Zhi-Fang, 1989, *Astrogeological Events in China*, Scottish Academic Press, Edinburgh.

Yeats, R. S., and Lawrence, R. D., 1984, Tectonics of the Himalayan thrust belt in northern Pakistan, in: *Marine Geology and Oceanography of Arabian Sea* (B. U. Haq and J. D. Milliman, eds.), Van Reinhold, New York, pp. 177–198.

Zhang Qinwen, Xu Daoyi, Sun Hing, Yang Zhengzhong, and Chai Zhifang, 1987, The rare event at the Precambrian–Cambrian boundary and the stratigraphic position of this boundary, *Mod. Geol.* **11**:69–77.

Zhuravleva, I. T., Repina, L. N., and Rozanov, A. Yu., 1990, Stage subdivision of the Lower Cambrian, in: *Abstracts of the Third International Symposium on the Cambrian System, Novosibirsk, USSR*, pp. 178–179.

Ziegler, A. M., and McKerrow, W. S., 1975, Silurian marine red beds, *Am. J. Sci.* **275**:31–56.

Chapter 16

The Macroevolution of Phyla

JAMES W. VALENTINE

1. Introduction

Macroevolution is taken to deal with the origins of novel branches of the tree of life at levels above that of species, and with the processes that produce and regulate changes within those branches as they wax and wane through geologic time. This stands in contrast to microevolution, which deals with the processes that produce and regulate heritable change within populations and species, and that can lead to speciation.

A phylum has had to originate as a founding species by definition, and thus via microevolutionary processes. The microevolutionary questions concern how and why such a speciation occurred: what genetic, ecologic, or other features in the population biology of the lineage conspired to produce the new species. There are macroevolutionary questions as well; they concern how and why a particular

JAMES W. VALENTINE • Department of Integrative Biology and Museum of Paleontology, University of California, Berkeley, California 94720.

Origin and Early Evolution of the Metazoa, edited by Jere H. Lipps and Philip W. Signor. Plenum Press, New York, 1992.

new species founded a higher taxon, such as an entire phylum, rather than being just another species within an extant taxon. If the answers indicate that higher taxa arise as simple extrapolations of the microevolutionary processes that we now understand, then while macroevolution would still have important questions of its own—what decides the fates of all the separate branches, why the tree has the architecture that we find, and so on—it would have little to add to our understanding of microevolution. If on the other hand the species that founded phyla have originated under circumstances that are special by modern standards, perhaps under modes or tempos or in environmental situations that do not occur at present, then those events should be of interest to all evolutionists.

Phyla are taken to be monophyletic branches or clades (they may be, for present purposes, paraphyletic) that have evolved unique body plans. Many of the features that characterize these body plans can be regarded as basic architectural "decisions" associated with evolutionary responses to adaptive opportunities, and as engineering solutions to the problems that they posed. These features include symmetry, number and developmental pattern of germ layers, presence and pattern of body cavities and their developmental sources, the sorts of skeletons present—tissue, fluid (hemocoelic or coelomic, hydrostatic or hydraulic), or rigid (internal or external, agglutinated, organic, or mineralized)—and the presence of more specialized features that complement the basic architecture, such as introverts, tentacles, lophophores, or radulae. Worked out by generations of morphologists, evaluations of likely patterns of appearance of these characters have led to groupings of some phyla that most authors believe to be natural ones, but have not led to any consensus on the details of the phylogenetic tree. Despite the addition of molecular evidence, the branching pattern of the phyla remains obscure.

The body plans of the various phyla have been much admired as exquisite engineering achievements and it is sometimes suggested that nearly or quite all of the different architectures that are possible for metazoan bodies have been explored by evolution. At one time it was not uncommon for this idea to be taken even further and used to suggest that the number of phyla must therefore be limited essentially to those that we know—a sort of variation of the doctrine of plenitude. Arguments involving this plenitude of design are not widely supported at present, however. Evolution has often explored a given adaptive opportunity a number of times beginning from different ancestral organisms (flight and sight are classic examples), and although this results in functional and "engineering" convergences, the morphological details and sometimes the biology of the resulting forms can be quite different. It is in keeping with experience at other levels of similarity that the architectural principles of body plans that are particularly well-suited to some potential mode of life and that are reasonably accessible evolutionarily might be employed repeatedly.

There are thus two key features that define phyla, uniqueness of body plan and independence of evolutionary history, and they are closely related in that the unique features must distance the morphology of a phylum from its *ancestors*, but not necessarily from other phyla. Phyla may resemble each other closely, but as long as they had different ancestors and so are independently novel achievements, they are valid phyla. The number of phyla is not limited by the number of possible body plans; the number of phyla reflects the number of times that lineages have produced descendants with body plans that are distinct from the ancestral plan. Just how distinctive a descendant clade should be to be ranked as a phylum is a question best approached historically.

2. Concepts of the Phylum

Early naturalists grouped animals according to their similarities with few guiding principles except, commonly, that of the Aristotelian *scala naturae*. As the number of animal types known to western naturalists became very large, particularly following the voyages of exploration of the fifteenth and sixteenth centuries, hierarchical systems of classification were often used to achieve some semblance of order. Linnaeus (1758) employed a five-tiered hierarchy in the tenth edition of *Systema Naturae:* Kingdom (*Regnum*), Class (*Classis*), Order (*Ordo*), Genus, and Species. Additional hierarchical levels were added to accommodate continued growth in the numbers and kinds of organisms known and to better express their patterns of resemblance. Cuvier (1812) used *embranchement* for the level that we now call the *phylum*, a term introduced by Haeckel (1866) from the greek *phylon*, a tribe or race. In the familiar ironic twist of etymological logic, *phyla* might be better used for *species* and *species* for *phyla*.

The few major animal groupings of these early naturalists (Linnaeus used four classes, Cuvier four embranchements of four classes each) began to be split up as distinctive anatomical plans of invertebrates that were lumped together under the early classifications were revealed by comparative morphologists. With the rise of evolutionary science in the nineteenth century, especially following publication of the *Origin of Species* (Darwin, 1859), it became generally understood that morphological resemblances suggested patterns of descent. While the increasing morphological information permitted the establishment of distinctive groupings of organisms such as phyla, it did not provide any very clear picture of their interrelationships, however. There seemed to be gaps in the family tree; the major branches could not be joined together by morphological intermediates, living or fossil; the phyla had no known ancestors.

Lacking empirical data, one had to join the branches from theoretical considerations. This state of affairs has led to a large number of phylogenetic hypotheses. Developmental evidence has played an important role in these speculations, partly because of recapitulationist theories, but also because it is plausible that phylogenetically conservative features are most likely to be found in early development. Many phylogenetic hypotheses attempt to infer the evolutionary history of the characterizing features of body plans, such as symmetries or various coelomic architectures, and some have concentrated upon hypothetical body plans that contain features believed to represent likely beginnings from which evolutionary divergence could produce the characters actually observed in extant forms. Some of these putative body plans have been *tours de force*, and some remain as plausible ancestral types, but the pattern of branching of the phyla remains hypothetical. Particularly trenchant accounts of the history of the concepts of many phyla, and of some of the various theories that have been entertained as guides to their relationships, are found in Hyman (1940–1959) [the historical introductions to each phylum, and especially Hyman (1950), Vol. 2, Chapter 9, and (not to be missed), Hyman (1959), Vol. 5, pp. 750–754].

The relationship between the concepts of phyla and the branching pattern of the phylogenetic tree can be illustrated by the example of the lophophorate invertebrates, which consist of three living groups, phoronids, brachiopods, and bryozoans. These groups are usually treated as separate phyla (Parker, 1982; Margulis and Schwartz, 1982; Barnes, 1984). The body plan of the last nonphoronid, nonbrachiopod, and nonbryozoan ancestor(s) of each of these respective groups is not known, nor is the body plan of their last common nonlophophorate ancestor; their interrelationships are hypothetical, and they may not form a monophyletic clade. Thus their treatment as separate phyla is prudent. Nevertheless, it appears that their body architectures can be derived from a (hypothetical) common lophophorate ancestor. It has been suspected by a number of workers that the brachiopods may be polyphyletic, articulates and inarticulates at least having separate nonbrachiopod ancestry. A. D. Wright (1979) has postulated that brachiopods have arisen from infaunal, somewhat phoronidlike ancestors seven times. If, in this case, the brachiopod and phoronid body plans are distinct enough to merit recognition as separate phyla, then the brachiopods become an assemblage of seven phyla, with patterns of parallel or convergent evolution. Although this idea has been questioned (Valentine, 1981; and especially Rowell and Caruso, 1985), a polyphyletic brachiopod origin cannot be entirely dismissed. Bryozoa may be polyphyletic also. At the other extreme, all the lophophorate groups are sometimes united as a single phylum [for example, as the Lophophorata (Emig, 1978)]. Such a lumping would relegate the problems of polyphyly among lophophorates to lower ranks but would not dismiss them. Without some knowledge of the body plans of the ancestors of the lophophorate "phyla" and (in this example) of the various "classes" and "orders," we are not in a position to evaluate the rankings that they merit.

Somewhat similar problems occur with taxa other than lophophorates. The arthropod taxa are split into two to several phyla by some authorities on the ground that their last common ancestors were probably not arthropods (Tiegs and Manton, 1958; Cisne, 1974; Manton and Anderson, 1979), contentions that could be settled if we knew or could reconstruct with some certainty the characters of the ancestors of the clades involved.

Clearly, any final assessment of phyla must rest upon two factors. One is an understanding of the body plan of ancestors. This factor would be definitive if our knowledge were complete enough. The second is an understanding of the major branching pattern of the phylogenetic tree. This factor would seem to be within our eventual grasp, at least for many clades, although even if completely understood, it might not remove all ambiguities, for the branching patterns and the origin of the body plans that define the phyla may be discordant. The number of living animal phyla usually recognized is in the low 30s. Three recent synopses of animal classification employ 30 (Parker, 1982), 32 (Margulis and Schwartz, 1982), and 34 (Barnes, 1984). An additional living phylum unknown to science at the time that those synopses were assembled has now been described [the Loricifera (Kristensen, 1983)].

3. Fossils and Phyla

3.1. Durably Skeletonized Body Fossils

The history of the early appearance of many phylum-level groups in the fossil record is detailed in preceding chapters, so that it is necessary to give here only a brief summary from the standpoint of phyla in general. For phyla with durable skeletons, and thus the better fossil records, the earliest fossils commonly represent extinct classes or orders. However, there is no hint in the fossil skeletons that their body plans were different from those of their living representatives. To be sure they are primitive with respect to living taxa and must have differed from them in some important ways. For example, the minute sizes of early brachiopods and bivalve mollusks indicate that they lived under conditions of low Reynolds numbers, and thus they may have functioned somewhat differently than their larger descendants. The skeletal plans of the early members of phyla do not converge upon each other, though, nor do they display definitive indications of ancestral body plans. So far as we can tell, the early members of the skeletonized phyla had acquired their characteristic body plans by the time they first appear.

Most of the durably skeletonized living phyla appear in the Early or Middle Cambrian: brachiopods and mollusks and possibly an arthropod phylum in the Tommotian, and echinoderms and trilobites, later in Lower Cambrian rocks. Graptolithines from the Middle Cambrian probably represent hemichordates. In Lower Ordovician rocks, bryozoan skeletons make their first appearance. The early bryozoans belong to taxa that require durable skeletons for their particular body subplans, but it is possible that soft-bodied bryozoans of more generalized forms were present earlier. In addition to the more or less durable remains assignable to several living phyla, a wide variety of small mineralized skeletons and skeletal elements ("small shelly fossils") and of agglutinated tests, not obviously referable to living phyla, occur in the fossil record, chiefly in Early Cambrian rocks (Rozanov, 1986; Brasier, 1989b; and references therein). There are over two dozen distinctive groups of such fossils that appear in latest Vendian to earliest Cambrian time, and the presence of still other fossils that are poorly preserved or still poorly known indicates that the full range of distinctive, durably skeletonized clades present during this interval has not yet been characterized.

3.2. Soft-Bodied Body Fossils

Soft-bodied phyla fossilize only under unusual circumstances and naturally furnish only a spotty picture of their history. Their pattern of appearance probably has more to do with the temporal distribution of Lagerstätten and with the environments that happen to be represented thereby than with their times or order of origin. Some phyla are not known as fossils at all. Most of these are small soft-bodied forms. The Phylum Platyhelminthes is moderately diverse today and is placed at the base of bilaterian evolution in most phylogenetic models. If this assessment is correct, then platyhelminths (or something very like them) have been extant at least since the Vendian (and possibly much earlier). Platyhelminths, however, possess no substantiated fossil representatives at least until the Miocene (Ruiz and Lindberg, 1989). The earliest well-defined record of ctenophores is from the Lower Devonian (G. D. Stanley and W. Stürmer, 1983); they do not reappear as fossils. The lack of clear Early Paleozoic records of ctenophores tell us as little about their time of origin as does their absence in post-Devonian rocks. We have little direct evidence as to when most soft-bodied phyla arose.

The soft-bodied fauna of the Late Proterozoic consists of relatively few body plans, chiefly appearing to be cnidarian or of cnidarian-grade forms (see Fedonkin, this volume, Chapter 4). There are also forms that have been considered to be either diploblastic or triploblastic. *Tribrachidium*, once suggested to be an ancestral echinoderm (Glaessner, 1962) has been placed with cnidarian-grade forms (Fedonkin, 1982, 1985a, b, d), which seems the more likely affinity. Another soft-bodied form with a symmetry resembling edrioasteroids, *Arakua*, remains as a possible echinoderm ancestor (Gehling, 1987). *Dickinsonia* is a flattened, ovate–elongate form with many divisions or "segments"; it may reach a length of 1 m (Runnegar, 1982a). *Dickinsonia* was first assigned to the Cnidaria (Sprigg, 1947), with which it may well be allied. It has also been considered to be a flatworm (Palij *et al.*, 1979; Conway Morris, 1981) and an annelid (Runnegar, 1982a; Glaessner, 1984) and has been assigned to an extinct phylum, the Proarticulata (Fedonkin, 1985b); its position remains enigmatic.

Other Late Proterozoic forms include the Vendomiidae, which are small and strongly segmented with few divisions, with either alternating or opposing segments. Vendomiids have been assigned to the Arthropoda (Glaessner, 1984) and the Proarticulata (Fedonkin, 1985b). Sprigginids, segmented forms with crescentic anteriors, bilobed axial ridges, and opposing lateral processes, have been assigned to annelids (Glaessner, 1984), arthropods (Fedonkin, 1985b), and a phylum of their own, Annulipodia (Birkett-Smith, 1981). Other possible bilaterians from the Late Proterozoic are morphologically even more enigmatic as well as phylogenetically obscure. Indeed, most of the Late Proterozoic fauna of body fossils have been suggested to belong to a clade of animals that arose independently of the rest of the Metazoa and thus that contains no cnidarians or bilaterians (Seilacher, 1984). Minute cylindrical, annulated fossils from Anhui, eastern China (Chen, 1988), while poorly dated, are evidently Late Proterozoic; if metazoans, they may represent pseudocoelomates.

Unusually well-preserved soft-bodied assemblages are known from the Lower Cambrian (Atdabanian) Chengjiang Formation of Yunnan, China, and from the Middle Cambrian Stephen Formation, including the famous Burgess Shale of British Columbia and its lateral equivalents to the south. A few soft-bodied genera are shared between the Chinese and Canadian localities. The Burgess Shale is particularly rich in body plans (Conway Morris *et al.*, 1982; Whittington, 1985; and references therein). Body fossils of worms of uncertain affinities, possibly including annelids, appear in the Chengjiang Formation [for example, *Fascivermis* (Hou and Chen, 1989)], whereas priapulids and annelids are first definitively represented in the Burgess Shale. Early and Middle Cambrian arthropods are represented by a wide variety of primitive forms with nonmineralized skeletons that are otherwise unknown (see, for example, Briggs, 1983; Whittington, 1985; Hou, 1987a, b). The Stephen Formation, containing the Burgess Shale, has also yielded what are evidently the earliest crustaceans (Briggs, 1978) and chelicerates (Briggs and Collins, 1988) yet known.

Xenusion, from boulders of the Lower Cambrian Kalmarsund Sandstone of the Baltic, and *Aysheaia*, from the Burgess Shale, may represent the marine clade from which the Onychophora arose (Dzik and Krumbiegel, 1989). Two Lower Cambrian (Atdabanian) forms from China, *Luolishania* from the Chenjiang Formation (Hou and Chen, 1989) and an unnamed "armored lobopod" from the Chiungchussu Formation (Ramskold and Hou, 1991), have series of annulated appendages that are overlain by spiny plates; these forms may be related to the onychopohora. Still other rather enigmatic Cambrian forms, which do not appear to be onychopohorans, may nevertheless belong to a lobopod alliance. One of these is *Microdictyon*, long known as a small shelly fossil represented by honeycomb-patterned plates in Tommotian and Atdabanian rocks. These plates have now been found to belong with a vermiform soft-bodied form with tubular limbs discovered in the Chenjiang Fauna (Chen *et al.*, 1989); the plates overlie the limbs as in the preceding forms. Another form is *Hallucigenia* (Conway Morris, 1977) from the Burgess Shale. As noted by Ramskold and Hou (1991), *Hallucigenia* can be reconstructed as possessing a series of limbs with overlying spines, and the various plates and spines in all these forms could conceivably be homologous.

Pikaia, from the Burgess Shale, is probably a cephalochordate (Conway Morris, 1979a). One Burgess Shale fossil, *Wiwaxia*, has an association of spines and scales that can be reconstructed as an extensive dermal skeleton (Conway Morris, 1985). Using *Wiwaxia* as guide, sclerites from among the small shelly fossils of the Tommotian and Atdabanian were reconstructed as forming a similar skeleton, *Halkieria* (Bengston and Conway Morris, 1984). Subsequently, articulated halkieriids were discovered in Greenland (Conway Morris and Peel, 1990) and are similar in general form to the reconstructions, though they bear two cap-shaped shells of a sort that were not previously known to be associated with spine-like sclerites. *Wiwaxia* has been interpreted as a phylodocid polychaete by Butterfield (1990); whether this assignment is correct, and if so what effect it might have on the halkieriids, is not yet clear. Still other body plans represented in the Cambrian differ from living phyla and do not appear to be derivative subtaxa of any of them (*Amiskwia*, *Nectocaris*, etc.).

3.3. Trace Fossils

Late Proterozoic trace fossils are proving to be of major significance in interpreting the early pattern of metazoan radiation (Seilacher, 1956; Crimes, 1974, 1989). Traces record the living activities of animals, especially locomotion, dwelling, and feeding. The ability to perform various of these activities depends very much upon the body plan of the animals, so that the traces provide important clues to the grades of construction that were evolved. For example, macroscopic penetrating vertical burrows (*Skolithos*, *Arenicolites*) imply the presence of a hydrostatic skeleton. Although diploblastic forms are capable of forming penetrating burrows, and other body plans capable of such activity can be hypothesized, common *Skolithos* and *Arenicolites* are most likely to be formed by triploblastic coelomate worms or arthropods. Such burrows were at one time reported from Late Proterozoic rocks and it was therefore postulated that coelomic body plans had appeared at that time (Valentine, 1975, 1977). However, as dating has improved, most of the clear-cut and well-dated penetrating burrows have proven to be of Tommotian age (or younger). Subsequent careful studies of trace fossil assemblages from Late Proterozoic rocks indicate that burrowing at that time was chiefly confined to shallow or semi-infaunal and largely horizontal patterns (for example, Fedonkin, 1977, 1985c; Crimes, 1987; Narbonne and Myrow, 1988; Crimes, this volume, Chapter 6). Evidently the deepest bioturbation of Vendian sediments occurs in the latest and shallowest facies and does include some small and rare penetrating burrows (Fedonkin, 1985b, Crimes, 1989). Some of the sediments with penetrating burrows, attributed by some authors to late Vendian, are believed to be Tommotian on other grounds [see Crimes (1989) and Brasier (1989a) on the age of the Jees Member of the Hartshill Formation Nuneaton sequence, England]. The essentially surficial or at least horizontal Late Proterozoic trails were earlier attributed to such living phyla as annelids and mollusks (Fedonkin, 1977), but contemporary body fossils of the trace-makers are lacking and the presence of the body plans characteristic of those living phyla has not been demonstrated. Traces suggestive of the scratch marks made by arthropods are reported from Late Proterozoic Ediacaran beds (Jenkins *et al.*, 1983). The horizontal trace fossils do suggest that moderately large bodies capable of prolonged creeping had evolved. These organisms were too large for purely ciliary locomotion, and at any rate created grooves in the sediments, implying locomotion involving body-wall musculature antagonized by a hydrostatic skeleton. While such a skeleton could have been tissue-based, it is far more likely to have been hemocoelic (Valentine, 1989) or possibly coelomic.

By contrast with the Late Proterozoic, Cambrian trace fossils become more diverse and indicate deeper burrowing, with vertical penetration of the substrate becoming more common and what appear to be vertical dwelling burrows appearing (Crimes, 1974, 1987; Fedonkin, 1977; Alpert, 1977; Glaessner, 1984). In well-studied sections, Tommotian bioturbation is rare even on the inner shelf, while in Atdabanian and later Lower Cambrian rocks it is common and deeper [averaging 2–3 cm (Droser and Bottjer, 1988a)]. Vertical burrows are only a few centimeters deep and are rare in the earliest Cambrian, increasing to 15 cm in depth with the appearance of presumed trilobite trace fossils and becoming common and much deeper with the appearance of trilobite body fossils (Droser and Bottjer, 1988b).

3.4. Summary of Early Appearance of Phyla

The fossil record is thus helpful but not precise about the timing of the origin of phyla. Multicellular organisms may have originated almost any time after the appearance of eukaryotes, so far as the record is concerned, and were certainly present to found the Ediacaran and correlative faunas. The bulk of Ediacaran fossils are best interpreted as diploblastic. Undoubted coelomate phyla first appear in the Tommotian, in association with a burst of durably skeletonized forms that probably represent many extinct higher taxa. That these myriad novel forms represent a time of evolutionary innovation of new body plans (Valentine, 1977; Runnegar, 1982b; Valentine and Erwin, 1987) and not simply the acquisition of biomineralization by previously existing types (Lowenstam and Margulis, 1980; Glaessner, 1984) is strongly suggested by several lines of evidence. For example, the body plans of brachiopods were probably coevolved with a mineralized exoskeleton (Cloud, 1949; Valentine, 1986), so that the appearance of inarticulate brachiopod skeletons in the early Tommotian should closely coincide with the origin of that group, to the limits of resolution of the fossil record. Similarly, the Tommotian shelly faunas include agglutinated as well as biomineralized skeletons. As agglutinated skeletons do not require sophisticated biomineralization physiologies, their appearance suggests that the organisms that formed them evolved during this Cambrian diversification—that it was new kinds of organisms, and not old lineages newly donning skeleton-armor, that appeared (Signor, 1988). Furthermore, the types of trace fossils and the extent of bioturbation found in early Cambrian rocks imply the appearance of new invertebrate behaviors that can be attributed to the rise of novel body plans, including forms not represented by contemporaneous fossils.

In our few samples of the soft-bodied biota of the Cambrian, we find numerous body plans, many of which have not survived. Their origins seem best referred to the Lower Cambrian diversifications recorded by durable skeletons and trace fossils. All evidence points to a high rate of evolution of new body plans across the Precambrian–Cambrian transition, among which many living phyla are counted. Some of the body plans appearing during that interval required or at least involved durable skeletons, biomineralized or agglutinated, and the appearance of these diverse skeletal forms created the Precambrian–Cambrian boundary and the onset of the Phanerozoic as traditionally defined. The event that is recorded then, however, involved far more major animal groups than just the durably skeletonized living phyla, the appearance of which is only symptomatic of a major radiation of metazoan body plans (Valentine, 1986).

4. Phylogenies of Phyla

4.1. Conventional Phylogenies

The recognition and understanding of the origins of phyla depend in part upon knowledge of their phylogenetic tree. It is a fair statement that the developmental and morphological evidence has gone far toward establishing natural taxa, although all questions are not settled, but has not been very effective in establishing the branching pattern among the groups. As morphological investigations become technically more sophisticated, involving, for example, comparative ultrastructural studies, the usual result is that they confirm the validity of previous taxonomic separations, and if anything emphasize the distinctiveness of the groups involved (e.g., Conway Morris *et al.*, 1985). By emphasizing particular morphological features it is possible to erect models of branching patterns that are consistent for those features, but there has not been general agreement as to which characters are decisive. Comparisons of trees that depict arrangements of phyla proposed by a variety of workers have been made by Clark (1964) and Salvini-Plawen (1982); new phylogenetic schemes continue to appear (e.g., Nielsen, 1985; Bergstrom, 1985; and Fig. 2, herein).

Historically, theories of the origins and evolution of the coelom have played a major role in hypotheses of phylogenetic relationships [for a more thorough history see Vols. 2 (1951) and 5 (1959) of Hyman (1940–1959), Hartman (1963) and Clark (1964)]. One theory, which has a number of variants, derives the coelom from outpocketings of the gut of cnidarians or, in some versions, from the common ancestor of cnidarians and coelomates. This, the enterocoel theory, was first given a thorough treatment by Sedgwick (1884) and has been developed in a variety of ways (Marcus, 1958; Remane, 1963; Jagersten, 1972). Some workers, such as Vandebroek (1952), suggest independent derivation of the coelom in different groups of phyla but from different cnidarian ancestors, while others (e.g., Beklemishev, 1969) favor polyphyletic coelomic origins not only from different cnidarian lineages, but also *de novo* in some superphyletic groupings. In these schemes at least one of the phylogenetic branches proceeds from cnidarians to coelomates and then to acoelomate bilaterians, which must involve loss of the coelom, the circulatory system, and other simplifications. There are many difficulties with such phylogenies (for example, Hyman, 1940–1959, Vol. 5; Hartman, 1963; Clark, 1964), partly because there is no explanation for the origin of mesoderm, and the reduction of the coelom to account for triploblastic acoelomates is *ad hoc* (though see Rieger, 1980).

A contrasting group of theories derive the coelom from an acoelomate bilaterian. The coelom thus originates as some space within mesodermal tissues, and speculations as to the original function of this space have included the following: ducts or lodging for nephridia (Lankester, 1874); reservoirs for metabolic wastes (Snodgrass, 1938); and gonadial sacs [the gonocoel theory (Bergh, 1885; Meyer, 1890; Lang, 1903)], a view that is probably the most widely held today (see Goodrich, 1945). It has also been suggested that the coelom arose *de novo* via schizocoely as a hydrostatic skeleton (Sarvaas, 1933; Livanov, 1955). The usual evolutionary sequence of body plans envisioned in this group of theories is from flatworms (which may or may not have arisen from diploblastic forms) to a coelomate, from which one branch (protostomes) leads to annelids and then arthropods, while another branch leads to the more derived deuterostomes. Lophophorates are commonly placed near this major branch point, sometimes on one branch, sometimes on the other.

In nearly all phylogenetic models (Hyman, 1940–1959, Vol. 1; Hadzi, 1963; Salvini-Plawen, 1969; Jagersten, 1972), arthropods are derived from annelids, and in some schemes (Gutman, 1981; Salvini-Plawen, 1982), the coelom first arises in a segmented annelid or protoannelid and is then modified in derivative phyla. Arthropodization of the protoannelid is visualized as proceeding by increased sclerotization of the cuticle and replacement of skeletal functions of the coelom chiefly by this exoskeleton, with consequent restriction of adult coelomic spaces to ducts (see Gasshoff, 1981). Those hydrostatic functions still required are transferred to the hemocoel [which develops from the blastocoel in annelids and is thus a pseudocoel (Anderson, 1973)]. However, Manton has produced a series of findings (summarized in Manton, 1977) that are difficult to reconcile with this proposed evolutionary pathway. The annelid parapodium depends upon a coelomic hydrostatic skeleton and aciculae, neither of which is found in the limbs of arthropods. Manton divided living arthropods into three major groups, the Uniramia (onychophorans, myriapods, and hexapods), Crustacea, and Chelicerata. Arthropod limbs are either hemocoelic lobopodia and their likely derivatives (Uniramia), which have nothing in common with the mechanisms or gait of annelid parapodia, or involve jointed exoskeletons with intrinsic musculature and sometimes with hemocoelic contributions, which also cannot be foreshadowed by parapodia (Crustacea, Chelicerata). Furthermore, the structure and mode of operation of the limbs and jaws in each of the major arthropod groups indicated to Manton a separate origin for each. Thus the last common ancestor of the arthropods antedates the evolution of jaws and of limbs. Such an ancestor would not have an arthropod body plan, so that the three major groups qualify as phyla. A plausible uniramian ancestor, in Manton's view, would be a segmented lobopodial worm with a hemocoelic hydrostatic skeleton. The capacious "eucoelom" of the annelids was in this view not replaced by a hemocoel in the arthropods—their common ancestor simply never had one.

4.2. Molecular Phylogenies

Following the stimulating paper by Zuckerkandl and Pauling (1965), there have been concerted attempts to investigate phylogenies from molecular evidence. For example, the amino acid sequences of some large protein molecules appear to be highly conserved within taxa but to vary somewhat between taxa; some protein structures vary among closely related forms and others only among distantly related ones. A likely explanation for this pattern is that the DNA base sequence coding for these proteins diverged once the taxa had branched, and that the more different the molecules, the longer since the branching. If this were strictly true, and if given molecules changed monotonically at a constant rate, then the order of branching of clades could be unambiguously determined from the differences between appropriate protein molecules. The changes would be clocklike, and differences among protein structures would date the times of their divergence. Molecules that vary significantly among closely-related taxa must change relatively rapidly (fast-clock molecules), while those that vary only among distantly-related taxa change much more slowly (slow-clock molecules). In phylogenetic analyses of distantly related organisms, study of slow-clock protein and ribosomal ribonucleic acid (rRNA) molecules have proven to be useful.

There have been too few studies involving large samples of animal phyla to support a comparative evaluation of the results. Mammalian orders are the highest animal taxa for which molecular phylogenies have been constructed for a number of proteins. Wyss *et al.* (1987) have evaluated published patterns of the amino acid sequence differences among species belonging to separate mammalian orders by techniques of cladistic analysis. Figure 1 depicts back-to-back cladograms for 19 orders as defined by two different proteins, myoglobin and alpha crystallin A. If the proteins had each diverged steadily following each branching event, the cladograms would form mirror images (with the exception that five more orders were studied by alpha crystallin A) and no branches would cross. Instead, the inferred lines of descent are quite different in the two

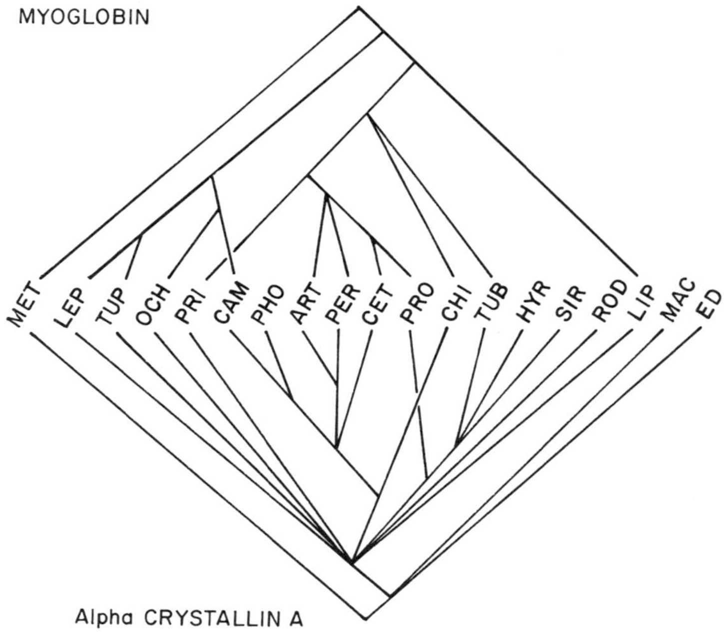

Figure 1. Cladograms of mammalian orders based on analysis of sequences of myoglobin (upper cladogram) and alpha crystallin A (lower cladogram). The relationships inferred from these proteins are rather different from each other, indicating homoplasies within the protein sequences. From left to right the orders are: MET, Metatheria; LEP, Leporidae; TUP, Tupaiidae; OCH, Ochotonidae; PRI, Primates; CAM, Camelidae; PHO, Pholidota; ART, Artiodactyla; PER, Perissodactyla; CET, Cetacea; PRO, Probiscidea; CHI, Chiroptera; TUB, Tublidentata; HYR, Hyracoidea; SIR, Sirenia; ROD, Rodentia; LIP, Lipotyphla; MAC, Macroscelidia; ED, Edentata. [After Wyss *et al.* (1987).]

cladograms. Tubulidentates (TUB; aardvarks) are allied to the Chiroptera (CHI; bats) and other groups according to the myoglobin phylogeny, but are close to the Hyracoidea (HYR; hyraxes) and Sirenia (SIR; manatees and dugongs) according to the alpha crystallin A patterns. Clearly, the sequence changes have not been monotonically divergent, but involve homoplasies (convergences and parallelisms). As the DNA sequences that underlie amino acid sequences have only four character states, reversals (back mutations) should not be uncommon (Fitch and Margoliash, 1967).

A consistency index (CI) that helps in evaluating cladograms was developed by Kluge and Farris (1969); it is a sort of proportion of all character states studied to the number of character states that must have changed. The values of CI may range from 0 (no homoplasies required) to 1. Wyss *et al.* (1987) calculated CIs for 11 cladograms of mammalian orders (Table I); seven of the cladograms were based on proteins, and four upon morphological features. There is little difference in the consistency of branching patterns evaluated by amino acid sequences from those evaluated by the more conventional morphological criteria. There is thus no assurance that the phylogenies inferred from molecular techniques are more accurate than the phylogenies constructed by comparative morphology. The patterns of branching of the taxa are difficult to resolve no matter what technique is used.

TABLE I. Consistency Indexes (CI) for Cladograms of Mammalian Orders Based on Several Protein Sequences and on Several Morphological Evaluations[a]

Cladogram	Interordinal CI
Alpha crystallin A	
McKenna (1987)	0.70 (19)
DeJong (1982)	0.72 (18)
Wyss *et al.* (1987)	0.63 (17)
Myoglobin	
McKenna (1987)	0.50 (22)
Wyss *et al.* (1987)	0.59 (19)
Alpha hemoglobin	0.49 (18)
Beta hemoglobin	0.60 (27)
Morphology	
Novacek (1986)	0.32–0.48 (104–46)
Shoshani (1986)	0.69 (99)
Novacek and Wyss (1986)	0.82 (68)

[a]The higher the CI, the fewer hemoplasies are required to explain the cladogram. The numbers in parentheses are the number of synapomorphies used in the index. After Wyss *et al.* (1987).

In addition to distortions resulting from homoplasies, molecular trees reconstructed from identical data but by different computational methods may be significantly different from each other (for example, Felsenstein, 1978, 1982, 1988; Lake, 1988). A major cause of such differences is that some "treeing" methods are sensitive to differences in the rates of evolution of molecular sequences among lineages. If rates have been very uneven, some treeing methods tend to indicate a close relationship between taxa because they share a common (slow) evolutionary pace rather than that they are actually closely allied. There has long been morphologically-based evidence that rates of evolution have indeed varied significantly, perhaps up to an order of magnitude, among different major animal taxa (Simpson, 1944, 1953; S. M. Stanley, 1979), and there is evidence of variation in neutral mutation rates of at least a factor of five on the molecular level as well (Britten, 1986). The biases resulting from evolutionary rate differentials in different branches of the phylogenetic tree become increasingly serious the earlier the time of branching.

Another difficulty in using molecular data to infer branching patterns in the remote past is that it is necessary to use slowly-evolving molecules that still retain some evidence of their common ancestry. If a series of branchings have occurred within a narrow window of time in some remote period, as appears to have occurred with many of the animal phyla, the molecules may have changed so slowly that they cannot discriminate among the actual sequence of branching events.

A final reservation in interpreting molecular phylogenies of the phyla, which arises even if the inferred branching pattern is correct, is that the divergences indicated by molecules do not necessarily record the origins of the body plans of phyla. A divergence within flatwormlike forms, for example, could conceivably have occurred tens or even hundreds of millions of years prior to the origin of the novel body plans that marked the actual origin of new phyla from separate flatworm stocks.

The most concerted attempts at molecular phylogenies of animal phyla published to date are based on 18S rRNA (Field *et al.*, 1988; Lake, 1989). These molecules have advantages over most proteins for phylogenetic research, for they are found in all phyla (and kingdoms), are not directly subjected to selection associated with different modes of life or environmental features, and contain highly conserved sequences (Woese, 1987). Field *et al.* (1988) used a distance matrix treeing method, which is somewhat sensitive to rate differences. In their tree, cnidarians arise independently of other animals [although see comments in Nielsen *et al.*, (1989)]. Of the remaining groups that they studied, flatworms appear as the earliest bilaterians, followed by the branching of four groups of metazoans. The arthropods form one group, their branch lying deepest in the tree to suggest an early origin. Echinoderms, chordates, and a group termed "eucoelomate protostomes" (annelids, vestimentiferans, mollusks, and brachiopods were studied) form the three other groups. The timing of appearances among and within these groups was difficult to resolve.

Lake (1989) has retreed the rRNA molecular data using a method, termed evolutionary parsimony (Lake, 1988), that is relatively insensitive to evolutionary rate differences. In this new tree metazoans appear to be monophyletic, with the cnidaria as the earliest members. The rRNA sequences of flatworms that have been studied thus far are rather "fast" and their placement not quite significant at the 0.05 level, but a position at the base of the bilateria is nevertheless plausible and certainly not contradicted by the data. The deuterostomes studied—echinoderms, urochordates, and chordates—cluster together, as do arthropodan phyla, while the remaining groups—annelids, vestimentiferans, sipunculans, mollusks, and brachiopods—form a third, somewhat loose cluster arising from the arthropods, with annelids as their deepest branch. In this tree, the main bilaterian branching may have occurred either between the deuterostomes and the other phyla or at any of several places within the arthropodan phyla. If we assume that the division between deuterostomes and protostomes has phylogenetic importance, then it is most plausible to root the higher invertebrates between them, with each branch descending from flatworms but along different paths.

4.3. Fossils and Phylogeny

The architectures of the various phyla are usually suited for and presumably adapted to particular sorts of locomotory functions or, if the group is sessile, to maintaining and functioning at a living site, with suitable modifications or constraints for growth, reproduction, and other activities to be sure. Primitive locomotory and site adaptations are probably largely in support of trophic strategies.

Analyses of the functioning of body designs of phyla or classes is usually made in terms of locomotor and trophic functions, and it is in these contexts that the phylogenetic significance of various skeletal designs, hydrostatic to rigid, can best be evaluated. The relations between design and function in invertebrate locomotion and feeding have been explored by a large number of investigators, for which relevant reviews include Clark (1964), Gray (1968), Trueman (1975), Manton (1977), and Elder and Trueman (1980).

One of the important keys to using the fossil record to aid in the interpretation of the rise of body plans is the timing of appearance of penetrating vertical burrows, in latest Proterozoic or earliest Cambrian time, relative to the appearance of body fossils at various grades of construction. The first such burrows may indicate a maximum age for the evolutionary origin of the annelid-style coelom, and thus the full-fledged annelidan body plan (Valentine, 1975). Whatever animal designs are found in earlier or contemporaneous deposits most probably belong either to ancestral or collateral branches to those of annelids—not to descendants. Of the Late Proterozoic body fossils, the sprigginids are least obscure morphologically and most suggestive of possible early metazoan phylogeny. Sprigginids, with their anteroposterior differentiation, "head" shields, and apparent dorsoventral differentiation in the "head" region, have every appearance of vagile bilaterians. They are large enough that they must have been vascularized, but there is no indication that they burrowed, and evidently they were primarily creepers. There is no evidence that they possessed a coelomic hydrostatic skeleton. Their annulated trunk suggests segmentation of some sort. There is no indication of legs, only short lateral extensions of the body that appear aligned with the trunk segments. These extensions are closely packed and perhaps not separated (Conway Morris, 1979b). The cuticle is inferred to have been lightly sclerotized, to account for the preservation as concave external molds, but to have retained considerable flexibility (Seilacher, 1984).

Such an organism has a body plan that is distinct from those of living phyla, but it is possible to suggest a possible locomotory mode by reference to the principles that operate in living forms (Valentine, 1989). Larger flatworms commonly employ retrograde pedal waves, which can be considered a form of peristalsis (Clark, 1964). In some flatworms the waves are restricted to the lateral margins of the body, where, aided by mucus, they form *points d'appui*, the body of the worm flowing forward, aided by ciliary activity (Pantin, 1950). Sprigginids are larger than these flatworms and though pedal waves might occur across their ventral surface, it is reasonable that they would possess a well-developed accessory gripping mechanism such as provided by rugae, chaetae, parapodia, and other features in various extant worms (see Elder, 1980). The sprigginid lateral segment extensions are obvious candidates to function in that manner. Furthermore, sprigginid fossils commonly display a lateral flexure. In elongated extant forms, such as chilopod uniramians and phyllodocid annelids (Clark, 1964; Gray, 1968), lateral flexures aid locomotion by moving lateral appendages forward around the outside of the flexure while those on the inside grip the substrate. Sections of lateral sprigginid margins could have been lifted off the substrate during an advancing flexure by contraction of transverse muscles similar to those that raise the body margin in flatworms. Turgor in the segments, and in the body in general, was presumably provided by the hemocoel. Perhaps this animal employed pedal waves on its lateral sections for slow locomotion and supplemented or replaced them by lateral undulatory locomotion for more speed or power (Gray, 1968).

Organisms with the general body plan just described, not necessarily sprigginids themselves, would make plausible ancestors for the arthropod phyla (Valentine, 1989). Uniramian lobopodia would be easily derived from the postulated hemocoelic hydrostatic skeleton, modified for somewhat ventrally-directed hemocoelic podia. Crustacean and chelicerate limbs could also be evolved from such a simple ancestral type through progressive sclerotization and jointing as appropriate musculature appeared. As the limbs of these two groups function differently, they need not have arisen from a common jointed-limb ancestor (Manton 1977), but could have evolved independently from a segmented heomocoelic form, developing distinctive approaches to the exploitation of the potential of jointed appendages. Thus the uniramians, crustaceans, and chelicerates might have evolved independently from ancestors with different body plans, and would qualify as phyla (Tiegs and Manton, 1958; Anderson, 1973; Manton, 1977). Trilobites have also been suspected of an independent origin from a nonarthropod (Tiegs and Manton, 1958). A large number of the many arthropod types known from the Cambrian, particularly from the Burgess Shale (Whittington, 1979; Briggs, 1983), cannot be assigned to extant clades (Briggs and Fortey, this volume, Chapter 10). Some (or all) may have radiated from primitive stocks with jointed appendages, from among which arose the lines leading to the major extant arthropod groups, in which case those forms could be considered as classes or orders of arthropods; or they might be considered to be separate phyla if their last common ancestor is judged not to be an arthropod. However many arthropod phyla there may be, they may never have possessed coelomic hydrostatic skeletons.

Arthropods do have spaces that are coelomic in the sense that they originate within mesoderm. Some arthropods possess mesoderm-lined cavities within their larval segments. These spaces disappear during development, some being occluded by the growth of muscles from their walls, some becoming confluent with the hemocoel (which otherwise develops as a pseudocoel), and some eventually forming genital ducts (Anderson, 1973). Whether these spaces ever have a mechanical function in the minute larvae is uncertain. There is no reason to believe that they ever contributed to an adult hydrostatic skeleton in the ancestors of arthropods, although they may have been present in ancestral larvae and have served some function therein, or they may simply have abetted the developmental topology. Ducts and sacs within the mesoderm have obvious functional significance as lodgings for organs or in transporting their products (such as gametes), and may well have evolved to serve precisely those purposes and no others.

The two ancestral types from which annelids are derived in most phylogenetic hypotheses are cnidarians and flatworms (Fig. 2). In flatworms we have a body plan that is sometimes seriated and from which larger, vascularized forms may have been derived. Narrow-bodied worms of this grade may have evolved the sort of segmented body imputed to sprigginids, with the serial internal structure becoming associated with the segments. To achieve an annlidan body plan, coelomic spaces must develop within adult segments, whatever their original source. If the larvae of this ancestral annelid possessed intramesodermal spaces such as annelid and arthropod larvae do, for some functional or developmental reason, these may have given rise to the annelid coelom. At any rate, the annelid coelom may have begun in a rather flattened segmented worm, presumably to enhance locomotion, possibly for more efficient semi-infaunal plowing. It was presumably for penetrating burrowing that a round body and the more capacious annelid eucoelom evolved (Clark, 1964), evidently not until the latest Proterozoic at least, and possibly not until well after the beginning of Cambrian time, if the earliest Cambrian burrows were made by forms with other body plans.

As undoubted coelomate organisms are known from the earliest Cambrian, and as they include oligomerous forms (with regionated but not segmented coeloms) such as the brachiopods, it seems very likely that the oligomerous coelom arose independently of the metameric one. Some authorities have concluded from developmental and morphological evidence that the brachiopod coelom originated *de novo* (Beklemishev, 1969). A possible origin for the large nonmetameric coelom is as a hydrostatic skeleton employed in horizontal creeping via direct peristaltic waves. As Elder (1980) has emphasized, direct peristaltic waves require that the volume of the coelomic cavity be altered locally, and as the volume of truly segmental coelomic compartments must remain constant, metameric architecture is not suitable for direct peristalsis, although it is an efficient and stable system for retrograde peristaltic locomotion. Large unsegmented coelomic spaces may be employed either in direct or retrograde peristaltic waves. Direct peristalsis is associated with locomotion in soft substrates such as flocculent muds and in the movement of some annelids (those with obsolete septa and therefore unsegmented body coeloms) within their burrows (Elder, 1972, 1980), rather than in burrowing in firm sediments. Perhaps small amerous worms (lacking coelomic partitions) developed a coelom for semi-infaunal, largely horizontal progression via direct peristalsis. Unless the coelomic cavity was quite asymmetrical with respect to an anteroposterior axis, these worms would have been round-bodied. According to the molecular phylogeny (Lake, 1989) such worms were either protoannelids or perhaps interstitial unsegmented annelids (Rieger, 1980). From such an ancestor, a phoronidlike form can be derived, adapted to a somewhat sessile existence on firm substrates and feeding upon suspended or resuspended particles via tentaculate body extensions. Septation was developed to separate the trunk functions (locomotion, support, visceral sacs) from those of the feeding tentacles. Brachiopods, bryozoa, and large burrowing phoronids could be descended from such an ancestral type (Valentine, 1981). Presumably the amerous worm ancestral to lophophorates would have descended in turn from vascularized forms derived from flatworms.

Deuterostomes may also have been derived from a similar amerous ancestor. It is likely that the organisms that led to the living deuterostome phyla possessed a trimeric architecture (Clark, 1964), which may or may not have been present at some stage in lophophorate evolution. Lophophorates and deuterostomes share some developmental features (Zimmer, 1964), but there are few developmental clues as to whether their coelomic cavities are homologous or were independently derived, although various rather inconspicuous cavities in lophophorates have been identified as protocoels. Mollusks have been interpreted by several workers as having arisen independently of other coelomates from flatworm ancestors (Stasek, 1972; Salvini-Plawen, 1980). It is possible that they are polyphyletic (Stasek, 1972). Bergstrom (1985) has suggested a phylogenetic model in which many phyla have branched from a conservative ancestral stock that began as flatwormlike and terminated as mollusks, with the deuterostome phyla branching nearest the molluscan end and being the most derived group. The molecular evidence, however, suggests that mollusks have branched from a protoannelidan or desegmented annelidan stock, much as brachiopods, and probably earlier (Field *et al.*, 1988; Lake, 1989).

Dzik and Krumbiegel (1989) have made the interesting suggestion that the derivation of lobopodal organisms and eventually of jointed-limbed arthropods was from pseudocoelomate worms, such as the priapulids, that employed direct peristaltic locomotion. In such an evolutionary scheme, the arthropods are descended from round-bodied worms that burrowed horizontally. A rounded body is thus not associated with penetrating vertical burrowing, and the roundness of onychophorans is simply ancestral, with surficial locomotory devices—gripping protuberances and eventually lobopods—appearing as the group radiated into the epifauna. This scenario also produces an arthropod ancestor—a lobopod—that has no coelomate ancestors but that evolved a complicated hydrostatic hemocoel from a simple hydrostatic pseudocoel. Thus the early lobopods must antedate the sprigginids and other flat-bodied members of the arthropod alliance and the burrowing pseudocoelomate ancestor must antedate the lobopods. It is not clear from trace fossil evidence that such a sequence of activities and body types can be documented, but it seems possible. Another possibility is that the flat- and round-bodied arthropod types have arisen from separate hemocoelic-grade stocks. Improvement in our understanding of the temporal sequence of various body and trace fossil types may eventually disprove or corroborate one or the other of the many phylogenetic hypotheses.

4.4. Summary of the Phylogeny of Phyla

Figure 2 depicts a tree of phyla that attempts to reconcile their patterns of developmental and morphological resemblance and the pattern of appearance of body and trace fossil types with the 18S rRNA molecular tree. The vertical lines are living (labeled) or extinct (short unlabeled or numbered) phyla; their lengths do not imply evolutionary rates. Horizontal lines indicate possible relationships, but their lengths have no meaning except to space phyla out on the page, and their relative heights do not indicate the relative times of origin of phyla proceeding from them except when constrained by a branching sequence. The skeletal coelom is interpreted as polyphyletic, arising in this diagram at least three times—in deuterostomes, annelid–lophophorates, and mollusks (cephalopods). The early history of the lineages is almost certainly more complicated than shown, with many more extinct body plans both in ancestral and collateral positions to the phyla depicted. Pseudocoelomate phyla are not depicted. Based solely on a sort of parsimony of body plan, the lophophorates would seem better placed between the arthropods and the deuterostomes, and thus in a position, common to many phylogenetic models, somewhat intermediate between the two great invertebrate alliances. However, they are relegated to a position within the protostomes because of the molecular evidence. It is expected that this tree will become obsolescent as have its predecessors, but it is hoped that the incorporation of fossil evidence may prove useful to future workers.

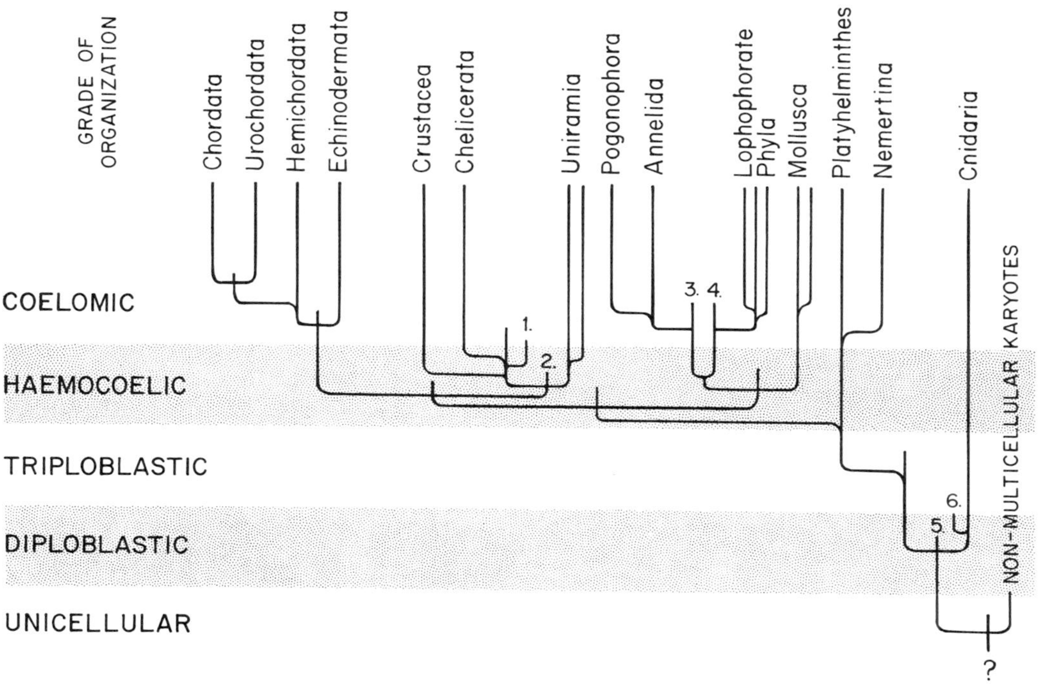

Figure 2. A phylogenetic model of some animal phyla that takes into account grades of organization (horizontal bands) and the sequence of appearance of organisms in the fossil record. Vertical lines represent phyla; living phyla are indicated by long lines and are named; extinct phyla are indicated by short lines and are unnamed or numbered. Horizontal lines indicate the possible branching patterns; their lengths are merely to separate them on the page; their relative heights do not imply relative timing of origin except as indicated by branching sequences. (1) Trilobita, (2) a sprigginid-grade phylum, (3) a narrow-bodied, crawling worm phylum with seriated organs and segmented longitudinal body musculature, (4) a narrow-bodied creeping worm phylum, presumably minute, that has suppressed seriation and is not segmented, (5, 6) diploblastic groups of the Late Proterozoic.

5. Modes of Origin of Phyla

5.1. Summary of the Problems

The fact that phyla and other major taxa remain cryptogenetic whether studied from the perspectives of comparative developmental and/or adult morphology, of molecular evolution, or of the fossil record, seems unlikely to be a fluke. Even though we may eventually learn the branching patterns of the phyla, it would appear that there is something about the origin of phyla that makes this a particularly difficult task; the evolutionary processes involved have obscured their tracks. Although workers using each sort of evidence can cite special circumstances to explain their difficulties, it is possible that the problems may arise from a common source. A feature that is conspicuously common to explanations arising from each of these various approaches to the major branching patterns is that the branches originated relatively abruptly, and that many of the branches originated within a narrow window of geologic time [for a review of our knowledge of the age of important datums in the Late Proterozoic–Early Cambrian geologic record see Cowie and Harland (1989)]. In considering models of rapid evolution, two aspects have been stressed by different workers. One is an internal aspect, explaining rapid evolution by considering what mechanisms might account for large, rapid genome changes. The other aspect is external, concentrating on environmental stresses or opportunities, and the sorts of selection required to achieve novelties rapidly. Granted that each of these features is important, there is still a third, historical aspect: the evolution of the proper grade of complexity to exploit those mechanisms and opportunities.

5.2. Internal Factors

Two end-members of genomic change for rapid evolution are (1) numerous small effects at high rates of successful mutations and recombinations (e.g., Simpson, 1944), and (2) few large effects from one or a few successful mutations (e.g., Goldschmidt, 1940). The effects of population size and structure on evolutionary rates under selective or drift regimes have been explored by many geneticists; especially rapid change may occur, for example, in populations that are small and under the influence of drift or are large and with subpopulations under the influence of strong selection (S. Wright, 1982b). Whether the stem populations of phyla answered to either of these descriptions is not known. Presumably many species have experienced each of these situations during the Phanerozoic Era without producing novel phyla; there must be other factors above and beyond these conditions for rapid evolution of body plans.

Rapid evolution via many small genetic changes requires a high degree of genetic variability and/or a high mutation rate. That mutation rates were much higher in very early organisms than in contemporary ones is a possibility. The demonstration by Britten (1986) that rates of DNA sequence change at neutral sites differ by a factor of five among living organisms, and tend to decline in groups that evolved most recently, suggests that rates may have been much higher still in primitive lineages. The causes and mechanisms of the decline in rate of sequence change are not certainly known, but variations in the efficiency of DNA replication and repair, perhaps partly residing in variation in the structure of DNA polymerase and in other molecules associated with these processes, are likely candidates (Goodman *et al.*, 1984; Britten, 1986). The possibility of elevated mutation rates in primitive organisms is of special interest because the chances of achieving independent mutations simultaneously (or at least contemporaneously) rises faster than the mutation rate (Woese *et al.*, 1985; Endler and McLellan, 1988). Thus the chances of evolutionary change when two or more mutations are required to raise fitness are improved disproportionately.

Buss (1987) has pointed out that in some invertebrate groups, cells may be recruited into the germ line relatively late in ontogeny. In such lineages the chances of mutations entering the gene pool are greatly increased, for the cells so recruited may have undergone great numbers of divisions and have a relatively high likelihood of having acquired some mutations. If the growth of such a variant cell type enhanced the adaptation of an ontogeny, then natural selection at the organismic level would also favor it; indeed the fixation of mutations for such variant cell lines would result in the evolution of differentiation. Those animal clades that exhibit late access of cell lineages to the germ line today, and thus have the most opportunity to reproduce developmental innovations in cell lines, tend to be those that have radiated into the most body subplans and that have become most diverse, though there are a number of exceptions (Buss, 1987, 1988). With higher inherent mutation rates, the evolution of novel cells and tissues and the exploration of distinctive developmental geometries could have been even more rapid and may have led to successful novelties in primitive metazoan lineages.

The notion that mutations with major effects have played evolutionary roles has had a mixed history. An argument against the contribution of major mutations is that they are almost always disruptive, so that useful ones would be exceedingly rare and likely to be infertile when crossed with parental genotypes. However, heterochrony or heterotopy may commonly be associated with the origin of novelty (Gould, 1977; Arthur, 1984, 1988; McKinney and McNamara, 1990) and such a process, involving the switching of the timing, location, or dosage of genes, perhaps of blocks of genes that are already coadapted, might not be very disruptive of developmental processes. If the changes enhance fitness, they may be accrued through relatively small steps, but with each step involving a shift in the expression of a large number of genes. Finally, there is a possibility that a horizontal transfer of mutations may occur, perhaps via retroviruses, bypassing normal reproduction and therefore evading the problem of fitness of mutant and parental generations. The role of any such mechanisms in the origin of higher taxa remains highly speculative.

5.3. External Factors

The usual role envisaged for external factors is in determining the timing, rather than the pace or extent, of the early metazoan radiations (Signor and Lipps, this volume, Chapter 1). An important exception is found in speculation regarding the role of the breadth of environmental opportunity. There is little doubt but that these radiations occurred in a biosphere in which relatively few metazoan species existed. On the other hand, higher taxa were appearing at an exceptional rate. This situation led to an estimate that at some times during the Precambrian–Cambrian transition, every fortieth species founded a new class or phylum (Valentine and Erwin, 1983). Whatever the accuracy of such a rough guess, it seems clear that evolution was proceeding in an altogether different biotic environment at that time, structurally as well as compositionally, one in which adaptive space was much less densely occupied. For large tracts of adaptive space, no very specialized adaptive types yet existed. As adaptive space filled, appropriate body types that were evolutionarily accessible were produced. The lack of appropriate body plans would appear to be a critical point, for open adaptive space alone does not appear to be a sufficient condition for the origin of new body plans. The marine Permian–Triassic extinctions removed so many lineages that marine adaptive space in the Early Triassic may have been as poorly or even more poorly inhabited than in Early Cambrian time, yet no phyla or classes were produced during the subsequent rediversifications (Valentine, 1969; Erwin *et al.*, 1987). Evidently, the large variety of body plans represented among the depauperate Triassic fauna provided an adequate source of adaptive types, modification of these plans enabling the surviving lineages to reoccupy adaptive space, and precluding the origin of new phyla.

5.4. Historical Factors: The Evolution of Precursory Body Plans

A utilitarian way of judging the complexity of a metazoan body plan is by the number of cell types differentiated during its construction (Bonner, 1988). As each cell type must evolve from a preexisting one, the evolution of differentiation will result in a branching, treelike pattern of cell descent (Kauffman, 1987). Such trees exist in principle for the cells appearing during the ontogeny of a particular organism, and for all the cell types within, say, all bilaterians. The increasing numbers of cell types present in organisms of more complex grades of construction are presumably associated with increasing numbers of structural genes and especially with disproportionate increases in gene regulatory complexity. Kauffman (1987) has described a model regulatory system in which the number of regulatory genes increases as about the square of the number of cell types. Perhaps the regulatory unit need not be a gene per se, but may be an epigenetic event or agent. Whatever the ratio of genes to cell types may be, the genome must have grown significantly, and epigenetic interactions increased, to encode or incorporate the additional information required during evolutionary trends toward increased complexity.

Bonner (1988) has given rough estimates of the number of cell types of some multicellular organisms, using a "lumper's" approach to characterizing the types. Sponges and cnidarians were recorded as having from about 9–12 cell types. No estimates were then available for flatworms or nemertines. Data on turbellarian flatworms summarized recently by Rieger *et al.* (1990), however, suggest that a "lumper's" count of their cell type numbers would range between 20 and 30 or so types, varying among orders, and similar work on nemertines (Turbeville, 1990) suggests cell-type numbers in the range of 35–40. These numbers fall nicely between Bonner's estimates for cnidarians and for higher invertebrates, for a notable feature of higher invertebrates is that they display a plateau of complexity or at least of cell-type number. Mollusks, annelids, arthropods, and echinoderms all have body plans that can be achieved by about 55 cell types. Although one might suspect that, for example, arthropod architectures are necessarily more complicated than annelidan ones, this impression is evidently conveyed by floor plans and not building blocks. The entire sequence of the evolution of many new cell types—the origin of the necessary variation, the subsequent divergence of parental and daughter types, the integration of new cells into the tissues and organs to which they contribute, with all the elaboration of the genetic regulatory machinery that is implied—would seem to be a relatively gradual process. Once the level of complexity of genomes, cells, tissues and organs had reached a point at which the regulatory capacity to produce complex organisms was available, however, the elaboration of large numbers of higher invertebrate body-plan geometries became possible. Such a radiation need not have required vast stretches of time, for it chiefly involved "only" a repatterning of regulatory links and pathways, not the invention of new levels of control (see Britten and Davidson, 1971; Valentine and

Campbell, 1975). The Precambrian–Cambrian transition may represent just such a stage in evolutionary history.

6. Summary and Conclusions

As is clear from the earlier contributions in this volume, there have been during the last few decades a large number of important discoveries, descriptions, and interpretations of fossils, previously unknown or unappreciated, representing early metazoans and providing new evidence of the events that led to the establishment of the great clades that have dominated during the Phanerozoic. The record that we have is not very supportive of models that posit a long period of the evolution of metazoan phyla in the Proterozoic. It is certainly possible that small, soft-bodied, multicellular animals have a long undiscovered history, but a metazoan origin within a few tens of millions of years of the beginning of the Cambrian is also consistent with available evidence, especially considering the uncertainties in geologic dating of the various fossiliferous units. Whenever metazoans first arose, they evolved to the point that they began to leave trace and body fossils only very late in the Proterozoic. The trace fossils indicate small or at least narrow-bodied forms capable of furrowing sedimentary substrates, while the body fossils include what appear to be segmented, broadly flattened forms, probably hemocoelic, that can be interpreted as surface creepers but were probably not responsible for the furrowed trails. As annelid ancestors were among the Late Proterozoic forms, it is possible that some of the narrow-bodied forms, perhaps flattened and vascularized but lacking hydrostatic coeloms, were also segmented. Round-bodied nonsegmented forms, perhaps pseudocoelomate, may be responsible for shallow horizontal burrows and may have a place in the ancestry of living phyla. Additionally, the ancestors of various oligomerous phyla and of mollusks may have been found among the small vagile worm fauna, together with flatworms or protoflatworms.

It is thus possible to picture the Late Proterozoic fauna as poised to produce the somewhat more complex body plans of the Cambrian, some of which came to dominate the marine benthos until the present time. This picture involves the postulation of a grade of metazoan organization that is now extinct—triploblastic hemocoelic segmented forms that lacked a eucoelom, although they may have possessed intramesodermal ducts and sacs that would qualify as coelomic by most definitions. Whether such spaces were developmental forerunners of hydrostatic coeloms is unclear; they may have been in some cases and not in others. These extinct forms, together with other vermiform animals that probably represent nemertine or pseudocoelomate grades, are hypothesized to have been moderately complex, but with their construction involving something fewer than, say, 50 cell types. Near the Precambrian–Cambrian boundary, then, many of these forms began to produce descendants of a coelomate grade, with all of the additional complexities and architectural improvements that were then made possible, including those associated with significant increases in body size now permitted and exploited by several lineages radiating independently. Although this radiation may rapidly have produced novel body plans within some lineages, it was not perfectly contemporaneous among lineages, although restricted to a few tens of millions of years in latest Proterozoic and Cambrian time. Some of the new body plans involved durable skeletons, being for some modes of life an improvement on body walls shaped by musculature and body fluids. The advent of these durable skeletons created the Precambrian–Cambrian boundary and the conventional fossil record of skeletons appeared.

Quite a few novel body plans, both soft-bodied and durably skeletonized, were evolved—far more than would have been necessary if one were designing just enough adaptive types to occupy the major available habitats. Early in the radiation, new body plans seem to have been appearing faster than established body plans were modified, even though modified subplans of established phyla could have invaded many of the empty adaptive zones, and indeed did so eventually.

To conclude, it may be that the early metazoan radiation occurred in lineages with high mutation rates, in which novel cell differentiates led to new developmental patterns that were available for incorporation into ontogeny, and that had attained a level of structural complexity permitting the further elaboration of what are now regarded as higher invertebrates, and finally that were presented with a broad range of adaptive opportunities. If there had not been such a concomitance of conditions, a radiation should still have occurred, but perhaps at a measured pace and without the production of so many distinctive body plans.

ACKNOWLEDGMENTS. It is a pleasure to thank the members of the Precambrian–Cambrian Transition Consortium for their advice and encouragement. James A. Lake (University of California, Los Angeles) provided trees indicated by the technique of evolutionary parsimony in advance of their publication. Valuable discussions and suggestions were tendered by Tracy McLellan (University of California, Santa Barbara), Simon Conway Morris (Cambridge University), Michael LaBarbara (University of Chicago), and Mary Droser (Oberlin College). Research from which this chapter was chiefly drawn was funded by grants EAR 84-17011 and EAR 87-21192 from the National Science Foundation.

References

Alpert, S. P., 1977, Trace fossils and the basal Cambrian boundary, *Geol. J. Spec. Issue* **9**:1–8.

Anderson, D. T., 1973, *Embryology and Phylogeny of Annelids and Arthropods*, Pergamon Press, Oxford.

Arthur, W., 1984, *Mechanisms of Morphological Evolution*, Wiley, New York.

Arthur, W., 1988, *A Theory of the Evolution of Development*, Wiley, Chichester, England.

Barnes, R. S. K. (ed.), 1984, *A Synoptic Classification of Living Organisms*, Sinauer, Sunderland, Massachusetts.

Beklemishev, W. N., 1969, *Principles of Comparative Anatomy of Invertebrates*, Chicago University Press, Chicago.

Bengston, S., and Conway Morris, S., 1984, A comparative study of Lower Cambrian *Halkieria* and Middle Cambrian *Wiwaxia*, *Lethaia* **17**:307–329.

Bergh, R. S., 1985, Die Exkretionsorgane der Wurmer, *Kosmos* (Lvov) **17**:97–122.

Bergstrom, J., 1985, Metazoan evolution—A new model, *Zool. Scripta* **15**:189–200.

Birket-Smith, S. J. R., 1981, A reconstruction of the Pre-Cambrian *Spriggina*, *Zool. Jahrb. Anat.* **105**:237–258.

Bonner, J. T., 1988, *The Evolution of Complexity by Means of Natural Selection,* Princeton University Press, Princeton, New Jersey.

Brasier, M. D., 1989a, Sections in England and their correlation, in: *The Precambrian–Cambrian Boundary* (J. W. Cowie and M. D. Brasier, eds.), Clarendon Press, Oxford, pp. 82–104.

Brasier, M. D., 1989b, Biostratigraphy of the earliest skeletal biotas, in: *The Precambrian–Cambrian Boundary* (J. W. Cowie and M. D. Brasier, eds.), Clarendon Press, Oxford, pp. 117–165.

Briggs, D. E. G., 1978, The morphology, mode of life, and affinities of *Canadaspis perfecta* (Crustacea: Phyllocarida), Middle Cambrian, Burgess Shale, British Columbia, *Philos. Trans. R. Soc. Lond. B* **281:**439–487.

Briggs, D. E. G., 1983, Affinities and early evolution of the Crustacea: The evidence of the Cambrian fossils, in *Crustacean Phylogeny* (F. R. Schram, ed.), Balkema, Rotterdam, pp. 1–29.

Briggs, D. E. G., and Collins, D., 1988, A Middle Cambrian chelicerate from Mount Stephen, British Columbia, *Palaeontology* **31:**779–798.

Britten, R. J., 1986, Rates of DNA sequence evolution differ between taxonomic groups, *Science* **231:**1393–1398.

Britten, R. J., and Davidson, E. H., 1971, Repetetive and non-repetetive DNA sequences and a speculation on the origins of evolutionary novelty, *Q. Rev. Biol.* **46:**111–133.

Buss, L. W., 1987, *The Evolution of Individuality,* Princeton University Press, Princeton, New Jersey.

Buss, L. W., 1988, Diversification and germ-line determination, *Paleobiology* **14:**313–321.

Butterfield, N. J., 1990, A reassessment of the enigmatic Burgess Shale fossil *Wiwaxia corrugata* (Matthew) and its relationship to the polychaete *Canadia spinosa, Paleobiology* **16:**287–303.

Chen, J.-Y., 1988, Precambrian metazoans of the Huai River drainage area (Anhui, E. China): Their taphonomic and ecological evidence, *Senckenb. Lethaea* **69:**189–215.

Chen, J.-Y., Hou, X.-G., and Lu, H.-Z., 1989, Early Cambrian netted scale-bearing worm-like sea animal, *Acta Palaeontol. Sinica* **28:**1–16.

Cisne, J. L., 1974, Trilobites and the origin of arthropods, *Science* **186:**13–18.

Clark, R. B., 1964, *Dynamics in Metazoan Evolution,* Clarendon Press, Oxford.

Cloud, P. E., 1949, Some problems and patterns of evolution exemplified by fossil invertebrates, *Evolution* **2:**322–350.

Conway Morris, S., 1977, A new metazoan from the Cambrian Burgess Shale of British Columbia, *Palaeontology* **20:**623–640.

Conway Morris, S., 1979a, The Burgess Shale (Middle Cambrian) fauna, *Annu. Rev. Ecol. Syst.* **10:**327–349.

Conway Morris, S., 1979b, Middle Cambrian polychaetes from the Burgess Shale of British Columbia, *Philos. Trans. R. Soc. Lond. B* **285:**227–274.

Conway Morris, S., 1981, Parasites and the fossil record, *Parasitology* **82:**489–509.

Conway Morris, S., 1985, The Middle Cambrian metazoan *Wiwaxia corrugata* (Matthew) from the Burgess Shale and *Ogygopsis* Shale, British Columbia, Canada, *Philos. Trans. R. Soc. Lond. B* **307:**507–582.

Conway Morris, S., and Peel, J. S., 1990, Articulated halkieriids from the Lower Cambrian of north Greenland, *Nature* **345:**802–805.

Conway Morris, S., Whittington, H. B., Briggs, D. E. G., Hughes, C. P., and Bruton, D. L., 1982, *Atlas of the Burgess Shale,* Palaeontological Association, London.

Conway Morris, S., George, J. D., Gibson, R., and Platt, H. M. (eds.), 1985, *The Origins and Relationships of Lower Invertebrates,* Clarendon Press, Oxford.

Cowie, J. W., and Harland, W. B., 1989, Chronometry, in: *The Precambrian–Cambrian Boundary* (J. W. Cowie and M. D. Brasier, eds.), Clarendon Press, Oxford, pp. 186–198.

Crimes, T. P., 1974, Colonisation of the early ocean floor, *Nature* **248:**328–330.

Crimes, T. P., 1987, Trace fossils and correlation of Late Precambrian and Early Cambrian strata, *Geol. Mag.* **124:**97–119.

Crimes, T. P., 1989, Trace fossils, in: *The Precambrian–Cambrian Boundary* (J. W. Cowie and M. D. Brasier, eds.), Clarendon Press, Oxford, pp. 166–185.

Cuvier, G., 1812, Sur un nouveau rapprochement à etablir entre les classes qui composent le regne animal, *Ann. Mus. Hist. Nat.* **19:**73–84.

Darwin, C. R., 1859, *On the Origin of Species,* Murray, London.

De Jong, W., 1982, Eye lens proteins and vertebrate phylogeny, in: *Macromolecular Sequences in Systematics and Evolutionary Biology* (M. Goodman, ed.), Plenum, New York, pp. 75–114.

Droser, M. L., and Bottjer, D. J., 1988a, Trends in depth and extent of bioturbation in Cambrian carbonate marine environments, western United States, *Geology* **16:**233–236.

Droser, M. L., and Bottjer, D. J., 1988b, Trends in extent and depth of Early Paleozoic bioturbation in the Great Basin (California, Nevada and Utah), in: *This Extended Land, Geological Journeys in the Southern Basin and Range* (D. L. Weide and M. L. Faber, eds.), Geological Society of America Cordilleran Section, Las Vegas, Nevada, pp. 123–135.

Dzik, J., and Krumbiegel, G., 1989, The oldest "onychoporan" *Xenusion:* A link connecting phyla?, *Lethaia* **22:**169–181.

Elder, H. Y., 1972, Connective tissues, body wall structure, and their significance for the polychaete *Polyphysia crassa* (Lipobranchius Jeffreysi) (Oersted), *J. Mar. Biol. Assoc. U. K.* **52:**747–764.

Elder, H. Y., 1980, Peristaltic mechanisms, in: *Aspects of Animal Movement* (H. Y. Elder and E. R. Trueman, eds.), Cambridge University Press, Cambridge, pp. 71–92.

Elder, H. Y., and Trueman, E. R., 1980, *Aspects of Animal Movement,* Cambridge University Press, Cambridge.

Emig, C. C., 1978, Un nouvel embranchment: Les lophophorata, *Bull. Soc. Zool. Fr.* **102:**341–344.

Endler, J. A., and McLellan, T., 1988, The processes of evolution: Toward a newer synthesis, *Annu. Rev. Ecol. Syst.* **19:**395–421.

Erwin, D. H., Valentine, J. W., and Sepkoski, J. J., Jr., 1987, A comparative study of diversification events: The early Paleozoic versus the Mesozoic, *Evolution* **41:**1177–1186.

Fedonkin, M. A., 1977, Precambrian–Cambrian ichnocoenoses of the East-European Platform, in: Trace Fossils 2 (T. P. Crimes and J. C. Harper, eds.), *Geol. J. Spec. Issue* **9:**183–194.

Fedonkin, M. A., 1982, Precambrian soft-bodied fauna and the earliest radiation of invertebrates, in: *Proceedings. Third North American Paleontological Convention,* Vol. 1, pp. 165–167.

Fedonkin, M. A., 1985a, Nonskeletal fauna of the Vendian: Promorphological analysis, in: *The Vendian System,* Vol. 1, *Paleontology* (B. S. Sokolov and A. B. Ivanovskii, eds.), Nauka, Moscow, pp. 18–69 [in Russian].

Fedonkin, M. A., 1985b, Systematic description of Vendian Metazoa, in: *The Vendian System,* Vol. 1, *Paleontology* (B. S. Sokolov and A. B. Ivanovskii, eds.), Nauka, Moscow, pp. 70–106 [in Russian].

Fedonkin, M. A., 1985c, Paleoichnology of Vendian Metazoa, in: *The Vendian System,* Vol. 1, *Paleontology* (B. S. Sokolov and A. B. Ivanovskii, eds.), Nauka, Moscow, pp. 112–117 [in Russian].

Fedonkin, M. A., 1985d, Precambrian metazoans: The problems of preservation, systematics and evolution, *Philos. Trans. R. Soc. Lond. B* **311:**27–45.

Felsenstein, J., 1978, Cases in which parsimony or compatibility methods will be positively misleading, *Syst. Zool.* **27:**401–410.

Felsenstein, J., 1982, Numerical methods for inferring evolutionary trees, *Q. Rev. Biol.* **57:**379–404.

Felsenstein, J., 1988, Perils of molecular introspection, *Nature* **335:**118.

Field, K. G., Olsen, G. J., Lane, D. J., Giovannoni, S. J., Ghiselin, M. T., Raff, E. C., Pace, N. R., and Raff, R. A., 1988, Molecular phylogeny of the animal kingdom, *Science* **239:**748–753.

Fitch, W. M., and Margoliash, E., 1967, Construction of phylogenetic trees, *Science* **155:**279–284.

Gasshoff, M., 1981, Arthropodisierung als biomechanischer Prozess und die Entstehung der Trilobiten-Konstruktion, *Palaeontol. Z.* **55:**219–235.

Gehling, J. G., 1987, Earliest known echinoderm—A new Ediacaran fossil from the Pound Subgroup of South Australia, *Alcheringa* **11:**337–345.

Glaessner, M. F., 1962, Precambrian fossils, *Biol. Rev.* **37:**467–494.

Glaessner, M. F., 1984, *The Dawn of Animal Life: A Biohistorical Study,* Cambridge University Press, Cambridge.

Goldschmidt, R., 1940, *The Material Basis of Evolution,* Yale University Press, New Haven, Connecticut.

Goodman, M., Koop, B. F., Czelusniak, J., Weiss, M. L., and Slightom, J. L., 1984, The η-globin gene: Its long evolutionary history in the β-globin gene family of mammals, *J. Mol. Biol.* **180:**803–823.

Goodrich, E. S., 1945, The study of nephridia and genital ducts since 1895, *Q. J. Microscop. Sci.* **86:**113–392.

Gould, S. J., 1977, *Ontogeny and Phylogeny,* Belknap Press, Cambridge, Massachusetts.

Gray, J., 1968, *Animal Locomotion,* Weidenfeld and Nicolson, London.

Gutmann, W. F., 1981, Relationships between invertebrate phyla based on functional–mechanical analysis of the hydrostatic skeleton, *Am. Zool.* **21:**63–81.

Hadzi, J., 1963, *The Evolution of the Metazoa,* Pergamon, Oxford.

Haeckel, E., 1866, *Generelle Morphologie der Organismen*, Vol. 2. Georg Reimer, Berlin.
Hartman, W. D., 1963, A critique of the enterocele theory, in: *The Lower Metazoa* (E. C. Dougherty, ed.), University of California Press, Berkeley, pp. 55–77.
Hou, X.-G., 1987a, Two new arthropods from Lower Cambrian, Chengjiang, eastern Yunnan, *Acta Palaeontol. Sinica* **26**:236–256.
Hou, X.-G., 1987b, Three new large arthropods from Lower Cambrian, Chengjiang, eastern Yunnan, *Acta Palaeontol. Sinica* **26**:272–285.
Hou, X.-G., and Chen, J.-Y., 1989, Early Cambrian tentacled worm-like animals *Facivermis* gen. nov. from Chengjiang, Yunnan, *Acta Palaeontol. Sinica* **28**:32–41.
Hou, X.-G., Chen, J.-Y., and Lu, H.-Z., 1989, Early Cambrian new arthropods from Chengjian, Yunnan, *Acta Palaeontol. Sinica* **28**:42–57.
Hyman, L. H., 1940–1959, *The Invertebrates*, Vols. 1–5, McGraw-Hill, New York.
Jagersten, G., 1972, *Evolution of the Metazoan Life Cycle*, Academic Press, London.
Jenkins, R. J. F., Ford, C. H., and Gehling, J. G., 1983, The Ediacara Member of the Rawnsley Quartzite: The context of the Ediacara assemblage (late Precambrian, Flinders Ranges), *J. Geol. Soc. Aust.* **30**:101–119.
Kauffman, S. A., 1987, Self-organization, selective adaptation, and its limits: A pattern of inference in evolution and development, in: *Neutral Models in Biology* (M. H. Nitecki and A. Hoffman, eds.), Oxford University Press, Oxford, pp. 56–89.
Kluge, A. G., and Farris, S., 1969, Quantitative phyletics and the evolution of anurans, *Syst. Zool.* **18**:1–32.
Kristensen, R. M., 1983, Loricifera, a new phylum with Aschelminthes characters from the meiobenthos, *Z. Zool. Syst. Evolutionsforsch.* **21**:163–180.
Lake, J. A., 1988, Origin of the eukaryotic nucleus determined by rate-invariant analysis of rRNA sequences, *Nature* **331**:184–186.
Lake, J. A., 1989, Origin of the multicellular animals, in: *The Hierarchy of Life* (B. Fernholm, K. Bremer, and H. Jornvall, eds.), Elsevier, Amsterdam, pp. 273–278.
Lang, A., 1903, Beitrage zu einer Trophocoltheorie, *Jena Z. Naturwiss.* **38**:1–373.
Lankester, E. R., 1874, Observations on the development of the pond snail (*Lymnaea stagnalis*), and on the early stages of other Mollusca, *Q. J. Microscop. Sci.* **14**:365–391.
Linnaeus, C., 1758, *Systema Naturae*, 10th ed., Laurentii Salvii, Holmiae [Facsimile edition, British Museum (Natural History) (1956)].
Livanov, N. A., 1955, *Paths of Evolution in the Animal World. Analysis of Organization of Principle of Phyla Multicellular Animals*, Moscow.
Lowenstam, H. A., and Margulis, L., 1980. Evolutionary prerequisites for early Phanerozoic calcareous skeletons, *Biosystems* **12**:27–41.
Manton, S. M., 1977, *The Arthropoda*, Oxford University Press, Oxford.
Manton, S. M., and Anderson, D. T., 1979, Polyphyly and the evolution of the arthropods, in: *The Origin of Major Invertebrate Groups* (M. R. House, ed.), Academic Press, New York, pp. 269–321.
Marcus, E., 1958, On the evolution of the animal phyla, *Q. Rev. Biol.* **33**:24–58.
Margulis, L., and Schwartz, K. V., 1982, *Five Kingdoms; An Illustrated Guide to the Phyla of Life on Earth*, Freeman, San Francisco.
McKenna, M. C., 1987, Molecular and morphologic analysis of high-level mammalian interrelationships, in: *Molecules and Morphology* (C. Patterson, ed.), Cambridge University Press, Cambridge, pp. 55–93.
McKinney, M. L., and McNamara, K. J., 1990, *Heterochrony: The Evolution of Ontogeny*, Plenum, New York.
Meyer, E., 1890, Die Abstammung der Anneliden. Der Ursprung der Metamerie und die Bedeutung des Mesoderms, *Biol. Zentralbl.* **10**:296–308.
Narbonne, G. M., and Myrow, P., 1988, Trace fossil biostratigraphy in the Precambrian–Cambrian boundary interval, in: *Trace Fossils, Small Shelly Fossils, and the Precambrian–Cambrian Boundary* (E. Landing, G. M. Narbonne, and P. Myrow, eds.), New York State Museum Bulletin 463, pp. 72–76.
Nielsen, C., 1985, Animal phylogeny in the light of the trochaea theory, *Biol. J. Linn. Soc.* **25**:243–299.
Nielsen, C., Walker, W. F., Bode, H. R., Steele, R. E., Field, K. A., Olsen, G. J., Giovannoni, S. J., Raff, E. C., Pace, N. R., and Raff, R. A., 1989, Phylogeny and molecular data, *Science* **243**:548–551.

Novacek, M. J., 1986, The skull of lepticid insectivorans and the higher-level classification of eutherian mammals, *Bull. Am. Museum Nat. Hist.* **183:**1–112.

Novacek, M. J., and Wyss, A. R., 1986, Higher-level relationships of the recent eutherian orders: Morphological evidence, *Cladistics* **2:**257–287.

Palij, V. M., Posti, E., and Fedonkin, M. A., 1979, Soft-bodied Metazoa and trace fossils of Vendian and Lower Cambrian, in: *Upper Precambrian and Cambrian Paleontology of East-European Platform* (B. M. Keller and A. Yu. Rozanov, eds.), USSR Academy of Sciences, Moscow, pp. 49–82 [in Russian].

Pantin, C. F. A., 1950, Locomotion in British terrestrial nemertines and planarians: With a discussion on the identity of *Rhynchodemus bilineatus* (Mecnikow) in Britain and on the name *Fasciola terrestris* O. F. Muller, *Proc. Linn. Soc. Lond.* **162:**23–37.

Parker, S. P. (ed.), 1982, *Synopsis and Classification of Living Organisms*, McGraw-Hill, New York.

Ramskold, L., and Hou, X., 1991, New Early Cambrian animal and onychophoran affinities of enigmatic metazoans, *Nature* **351:**225–228.

Remane, A., 1963, The enterocelic origin of the celom, in: *The Lower Metazoa* (E. C. Dougherty, ed.), University of California Press, Berkeley, pp. 78–90.

Rieger, R., 1980, A new group of interstitial worms, Lobatocerebridae nov. fam. (Annelida) and its significance for metazoan phylogeny, *Zoomorphologie* **95:**41–84.

Reiger, R. M., Tyler, S., Smith, J. P. S., III, and Rieger, G. E., 1990, Platyhelminthes: Turbellaria, in: *Microscopic Anatomy of Invertebrates, Vol. 3* (F. W. Harrison and B. J. Bogitsh, eds.), Wiley-Liss, New York, pp. 7–140.

Rowell, A. J., and Caruso, N. E., 1985, The evolutionary significance of *Nisusia sulcata*, an early articulate brachiopod, *J. Paleontol.* **59:**1227–1242.

Rozanov, A. Yu., 1986, Problematica of the Early Cambrian, in: *Problematic Fossil Taxa* (A. Hoffman and M. H. Nitecki, eds.), Oxford University Press, Oxford, pp. 87–96.

Ruiz, G., and Lindberg, D. R., 1989, A fossil record for trematodes: extent and potential uses, *Lethaia* **22:**431–438.

Runnegar, B., 1982a, Oxygen requirements, biology and phylogenetic significance of the late Precambrian worm *Dickinsonia*, and the evolution of the burrowing habit, *Alcheringa* **6:**223–239.

Runnegar, B., 1982b, The Cambrian explosion: Animals or fossils?, *J. Geol. Soc. Aust.* **29:**395–411.

Salvini-Plawen, L. von, 1969, Solenogastres und Caudofoveata (Mollusca: Aculifera): Organization und phylogenetische Bedeutung, *Malacologia* **9:**191–216.

Salvini-Plawen, L. von, 1980, A reconsideration of systematics in the Mollusca (phylogeny and higher classification), *Malacalogia* **19:**249–278.

Salvini-Plawen, L. v., 1982, A paedomorphic origin of the oligomerous animals?, *Zool. Scripta* **11:**77–81.

Sarvaas, A. E. du M., 1933, La Theorie du Coelome, Thesis, University of Utrecht.

Sedgwick, A., 1984, On the origin of metameric segmentation and some other morphological questions, *Q. J. Microscop. Sci.* **24:**43–82.

Seilacher, A., 1956, Der Beginn des Kambriums als biologische Wende, *Neues Jahrb. Geol. Palaeontol.* **103:**155–180.

Seilacher, A., 1984, Late Precambrian and Early Cambrian Metazoa: Preservational or real extinctions?, in: *Patterns of Change in Earth Evolution* (H. D. Holland and A. F. Trendall, eds.), Springer-Verlag, Berlin, pp. 159–168.

Shoshani, J., 1986, Mammalian phylogeny: Comparison of morphological and molecular results, *Mol. Biol. Evol.* **3:**222–242.

Signor, P. W., 1988, The Precambrian–Cambrian metazoan radiation: Significance of earliest Cambrian agglutinated skeletons, *Geol. Soc. Am. Abstr. Progr.* **20:**A104.

Simpson, G. G., 1944, *Tempo and Mode in Evolution*, Columbia University Press, New York.

Simpson, G. G., 1953, *The Major Features of Evolution*, Columbia University Press, New York.

Snodgrass, R. E., 1938, Evolution of the Annelida, Onychophora and Arthropoda, *Smithson. Misc. Collect.* **97:**1–159.

Sprigg, R. C., 1947, Early Cambrian (?) jellyfishes from the Flinders Ranges, South Australia, *Trans. R. Soc. S. Aust.* **71:**212–224.

Stanley, G. D., and Stürmer, W., 1983, The first fossil ctenophore from the Lower Devonian of West Germany, *Nature* **303:**518–520.

Stanley, S. M., 1979, *Macroevolution, Pattern and Process*, Freeman, San Francisco.

Stasek, C. R., 1972, The molluscan framework, in: *Chemical Zoology 7* (M. Florkin and B. J. Scheer, eds.), Academic Press, New York, pp. 1–44.
Tiegs, O. W., and Manton, S. M., 1958, The evolution of the Arthropoda, *Biol. Rev.* **33:**255–337.
Trueman, E. R., 1975, *The Locomotion of Soft-Bodied Animals*, Edward Arnold, London.
Turbeville, J. M., 1990, Nemertinea, in: *Microscopic Anatomy of Marine Invertebrates, Vol. 3* (F. W. Harrison and B. J. Bogitsh, eds.), Wiley-Liss, New York, pp. 285–328.
Turbeville, J. M., and Ruppert, E. E., 1985, Comparative ultrastructure and the evolution of nemertines, *Am. Zool.* **25:**53–71.
Valentine, J. W., 1969, Patterns of taxonomic and ecological structure of the shelf benthos during Phanerozoic time, *Palaeontology* **12:**684–709.
Valentine, J. W., 1975, Adaptive strategy and the origin of grades and ground plans, *Am. Zool.* **15:**391–404.
Valentine, J. W., 1977, General patterns of metazoan evolution, in: *Patterns of Evolution* (A. Hallam, ed.), Elsevier, Amsterdam, pp. 27–57.
Valentine, J. W., 1981, The lophophorate condition, in: *Lophophorates* (J. T. Dutro and R. S. Boardman, eds.), University of Tennessee Department of Geological Sciences, Studies in Geology, Vol. 5, pp. 190–204.
Valentine, J. W., 1986, Fossil record of the origin of bauplane and its implications, in: *Patterns and Processes in the History of Life* (D. M. Raup and D. Jablonski, eds.), Springer-Verlag, Berlin, pp. 209–222.
Valentine, J. W., 1989, Bilaterians of the Precambrian–Cambrian transition and the Annelid–Arthropod relationship, *Proc. Natl. Acad. Sci. USA* **86:**2272–2275.
Valentine, J. W., and Campbell, C. A., 1975, Genetic regulation and the fossil record, *Am. Sci.* **63:**673–680.
Valentine, J. W., and Erwin, D. H., 1983, Patterns of diversification of higher taxa: A test of macroevolutionary paradigms, in: Modalities, rythmes et mecanismes de l'evolution biologique: Gradualism phyletique ou equilibres ponctues? (J. Chaline, ed.), *Colloq. Int. CNRS* **330:**219–223.
Valentine, J. W., and Erwin, D. H., 1987, Interpreting great developmental experiments: The fossil record, in: *Development as an Evolutionary Process* (R. A. Raff and E. C. Raff, eds.), Liss, New York, pp. 71–107.
Vandebroek, G., 1952, La classification generale des Metazoaires superieurs et les recentes données embryologiques, *Ann. Soc. Zool. Belg.* **83:**131–142.
Whittington, H. B., 1979, Early arthropods, their appendages and relationships, in: *The Origin of Major Invertebrate Groups* (M. R. House, ed.), Academic Press, New York, pp. 253–268.
Whittington, H. B., 1985, *The Burgess Shale,* Yale University Press, New Haven, Connecticut.
Woese, C. R., 1987, Bacterial evolution, *Microbiol. Rev.* **51:**221–271.
Woese, C. R., Stackebrandt, E., and Ludwig, W., 1985, What are mycoplasmas: The relationship of tempo and mode in bacterial evolution, *J. Molec. Evol.* **21:**305–316.
Wright, A. D., 1979, Brachiopod radiation, in: *The Origin of Major Invertebrate Groups* (M. R. House, ed.), Academic Press, New York, pp. 235–252.
Wright, S., 1982a, Character change, speciation and higher taxa, *Evolution* **36:**427–443.
Wright, S., 1982b, The shifting balance theory and macroevolution, *Annu. Rev. Genet.* **16:**1–19.
Wyss, A. R., Novacek, M. J., and McKenna, M. C., 1987, Amino acid sequence versus morphological data and the interordinal relationships of mammals, *Mol. Biol. Evol.* **4:**99–116.
Zimmer, R. L., 1964, Reproductive biology and development of Phoronida, Ph.D. thesis, University of Washington, Seattle, Washington.
Zuckerkandl, E., and Pauling, L., 1965, Molecules as documents of evolutionary history, *J. Theoret. Biol.* **8:**357–366.

Taxonomic Index

Subject Index